The Human Body in Health & Disease

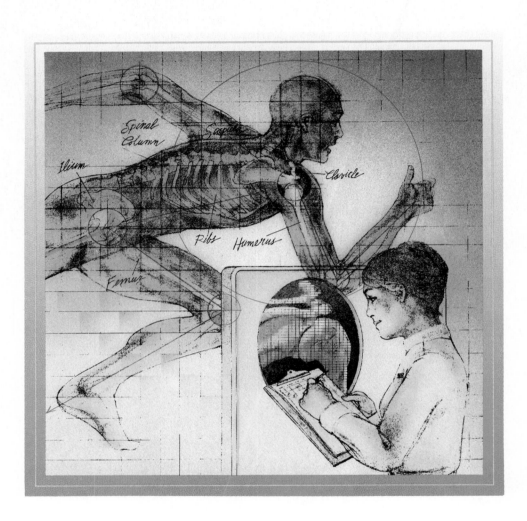

THE HUMAN BODY
—IN—
Health & Disease

GARY A. THIBODEAU, PhD

Chancellor and Professor of Biology,
University of Wisconsin—River Falls,
River Falls, Wisconsin

KEVIN T. PATTON, PhD

Professor, Department of Life Sciences,
St. Charles County Community College,
St. Peters, Missouri

with 485 illustrations, including 445 in color

Mosby
Year Book

St. Louis Baltimore Boston Chicago London Philadelphia Sydney Toronto

Mosby Year Book
Dedicated to Publishing Excellence

Publisher Edward F. Murphy
Editor Deborah Allen
Assistant Editor Laura J. Edwards
Project Manager Carol Sullivan Wiseman
Senior Production Editor Pat Joiner
Book Designer Diane Beasley
Cover Art by Gary Kaemmer
Illustrations by Ernest W. Beck (Lake Forest, Illinois)
　　　　　　　 Joan M. Beck (Minneapolis, Minnesota)
　　　　　　　 Rolin Graphics (Minneapolis, Minnesota)

Credits for other illustrations and photos used by
permission appear after the glossary.

Printed in the United States of America

Mosby–Year Book, Inc.
11830 Westline Industrial Drive
St. Louis, Missouri 63146

Library of Congress Cataloging in Publication Data
Thibodeau, Gary A.
　　The human body in health & disease / Gary A. Thibodeau, Kevin T. Patton.
　　　p.　cm.
　　Includes index.
　　ISBN 0-8016-6412-8 (soft cover). — ISBN 0-8016-6413-6
（hardbound)
　　1. Human physiology.　　2. Human anatomy.　　3. Physiology,
Pathological.　　I. Patton, Kevin T.　　II. Title.　　III. Title: The
human body in health and disease.
　　[DNLM　1. Anatomy.　　2. Pathology.　　3. Physiology.　　QT 104 T427h]
QP34.5.T495　1991
612—dc20
DNLM/DLC
for Library of Congress　　　　　　　　　　　　　　　　　91-29836
　　　　　　　　　　　　　　　　　　　　　　　　　　　　　　CIP

　　　95 96　C/CX/VH　9 8 7 6 5 4 3

Preface

This book is about the human body. *The Human Body in Health & Disease* serves as a guide for students of the health professions beginning their exploration of human structure and function and the basic mechanisms of disease. It not only presents introductory material on the elegance and efficiency of the human body, but it also presents pathologic conditions associated with each body system. To truly understand the human body, knowledge of both normal and abnormal aspects of human biology is essential.

As we approach the new century, our understanding of the human body is increasing at an explosive rate. Almost daily, new discoveries cause scientists to overturn old, established hypotheses and replace them with new concepts. Recent advances in medicine, biotechnology, biochemistry, immunology, neuroendocrinology, molecular genetics and other fields are overwhelming. This explosion of new understanding presents instructors with the dilemma of selecting the most appropriate information to present in introductory but nonetheless rigorous courses. *The Human Body in Health & Disease* has been carefully designed to present up-to-date information that is both accurate and "user friendly."

During the production of this book, each decision regarding the selection, sequencing, or method of presentation of information was evaluated by teachers actually working in the field—teachers currently assisting students in their learning of human structure, function, and disease for the first time. The result is a text that students will read—one designed to help the teacher teach and the student learn. It is particularly suited to introductory courses about the human body in nursing and allied health programs. *The Human Body in Health & Disease* emphasizes concepts required for entry into more advanced courses, completion of professional licensing examinations, and success in a practical work-related environment.

Special Features

UNIFYING THEMES

Anatomy, physiology, and introductory pathology encompass a body of knowledge that, because of its sheer magnitude, can easily discourage and overwhelm the beginning student. There is no question, however, that competency in these fields is essential for student success in almost every clinical or advanced course in a health-related or human science curriculum. If a book is to be successful as a teaching tool in such a complex and important learning environment, it must assist and complement the efforts of instructor and student. It must help unify information, stimulate critical thinking, and motivate students to master a new vocabulary as they learn about the "connectedness" of human structure and function and the "disjointedness" of human disease.

The Human Body in Health & Disease is dominated by two major unifying themes: the *complementarity of structure and function* and *homeostasis*. In every chapter, the student is shown how organized anatomical structures serve unique and specialized functions. Repeated emphasis of this principle encourages students to integrate otherwise isolated factual information into a cohesive and understandable whole. This integration of knowledge is further developed in each chapter as the breakdown of normal integration of form and function is identified as the basis for many disease processes. As a result, anatomy, physiology, and pathology emerge as living and dynamic topics of personal interest and importance to the student. The integrating principle of homeostasis is used to show how the "normal" interaction of structure and function is achieved and maintained by dynamic, counterbalancing forces within the body. Failures of homeostasis are shown as basic mechanisms of disease—a concept that reinforces understanding of the regulatory systems of the human body.

ORGANIZATION AND CONTENT

The 23 chapters of *The Human Body in Health & Disease* present the core concepts of anatomy, physiology, and pathology most important for introductory students. The selection of appropriate information in these disciplines was done with careful consideration—and in consultation with many scientists and educators. Our goal was to eliminate the confusing mix of nonessential and overly specialized material that unfortunately accompanies basic information in many introductory textbooks. Information is presented so that students know and understand what is *important*.

An equally important goal for us in designing this text was to present information in a conceptual framework on which the student can build an understanding of the human body. Rather than simply listing a set of facts, each chapter outlines the broad concepts that allow students to relate the facts to one another in a meaningful way. This approach is especially apparent in the passages that deal with human disease and related topics. Most anatomy and physiology books present diseases as a disjointed list of definitions or descriptions at the end of each chapter—almost as an afterthought. Our book, however, presents disease conditions within a framework that facilitates a more complete understanding of the *process* of disease and allows the student to compare and contrast related disorders easily.

The sequence of chapters in the book follows that most commonly used in courses taught at the undergraduate level. Basic concepts of human biology—anatomy, physiology, cytology, histology, and pathology—are presented in Chapters 1 through 4. Chapters 5 through 23 present material on more specialized topics, such as individual organs or systems, the senses (Chapter 9), immu-

nity (Chapter 14), or genetics and genetic disease (Chapter 23). Each chapter is self-contained, so instructors have the flexibility to alter the sequence of material to fit personal teaching preferences or the special content or time constraints of courses.

Instructors who teach courses with less emphasis on concepts of pathology may wish to examine an alternate text with a similar instructional design: *Structure & Function of the Body*, ninth edition, also available from Mosby–Year Book, Inc.

PEDAGOGICAL FEATURES

The Human Body in Health & Disease is a student-oriented book. Written in a very readable style, it has numerous pedagogical aids that maintain interest and motivation. Every chapter contains the following elements that facilitate learning by improving retention and also encourage integration of essential concepts:

Chapter Outline: An overview outline introduces each chapter and enables the student to preview the content and direction of the chapter at the major concept level before the detailed reading.

Chapter Objectives: Each chapter opens with a short list of measurable learning objectives for the student. Each objective clearly identifies for the student, before he or she reads the chapter, what the key goals should be and what information should be mastered.

Key Terms and Pronunciation Guide: Key terms, when introduced and defined in the body of the text, are identified in **boldface** to highlight their importance. A pronunciation guide follows each new term that students may find difficult to pronounce correctly.

Boxed Inserts and Essays: Brief boxed inserts or longer essays appear in every chapter. These inserts include information ranging from clinical applications to the latest developments in research to exercise and fitness applications. Such information generates interest in the topic at hand and helps motivate students as they progress through the course. All boxed material is highlighted and marked with an easily recognized symbol so students can see at a glance if the box contains clinical, fitness, or general information.

Illustrations: A major strength of *The Human Body in Health & Disease* is the exceptional quality, accuracy, and beauty of the illustration program. The truest test of any illustration is how effectively it can complement and strengthen written information found in the text and how successfully it can be used by the student as a learning tool. Extensive use has been made of full-color illustrations and diagrams, micrographs, medical images, and photographs of live subjects and cadavers.

Outline Summary: Extensive and detailed end-of-chapter summaries in outline format provide excellent guides for students as they review the text when preparing for examinations. Many students may also find such detailed guides useful as a chapter preview in conjunction with the chapter outline.

Word Lists: Two word lists appear at the end of each chapter. The first is a list of new terms related to basic, normal anatomy and physiology. The second is a brief list of new terms related to diseases and other clinical topics. These lists organize essential terminology so that students can study it more easily.

Chapter Tests: Objective Chapter Test questions are included at the end of each chapter. They serve as quick checks for the recall and mastery of important subject matter. They are also designed as aids to increase the retention of information. Answers to all Chapter Test questions are provided at the end of the text.

Review Questions: Subjective review questions at the end of each chapter allow students to use a narrative format to discuss concepts and synthesize important chapter information for review by the instructor. The answers to these review questions are available in an Instructor's Resource Manual.

Clinical Applications: Each chapter ends with a few case studies or other application questions that ask students to apply their knowledge of the human body to specific, practical problems. The questions range from simple applications to moderately complex problem-solving items. Complete narrative answers to each application question appear in the back of the book so that students can verify their answers or find the answer to a question that stumps them.

Additional learning and study aids at the end of the book include the following:

Chemistry Appendix: Recognizing that some students may need to review basic chemistry, we have added a fully illustrated *Chemistry of Life* appendix. This section discusses in simple, straightforward terms the concepts needed to understand basic anatomy and physiology with *concept summaries* to emphasize key points. As an appendix, this information will be readily available to students who need it, but it won't burden the text itself with unnecessary detail for those who don't. Page references within the chapters point students exactly to the appropriate coverage in the Appendix.

Pathological Conditions: A series of tables in this appendix summarize specific pathological conditions by characteristic. The tables serve as a mini-reference tool to supplement material presented in the chapters of the text. Summary tables include:

Leading health problems
Viral conditions
Bacterial conditions
Fungal conditions
Conditions caused by protozoa
Conditions caused by pathogenic animals
Conditions caused by physical agents
Endocrine conditions
Autoimmune conditions
Deficiency diseases
Genetic conditions

Medical Terminology: A list of word parts commonly used in terms related to medicine and pathology is given along with tips on dissecting complex terms to determine their meanings. Many of these word parts are also used within chapters to emphasize how knowledge of medical terminology can help in learning basic concepts.

Clinical Laboratory Values: Commonly observed values for human body content and physiological condition are listed along with their normal ranges. Normal and pathological blood, urine, and other values are summarized.

This information supplements related information presented in appropriate chapters.

Common Medical Abbreviations: A brief list of abbreviations and acronyms commonly used by nurses, medical specialists, pharmacists, and other health care professionals is presented to assist students in mastering relevant terminology.

Glossary: An extensive listing of key terms, pronunciations, and definitions serves as a handy reference for students as they progress through the course.

Index: A comprehensive index aids in locating information anywhere in the book quickly and easily.

Supplements

The supplements package has been carefully planned and developed to assist instructors and to enhance their use of the text. Each supplement, including the test items and study guide, has been thoroughly reviewed by many of the same instructors who reviewed the text.

INSTRUCTOR'S RESOURCE MANUAL AND TEST BANK

The Instructor's Resource Manual and Test Bank, prepared by Judith Diehl of Reid State Technical College, provides text adopters with substantial support in teaching from the text. The following features are included in every chapter:

Sample lecture outlines based on student objectives

Transparency masters for duplication or overhead projection

Outlines and **worksheets** for class demonstrations

Suggestions for student activities and assignments

Sources of audiovisual support

Information on current topics for distribution to students

Answers to the Chapter Tests in the textbook

The Instructor's Resource Manual also contains a comprehensive Test Bank. A variety of 30 to 50 objective questions, including some in the NCLEX

format, are provided for each chapter in the text. The answers are also provided.

TRANSPARENCY ACETATES

A set of full-color transparency acetates—all with large, easy-to-read labels—is available to adopters of the text for use as a teaching aid.

STUDY GUIDE

The Study Guide, written by Linda Swisher of Sarasota County Vocational Technical Center, provides students with additional self-study aids, including chapter overviews, topic reviews, review questions keyed to specific text pages, application and labeling exercises, and answers to all the questions in the Study Guide. These learning aids have been specially designed to prepare students for class discussion and examinations.

A WORD OF THANKS

Many people have contributed to the development of *The Human Body in Health & Disease*. We extend our thanks and deep appreciation to the various students and instructors who have provided us with helpful insights on how such a book should be designed and what information should be included. In particular, the assistance and expertise of Anne Lilly, RN, MS, and Judith Diehl, RN, MA, has helped ensure that the technical information presented in this book will be as current and relevant as possible to students entering the medical field.

A specific "thank you" goes to the following instructors and scientists who critiqued the drafts of this book. Their invaluable comments were instrumental in the development of this new book.

Bert Atsma,
*Union County College,
Cranford, New Jersey*

Virginia Clevenger,
*Mercer County Vocational
School, Trenton, New Jersey*

Mentor David,
*Barton County Community
College, Great Bend, Kansas*

Judy Fair,
*Sandusky School of Practical
Nursing, Sandusky, Ohio*

Beulah Hoffman,
*Indiana Vocational Technical
College, Terre Haute, Indiana*

Marilyn Hunter,
*Daytona Beach Community
College, Daytona Beach,
Florida*

Kathy Korona,
*Community College of
Allegheny County,
West Mifflin, Pennsylvania*

Donna Silsbee,
*SUNY Institute of Technology,
Utica, New York*

Shirley Yeargin,
*Rend Lake College,
Ina, Illinios*

At Mosby–Year Book, Inc., thanks are due all who have worked with us on this project. We wish especially to acknowledge the support and effort of our editor, Deborah Allen; assistant editor, Laura J. Edwards; project manager, Carol Sullivan Wiseman; senior production editor, Pat Joiner; and designer, Diane Beasley, all of whom were instrumental in bringing this book to successful completion.

Gary A. Thibodeau
Kevin T. Patton

Contents

Detailed Contents

An Introduction to the Structure and Function of the Body

CHAPTER

1

Outline

Objectives

After you have completed this chapter, you should be able to:

1 Define the terms *anatomy*, *physiology*, and *pathology*.

2 List and discuss in order of increasing complexity the levels of organization of the body.

3 Define the term *anatomical position*.

4 Discuss and contrast the axial and the appendicular subdivisions of the body. Identify a number of specific anatomical regions in each area.

5 List the nine abdominal regions and the abdominal quadrants.

6 List and define the principal directional terms and sections (planes) used in describing the body and the relationship of body parts to one another.

7 List the major cavities of the body and the subdivisions of each.

8 Explain the meaning of the term *homeostasis* and give an example of a typical homeostatic mechanism.

There are many wonders in our world, but none is more wondrous than the human body. This is a textbook about that incomparable structure. It deals with two very distinct and yet interrelated sciences: **anatomy** and **physiology.** As a science, anatomy is often defined as the study of the structure of an organism and the relationships of its parts. The word *anatomy* is derived from two Greek words that mean "a cutting up." Anatomists learn about the structure of the human body by cutting it apart. This process, called **dissection,** is still the principal technique used to isolate and study the structural components or parts of the human body. Physiology is the study of the functions of living organisms and their parts. It is a dynamic science that requires active experimentation. In the chapters that follow, you will see that anatomical structures seem designed to perform specific functions. Each has a particular size, shape, form, or position in the body because it is intended to perform a unique and specialized activity.

Although an understanding of the normal structure and function of the body is important, it is also important to know the mechanisms of **disease.** Disease conditions result from abnormalities of body structure or function that prevent the body from maintaining the internal stability that keeps us alive and healthy. **Pathology,** the scientific study of disease, uses principles of anatomy and physiology to determine the nature of particular diseases. The term *pathology* comes from *pathos,* the Greek word for "disease." Chapter 4 gives an overview of the basic mechanisms of disease, such as infection and cancer. Throughout the rest of this textbook, explanations of normal structure and function are supplemented by discussions of related disease processes. By knowing the structure and function of the healthy body, you will be better prepared to understand what can go wrong to cause disease. At the same time, a knowledge of disease states will enhance your understanding of normal structure and function.

Structural Levels of Organization

Before you begin the study of the structure and function of the human body and its many parts, it is important to think about how those parts are organized and how they might logically fit together into a functioning whole. Examine Figure 1-1. It illustrates the differing levels of organization that influence body structure and function. Note that the levels of organization progress from the least complex (chemical level) to the most complex (body as a whole).

Organization is one of the most important characteristics of body structure. Even the word *organism,* used to denote a living thing, implies organization.

Although the body is a single structure, it is made up of billions of smaller structures. Atoms and molecules are often referred to as the **chemical level** of organization (see Appendix A). The existence of life depends on the proper levels and proportions of many chemical substances in the cytoplasm of cells. Many of the physical and chemical phenomena that play important roles in the life process will be reviewed in the next chapter. Such information provides an understanding of the physical basis for life and for the study of the next levels of organization so important in the study of anatomy and physiology—cells, tissues, organs, and systems.

Cells are considered to be the smallest "living" units of structure and function in our body. Although long recognized as the simplest units of living matter, cells are far from simple. They are extremely complex, a fact you will discover in Chapter 2.

Tissues are somewhat more complex than cells. By definition a tissue is an organization of many similar cells that act together to perform a common function. Cells are held together and surrounded by varying amounts and varieties of gluelike, nonliving intercellular substances.

Organs are more complex than tissues. An organ is a group of several different kinds of tissues arranged so that they can together perform a special function. For instance, the lungs shown in Figure 1-1 are an example of organization at the organ level.

Systems are the most complex units that make up the body. A system is an organization of varying numbers and kinds of organs arranged so that they can together perform complex functions for the body. The organs of the respiratory system shown in Figure 1-1 permit air to enter the body

FIGURE **1-1** Structural Levels of Organization in the Body.

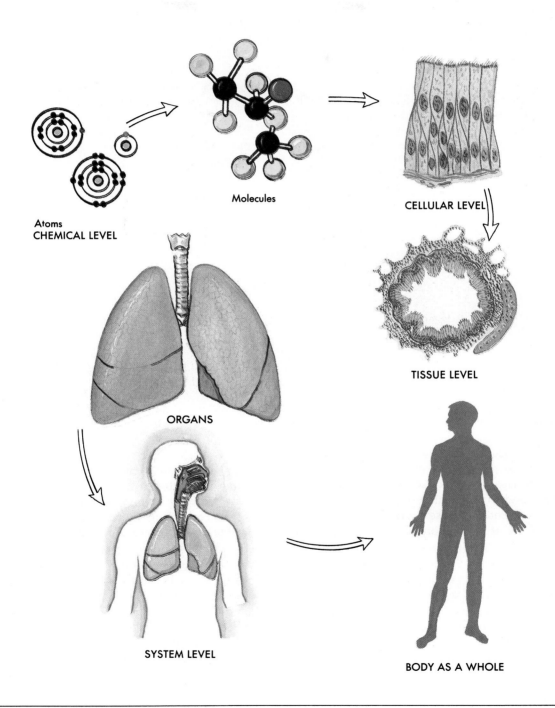

Molecules

Atoms
CHEMICAL LEVEL

CELLULAR LEVEL

TISSUE LEVEL

ORGANS

SYSTEM LEVEL

BODY AS A WHOLE

▼   **FIGURE 1-2 Body Cavities.** Location and subdivisions of the dorsal and ventral body cavities as viewed from the front (anterior) and from the side (lateral).

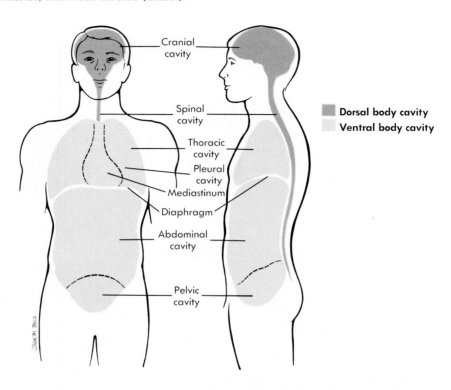

and travel to the lungs, where the eventual exchange of oxygen and carbon dioxide occurs. Organs of the respiratory system include the nose, the windpipe or trachea, and the complex series of bronchial tubes that permit passage of air into the lungs.

The **body as a whole** is all the atoms, molecules, cells, tissues, organs, and systems that you will study in subsequent chapters of this text. Although capable of being dissected or broken down into many parts, the body is a unified and complex assembly of structurally and functionally interactive components, each working together to ensure healthy survival.

Contrary to its external appearance, the body is not a solid structure. It is made up of open spaces or cavities that in turn contain compact, well-ordered arrangements of internal organs. The two major body cavities are called the **ventral** and

dorsal body cavities. The location and outlines of the body cavities are illustrated in Figure 1-2. The ventral cavity includes the **thoracic cavity,** a space that you may think of as your chest cavity. Its midportion is a subdivision of the thoracic cavity, called the **mediastinum,** and its other subdivisions, called the right and left **pleural cavities.** The ventral cavity in Figure 1-2 is broken down into an **abdominal cavity** and a **pelvic cavity.** Actually, they form only one cavity, the **abdominopelvic cavity,** because no physical partition separates them. In Figure 1-2 a dotted line shows the point of separation between the abdominal and pelvic subdivisions. Notice, however, that an actual physical partition, represented in the figure as a white band, separates the thoracic cavity from the abdominal cavity. This muscular partition is the **diaphragm.** It is dome-shaped and is the most important muscle for breathing.

FIGURE 1-3 Organs of the Thoracic and Abdominal Cavities. A view from the front.

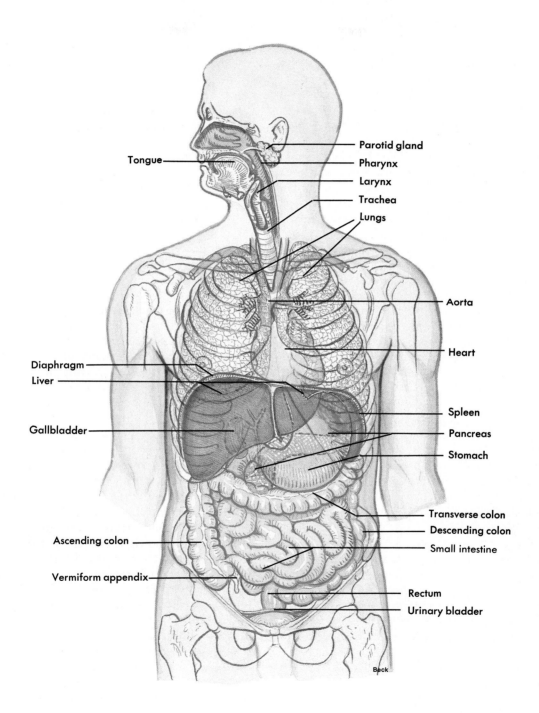

The dorsal cavity shown in Figure 1-2 includes the space inside the skull that contains the brain; it is called the **cranial cavity.** The space inside the spinal column is called the **spinal cavity;** it contains the spinal cord. The cranial and spinal cavities are **dorsal cavities,** whereas the thoracic and abdominopelvic cavities are called **ventral cavities.**

Some of the organs in the largest body cavities are visible in Figure 1-3 and are listed in Table 1-1. For example, the figure shows the trachea, aorta, and heart in the mediastinal portion of the thoracic cavity and the lungs in the pleural portions. Observe the organs shown in the abdominal cavity, including the liver, gallbladder, stomach, spleen, pancreas, small intestine, and parts of the large intestine, including the ascending, transverse, and descending colons. The sigmoid colon, rectum, urinary bladder, and reproductive organs lie in the pelvic portion of the abdominopelvic cavity. Although no specific anatomical structure separates the abdominal and pelvic portions of the abdominopelvic cavity, the area above the hipbones is considered to be abdominal, and the area below the hipbones is considered pelvic. Find each body cavity in a model of the human body if you have access to one. Try to identify the organs in each cavity, and try to visualize their locations in your own body. Study Figures 1-2 and 1-3.

The structure of the body changes in many ways and at varying rates during a lifetime. Before young adulthood, it develops and grows; after young adulthood, it gradually undergoes degenerative changes. With advancing age, there is a generalized decrease in size or a wasting away of many body organs and tissues that affects the structure and function of many body areas. This degenerative process is called **atrophy.** Nearly every chapter of this book will refer to a few of these changes.

Some Words Used in Describing Body Structures

When studying the body, it is often helpful to know where an organ is in relation to other structures. The following terms are used in describing relative positions:

1 Superior and **inferior** (Figure 1-4)—*superior* means "toward the head," and *inferior* means "toward the feet." *Superior* also means "upper" or "above," and *inferior* means "lower" or "below." For example, the lungs are superior to the diaphragm, whereas the stomach is inferior to it.

2 Anterior and **posterior** (Figure 1-4)—*anterior* means "front" or "in front of"; *posterior* means "back" or "in back of." In humans who walk in an upright position, *ventral* (toward the belly) can be used in place of anterior, and *dorsal* (toward the back) can be used for posterior. For example, the nose is on the anterior surface of the body, and the shoulder blades are on its posterior surface.

3 Medial and **lateral** (Figure 1-4)—*medial* means "toward the midline of the body"; *lateral* means "toward the side of the body or away from its midline." For example, the great toe is at the medial side of the foot, and the little toe is at its lateral side. The heart lies medial to the lungs, and the lungs lie lateral to the heart.

4 Proximal and **distal** (Figure 1-4)—*proximal* means "toward or nearest the trunk of the body, or nearest the point of origin of one of its parts"; *distal* means "away from or farthest from the trunk or the point of origin of a body part." For example, the elbow lies at the proximal end of the lower arm, whereas the hand lies at its distal end.

TABLE 1-1   Body Cavities

BODY CAVITY	ORGANS
Ventral Body Cavity	
Thoracic cavity	
Mediastinum	Trachea, heart, blood vessels
Pleural cavities	Lungs
Abdominopelvic cavity	
Abdominal cavity	Liver, gallbladder, stomach, spleen, pancreas, small intestine, parts of large intestine
Pelvic cavity	Lower (sigmoid) colon, rectum, urinary bladder, reproductive organs
Dorsal Body Cavity	
Cranial cavity	Brain
Spinal cavity	Spinal cord

▼ **FIGURE 1-4** Directions and Planes of the Body.

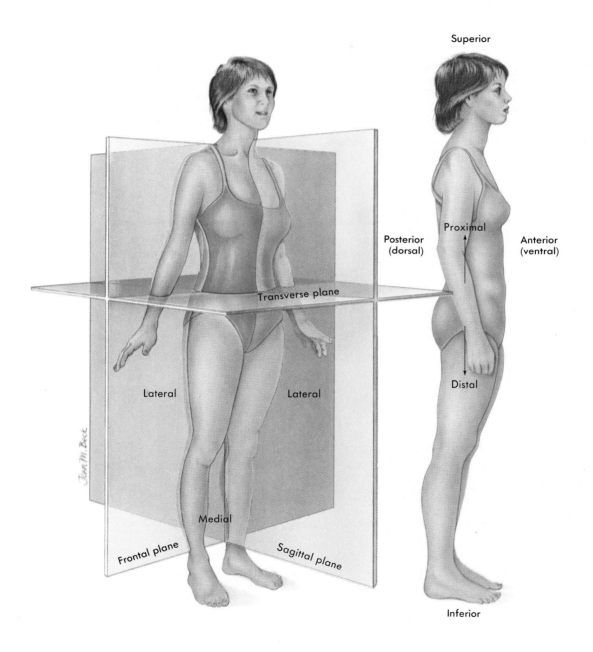

5 Superficial and **deep**—*superficial* means nearer the surface; *deep* means farther away from the body surface. For example, the skin of the arm is superficial to the muscles below it, and the bone of the upper arm is deep to the muscles that surround and cover it.

6 Supine and **prone**—*supine* and *prone* are terms used to describe the body lying in a horizontal position. In the supine position the body is lying face upward, and in the prone position the body is lying face downward.

7 Abdominal regions—to make it easier to locate abdominal organs, anatomists have divided the abdomen into the nine regions shown in Figure 1-5 and defined them as follows:

a *Upper abdominal regions*—the right and left hypochondriac regions and the epigastric region lie above an imaginary line across the abdomen at the level of the ninth rib cartilages

b *Middle regions*—the right and left lumbar regions and the umbilical region lie below an imaginary line across the abdomen at the level of the ninth rib cartilages and above an imaginary line across the abdomen at the top of the hipbones

c *Lower regions*—the right and left iliac (or inguinal) regions and the hypogastric region lie below an imaginary line across the abdomen at the top of the hip-bones

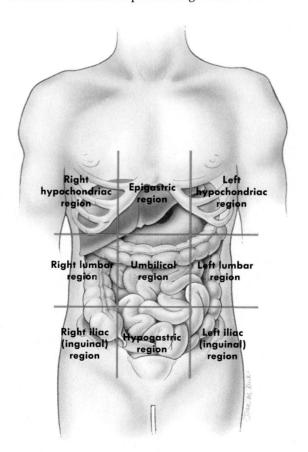

▼ **FIGURE 1-5 The Nine Regions of the Abdomen.** The most superficial organs are shown.

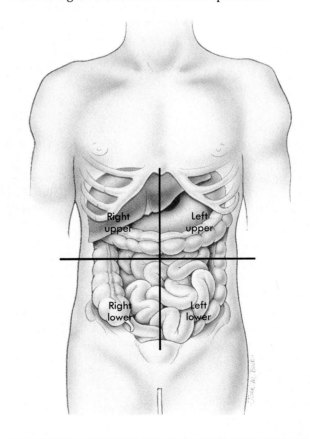

▼ **FIGURE 1-6 Division of the Abdomen into Four Quadrants.** Diagram showing relationship of internal organs to the four abdominal quadrants.

Another and perhaps easier way to divide the abdomen is shown in Figure 1-6. This method is frequently used by health professionals and is useful for locating pain or describing the location of a skin lesion or abdominal tumor. As you can see in Figure 1-6, midsagittal and transverse planes, which are described in the next section, pass through the navel (umbilicus) and divide the abdomen into **four quadrants:** right upper or superior, right lower or inferior, left upper or superior, and left lower or inferior.

Planes or Body Sections

To facilitate the study of individual organs or the body as a whole, it is often useful to subdivide or "cut" it into smaller segments. To do this, body planes or sections have been identified by special names. Read the following definitions and identify each term in Figure 1-4.

1 Sagittal—a sagittal cut or section is a lengthwise plane running from front to back. It divides the body or any of its parts into right and left sides. The sagittal plane shown in Figure 1-4 divides the body into two *equal halves.* This unique type of sagittal plane is called a **midsagittal plane.**

2 Frontal—a frontal *(coronal)* plane is a lengthwise plane running from side to side. As you can see in Figure 1-4, a frontal plane divides the body or any of its parts into anterior and posterior (front and back) portions.

3 Transverse—a transverse plane is a horizontal or crosswise plane. Such a plane (Figure 1-4) divides the body or any of its parts into upper and lower portions.

Anatomical Position

Discussions about the body, the way it moves, its posture, or the relationship of one area to another assume that the body as a whole is in a position called the **anatomical position.** In this reference position (Figure 1-7) the body is in an erect or standing posture with the arms at the sides and palms turned forward. The head and feet also point forward. The anatomical position is a reference position that gives meaning to the directional terms used to describe body parts and regions.

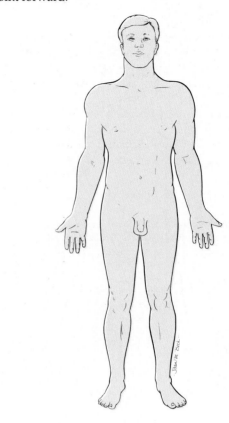

FIGURE 1-7 Anatomical Position. The body is in an erect or standing posture with the arms at the sides and the palms forward. The head and feet also point forward.

Body Regions

To recognize an object, you usually first notice its overall or generalized structure and form. For example, a car is recognized as a car before the specific details of its tires, grill, or wheel covers are noted. Recognition of the human form also occurs as you first identify overall shape and basic outline. However, for more specific identification to occur, details of size, shape, and appearance of individual body areas must be described. Individuals differ in overall appearance because specific body areas such as the face or torso have unique identifying characteristics. Detailed descriptions of the human form require that specific

FIGURE 1-8 Axial and Appendicular Divisions of the Body. Specific body regions are labeled.

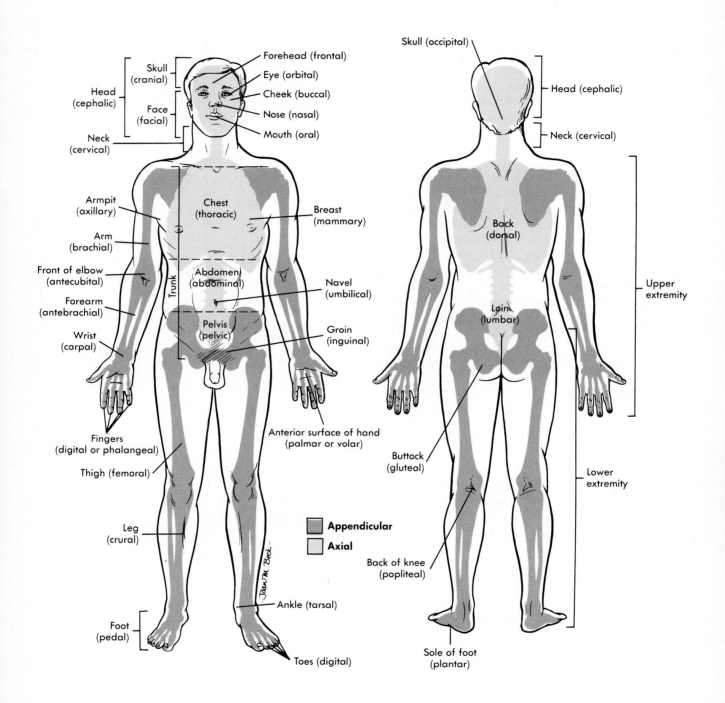

regions be identified and appropriate terms be used to describe them.

The ability to identify and correctly describe specific body areas is particularly important in the health sciences. For a patient to complain of pain in the head is not as specific and therefore not as useful to a physician or nurse as a more specific and localized description. Saying that the pain is facial provides additional information and helps to more specifically identify the area of pain. By using correct anatomical terms such as

forehead, cheek, or chin to describe the area of pain, attention can be focused even more quickly on the specific anatomical area that may need attention. Familiarize yourself with the more common terms used to describe specific body regions identified in Figure 1-8 and listed in Table 1-2.

The body as a whole can be subdivided into two major portions or components: **axial** and **appendicular.** The axial portion of the body consists of the head, neck, and torso or trunk; the

⬇ **TABLE 1-2** Descriptive Terms for Body Regions

BODY REGION	AREA OR EXAMPLE	BODY REGION	AREA OR EXAMPLE
Abdominal (ab-DOM-in-al)	Anterior torso below diaphragm	**Facial, cont'd**	
Antebrachial (an-te-BRAY-kee-al)	Forearm	**Zygomatic** (zye-go-MAT-ik)	Cheek
Antecubital (an-te-KYOO-bi-tal)	Depressed area just in front of elbow	**Femoral** (FEM-or-al)	Thigh
Axillary (AK-si-lair-ee)	Armpit	**Gluteal** (GLOO-tee-al)	Buttock
Brachial (BRAY-kee-al)	Arm	**Inguinal** (ING-gwi-nal)	Groin
Buccal (BUK-al)	Cheek	**Lumbar** (LUM-bar)	Lower back between ribs and pelvis
Carpal (CAR-pal)	Wrist	**Mammary** (MAM-er-ee)	Breast
Cephalic (se-FAL-ik)	Head	**Navel** (NAY-val)	Area around navel or umbilicus
Cervical (SER-vi-kal)	Neck	**Occipital** (ok-SIP-i-tal)	Back of lower skull
Cranial (KRAY-nee-al)	Skull	**Palmar** (PAHL-mar)	Palm of hand
Crural (KROOR-al)	Leg	**Pedal** (PEED-al)	Foot
Cubital (KYOO-bi-tal)	Elbow	**Pelvic** (PEL-vik)	Lower portion of torso
Cutaneous (kyoo-TANE-ee-us)	Skin (or body surface)	**Perineal** (pair-i-NEE-al)	Area (perineum) between anus and genitals
Digital (DIJ-i-tal)	Fingers or toes	**Plantar** (PLAN-tar)	Sole of foot
Dorsal (DOR-sal)	Back	**Popliteal** (pop-li-TEE-al)	Area behind knee
Facial (FAY-shal)	Face	**Supraclavicular** (soo-prah-cla-VIK-yoo-lar)	Area above clavicle
Frontal (FRON-tal)	Forehead	**Tarsal** (TAR-sal)	Ankle
Nasal (NAY-zal)	Nose	**Temporal** (TEM-po-ral)	Side of skull
Oral (OR-al)	Mouth	**Thoracic** (thor-AS-ik)	Chest
Orbital or **ophthalmic** (OR-bi-tal or op-THAL-mik)	Eyes	**Volar** (VO-lar)	Palm or sole

AUTOPSY

Knowledge of human anatomy is important in an **autopsy** (AW-top-see) or postmortem examination. The term *autopsy* comes from the Greek words *auto* (meaning "self") and *opsis* (meaning "view"). Autopsies are procedures in which a human body is examined after death to accurately determine the cause of death, to confirm the accuracy of diagnostic tests, to discover previously undetected problems, and to assess the effectiveness of surgeries or other treatments. Medical and allied health students often attend autopsies to improve their knowledge of human anatomy.

Autopsies are usually performed in three stages. In the first stage, the exterior of the body is examined for abnormalities such as wounds or scars from injuries or surgeries. In the second stage, the ventral body cavity is opened by a deep, Y-shaped incision. The arms of the Y start at the anterior surface of the shoulders and join at the inferior point of the breastbone (sternum) to form a single cut that extents to the pubic area. After the rib cage is sawn through, the walls of the thoracic and abdominopelvic cavities can be opened like hinged doors to expose the internal organs. The second stage of the autopsy includes careful dissection of many or all of the internal organs. If the brain is to be examined, a portion of the skull must be removed. The face, arms, and legs are usually not dissected unless there is a specific reason for doing so. After the organs are returned to their respective body cavities, and the body is sewn up, the third phase of the autopsy begins. It is a microscopic examination of tissues collected during the first two phases. Tests to analyze the chemical content of body fluids or to determine the presence of infectious organisms may also be performed.

---

appendicular portion consists of the upper and lower extremities. Each major area is subdivided as shown in Figure 1-8. Note, for example, that the torso is composed of thoracic, abdominal, and pelvic areas, and the upper extremity is divided into arm, forearm, wrist, and hand components. Although most terms used to describe gross body regions are well understood, misuse is common. The word *leg* is a good example: it refers to the area of the lower extremity between the knee and ankle and *not* to the entire lower extremity.

▼ *Some Basic Facts About Body Functions*

Although they may have very different structures, all living organisms have the following basic facts in common:

1 Maintaining mechanisms that ensure survival of the body and success in propagating its genes through its offspring is the primary business of life.

2 Survival depends on the body's maintaining or restoring homeostasis. **Homeostasis** is the relative constancy of the internal environment. More specifically, homeostasis means that the chemical composition, the volume, and certain other characteristics of blood and interstitial fluid (fluid around cells) remain within narrow limits called *normal ranges*. Maintaining homeostasis means that the cells of the body are in an environment that meets their needs and permits them to function normally under changing conditions.

3 Homeostasis depends on the body's ceaselessly performing many activities. It must continually respond to changes in its environment, exchange materials between its environment and its cells, metabolize food, and control all its diverse activities.

To accomplish this self-regulation, a highly complex and integrated communication control system or network is required. One type of network, called a **feedback control loop,** is shown in Figure 1-9. Different networks in the body control diverse functions such as blood carbon dioxide

EXERCISE PHYSIOLOGY

Exercise physiologists study the effects of exercise on the body organ systems. They are especially interested in the complex control mechanisms that preserve or restore homeostasis during or immediately after periods of strenuous physical activity. Exercise, defined as any significant use of skeletal muscles, is a normal activity with beneficial results. However, exercise disrupts homeostasis. For example, when muscles are worked, the core body temperature rises and blood CO_2 levels increase. These and many other body functions quickly deviate from "normal ranges" that exist at rest. Complex control mechanisms must then "kick in" to restore homeostasis.

As a scientific discipline, exercise physiology attempts to explain many body processes in terms of how they maintain homeostasis. Exercise physiology has many practical applications in therapy and rehabilitation, athletics, occupational health, and general wellness. This specialty concerns itself with the function of the whole body, not just one or two body systems.

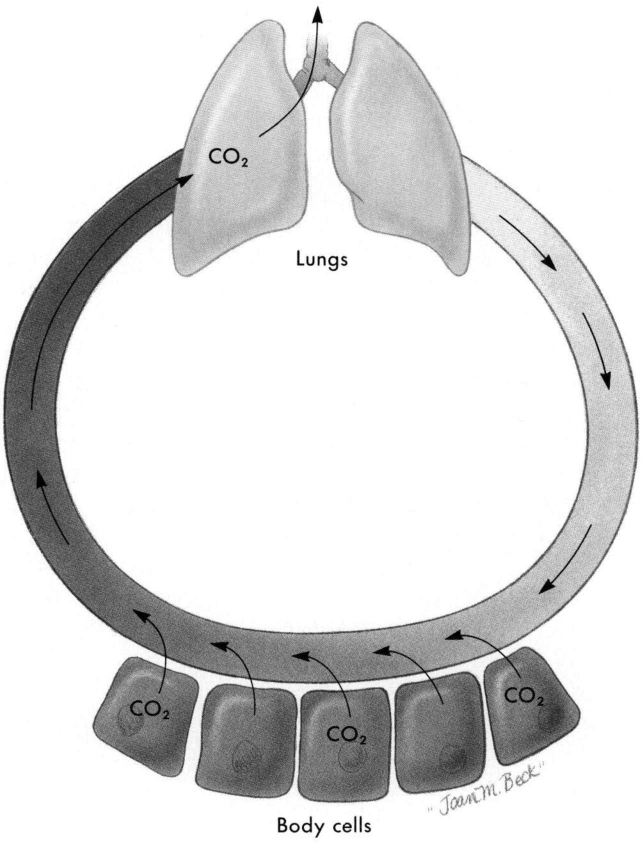

FIGURE 1-9 Negative Feedback Loop. Homeostasis of blood carbon dioxide level is maintained by a negative feedback loop. Because carbon dioxide (CO_2) is constantly being produced by body cells, a specific homeostatic mechanism is required to maintain acceptable levels in blood and the fluid around cells. In the absence of homeostasis, body CO_2 levels quickly rise to lethal levels.

levels, temperature, and heart and respiratory rates. Homeostatic control mechanisms are **negative** or **positive feedback loops.** The most important and numerous of the homeostatic control mechanisms involve negative feedback loops.

Negative feedback loops are stabilizing mechanisms. The example of a negative feedback loop shown in Figure 1-9 illustrates the system used to maintain homeostasis of blood carbon dioxide (CO_2) concentration. As blood CO_2 increases, the respiratory rate increases to permit CO_2 to exit the body in increased amounts through expired air. Without this homeostatic mechanism, body CO_2 levels rapidly rise to toxic levels, and death results.

Although not common, positive feedback loops exist in the body and are involved in normal function. Positive feedback control loops are stimulatory. Instead of opposing a change in the internal environment and causing a "return to normal," positive feedback loops amplify or reinforce the change that is occurring. This type of feedback loop causes an ever-increasing rate of events to occur until something stops the process. Examples of positive feedback loops include the events that lead to the birth of a baby or the formation of a blood clot. Examples of both negative and positive feedback mechanisms involved in homeostasis will be discussed throughout the text.

4 All body functions are ultimately cell functions.

5 Body functions are related to age. During childhood, body functions gradually become more and more efficient and effective. They operate with maximum efficiency and effectiveness during young adulthood. During late adulthood and old age, they gradually become less and less efficient and effective. Changes and functions occurring during the early years are called *developmental processes;* those occurring after young adulthood are called *aging processes.* In general, developmental processes improve functions; aging processes usually diminish them.

OUTLINE SUMMARY

STRUCTURAL LEVELS OF ORGANIZATION
(Figure 1-1)

A Organization is an outstanding characteristic of body structure

B The body is a unit constructed of the following smaller units:

  1 Cells—the smallest structural units; organizations of various chemicals

  2 Tissues—organizations of similar cells

  3 Organs—organizations of different kinds of tissues

  4 Systems—organizations of many different kinds of organs

C The presence of the following cavities is a prominent feature of body structure (Figure 1-2):

  1 Ventral cavity

    a Thoracic cavity

      (1) Mediastinum—midportion of thoracic cavity; heart, and trachea located in mediastinum

      (2) Pleural cavities—right lung located in right pleural cavity, left lung in left pleural cavity

    b Abdominopelvic cavity

      (1) Abdominal cavity contains stomach, intestines, liver, gallbladder, pancreas, and spleen

      (2) Pelvic cavity contains reproductive organs, urinary bladder, and lowest part of intestine

  2 Dorsal cavity

    a Cranial cavity contains brain

    b Spinal cavity contains spinal cord

SOME WORDS USED IN DESCRIBING BODY STRUCTURES

A Superior—toward the head, upper, above
Inferior—toward the feet, lower, below

B Anterior—front, in front of (same as ventral in humans)
Posterior—back, in back of (same as dorsal in humans)

C Medial—toward the midline of a structure
Lateral—away from the midline or toward the side of a structure

D Proximal—toward or nearest the trunk, or nearest the point of origin of a structure
Distal—away from or farthest from the trunk, or farthest from a structure's point of origin

E Superficial—nearer the body surface
Deep—farther away from the body surface

F Supine—lying face up
Prone—lying face down

G Abdominal regions (Figures 1-5 and 1-6)

  1 Nine regions of abdomen

  2 Four quadrants of abdomen

PLANES OR BODY SECTIONS (Figure 1-4)

A Sagittal plane—lengthwise plane that divides a structure into right and left sections

B Midsagittal—sagittal plane that divides the body into two equal halves

C Frontal (coronal) plane—lengthwise plane that divides a structure into anterior and posterior sections

D Transverse plane—horizontal plane that divides a structure into upper and lower sections

ANATOMICAL POSITION (Figure 1-7)

Standing erect with the arms at the sides and palms turned forward

BODY REGIONS (Figure 1-8)

A Axial region—head, neck, and torso or trunk

B Appendicular region—upper and lower extremities

SOME BASIC FACTS ABOUT BODY FUNCTIONS

A Survival of the individual and of the human species is the body's most important business

B Survival depends on the maintenance or restoration of homeostasis (relative constancy of the internal environment; Figure 1-9); the body uses negative feedback loops and, less often, positive feedback loops to maintain or restore homeostasis

C Homeostasis depends on never-ceasing activities: response to changes in the external and internal environment, exchange of materials between the environment and cells, metabolism, and control

D All body functions are ultimately cell functions

E Body functions are related to age; peak efficiency is during young adulthood, diminishing efficiency occurs after young adulthood

NEW WORDS

abdominal quadrants (4)
abdominal regions (9)
anatomical position
anatomy
atrophy
cavities
  abdominal
  cranial
  pelvic
  pleural
  spinal
  thoracic

directional terms
  superior
  inferior
  anterior
  posterior
  ventral
  dorsal
  medial
  lateral
  proximal
  distal
  superficial
  deep

homeostasis
mediastinum
organization
  (structural levels)
  chemical
  cellular
  tissue
  organ
  system

pathology
physiology
planes of section
  sagittal
  midsagittal
  frontal
  transverse
prone
supine

CHAPTER TEST

1. The study of the structure of an organism and the relationship of its parts is called _____ ; the study of the functions of that organism is called _____ ; the study of disease is called _____ .
2. An organization of many similar cells that together perform a common function is called a _____ .
3. The term _____ means "toward the side of the body."
4. A frontal plane divides the body or any of its parts into _____ and _____ portions.
5. The body as a whole can be subdivided into two major portions: _____ and _____ .
6. The relative constancy of the body's internal environment is described by the term _____ .

Match the body area in column B with the area or example in column A. (Only one answer is correct.)

COLUMN A

7. _____ Skull
8. _____ Groin
9. _____ Breast
10. _____ Sole of foot
11. _____ Chest
12. _____ Wrist
13. _____ Fingers or toes
14. _____ Arm pit
15. _____ Mouth
16. _____ Buttock

COLUMN B

a. Mammary
b. Thoracic
c. Digital
d. Carpal
e. Inguinal
f. Axillary
g. Gluteal
h. Cranial
i. Oral
j. Plantar

17. The mediastinum is a subdivision of the:
    a. Thoracic cavity
    b. Pleural cavity
    c. Abdominal cavity
    d. Pelvic cavity
18. Which of the following is an example of a lower abdominal region?
    a. Epigastric region
    b. Umbilical region
    c. Right hypochondriac region
    d. Hypogastric region
19. Which of the following represents the least complex of the structural levels of organization?
    a. System
    b. Tissue
    c. Cell
    d. Organ
20. The diaphragm separates the:
    a. Right and left pleural cavities
    b. Abdominal and pelvic cavities
    c. Thoracic and abdominal cavities
    d. Mediastinum and pleural cavities

REVIEW QUESTIONS

1 Name the four kinds of structural units of the body. Define each briefly.
2 In what cavity could you find each of the following?
    brain
    esophagus
    gallbladder
    heart
    liver
3 In one word, what is the one dominant function of the body or of any living thing?
4 Explain briefly what *homeostasis* means.

5 What do *proximal* and *distal* mean?
6 On what surface of the foot are the toenails located?
7 What structures lie lateral to the bridge of the nose?
8 Which joint—hip or knee—lies at the distal end of the thigh?
9 Define *anatomical position.*
10 List the major subdivisions of the axial and appendicular areas of the body.
11 List and define the four major planes of section.
12 Identify and discuss the mechanisms of negative and positive feedback loops in homeostasis.

CLINICAL APPLICATIONS

1 Mrs. Miller was referred to the clinic by her regular physician to have a mole on her skin examined. The referral form states that the mole is on the left upper quadrant of her abdomen. Describe its location. In preparing Mrs. Miller for the examination, how would you position her? Should you ask her to assume a supine or prone position? During the examination, the physician notices that the referral form states that Mrs. Miller has a similar mole in the occipital region. What position should she assume so that the physician can examine this mole?

2 Mr. Sanchez has just severed the distal tip of a digit on his upper extremity. Where was his injury? As blood poured out of the injured tissue, his blood pressure dropped. His heart pumped faster to restore normal pressure. What effect would this response have on Mr. Sanchez's homeostasis? Would such a response be an example of negative or positive feedback?

Cells and Tissues

C H A P T E R

2

Objectives

After you have completed this chapter, you should be able to:

1 Identify and discuss the basic structure and function of the three major components of a cell.

2 List and briefly discuss the functions of the primary cellular organelles.

3 Compare the major passive and active transport processes that act to move substances through cell membranes.

4 Compare and discuss DNA and RNA and their function in protein synthesis.

5 Discuss the stages of mitosis and explain the importance of cellular reproduction.

6 Explain how epithelial tissue is grouped according to shape and arrangement of cells.

7 List and briefly discuss the major types of connective and muscle tissue.

8 List the three structural components of a neuron.

*A*bout 300 years ago Robert Hooke looked through his microscope—one of the very early, somewhat primitive ones—at some plant material. What he saw must have surprised him. Instead of a single magnified piece of plant material, he saw many small pieces. Because they reminded him of miniature prison cells, that is what he called them—cells. Since Hooke's time, thousands of individuals have examined thousands of plant and animal specimens and found them all, without exception, to be composed of cells. This fact, that cells are the smallest structural units of living things, has become the foundation of modern biology. Many living things are so simple that they consist of just one cell. The human body, however, is so complex that it consists not of a few thousand or millions or even billions of cells but of many trillions of them. This chapter discusses cells first and then tissues.

Cells

SIZE AND SHAPE

Human cells are microscopic in size; that is, they can be seen only when magnified by a microscope. However, they vary considerably in size. An ovum (female sex cell) for example, has a diameter of a little less than 1000 micrometers (about $1/25$ of an inch), whereas red blood cells have a diameter of only 7.5 micrometers. Cells differ even more notably in shape than in size. Some are flat, some are brick shaped, some are threadlike, and some have irregular shapes.

COMPOSITION

Cells contain **cytoplasm** (SI-to-plazm), or "living matter," a substance that exists only in cells. The term *cyto-* is a combining form from the Greek and denotes a relationship to a cell. Each cell in the body is surrounded by a thin membrane, the **plasma membrane.** This membrane separates the cell contents from the dilute salt water solution called **tissue fluid** that bathes every cell in the body. A small, circular body called the **nucleus** (NOO-kle-us) and numerous specialized structures called **organelles** (or-gan-ELZ), which will be described in subsequent sections, are contained within the cytoplasm of each cell.

Important information related to body composition is included in Appendix A. You are encour-

aged to review this material, which includes a discussion of the chemical elements and compounds important to body structure and function.

STRUCTURAL PARTS

The three main parts of a cell are:

1 Plasma membrane
2 Cytoplasm
3 Nucleus

Plasma Membrane

As the name suggests, the **plasma membrane** is the membrane that encloses the cytoplasm and forms the outer boundary of the cell. It is an incredibly delicate structure—only about 7 nm (nanometers) or 3/10,000,000 of an inch thick! Yet it has a precise, orderly structure (Figure 2-1). Two layers of phosphate-containing fat molecules called **phospholipids** form a fluid framework for the plasma membrane. Note in Figure 2-1 that protein molecules lie at both outer and inner surfaces of this framework, and many extend all the way through it. Cholesterol is also a component of the plasma membrane.

Despite its seeming fragility, the plasma membrane is strong enough to keep the cell whole and intact. It also performs other life-preserving functions for the cell. It serves as a well-guarded gateway between the fluid inside the cell and the fluid around it. Certain substances move through it, but it bars the passage of others. The plasma membrane even functions as a communication device. How? Some protein molecules on the membrane's outer surface serve as receptors for certain other molecules when the other molecules contact them. In other words, certain molecules bind to certain receptor proteins. For example, some hormones (chemicals secreted into blood from ductless glands) bind to membrane receptors, and a change in cell functions follows. We might therefore think of such hormones as chemical messages, communicated to cells by binding to their cytoplasmic membrane receptors. A majority of hormones are themselves composed of proteins.

The plasma membrane also identifies a cell as coming from one particular individual. Its surface proteins serve as positive identification tags because they occur only in the cells of that indi-

▼ **FIGURE 2-1 Structure of the Plasma Membrane.** Note that protein molecules may penetrate completely through the two layers of phospholipid molecules.

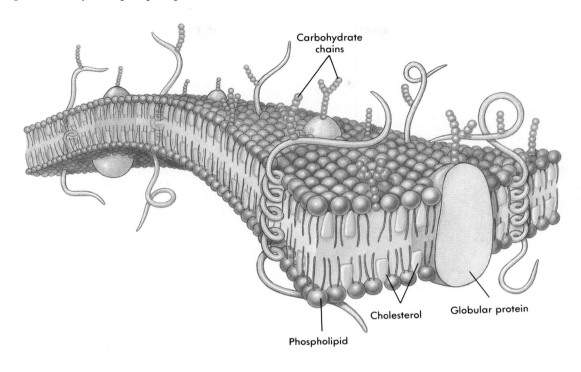

Carbohydrate chains

Phospholipid

Cholesterol

Globular protein

METRIC SYSTEM

Scientists, many government agencies, and increasing numbers of American industries are using or moving toward the conversion of our system of English measurements to the metric system. The metric system is a decimal system in which measurement of length is based on the *meter* (39.37 inches) and weight or mass is based on the *gram* (about 454 grams equal a pound).

A micrometer is one millionth of a meter. (*Micron* is another name for micrometer.) In the metric systems the units of length are as follows:

1 meter (m) = 39.37 inches
1 centimeter (cm) = 1/100 m
1 millimeter (mm) = 1/1000 m
1 micrometer (μm)
    or micron (μ) = 1/1,000,000 m
1 nanometer (nm) = 1/1,000,000,000 m
1 Angstrom (Å) = 1/10,000,000,000 m

Approximately equal to 1 inch:
2.5 cm
25 mm
25,000 μm
25,000,000 nm
250,000,000 Å

FIGURE 2-2 General Characteristics of the Cell. A, Artist's interpretation of cell structure. **B,** A cell as seen under an electron microscope. Note in both the many mitochondria, popularly known as the "power plants of the cell." Note, too, the innumerable dots bordering the endoplasmic reticulum. These are ribosomes, the cell's "protein factories."

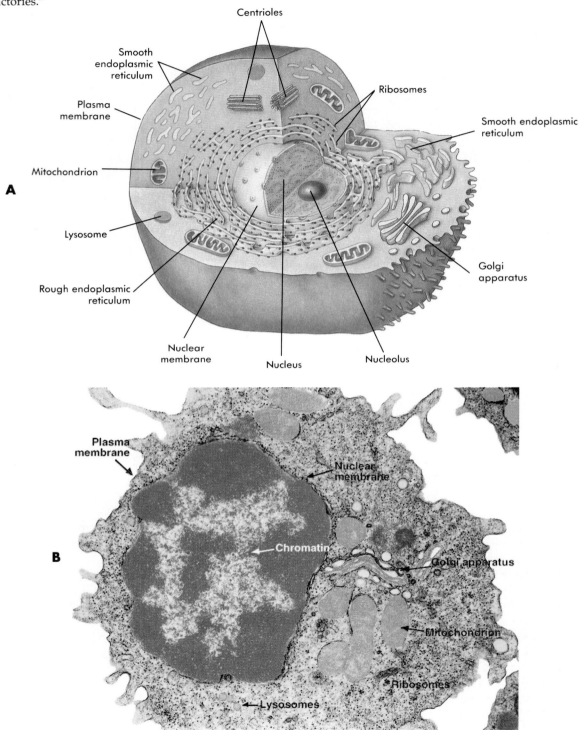

vidual. A practical application of this fact is made in *tissue typing,* a procedure performed before an organ from one individual is transplanted into another. Carbohydrate chains attached to the surface of cells may play a role in the identification of cell types.

Cytoplasm

Cytoplasm is the specialized living material of cells. It lies between the plasma membrane and the nucleus, which can be seen in Figure 2-2 as a round or spherical structure in the center of the cell. Numerous small structures are located in the cytoplasm. As a group, they are called **organelles** (that is, little organs), an appropriate name because they function like organs function for the body.

Look again at Figure 2-2. Notice how many different kinds of structures you can see in the cytoplasm of this cell. A little more than a generation ago, almost all of these organelles were unknown. They are so small that they are invisible even when magnified 1000 times by a light microscope. Electron microscopes brought them into view by magnifying them many thousands of times. We shall briefly discuss the following organelles, which are found in cytoplasm (see also Table 2-1):

1 Endoplasmic reticulum
2 Ribosomes
3 Mitochondria
4 Lysosomes
5 Golgi apparatus
6 Centrioles
7 Cilia
8 Flagella
9 Nucleus
10 Nucleolus

Endoplasmic reticulum. The **endoplasmic reticulum** (en-doe-PLAZ-mik ree-TIK-yoo-lum) **(ER)** is a system of membranes forming a network of many connecting sacs and canals that wind tortuously through a cell's cytoplasm, all the way from its plasma membrane to its nucleus. The tubular passageways or canals in the ER carry the proteins and other substances through the cytoplasm of the cell from one area to another area. There are two types of ER: *smooth* and *rough.* Smooth ER is found in cells that handle or manufacture fatty substances, and rough ER is found in

---

TABLE 2-1 Some Major Cell Structures and Their Functions

CELL STRUCTURE	FUNCTIONS
Plasma membrane	Serves as the boundary of the cell; protein and carbohydrate molecules on outer surface of plasma membrane perform various functions; for example, they serve as markers that identify cells of each individual or as receptor molecules for certain hormones
Endoplasmic reticulum (ER)	Ribosomes attached to rough ER synthesize proteins; smooth ER synthesizes lipids and certain carbohydrates
Ribosomes	Synthesize proteins; a cell's "protein factories"
Mitochondria	ATP synthesis; a cell's "powerhouses"
Lysosomes	A cell's "digestive system"
Golgi apparatus	Synthesizes carbohydrate, combines it with protein, and packages the product as globules of glycoprotein
Centrioles	Function in cell reproduction
Cilia	Short, hairlike extensions on the free surfaces of some cells capable of movement
Flagella	Single and much larger projections of cell surfaces than cilia; the only example in humans is the "tail" of a sperm cell
Nucleus	Dictates protein synthesis, thereby playing an essential role in other cell activities, namely, active transport, metabolism, growth, and heredity
Nucleoli	Play an essential role in the formation of ribosomes

SCREENING DONATED ORGANS AND TISSUES

Tissue typing is a screening process in which cell markers in a donated organ or tissue are identified so that they can be matched to donors with similar cell markers. Cell markers are specific protein molecules (called *antigens* [see Chapter 14]) on the surface of plasma membranes. If the cell markers in donated tissue are different than those in the recipient's normal tissue the recipient's immune system will recognize the tissue as foreign. When the immune system mounts a significant attack against the donated tissue, a *rejection reaction* occurs. The inflammation and tissue destruction that occurs in a rejection reaction not only destroys or "rejects" the donated tissue but may also threaten the life of the recipient. Although drugs such as cyclosporine can be used to inhibit the immune system's attack against donated tissue, cross-matching of tissues by their cell markers is the primary method of preventing rejection reactions.

If you know your blood type, you already know some of your tissue markers. In the ABO system (see Chapter 11), type A blood has the A marker, type B blood has the B marker, type AB blood has both A and B markers, and type O blood has neither A nor B markers. It is important to type and cross-match blood before a blood transfusion takes place to prevent a rejection reaction that could kill the recipient. The American Red Cross and other agencies that coordinate procurement of organs and tis-

sues for transplant are developing computer networks to monitor availability of organs with specific cell markers. Such high-speed computer networks allow physicians to immediately locate organs or tissues for emergency transplants or transfusions.

Another procedure used to screen potential donor organs and tissues involves checking for the presence of infectious agents, especially viruses. Because viruses are difficult to find, most screening tests screen for the presence of specific antibodies. Antibodies are protein molecules produced by some white blood cells on exposure to a virus or other infectious agent. Each type of virus triggers production of a specific kind of antibody, so the presence of a specific antibody type means that the tissue may have the corresponding virus. For example, a test called *ELISA (enzyme-linked immunoabsorbent assay)* is used to test for the presence of antibodies produced in response to HIV (human immunodeficiency virus). HIV causes *acquired immune deficiency syndrome (AIDS)*, a fatal disease that can be transmitted through HIV-contaminated tissues or body fluids. Routine screening for hepatitis-B antigen and other viral antibodies is also done by most tissue and blood banks. Because there is some lag time between infection by a virus and the resulting production of antibodies, such screening tests may fail to identify a virus-contaminated tissue from a recently infected donor.

cells that manufacture the proteins. In both cases the ER functions as a miniature circulatory system for the cell.

Ribosomes. Organelles called **ribosomes** (RI-bo-sohms), shown as dots in Figure 2-2, *A*, are attached to and are part of the rough ER. Although these vitally important organelles are usually attached to ER, they may also be free in the cytoplasm. Ribosomes perform a very com-

plex function; they make enzymes and other protein compounds. Their nickname, "protein factories," indicates this function. Because the cells of the pancreas, which secrete digestive enzymes, are rich in ribosomes, they are the cell type that has been researched the most to discover how ribosomes work.

Mitochondria. The **mitochondria** (my-toe-KON-dree-ah) are another kind of organelle in all cells. Observe their appearance in Figure 2-2, *B*, where they are magnified thousands of times. Mitochondria are so tiny that a lineup of 15,000 or more of them would fill a space only about 2.5 cm or 1 inch long. Two membranous sacs, one inside the other, compose a mitochondrion. The inner membrane forms folds that look like miniature incomplete partitions. Within a mitochondrion's fragile walls, complex, energy-releasing chemical reactions occur continuously. Because these reactions supply most of the power for cellular work, mitochondria have been nicknamed the cell's "power plants." The survival of cells and therefore of the body depends on mitochondrial chemical reactions. Enzymes (chemical catalysts), which are found in mitochondrial walls and matrixes, use oxygen to break down glucose and other nutrients to release energy required for cellular work. The process is called *aerobic* or *cellular respiration.*

Lysosomes. The **lysosomes** (LYE-so-sohms) are membranous-walled organelles that in their active stage look like small sacs, often with tiny particles in them (Figure 2-2). Because lysosomes contain chemicals (enzymes) that can digest food compounds, one of their nicknames is "digestive bags." Lysosomal enzymes can also digest substances other than foods. For example, they can digest and thereby destroy microbes that invade the cell. Thus lysosomes can protect cells against destruction by microbes. Yet, paradoxically, lysosomes sometimes kill cells instead of protecting them. If their powerful enzymes escape from the lysosome sacs into the cytoplasm, they kill the cell by digesting it. This fact has earned lysosomes their other nickname, which is "suicide bags."

Golgi Apparatus. The **Golgi** (GOL-jee) **apparatus** consists of tiny sacs stacked on one another near the nucleus. It makes certain carbohydrate compounds, combines them with certain protein molecules, and packages the product in neat little globules. Then these globules move slowly outward to and through the cell membrane. Once outside the cell, they break open, and their contents spill out. An example of a Golgi apparatus product is the slippery substance called mucus. If we wanted to nickname the Golgi apparatus, we might call it the cell's "carbohydrate-producing and -packaging factory."

Centrioles. The **centrioles** (SEN-tree-olz) are paired organelles. Two of these rod-shaped struc-

tures exist in every cell. They are arranged so that they lie at right angles to each other (Figure 2-2, *A*). Each centriole is composed of fine tubules that play an important role during cell division.

Cilia. **Cilia** (SIL-ee-ah) are extremely fine, almost hairlike extensions on the exposed or free surfaces of some cells. Cilia are organelles capable of movement. One cell may have a hundred or more cilia capable of moving together in a wavelike fashion over the surface of a cell. They often have highly specialized functions. For example, by moving as a group in one direction, they propel mucus upward over the cells that line the respiratory tract.

Flagella. A **flagellum** (flah-JEL-um) is a single projection extending from the cell surface. Flagella are much larger than cilia. In the human, the only example of a flagellum is the "tail" of the male sperm cell. Propulsive movements of the flagellum make it possible for sperm to "swim" or move toward the ovum after they are deposited in the female reproductive tract (Figure 2-3).

Nucleus

Viewed under a light microscope, the **nucleus** of a cell looks like a very simple structure—just a

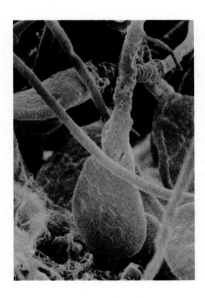

FIGURE 2-3 Human Sperm. Note the tail-like flagellum on each sperm cell.

small sphere in the central portion of the cell. However, its simple appearance belies the complex and critical role it plays in cell function. The nucleus ultimately controls every organelle in the cytoplasm. It also controls the complex process of cell reproduction. In other words, the nucleus must function properly for a cell to accomplish its normal activities and be able to duplicate itself.

Note that the cell nucleus in Figure 2-2 is surrounded by a **nuclear membrane.** The membrane encloses a special type of cytoplasm in the nucleus called **nucleoplasm.** Nucleoplasm contains a number of specialized structures; two of the most important are shown in Figure 2-2. They are the **nucleolus** (noo-KLEE-oh-lus) and the **chromatin** (KRO-mah-tin) **granules.**

Nucleolus and Chromosomes. The nucleolus is critical in protein formation because it "programs" the formation of ribosomes in the nucleus. The ribosomes then migrate through the nuclear membrane into the cytoplasm of the cell and produce proteins. Chromatin granules in the nucleus are threadlike structures made of proteins and hereditary material called **DNA** or **deoxyribonucleic** (dee-OK-see-rye-bo-noo-KLEE-ik) **acid.** DNA is the genetic material often described as the chemical "blueprint" of the body. It determines everything from sex to body build and hair color in every human being. During cell division, DNA molecules become tightly coiled. They then look like short, rodlike structures and are called **chromosomes.** The importance and function of DNA will be explained in greater detail in the section on cell reproduction later in this chapter.

CELL FUNCTIONS

Every human cell performs certain functions; some maintain the cell's survival, and others help maintain the body's survival. In many instances, the number and type of organelles allow cells to differ dramatically in terms of their specialized functions. For example, cells that contain large numbers of mitochondria, such as heart muscle cells, are capable of sustained work. Why? Because the numerous mitochondria found in these cells supply the necessary energy required for rhythmic and ongoing contractions. Movement of the flagellum of a sperm cell is another example of the way a specialized organelle has a specialized function. The sperm's flagellum propels it through the reproductive tract of the female, thus increasing the chances of successful fertilization. This is how and why organizational structure at the cellular level is so important for function in living organisms. There are examples in every chapter of the text to illustrate how structure and function are intimately related at every level of body organization.

Movement of Substances Through Cell Membranes

The cytoplasmic membrane in every healthy cell separates the contents of the cell from the tissue fluid that surrounds it. At the same time the membrane must permit certain substances to enter the cell and allow others to leave. Heavy traffic moves continuously in both directions through cell membranes. Molecules of water, foods, gases, wastes, and many other substances stream in and out of all cells in endless procession. A number of processes allow this mass movement of substances into and out of cells. These transport processes are classified under two general headings:

1 Passive transport processes
2 Active transport processes

As implied by their name, active transport processes require the expenditure of energy by the cell, and passive transport processes do not. The energy required for active transport processes is obtained from a very important chemical substance called **adenosine triphosphate** (ah-DEN-o-sen tri-FOS-fate) or **ATP.** ATP is produced in the mitochondria from nutrients and is capable of releasing energy that in turn enables the cell to work. For active transport processes to occur, the breakdown of ATP and use of the related energy is required.

The details of active and passive transport of substances across cell membranes is much easier to understand if you keep in mind the following two key facts: (1) in passive transport processes, no cellular energy is required to move substances from a high concentration to a low concentration; and (2) in active transport processes, cellular energy is required to move substances from a low concentration to a high concentration.

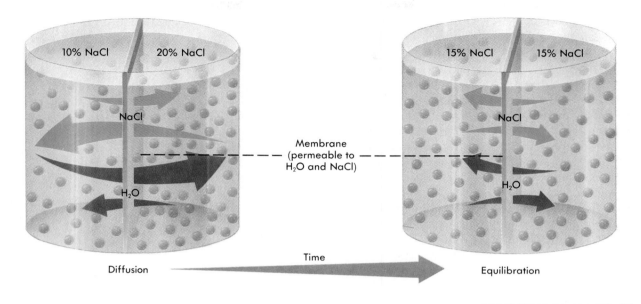

FIGURE 2-4 Diffusion. Note that the membrane is permeable to salt (NaCl) and water and that it separates a 10% NaCl solution from a 20% NaCl solution. The container on the left shows the two solutions separated by the membrane at the start of diffusion. The container on the right shows the result of diffusion after time.

PASSIVE TRANSPORT PROCESSES

The primary passive transport processes that move substances through the cell membranes include the following:

1 Diffusion
   a Osmosis
   b Dialysis
2 Filtration

Scientists describe the movement of substances in passive systems as going "down a concentration gradient." This means that substances in these passive systems move from a region of high concentration to a region of low concentration until they reach equal proportions on both sides of the membrane.

Diffusion

Diffusion is a good example of a passive transport process. Diffusion is the process by which substances scatter themselves evenly throughout an available space. The system does not require additional energy for this movement. To demonstrate diffusion of particles throughout a fluid,

perform this simple experiment the next time you pour yourself a cup of coffee or tea. Place a cube of sugar on a teaspoon and lower it gently to the bottom of the cup. Let it stand for 2 or 3 minutes, and then, holding the cup steady, take a sip off the top. It will taste sweet. Why? Because some of the sugar molecules will have diffused from the area of high concentration near the sugar cube at the bottom of the cup to the area of low concentration at the top of the cup.

The process of diffusion is shown in Figure 2-4. Note that both substances diffuse rapidly through the membrane in both directions. However, as indicated by the red arrows, more sodium chloride moves out of the 20% solution, where the concentration is higher, into the 10% solution, where the concentration is lower, than in the opposite direction. This is an example of movement down a concentration gradient. Simultaneously, more water moves from the 10% solution, where there are more water molecules, into the 20% solution, where there are fewer water molecules. This is also an example of movement down a concentration gradient. Water moves

TONICITY

A salt (NaCl) solution is said to be **isotonic** (*iso* = equal) if it contains the same concentration of salt normally found in a living red blood cell. A 0.9% NaCl solution is isotonic; that is, it contains the same level of NaCl as found in red cells. A solution that contains a higher level of salt (above 0.9%) is said to be **hypertonic** (*hyper* = above) and one containing less (below 0.9%) is **hypotonic** (*hypo* = below). With what you now know about filtration, diffusion, and osmosis, can you predict what would occur if red blood cells were placed in isotonic, hypotonic, and hypertonic solutions?

Examine the figure. Note that red blood cells placed in isotonic solution remain unchanged because there is no effective difference in salt or water concentrations. The movement of water into and out of the cells is about equal. This is not the case with red

cells placed in hypertonic salt solution; they immediately lose water from their cytoplasm into the surrounding salty solution, and they shrink. This process is called **crenation.**

The opposite occurs if red cells are placed in a hypotonic solution; they swell as water enters the cell from the surrounding dilute solution. Eventually the cells break or **lyse,** and the hemoglobin they contain is released into the solution.

Hypotonic solution (cells lyse)   Isotonic solution   Hypertonic solution (cells crenate)

from high to low concentration. The result? Equilibration of the concentrations of the two solutions after an interval of time. From then on, equal amounts of salt will diffuse in both directions, as will equal amounts of water.

Osmosis and dialysis. Osmosis (os-MO-sis) and **dialysis** (dye-AL-i-sis) are specialized examples of diffusion. In both cases, diffusion occurs across a **selectively permeable membrane.** The plasma membrane of a cell is said to be selectively permeable because it permits the passage of certain substances but not others. This is a necessary property if the cell is to permit some substances, such as nutrients, to gain entrance to the cell while excluding others. Osmosis is the diffusion of *water* across a selectively permeable membrane. However, in the case of dialysis, substances called **solutes,** which are dissolved particles in water, move across a selectively permeable membrane by diffusion.

Filtration

Filtration is the movement of water and solutes through a membrane because of a greater push-

ing force on one side of the membrane than on the other side. The force is called *hydrostatic pressure,* which is simply the force or weight of a fluid pushing against some surface. A principle about filtration that is of great physiological importance is that it always occurs *down* a hydrostatic pressure gradient. This means that, when two fluids have unequal hydrostatic pressures and are separated by a membrane, water and diffusible solutes or particles (those to which the membrane is permeable) filter out of the solution that has the higher hydrostatic pressure into the solution that has the lower hydrostatic pressure. Filtration is the process responsible for urine formation in the kidney; wastes are filtered out of the blood into the kidney tubules because of a difference in hydrostatic pressure.

ACTIVE TRANSPORT PROCESSES

Active transport is the uphill movement of a substance through a living cell membrane. *Uphill* means "up a concentration gradient" (that is, from a lower to a higher concentration). The energy required for moving a substance up its concentration gradient is obtained from ATP. Because the formation and breakdown of ATP requires complex cellular activity, active transport mechanisms can take place only through living membranes.

Permease System

A specialized cellular component called the **permease** (PER-mee-ase) **system** makes possible a number of active transport mechanisms. Permease is a protein complex in the cell membrane that uses energy from ATP to actively move many substances across cell membranes *against* their concentration gradients. Many permease-driven transport systems are called **pumps,** an appropriate term because it suggests that active transport moves a substance in an uphill direction just as a water pump, for example, moves water uphill. Permeases are very specific, and different permease pumps are required to move different substances. Many permease pumps are "coupled" to one another so that two or more different substances, such as sodium and potassium ions, may be simultaneously moved in opposite directions through the cell membrane (sodium-potassium pump). The sodium pump moves sodium through the plasma membrane from the inside of the cell, where sodium concentration is low, to the outside of the cell, where sodium concentration is high. Such a pump is required to remove sodium from the inside of a nerve cell after it has

rushed in as a result of the passage of a nerve impulse. In addition to pumping sodium and potassium ions, other specific permease pumps transport certain sugars and important amino acids into cells.

Phagocytosis and Pinocytosis

Phagocytosis (fag-o-sye-TOE-sis) is another example of how a cell can use its active transport mechanism to move an object or substance through the plasma membrane and into the cytoplasm. The term *phagocytosis* comes from a Greek word meaning "to eat." The word is appropriate because the process permits a cell to engulf and literally "eat" foreign material (Figure 2-5). Certain white blood cells destroy bacteria in the body by phagocytosis. During this process the cell membrane forms a pocket around the material to be moved into the cell and, by expenditure of energy from ATP, the object is moved to the interior of the cell. Once inside the cytoplasm, the bacterium (shown in Figure 2-5) fuses with a lysosome and is destroyed. **Pinocytosis** (pin-o-sye-TOE-sis) is an active transport mechanism used to incorporate fluids or dissolved substances into cells. Again, the term is appropriate because it comes from the Greek word meaning "drink."

CELL TRANSPORT AND DISEASE

Considering the importance of active and passive transport processes to cell survival, you can imag-

---

FIGURE 2-5 Phagocytosis of Bacteria by a White Blood Cell. Phagocytosis is an active transport mechanism that requires expenditure of energy. Note how an extension of cytoplasm envelops the bacteria, which are drawn through the cell membrane and into the cytoplasm.

ine the problems that arise when one of these processes fails. Several very severe diseases result from damage to cell transport processes. **Cystic fibrosis (CF),** for example, is an inherited condition in which chloride ion (Cl^-) pumps in the plasma membrane are missing. Because chloride ion transport is altered, sweat, pancreatic, and mucous glands produce secretions that are very salty—and often thick. Abnormally thick mucus in the lungs impairs normal breathing and often leads to recurring lung infections. Figure 2-6 shows a child with CF next to a normal child of the same age. Because of the breathing, digestive, and other problems caused by the disease, the affected child has not developed normally. Thick pancreatic secretions may plug the duct leading from the pancreas and prevent important digestive juices from flowing into the intestines. Recent evidence suggests that **Duchenne muscular dystrophy (DMD),** another inherited lethal muscle disease, results from "leaky" membranes in muscle cells. Calcium (Ca^{++}) enters affected muscle cells through the leaky membranes and triggers chemical reactions that destroy the muscle, causing life-threatening paralysis (see Chapter 7).

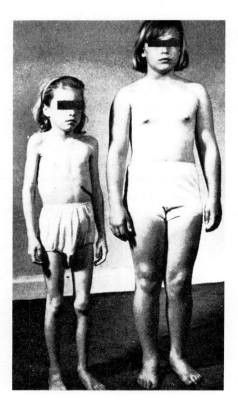

FIGURE 2-6 Cystic Fibrosis. Even though both children are the same age, the child with cystic fibrosis *(left)* is smaller and thinner than the normal child *(right).* In cystic fibrosis, the absence of chloride ion pumps causes thickening of some glandular secretions. Because thickened secretions block airways and digestive ducts, children born with this disease become weakened—often dying before adulthood.

Cell Reproduction and Heredity

All human cells that reproduce do so by a process called **mitosis** (my-TOE-sis). During this process a cell divides to multiply; one cell divides to form two cells. Cell reproduction and ultimately the transfer of heritable traits is closely tied to the production of proteins. Two *nucleic acids,* **ribonucleic acid** or **RNA** in the cytoplasm and **deoxyribonucleic acid** or **DNA** in the nucleus play crucial roles in protein synthesis.

DNA MOLECULE AND GENETIC INFORMATION

Chromosomes, which are composed largely of DNA, make heredity possible. The "genetic information" contained in DNA molecules ultimately determines the transmission and expression of heritable traits such as skin color and blood group from each generation of parents to their children.

Structurally, the DNA molecule resembles a long, narrow ladder made of a pliable material. It is twisted round and round its axis taking on the shape of a double helix. Each DNA molecule is made of many smaller units, namely, a sugar, bases, and phosphate units. The bases are adenine, thymine, guanine, and cytosine. As you can see in Figure 2-7, *A,* each step in the DNA ladder consists of a pair of bases. Only two combinations of bases occur, and the same two bases invariably pair off with each other in a DNA molecule. Adenine always binds to thymine, and cytosine always binds to guanine. This characteristic of DNA structure is **complementary base pairing.**

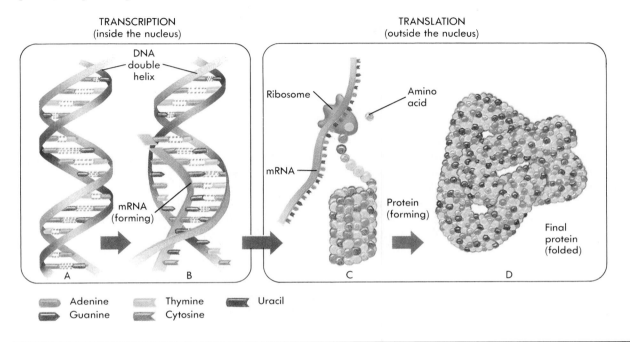

FIGURE 2-7 **Protein Synthesis. A,** The DNA molecule contains condons that represent a sequence of amino acids. **B,** During transcription, the DNA code is "transcribed" as an mRNA molecule forms. **C,** During translation, the mRNA code is "translated" at the ribosome and the proper sequence of amino acids is assembled. The amino acid strand coils or folds as it is formed. **D,** The coiled amino acid strand folds again to form a protein molecule with a specific, complex shape.

Another important fact about DNA structure is that the sequence of its base pairs is not the same in all DNA molecules, although the base pairs are the same. This fact has tremendous functional importance because it is the sequence of base pairs that determines heredity. In fact, that is what a *gene* is: a specific sequence of base pairs in a chromosome. Each gene directs the synthesis of one kind of protein molecule that may function, for example, as an enzyme, a structural component of a cell or a specific hormone. In humans having 46 chromosomes in each body cell, the nuclear DNA represents a total genetic information package or **genome** (JEE-nohm) of over 3 *billion* base pairs separated into the 23 pairs of chromosomes received from each parent. Is it any wonder, then, with such vast amounts of genetic information in each of our cells, that no two of us inherit exactly the same traits?

Genetic Code

How do genes bring about heredity? There is, of course, no short and easy answer to that question. We know that the genetic information contained in each gene is capable of "directing" the synthesis of a specific protein. The unique sequence of a thousand or so base pairs in a gene determines the sequence of specific building blocks required to form a particular protein. This store of information in each gene is called the *genetic code*. In summary, the coded information in genes control protein and enzyme production, enzymes facilitate cellular chemical reactions, and cellular chemical reactions determine cell structure and function and therefore heredity.

RNA Molecules and Protein Synthesis

DNA with its coded genetic information is contained in the nucleus of the cell. Protein synthesis, however, occurs in ribosomes and on ER. Another specialized nucleic acid, ribonucleic acid or RNA, transfers this genetic information from the nucleus to the cytoplasm.

Both RNA and DNA are composed of four bases, a sugar, and phosphate. RNA, however, is a single- rather than a double-stranded molecule, and it contains a different sugar and base component. The base uracil replaces thymine.

The process of transferring genetic information from the nucleus into the cytoplasm where proteins are actually produced requires completion of two specialized steps called **transcription** and **translation.**

Transcription. During *transcription* the double-stranded DNA molecule separates or unwinds, and a special type of RNA called **messenger RNA** or **mRNA** is formed (Figure 2-7, *B*). Each strand of mRNA is a duplicate or copy of a particular gene sequence along one of the newly separated DNA spirals. The messenger RNA is said to have been "transcribed" from its DNA mold or template. The mRNA molecules pass from the nucleus to the cytoplasm to direct protein synthesis in the ribosomes and ER.

Translation. *Translation* is the synthesis of a protein by ribosomes, which use the information contained in an mRNA molecule to direct the choice and sequencing of the appropriate chemical building blocks (Figure 2-7, *C*).

CELLS, GENETICS, AND DISEASE

Many other diseases have a cellular basis; that is, they are basically cell problems even though they may affect the entire body. Because individual cells are members of an interacting "community" of cells, it is no wonder that a problem in just a few cells can have a "ripple effect" that influences the entire body. Most of these cell problems can be traced to abnormalities in the DNA itself or in the process by which DNA information is transcribed and translated into proteins.

In individuals with inherited diseases, abnormal DNA from one or both parents may cause production of dysfunctional proteins in certain cells or prevent a vital protein from being synthesized. For example, DNA may contain a mistake in its genetic code that prevents production of normal blood-clotting proteins. Absence of these essential proteins results in excessive, uncontrol-

lable bleeding—a condition called *hemophilia* (see Chapter 11). Chemical or mechanical irritants, radiation, bacteria, viruses, and other factors can directly damage DNA molecules and thus disrupt a cell's normal function. For example, the virus that causes *acquired immune deficiency syndrome (AIDS)* eventually inserts its own genetic codes into the DNA of certain cells. The viral codes trigger synthesis of viral molecules, detouring raw materials intended for use in building normal human products. This does two things: it prevents human white blood cells from performing their normal functions and it provides a mechanism by which the virus can reproduce itself and spread to other cells. When enough cells of the human immune system are affected, they can no longer protect us from infections and cancer—a condition that eventually leads to death.

The genetic basis for disease discussed briefly in Chapter 4 is fully explained in Chapter 23.

DNA Replication

DNA molecules possess a unique ability that no other molecules in the world have. They can make copies of themselves, a process called **DNA replication.** Before a cell divides to form two new cells, each DNA molecule in its nucleus forms another DNA molecule just like itself. When a DNA molecule is not replicating, it has the shape of a tightly coiled double helix. As it begins replication, short segments of the DNA molecule uncoil and the two strands of the molecule pull apart between their base pairs. The separated strands therefore contain unpaired bases. Each unpaired base in each of the two separated strands attracts its complementary base (in the nucleoplasm) and binds to it. Specifically, each adenine attracts and binds to a thymine, and each cytosine attracts and binds to a guanine. These steps are repeated over and over throughout the length of the DNA molecule. Thus each half of a DNA molecule becomes a whole DNA molecule identical to the original DNA molecule.

STAGES OF CELL DIVISION

A cell is ready to reproduce itself after the DNA molecules have duplicated themselves. The process of cell division involves the division of the nucleus (mitosis) and the cytoplasm. After the process is complete, two daughter cells result; both have the same genetic material as the cell that preceded them. As you can see in Figure 2-8, the specific and visible stages of cell division are preceded by a period called **interphase** (IN-ter-faze). During interphase the cell is said to be "resting." However, it is resting only from the standpoint of reproductive activity. In all other aspects it is exceedingly active. During interphase and just before mitosis begins, the DNA of each chromosome replicates itself.

The stages of mitosis are listed in Table 2-2 with a brief description of the changes that occur during each stage.

Prophase

Look at Figure 2-8 and note the changes that identify the first stage of mitosis, prophase (PRO-faze). The chromatin becomes "organized." Chromosomes in the nucleus have formed two strands called **chromatids** (KRO-mah-tids). Note that the two chromatids are held together by a beadlike structure called the **centromere** (SEN-tro-meer). In the cytoplasm the centrioles are moving away from each other as a network of tubules called **spindle fibers** forms between them. These spindle fibers serve as "guidewires" and assist the chromosomes to move toward opposite ends of the cell later in mitosis.

FIGURE 2-8 Mitosis. For simplicity, only four chromosomes are shown in the diagram.

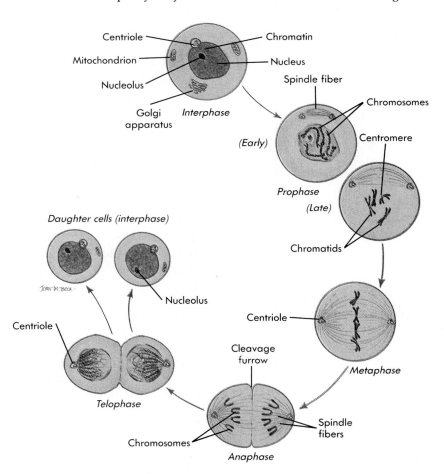

Metaphase

By the time metaphase (MET-ah-faze) begins, the nuclear membrane and nucleolus have disappeared. Note in Figure 2-8 the chromosomes have aligned themselves across the center of the cell. Also, the centrioles have migrated to opposite ends of the cell, and spindle fibers are attached to each chromatid.

Anaphase

As anaphase (AN-ah-faze) begins, the beadlike centromeres, which were holding the paired chromatids together, break apart. As a result, the individual chromatids, identified once again as chromosomes, move away from the center of the cell. Movement of chromosomes occurs along spindle fibers toward the centrioles. Note in Figure 2-8 that chromosomes are being pulled to opposite ends of the cell. A **cleavage furrow** that begins to divide the cell into two daughter cells can be seen for the first time at the end of anaphase.

Telophase

During telophase (TEL-o-faze) cell division is completed. Two nuclei appear, and chromosomes become less distinct and appear to break up. As the nuclear membrane forms around the chromatin, the cleavage furrow completely divides the cell into two parts. Before division is complete, each nucleus is surrounded by cytoplasm in which organelles have been equally distributed. By the end of telophase, two separate daughter cells, each having identical genetic characteristics, are formed. Each cell is fully functional and will perhaps itself undergo mitosis in the future.

Results of Cell Division

Mitosis results in the production of identical new cells. In the adult, mitosis replaces cells that have become less functional with age or have been damaged or destroyed by illness or injury. During early development, unspecialized cells *differentiate* to become specialized cells. In other words, primitive embryonic cells become muscle cells, nerve cells, and so on. During subsequent periods of body growth, mitosis allows groups of similar cells to develop into **tissues**.

CHANGES IN CELL GROWTH AND REPRODUCTION

Cells have the ability to adapt to changing conditions. Cells may alter their size, reproductive rate, or other characteristics to adapt to changes in the internal environment. Such adaptations usually allow cells to work more efficiently. However, sometimes cells alter their characteristics abnormally—decreasing their efficiency and threatening the health of the body. Common types of changes in cell growth and reproduction are summarized below and in Table 2-3.

Cells may respond to changes in function, hormone signals, or availability of nutrients by increasing or decreasing in size. The term **hypertrophy** (hye-PER-tro-fee) refers to an increase in cell size, and the term **atrophy** (AT-ro-fee) refers to a decrease in cell size. Either type of adaptive change can occur easily in muscle tissue. When a person continually uses muscle cells to pull against heavy resistance, as in weight training, the cells respond by increasing in size. Bodybuilders thus increase the size of their muscles by hypertrophy—increasing the size of muscle cells. Atrophy often occurs in underused muscle cells.

TABLE 2-2   Stages of Cell Division

STAGE	CHARACTERISTICS
Prophase	The chromatin condenses into visible chromosomes.
	Chromatids become attached at the centromere.
	Spindle fibers appear.
Metaphase	The nucleolus and nuclear membrane have disappeared.
	Spindle fibers attach to each chromatid.
	Chromosomes align across the center of the cell.
Anaphase	Centromeres break apart.
	Chromosomes move away from the center of the cell.
	The cleavage furrow appears.
Telophase	The nuclear membrane and both nuclei appear.
	The cytoplasm and organelles divide equally.
	The process of cell division is completed.

For example, when a broken arm is immobilized in a cast for a long period, muscles that move the arm often atrophy. Because the muscles are temporarily out of use, muscle cells decrease in size. Atrophy may also occur in tissues whose nutrient or oxygen supply is diminished.

Sometimes cells respond to changes in the internal environment by increasing their rate of reproduction—a process called **hyperplasia** (hye-per-PLAY-zha). The word *-plasia* comes from a Greek word that means "formation"—referring to formation of new cells. Because *hyper-* means "excessive," *hyperplasia* means excessive cell reproduction. Like hypertrophy, hyperplasia causes an increase in the size of a tissue or organ. However, hyperplasia is an increase in the *number of cells* rather than an increase in the size of each cell. A common example of hyperplasia occurs in the milk-producing glands of the female breast during pregnancy. In response to hormone signals, the glandular cells reproduce rapidly, preparing the breast for nursing.

If the body loses its ability to control mitosis, abnormal hyperplasia may occur. The new mass of cells thus formed is a tumor or **neoplasm** (NEE-o-plazm). Many neoplasms also exhibit a characteristic called **anaplasia** (an-ah-PLAY-zha). Anaplasia is a condition in which cells change in orientation to each other and fail to mature normally; that is, they fail to *differentiate* into a specialized cell type. Neoplasms may be relatively harmless growths called *benign* (be-NINE) tumors. If tumor cells can break away and travel through the blood or lymphatic vessels to other parts of the body (Figure 2-9), the neoplasm is a *malignant* (mah-LIG-nant) *tumor* or **cancer.** Neoplasms are discussed further in Chapter 4.

TABLE 2-3 Alterations in Cell Growth and Reproduction

TERM	DEFINITION
Changes in Growth of Individual Cells	
Hypertrophy	Increase in size of individual cells
Atrophy	Decrease in size of individual cells
Changes in Cell Reproduction	
Hyperplasia	Increase in cell reproduction
Anaplasia	Production of abnormal, undifferentiated cells

FIGURE 2-9 Cancer. This abnormal mass of proliferating cells in the lining of lung airways is a malignant tumor—lung cancer. Notice how some cancer cells are leaving the tumor and entering the blood and lymph vessels.

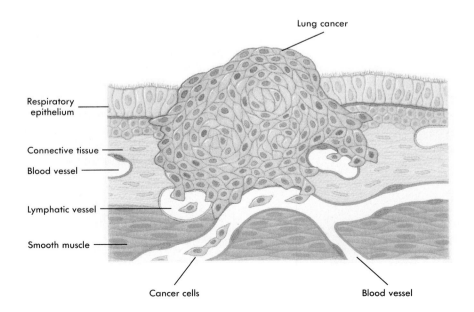

Lung cancer

Respiratory epithelium

Connective tissue

Blood vessel

Lymphatic vessel

Smooth muscle

Cancer cells

Blood vessel

Tissues

The four main kinds of tissues that compose the body's many organs follow:

1 Epithelial tissue
2 Connective tissue
3 Muscle tissue
4 Nervous tissue

Tissues differ from each other in the size and shape of their cells, in the amount and kind of material between the cells, and in the special functions they perform to help maintain the body's survival. In Table 2-4, you will find a listing of the four major tissues and the various sub-

types of each. The table also includes examples of the location of the tissues and a primary function of each tissue type.

EPITHELIAL TISSUE

Epithelial (ep-i-THEE-lee-al) **tissue** covers the body and many of its parts. It also lines various parts of the body. Because epithelial cells are packed close together with little or no intercellular material between them, they form continuous sheets that contain no blood vessels. Examine Figure 2-10. It illustrates how this large group of tissues can be subdivided according to the **shape** and **arrangement** of the cells found in each type.

FIGURE 2-10 Classification of Epithelial Tissues. The tissues are classified according to the shape and arrangement of cells.

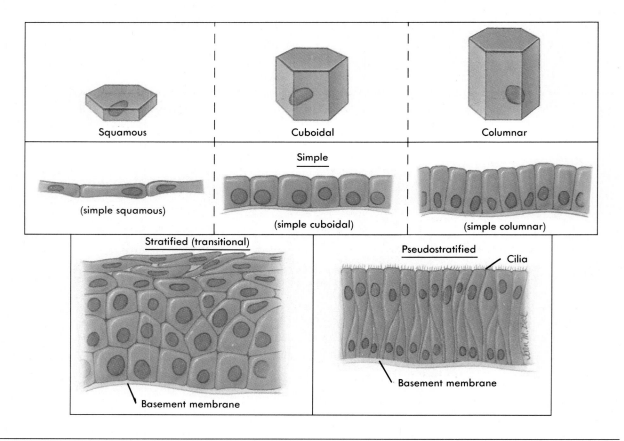

TISSUE	LOCATIONS	FUNCTIONS
Epithelial		
Simple squamous	Alveoli of lungs	Absorption by diffusion of respiratory gases between alveolar air and blood
	Lining of blood and lymphatic vessels	Absorption by diffusion, filtration, and osmosis
Stratified squamous	Surface of lining of mouth and esophagus	Protection
	Surface of skin (epidermis)	Protection
Simple columnar	Surface layer of lining of stomach, intestines, parts of respiratory tract	Protection; secretion; absorption
Stratified transitional	Urinary bladder	Protection
Pseudostratified	Surface of lining of trachea	Protection
Connective (most widely distributed of all tissues)		
Areolar	Area between other tissues and organs	Connection
Adipose (fat)	Area under skin	Protection
	Padding at various points	Insulation; support; nutrient reserve
Dense fibrous	Tendons; ligaments; fascia; scar tissue	Flexible but strong connection
Bone	Skeleton	Support; protection
Cartilage	Part of nasal septum; area covering articular surfaces of bones; larynx; rings in trachea and bronchi	Firm but flexible support
	Disks between vertebrae	Withstand pressure
	External ear	Flexible support
Blood	Blood vessels	Transportation
Hemopoietic	Red bone marrow	Blood cell formation
Muscle		
Skeletal (striated voluntary)	Muscles that attach to bones	Maintenance of posture; movement of bones
	Eyeball muscles	Eye movements
	Upper third of esophagus	First part of swallowing
Cardiac (striated involuntary)	Wall of heart	Contraction of heart
Smooth (nonstriated involuntary or visceral)	Walls of tubular viscera of digestive, respiratory, and genitourinary tracts	Movement of substances along respective tracts
	Walls of blood vessels and large lymphatic vessels	Changing of diameter of vessels
	Ducts of glands	Movement of substances along ducts
	Intrinsic eye muscles (iris and ciliary body)	Changing of diameter of pupils and shape of lens
	Arrector muscles of hairs	Erection of hairs (goose pimples)
Nervous		
	Brain; spinal cord; nerves	Irritability; conduction

Shape of Cells

If classified according to *shape,* epithelial cells are:

1 Squamous (flat and scalelike)
2 Cuboidal (cube shaped)
3 Columnar (higher than they are wide)

Arrangement of Cells

If categorized according to *arrangement* of cells, epithelial tissue can be classified as the following:

1 Simple (a single layer of cells of the same shape)
2 Stratified (many layers of cells of the same shape)
3 Transitional (several layers of cells of differing shapes)

Several types of epithelium are described in the paragraphs that follow and are illustrated in Figures 2-11 to 2-14.

Simple Squamous Epithelium

Simple squamous (SKWAY-mus) **epithelium** consists of a single layer of very thin and irregularly shaped cells. Because of its structure, substances can readily pass through simple squa-mous epithelial tissue, making absorption its special function. Absorption of oxygen into the blood, for example, takes place through the simple squamous epithelium that forms the tiny air sacs in the lungs (Figure 2-11).

Stratified Squamous Epithelium

Stratified squamous epithelium (Figure 2-12) consists of several layers of closely packed cells, an arrangement that makes this tissue a specialist at protection. For instance, stratified squamous epithelial tissue protects the body against invasion by microorganisms. Most microbes cannot work their way through a barrier of stratified squamous tissue such as that which composes the surface of skin and of mucous membranes.

One way of preventing infections, therefore, is to take good care of your skin. Don't let it become cracked from chapping, and guard against cuts and scratches.

Simple Columnar Epithelium

Simple columnar epithelium can be found lining the inner surface of the stomach, intestines, and some areas of the respiratory and reproductive tracts. Note in Figure 2-13 that the simple colum-

FIGURE 2-11 Simple Squamous Epithelium. Photomicrograph of lung tissue shows thin simple squamous epithelium lining the alveolar air sacs.

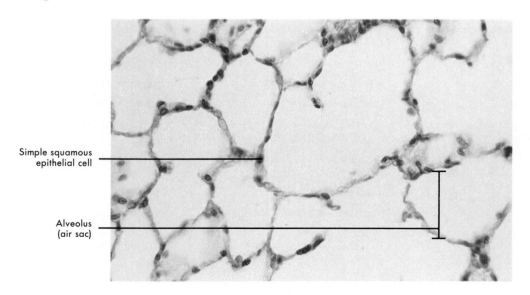

Simple squamous
epithelial cell

Alveolus
(air sac)

FIGURE 2-12 Stratified Squamous Epithelium. A, Photomicrograph. **B,** Sketch of the photomicrograph. Note the many layers of epithelial cells that have been stained yellow.

A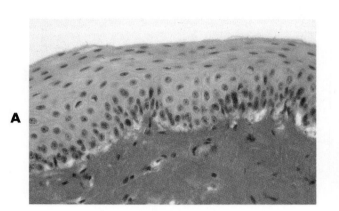

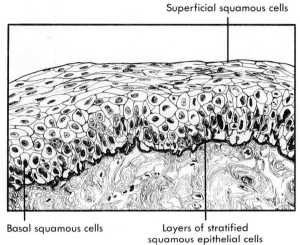
B

Superficial squamous cells

Basal squamous cells

Layers of stratified squamous epithelial cells

FIGURE 2-13 Simple Columnar Epithelium. A, Photomicrograph. **B,** Sketch of the photomicrograph. Note the goblet or mucus-producing cells that are present.

Goblet cell

A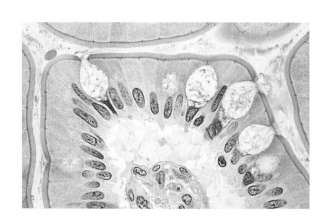

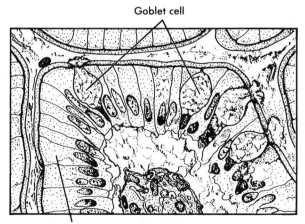
B

Columnar epithelial cell

▼   **FIGURE 2-14 Stratified Transitional Epithelium. A,** Photomicrograph of tissue lining the urinary bladder wall. **B,** Sketch of the photomicrograph. Note the many layers of epithelial cells of various shapes.

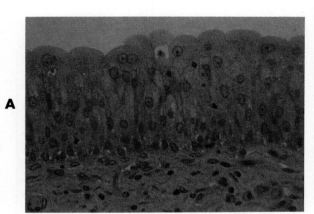

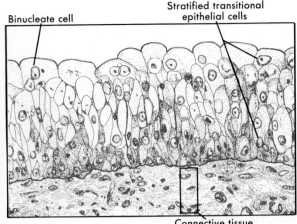

A   B

nar cells are arranged in a single layer lining the inner surface of the colon or large intestine. These epithelial cells are higher than they are wide, and the nuclei are located toward the bottom of each cell. The "open spaces" between the cells are specialized **goblet cells** that produce mucus. The regular columnar-shaped cells specialize in absorption.

Stratified Transitional Epithelium

Stratified transitional epithelium is typically found in body areas subjected to stress and must be able to stretch, such as the wall of the urinary bladder. In many instances, up to 10 layers of cuboidal-shaped cells of varying sizes are present in the absence of stretching. When stretching occurs, the epithelial sheet expands, the number of cell layers decreases, and cell shape changes from cuboidal to squamous (flat) in appearance. The fact that transitional epithelium has this ability keeps the bladder wall from tearing under the pressures of stretching. Stratified transitional epithelium is shown in Figures 2-10 and 2-14.

Pseudostratified Epithelium

Pseudostratified epithelium, illustrated in Figure 2-10 is typical of that which lines the trachea or windpipe. Look carefully at the illustration and note that each cell actually touches the basement membrane. Although the epithelium in Figure 2-10 (pseudostratified) appears to be two cell layers thick, it is not. This is the reason it is called *pseudo* (or false) stratified epithelium. The cilia that extend from the cells are capable of moving in unison. In doing so, they move mucus along the lining surface of the trachea thus affording protection against entry of dust or foreign particles into the lungs.

Glandular Epithelium

Glandular epithelium differs from the membranous and sheetlike arrangement of epithelial cells. Instead of occurring in protective coverings or linings, glandular epithelial cells, generally cuboidal in shape, are specialized for secretory activity. These specialized cells may function alone or in clusters or groups of secretory cells commonly called **glands.** Glands of the body may be classified as **exocrine** if they release their secretion through a duct or as **endocrine** if they release their secretion directly into the bloodstream. Examples of glandular secretions include saliva produced by the salivary glands, digestive juices, sweat or perspiration, and hormones such as those secreted by the pituitary or the thyroid glands.

CONNECTIVE TISSUE

Connective tissue is the most abundant and widely distributed tissue in the body. It also exists in more varied forms than any of the other tissue types. It is found in skin, membranes, muscles, bones, nerves, and all internal organs. Connective tissue exists as delicate, paper-thin webs that hold internal organs together and give them shape. It also exists as strong and tough cords, rigid bones, and even in the form of a fluid: blood.

The functions of connective tissue are as varied as its structure and appearance. It connects tissues to each other and forms a supporting framework for the body as a whole and for its individual organs. As blood, it transports substances throughout the body. Several other kinds of connective tissue function to defend us against microbes and other invaders.

Connective tissue differs from epithelial tissue in the arrangement and variety of its cells and in the amount and kinds of intercellular material, called **matrix,** found between its cells. Besides the relatively few cells embedded in the matrix of most types of connective tissue, varying numbers and kinds of fibers are also present. The structural quality and appearance of the matrix and fibers determine the qualities of each type of connective tissue. The matrix of blood, for example, is a liq-uid, but other types of connective tissue, such as cartilage, have the consistency of firm rubber. The matrix of bone is hard and rigid, although the matrix of connective tissues such as tendons and ligaments is strong and flexible.

The following list identifies major types of connective tissue in the body. Photomicrographs and sketches of several are also shown.

1 Areolar connective tissue
2 Adipose or fat tissue
3 Fibrous connective tissue
4 Bone
5 Cartilage
6 Blood
7 Hemopoietic tissue

Areolar and Adipose Connective Tissue

Areolar (ah-REE-o-lar) **connective tissue** is the most widely distributed of all connective tissue types. It is the "glue" that gives form to the internal organs. It consists of delicate webs of fibers and of a variety of cells embedded in a loose matrix of soft, sticky gel.

Adipose (AD-i-pose) or **fat tissue** is specialized to store lipids. In Figure 2-15, numerous spaces have formed in the tissue so that large quantities of fat can accumulate inside cells.

FIGURE 2-15 Adipose Tissue. A, Photomicrograph. **B,** Sketch of the photomicrograph. Note the large storage spaces for fat inside the adipose tissue cells.

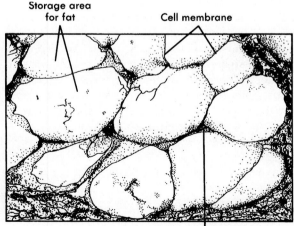

▼ **FIGURE 2-16 Dense Fibrous Connective Tissue. A,** Photomicrograph of tissue in the tendon. **B,** Sketch of the photomicrograph. Note the multiple layers of flattened collagenous fibers arranged in parallel rows.

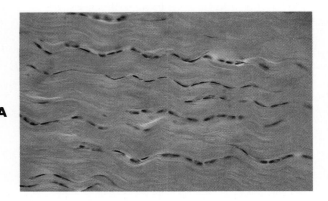

A

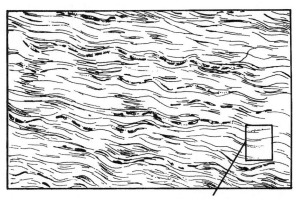

B

Collagen fibers

▼ **FIGURE 2-17 Bone Tissue. A,** Photomicrograph of dried, ground bone. **B,** Sketch of the photomicrograph. Many wheel-like structural units of bone, known as *Haversian systems,* are apparent in this section.

Haversian system

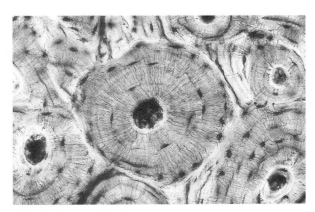

A

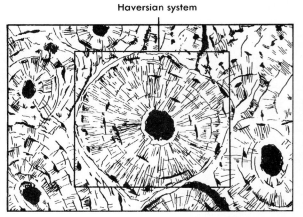

B

Fibrous Connective Tissue

Fibrous connective tissue (Figure 2-16) consists mainly of bundles of strong, white **collagen** fibers arranged in parallel rows. This type of connective tissue composes tendons. It provides great strength and nonstretchability, desirable characteristics for these structures that anchor our muscles to our bones.

Bone and Cartilage

Bone is one of the most highly specialized forms of connective tissue. The matrix of bone is hard and calcified. It forms numerous structural building blocks called **Haversian** (ha-VER-shan) **systems.** When bone is viewed under a microscope, we can see these circular arrangements of calcified matrix and cells that give bone its characteristic appearance (Figure 2-17). Bones are a storage area for calcium and provide support and protection for the body.

Cartilage differs from bone in that its matrix is the consistency of a firm plastic or gristlelike gel. Cartilage cells, which are called **chondrocytes** (KON-dro-sites), are located in many tiny spaces distributed throughout the matrix (Figure 2-18).

Blood and Hemopoietic Tissue

Because its matrix is liquid, blood is perhaps the most unusual form of connective tissue. It has transportational and protective functions in the body. Red and white blood cells are the cell types common to blood (Figure 2-19).

Hemopoietic (hee-mo-poy-ET-ik) **tissue** is the connective tissue found in the red marrow cavities of bones and in organs such as the spleen, tonsils, and lymph nodes. This specialized connective tissue is responsible for the formation of blood cells and lymphatic system cells important in our defense against disease.

MUSCLE TISSUE

There are three kinds of muscle tissue: **skeletal, cardiac,** and **smooth.** Skeletal or striated muscle is called *voluntary* because willed or voluntary control of skeletal muscle contractions is possible. Note in Figure 2-20 that, when viewed under a microscope, skeletal muscle is characterized by many cross striations and many nuclei per cell. Individual cells are long and threadlike and are often called *fibers.* Skeletal muscles are attached to bones and when contracted produce voluntary and controlled body movements.

FIGURE 2-18 Cartilage. A, Photomicrograph. **B,** Sketch of the photomicrograph. Note the chondrocytes distributed throughout the gel-like matrix.

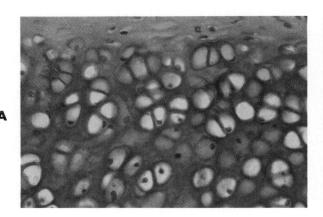
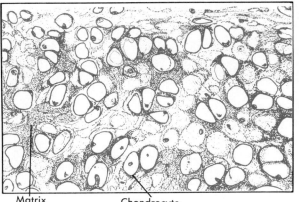

A

B

Matrix   Chondrocyte

▼ **FIGURE 2-19 Blood. A,** Photomicrograph of a human blood smear. **B,** Sketch of the photomicrograph. This smear shows two white blood cells surrounded by a number of smaller red blood cells.

A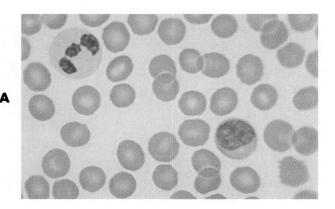

B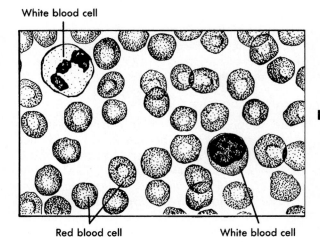

▼ **FIGURE 2-20 Skeletal Muscle. A,** Photomicrograph. **B,** Sketch of the photomicrograph. Note the striations of the muscle cell fibers in longitudinal section.

A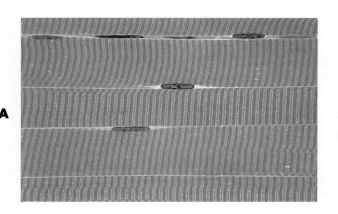

B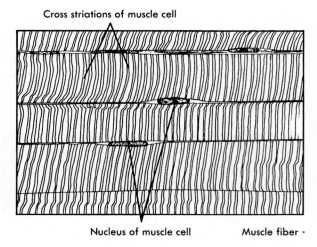

Cardiac muscle forms the walls of the heart, and the regular but involuntary contractions of cardiac muscle produce the heartbeat. Under the light microscope (Figure 2-21), cardiac muscle fibers have cross striations (like skeletal muscle) and unique dark bands called *intercalated disks.*

Cardiac muscle fibers branch and reform to produce an interlocking mass of contractile tissue.

Smooth (visceral) muscle is said to be involuntary because it is not under conscious or willful control. Under a microscope (Figure 2-22), smooth muscle cells are seen as long, narrow

▽ **FIGURE 2-21 Cardiac Muscle. A,** Photomicrograph. **B,** Sketch of the photomicrograph. The dark bands, called *intercalated disks*, which are characteristic of cardiac muscle, are easily identified in this tissue section.

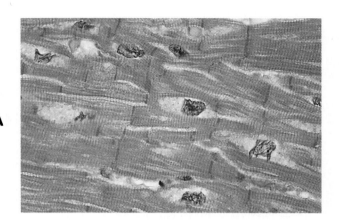

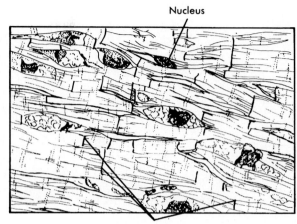

A

B

Nucleus

Intercalated disk

▽ **FIGURE 2-22 Smooth Muscle. A,** Photomicrograph, longitudinal section. **B,** Sketch of the photomicrograph. Note the central placement of nuclei in the spindle-shaped smooth muscle fibers.

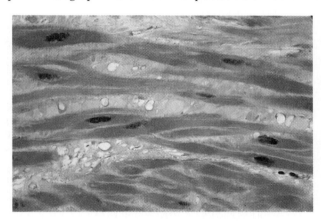

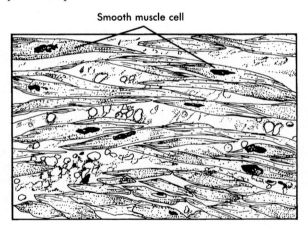

A

B

Smooth muscle cell

fibers but not nearly as long as skeletal or striated fibers. Individual smooth muscle cells appear smooth (that is, without cross striations) and have only one nucleus per fiber. Smooth muscle helps form the walls of blood vessels and hollow organs such as the intestines and other tube-shaped structures in the body. Contractions of smooth (visceral) muscle propel food material through the digestive tract and help regulate the diameter of blood vessels. Contraction of smooth muscle in the tubes of the respiratory system such as the bronchioles in the lungs can impair breathing

FIGURE 2-23 **Nervous Tissue. A,** Photomicrograph of neurons in a smear of the spinal cord. **B,** Sketch of the photomicrograph. Both neurons in this slide show characteristic cell bodies and multiple cell processes.

A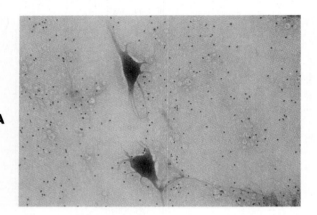

B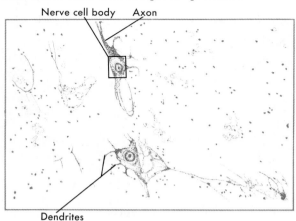

and result in asthma attacks and labored respiration.

Muscle cells are the movement specialists of the body. They have a higher degree of contractility (ability to shorten or contract) than any other tissue cells. Unfortunately, injured muscle cells are often slow to heal and are frequently replaced by scar tissue if injured.

NERVOUS TISSUE

The basic function of **nervous tissue** is rapid communication between body structures and control of body functions. Actual nervous tissue consists of nerve cells or **neurons** (NOO-rons), which are the functional or conducting units of the system, and special connecting and supporting cells called **neuroglia** (noo-ROG-lee-ah).

All neurons are characterized by a **cell body** and two types of processes: one **axon,** which transmits a nerve impulse away from the cell body, and one or more **dendrites** (DEN-drites), which carry impulses toward the cell body. Both neurons in Figure 2-23 have many dendrites extending from the cell body.

TISSUE REPAIR

When damaged by mechanical or other injuries, tissues have varying capacity to repair themselves. Damaged tissue will regenerate or be replaced by tissue we know as scars. Tissues usually repair themselves by allowing the phagocytic cells to remove dead or injured cells, then filling in the gaps that are left. This growth of new tissue is called **regeneration.**

Epithelial and connective tissues have the greatest capacity to regenerate. When a break in an epithelial membrane occurs, as in a cut, cells quickly divide to form daughter cells that fill the wound. In connective tissues, cells that form collagen fibers become active after an injury and fill in a gap with an unusually dense mass of fibrous connective tissue. If this dense mass of fibrous tissue is small, it may be replaced by normal tissue later. If the mass is deep or large, or if cell damage was extensive, it may remain as dense fibrous mass called a **scar.** An unusually thick scar that develops in the lower layer of the skin, such as that shown in Figure 2-24, is called a **keloid** (KEE-loyd).

TISSUES AND FITNESS

Achieving and maintaining an ideal body weight is a health-conscious goal. However, a better indicator of health and fitness is **body composition.** Exercise physiologists make an assessment of body composition to identify the percentage of the body made of lean tissue and the percentage made of fat. A person with a low body weight may still have a high ratio of fat to muscle, an unhealthy condition. In such a case the individual is "underweight" but "overfat." In other words, fitness depends more on the percentage and ratio of specific tissue types than the overall amount of tissue present.

Therefore one goal of a good fitness program is a desirable body-fat percentage. For men, the ideal is 15% to 18%, and for women, the ideal is 20% to 22%. Because fat contains stored energy (measured in calories), a low-fat percentage means a low-energy reserve. High body-fat percentages are associated with several life-threatening conditions, including cardiovascular disease. A balanced diet and an exercise program ensure that the ratio of fat to muscle tissue stays at a level appropriate for maintaining homeostasis.

FIGURE 2-24 Keloid. Keloids are thick scars that form in the lower layer of the skin in predisposed individuals. This photograph shows keloids that formed at suture marks after surgery.

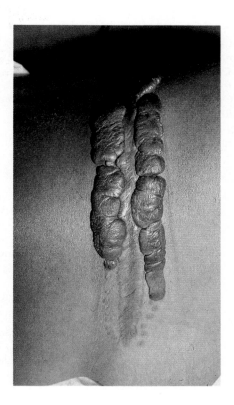

Muscle tissue, on the other hand, has a very limited capacity to regenerate and thus heal itself. Damaged muscle is often replaced with fibrous connective tissue instead of muscle tissue. When this happens, the organ involved loses some or all of its ability to function.

Like muscle tissue, nerve tissue also has a limited capacity to regenerate. Neurons outside the brain and spinal cord can sometimes regenerate but very slowly and only if certain neuroglia are present to "pave the way." In the normal adult brain and spinal cord, neurons do not grow back when injured. Thus brain and spinal cord injuries nearly always result in permanent damage. Fortunately, the discovery of *nerve growth factors* produced by neuroglia offers the promise of treating brain damage by stimulating release of these factors.

OUTLINE SUMMARY

CELLS

A Size and shape
  1 Human cells vary considerably in size
  2 All are microscopic
  3 Cells differ notably in shape
B Composition
  1 Cytoplasm containing specialized organelles surrounded by a plasma membrane
  2 Organization of cytoplasmic substances important for life
C Structural parts
  1 Plasma membrane (Figure 2-1)
    a Forms outer boundary of cell
    b Thin, two-layered membrane of phospholipids containing proteins
    c Is selectively permeable
  2 Cytoplasm (Figure 2-2)
    a Organelles
      (1) Endoplasmic reticulum (ER)
        (a) Network of connecting sacs and canals
        (b) Carry substances through cytoplasm
        (c) Types are rough and smooth
        (d) Smooth ER handles fatty substances
        (e) Rough ER handles proteins
      (2) Ribosomes
        (a) May attach to rough ER or lie free in cytoplasm
        (b) Manufacture proteins
        (c) Often called *protein factories*
      (3) Mitochondria
        (a) Composed of inner and outer membranes
        (b) Involved with energy-releasing chemical reactions
        (c) Often called *power plants* of the cell
      (4) Lysosomes
        (a) Membranous-walled organelles
        (b) Contain digestive enzymes
        (c) Have protective function (eat microbes)
        (d) Often called *suicide bags*
      (5) Golgi apparatus
        (a) Collection of small sacs near nucleus
        (b) Manufacture carbohydrate secretion products
        (c) Called *carbohydrate-producing and -packaging factory*
      (6) Centrioles
        (a) Paired organelles
        (b) Lie at right angles to each other near nucleus
        (c) Function in cell reproduction.

      (7) Cilia
        (a) Fine, hairlike extensions found on free or exposed surfaces of some cells
        (b) Capable of moving in unison in a wavelike fashion
      (8) Flagella
        (a) Single projections extending from cell surfaces
        (b) Much larger than cilia
        (c) "Tails" of sperm cells only example of flagella in humans
  3 Nucleus
    a Controls cytoplasmic organelles
    b Controls all steps in cell reproduction
    c Component structures include nuclear membrane, nucleoplasm, nucleolus, and chromatin granules
    d Chromosomes contain DNA
D Cell functions
  1 Regulation of life processes
  2 Survival of species through reproduction of the individual
  3 Relationship of structure to function apparent in number and type of organelles seen in different cells
    a Heart muscle cells contain many mitochondria required to produce adequate energy needed for continued contractions
    b Flagellum of sperm cell gives motility, allowing movement of sperm through female reproductive tract, thus increasing chances for fertilization (Figure 2-3)

MOVEMENTS OF SUBSTANCES THROUGH CELL MEMBRANES

A Passive transport processes—do not require added energy and result in movement "down a concentration gradient"
  1 Diffusion (Figure 2-4)
    a Substances scatter themselves evenly throughout an available space
    b It is unnecessary to add energy to the system
    c Movement is from high to low concentration
    d Osmosis and dialysis are specialized examples of diffusion across a selectively permeable membrane
    e Osmosis is diffusion of water
    f Dialysis is diffusion of solutes

2 Filtration
  a Movement of water and solutes caused by hydrostatic pressure on one side of membrane
  b Responsible for urine formation
B Active transport processes—occur only in living cells; movement of substances is "up the concentration gradient"; requires energy from ATP
  1 Permease systems
    a Permease is protein complex in cell membrane
    b Permease systems use energy from ATP to move substances across cell membranes against their concentration gradients
    c Permease-driven transport systems often called *pumps;* examples include sodium-potassium pump and movement of sugars and amino acids
  2 Phagocytosis and pinocytosis
    a Both are active transport mechanisms
    b Phagocytosis is a protective mechanism often used to destroy bacteria (Figure 2-5)
    c Pinocytosis is used to incorporate fluids or dissolved substances into cells
C Several severe diseases result from damage to cell transport processes
  1 Cystic fibrosis, characterized by abnormally thick secretions in the airways and digestive ducts, results from improper Cl^- transport.
  2 Duchenne muscular dystrophy, a fatal disease characterized by progressive paralysis, results from improper Ca^{++} transport

CELL REPRODUCTION

A DNA structure—large molecule shaped like a spiral staircase; sugar (deoxyribose), and phosphate units compose sides of the molecule; base pairs (adenine-thymine or guanine-cytosine) compose "steps"; base pairs always the same but sequence of base pairs differs in different DNA molecules; a gene is a specific sequence of base pairs; genes dictate formation of enzymes and other proteins by ribosomes, thereby indirectly determining a cell's structure and functions; in short, genes are heredity determinants (Figure 2-6)
B Genetic code
  1 Genetic information—stored in base-pair sequences on genes—expressed through protein synthesis
  2 RNA molecules and protein synthesis
    a DNA—contained in cell nucleus
    b Protein synthesis—occurs in cytoplasm, thus genetic information must pass from the nucleus to the cytoplasm
    c Process of transferring genetic information from nucleus to cytoplasm where proteins are produced requires completion of *transcription* and *translation* (Figure 2-7).

3 Transcription
    a Double-stranded DNA separates to form messenger RNA or mRNA
    b Each strand of mRNA duplicates a particular gene sequence from a segment of DNA
    c mRNA molecules pass from the nucleus to the cytoplasm where they direct protein synthesis in ribosomes and ER
  4 Translation
    a Involves synthesis of proteins in cytoplasm by ribosomes
    b Requires use of information contained in mRNA
C Abnormal DNA that is inherited, or that results from damage from viruses or other factors, is often the basis of disease
D DNA replication—process by which each half of a DNA molecule becomes a whole molecule identical to the original DNA molecule; precedes mitosis
E Mitosis—process in cell division that distributes identical chromosomes (DNA molecules) to each new cell formed when the original cell divides; enables cells to reproduce their own kind; makes heredity possible
F Stages of mitosis (Figure 2-8)
  1 Prophase—first stage
    a Chromatin granules become organized
    b Chromatids appear
    c Centrioles move away from nucleus
    d Spindle fibers appear
  2 Metaphase—second stage
    a Nuclear membrane disappears
    b Nucleus disappears
    c Chromosomes align across center of cell
    d Centrioles move to opposite ends of cell
    e Spindle fibers attach themselves to each chromatid
  3 Anaphase—third stage
    a Centromeres break apart
    b Separated chromatids now called *chromosomes*
    c Chromosomes are pulled to opposite ends of cell
    d Cleavage furrow develops
  4 Telophase—fourth stage
    a Cell division is completed
    b Nuclei appear in daughter cells
    c Nuclear membranes and nucleoli appear
    d Cytoplasm is divided
    e Daughter cells become fully functional
G Changes in cell growth and reproduction
  1 Changes in growth of individual cells
    a Hypertrophy—increase in size of individual cells, increasing size of tissue
    b Atrophy—decrease in size of individual cells, decreasing size of tissue

2 Changes in cell reproduction

    a Hyperplasia—increase in cell reproduction, increasing size of tissue

    b Anaplasia—production of abnormal, undifferentiated cells

    c Uncontrolled cell reproduction results in formation of a benign or malignant neoplasm (tumor)

TISSUES

A Epithelial tissue

  1 Covers body and lines body cavities

  2 Cells packed closely together with little matrix

  3 Classified by shape of cells (Figure 2-10)

    a Squamous

    b Cuboidal

    c Columnar

  4 Classified by arrangement of cells

    a Simple

    b Stratified

    c Transitional

  5 Simple squamous epithelium (Figure 2-11)

    a Single layer of scalelike cells

    b Absorption is function

  6 Stratified squamous epithelium (Figure 2-12)

    a Several layers of closely packed cells

    b Protection is primary function

  7 Simple columnar epithelium (Figure 2-13)

    a Columnar cells arranged in a single layer

    b Line stomach and intestines

    c Contain mucus-producing goblet cells

    d Specialized for absorption

  8 Stratified transitional epithelium (Figure 2-14)

    a Found in body areas, such as urinary bladder, that stretch

    b Up to 10 layers of cuboidal-shaped cells

  9 Pseudostratified epithelium

    a Each cell touches basement membrane

    b Lines the trachea

 10 Glandular epithelium

    a Specialized for secretory activity

    b Cuboidal cells grouped into glands

    c May secrete into ducts, directly into blood, and on body surface

    d Examples include saliva, digestive juice, and hormones

B Connective tissue

  1 Most abundant tissue in body

  2 Most widely distributed tissue in body

  3 Multiple types, appearances, and functions

  4 Relatively few cells in intercellular matrix

  5 Types

    a Areolar—glue that holds organs together

    b Adipose (fat)—lipid storage is primary function (Figure 2-15)

    c Fibrous—strong fibers; example is tendon (Figure 2-16)

    d Bone—matrix is calcified; function in support and protection (Figure 2-17)

    e Cartilage—chondrocyte is cell type (Figure 2-18)

    f Blood—matrix is fluid; function is transportation (Figure 2-19)

C Muscle tissue (Figures 2-20 to 2-22)

  1 Types

    a Skeletal—attaches to bones; also called *striated* or *voluntary;* control is voluntary; striations apparent when viewed under a microscope (Figure 2-20)

    b Cardiac—also called *striated involuntary;* composes heart wall; ordinarily cannot control contractions (Figure 2-21)

    c Smooth—also called *nonstriated (visceral)* or *involuntary;* no cross striations; found in blood vessels and other tube-shaped organs (Figure 2-22)

D Nervous tissue (Figure 2-23)

  1 Cell types

    a Neurons—conducting cells

    b Neuroglia—supportive and connecting cells

  2 Neurons

    a Cell components

      (1) Cell body

      (2) Axon (one) carries nerve impulse away from cell body

      (3) Dendrites (one or more) carry nerve impulse toward the cell body

  3 Function—rapid communication between body structures and control of body functions

E Tissue repair—usually accomplished by means of regeneration of tissue

  1 Epithelial and connective tissues regenerate easily

  2 Muscle and nervous tissues have very limited abilities to repair themselves

NEW WORDS

adenosine triphosphate
  (ATP)
adipose
areolar
axon
centriole
centromere
chondrocyte
chromatid
chromatin
collagen
columnar
crenation
cuboidal
cytoplasm
dendrite

deoxyribonucleic acid
  (DNA)
goblet cell
Haversian system
hemopoietic
hypertonic
hypotonic
interphase
lyse
matrix
mitosis
  prophase
  metaphase
  anaphase
  telophase

neuroglia
neuron
nucleoplasm
organelle
  cilia
  endoplasmic
    reticulum (ER)
  flagellum
  Golgi apparatus
  lysosome
  mitochondria
  nucleoli
  nucleus
  plasma membrane
  ribosome

ribonucleic acid (RNA)
spindle fiber
squamous
transcription
translation
transport
  dialysis
  diffusion
  filtration
  osmosis
  phagocytosis
  pinocytosis

Diseases and Other Clinical Terms

cancer
cystic fibrosis (CF)

Duchenne muscular
  dystrophy (DMD)

keloid
neoplasm

regeneration
scar

CHAPTER TEST

1. Which of the following groups represents the three principal parts of a cell?
   a. Plasma membrane, nucleus, cytoplasm
   b. Ribosomes, plasma membrane, mitochondria
   c. Lysosomes, centrioles, plasma membrane
   d. Nucleus, Golgi apparatus, mitochondria
2. Which of the following is considered an active transport process?
   a. Osmosis
   b. Permease system
   c. Filtration
   d. Dialysis
3. Deoxyribonucleic acid or DNA:
   a. Is exactly the same as RNA
   b. Contains the compound uracil
   c. Is sometimes called the "heredity molecule"
   d. Is not involved with genes or chromosomes

4. The stage of mitosis in which the chromosomes align themselves across the center of the cell is called:
   a. Prophase
   b. Anaphase
   c. Telophase
   d. Metaphase
5. Which of the following terms describes flat, platelike cells that permit substances to readily pass through them?
   a. Squamous
   b. Stratified
   c. Transitional
   d. Columnar
6. Haversian systems are associated with which of the following?
   a. Adipose or fat tissue
   b. Bone
   c. Fibrous connective tissue
   d. Cartilage
7. Which of the following types of muscle is considered voluntary?
   a. Cardiac
   b. Smooth
   c. Visceral
   d. Skeletal

Select the most appropriate answer in column B for each item in Column A. (Only one answer is correct.)

COLUMN A	COLUMN B
8. _____ Cellular organelles	a. Osmosis
9. _____ Example of diffusion	b. Blood
10. _____ Active transport mechanism	c. Mitochondria
11. _____ Characteristic of DNA	d. Anaphase
12. _____ Stage of mitosis	e. Complementary base pairing
13. _____ Type of connective tissue	f. Pinocytosis
14. _____ Composed of epithelial tissue	g. Dendrite
15. _____ Component of a nerve cell	h. Glands

16. The nickname, "protein factories," is often used to describe _____ .
17. Substances move "up a concentration gradient" in _____ transport processes.
18. The first stage of mitosis is called _____ .
19. Absorption, protection, and secretion are important functions of _____ tissue.
20. No blood vessels are found in _____ tissue.

REVIEW QUESTIONS

1 One inch is equal to approximately how many centimeters? How many millimeters? How many micrometers?

2 Identify the three main parts of a cell.

3 Describe the structure and functions of the plasma membrane.

4 What and where are the following? What functions do they perform?

    nucleolus

    endoplasmic reticulum

    ribosomes

    chromatin

5 Discuss the importance of the following three chemical substances: DNA, RNA, and ATP. Where in the cell would you expect to find the largest quantity of each?

6 Discuss the genetic code and its relationship to protein synthesis.

7 What is a gene? How does a gene differ from a chromosome?

8 Compare and contrast active and passive transport systems. List by name the major active and passive transport processes.

9 What actually moves during osmosis? During dialysis? What is required for both osmosis and dialysis to occur that is not required for diffusion?

10 How is ATP involved in active transport processes?

11 Name two diseases caused by improper cell transport.

12 List the stages of mitosis and briefly describe what occurs during each period.

13 When does DNA replication occur with respect to mitosis?

14 Identify the stage of mitosis that could be described as "prophase in reverse."

15 Explain how epithelial tissue can be classified according to the shape and arrangement of the cells.

16 Compare the matrix found in bone, areolar connective tissue, and blood.

17 What is the most widely distributed tissue in the body? What tissue exists in more varied forms than any other tissue type? In what type of tissue is appearance most determined by the nature of the matrix?

18 Compare and contrast the three major types of muscle tissue.

19 Identify the two basic types of cells found in nervous tissue. What is the difference between an axon and a dendrite?

20 Compare and contrast tissue repair in epithelial, connective, muscle, and nervous tissues.

CLINICAL APPLICATIONS

1 One form of the inherited condition *glycogen storage disease*—a form called Pompe's disease—results in accumulation of excessive glycogen in cells of the heart, liver, and other organs. The accumulation of glycogen can disrupt cell function, causing heart and other problems that can progress to death. Glycogen is a large carbohydrate molecule formed by linking numerous glucose molecules into a branched chain (see Appendix A). Glycogen formation is normal in the affected cells. The accumulation of excessive glycogen results from the failure of enzymes that are supposed to break apart the glycogen so that the cell can use the glucose subunits. In what organelle would you expect to find these glycogen-digesting enzymes? Can you explain how the presence of nonfunctional enzymes could have been inherited?

2 Malignant tumors are sometimes treated with drugs that halt mitosis, and thus stop the production of new cancer cells. Two such drugs, *vincristine sulfate* and *vinblastine sulfate*, interfere with the formation of spindle fibers. How could this action halt mitosis? Antibiotics such as *mitomycin C* and inorganic compounds such as *cis-platinum* can also be used to stop the growth of tumors. These drugs interfere with DNA synthesis in treated cells. How could this action halt mitosis?

3 Lauren is a 2-year-old girl with cystic fibrosis (CF). Because Lauren has this condition, her mother frequently turns her over, cups her hand, and quickly but firmly pats her sharply on the back between the shoulder blades. How could this help Lauren's condition?

Organ Systems of the Body

C H A P T E R

3

Objectives

After you have completed this chapter,
you should be able to:

1 Define and contrast the terms *organ* and *organ system*.

2 List the 11 major organ systems of the body.

3 Identify and locate the major organs of each major organ system.

4 Briefly describe the major functions of each major organ system.

5 Identify and discuss the major subdivisions of the reproductive system.

6 Describe current approaches to organ replacement.

The words *organ* and *system* were discussed in Chapter 1 as having special meanings when applied to the body. An **organ** is a structure made up of two or more kinds of tissues organized in such a way that the tissues can together perform a more complex function than can any tissue alone. A **system** is a group of organs arranged in such a way that they can together perform a more complex function than can any organ alone. This chapter gives an overview of the 11 major organ systems of the body.

In the chapters that follow, the presentation of information on individual organs and an explanation of how they work together to accomplish complex body functions will form the basis for the discussion of each organ system. For example, a detailed description of the skin as the primary organ of the integumentary system will be covered in Chapter 5, and information on the bones of the body as organs of the skeletal system will be presented in Chapter 6. A knowledge of individual organs and how they are organized into groups makes much more meaningful the understanding of how a particular organ system functions as a unit in the body.

When you have completed your study of the major organ systems in the chapters that follow, it will be possible to view the body not as an assembly of individual parts but as an integrated and functioning whole. This chapter names the systems of the body and the major organs that compose them, and it briefly describes the functions of each system. It is intended to provide a basic "road map" to help you anticipate and prepare for the more detailed information that follows in the remainder of the text.

Organ Systems of the Body

In contrast to cells, which are the smallest structural units of the body, organ systems are its largest and most complex structural units. The 11 major organ systems that compose the human body are listed below.

1 Integumentary
2 Skeletal
3 Muscular
4 Nervous
5 Endocrine
6 Circulatory
7 Lymphatic
8 Respiratory
9 Digestive
10 Urinary
11 Reproductive
   a Male subdivision
   b Female subdivision

Examine Figure 3-1 to find a diagrammatical listing of the body systems and the major organs in each. In addition to the information contained in Figure 3-1, each system is presented in visual form in Figures 3-2 through 3-13. Visual presentation of material is often useful in understanding the interrelationships that are so important in anatomy and physiology.

INTEGUMENTARY SYSTEM

Note in Figure 3-2 that the skin is the largest and most important organ in the **integumentary** (in-teg-yoo-MEN-tar-ee) **system.** Its weight in most adults is 20 pounds or more, accounting for about 16% of total body weight and making it the body's heaviest organ. The integumentary system includes the skin and its **appendages,** which include the hair, nails, and specialized sweat- and oil-producing glands. In addition, a number of microscopic and highly specialized sense organs are embedded in the skin. They permit the body to respond to pain, pressure, touch, and changes in temperature.

The integumentary system is crucial to survival. Its primary function is *protection*. The skin protects underlying tissue against invasion by harmful bacteria, bars entry of most chemicals, and minimizes the chances of mechanical injury to underlying structures. In addition, the skin regulates body temperature by sweating, synthesizes important chemicals and hormones, and functions as a sophisticated sense organ.

SKELETAL SYSTEM

The sternum or breastbone, the humerus, and the femur shown in Figure 3-3 are examples of the 206 individual organs (bones) found in the **skeletal system.** The system includes not only bones but also related tissues such as cartilage and ligaments that together provide the body with a rigid

FIGURE 3-1 Body Systems and Their Organs.

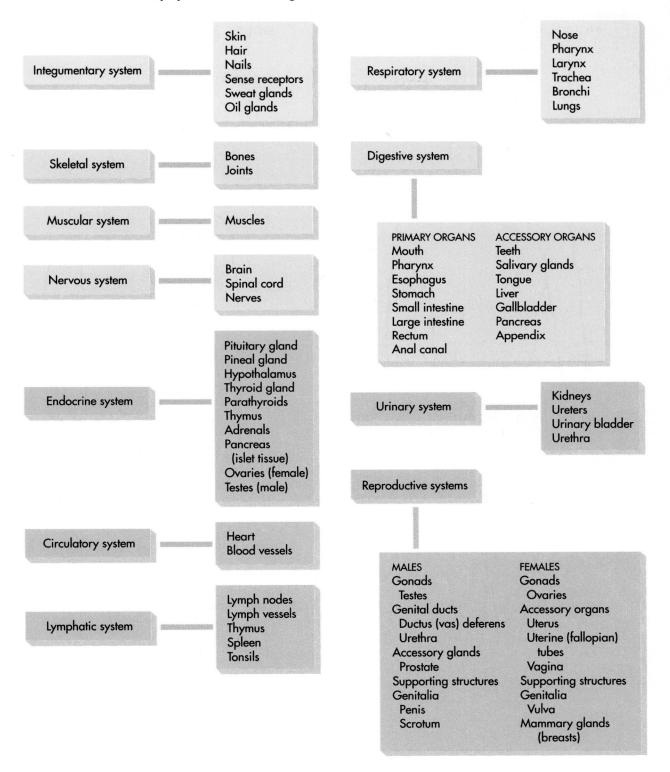

Integumentary system
- Skin
- Hair
- Nails
- Sense receptors
- Sweat glands
- Oil glands

Skeletal system
- Bones
- Joints

Muscular system
- Muscles

Nervous system
- Brain
- Spinal cord
- Nerves

Endocrine system
- Pituitary gland
- Pineal gland
- Hypothalamus
- Thyroid gland
- Parathyroids
- Thymus
- Adrenals
- Pancreas (islet tissue)
- Ovaries (female)
- Testes (male)

Circulatory system
- Heart
- Blood vessels

Lymphatic system
- Lymph nodes
- Lymph vessels
- Thymus
- Spleen
- Tonsils

Respiratory system
- Nose
- Pharynx
- Larynx
- Trachea
- Bronchi
- Lungs

Digestive system

PRIMARY ORGANS	ACCESSORY ORGANS
Mouth	Teeth
Pharynx	Salivary glands
Esophagus	Tongue
Stomach	Liver
Small intestine	Gallbladder
Large intestine	Pancreas
Rectum	Appendix
Anal canal	

Urinary system
- Kidneys
- Ureters
- Urinary bladder
- Urethra

Reproductive systems

MALES	FEMALES
Gonads	Gonads
Testes	Ovaries
Genital ducts	Accessory organs
Ductus (vas) deferens	Uterus
Urethra	Uterine (fallopian)
Accessory glands	tubes
Prostate	Vagina
Supporting structures	Supporting structures
Genitalia	Genitalia
Penis	Vulva
Scrotum	Mammary glands
	(breasts)

FIGURE 3-2 Integumentary System.

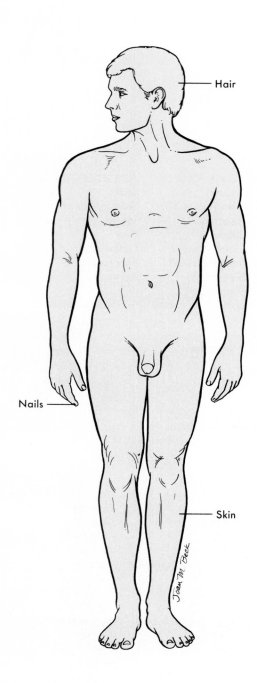

FIGURE 3-3 Skeletal System.

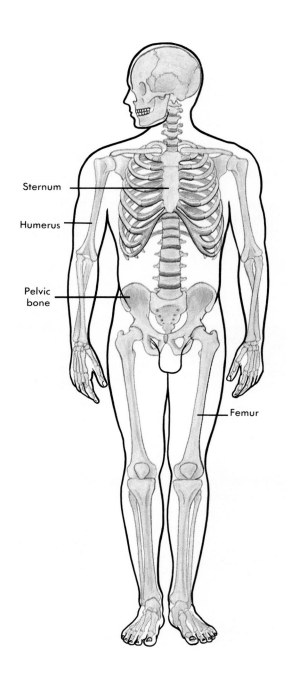

framework for support and protection. In addition, the skeletal system, through the existence of **joints** between bones, makes possible the movements of body parts. Without joints, we could make no movements; our bodies would be rigid, immobile hulks. Bones also serve as storage areas for important minerals such as calcium and phosphorus. The formation of blood cells in the red marrow of certain bones is another crucial function of the skeletal system.

MUSCULAR SYSTEM

Individual skeletal muscles are the organs of the **muscular system.** In addition to **voluntary** or **skeletal** muscles, which have the ability to contract when stimulated and are under conscious control, the muscular system also contains **smooth** or **involuntary** muscles found in organs such as the stomach and small intestine. The third type of muscle tissue is the **cardiac muscle** of the heart. Muscles not only produce movement and maintain body posture but also generate the heat required for maintaining a constant core body temperature.

The tendon labeled in Figure 3-4 represents how muscles attach to bones. When stimulated by a nervous impulse, muscle tissue shortens or contracts. Voluntary movement of the body occurs when skeletal muscles contract because of the way muscles are attached to bones and the way bones articulate or join together with one another in joints.

NERVOUS SYSTEM

The brain, spinal cord, and nerves are the organs of the **nervous system.** As you can see in Figure 3-5, nerves extend from the brain and spinal cord to every area of the body. The extensive networking of the components of the nervous system makes it possible for this complex system to perform its primary functions. These include the following:

1 Communication between body functions
2 Integration of body functions
3 Control of body functions
4 Recognition of sensory stimuli

These functions are accomplished by specialized signals called **nerve impulses.** In general, the functions of the nervous system result in rapid

FIGURE 3-4 Muscular System.

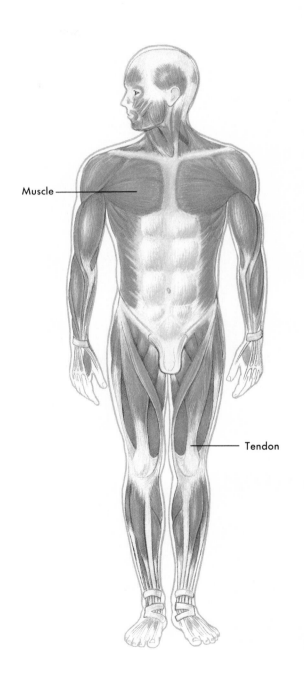

Muscle

Tendon

FIGURE 3-5 Nervous System.

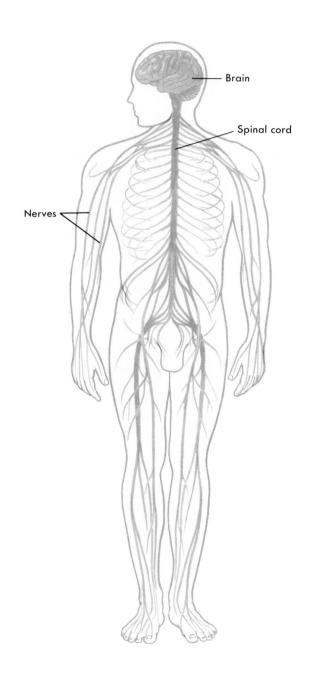

Brain

Spinal cord

Nerves

activity that lasts usually for a short duration. For example, we can chew our food normally, walk, and perform coordinated muscular movements only if our nervous system functions properly. The nerve impulse permits the rapid and precise control of diverse body functions. Other types of nerve impulses cause glands to secrete fluids. In addition, elements of the nervous system can recognize certain **stimuli** (STIM-yoo-lye), such as heat, light, pressure, or temperature, that affect the body. Nervous impulses may then be generated to convey this information to the brain, where it can be analyzed and action can be initiated.

ENDOCRINE SYSTEM

The **endocrine system** is composed of specialized glands that secrete chemicals known as **hormones** directly into the blood. Sometimes called *ductless glands,* the organs of the endocrine system perform the same general functions as the nervous system: communication, integration, and control. The nervous system provides rapid, brief control by fast-traveling nerve impulses. The endocrine system provides slower but longer-lasting control by hormone secretion; for example, secretion of growth hormone controls the rate of development over long periods of gradual growth.

In addition to controlling growth, hormones are the main regulators of metabolism, reproduction, and many other body activities. They play important roles in fluid and electrolyte balance, acid-base balance, and energy metabolism.

As you can see in Figure 3-6 the endocrine glands are widely distributed throughout the body. The **pituitary** (pi-TOO-i-TAIR-ee) **gland, pineal** (PIN-e-al) **gland,** and **hypothalamus** (hi-po-THAL-ah-mus) are located in the skull. The **thyroid** (THY-roid) and **parathyroid** (PAIR-ah-THY-roid) **glands** are in the neck, and the **thymus** (THY-mus) **gland** is in the thoracic cavity, specifically in the mediastinum (see p. 4). The **adrenal** (ah-DRE-nal) **glands** and **pancreas** (PAN-kree-as) are found in the abdominal cavity. Note in Figure 3-6 that the ovaries in the female and the testes in the male also function as endocrine glands.

CIRCULATORY SYSTEM

The **circulatory system** consists of the heart, which is a muscular pumping device as shown in Figure 3-7 and a closed system of vessels made up

FIGURE 3-6 Endocrine System.

FIGURE 3-7 Circulatory System.

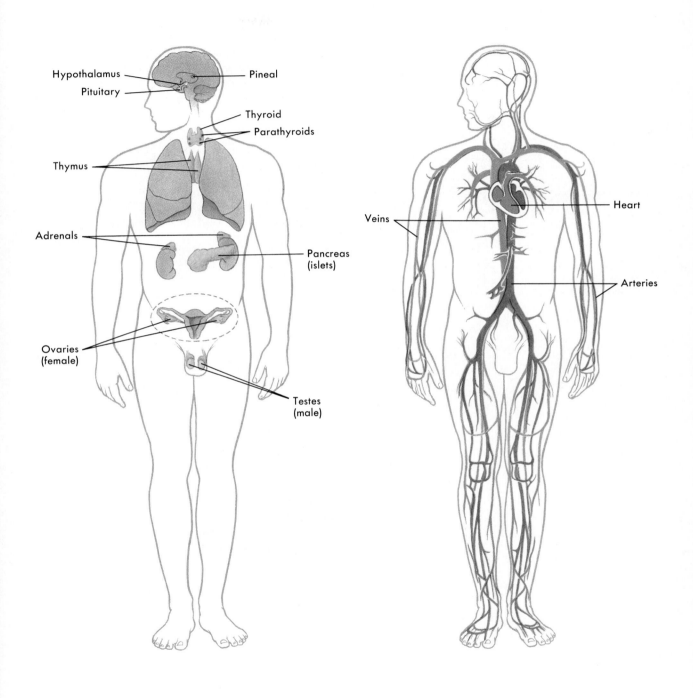

FIGURE **3-6** Endocrine System.

FIGURE **3-7** Circulatory System.

FIGURE **3-8** Lymphatic System.

FIGURE **3-9** Respiratory System.

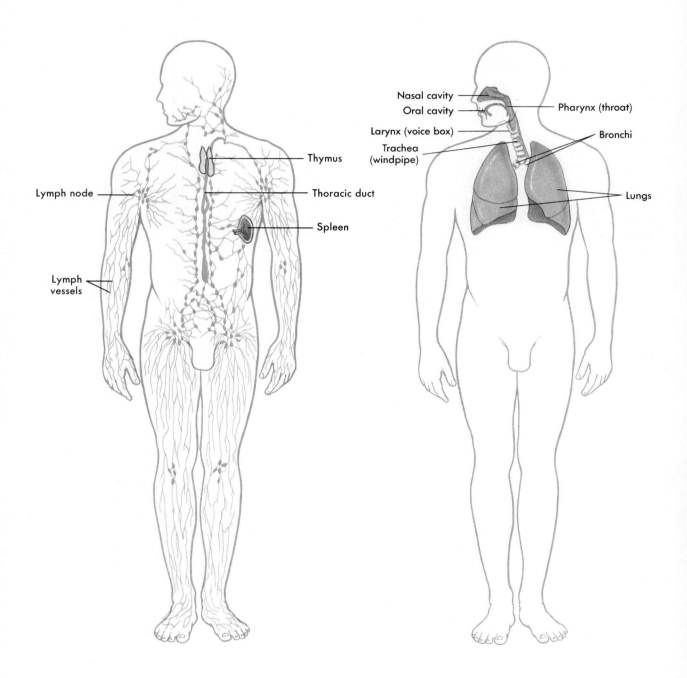

Thymus

Lymph node

Thoracic duct

Spleen

Lymph
vessels

Nasal cavity

Oral cavity

Pharynx (throat)

Larynx (voice box)

Bronchi

Trachea
(windpipe)

Lungs

of **arteries, veins,** and **capillaries.** As the name implies, blood contained in the circulatory system is pumped by the heart around a closed circle or circuit of vessels as it passes through the body. The term **cardiovascular** refers to the system of the heart and blood vessels.

The primary function of the circulatory system is *transportation*. The need for an efficient transportation system in the body is critical. Transportation needs include continuous movement of oxygen and carbon dioxide, nutrients, hormones, and other important substances. Wastes produced by the cells are released into the bloodstream on an ongoing basis and are transported by the blood to the excretory organs. The circulatory system also helps regulate body temperature by distributing heat throughout the body and by assisting in retaining or releasing heat from the body by regulating blood flow near the body surface. Certain cells of the circulatory system can also become involved in the defense of the body or immunity.

LYMPHATIC SYSTEM

The **lymphatic system** is composed of **lymph nodes, lymphatic vessels,** and specialized lymphatic organs such as the **tonsils, thymus,** and **spleen.** Note that the thymus in Figure 3-8 functions as an endocrine and as a lymphatic gland. Instead of containing blood, the lymphatic vessels are filled with a whitish, watery fluid that contains lymphocytes, proteins, and some fatty molecules. No red blood cells are present. The lymph is formed from the fluid around the body cells and diffuses into the lymph vessels. However, unlike blood, lymph does not circulate repeatedly through a closed circuit or loop of vessels. Instead, lymph flowing through lymphatic vessels eventually enters the circulatory system by passing through large ducts, including the **thoracic duct** shown in Figure 3-8, which in turn connect with veins in the upper area of the thoracic cavity. Collections of lymph nodes can be seen in the axillary (armpit) and in the inguinal (groin) areas of the body in Figure 3-8. The formation and the movement of lymph is discussed in Chapter 14.

The functions of the lymphatic system include movement of fluids and certain large molecules from the tissue spaces around the cells and movement of fat-related nutrients from the digestive tract back to the blood. The lymphatic system is also involved in the functioning of the immune system, which plays a critical role in the defense mechanism of the body against disease.

RESPIRATORY SYSTEM

The organs of the **respiratory system** include the nose, **pharynx** (FAIR-inks), **larynx** (LAR-inks), **trachea** (TRAY-kee-ah), **bronchi** (BRON-ki), and lungs (Figure 3-9). Together these organs permit the movement of air into the tiny, thin-walled sacs of the lungs called **alveoli** (al-VE-o-li). In the alveoli, oxygen from the air is exchanged for the waste product, carbon dioxide, which is carried to the lungs by the blood so that it can be eliminated from the body.

DIGESTIVE SYSTEM

The organs of the **digestive system** (Figure 3-10) are often separated into two groups: the *primary organs* and the *secondary* or *accessory organs* (Figure 3-1). They work together to ensure proper digestion and absorption of nutrients. The primary organs include the mouth, pharynx, esophagus, stomach, small intestine, large intestine, rectum, and anal canal. The accessory organs of digestion include the teeth, salivary glands, tongue, liver, gallbladder, pancreas, and appendix.

The primary organs of the digestive system form a tube, open at both ends, called the **gastrointestinal** (GAS-tro-in-TES-ti-nal) or **GI tract.** Food that enters the tract is digested, its nutrients are absorbed, and the undigested residue is eliminated from the body as waste material called **feces** (FEE-seez). The accessory organs assist in the mechanical or chemical breakdown of ingested food. The appendix, although classified as an accessory organ of digestion and physically attached to the digestive tube, is not functionally important in the digestive process. However, inflammation of the appendix, called **appendicitis** (ah-PEN-di-SYE-tis) is a very serious clinical condition and frequently requires surgery.

URINARY SYSTEM

The organs of the **urinary system** include the **kidneys, ureters** (u-REE-ters), **bladder,** and **urethra** (yoo-RE-thrah).

▼ **FIGURE 3-10** Digestive System.   ▼ **FIGURE 3-11** Urinary System.

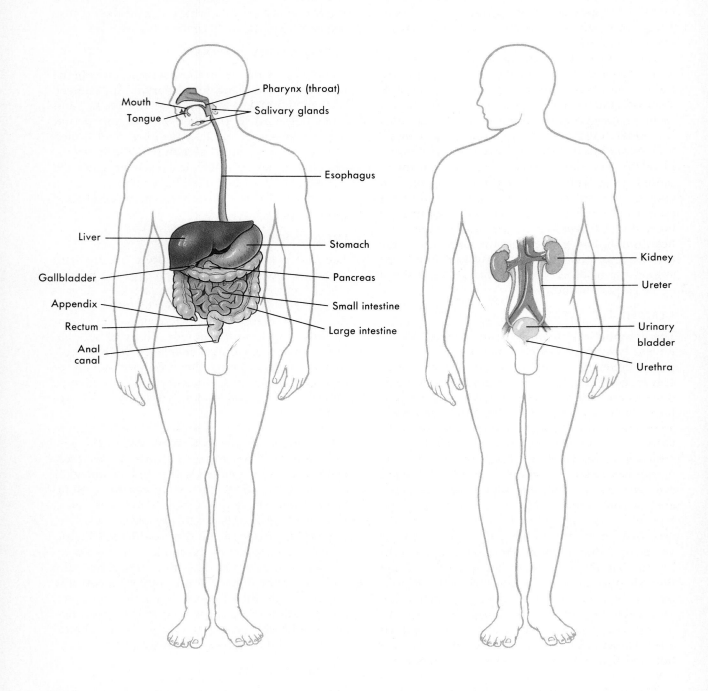

The kidneys (Figure 13-11) "clear" or clean the blood of the waste products continually produced by the metabolism of food-stuff in the body cells. The kidneys also play an important role in maintaining the electrolyte, water, and acid-base balances in the body.

The waste product produced by the kidneys is called **urine** (YOOR-in). After it is produced by the kidneys, it flows out of the kidneys through the ureters into the urinary bladder, where it is stored. Urine passes from the bladder to the outside of the body through the urethra. In the male the urethra passes through the penis, which has a double function; it transports urine and semen or seminal fluid. Therefore it has urinary and reproductive purposes. In the female the urinary and reproductive passages are completely separate, so the urethra performs only a urinary function.

Other organs besides those of the urinary system are also involved in the elimination of body wastes. Undigested food residues and metabolic wastes are eliminated from the intestinal tract as feces, and the lungs rid the body of carbon dioxide. The skin also serves an excretory function by eliminating water and some salts in sweat.

REPRODUCTIVE SYSTEM

The normal function of the **reproductive system** is different from the normal function of other organ systems of the body. The proper functioning of the reproductive systems ensures survival, not of the individual but of the species—the human race. In addition, production of the hormones that permit the development of sexual characteristics occurs as a result of normal reproductive system activity.

Male Reproductive System

The male reproductive structures shown in Figure 3-12 include the **gonads** (GO-nads) called **testes** (TES-teez), which produce the sex cells or **sperm;** one of the important **genital ducts** called the **vas deferens** (vas DEF-er-enz); and the **prostate** (PROSS-tate), which is classified as an **accessory gland** in the male. The **penis** (PEE-nis) and **scrotum** (SKRO-tum) are **supporting structures** and together are known as the **genitalia** (JEN-i-tail-yah). The urethra, which is identified in Figure 3-11 as part of the urinary system, passes through

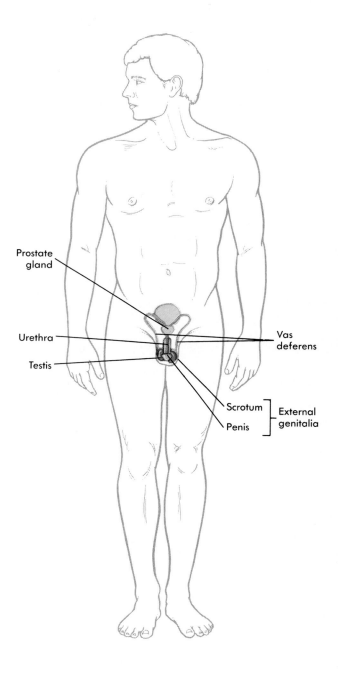

FIGURE 3-12 Male Reproductive System.

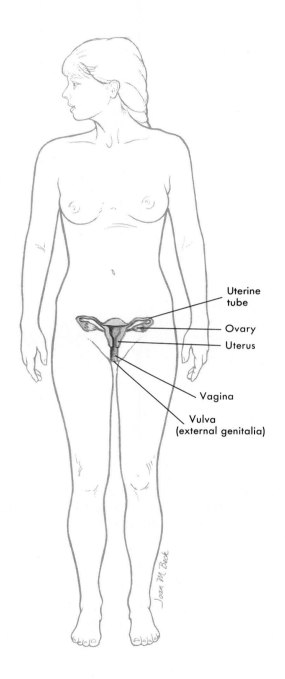

FIGURE **3-13** **Female Reproductive System.**

Uterine tube

Ovary

Uterus

Vagina

Vulva
(external genitalia)

the penis. It serves as a genital duct that carries sperm to the exterior and as a passageway for the elimination of urine. Functioning together, these structures produce, transfer, and ultimately introduce sperm into the female reproductive tract, where fertilization can occur. Sperm produced by the testes travels through a number of genital ducts, including the vas deferens, to exit from the body. The prostate and other accessory glands, which add fluid and nutrients to the sex cells as they pass through the ducts and the supporting structures (especially the penis), permit transfer of sex cells into the female reproductive tract.

Female Reproductive System

The female **gonads** are the **ovaries.** The **accessory organs** shown in Figure 3-13 include the **uterus** (YOO-ter-us), **uterine** (YOO-ter-in) or **fallopian tubes,** and the **vagina** (vah-JYE-nah). In the female the term **vulva** (VUL-vah) is used to describe the external genitalia. The breasts or **mammary glands** are also classified as external accessory sex organs in the female.

The reproductive organs in the female produce the sex cells or **ova,** receive the male sex cells (sperm), permit fertilization and transfer of the sex cells to the uterus, and allow for the development, birth, and nourishment of offspring.

As you study the more detailed structure and function of the organ systems in the chapters that follow, always relate the system and its component organs to the body as a whole. No one body system functions entirely independently of other systems. Instead, you will find that they are structurally and functionally interrelated and interdependent.

Organ Replacement

As we all know, disease and injury are sometimes unavoidable. It is common therefore to suffer damage that renders an organ incapable of proper function. By definition, a nonvital organ is not required for life to continue—a vital organ is. If a *nonvital organ* is damaged, our health may be in some peril, but even permanent loss of that organ's function will not result in death. For example, we can survive easily without the use of our spleen, appendix, and tonsils. We can also survive, although less easily, without the use of our eyes, arms, and legs. However, if the functions of a *vital organ* are lost, we are in immediate danger of dying. For example, when the heart or brain cease to function, death results.

Over the past few decades, health science professionals have made great advances in the ability to replace lost or damaged organs. In the case of nonvital organs, these techniques have improved the quality of life for many patients. In the case of vital organs, these techniques have extended life.

ARTIFICIAL ORGANS

A nonvital organ, especially, is often successfully replaced or enhanced by an artificial organ or **prosthesis** (pros-THEE-sis). Figure 3-14 shows examples of many types of prostheses now in use or in development. Crude artificial limbs have been used for centuries, but the availability of new materials and advanced engineering has made more-efficient types possible. For example, new computer-assisted arm and hand replacements can manipulate small objects with amazing dexterity. Artificial sense organs have even been able to restore sight to the blind and hearing to the deaf. For example, many people suffering from deafness have had their hearing partially restored by "artificial ears" called *cochlear* (KO-klee-er) *implants.* In cochlear implants, a miniature microphone is surgically implanted under the skin near the outer ear and wired to an electrode in the inner ear or *cochlea.* Sound picked up by the microphone is converted to electrical signals that are relayed directly to the auditory nerve in the cochlea.

Unfortunately, the use of artificial organs to replace vital organs has been less successful. One of the earliest devices to augment vital functions was the "artificial kidney" or *dialysis machine* (see Chapter 18). Dialysis machines pump blood through permeable tubes in an external apparatus, allowing waste products to diffuse out of the blood and into a saltwater bath. Patients with kidney failure must be "hooked up" to the machine for many hours at a time. Dialysis machines extend the lives of kidney failure victims for long periods, perhaps years. However, because they cannot perform *all* the functions of the kidney, they are considered only short-term remedies. Artificial hearts have briefly extended the lives of

PAIRED ORGANS

Have you ever wondered what advantage there might be in having two kidneys, two lungs, two eyes, and two of many other organs? Although the body could function well with only one of each, most of us are born with a pair of these organs. For paired organs that are vital to survival, such as the kidneys, this arrangement allows for the accidental loss of one organ without immediate threat to the survival of the individual. Athletes who have lost one vital paired organ through injury or disease are often counseled against participating in events that carry the risk of damaging the remaining organ. If the second organ is damaged total loss of a vital function, such as sight, or even death may result.

some patients, but most artificial means of pumping blood are cumbersome and inefficient. They involve extensive surgery to implant them and require large, external motors that often prevent the patient from ever moving out of a hospital bed. A new device called the *Hemopump,* shown in Figure 3-14, requires only a small external motor and can be inserted through a tiny incision in the artery in the left leg. But even the small and efficient Hemopump is not considered adequate for long-term use. Because of their inefficiency, artificial replacements for vital organs are generally used only to ensure survival until a more permanent solution can be found.

ORGAN TRANSPLANTATION

One approach that offers the hope of a permanent solution to loss of vital organ function is *organ transplantation.* In this technique, a normal living organ from a donor is surgically transplanted into the recipient. Kidney, liver, pancreas, and heart transplants are now done at many hospitals throughout the world. When a new organ is transplanted into the body, the old organ may or may not be removed. For example, failed kidneys are often left in place at the back of the ventral

▼ **FIGURE 3-14 Examples of Prostheses.** Damaged organs or tissues can often be replaced or repaired by using artificial materials or devices.

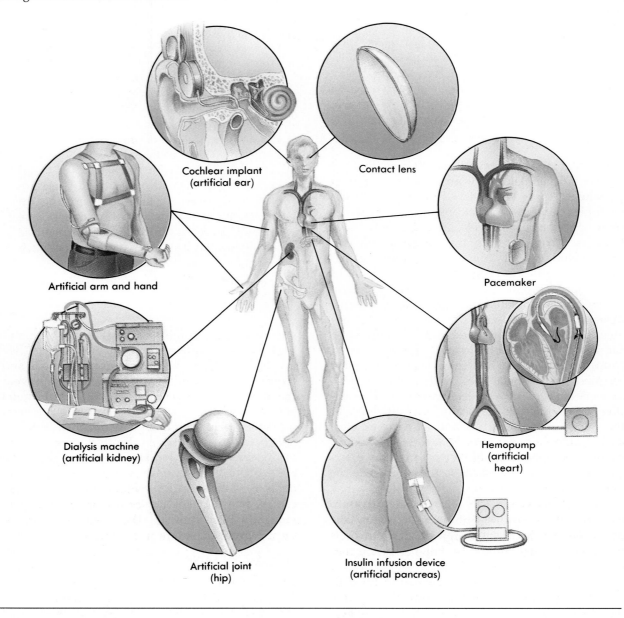

Cochlear implant
(artificial ear)

Contact lens

Artificial arm and hand

Pacemaker

Dialysis machine
(artificial kidney)

Hemopump
(artificial
heart)

Artificial joint
(hip)

Insulin infusion device
(artificial pancreas)

body cavity. As Figure 3-15 shows, the "new" kidney is nestled far below it in the curve of the pelvic bone, where it is attached to major blood vessels and to the bladder. Using this strategy, the trauma of removing the damaged kidneys is avoided and the transplanted kidney can still process blood efficiently.

Despite its many successes, organ transplantation has some problems. One is that a recipient's immune system often rejects transplanted organs.

▼ **FIGURE 3-15 Kidney Transplantation.** In kidney transplantations, the diseased organs are left in place, and the donated organ is nestled in another part of the body.

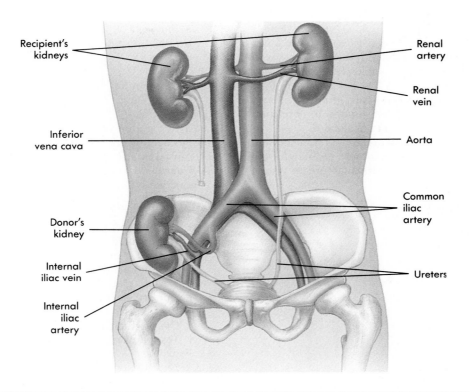

Some **immunosuppressive** drugs that suppress the immune system and inhibit rejection reactions also pose the risk of severe infection. Cyclosporine is an immunosuppressive drug that solves this problem to some degree by suppressing rejection reactions without severely inhibiting infection control. Development of new drugs and better tissue-typing procedures offer the hope of reducing organ rejection problems.

Another way to solve the rejection problem is to build "new" organs from a patient's own tissues. For example, in a method called *free-flap surgery,* pieces of tissue from one part of the body are surgically remodelled and then grafted to a new part of the body. After cancerous breasts are removed, "new" breasts can be formed from skin and muscle tissue taken from the thighs, buttocks, or abdomen. Parts of the intestine can be used to repair the urinary bladder. Toes can even be transplanted to the hand to replace missing fingers. The advantage to using a patient's tissues is that the possibility of rejection is eliminated.

Another major problem with organ transplants is the limited availability of donor organs. One solution is to eliminate the need for human donors. Researchers are now working on a variety of methods by which new organs or tissues can be "grown" in a tissue culture or in a patient's body. For example, it is hoped that normal liver cells can be safely removed from a healthy donor and implanted in a plastic sponge that is placed in the recipient's body. The transplanted cells may then reproduce and form a mass that provides some liver function in the recipient. Researchers are also culturing colonies of healthy nervous tissue in laboratory dishes in the hope that it can someday be used to repair damaged sections of the brain or spinal cord.

OUTLINE SUMMARY

DEFINITIONS AND CONCEPTS

A Organ—a structure made up of two or more kinds of tissues organized in such a way that they can together perform a more complex function than can any tissue alone

B Organ system—a group of organs arranged in such a way that they can together perform a more complex function than can any organ alone

C A knowledge of individual organs and how they are organized into groups makes more meaningful the understanding of how a particular organ system functions as a whole

ORGAN SYSTEMS

A Integumentary system
 1 Structure—organs
  a Skin
  b Hair
  c Nails
  d Sense receptors
  e Sweat glands
  f Oil glands
 2 Functions
  a Protection
  b Regulation of body temperature
  c Synthesis of chemicals and hormones
  d Sense organ

B Skeletal system
 1 Structure
  a Bones
  b Joints
 2 Functions
  a Support
  b Movement (with joints and muscles)
  c Storage of minerals
  d Blood cell formation

C Muscular system
 1 Structure
  a Muscles
   (1) Voluntary or striated
   (2) Involuntary or smooth
   (3) Cardiac
 2 Functions
  a Movement
  b Maintenance of body posture
  c Production of heat

D Nervous system
 1 Structure
  a Brain
  b Spinal cord
  c Nerves
 2 Functions
  a Communication
  b Integration
  c Control
  d Recognition of sensory stimuli
 3 System functions by production of nerve impulses caused by stimuli such as heat and pressure
 4 Control is fast-acting and of short duration

E Endocrine system
 1 Structure
  a Pituitary gland
  b Pineal gland
  c Hypothalamus
  d Thyroid gland
  e Parathyroid glands
  f Thymus gland
  g Adrenal glands
  h Pancreas
  i Ovaries (female)
  j Testes (male)
 2 Functions
  a Secretion of special substances called *hormones* directly into the blood
  b Same as nervous system—communication, integration, control
  c Control is slow and of long duration
  d Examples of hormone regulation: growth, metabolism, reproduction, fluid and electrolyte balance

F Circulatory system
 1 Structure
  a Heart
  b Blood vessels
 2 Functions
  a Transportation
  b Immunity (body defense)

G Lymphatic system
 1 Structure
  a Lymph nodes
  b Lymph vessels
  c Thymus
  d Spleen
  e Tonsils
 2 Functions
  a Transportation
  b Immunity (body defense)

H Respiratory system
 1 Structure

a Nose
b Pharynx
c Larynx
d Trachea
e Bronchi
f Lungs
2 Functions
  a Exchange of waste gas (carbon dioxide) for oxygen in the lungs
  b Area of gas exchange in the lungs called *alveoli*
I Digestive system
 1 Structure
  a Primary organs: mouth, pharynx, esophagus, stomach, small intestine, large intestine, rectum, anal canal
  b Accessory organs: teeth, salivary glands, tongue, liver, gallbladder, pancreas, appendix
 2 Functions
  a Mechanical and chemical breakdown (digestion) of food
  b Absorption of nutrients
  c Undigested waste product that is eliminated is called *feces*
  d Appendix is a structural but not a functional part of digestive system
  e Inflammation of appendix is called *appendicitis*
J Urinary system
 1 Structure
  a Kidneys
  b Ureters
  c Urinary bladder
  d Urethra
 2 Functions
  a "Clearing" or cleaning blood of waste products—waste product excreted from body is called *urine*
  b Electrolyte balance
  c Water balance
  d Acid-base balance
  e In male, urethra has urinary and reproductive functions
K Reproductive system
 1 Structure
  a Male
   (1) Gonads—testes
   (2) Genital ducts—vas deferens, urethra
   (3) Accessory gland—prostate
   (4) Supporting structures—genitalia (penis and scrotum)
  b Female
   (1) Gonads—ovaries
   (2) Accessory organs—uterus, uterine (fallopian) tubes, vagina
   (3) Supporting structure—genitalia (vulva), mammary glands (breasts)
 2 Functions
  a Survival of species
  b Production of sex cells (male: sperm; female: ova)
  c Transfer and fertilization of sex cells
  d Development and birth of offspring
  e Nourishment of offspring
  f Production of sex hormones

ORGAN REPLACEMENT

A Loss of function in nonvital organs is not immediately life-threatening; loss of function in vital organs is immediately life-threatening.
B Loss of function in organs can be treated by organ replacement.
 1 Artificial organs (prostheses)
 2 Organ transplantation
 3 Free-flap surgery

NEW WORDS

Review the names of organ systems and individual organs in Figures 3-1 and 3-2.

appendix	gastrointestinal (GI)	hormone	nerve impulse
cardiovascular	tract	integumentary	stimuli
endocrine	genitalia	lymphatic	urine
feces			

Diseases And Other Clinical Terms

appendicitis	dialysis	immunosuppressive
cochlear implant	Hemopump	prosthesis

CHAPTER TEST

1. The largest and most important organ of the integumentary system is the _____.
2. The 206 individual organs of the skeletal system are called _____.
3. The four primary functions of the nervous system are _____, _____, _____, and _____.
4. The organs of the endocrine system secrete substances called _____ into the blood.
5. The thymus and spleen are classified as _____ organs.
6. The waste product produced by the kidneys is called _____ .
7. The tongue, liver, and gallbladder are classified as _____ organs of digestion.
8. Oxygen from the air is exchanged for the waste product carbon dioxide in thin-walled sacs in the lungs called _____.

Select the most appropriate answer in column B for each item in column A. (Only one answer is correct.)

COLUMN A

9. _____ Appendage of skin
10. _____ Storage site for calcium
11. _____ Function to generate heat
12. _____ Permit rapid communication
13. _____ Important in body defense mechanism
14. _____ Undigested residue in GI tract
15. _____ Female external genitalia
16. _____ Male gonads

COLUMN B

a. Lymphatic system
b. Muscles
c. Feces
d. Hair
e. Vulva
f. Testes
g. Bones
h. Nerve impulses

REVIEW QUESTIONS

1 Give brief definitions of the terms *organ* and *organ system.*
2 List the 11 major organ systems of the body.
3 Discuss the structure and generalized functions of the integumentary system.
4 Explain the functional interaction that occurs between the skeletal, muscular, and nervous systems.
5 Compare the generalized functions of the nervous and endocrine systems. How are they similar? How do they differ?
6 What is the relationship between a stimulus and a nerve impulse?

7 Define the term *hormone.*
8 List and discuss the structural and functional components of the circulatory system.
9 Identify the waste products associated with the urinary and digestive systems.
10 List the primary and accessory organs of the digestive system.
11 Compare and contrast the structure and function of the male and female reproductive systems.
12 Describe these approaches to organ replacement: artificial organs and organ transplantation. List the advantages and disadvantages of these approaches.

CLINICAL APPLICATIONS

1 Tommy has been diagnosed as having irreversible kidney failure. Which system of the body is involved in this condition? What functions has Tommy lost? What options do his physicians have in treating Tommy's condition?

2 Mr. Davidson was referred to a *urologist* for diagnosis and treatment of an obstruction in his urethra. What bodily functions may be affected by Mr. Davidson's conditions?

Mechanisms of Disease

CHAPTER

4

Objectives

*After you have completed this chapter,
you should be able to:*

1 Define the terms *health* and *disease*.

2 List and describe the basic mechanisms of disease and risk factors associated with disease.

3 List and describe five categories of pathogenic organisms and explain how they cause disease.

4 Distinguish between the terms *benign* and *malignant* as they apply to tumors.

5 Describe the pathogenesis of cancer.

6 Outline the events of the inflammatory response and explain its role in disease.

*T*he title of this book uses the words *health* and *disease*. We use these words all the time, but what do they really mean? For the purposes of scientific study, health is physical, mental, and social well-being—not merely the absence of disease. Disease can be described as an abnormality in body function that threatens well-being. Named diseases are specific structural or functional abnormalities. In this chapter, we will explore some basic ideas about disease, especially how it disrupts normal function.

Studying Disease

DISEASE TERMINOLOGY

Everyone is interested in *pathology*—the study of disease. Researchers want to know the scientific basis of abnormal conditions. Health practitioners want to know how to prevent and treat a wide variety of diseases. When we suffer from the inevitable cold or something more serious, we want to know what is going on and how best to deal with it. Pathology has its own terminology, as in any specialized field. Most of these terms are derived from Latin and Greek word parts. For example, *patho-* comes from the Greek word for "disease" *(pathos)* and is used in many terms, including *pathology* itself. If you are unfamiliar with word parts commonly used in medical science, refer to Appendix C.

Disease conditions are usually *diagnosed* or identified by signs and symptoms. **Signs** are objective abnormalities that can be seen or measured by someone other than the patient, whereas **symptoms** are the subjective abnormalities felt only by the patient. Although *sign* and *symptom* are distinct terms, we often use them interchangeably. A **syndrome** (SIN-drome) is a collection of different signs and symptoms, usually with a common cause, that presents a clear picture of a pathological condition. When signs and symptoms appear suddenly, persist for a short time, then disappear, we say that the disease is **acute.** On the other hand, diseases that develop slowly and last for a long time (perhaps for life) are labeled **chronic** diseases. The term *subacute* refers to diseases with characteristics somewhere between acute and chronic.

The study of all factors involved in causing a disease is its **etiology** (ee-tee-OL-o-jee). The etiol-ogy of a skin infection often involves a cut or abrasion and subsequent invasion and growth of a bacterial colony. Diseases with undetermined causes are said to be **idiopathic** (id-ee-o-PATH-ik). **Communicable** diseases can be transmitted from one person to another.

The term *etiology* refers to the theory of a disease's cause, but the actual pattern of a disease's development is called its **pathogenesis** (path-o-JEN-es-is). The common cold, for example, begins with a *latent* or "hidden" stage during which the cold virus establishes itself in the patient. No signs of the cold are yet evident. In infectious diseases, the latent stage is also called **incubation.** The cold may then manifest itself as a mild nasal drip, triggering a few sneezes. It then progresses to its full fury and continues for a few days. After the cold has run its course, *convalescence* or recovery occurs. During this stage, body functions return to normal. Some chronic diseases, such as cancer, exhibit a temporary reversal that seems to be a recovery. Such reversal of a chronic disease is called a **remission.** If a remission is permanent, we say that the person is "cured."

PATTERNS OF DISEASE

Epidemiology (ep-id-ee-mee-OL-o-jee) is the study of the occurrence, distribution, and transmission of diseases in humans. Epidemiologists are physicians or medical scientists who study patterns of disease occurrence in specific groups of people. For example, a hospital may employ a staff epidemiologist who is responsible for infection-control programs within the hospital. Many governments and other agencies employ epidemiologists who track the spread of disease through a local community or even the world.

A disease that is native to a local region is called an **endemic** disease. If the disease spreads to many individuals at the same time, the situation is called an **epidemic. Pandemics** are epidemics that affect large geographic regions, perhaps spreading throughout the world. Because of the speed and availability of modern air travel, pandemics are more common than they once were. Almost every flu season, we see a new strain of influenza virus quickly spreading from continent to continent.

Tracking the cause of a disease and its pattern of spread through a population can be very difficult. One reason is that there are so many different factors involved in the spread of disease. Nutrition, age, gender, sanitation practices, and socioeconomic conditions may play a role in the spread of disease. Infectious agents, for example, can spread quickly and easily through an unsanitary water supply. Likewise, accumulation of untreated sewage or garbage can harbor disease-causing organisms or chemicals. Infectious agents or other contaminants in food may also spread disease to a large number of people. Crowded conditions may also play a role in spreading disease because more people come in close contact with one another. In crowded regions with poor sanitation and food-handling practices, disease may spread quickly.

The pattern of a disease's spread may be difficult to explain because of the different kinds of agents that may cause disease. For example, imagine that the majority of students in your class became ill with headaches and nausea (sick stomach) at about the same time. One would have to investigate many possible causes and modes of transmission before an explanation could be offered. Is it food poisoning? Is it an outbreak of the "flu" or another virus? Is the water supply at the drinking fountain contaminated? Is there a leak of toxic fumes in the building? Is there radioactive material nearby? Because any of these can cause the situation that is described, a thorough investigation is needed to distinguish the *causal* relationships from the *coincidental* relationships. Only then can a reasonable answer be proposed.

Epidemiologists study the spread of disease so that ways of stopping it can be found. The two most obvious strategies for combatting disease are *prevention* and *therapy*. Therapy or treatment of diseases was perhaps the first strategy used by humans to fight disease. The continued search for therapeutic treatments for almost all known diseases is evidence that we still value this strategy. However, we have always known that an even more effective disease-fighting strategy is prevention. Only recently have we understood many diseases well enough to know how to prevent them. Although the war on human disease will probably never end, we have had some dramatic successes. The often fatal viral infection *smallpox* once caused catastrophic epidemics but is now

FIGURE 4-1 The Last Smallpox Victim. Ali Maow Maalin of Somalia contracted the last known case of smallpox in 1977. Successful disease prevention techniques have completely eradicated a disease that once killed millions.

gone because of successful prevention strategies such as worldwide *vaccination* and education (Figure 4-1).

Pathophysiology

MECHANISMS OF DISEASE

Pathophysiology is the organized study of the underlying physiological processes associated with disease. Pathophysiologists attempt to understand the mechanisms of a disease and its pathogenesis. Although pathophysiologists uncover information that leads to strategies of prevention and treatment, developing and applying these strategies is left to other professionals.

Many diseases are best understood as disturbances of homeostasis, the relative constancy of the body's internal environment. If homeostasis is disturbed, a variety of negative-feedback mechanisms usually returns the body to normal. When a disturbance goes beyond the normal fluctuation of everyday life, a disease condition exists. In acute conditions, the body recovers its homeostatic balance quickly. In chronic diseases, a normal state of balance may never be restored. If the disturbance keeps the body's internal environment

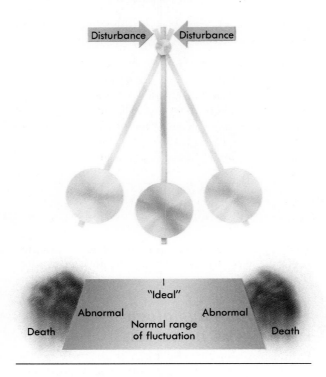

FIGURE 4-2 Model of Homeostatic Balance. Movement of the pendulum represents the fluctuation of a physiological variable, such as body temperature or blood pressure, around an "ideal" value. Disturbance in either direction could move the variable into the "abnormal" range, the disease state. If the disturbance is so extreme that it goes "off the scale" or outside the tolerable range, death results.

too far from normal for too long, death may result (Figure 4-2).

Disturbances to homeostasis and the body's responses are the basic mechanisms of disease. Because of their variety, the following disease mechanisms can be categorized for easier study:

1 Genetic mechanisms—Altered or *mutated* genes can cause abnormal proteins to be made. These abnormal proteins often do not perform their intended function, resulting in the absence of an essential function. On the other hand, such proteins may perform an abnormal, disruptive function. Either case may be a threat to the constancy of the body's internal environment. The basis for genetic diseases is discussed in Chapter

23, and important genetic conditions are summarized in Appendix B.

2 Pathogenic organisms—Many important disorders are caused by *pathogenic* (disease-causing) organisms that damage the body in some way. An organism that lives in or on another organism to obtain its nutrients is called a **parasite.** The presence of microscopic or larger parasites may interfere with normal body functions of the *host* and cause disease. Besides parasites, there are organisms that poison or otherwise damage the human body to cause disease. Some of the major pathogenic organisms are listed later in this chapter and in Appendix B.

3 Tumors and cancer—Abnormal tissue growths or *neoplasms* can cause a variety of physiological disturbances, such as those described later in this chapter.

4 Physical and chemical agents—Agents such as toxic or destructive chemicals, extreme heat or cold, mechanical injury, and radiation can affect the normal homeostasis of the body. Examples of pathological conditions caused by physical agents are summarized in Appendix B.

5 Malnutrition—Insufficient or imbalanced intake of nutrients causes a variety of diseases, as outlined in Chapter 16 and Appendix B.

6 Autoimmunity—Some diseases result from the immune system attacking the body *(autoimmunity)* or from mistakes or overreactions of the immune response. Autoimmunity, literally "self-immunity," is discussed in Chapter 14 with other immune system disturbances. Examples of autoimmune conditions are listed in Appendix B.

7 Inflammation—The body often responds to disturbances with the *inflammatory response.* The inflammatory response is a normal mechanism that usually speeds recovery from an infection or injury. However, when the inflammatory response occurs at inappropriate times or is abnormally prolonged or severe, normal tissues may become damaged. Thus some disease symptoms are *caused by* the inflammatory response.

8 Degeneration—By means of many still unknown processes, tissues sometimes break apart or *degenerate.* Although a normal consequence of aging, degeneration of one or more tissues resulting from disease can occur at any time. The degeneration of tissues associated with aging is discussed in Chapter 22.

RISK FACTORS

Other than direct causes or disease mechanisms, certain *predisposing conditions* may make the development of a disease more likely to occur. Usually called **risk factors,** they often do not actually cause a disease but may put one "at risk" for developing it. Some of the major types of risk factors follow:

1 Genetic factors—There are several types of genetic risk factors. Sometimes, an inherited trait puts a person at a greater than normal risk for developing a specific disease. For example, light-skinned people are more at risk for developing certain forms of skin cancer than dark-skinned people. This occurs because light-skinned people have less pigment in their skin to protect them from cancer-causing ultraviolet radiation (see Chapter 5). Membership in a certain ethnic group or *gene pool* involves the "risk" of inheriting a disease-causing gene that is common in that gene pool. For example, certain Africans and their descendants are at a greater-than-average risk of inheriting *sickle-cell anemia*—a deadly blood disorder.

2 Age—Biological and behavioral variations during different phases of the human life cycle put us at greater risk for developing certain diseases at certain times in life. For example, middle ear infections are more common in infants than in adults because of the difference in ear structure at different ages.

3 Lifestyle—The way we live and work can put us at risk for some diseases. People whose work or personal activity puts them in direct sunlight for long periods have a greater chance of developing skin cancer because they are in more frequent contact with ultraviolet radiation from the sun. Some researchers believe that the high-fat, low-fiber diet common among people in the "developed" nations increases their risk of developing certain cancers.

4 Stress—Physical, psychological, or emotional stress can put one at risk of developing problems such as chronic high blood pressure (hypertension), peptic ulcers, and headaches. Conditions caused by psychological factors are sometimes called *psychogenic* ("mind-caused") disorders.

5 Environmental factors—Although environmental factors such as climate and pollution can cause injury or disease, some environmental situations simply put us at greater risk for getting certain diseases. For example, because some parasites survive only in tropical environments, we are not at risk if we live in a temperate climate.

6 Preexisting conditions—A preexisting disease, such as an infection, can adversely affect our capacity to defend ourselves against further attack. Thus a *primary* (preexisting) condition can put a person at risk of developing a *secondary* condition. For example, blisters from a preexisting burn may break open and thus increase the risk of a bacterial infection of the skin.

Risk factors can combine, increasing a person's chances of developing a specific disease even more. For example, a light-skinned person can add to the genetic risk of developing skin cancer by spending a large amount of time in the sun without skin protection—a lifestyle risk. As you may have guessed, many of these categories of risk factors overlap. For example, stress can be a component of lifestyle, or it could be a preexisting condition. Sometimes a high-risk group is identified by epidemiologists, but the exact risk mecha-

CENTERS FOR DISEASE CONTROL

Epidemiology is a major concern of the Centers for Disease Control (CDC), a branch of the U.S. Public Health Service. Scientists at CDC headquarters in Atlanta, Georgia, continuously track the incidence and spread of disease in this country and worldwide. Much of this information is published in the *Morbidity and Mortality Weekly Report (MMWR).* Sent to physicians and other health professionals, this report provides recent information on disease rates in specific populations **(morbidity)** and the numbers of deaths caused by specific diseases **(mortality).** Much of the information in the MMWR concerns *notifiable diseases*—diseases that physicians must report to the U.S. Public Health Service. Gonorrhea, measles, AIDS, and tetanus are examples of notifiable diseases.

nism is uncertain. For example, a high incidence of heart disease in a small ethnic group may point to a genetic risk factor but could also result from some aspect of a shared lifestyle.

Risk factors for many deadly diseases can be avoided. Risk of heart disease, cancer, infections, and other types of disease can be decreased by making informed choices about lifestyle, stress management, the environment, and treatment of preexisting conditions.

▼ *Pathogenic Organisms*

TYPES OF ORGANISMS

Many kinds of organisms can cause disease in humans. Even humans can cause human disease through accidental or intentional injury. In pathophysiology, the pathogenic organisms most often

▼ **FIGURE 4-3 HIV.** The human immunodeficiency virus or HIV (blue in this electron micrograph), which is released from infected white blood cells, soon spreads over neighboring cells, infecting them in turn. The individual viruses are very small; more than 200 million would fit on the period at the end of this sentence.

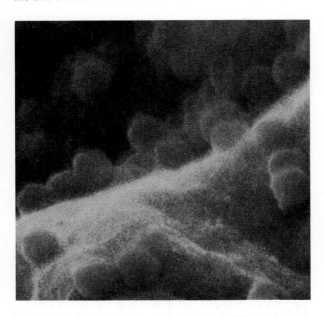

studied are microscopic or just barely visible to the unaided eye. Microscopic organisms, also called **microbes,** include *bacteria, fungi,* and *protozoa.* Larger organisms, the pathogenic *animals,* are also medically important. The smallest of all pathogens, microscopic nonliving particles called *viruses,* lead our list of important disease-causing agents.

Viruses

Viruses are intracellular parasites that consist of a nucleic acid (DNA or RNA) core surrounded by a protein coat and sometimes a lipoprotein envelope. Although they are not technically living organisms, viruses have a genetic code and, like living organisms, they multiply. They invade cells and insert their own genetic code into the host cell's genetic code, causing the cell to produce viral DNA or RNA and protein coats. They thus pirate the host cell's nutrients and organelles to produce more virus particles. New viruses may leave the cell to infect other cells by exocytosis or by bursting the cell membrane (Figure 4-3).

The symptoms of viral infections may not appear right away. The viral genetic code may not become active for some time, or viral multiplication may not immediately cause significant cellular damage. In any case, the effects of the intracellular viral parasite eventually take their toll and thus produce symptoms of disease.

Viruses are a very diverse group, as illustrated in Figure 4-4. They are usually classified according to their shape, DNA or RNA content, and their method of multiplying. Some examples of medically important viruses are listed in Table 4-1. Many of these and other viral diseases are discussed in detail in later chapters.

Bacteria

A **bacterium** is a tiny, primitive cell without a nucleus. Bacteria produce disease in a variety of ways. They can secrete toxic substances that damage human tissues, they may become parasites inside human cells, or they may form colonies in the body that disrupt normal human function. Like viruses, bacteria are also a diverse group of *pathogens* ("disease-producers"). There are several ways to classify bacteria:

FIGURE **4-4 Diversity of Pathogenic Viruses.** Some viruses are relatively large, others are extremely tiny.

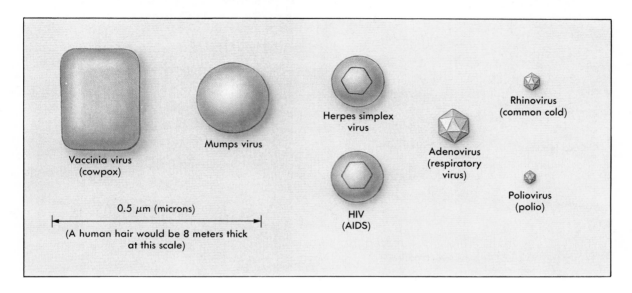

TABLE **4-1**   Examples of Pathogenic Viruses

VIRAL TYPE	VIRUS	DISEASES CAUSED
DNA	Human papilloma virus (HPV)	Warts
	Hepatitis B virus	Hepatitis (viral liver infection)
	Herpes simplex 1 and 2	Fever blisters and genital herpes
	Epstein-Barr virus (EBV)	Mononucleosis
RNA	Influenza A, B, and C	Various influenza infections
	Human immunodeficiency virus (HIV)	Acquired immune deficiency syndrome (AIDS)
	Paramyxovirus	Measles, mumps, and parainfluenza
	Rhinovirus	Common cold and upper respiratory infections

See Table 2 in Appendix B for a list of viral diseases and their descriptions.

▼ **FIGURE 4-5 Major Groups of Pathogenic Bacteria.** Bacteria can be classified according to size and shape.

▼ **FIGURE 4-6 Streptococci Bacteria.** As this scanning electron micrograph shows, individual spherical bacteria (cocci) may adhere to form chains.

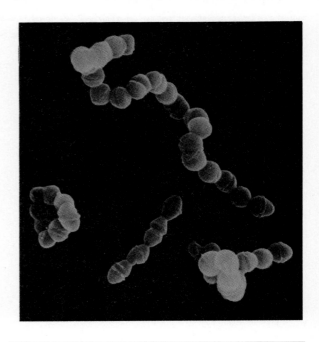

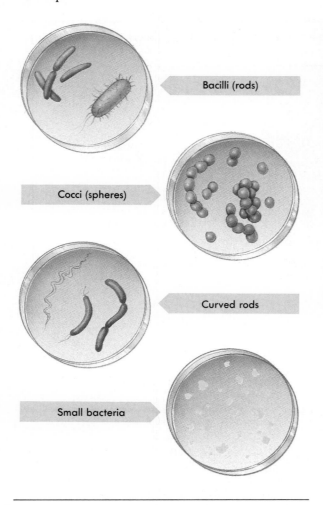

1 Function—For example, bacteria can be categorized as *aerobic* (requiring oxygen for their metabolism) or *anaerobic* (requiring an absence of oxygen).

2 Staining properties—Different bacteria stain differently, depending on the compounds in their walls. For example, *gram-positive* bacteria are stained purple by Gram's staining technique, whereas *gram-negative* bacteria are not.

3 Shape and size—Bacteria are most commonly classified by their varied shapes (Figure 4-5). Medically significant bacteria range in size from under 0.5 μm to more than 5 μm, making size a useful characteristic for classification. The unit μ*m* represents *micrometers* or *microns*, one-millionth of a meter. Some major groupings based on shape and size follow:

a Bacilli (ba-SIL-eye)—These are large, rod-shaped cells found singly or in groups.

b Cocci (KOKS-eye)—These large, round bacteria are found singly, in pairs *(diplococci)*, in strings *(streptococci)* as shown in Figure 4-6, or in clusters *(staphylococci)*.

c Curved or spiral rods—These can be curved rods arranged singly or in strands, or they can be large curved or spiral cells or cell colonies.

d Small bacteria—These round or oval bacteria are so small that some of them were once thought to be viruses. They can only reproduce inside other living cells, so they are sometimes called *obligate parasites*. **Rickettsia** (ri-KET-see-ah) and **Chlamydia** (kla-MID-ee-ah) are two types of small bacteria.

▼ **TABLE 4-2** Examples of Pathogenic Bacteria

STRUCTURAL CLASSIFICATION	GRAM STAIN CLASSIFICATION	BACTERIUM	DISEASES CAUSED
Bacilli (rods)	Gram-positive	*Bacillus* organisms	Anthrax and gastroenteritis
	Gram-positive	*Clostridium* organisms	Botulism, tetanus, and soft tissue infections
	Gram-negative	*Enterobacteria* organisms	*Salmonella* diseases and gastroenteritis
	Gram-negative	*Pseudomonas* organisms	External otitis (swimmer's ear), endocarditis, and pulmonary infections
Cocci (spheres)	Gram-positive	*Staphylococcus* organisms	Staphylococci infections, food poisoning, urinary tract infections, and toxic shock syndrome
	Gram-positive	*Streptococcus* organisms	Throat infections, pneumonia, sinusitis, otitis media, rheumatic fever, and dental caries
	Gram-negative	*Neisseria* organisms	Meningitis, gonorrhea, and pelvic inflammatory disease
Curved or spiral rod	Gram-negative	*Vibrio* organisms	Cholera, gastroenteritis, and wound infections
	Gram-negative	*Campylobacter* organisms	Diarrhea
	Gram-negative	Spirochetes	Syphilis and Lyme disease
Small bacterium	Gram-negative	*Rickettsia* organisms	Rocky Mountain spotted fever and Q fever
	Gram-negative	*Chlamydia* organisms	Genital infections, lymphogranuloma venereum, conjunctivitis, and parrot fever

See Table 3 in Appendix B for a list of bacterial diseases and their descriptions.

Table 4-2 summarizes bacterial types and some of the diseases each group causes.

Some bacteria can produce resistant forms called **spores** under adverse environmental conditions. Spores are resistant to chemicals, heat, and dry conditions. When environmental conditions become more suitable for activities such as reproduction, the spores revert back to the active form of bacterium. Although advantageous for the bacterium, this strategy often makes it difficult for humans to destroy pathogenic bacteria.

Fungi

Fungi (FUN-jye) are a group of simple organisms similar to plants but without chlorophyll (green pigment). Without chlorophyll, pathogenic fungi cannot produce their own food, so they must consume or parasitize other organisms. Most pathogenic fungi parasitize tissue on or near the skin or mucous membranes, as in athlete's foot and vaginal yeast infections. A few systemic (body-wide) fungal infections, such as San Joaquin fever, can disrupt the entire body. Figures 4-7

▼ **FIGURE 4-7 Major Groups of Pathogenic Fungi: Yeasts and Molds.**

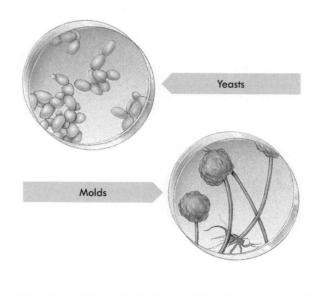

Yeasts

Molds

▼ **FIGURE 4-8** **Examples of Pathogenic Fungi.**
A, Scanning electron micrograph of yeast cells. Yeasts
commonly infect the urinary and reproductive tract. **B,**
This electron micrograph shows *Aspergillus* organisms,
a mold that can infect different parts of the body where
it forms characteristic "fungus balls."

A **B**

▼ **TABLE 4-3** Examples of Pathogenic
Fungi

FUNGUS	DISEASES CAUSED
Candida organisms	Thrush and mucous membrane infections (including vaginal yeast infections)
Epidermophyton and *Microsporum* organisms	Tinea infections: ringworm, jock itch, and athlete's foot
Histoplasma organisms	Histoplasmosis
Aspergillus organisms	Aspergillosis and pneumonia
Coccidioides organisms	Coccidioidomycosis (San Joaquin fever)

See Table 4 in Appendix B for a list of mycotic diseases and their
descriptions.

and 4-8 shows that *yeasts* are small, single-celled
fungi and *molds* are large, multicellular fungi.
Fungal or **mycotic** (my-KOT-ik) infections often
resist treatment, so they can become quite serious.
Table 4-3 lists some of the important pathogenic
fungi and the diseases that they cause.

Protozoa

Protozoa (pro-toe-ZO-ah) are *protists*, one-celled
organisms larger than bacteria and whose DNA is
organized in a nucleus. Figure 4-9 illustrates some
of the pathogenic protozoa. Protozoa can infest
human fluids and cause disease by parasitizing
cells or directly destroying them. The major
groups of protozoa include:

1 Amebas (ah-ME-bahs)—Large cells of changing shape, amebas extend their membranes to
form *pseudopodia* ("false feet") that pull them
along.

2 Flagellates (FLAJ-el-ates)—These protozoa
are similar to amebas but move by wiggling long,
whiplike extensions called **flagella.** Figure 4-10
shows *Giardia lamblia*, a common flagellate protozoan that causes diarrhea.

3 Ciliates (SILL-ee-ates)—Ciliates are protozoa
that move by means of many short, hairlike projections called **cilia.**

4 Sporozoa (spor-oh-ZO-ah)—Also called *coccidia,* the sporozoa have unusual organelles at
their tips that allow them to enter host cells. They
often oscillate between two different hosts, having two different stages in their life cycle. The
sporozoa that cause malaria exhibit this pattern.

Table 4-4 lists some of the major disease-causing
protozoa.

Pathogenic Animals

Sometimes called *metazoa* (met-a-ZO-a), these
pathogenic animals are large, multicellular organisms. Animals can cause disease by parasitizing
humans or causing injury in some other way.
Figure 4-11 illustrates some animals that cause
disease. The major groups of pathogenic animals
include the following:

1 Nematodes (nem-a-TOADS)—Also called
roundworms, these large parasites infest a variety
of different human tissues. They are often transmitted by food or by flies that bite.

FIGURE 4-9 Major Groups of Pathogenic Protozoa. Protozoa are complex, one-celled organisms.

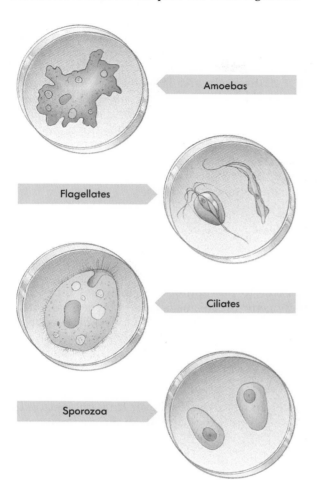

FIGURE 4-10 Flagellate Protozoans. This scanning electron micrograph shows several individual *Giardia lamblia*. Notice the numerous flagella (long, hairlike extensions) in each cell.

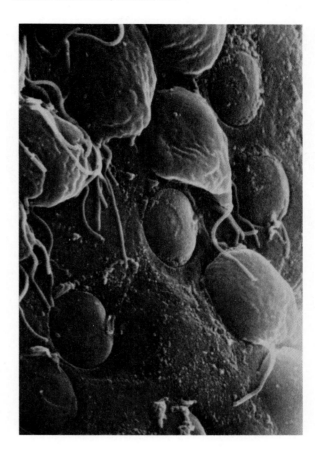

TABLE 4-4 Examples of Pathogenic Protozoa

CLASSIFICATION	PROTOZOAN	DISEASES CAUSED
Ameba	*Entamoeba* organisms	Diarrhea, amebic dysentery, and liver and lung infections
Flagellate	*Giardia* organisms	Giardiasis, diarrhea, and malabsorption syndrome
	Trichomonas organisms	Trichomoniasis, vaginitis, and urinary tract infections
Ciliate	*Balantidium* organisms	Gastrointestinal disturbances, including pain, nausea, and anorexia
Sporozoan (coccidium)	*Isospora* organisms	Isosporiasis infection of gastrointestinal tract, diarrhea, and malabsorption syndrome
	Plasmodium organisms	Malaria
	Toxoplasma organisms	Toxoplasmosis and congenital damage to fetus

See Table 5 in Appendix B for a list of diseases caused by protozoa.

▽ **FIGURE 4-11 Examples of Pathogenic Animals.**

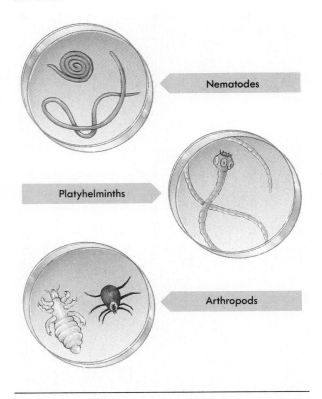

2 Platyhelminths   (plat-ee-HEL-minths)—Otherwise known as *flatworms* and *flukes*, these large parasites can infest several different human organs. The *Schistosoma* flukes shown in Figure 4-12 cause "snail fever" or *schistosomiasis.*

3 Arthropods (ARTH-ro-pods)—The arthropod group includes parasitic *mites, ticks, lice,* and *fleas.* Also included are biting or stinging *wasps, bees, mosquitoes,* and *spiders.* All are capable of causing injury or infestation themselves but can also carry other pathogenic organisms. An organism that spreads disease to other organisms is called a **vector** of the disease.

Table 4-5 summarizes some of the major health problems associated with selected pathogenic animals.

PREVENTION AND CONTROL

The key to preventing many diseases caused by pathogenic organisms is to stop them from entering the human body. This sounds simple enough but is often very difficult to accomplish. The following is a partial list of the ways in which pathogens can spread:

1 Person-to-person contact—Small pathogens can often be carried in the air from one person to another. Direct contact with an infected person or

▽ **TABLE 4-5   Examples of Pathogenic Animals**

CLASSIFICATION	ANIMAL	DISEASES CAUSED
Nematode	*Ascaris* organisms	Intestinal roundworm infestation, gastrointestinal obstruction, and bronchial damage
	Enterobius organisms	Pinworm infestation of the lower gastrointestinal tract, rectal itching, and insomnia
	Trichinella organisms	Trichinosis, fever, and muscle pain
Platyhelminth	*Schistosoma* organisms	Schistosomiasis (snail fever)
	Fasciola organisms	Liver fluke infestation
	Taenia organisms	Pork and beef tapeworm infestation
Arthropod	*Arachnida* organisms	Infestation by mites and ticks; toxic bites by spiders, scorpions; and transmission of other pathogens
	Insecta	Infestation by fleas and lice; toxic bites by wasps, mosquitoes, and bees; and transmission of other pathogens; ticks (Lyme disease)

See Table 6 in Appendix B for a list of diseases caused by pathogenic animals.

with contaminated materials handled by the infected person is a common mode of transmission. The *rhinovirus* that causes the common cold is often transmitted in these ways. Some viruses, such as those that cause *hepatitis B* and *AIDS* are transmitted when infected blood, semen, or another body fluid enters a person's blood stream. Preventing the spread of these diseases often involves educating people about avoiding certain types of contact with individuals known or suspected of carrying the disease. Another strategy, called **aseptic** (a-SEP-tik) **technique,** involves killing or disabling pathogens on surfaces before they can spread to other people. Table 4-6 summarizes the major approaches taken when using aseptic technique.

2 Environmental contact—Many pathogens are found throughout the local environment—in food, water, soil, and on assorted surfaces. Under normal conditions, these pathogens infect only individuals who happen to come across them or who are already weakened by some other condition. If improper sanitation practices create an environment that promotes increased growth and spread of pathogens, an epidemic could result. Disease caused by environmental pathogens can often be prevented by avoiding contact with certain materials and by maintaining safe sanitation practices.

3 Opportunistic invasion—Some potentially pathogenic organisms are found on the skin and mucous membranes of nearly everyone. How-

ever, they do not cause disease until they have the opportunity. That is, they do not create a problem until and unless conditions change or they enter the body's internal environment. For example, the fungi that cause athlete's foot are often present on the skin of people who do not have symptoms of this infection. Only when the skin is kept warm and moist for prolonged periods can the fungus reproduce and create an infection. Preventing

▾ **FIGURE 4-12 Platyhelminths.** This light micrograph shows a male and female *Schistosoma* fluke mating in the human bloodstream (the male is the larger of the two).

▾ **TABLE 4-6** Common Aseptic Methods that Prevent the Spread of Pathogens*

METHOD	ACTION	EXAMPLES
Sterilization	Destruction of all living organisms	Pressurized steam bath, extreme temperature, or radiation used to sterilize surgical instruments and garments or other surfaces
Disinfection	Destruction of most or all pathogens on inanimate objects but not necessarily all harmless microbes	Chemicals such as iodine, chlorine, alcohol, phenol, and soaps
Antisepsis	Inhibition or inactivation of pathogens	Chemicals such as alcohol, iodine, quaternary ammonium compounds (quats), and dyes
Isolation	Separation of potentially infectious people or materials from noninfected people	Quarantine of affected patients; protective apparel worn while giving treatments; and sanitary transport, storage, and disposal of body fluids, tissues, and other materials

*Spores (special bacterial forms) may resist methods that would ordinarily limit active bacterial cells.

opportunistic infection involves avoiding conditions that could promote infections. Changes in the pH (acidity), moisture, temperature, or other characteristics of skin and mucous membranes often promote opportunistic infections. Cleansing and aseptic treatment of accidental or surgical wounds can also prevent these infections.

4 Transmission by a vector—As stated previously, a vector is an arthropod that acts as a carrier of a pathogenic organism. For example, the spirochete bacterium that causes *Lyme disease* is not usually transmitted directly from human to human. Instead, a vector such as the deer tick carries it from one person to another or between animals and humans. The most effective way to stop such diseases from spreading is a combination of reducing the population of vectors and reducing the number of contacts with vectors. *Malaria,* still a major killer in some parts of the world, was virtually eliminated from North America in this way. Many mosquitoes that transmit the malaria organism were destroyed with pesticides while people were educated about ways to prevent mosquito bites. Consistent use of both strategies resulted in the collapse of the pathogen population in the vector and host.

A prevention strategy that has worked with some bacterial and viral pathogens has been the **vaccine.** A vaccine is a killed or attenuated (weakened) pathogen that is given to a person to stimulate immunity. Vaccination is a way to stimulate a person's own immune system to develop resistance to a particular pathogen. More discussion of vaccination and other immune-system strategies of disease prevention is found in Chapter 14.

After an infection has begun, there are several ways to treat the patient and attempt to gain control of the disease. One common approach is the use of chemicals to destroy pathogens or inhibit their growth. **Antibiotics,** for example, are compounds produced by certain living organisms that kill or inhibit pathogens. *Penicillin* produced by a fungus and *streptomycin* produced by a bacterium are well-known antibiotics. A few synthetic chemicals are now used to treat bacterial and viral infections. Among the most well-known antiviral agents are *acyclovir (ACV)* for treating herpes infections and *azidothymidine (AZT)* used to treat AIDS. These antiviral agents do not stop infections but merely inhibit viral reproduction and thus slow down the progression of viral diseases.

A NEW MOSQUITO VECTOR

Several species of mosquitoes commonly found in the United States carry deadly pathogens such as those that cause yellow fever and encephalitis (brain inflammation). However, a new vector has recently been added to the list. The *Asian tiger mosquito* arrived in Texas in 1985 in a batch of old tires from Japan. It has spread rapidly to many areas in the United States, especially in the south and midwest. The Asian tiger mosquito is a much more aggressive biter than other mosquitoes, making it a potentially more dangerous vector. It is already known to be vector of several deadly Asian diseases and is capable of carrying mosquito-borne pathogens native to North America. Only time will tell how important a role this new arrival will play in the transmission of disease in this country.

Tumors and Cancer

NEOPLASMS

The term **neoplasm** literally means "new matter" and refers to an abnormal growth of cells. Also called **tumors,** neoplasms can be distinct lumps of abnormal cells or, in blood tissue, can be diffuse. Neoplasms are often classified as **benign** or **malignant** (Table 4-7). Benign tumors remain localized within the tissue from which they arose. Malignant tumors tend to spread to other regions of the body. Another term for a malignant tumor is **cancer.**

Benign tumors are called that because they do not spread to other tissues and they usually grow very slowly. Their cells are often well differentiated, unlike the undifferentiated cells typical of malignant tumors. Cells in a benign tumor tend to stay together, and they are often surrounded by a capsule of dense tissue. Benign tumors are usually not life threatening but can be if they disrupt the normal function of a vital organ (Figure 4-13).

Malignant tumors, on the other hand, are not encapsulated and do not stay in one place. Their cells tend to fall away from the original neoplasm

LABORATORY IDENTIFICATION OF PATHOGENS

Often the signs or symptoms of a disease caused by bacteria or other pathogens are enough for a health professional to make a diagnosis. So that the correct course of treatment is given, laboratory tests are often required to positively identify a pathogen.

Sometimes pathogens can be observed in specimens of blood, feces (stool), cerebrospinal fluid, mucus, urine, or other substances from the body. A portion of the specimen is smeared on a microscope slide and then stained. Certain stains color only certain types of cells. For example, only gram-positive bacteria retain the violet stain used in Gram's staining technique (Figure *A*). Gram-negative bacteria do not retain the violet stain but only a red counterstain (Figure *B*). Thus gram-positive (violet) bacteria can be distinguished from gram-negative (red) bacteria using Gram's method. The staining properties, shape, and size can sometimes be used to identify pathogens in specimen samples.

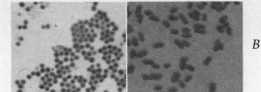

A   *B*

Pathogens are sometimes identified by growing **cultures** from specimens taken from a patient. Populations of bacteria can be grown only on certain *media* (liquid or *agar* gel containing nutrients). Thus pathogenic bacteria are often identified by the type of medium in which they grow best. For example, mucus swabbed from a sore throat and placed in a medium that contains blood may produce pinpoint-sized colonies of pathogenic streptococci bacteria. The streptococci bacteria that cause "strep throat" typically have a distinct, transparent ring around each colony. The rings result from hemolysis—bursting of red blood cells in the surrounding medium. Viruses can also be cultivated but only within living cells.

Some infections can be diagnosed on the basis of immunological tests that check for antibodies against a particular pathogen. If antibodies are found, it is assumed that the patient has been exposed to a pathogen; a large number of antibodies usually indicates an active infection. An example is the test for anti-HIV antibodies used to identify HIV infections. Recall from Chapter 2 that such tests are often used to screen donated tissues and organs for pathogenic organisms. A wide variety of different immunological tests are now available for bacterial and viral infections.

TABLE 4-7 Comparison of Benign and Malignant Tumors

CHARACTERISTIC	BENIGN TUMOR	MALIGNANT TUMOR
Rate of growth	Slow	Rapid
Structure	Encapsulated	Nonencapsulated (infiltrates surrounding tissue)
Pattern of growth	Expanding but not spreading to other tissues	Metastasizing (spreading) to other tissues
Cell type	Well differentiated (similar to normal tissue cells)	Undifferentiated (abnormal in structure and function)
Mortality rate	Low	High if condition remains untreated

FIGURE **4-13** **Types of Neoplasms. A,** Benign neoplasms (tumors) are usually encapsulated and grow slowly. **B,** Malignant neoplasms or cancers are not encapsulated. They grow rapidly, extending into surrounding tissues. Some cells metastasize, falling away and forming tumors in other parts of the body.

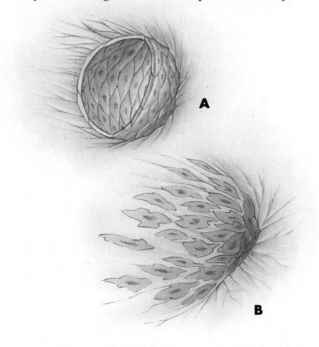

FIGURE **4-14** **Metastasis.** Abnormal cells from malignant tumors fall away from the original neoplasm, travel along lymphatic vessels, through which they can enter and exit easily. Malignant cells also travel through the blood stream, and burrow through a blood vessel wall to invade other tissues.

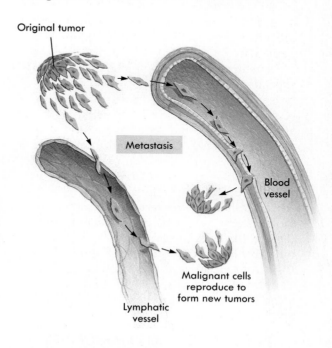

Original tumor

Metastasis

Blood vessel

Malignant cells reproduce to form new tumors

Lymphatic vessel

and may start new tumors in other parts of the body. For example, cells from malignant breast tumors usually form new (secondary) tumors in bone, brain, and lung tissues. The cells migrate by way of lymphatic or blood vessels. This manner of spreading is called **metastasis** (met-TAS-ta-sis). Cells that do not metastasize can spread another way: they grow rapidly and extend the tumor into nearby tissues. Malignant tumors may replace part of a vital organ with abnormal, undifferentiated tissue—a life-threatening situation (Figures 4-13 and 4-14).

Benign and malignant neoplasms are classified into subgroups depending on appearance and the location where they originate. Benign and malignant tumors can be divided into three types— epithelial, connective tissue, and miscellaneous tumors. Some examples of each follow:

1 Benign tumors that arise from epithelial tissues
  a **Papilloma** (pap-i-LO-mah)—This type of tumor forms a fingerlike projection, as in a wart.
  b **Adenoma** (ad-e-NO-mah)—This is a general term for benign tumors of glandular epithelium.
  c **Nevus** (NEE-vus)—Nevi include a variety of small, pigmented tumors of the skin, such as moles.
2 Benign tumors that arise from connective tissues
  a **Lipoma** (lip-O-mah)—A lipoma is a tumor arising from adipose (fat) tissue.
  b **Osteoma** (os-tee-O-mah)—This tumor involves bone tissues.
  c **Chondroma** (kon-DRO-mah)—Chondromas are tumors of cartilage tissue.

3 Malignant tumors that arise from epithelial tissues, generally called **carcinomas** (kar-sin-O-mahs)

   a Melanoma (mel-ah-NO-mah)—This type of cancer involves melanocytes, the pigment-producing cells of the skin.

   b Adenocarcinoma (ad-en-o-kar-sin-O-mah)—This is the general term for malignant tumors of glandular epithelium.

4 Malignant tumors that arise from connective tissues, generally called **sarcomas** (sar-KO-mahs)

   a Lymphoma (lim-FO-mah)—Lymphoma is a term used to describe a cancer of lymphatic tissue.

   b Osteosarcoma (os-tee-o-sar-KO-mah)—This term refers to a malignant tumor of bone tissue.

   c Myeloma (my-el-O-mah)—This is a type of malignant bone marrow tumor.

   d Fibrosarcoma (fy-bro-sar-KO-mah)—This is a general term used to describe cancers involving fibrous connective tissues.

Miscellaneous tumors do not fit either category. For example, an *adenofibroma* (ad-en-o-fy-BRO-mah) is a benign neoplasm formed by epithelial and connective tissues. Another example is **neuroblastoma** (noo-roh-blast-OH-mah), a malignant tumor that arises from nerve tissue.

Cancers can be further classified by their location. For example, malignant tumors may be labeled *skin cancer*, *stomach cancer*, or *lung cancer* according to the location of the affected tissues. The more common forms of cancer in the United States are listed in Table 4-8 and are described in later chapters.

CAUSES OF CANCER

The etiologies of various forms of cancer puzzle researchers no less today than 100 years ago. The more we know about how cancer develops, the more questions we have. Currently, the best answer to the question "What causes cancer?" is "Many different things." We know that cancer is a type of neoplasm, which means that it involves uncontrolled cell division. A process called **hyperplasia** (hye-per-PLAY-zha) produces too many cells. Also, abnormal, undifferentiated tumor cells are often produced by a process called

▽ TABLE 4-8   Major Forms of Cancer*
Lung cancer
Colorectal cancer
Breast cancer
Prostate cancer
Uterine cancer (including cervical cancer)
Urinary (bladder and kidney) cancer
Oral (lip, mouth, and throat) cancer
Pancreatic cancer
Leukemia (cancer of blood tissue)
Lymphoma (cancer of lymphatic tissue)
Ovarian cancer
Skin cancer

*By location.

anaplasia (an-a-PLAY-zha). Thus the mechanism of all cancers is a mistake or problem in cell division. We are uncertain of the cause of the abnormal cell division. Currently, the following factors are known to play a role:

1 **Genetic factors**—Over a dozen forms of cancer are known to be directly inherited, perhaps involving abnormal "cancer genes" called **oncogenes** (AHN-ko-jeens). The way in which oncogenes work is not yet clearly understood. Other cancers may develop primarily in those with genetic predispositions to specific forms of cancer. Cancers with known genetic risk factors include basal cell carcinoma (a type of skin cancer), breast cancer, and neuroblastoma (a cancer of nerve tissue).

2 **Carcinogens** (kar-SIN-o-jens)—Carcinogens ("cancer makers") are chemicals that affect genetic activity in some way, causing abnormal cell reproduction. Some carcinogens are **mutagens** (MYOOT-ah-jens) ("mutation makers"). Mutagens cause changes in a cell's DNA structure. Although many industrial products such as benzene are known to be carcinogens, a wide variety of natural vegetable and animal materials are also carcinogenic.

3 **Age**—Certain cancers are found primarily in young people (for example, leukemia) and others

▼ **TABLE 4-9** The Warning Signs of Cancer
Sores that do not heal
Unusual bleeding
A change in wart or mole
A lump or thickening in any tissue
Persistent hoarseness or cough
Chronic indigestion
A change in bowel or bladder function

primarily in older adults (for example, colon cancer). The age factor may result from changes in the genetic activity of cells over time or from accumulated effects of cell damage.

4 Environment—Exposure to damaging types of radiation or chronic mechanical injury can cause cancer. For example, sunlight can cause skin cancer, and breathing asbestos fibers can cause lung cancer. Also, exposure to high concentrations of certain metals such as nickel or chromium can cause tumors to develop.

5 Viruses—Several cancers have now been identified as having a viral origin. This makes sense because we know that viruses often change the genetic machinery of infected cells. For example, papilloma viruses have been blamed for some cases of cervical cancer in women.

PATHOGENESIS OF CANCER

Signs of cancer include those a person would expect of a malignant neoplasm—the appearance of abnormal, rapidly growing tissue. Cancer specialists, or *oncologists,* have summarized some major signs of early stages of cancer. These signs are listed in Table 4-9.

Early detection of cancer is important because in the early stages of development of primary tumors, before metastasis and the development of secondary tumors has begun, cancer is most treatable. Some methods currently used to detect the presence of cancer include the following:

1 Self-examination—Self-examination for the early signs of cancer is one method of detection.

For example, women are encouraged to perform a monthly breast self-examination. Likewise, men are encouraged to perform a monthly testicular self-examination. If an abnormality is found, it can be further investigated with one of the methods described later. Self-examination of the skin and other accessible organs or tissues is also recommended by cancer specialists.

2 Diagnostic imaging—A variety of methods are available for forming images of internal body organs to detect tumors without exploratory surgery. **Radiography** is the oldest and still the most widely used method of noninvasive imaging of internal body structures. Radiography is the use of *x rays* to form a still or moving picture of some of the internal tissues of the body. A *mammogram,* for example, is an x-ray photograph of a breast. Potentially cancerous lumps show up as small, white areas on the mammogram (Figure 4-15, *A*). **Computed tomography (CT)** scanning is a type of radiography in which x rays produce a cross-sectional image of body regions (Figure 4-15, *B*). **Magnetic resonance imaging (MRI)** is a type of scanning that uses a magnetic field to induce tissues to emit radio waves. Different tissues can be distinguished because each emits different signals. With MRI, tumors can then be visualized on a computer screen in cross sections similar to those produced in CT scanning (Figure 4-15, *C*). MRI is also sometimes called *nuclear magnetic resonance (NMR) imaging.* In **ultrasonography,** high-frequency sound waves can be reflected off internal tissues to produce images, or *sonograms,* of tumors (Figure 4-15, *D*).

3 Biopsy—After a neoplasm has been identified with one of the previously mentioned techniques, a biopsy of the tumor may be done. A biopsy is the removal and examination of living tissue. Microscopic examination of tumor tissue removed surgically or through a needle sometimes reveals whether the tissue is malignant or benign. A very simple, noninvasive type of biopsy used to detect some types of cancer involves simply scraping cells from an exposed surface and smearing them on a glass microscope slide. For example, the *Papanicolaou* (pap-ah-nik-o-LAH-oo) *test* or "Pap smear" is a common screening procedure in which cells from the neck of the uterus (cervix) are examined (see Chapter 21).

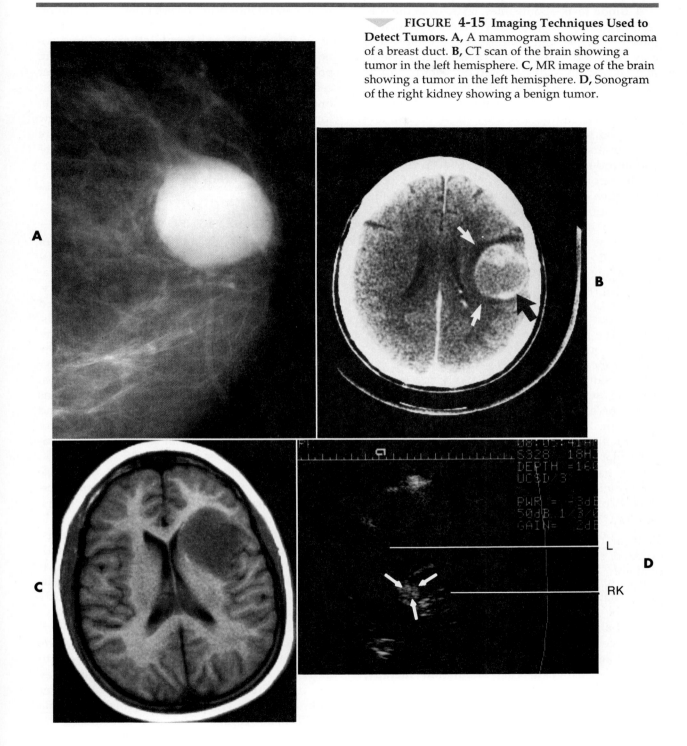

FIGURE **4-15** Imaging Techniques Used to Detect Tumors. **A,** A mammogram showing carcinoma of a breast duct. **B,** CT scan of the brain showing a tumor in the left hemisphere. **C,** MR image of the brain showing a tumor in the left hemisphere. **D,** Sonogram of the right kidney showing a benign tumor.

4 Blood tests—Sometimes, changes in the concentration of normal blood components, such as ions or enzymes, can indicate cancer. Cancer cells may also produce or trigger production of abnormal substances—substances often referred to as *tumor markers*. For example, bone cancer and other malignancies can elevate the blood concentration of calcium ions (Ca^{++}) above normal levels. Blood tests to help detect tumor markers of prostate and other cancers are being developed and introduced.

The information gained from these and other techniques can be used to *stage* and *grade* malignant tumors. Staging involves classifying a tumor based on its size and the extent of its spread. Grading is an assessment of what the tumor is likely to do based on the degree of cell abnormality. Grading is a useful basis for making a **prognosis** (prog-NO-sis) or statement of the probable outcome of the disease.

Without treatment, cancer usually results in death. The progress of a particular type of cancer depends on the type of cancer and its location. Many cancer patients suffer from **cachexia** (ka-KEK-see-ah), a syndrome involving loss of appetite, severe weight loss, and general weakness. The cause of cachexia in cancer patients is uncertain. A variety of anatomical or functional abnormalities may arise as a result of damage to particular organs. The ultimate causes of death in cancer patients include secondary infection by pathogenic microbes, organ failure, hemorrhage (blood loss), and in some cases, undetermined factors.

Of course, after cancer has been identified, every effort is made to treat it and thus prevent or delay its development. Surgical removal of cancerous tumors is sometimes done, but the probability that malignant cells have been left behind must be addressed. **Chemotherapy** (kee-moh-THAIR-ah-pee) or "chemical therapy" using *cytotoxic* ("cell-killing") compounds or *antineoplastic* drugs can be used after surgery to destroy any remaining malignant cells. **Radiation therapy,** also called *radiotherapy,* using destructive x-ray or gamma radiation may be used alone or with chemotherapy to destroy remaining cancer cells. Chemotherapy and radiation therapy may have severe side effects because normal cells are often killed with the cancer cells. **Laser** therapy, in which an intense beam of light destroys a tumor, is also sometimes performed with chemotherapy or radiation therapy. **Immunotherapy** (im-yoo-no-THAIR-ah-pee), a newer cancer treatment, bolsters the body's own defenses against cancer cells. Because viruses cause some types of cancer, oncologists hope that vaccines against certain forms of cancer will be developed. Although new and different approaches to cancer treatment are being investigated, many researchers are concentrating on improving existing methods.

Inflammation

INFLAMMATORY RESPONSE

The **inflammatory response** is a combination of processes that attempt to minimize injury to tissues, thus maintaining homeostasis. Inflammation may occur as a response to any tissue injury, including mechanical injuries such as cuts and burns or many other irritants such as chemicals, radiation, or toxins released by bacteria. Inflammation may also accompany specific immune system reactions (which are discussed in Chapter 14). First described by a Roman physician almost 2000 years ago, the inflammatory response has four primary signs—redness, heat, swelling, and pain. These signs are indicators of a complex process that is summarized in the following paragraphs and in Figures 4-16 and 4-17.

As tissue cells are damaged, they release **inflammation mediators** such as *histamine, prostaglandins,* and compounds called *kinins*. Some inflammation mediators cause blood vessels to dilate (widen), increasing blood volume in the tissue. Increased blood volume produces the redness and heat of inflammation. This response is important because it allows immune system cells (white blood cells) in the blood to travel quickly and easily to the site of injury.

Some inflammation mediators increase the permeability of blood vessel walls. This allows immune cells and other blood components to move out of the blood vessels easily where they can deal directly with injured tissue. As water leaks out of the vessel, tissue swelling or **edema** (ed-EE-mah) results. The pressure caused by edema triggers pain receptors, consciously alerting an individual of the damage. The excess fluid often has the beneficial effect of diluting the irri-

FIGURE **4-16** Inflammatory Response.

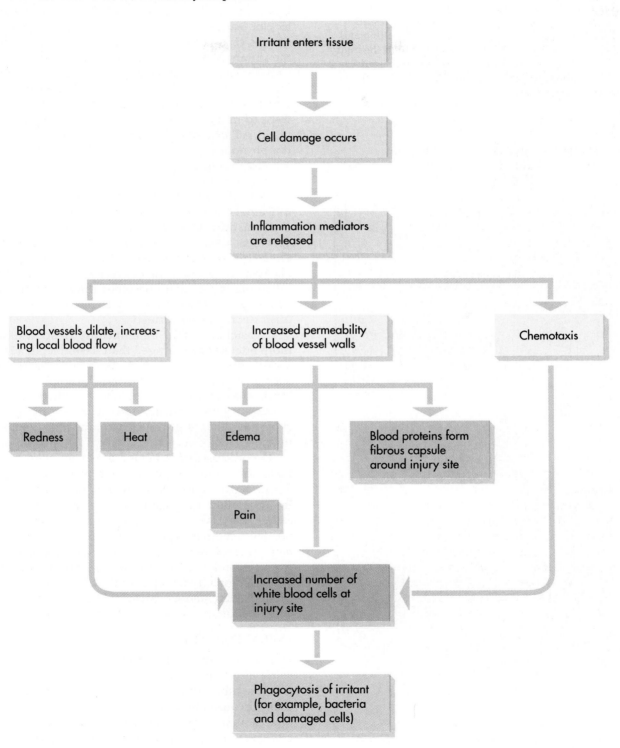

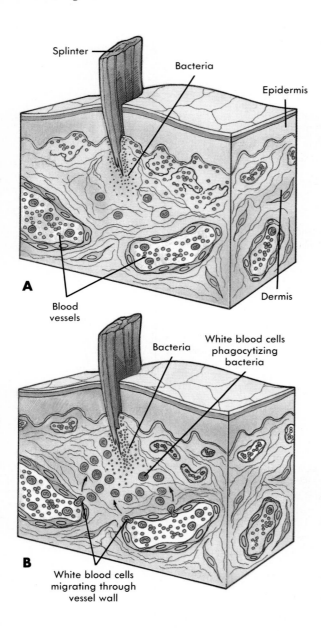

▽ **FIGURE 4-17 Typical Inflammatory Response to a Mechanical Injury. A,** A splinter damages tissue and carries bacteria. Blood vessels dilate and begin leaking fluids, causing swelling and redness. **B,** White blood cells are attracted to the injury site and begin to consume bacteria and damaged tissue cells. A fibrous capsule separates the injury site from surrounding tissue.

tant. The fluid that accumulates in inflamed tissue is called **inflammatory exudate** (EKS-yood-ate). Blood proteins that leak into tissue spaces begin to clot within a few minutes. The clot forms a fibrous capsule around the injury site, preventing the irritant from spreading to nearby tissues.

Inflammatory exudate is slowly removed by lymphatic vessels and is carried to lymph nodes, which act as filters. Bacteria and damaged cells trapped in the lymph nodes are acted on by white blood cells in each lymph node. In some cases, lymph nodes enlarge when they process a large amount of infectious material.

Inflammation mediators can also act as signals that attract white blood cells to the injury site. The movement of white blood cells in response to chemical attractants is called **chemotaxis** (kee-mo-TAK-sis). Once in the tissue, white blood cells often consume damaged cells and pathogenic bacteria by means of *phagocytosis*. When the inflammatory exudate becomes thick with white blood cells, dead tissue and bacterial cells, and other debris, **pus** is formed.

Occasionally, the inflammatory response is more intense or prolonged than desirable. In such a case, inflammation can be suppressed by drugs such as antihistamines or aspirin. Antihistamines block the action of histamine, as their name implies. Aspirin disrupts the body's synthesis of prostaglandins, a group of inflammation mediators.

The processes of inflammation eventually eliminate the irritant, and tissue repair can begin. Tissue repair is the replacement of dead cells with living cells. In a type of tissue repair called *regeneration*, the new cells are similar to those that they replace. Another type of tissue repair is *replacement*. In replacement, the new cells are different from those that they replace, resulting in a scar. Often, fibrous tissue replaces the old tissue, a condition called *fibrosis*. Most tissue repairs are a combination of regeneration and replacement.

INFLAMMATORY DISEASE

Although many inflammation events are *local*, some affect the entire body, producing *systemic* inflammation. Local inflammation occurs when damage caused by an irritant remains isolated in a limited area, as in a small cut that becomes infected. Systemic inflammation occurs when the irritant spreads widely through the body or when inflammation mediators cause changes throughout the body.

One example of a systemic (body-wide) manifestation of the inflammatory response is a **fever.** The irritant or inflammation mediators can cause the "thermostat" of the brain to reset at a higher temperature. Instead of the normal body temperature, the body achieves and maintains a new, higher temperature. Increased temperature often kills or inhibits pathogenic microbes. Some pathophysiologists also believe that the higher temperature enhances the activity of the immune system. Fevers usually subside or "break" after the irritant has been eliminated. Fevers can also be reduced by drugs that block the fever-producing agents.

The fever response in children and in the elderly often differs from that in the normal adult. Young children often develop very high tempera-

tures in mild infections compared with adults, sometimes causing *seizures.* Elderly people often have reduced or absent fever responses during infections, which may reduce their ability to resist the infectious agent.

Acute inflammation is an immediate, protective response that promotes elimination of an irritant and subsequent tissue repair. Occasionally, chronic inflammatory conditions occur. Chronic inflammation, whether local or systemic, is always damaging to affected tissues. Thus conditions involving chronic inflammation are classified as *inflammatory diseases.* Although some inflammatory diseases are caused by known pathogens or by an abnormal immune response (allergy or autoimmunity), the causes of many of them are uncertain. Inflammatory conditions such as arthritis, asthma, eczema, and chronic bronchitis are among the most common chronic diseases in the world.

OUTLINE SUMMARY

STUDYING DISEASE

A Disease terminology
 1 Health—physical, mental, and social well-being—not merely the absence of disease
 2 Disease—an abnormality in body function that threatens health
 3 Etiology—the study of the factors that cause a disease
 4 Idiopathic—refers to a disease with an unknown cause
 5 Signs and symptoms—the objective and subjective abnormalities associated with a disease
 6 Pathogenesis—the pattern of a disease's development
B Patterns of disease
 1 Epidemiology is the study of occurrence, distribution, and transmission of diseases in human populations
 2 Endemic diseases are native to a local region
 3 Epidemics occur when a disease affects many people at the same time
 4 Pandemics are widespread, perhaps global, epidemics
 5 Discovering the cause of a disease is difficult because many factors affect disease transmission
 6 Disease can be fought through prevention and therapy (treatment)

PATHOPHYSIOLOGY

A Mechanisms of disease
 1 Pathophysiology—the study of underlying physiological aspects of disease
 2 Genetic mechanisms
 3 Pathogenic organisms
 4 Tumors and cancer
 5 Physical and chemical agents
 6 Autoimmunity
 7 Inflammation
 8 Degeneration
B Risk factors (predisposing conditions)
 1 Genetic factors
 2 Age
 3 Lifestyle
 4 Stress
 5 Environmental factors
 6 Preexisting conditions

PATHOGENIC ORGANISMS

A Types of organisms
 1 Viruses (Table 4-1 and Figure 4-4)
   a Microscopic, intracellular parasites that consist of a nucleic acid core with a protein coat
   b Invade host cells and pirate organelles and raw materials
   c Classified by shape, nucleic acid type, and method of reproduction

2 Bacteria (Table 4-2 and Figure 4-5)
  a Tiny cells without nuclei
  b Secrete toxins, parasitize host cells, or form colonies
  c Classification
    (1) By function
      (a) Aerobic—require oxygen
      (b) Anaerobic—require no oxygen
    (2) By staining properties (composition of cell wall)
      (a) Gram-positive
      (b) Gram-negative
    (3) By shape and size
      (a) Bacilli—rod-shaped cells
      (b) Cocci—round cells
      (c) Curved or spiral rods
      (d) Small bacteria—obligate parasites
  d Spores—nonreproducing forms of bacteria that resist unfavorable environmental conditions
3 Fungi (Table 4-3 and Figure 4-7)
  a Simple organisms similar to plants but lack chlorophyll
  b Yeasts—small, single-celled fungi
  c Molds—large, multicellular fungi
  d Mycotic infections—often resist treatment
4 Protozoa (Table 4-4 and Figure 4-9)
  a Large, one-celled organisms having organized nuclei
  b May infest human fluids and parasitize or destroy cells
  c Major groups
    (1) Amebas—possess pseudopodia
    (2) Flagellates—possess flagella
    (3) Ciliates—possess cilia
    (4) Sporozoa (coccidia)—enter cells during one phase of a two-part life cycle; borne by vectors (transmitters) during the other phase
5 Pathogenic animals (Table 4-5 and Figure 4-11)
  a Large, complex multicellular organisms
  b Parasitize or otherwise damage human tissues or organs
  c Major groups
    (1) Nematodes—roundworms
    (2) Platyhelminths—flatworms and flukes
    (3) Arthropods
      (a) Parasitic mites, ticks, lice, fleas
      (b) Biting or stinging wasps, bees, mosquitoes, spiders
      (c) Are often vectors of disease

B Prevention and control
  1 Mechanisms of transmission
    a Person-to-person contact
      (1) Can be prevented by education
      (2) Can be prevented by using aseptic technique (Table 4-6)
    b Environmental contact
      (1) Can be prevented by avoiding contact
      (2) Can be prevented by safe sanitation practices
    c Opportunistic invasion
      (1) Can be prevented by avoiding changes in skin and mucous membranes
      (2) Can be prevented by cleansing of wounds
    d Transmission by a vector
      (1) Can be prevented by reducing the population of vectors and reducing contact with vectors
  2 Other prevention and treatment strategies
    a Vaccination—stimulates immunity
    b Chemicals—destroy or inhibit pathogens
      (1) Antibiotics—natural compounds derived from living organisms
      (2) Synthetic compounds (for example, ACV and AZT)

TUMORS AND CANCER

A Neoplasms (tumors)—abnormal growths of cells
  1 Benign tumors remain localized
  2 Malignant tumors spread, forming secondary tumors
  3 Metastasis—cells leave a primary tumor and start a secondary tumor at a new location (Figure 4-14)
  4 Classification of tumors
    a Benign, epithelial tumors
      (1) Papilloma—fingerlike projection
      (2) Adenoma—glandular tumor
      (3) Nevus—small, pigmented tumor
    b Benign, connective tissue tumors
      (1) Lipoma—adipose (fat) tumor
      (2) Osteoma—bone tumor
      (3) Chondroma—cartilage tumor
    c Carcinomas (malignant epithelial tumors)
      (1) Melanoma—involves melanocytes
      (2) Adenocarcinoma—glandular cancer
    d Sarcomas (connective tissue cancers)
      (1) Lymphoma—lymphatic cancer
      (2) Osteosarcoma—bone cancer
      (3) Myeloma—bone marrow tumor
      (4) Fibrosarcoma—cancer of fibrous tissue

B The causes of cancer—varied and still not clearly understood
   1 Cancer involves hyperplasia (growth of too many cells) and anaplasia (development of undifferentiated cells)
   2 Factors known to play a role in causing cancer
      a Genetic factors (for example, oncogenes—cancer genes)
      b Carcinogens—chemicals that alter genetic activity
      c Age
      d Injury—chronic exposure to it
      e Viruses
C Pathogenesis of cancer
   1 Early detection is important
   2 Methods of detecting cancers (Figure 4-15)
      a Self-examination
      b Diagnostic imaging—radiography (for example, mammogram and CT scan), magnetic resonance imaging (MRI), Ultrasonography
      c Biopsy (for example, Pap smear)
      d Blood tests
   3 Staging—classifying tumors by size and extent of spread
   4 Grading—assessing the likely pattern of a tumor's development
   5 Cachexia—syndrome including appetite loss, weight loss, and general weakness
   6 Causes of death by cancer—secondary infections, organ failure, hemorrhage, and undetermined factors

7 Treatments
   a Surgery
   b Chemotherapy (chemical therapy)
   c Radiation therapy (radiotherapy)
   d Laser therapy
   e Immunotherapy
   f New strategies (for example, vaccines)

INFLAMMATION

A The inflammatory response—reduces injury to tissues, thus maintaining homeostasis (Figures 4-16 and 4-17)
   1 Signs—redness, heat, swelling, and pain
   2 Inflammation mediators (histamine, prostaglandins, and kinins)
      a Some cause blood vessels to dilate, increasing blood volume (redness and heat)—white blood cells travel quickly to injury site
      b Some increase blood vessel permeability (causing swelling or edema and pain)—white blood cells move easily out of vessels, irritant is diluted, and exudate accumulates
      c Some attract white blood cells to injury site (chemotaxis)
B Inflammatory diseases
   1 Inflammation can be local or systemic (body-wide)
   2 Fever—high body temperature caused by a resetting of the body's "thermostat"—destroys pathogens and enhances immunity
   3 Chronic inflammation can constitute a disease itself because its causes damage to tissues

NEW WORDS

ameba	cocci	microbe	protozoa
arthropod	edema	nematode	pus
autoimmunity	flagellate	oncogene	risk factor
bacilli	fungi	parasite	spore
bacterium	inflammation	pathogenesis	sporozoa
chemotaxis	inflammation mediators	pathophysiology	vector
ciliate	inflammatory exudate	platyhelminth	virus

Diseases and Other Clinical Terms

acute	chondroma	immunotherapy	nevus
adenocarcinoma	chronic	incubation	osteoma
adenoma	communicable	lipoma	osteosarcoma
antibiotic	computed tomography	lymphoma	pandemic
aseptic technique	(CT)	magnetic resonance	papilloma
biopsy	endemic	imaging (MRI)	remission
cachexia	epidemic	melanoma	rickettsia
carcinogen	etiology	metastasis	syndrome
chemotherapy	fibrosarcoma	mutagen	ultrasonography
chlamydia	idiopathic	myeloma	vaccine

CHAPTER TEST

1. A disease with an unknown cause is said to be _____.
2. The pattern of development of a disease is its _____.
3. _____ is the study of all the factors that may have caused a disease.
4. _____ is the study of the underlying physiological aspects of disease.
5. Any microscopic organism can be called a(n) _____.
6. Rod-shaped bacterial cells are called _____.
7. An arthropod that carries an infectious disease from one organism to another is called a(n) _____.
8. A(n) _____ is an killed or attenuated pathogen given to a person to stimulate immunity.
9. _____ are compounds produced by living organisms that kill or inhibit pathogens.
10. An abnormal growth of cells can be called a tumor or _____.
11. A malignant tumor is often called _____.
12. _____ are substances that trigger the development of cancer.
13. A(n) _____ is a gene that is thought to be responsible for the growth of a malignant tumor.
14. _____ is the removal and examination of living tissue.
15. An inflammation _____ is a chemical that triggers the inflammation response.

Select the most correct answer from Column B for each statement in Column A. (Only one answer is correct.)

COLUMN A

16. _____ Immune system attack against one's own cells
17. _____ Predisposing condition
18. _____ Tiny, primitive cell without an organized nucleus
19. _____ Yeasts and molds
20. _____ Amebas, flagellates, ciliates, and sporozoa
21. _____ Nematodes, platyhelminths, and arthropods
22. _____ The use of chemicals to control disease
23. _____ Spreading of cancer
24. _____ Development of abnormal cells
25. _____ Swelling

COLUMN B

a. Anaplasia
b. Autoimmunity
c. Bacterium
d. Chemotherapy
e. Edema
f. Fungi
g. Metastasis
h. Pathogenic animals
i. Protozoa
j. Risk factor

REVIEW QUESTIONS

1 Distinguish between the terms *health* and *disease*.
2 Explain what is meant by these terms: *etiology, idiopathic, communicable, endemic, epidemic, pandemic, pathogenesis, incubation,* and *remission*.
3 What is the difference between a sign and a symptom? What is a syndrome?
4 List eight categories of disease mechanisms. Briefly explain how each may cause disease.
5 List and describe six risk factors, or predisposing conditions, of disease.
6 What are viruses? Explain how viruses can cause disease.
7 Describe bacteria and list three ways to classify them.
8 How do fungi cause disease?
9 Name and describe the four major groups of pathogenic protozoa.
10 What are pathogenic animals? Name and describe three major groups of pathogenic animals.
11 Describe four ways in which pathogenic microbes can be transmitted.
12 What is a vector? Give some examples.
13 Explain some of the approaches used to treat or prevent microbial diseases.
14 What is the difference between a benign and a malignant neoplasm? Can a benign neoplasm cause death?
15 How does cancer spread?
16 List some examples of benign and malignant tumor types.
17 Name five factors involved in causing cancer.
18 How can cancer be detected? Describe several current methods.
19 Describe the pathogenesis of cancer.
20 List the events of the inflammatory response.
21 How does each part of the inflammatory response help maintain homeostasis of the body?
22 What is a fever? What role does it play in the body's defense against disease?

CLINICAL APPLICATIONS

1 Without warning, Mr. Lee begins to feel sick. His most obvious symptom is a high fever. Within 24 hours, everyone in the Lee household also feels sick and has a high temperature. Before long, nearby households have the same experience—many people in the community are now sick. The local health department would probably call on what type of health professional to investigate this situation? Would the health professional label this situation an epidemic or a pandemic? If the symptoms are caused by a bacterial infection, list some ways the pathogen could have been transmitted to so many people within a short span of time.

2 Sandy is a nurse at the local university hospital. One of her patients has a severe staphylococcal infec-

tion. What would the pathogen responsible for this infection look like under a microscope? Sandy's patient is taking a newly developed antibiotic in the hope that it will cure the infection. Do you think that this drug is natural or synthetic?

3 Fred is a first-year medical student. He received a minor scrape during a basketball game on the parking lot outside his dorm. He has cleansed the wound and applied an antibiotic as a preventive measure. The affected area is red, swollen, and mildly painful. How do you explain these symptoms? Fred's roommate suggested applying an antiinflammatory drug to the wound, but Fred refuses. What advantage might Fred see in avoiding such treatment?

The Integumentary System and Body Membranes

Objectives

*After you have completed this chapter,
you should be able to:*

1 Classify, compare the structure of, and give examples of each type of body membrane.

2 Describe the structure and function of the epidermis and dermis.

3 List and briefly describe each accessory organ of the skin.

4 List and discuss the three primary functions of the integumentary system.

5 List and describe major skin disorders and infections.

6. Classify burns and describe how to estimate the extent of a burn injury.

In Chapter 1 the concept of progressive organization of body structures from simple to complex was established. Complexity in body structure and function progresses from cells to tissues and then to organs and organ systems. This chapter discusses the skin and its **appendages**—the hair, nails, and skin glands—as an organ system. This system is called the **integumentary system. Integument** (in-TEG-yoo-ment) is another name for the skin, and the skin itself is the principal organ of the integumentary system. The skin is one of a group of anatomically simple but functionally important sheetlike structures called **membranes.** This chapter will begin with classification and discussion of important body membranes. Study of the structure and function of the integument will follow. Ideally, you should study the skin and its appendages before proceeding to the more traditional organ systems in the chapters that follow to improve your understanding of how structure is related to function.

Classification of Body Membranes

The term **membrane** refers to a thin, sheetlike structure that may have many important functions in the body. Membranes cover and protect the body surface, line body cavities, and cover the inner surfaces of the hollow organs such as the digestive, reproductive, and respiratory passageways. Some membranes anchor organs to each other or to bones, and others cover the internal organs. In certain areas of the body, membranes secrete lubricating fluids that reduce friction during organ movements such as the beating of the heart or lung expansion and contraction. Membrane lubricants also decrease friction between bones in joints. There are two major categories or types of body membranes:

1 **Epithelial membranes,** composed of epithelial tissue and an underlying layer of specialized connective tissue
2 **Connective tissue membranes** composed exclusively of various types of connective tissue; no epithelial cells are present in this type of membrane

EPITHELIAL MEMBRANES

There are three types of epithelial tissue membranes in the body:

1 Cutaneous membrane
2 Serous membranes
3 Mucous membranes

Cutaneous Membrane

The **cutaneous** (kyoo-TAY-nee-us) **membrane** or **skin** is the primary organ of the integumentary system. It is one of the most important and certainly one of the largest and most visible organs. In most individuals the skin composes some 16% of the body weight. It fulfills the requirements necessary for an epithelial tissue membrane in that it has a superficial layer of epithelial cells and an underlying layer of supportive connective tissue. Its structure is uniquely suited to its many functions. The skin will be discussed in depth later in the chapter.

Serous Membranes

Like all epithelial membranes, a **serous** (SE-rus) **membrane** is composed of two distinct layers of tissue. The epithelial sheet is a thin layer of simple squamous epithelium. The connective tissue layer forms a very thin gluelike **basement membrane** that holds and supports the epithelial cells.

The serous membrane that lines body cavities and covers the surfaces of organs in those cavities is in reality a single, continuous sheet of tissue covering two different surfaces. The name of the serous membrane is determined by its location. Using this criterion results in two types of serous membranes; the first type lines body cavities, and the second type covers the organs in those cavities. The serous membrane, which lines the walls of a body cavity much like wallpaper covers the walls of a room, is called the **parietal** (pah-RYE-i-tal) **portion.** The other type of serous membrane, which covers the surface of organs found in body cavities, is called the **visceral** (VIS-er-al) **portion.**

The serous membranes of the thoracic and abdominal cavities are identified in Figure 5-1. In the thoracic cavity the serous membranes are called **pleura** (PLOOR-ah), and in the abdominal cavity, they are called **peritoneum** (pair-i-toe-NEE-um). Look again at Figure 5-1 to note the

▽ **FIGURE 5-1 Types of Body Membranes. A,** Epithelial membranes, including cutaneous membrane (skin), serous membranes (parietal and visceral pleura and peritoneum), and mucous membranes. **B,** Connective tissue membranes, including synovial membranes. See text for explanation.

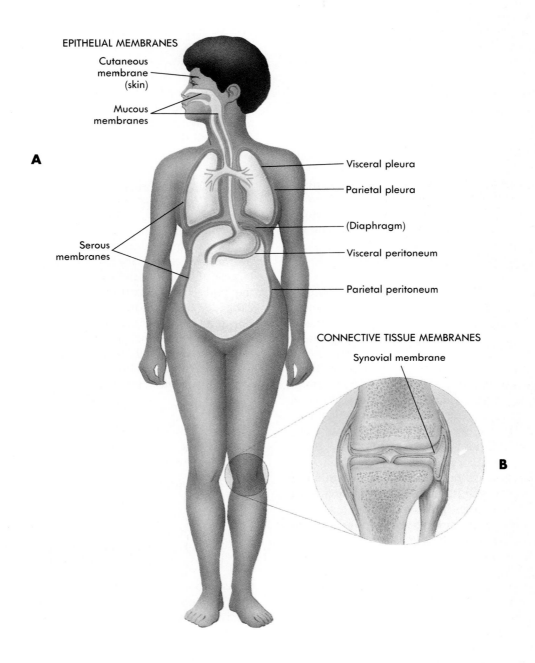

placement of the **parietal** and **visceral pleura** and the **parietal** and **visceral peritoneum.** In both cases the parietal layer forms the lining of the body cavity, and the visceral layer covers the organs found in that cavity.

Serous membranes secrete a thin, watery fluid that helps reduce friction and serves as a lubricant when organs rub against one another and against the walls of the cavities that contain them. **Pleurisy** (PLOOR-i-see) is a very painful pathological condition characterized by inflammation of the serous membranes (pleura) that line the chest cavity and cover the lungs. Pain is caused by irritation and friction as the lungs rub against the walls of the chest cavity. In severe cases the inflamed surfaces of the pleura fuse, and permanent damage may develop. The term **peritonitis** (pair-i-toe-NYE-tis) is used to describe inflammation of the serous membranes in the abdominal cavity. Peritonitis is sometimes a serious complication of an infected appendix.

Mucous Membranes

Mucous (MYOO-kus) **membranes** are epithelial membranes that line body surfaces opening directly to the exterior. Examples of mucous membranes include those lining the respiratory, digestive, urinary, and reproductive tracts. The epithelial component of a mucous membrane varies, depending on its location and function. In the esophagus, for example, a tough, abrasion-resistant stratified squamous epithelium is found. A thin layer of simple columnar epithelium covers the walls of the lower segments of the digestive tract.

The epithelial cells of most mucous membranes secrete a thick, slimy material called **mucus** that keeps the membranes moist and soft.

The term **mucocutaneous** (myoo-ko-kyoo-TAY-nee-us) **junction** is used to describe the transitional area that serves as a point of "fusion" where skin and mucous membranes meet. Such junctions lack accessory organs such as hair or sweat glands that characterize skin. These transitional areas are generally moistened by mucous glands within the body orifices or openings where these junctions are located. The eyelids, nasal openings, vulva, and anus have mucocutaneous junctions that may become sites of infection or irritation.

CONNECTIVE TISSUE MEMBRANES

Unlike cutaneous, serous, and mucous membranes, connective tissue membranes do not contain epithelial components. The **synovial** (si-NO-vee-al) **membranes** lining the spaces between bones and joints that move are classified as connective tissue membranes. These membranes are smooth and slick and secrete a thick and colorless lubricating fluid called **synovial fluid.** The membrane itself, with its specialized fluid, helps reduce friction between the opposing surfaces of bones in movable joints. Synovial membranes also line the small, cushionlike sacs called **bursae** (BER-see) that may be found between moving body parts.

The Skin

The brief description of the skin in Chapter 3 (see p. 56) identified it not only as the primary organ of the integumentary system but also as the largest and one of the most important organs of the body. Architecturally the skin is a marvel. Consider the incredible number of structures fitting into 1 square inch of skin: 500 sweat glands; over 1000 nerve endings; yards of tiny blood vessels; nearly 100 oil or sebaceous (se-BAY-shus) glands; 150 sensors for pressure, 75 for heat, 10 for cold; and millions of cells.

STRUCTURE OF THE SKIN

The skin or cutaneous membrane is a sheetlike organ composed of the following layers of distinct tissue (Figure 5-2):

1 The **epidermis** is the outermost layer of the skin. It is a relatively thin sheet of stratified squamous epithelium.
2 The **dermis** is the deeper of the two layers. It is thicker than the epidermis and is made up largely of connective tissue.

As you can see in Figure 5-2, the layers of the skin are supported by a thick layer of loose connective tissue and fat called **subcutaneous** (sub-kyoo-TAY-nee-us) **tissue.** Fat in the subcutaneous layer insulates the body from extremes of heat and cold. It also serves as a stored source of energy for the body and can be used as a food source if required. In addition, the subcutaneous tissue

FIGURE 5-2 Microscopic View of the Skin. The epidermis, shown in longitudinal section, is raised at one corner to reveal ridges, called *dermal papillae,* found in the dermis.

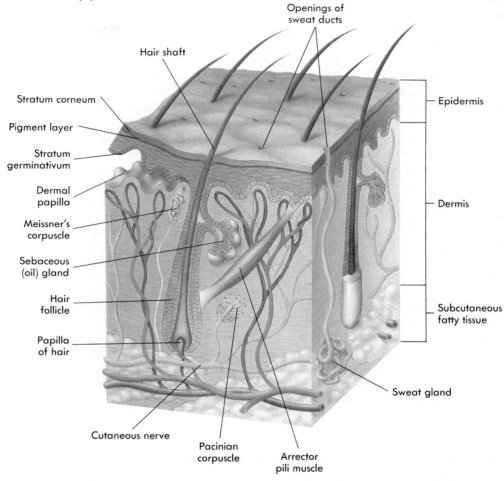

acts as a shock-absorbing pad and helps protect underlying tissues from injury caused by bumps and blows to the body surface.

Epidermis

The tightly packed epithelial cells of the epidermis are arranged in many distinct layers. The cells of the innermost layer, called the **stratum germinativum** (jer-min-ah-TY-vum), undergo mitosis and reproduce themselves (Figure 5-2). This ability is of critical clinical significance. It enables the skin to repair itself if it is injured. The self-repair-

ing characteristic of normal skin makes it possible for the body to maintain an effective barrier against infection, even when it is subjected to injury and normal wear and tear. As new cells are produced in the deep layer of the epidermis, they move toward the surface. As they approach the surface, the cytoplasm is replaced by one of nature's most unique proteins, a substance called **keratin** (KARE-ah-tin). Keratin is a tough, waterproof material that provides cells in the outer layer of the skin with a horny, abrasion-resistant, and protective quality. The tough outer layer of the epidermis is called the **stratum corneum** (KOR-nee-um). Cells filled with keratin are continually pushed to the surface of the epidermis. In the photomicrograph of the skin shown in Figure 5-3, many of the outermost cells of the stratum corneum have been dislodged. These dry, dead cells filled with keratin "flake off" by the thousands onto our clothes, our bathwater, and things we handle. Millions of epithelial cells reproduce daily to replace the millions shed—just one example of the work our bodies do without our knowledge, even when we seem to be resting.

Another cell layer of the epidermis identified in Figure 5-2 is the **pigment layer.** The term *pigment* comes from a Latin word meaning "paint." It is an appropriate name for the epidermal layer that gives color to the skin. The brown pigment **melanin** (MEL-ah-nin) is produced by specialized cells in the pigment layer. These cells are called **melanocytes** (MEL-ah-no-sites). The higher the concentration of melanin, the deeper the color of skin. The amount of melanin in your skin depends first on the skin color genes you have inherited. That is, heredity determines how dark or light your basic skin color is. However, other factors such as sunlight can modify this hereditary effect. Prolonged exposure to sunlight in light-skinned people darkens the exposed area because it leads to increased melanin deposits in the epidermis. If the skin contains little melanin, a change in color can occur if the volume of blood in the skin changes significantly or if the amount of oxygen in the blood is increased or decreased. In these individuals, increased blood flow to the skin or increased blood oxygen levels can cause a pink flush to appear. However, if blood oxygen levels decrease or if actual blood flow is reduced dramatically, the skin turns a bluish gray color—a condition called **cyanosis** (SYE-ah-NO-sis). In general, the less abundant the melanin deposits in the skin, the more visible the changes in color caused by the change in skin blood volume or oxygen level. Conversely, the richer the skin's pigmentation, the less noticeable such changes will be.

The cells of the epidermis are packed tightly together. They are held firmly to one another and to the dermis below by specialized junctions between the membranes of adjacent cells. If these specialized links, sometimes described as "spot welds," are weakened or destroyed, the skin falls apart. When this occurs because of burns, friction injuries, or exposure to irritants, **blisters** may result. The blisters shown in Figure 5-4 were caused by the irritant chemicals in poison ivy that caused cell injury and death.

The junction that exists between the thin epidermal layer of the skin above and the dermal layer below is called the **dermal-epidermal junc-**

FIGURE 5-3 Photomicrograph of the Skin. Many dead cells of the stratum corneum have flaked off from the surface of the epidermis. Note that the epidermis is very cellular. The dermis has fewer cells and more connective tissue.

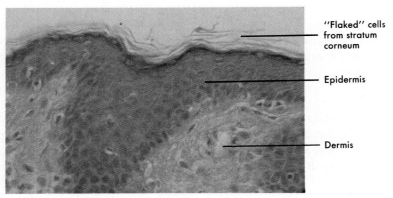

"Flaked" cells from stratum corneum

Epidermis

Dermis

▼ **FIGURE 5-4 Blisters Resulting from Contact with Poison Ivy.**

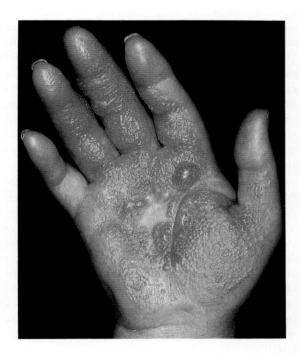

▼ **FIGURE 5-5 Skin Changes with Age.** As we age, the skin loses elasticity and forms wrinkles.

tion. The area of contact between dermis and epidermis "glues" them together and provides support for the epidermis, which is attached to its upper surface. Blister formation also occurs if this junction is damaged or destroyed. The junction is visible in Figure 5-2, which shows the epidermis raised on one corner to reveal the underlying dermis more clearly.

Dermis

The dermis is the deeper of the two primary skin layers and is much thicker than the epidermis. It is composed largely of connective tissue. Instead of cells being crowded close together like the epithelial cells of the epidermis, they are scattered far apart, with many fibers in between. Some of the fibers are tough and strong (collagen or white fibers), and others are stretchable and elastic (elastic or yellow fibers).

The upper region of the dermis is characterized by parallel rows of peglike projections called **der-**

mal papillae (pah-PIL-ee), which are visible in Figure 5-2. These upward projections are interesting and useful structural features. They form an important part of the dermal-epidermal junction that helps bind the two skin layers together. In addition, they form the ridges and grooves that make possible fingerprinting as a means of identification.

Observe these ridges on the tips of the fingers and on the skin covering the palms of your hands. Observe in Figure 5-2 how the epidermis follows the contours of the dermal papillae. These ridges develop sometime before birth. Their pattern is unique in each individual, and it never changes except to grow larger, which explains why our fingerprints or footprints positively identify us. Many hospitals identify newborn babies by footprinting them soon after birth.

The deeper area of the dermis is filled with a dense network of interlacing fibers. Most of the fibers in this area are collagen that gives toughness to the skin. However, elastic fibers are also present. These make the skin stretchable and elastic (able to rebound). As we age, the number of elastic fibers in the dermis decreases, and the amount of fat stored in the subcutaneous tissue is reduced. Wrinkles develop as the skin loses elasticity, sags, and becomes less soft and pliant (Figure 5-5).

▼ **FIGURE 5-6 Hair Follicle.** Relationship of a hair follicle and related structures to the epidermal and dermal layers of the skin.

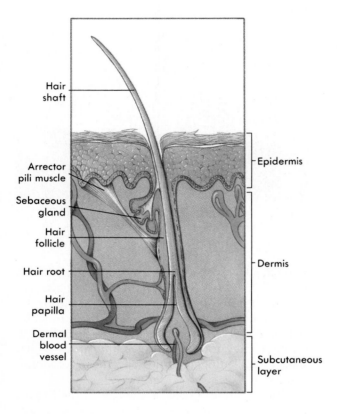

In addition to connective tissue elements, the dermis contains a specialized network of nerves and nerve endings to process sensory information such as pain, pressure, touch, and temperature. At various levels of the dermis, there are muscle fibers, hair follicles, sweat and sebaceous glands, and many blood vessels.

APPENDAGES OF THE SKIN

Hair

The human body is covered with millions of hairs. Indeed, at the time of birth most of the specialized structures called **follicles** (FOL-li-kuls) that are required for hair growth are already present. They develop early in fetal life and by birth are present in most parts of the skin. The hair of a newborn infant is extremely fine and soft; it is called **lanugo** (lah-NOO-go) from the Latin word meaning "down." In premature infants, lanugo may be noticeable over most of the body, but soon after birth the lanugo is lost and replaced by new hair that is stronger and more pigmented. Although only a few areas of the skin are hairless—notably the lips, the palms of the hands, and the soles of the feet—most body hair remains almost invisible. Hair is most visible on the scalp, eyelids, and eyebrows. The coarse hair that first appears in the pubic and axillary regions at the time of puberty develops in response to the secretion of sex hormones.

Hair growth begins when cells of the epidermal layer of the skin grow down into the dermis, forming a small tube called the **hair follicle.** The relationship of a hair follicle and its related structures to the epidermal and dermal layers of the skin is shown in Figure 5-6. Hair growth begins from a small, cap-shaped cluster of cells called the **hair papilla** (pah-PIL-ah), which is located at the base of the follicle. The papilla is nourished by a dermal blood vessel. Note in Figure 5-6 that part of the hair, namely the **root,** lies hidden in the follicle. The visible part of a hair is called the **shaft.**

As long as cells in the papilla of the hair follicle remain alive, new hair will replace any that is cut or plucked. Contrary to popular belief, frequent cutting or shaving does not make hair grow faster or become coarser. Why? Because neither process affects the epithelial cells that form the hairs, since they are embedded in the dermis.

Hair loss of any kind is called **alopecia** (al-o-PEE-sha). Some forms of alopecia, such as *male pattern baldness* are not diseases but are simply inherited traits. Alopecia may also be a normal consequence of aging (see Chapter 22). Rapid loss of hair in rounded patches, such as that seen in Figure 5-7, can occur without a known cause but is often associated with metabolic diseases such as thyroid disease and lupus erythematosus (see Appendix B). This rapid hair loss is usually followed by normal regrowth. Scalp infections, chemotherapy, and radiation sickness can also cause rapid hair loss.

A tiny, smooth (involuntary) muscle can be seen in Figure 5-6. It is called an **arrector pili** (ah-REK-tor PYE-lie) muscle. It is attached to the base of a dermal papilla above and to the side of a hair follicle below. Generally, these muscles contract only when we are frightened or cold. When contraction occurs, each muscle simultaneously pulls on its two points of attachment (that is, up on a hair follicle but down on a part of the skin). This produces little raised places, called *goose pimples,* between the depressed points of the skin and at the same time pulls the hairs up until they are more or less straight. The name *arrector pili* describes the function of these muscles; it is Latin for "erectors of the hair." We unconsciously recognize these facts in expressions such as "I was so frightened my hair stood on end."

Receptors

Receptors in the skin make it possible for the body surface to act as a sense organ, relaying messages to the brain concerning sensations such as touch, pain, temperature, and pressure. Receptors differ in structure from the highly complex to the very simple. Figure 5-8 shows enlarged views of a **Meissner's** (MIZE-ners) **corpuscle** and a **pacinian** (pah-SIN-ee-an) **corpuscle.** Look again at Figure 5-2 and find these receptors. The pacinian corpuscle is deep in the dermis. It is capable of detecting *pressure* on the skin surface.

The Meissner's corpuscle is generally located close to the skin surface. It is capable of detecting sensations of *light touch.* Both specialized receptors are widely distributed in skin. Additional receptors in the skin respond to other types of

▽ **FIGURE 5-7** Alopecia.

▽ **FIGURE 5-8 Skin Receptors.** Receptors are specialized nerve endings that make it possible for the skin to act as a sense organ. **A,** Meissner's corpuscle. **B,** Pacinian corpuscle. (See also Figure 5-2.)

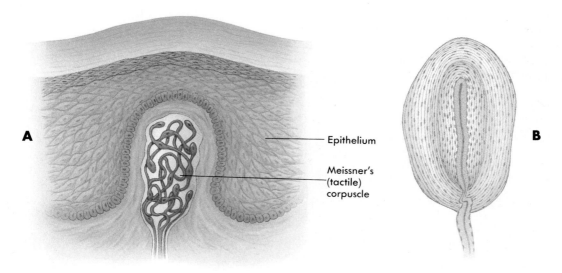

A

B

Epithelium

Meissner's (tactile) corpuscle

stimuli. For example, **free nerve endings** respond to pain, and receptors called **Krause's end bulbs** detect sensations of cold. Other receptors mediate sensations of heat, crude touch, and vibration.

Nails

Nails are classified as accessory organs of the skin and are produced by cells in the epidermis. They form when epidermal cells over the terminal ends of the fingers and toes fill with keratin and become hard and platelike. The components of a typical fingernail and its associated structures are shown in Figure 5-9. In this illustration the fingernail of the index finger is viewed from above and in sagittal section. (Recall that a sagittal section divides a body part into right and left portions.) Look first at the nail as seen from above. The visible part of the nail is called the **nail body.** The rest of the nail, namely, the **root,** lies in a groove and is hidden by a fold of skin called the

cuticle (KYOO-ti-kul). In the sagittal section you can see the nail root from the side and note its relationship to the cuticle, which is folded back over its upper surface. The nail body nearest the root has a crescent-shaped white area known as the **lunula** (LOO-nyoo-lah), or "little moon." You should be able to identify this area easily on your own nails; it is most noticeable on the thumbnail. Under the nail lies a layer of epithelium called the **nail bed,** which is labeled on the sagittal section in Figure 5-9. Because it contains abundant blood vessels, it appears pink in color through the translucent nail bodies. If blood oxygen levels drop and cyanosis develops, the nail bed will turn blue.

Skin Glands

The skin glands include the two varieties of **sweat** or **sudoriferous** (soo-doe-RIF-er-us) **glands** and the microscopic **sebaceous glands.**

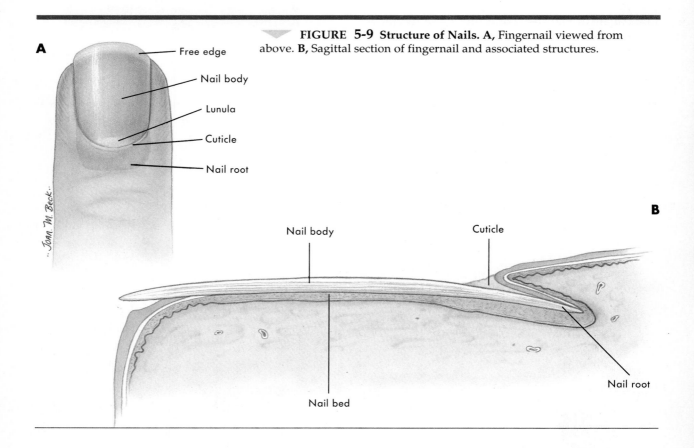

FIGURE 5-9 Structure of Nails. A, Fingernail viewed from above. **B,** Sagittal section of fingernail and associated structures.

A

Free edge

Nail body

Lunula

Cuticle

Nail root

Joan M. Beck.

B

Nail body

Cuticle

Nail root

Nail bed

Sweat (Sudoriferous) glands. Sweat or **sudoriferous glands** are the most numerous of the skin glands. They can be classified into two groups—**eccrine** (EK-rin) and **apocrine** (AP-o-krin)—based on type of secretion and location. **Eccrine sweat glands** are by far the more numerous, important, and widespread sweat glands in the body. They are quite small and with few exceptions are distributed over the total body surface. Throughout life, they produce a transparent watery liquid called **perspiration,** or **sweat.** Sweat assists in the elimination of waste products such as ammonia and uric acid. In addition to elimination of waste, sweat plays a critical role in helping the body maintain a constant temperature. Anatomists estimate that a single square inch of skin on the palms of the hands contains about 3000 eccrine sweat glands. With a magnifying glass you can locate the pinpoint-size openings on the skin that you probably call **pores.** The pores are outlets of small ducts from the eccrine sweat glands.

Apocrine sweat glands are found primarily in the skin in the armpit (axilla) and in the pigmented skin areas around the genitals. They are larger than the eccrine glands, and instead of watery sweat, they secrete a thicker milky secretion. The odor associated with apocrine gland secretion is not caused by the secretion itself. Instead, it is caused by the contamination and decomposition of the secretion by skin bacteria. Apocrine glands enlarge and begin to function at puberty.

Sebaceous glands. Sebaceous glands secrete oil for the hair and skin. Oil or sebaceous glands grow where hairs grow. Their tiny ducts open into hair follicles (Figure 5-6) so that their secretion, called **sebum** (SEE-bum), lubricates the hair and skin. Someone aptly described sebum as "nature's skin cream" because it prevents drying and cracking of the skin. Sebum secretion increases during adolescence, stimulated by the increased blood levels of the sex hormones. Frequently sebum accumulates in and enlarges some of the ducts of the sebaceous glands, forming white pimples. This sebum often darkens, forming a **blackhead.** Sebum secretion decreases in late adulthood, contributing to increased wrinkling and cracking of the skin.

Acne. The most common kind of acne, **acne vulgaris** (AK-nee vul-GAIR-is) (Figure 5-10), oc-

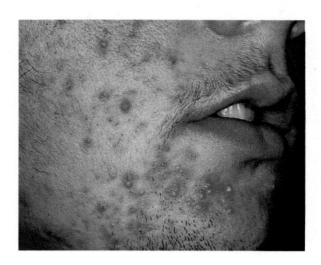

▼ **FIGURE 5-10** Acne.

curs most frequently during adolescence. This condition results from the more than fivefold increase in sebum secretion between the ages of 10 and 19. The oversecretion of sebum results in blockage of the sebaceous gland ducts with sebum, skin cells, and bacteria. The inflamed lesions that result are called *papules* (PAP-yools). Pus-filled pimples called **pustules** (PUS-tyools) often develop and then rupture, resulting in secondary infections in the surrounding skin. Formation of acne lesions can be minimized by careful cleansing of the skin and use of acne-controlling drugs. Many topical (external) creams used to treat acne contain *benzoyl peroxide,* a drug that causes drying and peeling of skin to remove sebaceous plugs and releases oxygen to inhibit anaerobic skin bacteria.

FUNCTIONS OF THE SKIN

The skin or cutaneous membrane serves three important functions that contribute to survival. The most important functions are:

1 Protection
2 Temperature regulation
3 Sense organ activity

Protection

The skin as a whole is often described as our "first line of defense" against a multitude of hazards. It protects us against the daily invasion of deadly microbes. The tough, keratin-filled cells of the stratum corneum also resist the entry of harmful chemicals and protect against physical tears and cuts. Because it is waterproof, **keratin** also protects the body from excessive fluid loss. Melanin in the pigment layer of the skin prevents the sun's harmful ultraviolet rays from penetrating the interior of the body.

Temperature Regulation

The skin plays a key role in regulating the body's temperature. Incredible as it seems, on a hot and humid day the skin can serve as a means for releasing almost 3000 calories of body heat—enough heat energy to boil over 20 liters of water! It accomplishes this feat by regulating sweat secretion and by regulating the flow of blood close to the body surface. When sweat evaporates from the body surface, heat is also lost. The principle of heat loss through evaporation is basic to many cooling systems. When increased quantities of blood are allowed to fill the vessels close to the skin, heat is also lost by radiation. Blood supply to the skin far exceeds the amount needed by the skin. Such an abundant blood supply primarily enables the regulation of body temperature.

Sense Organ Activity

The skin functions as an enormous sense organ. Its millions of nerve endings serve as antennas or receivers for the body, keeping it informed of changes in its environment. The specialized receptors shown in Figures 5-2 and 5-8 make it possible for the body to detect sensations of light touch (Meissner's corpuscles) and pressure (pacinian corpuscles). Other receptors make it possible for us to respond to the sensations of pain, heat, and cold.

Disorders of the Skin

Any disorder of the skin can be called a **dermatosis** (der-ma-TOE-sis), which simply means "skin condition." Many dermatoses involve inflammation of the skin, or **dermatitis** (der-ma-TIE-tis). Only a few of the many disorders of the skin are discussed here.

SKIN LESIONS

A **lesion** (LEE-zhun) is any measurable variation from the normal structure of a tissue. Lesions are not necessarily signs of disease but may be benign variations that do not constitute a disorder. For example, freckles are considered lesions but are not signs of disease.

Almost all diseases affecting the skin are discovered and diagnosed after observing the nature of the lesions present. Lighting the skin from the side with a penlight is a method used to determine the category of a lesion: elevated, flat, or depressed. Elevated lesions cast shadows outside their edges; flat lesions do not cast shadows, and depressed lesions cast shadows inside their edges. Important examples of each type of lesion are summarized in Table 5-1 on pp. 114 and 115.

Lesions are often distinguished by abnormal density of tissue or abnormal coloration. Overgrowth or deficient growth of skin cells, calcification, and edema can cause changes in skin density. Discoloration can result from overproduction or underproduction of skin pigments such as the increase in melanin seen in a mole. A decrease in blood flow or oxygen content can give the skin a bluish cast (cyanosis), whereas an increased blood flow or oxygen content can give a red or darker hue to the skin. Discoloration of the affected area is associated with most skin lesions.

BURNS

Burns constitute one of the most serious and frequent problems that affect the skin. Typically, we think of a burn as an injury caused by fire or by contact of the skin with a hot surface. However, overexposure to ultraviolet light (sunburn) or contact of the skin with an electric current or a harmful chemical such as an acid can also cause burns.

Estimating Body Surface Area

When burns involve large areas of the skin, treatment and the possibility for recovery depend in large part on the **total area involved** and the **severity of the burn.** The severity of a burn is determined by the depth of the injury, as well as by the amount of body surface area affected.

The **"rule of nines"** is one of the most frequently used methods of determining the extent of a burn injury. With this technique (Figure 5-11) the adult body is divided into 11 areas of 9% each, with the area around the genitals representing the additional 1% of body surface area. As you can see in Figure 5-11 on p. 116, in the adult 9% of the skin covers the head and each upper extremity, including front and back surfaces. Twice as much, or 18%, of the total skin area covers the front and back of the trunk and each lower extremity, including front and back surfaces. The formula for the rule of nines varies slightly for infants and children because their body proportions differ from the adult.

Classification of Burns

The classification system used to describe the severity of burns is based on the number of tissue layers involved. The most severe burns destroy not only layers of the skin and subcutaneous tissue but underlying tissues, as well.

First-degree burns. A **first-degree burn** (for example, a typical sunburn) causes minor discomfort and some reddening of the skin. Although the surface layers of the epidermis may peel in 1 to 3 days, no blistering occurs, and actual tissue destruction is minimal.

Second-degree burns. A **second-degree burn** involves the deep epidermal layers and always causes injury to the upper layers of the dermis. Although deep second-degree burns damage sweat glands, hair follicles, and sebaceous glands, complete destruction of the dermis does not occur. Blisters, severe pain, generalized swelling, and fluid loss characterize this type of burn. Scarring is common.

Third-degree burns. A **third-degree burn** is characterized by complete destruction of the epidermis and dermis. Tissue death extends below the primary skin layers into the subcutaneous tissue. Third-degree burns often involve underlying muscles and even bone. One distinction between second- and third-degree burns is that third-degree lesions are insensitive to pain immediately after injury because of the destruction of nerve endings. The fluid loss that results from third-degree burns is a very serious problem.

EXERCISE AND THE SKIN

Excess heat produced by the skeletal muscles during exercise increases the core body temperature far beyond the normal range. Because blood in vessels near the skin's surface dissipates heat well, the body's control centers adjust blood flow so that more warm blood from the body's core is sent to the skin for cooling. During exercise, blood flow in the skin can be so high that the skin takes on a redder coloration.

To help dissipate more heat, sweat production increases to as high as 3 L per hour during exercise. Although each sweat gland produces very little of this total, over 3 million individual sweat glands are found throughout the skin. Sweat evaporation is essential to keeping body temperature in balance, but excessive sweating can lead to a dangerous loss of fluid. Because normal drinking may not replace the water lost through sweating, it important to increase fluid consumption during and after any type of exercise to avoid **dehydration.**

	TABLE 5-1	Common Skin Lesions			
LESION	**DESCRIPTION**	**EXAMPLE**	**LESION**	**DESCRIPTION**	**EXAMPLE**
Elevated			**Elevated**—cont'd		
Papule	Firm, raised lesion (less than 1 cm in diameter)	Warts	Pustule	Elevated lesion filled with pus	Acne
Plaque	Large, raised lesion (greater than 1 cm in diameter)	Psoriasis	Crust	Scab; area with dried blood or exudate	Scrape wound
Vesicle	Thin-walled blister filled with fluid that is smaller than 1 cm (a vesicle larger than 1 cm is a *bulla*)	Second-degree burn	Wheal (hive)	Firm, raised area of irregular shape with a light center	Insect bite

TABLE 5-1—cont'd   Common Skin Lesions

LESION	DESCRIPTION	EXAMPLE	LESION	DESCRIPTION	EXAMPLE
Flat			**Depressed**—cont'd		
Macule	Area distinguished from surrounding skin by color	Freckle	Ulcer	Craterlike lesion caused by disintegration of skin	Bedsore or pressure sore

Depressed			Fissure	Linear crack or break from epidermis to dermis	Athlete's foot
Excoriation	Area in which epidermis is missing exposing the dermis	Scratch			

▼ **FIGURE 5-11 The "Rule of Nines."** Dividing the body into 11 areas of 9% each helps one to estimate the amount of skin surface burned in an adult.

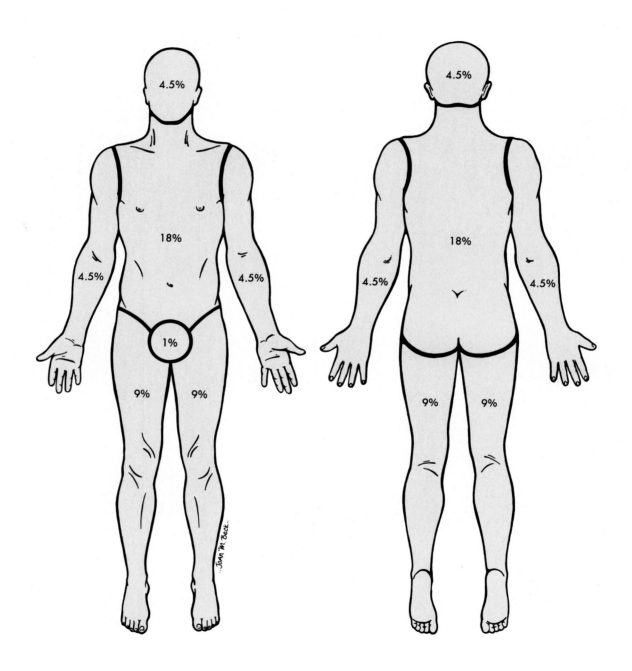

SKIN INFECTIONS

The skin is the first line of defense against microbes that might otherwise invade the body's internal environment. So the skin is a common site of infection. Viruses, bacteria, fungi, or larger parasites cause skin conditions such as those listed here. Refer to Appendix B for more information on these and other skin infections.

1 Impetigo (im-pe-TIE-go)—This highly contagious condition results from *staphylococcal* or *streptococcal* infection and occurs most often in young children. It starts as a reddish discoloration or **erythema** (er-ih-THEE-ma) but soon develops into vesicles and yellowish crusts (Figure 5-12, *A*). Occasionally, it becomes systemic (body-wide) and thus life threatening.

FIGURE 5-12 Common Skin Lesions. A, Impetigo. **B,** Tinea (ringworm). **C,** Common warts. **D,** Furuncle (boil). **E,** Scabies.

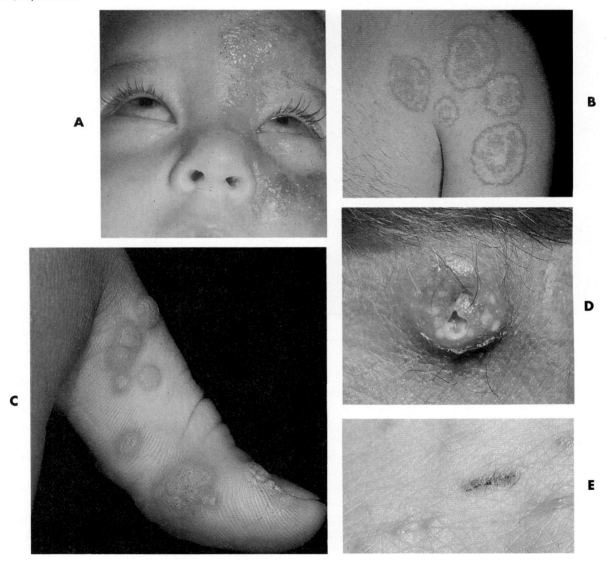

▼ **FIGURE 5-13 Decubitus Ulcer.** Also called *pressure sores* or *bedsores*, these lesions result from reduced blood flow in local areas of the skin.

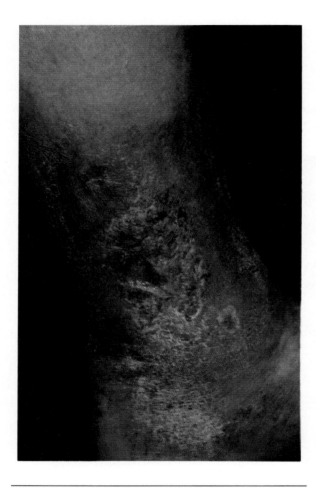

2 Tinea (TIN-ee-a)—*Tinea* is the general name for many different *mycoses* (fungal infections) of the skin. Ringworm, jock itch, and athlete's foot are classified as tinea. Signs of tinea include erythema, scaling, and crusting. Occasionally, fissures or cracks develop at creases in the epidermis. Figure 5-12, *B*, shows a case of ringworm, a tinea infection that typically forms a round rash that heals in the center to form a ring. Antifungal agents usually stop the acute infection. Recurrence can be avoided by keeping the skin dry because fungi require a moist environment to grow.

3 Warts—Caused by a papilloma virus, warts are a type of benign neoplasm of the skin. Some warts transform to become malignant. The nipplelike projections characteristic of this contagious condition are shown in Figure 5-12, *C*. Transmission of warts generally occurs through direct contact with lesions on the skin of an infected person. Warts can be removed by freezing, drying, laser therapy, or application of chemicals.

4 Boils—Also called **furuncles** (FUR-un-klz), boils are local staphylococci infections of hair follicles characterized by large, inflamed pustules (Figure 5-12, *D*). A group of untreated boils may fuse into even larger pus-filled lesions called **carbuncles** (KAR-bun-klz).

5 Scabies (SKAY-beez)—Scabies is a contagious skin condition caused by the itch mite *(Sarcoptes scabiei)*. Transmitted by skin-to-skin contact, as in sexual activity, the female mite digs under the hard stratum corneum and forms a short, winding burrow where she deposits her eggs (Figure 5-12, *E*). Young mites called *larvae* hatch out, forming tiny, red papules. After a month or so, a hypersensitivity reaction (see Chapter 14) may cause a rash characterized by erythema and numerous papules. As the name of the culprit indicates, infestation of the skin by itch mites causes intense itching. Excoriation that results from scratching the itchy infested areas may lead to secondary bacterial infections.

VASCULAR AND INFLAMMATORY SKIN DISORDERS

Every health care giver should be aware of the causes and nature of pressure sores or **decubitus** (de-KYOO-bit-us) **ulcers** (Figure 5-13). *Decubitus* means "lying down," a name that hints at a common cause of pressure sores: lying in one position for long periods. Also called *bedsores*, these lesions appear after blood flow to a local area of skin slows or is obstructed because of pressure on skin covering bony prominences such as the ankles. Ulcers form and infections develop because lack of blood flow causes tissue damage or death. Frequent changes in body position and soft support cushions help prevent decubitus ulcers.

A common type of skin disorder that involves blood vessels is **urticaria** (ur-ti-KAR-ee-a) or *hives*. This condition is characterized by raised red lesions called *wheals* caused by leakage of fluid from the skin's blood vessels. Urticaria is often associated with severe itching. Hypersensitivity or allergic reactions, physical irritants, and systemic diseases are common causes.

Scleroderma (skla-ro-DER-ma) is an autoimmune disease that affects the blood vessels and connective tissues of the skin. The name *scleroderma* comes from the word parts, *sclera,* which means "hard," and *derma,* which means "skin." Hard skin is a good description of the lesions characteristic of scleroderma. Scleroderma begins as an area of mild inflammation that later develops into a patch of yellowish, hardened skin. Scleroderma most commonly remains a mild, localized condition. Very rarely, localized scleroderma progresses to a systemic form, affecting large areas of the skin and other organs. Persons with advanced systemic scleroderma seem to be wearing a mask because skin hardening prevents them from moving their mouths freely. Both forms of scleroderma occur more commonly in women than in men.

Psoriasis (so-RYE-a-sis) is a chronic inflammatory disorder of the skin thought to have a genetic basis. This common skin problem is characterized by cutaneous inflammation accompanied by scaly plaques (Figure 5-14). The scales or plaques associated with psoriasis develop from an excessive rate of epithelial cell growth.

Eczema (EK-ze-ma) is the most common inflammatory disorder of the skin. This condition is characterized by inflammation often accompanied by papules, vesicles, and crusts. Eczema is not a distinct disease but rather a sign or symptom of an underlying condition. For example, an allergic reaction called *contact dermatitis* can progress to become eczematous. Poison ivy (Figure 5-4) is a form of contact dermatitis—occurring upon *contact* with chemicals coating the poison ivy plant.

SKIN CANCER

Of the many types of skin cancer, the most common are **squamous cell carcinoma, basal cell carcinoma,** and malignant **melanoma:**

1 Squamous cell carcinoma—This slow-growing malignant tumor of the epidermis is the most common type of skin cancer. Lesions typical of this form of skin cancer are hard, raised nodules that are usually painless (Figure 5-15, *A*). If not treated, squamous cell carcinoma will metastasize, invading other organs.

2 Basal cell carcinoma—Usually occurring on the upper face, this type of skin cancer is much less likely to metastasize than other types. This malignancy begins in cells at the base of the epidermis (the basal layer of stratum germinativum). Basal cell carcinoma lesions typically begin as papules that erode in the center to form a bleeding, crusted crater (Figure 5-15, *B*).

3 Melanoma—Malignant melanoma, the fastest increasing cancer in the United States today, is the most serious form of skin cancer; it causes death in about one in every four cases. This type of cancer sometimes develops from a pigmented nevus (mole) to become a dark, spreading lesion (Figure 5-15, *C*). Benign moles should be checked regularly for warning signs of melanoma because early detection and removal is essential in treating this rapidly spreading cancer. The "ABCD" rule of self-examination of moles is summarized in Table 5-2.

FIGURE 5-14 Psoriasis. Note the scaly plaques characteristic of this condition.

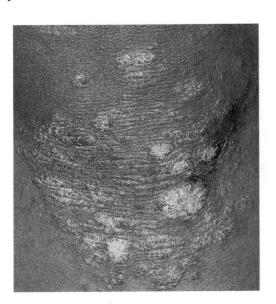

FIGURE **5-15** **Examples of Skin Cancer Lesions. A,** Squamous cell carcinoma. **B,** Basal cell carcinoma. **C,** Malignant melanoma. **D,** Kaposi's sarcoma.

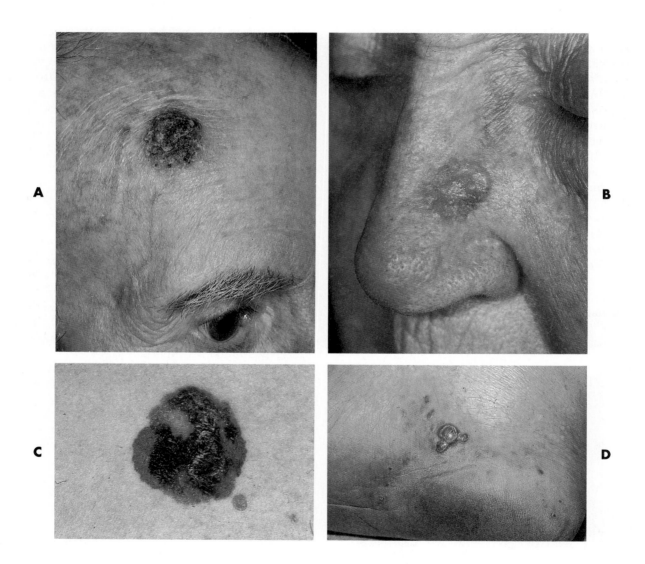

Although genetic predisposition also plays a role, many pathophysiologists believe that exposure to the sun's ultraviolet (UV) radiation is the most important factor in causing the common skin cancers. UV radiation damages the DNA in skin cells, causing the mistakes in mitosis that produce cancer. Skin cells have a natural ability to repair UV damage to the DNA, but in some people, this mechanism may not be able to deal with a massive amount of damage. People with the rare, inherited condition *xeroderma* (zee-roh-DERM-ah) *pigmentosum* cannot repair UV damage at all and almost always develop skin cancer.

One of the rarer skin cancers, **Kaposi's** (ka-PO-sees) **sarcoma,** has increased recently in some parts of the world. Once associated mainly with certain ethnic groups, a form of this cancer now appears in many cases of AIDS and other immune deficiencies. Kaposi's sarcoma, first appearing as purple papules (Figure 5-15, *D*), quickly spreads to the lymph nodes and internal organs. Some pathophysiologists believe that a virus or other agent, perhaps transmitted along with the HIV, is a possible cause of this cancer.

TABLE 5-2 Warning Signs of Malignant Melanoma

ABCD	RULE
Asymmetry	Benign moles are *symmetrical;* their halves are mirror images of each other. Melanoma lesions are asymmetrical or lopsided.
Border	Benign moles are outlined by a distinct border, but malignant melanomal lesions are often irregular or indistinct.
Color	Benign moles may be any shade of brown but are relatively evenly colored. Melanoma lesions tend to be unevenly colored, exhibiting a mixture of shades or colors.
Diameter	By the time a melanoma lesion exhibits characteristics A, B, and C, it is also probably larger than 6 mm (¼ inch).

SUNBURN AND SKIN CANCER

Burns caused by exposure to harmful UV radiation in sunlight are commonly called *sunburns.* As with any burn, serious sunburns can cause tissue damage and lead to secondary infections and fluid loss. Cancer researchers have recently theorized that blistering (second-degree) sunburns during childhood may trigger the development of malignant melanoma later in life. Some epidemiological studies show that adults who had more than two blistering sunburns before the age of 20 have a much greater risk of developing melanoma than someone who experienced no such burns. If this theory is true, it could explain the dramatic increase in skin cancer rates in the United States observed in recent years. Those who grew up as sunbathing and the resulting "suntans" became popular in the 1950s and 1960s are now, as adults, exhibiting melanoma at a much higher rate than in previous generations.

OUTLINE SUMMARY

CLASSIFICATION OF BODY MEMBRANES

A Classification of body membranes (Figure 5-1)
  1 Epithelial membranes—composed of epithelial tissue and an underlying layer of connective tissue
  2 Connective tissue membranes—composed exclusively of various types of connective tissue
B Epithelial membranes
  1 Cutaneous membrane—the skin
  2 Serous membranes—simple squamous epithelium on a connective tissue basement membrane
    a Types
      (1) Parietal—line walls of body cavities
      (2) Visceral—cover organs found in body cavities
    b Examples
      (1) Pleura—parietal and visceral layers line walls of thoracic cavity and cover the lungs
      (2) Peritoneum—parietal and visceral layers line walls of abdominal cavity and cover the organs in that cavity
    c Diseases
      (1) Pleurisy—inflammation of the serous membranes that line the chest cavity and cover the lungs
      (2) Peritonitis—inflammation of the serous membranes in the abdominal cavity that line the walls and cover the abdominal organs
  3 Mucous membranes
    a Line body surfaces that open directly to the exterior
    b Produce mucus, a thick secretion that keeps the membranes soft and moist
C Connective tissue membranes
  1 Do not contain epithelial components
  2 Produce a lubricant called *synovial fluid*
  3 Examples are the synovial membranes in the spaces between joints and in the lining of bursal sacs

THE SKIN

A Structure (Figure 5-2)—two primary layers called *epidermis* and *dermis*
  1 Epidermis
    a Outermost and thinnest primary layer of skin
    b Composed of several layers of stratified squamous epithelium
    c Innermost layer of cells continually reproduces, and new cells move toward the surface

d As cells approach surface, they are filled with a tough, waterproof protein called *keratin* and eventually flake off
    e Stratum corneum—outermost layer of keratin-filled cells
    f Pigment layer—epidermal layer that contains pigment cells called *melanocytes,* which produce the brown pigment melanin
    g Blisters caused by breakdown of union between cells or primary layers of skin
    h Dermal-epidermal junction—specialized area between two primary skin layers
  2 Dermis
    a Deeper and thicker of the two primary skin layers composed of connective tissue
    b Upper area of dermis characterized by parallel rows of peglike dermal papillae
    c Ridges and grooves in dermis form pattern unique to each individual (basis of fingerprinting)
    d Deeper areas of dermis filled with network of tough collagenous and stretchable elastic fibers
    e Number of elastic fibers decreases with age and contributes to wrinkle formation
    f Dermis also contains nerve endings, muscle fibers, hair follicles, sweat and sebaceous glands, and many blood vessels
B Appendages of the skin
  1 Hair (Figure 5-6)
    a Soft hair of fetus and newborn called *lanugo*
    b Hair growth requires epidermal tubelike structure called *hair follicle*
    c Hair growth begins from hair papilla
    d Hair root lies hidden in follicle and visible part of hair called *shaft*
    e Alopecia—hair loss
    f Arrector pili—specialized smooth muscle that produces "goose pimples" and causes hair to stand up straight
  2 Receptors (Figure 5-8)
    a Specialized nerve endings—make it possible for skin to act as a sense organ
    b Meissner's corpuscle—capable of detecting light touch
    c Pacinian corpuscle—capable of detecting pressure
  3 Nails (Figure 5-9)
    a Produced by epidermal cells over terminal ends of fingers and toes

b Visible part called *nail body*

c Root lies in a groove and is hidden by cuticle

d Crescent-shaped area nearest root called *lunula*

e Nail bed may change color with change in blood flow

4 Skin glands

  a Types

    (1) Sweat or sudoriferous

    (2) Sebaceous

  b Sweat or sudoriferous glands

    (1) Types

      (a) Eccrine sweat glands

        • Most numerous, important, and widespread of the sweat glands

        • Produce perspiration or sweat, which flows out through pores on skin surface

        • Function throughout life and assist in body heat regulation

      (b) Apocrine sweat glands

        • Found primarily in axilla and around genitalia

        • Secrete a thicker milky secretion quite different from eccrine perspiration

        • Breakdown of secretion by skin bacteria produces odor

  c Sebaceous glands

    (1) Secrete oil or sebum for hair and skin

    (2) Level of secretion increases during adolescence

    (3) Amount of secretion regulated by sex hormones

    (4) Sebum in sebaceous gland ducts may darken to form a blackhead

    (5) Acne vulgaris—inflammation of sebaceous gland ducts

C Functions of the skin

  1 Protection—first line of defense

    a Against infection by microbes

    b Against ultraviolet rays from sun

    c Against harmful chemicals

    d Against cuts and tears

  2 Temperature regulation

    a Skin can release almost 3000 calories of body heat per day

    (1) Mechanisms of temperature regulation

      (a) Regulation of sweat secretion

      (b) Regulation of flow of blood close to the body surface

  3 Sense organ activity

    a Skin functions as an enormous sense organ

    b Receptors serve as receivers for the body, keeping it informed of changes in its environment

DISORDERS OF THE SKIN (DERMATOSES)

A Skin lesions

  1 Elevated lesions—cast a shadow outside their edges

    a Papule—small, firm raised lesion

    b Plaque—large raised lesion

    c Vesicle—blister

    d Pustule—pus-filled lesion

    e Crust—scab

    f Wheal (hive)—raised, firm lesion with a light center

  2 Flat lesions—do not cast a shadow

    a Macule—flat, discolored region

  3 Depressed lesions—cast a shadow within their edges

    a Excoriation—missing epidermis, as in a scratch wound

    b Ulcer—craterlike lesion

    c Fissure—deep crack or break

B Burns

  1 Treatment and recovery or survival depend on total area involved and severity or depth of the burn

  2 Estimating body surface area using the "rule of nines" (Figure 5-11) in adults

    a Body divided into 11 areas of 9% each

    b Additional 1% of body surface area around genitals

  3 Classification of burns

    a First-degree burns—only surface layers of epidermis involved

    b Second-degree burns—involve the deep epidermal layers and always cause injury to the upper layers of the dermis

    c Third-degree burns—characterized by complete destruction of the epidermis and dermis

      (1) May involve underlying muscle and bone

      (2) Lesion is insensitive to pain because of destruction of nerve endings immediately after injury—intense pain is soon experienced

C Skin infections

  1 Impetigo—highly contagious staphylococci infection

  2 Tinea—fungal infection (mycosis) of the skin; several forms occur

  3 Warts—benign neoplasms caused by papillomavirus

  4 Boils—furuncles; staphylococci infection in hair follicles

D Vascular and inflammatory skin disorders

  1 Decubitus ulcers (bedsores) develop when pressure slows down blood flow to local areas of the skin

2 Urticaria or hives—red lesions caused by fluid loss from blood vessels

3 Scleroderma—disorder of vessels and connective tissue characterized by hardening of the skin; two types: localized and systemic

4 Psoriasis—chronic inflammatory condition accompanied by scaly plaques

5 Eczema—common inflammatory condition characterized by papules, vesicles, and crusts; not a disease itself but a symptom of an underlying condition

E Skin cancer

1 Three common types

a Squamous cell carcinoma—the most common type, characterized by hard, raised tumors

b Basal cell carcinoma—characterized by papules with a central crater; rarely spreads

c Melanoma—malignancy in a nevus (mole); the most serious type

2 The most important causative factor in common skin cancers is exposure to sunlight

3 Kaposi's sarcoma, characterized by purple lesions, is associated with AIDS and other immune deficiencies

$\mathcal{N}$EW WORDS

apocrine sweat gland
arrector pili
bursa
cutaneous
cuticle
depilatories
dermis
eccrine sweat gland
epidermis

follicle
keratin
Krause's end bulb
lanugo
lunula
Meissner's corpuscle
melanin
melanocyte

mucocutaneous junction
mucous membrane
mucus
pacinian corpuscle
papilla
parietal
peritoneum

pleura
sebaceous gland
serous
stratum corneum
subcutaneous
sudoriferous gland
synovial
visceral

Diseases and Other Clinical Terms

acne vulgaris
alopecia
blister
carbuncle
crust
cyanosis
decubitus ulcer
dehydration

dermatitis
dermatosis
eczema
excoriation
fissure
furuncle
impetigo
Kaposi's sarcoma

lesion
macule
mole
nevus
papule
peritonitis
plaque

pleurisy
psoriasis
pustule
scleroderma
urticaria
vesicle
wheal

CHAPTER TEST

1. There are two major categories of body membranes called _____ membranes and _____ membranes.
2. The membrane that lines the walls of a body cavity is called the _____ portion of the membrane, and the portion that covers the surface of organs found in body cavities is called the _____ portion of that membrane.
3. The connective tissue membranes that line joint spaces are called _____ membranes.
4. The principal organ of the integumentary system is the _____.
5. The skin is classified as a _____ membrane.
6. The two principal layers of the skin are called the _____ and the _____.
7. The tough waterproof material that provides cells in the outer layer of the skin with a protective quality is called _____.
8. The brown pigment that gives color to the skin is known as _____.
9. The part of a hair that lies hidden in the follicle is called the _____.
10. The pacinian corpuscle is capable of detecting _____ on the skin surface.
11. Sweat or sudoriferous glands are classified into two groups, _____ and _____, based on type of secretion and location.
12. Specialized glands that secrete oil for the hair and skin are known as _____ glands.
13. The "rule of nines" is used in the treatment and prognosis of _____.
14. _____ are pressure sores caused by reduced blood flow to local areas of the skin.
15. The most common type of skin cancer is _____ carcinoma.

Select the most correct answer from Column B for each statement in Column A. (Only one answer is correct.)

COLUMN A

16. _____ Another name for skin
17. _____ Outermost layer of skin
18. _____ Contains melanocytes
19. _____ Fine, soft hair of newborn
20. _____ Visible part of a hair
21. _____ Produce "goose pimples"
22. _____ Lined with synovial membrane
23. _____ Another name for hives
24. _____ Skin lesion caused by a shallow scratch
25. _____ Another name for a skin boil

COLUMN B

a. Lanugo
b. Arrector pili
c. Shaft
d. Integument
e. Pigment layer
f. Bursae
g. Epidermis
h. Furuncle
i. Urticaria
j. Excoriation

REVIEW QUESTIONS

1 Define the term *membrane* and discuss a variety of functions that membranes serve in the body.
2 Classify body membranes.
3 Discuss the two types of serous membranes and give examples of each.
4 Compare mucous and synovial membranes.
5 How do the terms *integument* and *integumentary system* differ in meaning?
6 Identify and compare the two main layers of the skin. How are these layers related to the subcutaneous layer?
7 List the appendages of the skin.
8 Discuss the three primary functions of the skin.
9 Classify the skin glands. Locate each type of gland and compare the secretions of each.
10 What are the two major factors that determine the treatment of and possibility for recovery from burn injuries?
11 Give examples of elevated, flat, and depressed skin lesions.
12 Classify and compare the three major types of burn injuries.
13 Compare and contrast boils (furuncles) and carbuncles.
14 Name and describe five vascular and inflammatory skin disorders.
15 List the three most common forms of skin cancer and explain the factors involved in their development.

CLINICAL APPLICATIONS

1 Dana has just been assigned to the Burn Unit at St. John Hospital. One patient in the unit has burns covering the lower half of each arm (front and back). How can Dana estimate the total percent of skin surface area affected by the burn? What should Dana's estimate be?

2 Uncle Ed, a light-skinned older gentleman, has just come from a visit to his physician with the news that the spot on his forehead is skin cancer. Of course, dark-skinned Aunt Gina is very upset. Before Uncle Ed informs the family what type of skin cancer he has, you examine the lesion and notice that it is a papule with an ulcer in the center. What type of skin cancer do you think Uncle Ed has? What do you know about this type of cancer that may help comfort Aunt Gina?

3 During your shift at the clinic, a young man presents himself with a red, scaly rash formed into rings. What is this patient's diagnosis likely to be? What causes this condition? How can he avoid this type of rash in the future?

The Skeletal System

Outline

Objectives

After you have completed this chapter, you should be able to:

1 Explain how bones are formed, how they grow, and how they are remodeled.

2 Discuss the microscopic structure of bone and cartilage, including the identification of specific cell types and structural features.

3 Identify the major anatomical structures found in a typical long bone and discuss bone formation and growth.

4 List and discuss the generalized functions of the skeletal system.

5 Identify the two major subdivisions of the skeleton and list the bones found in each area.

6 List and compare the major types of joints in the body and give an example of each.

7 Name and describe major disorders of bones and joints.

Boxed Essays

Epiphyseal Fracture
Palpable Bony Landmarks
The Knee Joint
Arthroscopy

The primary organs of the skeletal system, bones, lie buried within the muscles and other soft tissues, providing a rigid framework and support structure for the whole body. In this respect the skeletal system functions like steel girders in a building, but unlike steel girders, bones can be moved. Bones are also living organs. They can change and help the body respond to a changing environment. This ability of bones to change allows our bodies to grow and change as well.

Our study of the skeletal system will begin with a discussion of bone formation and growth. After information about the microscopic structure of skeletal tissues, we will classify bones and look at the specific characteristics of a "typical" bone that are representative of many other bones in the skeleton. With this information, the study of specific bones and the way they are assembled in the skeleton will be more meaningful. The chapter will end with a discussion of skeletal functions, joints or **articulations** (ar-tick-yoo-LAY-shuns), and an overview of skeletal disorders.

An understanding of how bones articulate with one another in joints and how they relate to other body structures provides a basis for understanding the functions of many other organ systems. Coordinated movement, for example, is possible only because of the way bones are joined to one another in joints and because of the way muscles are attached to those bones. In addition, knowledge of the placement of bones within the soft tissues assists you in locating and identifying other body structures that will be discussed in subsequent chapters of the text.

Bone Formation and Growth

When the skeleton begins to form in a baby before its birth, it consists not of bones but of cartilage and fibrous structures shaped like bones. Gradually these cartilage "models" become transformed into real bones when the cartilage is replaced with calcified bone matrix. This process of constantly "remodeling" a growing bone as it changes from a small cartilage model to the characteristic shape and proportion of the adult bone requires continuous activity by bone-forming cells called **osteoblasts** (OS-tee-o-blasts) and bone-resorbing cells called **osteoclasts** (OS-tee-o-clasts). The laying down of calcium salts in the gel-like matrix of the forming bones is an ongoing process. This calcification process is what makes bones as "hard as bone." The combined action of the osteoblasts and osteoclasts sculpts bones into their adult shapes (Figure 6-1). The process of "sculpting" by the bone-forming and bone-resorbing cells allows bones to respond to stress or injury by changing size, shape, and density. The stresses placed on certain bones during exercise increase the rate of bone deposition. For this reason, athletes or dancers may have denser, stronger bones than less active people.

Most bones of the body are formed from cartilage models as illustrated in Figure 6-1. This process is called **endochondral** (en-doe-KON-dral) **ossification** (os-i-fi-KAY-shun), meaning "formed in cartilage." A few flat bones are formed by another process in connective tissue membranes.

As you can see in Figure 6-1, a long bone grows and ultimately becomes "ossified" from small centers located in both ends of the bone, called **epiphyses** (e-PIF-i-sees), and from a larger center located in the shaft or the **diaphysis** (dye-AF-i-sis) of the bone. As long as any cartilage, called an **epiphyseal plate,** remains between the epiphyses and the diaphysis, growth continues. Growth ceases when all epiphyseal cartilage is transformed into bone. Physicians sometimes use this knowledge to determine whether a child is going to grow any more. They have an x-ray study performed on the child's wrist, and if it shows a layer of epiphyseal cartilage, they know that additional growth will occur. However, if it shows no epiphyseal cartilage, they know that growth has stopped and that the individual has attained adult height. Epiphyseal plates are visible in the long bone shown in Figure 6-1.

Microscopic Structure of Bone and Cartilage

The skeletal system contains two major types of connective tissue: **bone** and **cartilage.** Bone has a different appearance and texture, depending on its location. In Figure 6-2, *A,* the outer layer of bone is hard and dense. Bone of this type is called **dense** or **compact bone.** The porous bone in the

FIGURE 6-1 Endochondral Ossification. Bone formation begins with a cartilage model. Invasion by blood vessels and the combined action of osteoblast and osteoclast cells result in cavity formation, calcification, and ultimately the appearance of a mature bone. Note the epiphyseal plates, indications that this bone is not yet mature and that additional growth is possible.

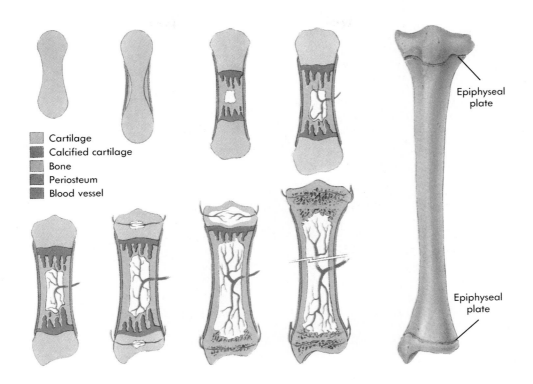

Cartilage
Calcified cartilage
Bone
Periosteum
Blood vessel

Epiphyseal plate

Epiphyseal plate

EPIPHYSEAL FRACTURE

The point of articulation between the epiphysis and diaphysis of a growing long bone is susceptible to injury if over-stressed—expecially in the young child or preadolescent athlete. In these individuals the epiphyseal plate can be separated from the diaphysis or epiphysis, causing an **epiphyseal fracture.** This x-ray study shows such a fracture in a young boy.

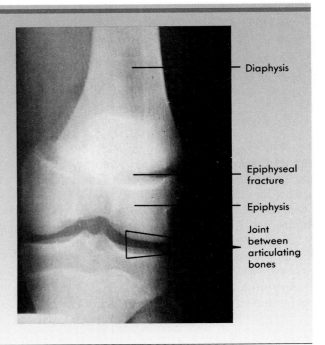

Diaphysis

Epiphyseal fracture

Epiphysis

Joint between articulating bones

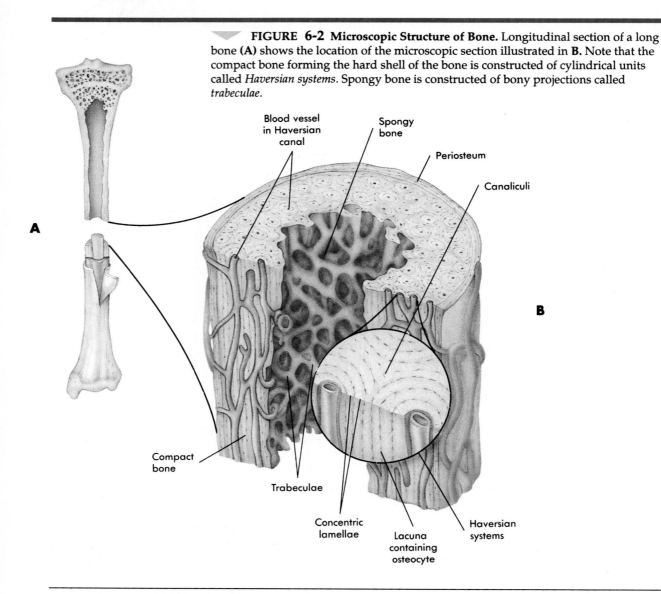

FIGURE 6-2 Microscopic Structure of Bone. Longitudinal section of a long bone **(A)** shows the location of the microscopic section illustrated in **B.** Note that the compact bone forming the hard shell of the bone is constructed of cylindrical units called *Haversian systems.* Spongy bone is constructed of bony projections called *trabeculae.*

end of the long bone is called **spongy bone.** As the name implies, spongy bone contains many spaces that may be filled with marrow. Compact or dense bone appears solid to the naked eye. Figure 6-2, *B,* shows the microscopic appearance of spongy and compact bone. The needlelike threads of spongy bone that surround a network of spaces are called **trabeculae** (trah-BEK-yoo-lee).

As you can see in Figures 6-2 and 6-3, compact or dense bone does not contain a network of open spaces. Instead, the matrix is organized into numerous structural units called **Haversian systems.** Each circular and tubelike Haversian system is composed of multiple layers of calcified matrix arranged like the rings of an onion. Each ring is called a **concentric lamella** (lah-MEL-ah). The rings or lamellae surround the **Haversian canal,** which contains a blood vessel.

Bones are not lifeless structures. Within their hard, seemingly lifeless matrix are many living

▼ **FIGURE 6-3 Compact Bone.** Photomicrograph shows Haversian system of organization.

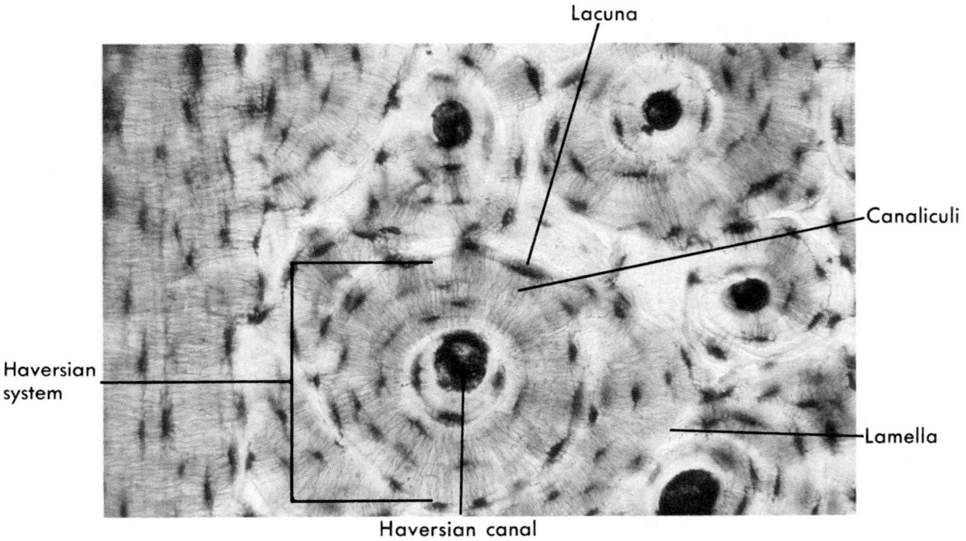

Lacuna

Canaliculi

Haversian
system

Lamella

Haversian canal

---

bone cells called **osteocytes** (OS-tee-o-sites). Osteocytes lie between the hard layers of the lamellae in little spaces called **lacunae** (lah-KOO-nee). In Figures 6-2, *B,* and 6-3, note that tiny passageways or canals called **canaliculi** (kan-ah-LIK-yoo-lye) connect the lacunae with one another and with the central canal in each Haversian system. Nutrients pass from the blood vessel in the Haversian canal through the canaliculi to the osteocytes. Note also in Figure 6-2, *B,* that numerous blood vessels from the outer **periosteum** (pair-ee-OS-tee-um) enter the bone and eventually pass through the Haversian canals.

Cartilage resembles and differs from bone. Like bone, it consists more of intercellular substance than of cells. Innumerable collagenous fibers reinforce the matrix of both tissues. However, in cartilage the fibers are embedded in a firm gel instead of in a calcified cement substance like they are in bone; hence cartilage has the flexibility of a firm plastic rather than the rigidity of bone. Note in Figure 6-4 that cartilage cells, called **chondrocytes** (kon-dro-sites), like the osteocytes of bone, are located in lacunae. In cartilage, lacunae are suspended in the cartilage

▼ **FIGURE 6-4 Cartilage Tissue.** Photomicrograph shows chondrocytes scattered throughout openings called *lacunae.*

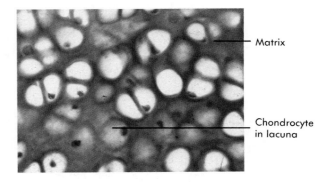

Matrix

Chondrocyte
in lacuna

matrix much like air bubbles in a block of firm gelatin. Because there are no blood vessels in cartilage, nutrients must diffuse through the matrix to reach the cells.

FIGURE 6-5 Longitudinal Section of a Long Bone.

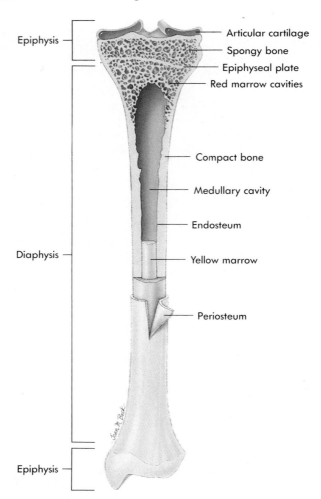

Epiphysis

Diaphysis

Epiphysis

Articular cartilage

Spongy bone

Epiphyseal plate

Red marrow cavities

Compact bone

Medullary cavity

Endosteum

Yellow marrow

Periosteum

Types of Bones

There are four types of bones. Their names suggest their shapes: *long* (for example, the humerus or upper arm bone), *short* (for example, the carpals or wrist bones), *flat* (for example, the frontal or skull bone), and *irregular* (for example, the vertebrae or spinal bones). Many important bones in the skeleton are classified as long bones, and all have several common characteristics. By studying a typical long bone, you can become familiar with the structural features of the entire group.

Structure of Long Bones

Figure 6-5 will help you learn the names of the main parts of a long bone. Identify each of the following:

1 **Diaphysis** or shaft—a hollow tube made of hard compact bone, hence a rigid and strong structure light enough in weight to permit easy movement
2 **Medullary cavity**—the hollow area inside the diaphysis of a bone; contains soft yellow bone marrow

3 **Epiphyses** or the ends of the bone—red bone marrow fills in small spaces in the spongy bone composing the epiphyses

4 **Articular cartilage**—a thin layer of cartilage covering each epiphysis; functions like a small rubber cushion would if it were placed over the ends of bones where they form a joint

5 **Periosteum**—a strong fibrous membrane covering a long bone except at joint surfaces, where it is covered by articular cartilage

6 **Endosteum**—a fibrous membrane that lines the medullary cavity

Functions of Bone

SUPPORT

Bones form the body's supporting framework much like steel girders form the supporting framework of modern buildings.

PROTECTION

Hard, bony "boxes" protect delicate structures enclosed within them. For example, the skull protects the brain. The breastbone and ribs protect vital organs (heart and lungs) and also a vital tissue (red bone marrow, the blood cell–forming tissue).

MOVEMENT

Muscles are anchored firmly to bones. As muscles contract and shorten, they pull on bones and thereby move them.

STORAGE

Bones play an important part in maintaining homeostasis of blood calcium, a vital substance required for normal nerve and muscle function. They serve as a safety-deposit box for calcium. When the amount of calcium in blood increases above normal, calcium moves out of the blood and into the bones for storage. Conversely, when blood calcium decreases below normal, calcium moves in the opposite direction. It comes out of storage in bones and enters the blood.

HEMOPOIESIS

The term **hemopoiesis** (hee-mo-poy-EE-sis) is used to describe the process of blood cell formation. It is a combination of two Greek words: *hemo*

(HEE-mo) meaning "blood" and *poiesis* (poy-EE-sis) meaning "to make." Blood cell formation is a vital process carried on in **red bone marrow.** As you can see in Figure 6-5, red marrow is found in the spongy type of bone located in the ends of some long bones. There is more red marrow in an infant's or child's body than in an adult's. As an individual ages, much of the red marrow is transformed into **yellow bone marrow,** which is an inactive fatty tissue. Yellow bone marrow is shown in the **medullary cavity** of the long bone in Figure 6-5.

Divisions of Skeleton

The human skeleton has two divisions: the **axial skeleton** and the **appendicular skeleton.** Bones of the skull, spine, and chest and the hyoid bone in the neck make up the axial skeleton. The appendicular skeleton consists of the bones of the upper extremities (shoulder, pectoral girdles, arms, wrists, and hands) and the lower extremities (hip, pelvic girdles, legs, ankles, and feet) (Table 6-1). Locate the various parts of the axial skeleton and the appendicular skeleton in Figure 6-6.

▼ **TABLE 6-1**   Main Parts of the Skeleton*

AXIAL SKELETON[†]	APPENDICULAR SKELETON[‡]
Skull	Upper Extremities
Cranium	Shoulder (pectoral) girdle
Ear bones	Arms
Face	Wrists
	Hands
Spine	Lower Extremities
Vertebrae	Hip (pelvic) girdle
Thorax	Legs
Ribs	Ankles
Sternum	Feet
Hyoid Bone	

*Total bones = 206.
[†]Total = 80 bones.
[‡]Total = 126 bones.

FIGURE 6-6 Human Skeleton. The axial skeleton is distinguished by its blue color. **A,** Anterior view.

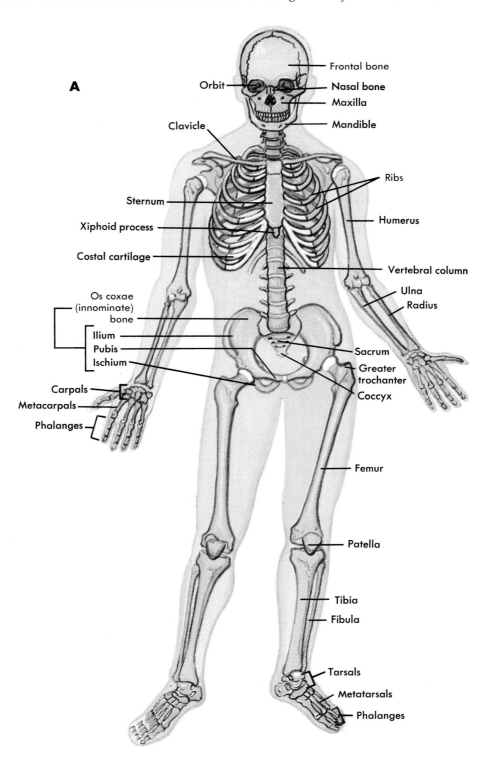

FIGURE 6-6, cont'd B, Posterior view.

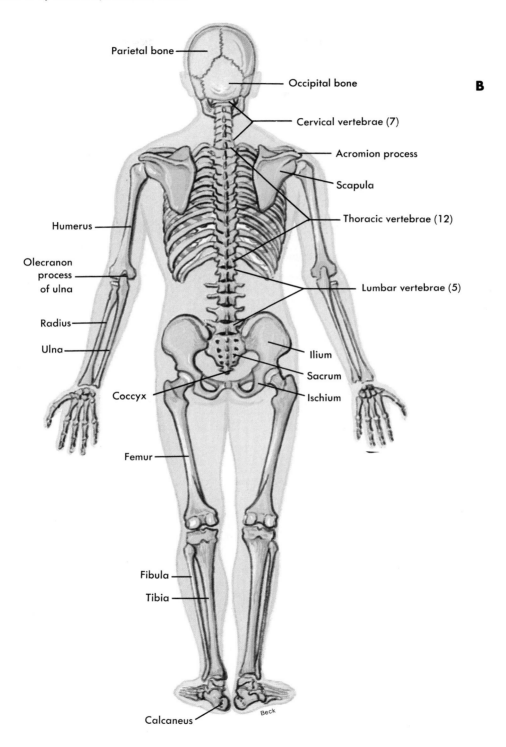

B

Parietal bone

Occipital bone

Cervical vertebrae (7)

Acromion process

Scapula

Thoracic vertebrae (12)

Humerus

Olecranon process of ulna

Lumbar vertebrae (5)

Radius

Ulna

Ilium

Sacrum

Coccyx

Ischium

Femur

Fibula

Tibia

Calcaneus

Beck

AXIAL SKELETON
Skull

The skull consists of 8 bones that form the **cranium,** 14 bones that form the **face,** and 6 tiny bones in the **middle ear.** You will probably want to learn the names and locations of these bones. These are given in Table 6-2. Find as many of them as you can on Figure 6-7. Feel their outlines in your own body where possible. Examine them on a skeleton if you have access to one.

"My sinuses give me so much trouble." Have you ever heard this complaint or perhaps uttered it yourself? **Sinuses** are spaces or cavities inside some of the cranial bones. Four pairs of them

▼ **TABLE 6-2**   Bones of the Skull

NAME	NUMBER	DESCRIPTION
Cranial Bones		
Frontal	1	Forehead bone; also forms front part of floor of cranium and most of upper part of eye sockets; cavity inside bone above upper margins of eye sockets (orbits) called *frontal sinus;* lined with mucous membrane
Parietal	2	Form bulging topsides of cranium
Temporal	2	Form lower sides of cranium; contain *middle* and *inner ear structures; mastoid sinuses* are mucosa-lined spaces in *mastoid process,* the protuberance behind ear; *external auditory canal* is tube leading into temporal bone; muscles attach to *styloid process*
Occipital	1	Forms back of skull; spinal cord enters cranium through large hole *(foramen magnum)* in occipital bone
Sphenoid	1	Forms central part of floor of cranium; pituitary gland located in small depression in sphenoid called *sella turcica (Turkish saddle);* muscles attach to *pterygoid process*
Ethmoid	1	Complicated bone that helps form floor of cranium, side walls and roof of nose and part of its middle partition (nasal septum—made up of the *vomer* and the *perpendicular plate*), and part of orbit; contains honeycomb-like spaces, the *ethmoid sinuses; superior* and *middle conchae* are projections of ethmoid bone; forms "ledges" along side wall of each nasal cavity
Face Bones		
Nasal	2	Small bones that form upper part of bridge of nose
Maxilla	2	Upper jawbones; also help form roof of mouth, floor, and side walls of nose and floor of orbit; large cavity in maxillary bone is *maxillary sinus*
Zygomatic	2	Cheek bones; also help form orbit
Mandible	1	Lower jawbone articulates with temporal bone at *condyloid process;* small anterior hole for passage of nerves and vessels is the *mental foramen*
Lacrimal	2	Small bones; help form medial wall of eye socket and side wall of nasal cavity
Palatine	2	Form back part of roof of mouth and floor and side walls of nose and part of floor of orbit
Inferior concha	2	Form curved "ledge" along inside of side wall of nose, below middle concha
Vomer	1	Forms lower, back part of nasal septum
Ear Bones		
Malleus	2	Malleus, incus, and stapes are tiny bones in middle ear cavity in temporal bone; *malleus* means "hammer"—shape of bone
Incus	2	*Incus* means "anvil"—shape of bone
Stapes	2	*Stapes* means "stirrup"—shape of bone

FIGURE 6-7 The Skull. A, Right side. **B,** Front.

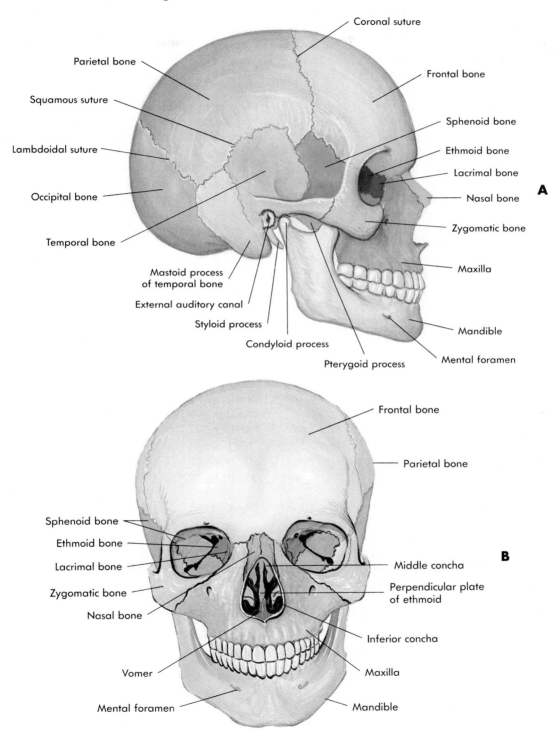

(those in the frontal, maxillary, sphenoid, and ethmoid bones) have openings into the nose and thus are referred to as **paranasal sinuses.** Sinuses give trouble when the mucous membrane that lines them becomes inflamed, swollen, and painful. For example, inflammation in the frontal sinus *(frontal sinusitis)* often starts from a common cold. The letters *-itis* added to a word mean "inflammation of" (see Appendix C).

Mastoiditis (mas-toy-DYE-tis), inflammation of the air spaces within the mastoid portion of the temporal bone, can produce very serious medical problems if not treated promptly. Locate the mastoid process in Figure 6-7, *A.* Infectious material sometimes finds its way into the mastoid air cells from middle ear infections. These air cells do not drain into the nose like the paranasal sinuses do. Thus, infectious material that accumulates may damage the thin, bony partition that separates the air cells from the brain. If this occurs, the infection may spread to the brain or the membranes covering the brain, a life-threatening situation. Chronic mastoiditis may be treated by surgically removing the affected tissue, including internal parts of the ear—rendering the individual deaf in the affected ear.

Note in Figure 6-7 that the two parietal bones, which give shape to the bulging topside of the skull, form immovable joints called **sutures** with several bones: the *lambdoidal suture* with the occipital bone, the *squamous suture* with the temporal bone and part of the sphenoid, and the *coronal suture* with the frontal bone.

You may be familiar with the "soft spots" on a baby's skull. These are six **fontanels,** or areas where ossification is incomplete at birth. They allow some compression of the skull during birth and may also be important in determining the position of the baby's head before delivery. The fontanels fuse to form sutures before a baby is 2 years old.

Spine (Vertebral Column)

The term *vertebral column* may conjure up a mental picture of the spine as a single long bone shaped like a column in a building, but this is far from true. The vertebral column consists of a series of separate bones or **vertebrae** connected in such a way that they form a flexible curved rod (see Figure 6-10, *A*). Different sections of the spine have different names: cervical region, thoracic region, lumbar region, sacrum, and coccyx. They are illustrated in Figure 6-8 and described in Table 6-3.

Although individual vertebrae are small bones, irregular in shape, they have several well-defined parts. Note, for example, in Figure 6-9, the body of the lumbar vertebra shown there, its spinous process (or spine), its two transverse processes, and the hole in its center, called the *vertebral foramen.* The superior and inferior articular processes permit limited and controlled movement between adjacent vertebrae. To feel the tip of the spinous process of one of your vertebrae, simply bend your head forward and run your fingers down the back of your neck until you feel a projection of bone at shoulder level. This is the tip of the seventh cervical vertebra's long spinous process. The seven cervical vertebrae form the supporting framework of the neck.

TABLE 6-3	Bones of the Vertebral Column	
NAME	**NUMBER**	**DESCRIPTION**
Cervical vertebrae	7	Upper seven vertebrae, in neck region; first cervical vertebra called *atlas;* second, *axis*
Thoracic vertebrae	12	Next twelve vertebrae; ribs attach to these
Lumbar vertebrae	5	Next five vertebrae; are in small of back
Sacrum	1	In child, five separate vertebrae; in adult, fused into one
Coccyx	1	In child, three to five separate vertebrae; in adult, fused into one

FIGURE 6-8 The Spinal Column. View shows the 7 cervical vertebrae, the 12 thoracic vertebrae, the 5 lumbar vertebrae, the sacrum, and the coccyx. **A,** Anterior view. **B,** Posterior view.

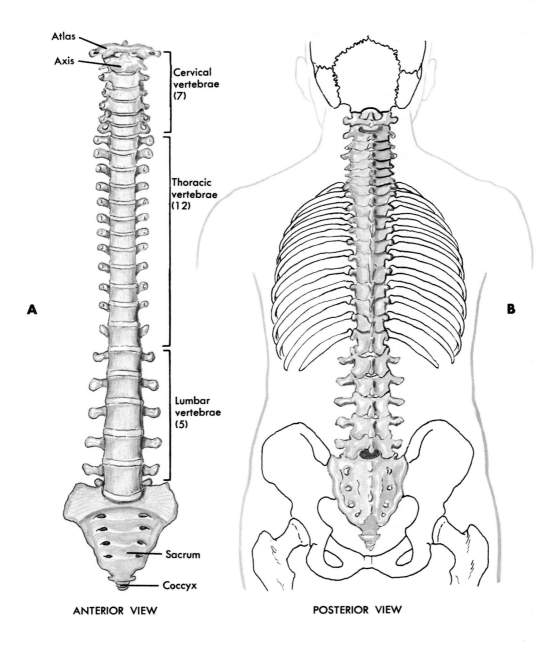

ANTERIOR VIEW

POSTERIOR VIEW

FIGURE 6-9 The Third Lumbar Vertebra. A, From above. **B,** From the side.

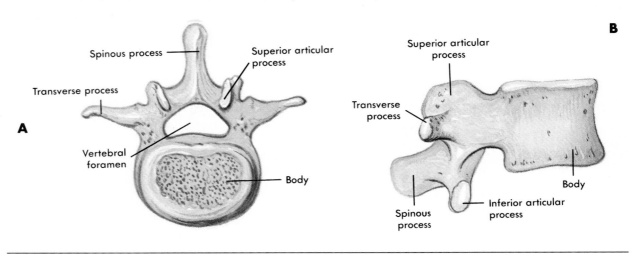

Have you ever noticed the four curves in your spine? Your neck and the small of your back curve slightly inward or forward, whereas the chest region of the spine and the lowermost portion curve in the opposite direction (Figure 6-10, *A*). The cervical and lumbar curves of the spine are called *concave curves,* and the thoracic and sacral curves are called *convex curves.* These two curves are not true, however, of a newborn baby's spine. It forms a continuous convex curve from top to bottom (Figure 6-10, *B*). Gradually, as the baby learns to hold up his or her head, a reverse or concave curve develops in the neck, (cervical region). Later, as the baby learns to stand, the lumbar region of his or her spine also becomes concave.

The normal curves of the spine have important functions. They give it enough strength to support the weight of the rest of the body. They also provide the balance necessary for us to stand and walk on two feet instead of having to crawl on all fours. A curved structure has more strength than a straight one of the same size and materials. (The next time you pass a bridge, look to see whether or not its supports form a curve.) Clearly the spine needs to be a strong structure. It supports the head balanced on top of it, the ribs and inter-

nal organs suspended from it in front, and the hips and legs attached to it below.

Poor posture or disease may cause the lumbar curve to become abnormally exaggerated, a condition known to some as *swayback* or **lordosis** (lor-DOE-sis). Abnormal thoracic curvature is **kyphosis** (ki-FO-sis) or "hunchback." Abnormal side-to-side curvature is **scoliosis** (sko-lee-O-sis). Sometimes these abnormal curvatures (Figure 6-11) interfere with normal breathing and other vital functions. Scoliosis is a relatively common condition that appears before adolescence, usually of unknown cause. Depending on the degree of lateral curvature and resulting deformity of individual vertebrae, various treatments may be applied. The traditional treatment for scoliosis is the use of a supportive brace called the *Milwaukee brace.* This brace is worn on the upper body for 23 hours a day for up to several years. A newer approach to straightening abnormal curvature is *transcutaneous* ("through-the-skin") *muscle stimulation.* In this method, muscles on one side of the vertebral column are electrically stimulated to contract and pull the vertebrae into a more normal position. If these methods fail, surgery may be used to correct scoliosis. Often, chips of bone from elsewhere in the skeleton are grafted to

FIGURE 6-10 **Curvature of the Spine. A,** The normal curves of the adult spine from the right lateral view. **B,** The spine of the newborn baby forms a continuous convex curve.

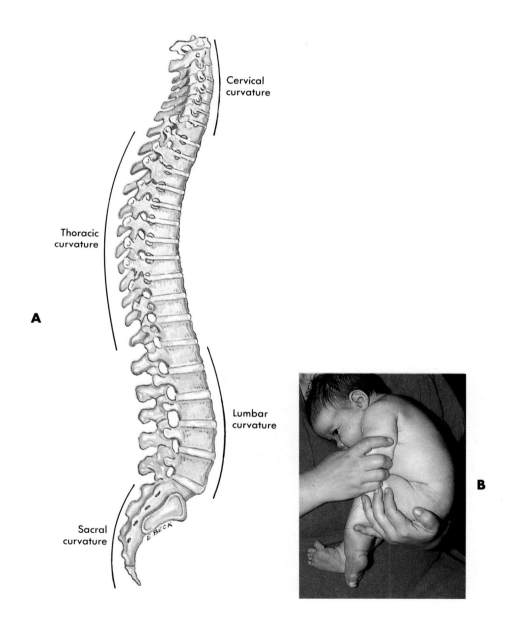

Cervical
curvature

Thoracic
curvature

A

Lumbar
curvature

Sacral
curvature

E BECK

B

deformed vertebrae to hold them in a more normal position. Internal metal rods and body casts hold the vertebral column straight until the vertebrae and bone chips heal and "fuse" into a strong, solid support.

Thorax

Twelve pairs of ribs, the sternum (breastbone), and the thoracic vertebrae form the bony cage known as the **thorax** or **chest.** Each of the twelve pairs of ribs is attached posteriorly to a vertebra. Also, all the ribs except the lower two pairs are attached to the sternum and so have anterior and posterior anchors. Look closely at Figure 6-12, and you can see that the first seven pairs of ribs (sometimes referred to as the *true ribs*) are attached to the sternum by costal cartilage. The eighth, ninth, and tenth pairs of ribs are attached to the cartilage of the seventh ribs and are sometimes called *false ribs.* The last two pairs of ribs, in contrast, are not attached to any costal cartilage but seem to float free in front, hence their descriptive name, *floating ribs* (Table 6-4).

APPENDICULAR SKELETON

Of the 206 bones that form the skeleton as a whole, 126 are contained in the appendicular sub-

---

FIGURE **6-11 Abnormal Spinal Curvatures. A,** Lordosis. **B,** Kyphosis. **C,** Scoliosis.

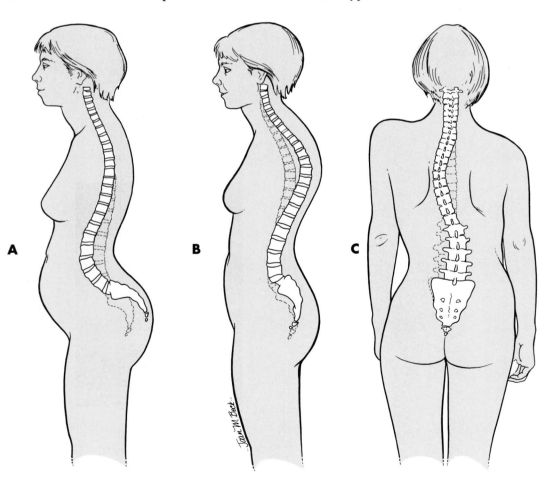

division. Look again at Figure 6-6 to identify the appendicular components of the skeleton. Note that the bones in the shoulder or pectoral girdle connect the bones of the arm, forearm, wrist, and hands to the axial skeleton of the thorax, and the hip or pelvic girdle connects the bones of the thigh, leg, ankle, and foot to the axial skeleton of the pelvis.

Upper Extremity

The **scapula** (SKAP-yoo-lah) or shoulder blade and the **clavicle** (KLAV-ik-kul) or collar bone compose the *shoulder* or *pectoral girdle*. This connects the upper extremity to the axial skeleton. The only direct point of attachment between bones occurs at the **sternoclavicular** (ster-no-klah-VIK-yoo-lar) **joint** between the clavicle and the

---

TABLE 6-4 Bones of the Thorax

NAME	NUMBER	DESCRIPTION
True ribs	14	Upper seven pairs; attached to sternum by *costal cartilages*
False ribs	10	Lower five pairs; lowest two pairs do not attach to sternum, therefore, called *floating ribs*; next three pairs attached to sternum by costal cartilage of seventh ribs
Sternum	1	Breastbone; shaped like a dagger; piece of cartilage at lower end of bone called *xiphoid process*

FIGURE 6-12 **Bones of the Thorax.** Rib pairs 1 through 7, the true ribs, are attached by cartilage to the sternum. Rib pairs 8 through 10, the false ribs, are attached to the cartilage of the seventh pair. Rib pairs 11 and 12 are called *floating ribs* because they have no anterior cartilage attachments.

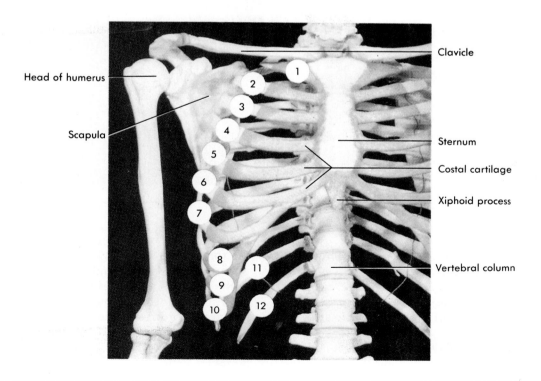

PALPABLE BONY LANDMARKS

Health professionals often identify externally palpable bony landmarks when dealing with the sick and injured. **Palpable bony landmarks** are bones that can be touched and identified through the skin. They serve as reference points in identifying other body structures.

There are externally palpable bony landmarks throughout the body. Many skull bones such as the zygomatic bone, can be palpated. The medial and lateral epicondyles of the humerus, the olecranon process of the ulna, and the styloid process of the ulna and the radius at the wrist can be palpated on the upper extremity. The highest corner of the shoulder is the acromion process of the scapula.

When you put your hands on your hips, you can feel the superior edge of the ilium called the *iliac crest.* The anterior end of the crest, called the *anterior superior iliac spine,* is a prominent landmark used often as a clinical reference. The medial malleolus of the tibia and the lateral malleolus of the fibula are prominent at the ankle. The calcaneus or heel bone is easily palpated on the posterior aspect of the foot. On the anterior aspect of the lower extremity, examples of palpable bony landmarks include the patella or knee-cap, the anterior border of the tibia or shin bone, and the metatarsals and phalanges of the toes. Try to identify as many of the externally palpable bones of the skeleton as possible on your own body. Using these as points of reference will make it easier for you to visualize the placement of other bones that cannot be touched or palpated through the skin.

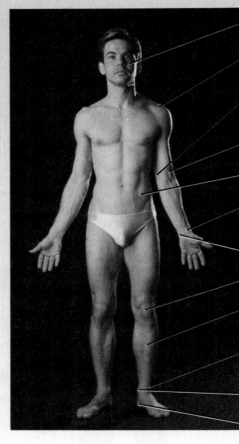

Zygomatic bone

Acromion process of scapula

Medial epicondyle of humerus

Lateral epicondyle of humerus

Iliac crest

Styloid process of radius

Styloid process of ulna

Patella

Anterior border of tibia

Lateral malleolus of fibula

Medial malleolus of tibia

Calcaneus

sternum or breastbone. As you can see in Figures 6-6 and 6-12, this joint is very small. Because the upper extremity is capable of a wide range of motion, great pressures can occur at or near the joint. As a result, fractures of the clavicle are very common.

The **humerus** (HYOO-mer-us) is the long bone of the arm and the second longest bone in the body. It is attached to the scapula at its proximal end and articulates distally with the two bones of the forearm at the elbow joint. The bones of the forearm are the **radius** and the **ulna.** The anatomy of the elbow is a good example of how structure determines function. Note in Figure 6-13 that the large bony process of the ulna, called the **olecranon** (o-LEK-rah-non) **process,** fits nicely into a large depression on the posterior surface of the humerus, called the **olecranon fossa.** This structural relationship makes possible movement at the joint.

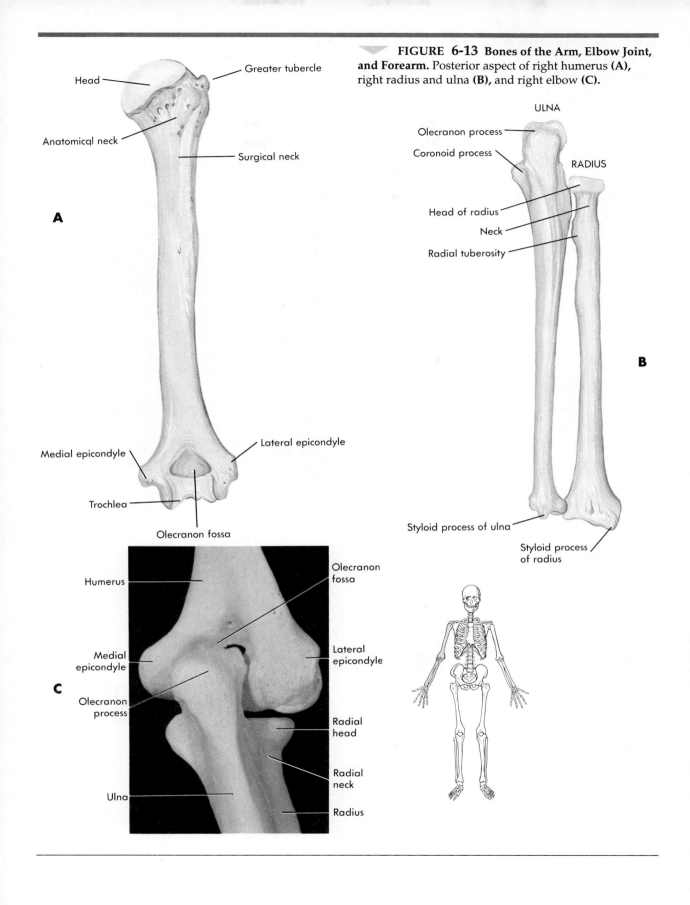

FIGURE 6-13 Bones of the Arm, Elbow Joint, and Forearm. Posterior aspect of right humerus **(A)**, right radius and ulna **(B)**, and right elbow **(C)**.

Head

Greater tubercle

Anatomical neck

Surgical neck

A

Medial epicondyle

Lateral epicondyle

Trochlea

Olecranon fossa

ULNA

Olecranon process

Coronoid process

RADIUS

Head of radius

Neck

Radial tuberosity

B

Styloid process of ulna

Styloid process of radius

Humerus

Olecranon fossa

Medial epicondyle

Lateral epicondyle

Olecranon process

Radial head

Ulna

Radial neck

Radius

C

FIGURE 6-14 **Bones of the Right Hand and Wrist.** There are 14 phalanges in each hand. Each of these bones is called a *phalanx*.

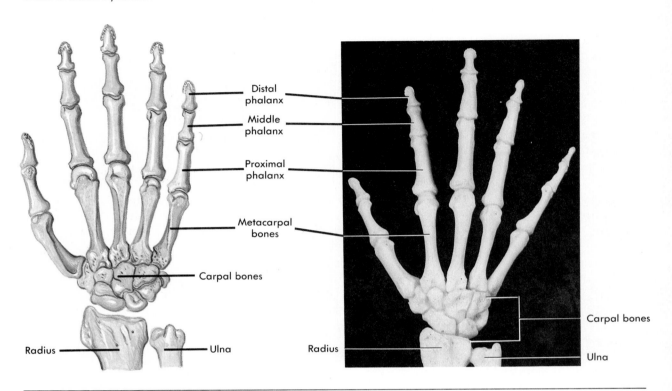

TABLE 6-5   Bones of the Shoulder Girdle and Upper Extremities

NAME	NUMBER	DESCRIPTION
Clavicle	2	Collarbones; only joints between shoulder girdle and axial skeleton are those between each clavicle and sternum *(sternoclavicular joints)*
Scapula	2	Shoulder blades; scapula plus clavicle forms *shoulder girdle; acromion process*—tip of shoulder that forms joint with clavicle; *glenoid cavity*—arm socket
Humerus	2	Upper arm bone (Muscles are attached to the *greater tubercle* and to the *medial* and *lateral epicondyles;* the *trochlea* articulates with the ulna; the *surgical neck* is a common fracture site.)
Radius	2	Bone on thumb side of lower arm (Muscles are attached to the *radial tuberosity* and to the *styloid process.*)
Ulna	2	Bone on little finger side of lower arm; *olecranon process*—projection of ulna known as elbow or "funny bone" (Muscles are attached to the *coronoid process* and to the *styloid process.*)
Carpal bones	16	Irregular bones at upper end of hand; anatomical wrist
Metacarpals	10	Form framework of palm of hand
Phalanges	28	Finger bones; three in each finger, two in each thumb

The radius and the ulna of the forearm articulate with each other and with the distal end of the humerus at the elbow joint. In addition, they also touch each another distally where they articulate with the bones of the wrist. In the anatomical position with the arm at the side and the palm facing forward, the radius runs along the lateral side of the forearm, and the ulna is located along the medial border.

The wrist and the hand have more bones in them for their size than any other part of the body—8 **carpal** (KAR-pal) or wrist bones, 5 **metacarpal** (met-ah-KAR-pal) bones that form the support structure for the palm of the hand, and 14 **phalanges** (fah-LAN-jeez) or finger bones—a total of 27 bones in all (Table 6-5). This structure is very important structurally. The presence of many small bones in the hand and wrist and the many movable joints between them makes the human hand highly dexterous. Some anatomists refer to the hand and wrist as the functional "reason" for the upper extremity. Refer to Figure 6-14 to see the relationships between the bones of the wrist and hand.

Lower Extremity

The *hip* or *pelvic girdle* connects the legs to the trunk. The hip girdle as a whole consists of two large **os coxae** or bones, one located on each side of the pelvis. These two bones, with the sacrum and coccyx behind, provide a strong base of support for the torso and connect the lower extremities to the axial skeleton. In an infant's body each os coxa consists of three separate bones—the **ilium** (ILL-ee-um), the **ischium** (IS-kee-um), and the **pubis** (PYOO-bis). These bones grow together

THE KNEE JOINT

The knee is the largest and most vulnerable joint. Because the knee is often subjected to sudden, strong forces during athletic activity, knee injuries are among the most common type of athletic injury. Sometimes, the articular cartilages on the tibia become torn when the knee twists while bearing weight. The ligaments holding the tibia and femur together can also be injured in this way. Knee injuries may also occur when a weight-bearing knee is hit by another person.

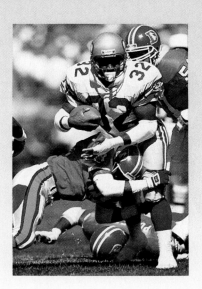

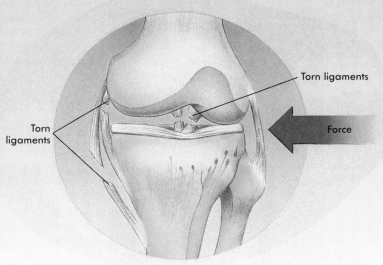

Torn ligaments

Torn ligaments

Force

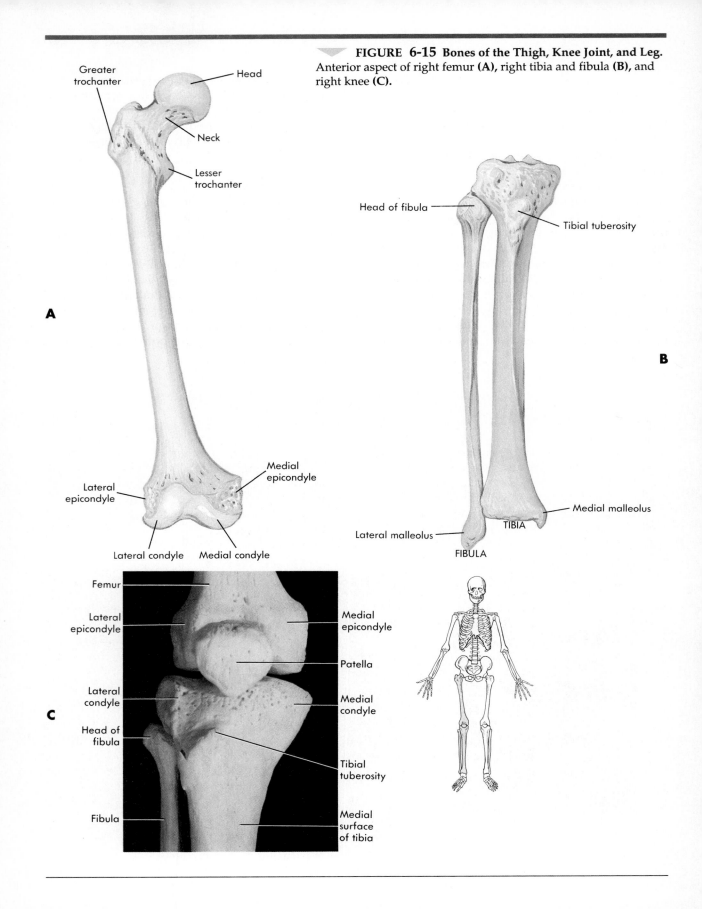

FIGURE **6-15 Bones of the Thigh, Knee Joint, and Leg.** Anterior aspect of right femur **(A)**, right tibia and fibula **(B)**, and right knee **(C)**.

A

Greater trochanter
Head
Neck
Lesser trochanter
Medial epicondyle
Lateral epicondyle
Lateral condyle
Medial condyle

B

Head of fibula
Tibial tuberosity
Medial malleolus
Lateral malleolus
TIBIA
FIBULA

C

Femur
Lateral epicondyle
Lateral condyle
Head of fibula
Fibula
Medial epicondyle
Patella
Medial condyle
Tibial tuberosity
Medial surface of tibia

to become one bone in an adult (Figures 6-6 and 6-18).

Just as the humerus is the only bone in the arm, the **femur** (FEE-mur) is the only bone in the thigh (Figure 6-15). It is the longest bone in the body and articulates proximally (toward the hip) with the os coxa in a deep, cup-shaped socket called the **acetabulum** (as-e-TAB-yoo-lum). The articulation of the head of the femur in the acetabulum is more stable than the articulation of the head of the humerus with the scapula in the upper extremity. As a result, dislocation of the hip occurs less often than does disarticulation of the shoulder. Distally, the femur articulates with the knee cap or **patella** (pah-TEL-ah) and the **tibia** or "shinbone." The tibia forms a rather sharp edge along the front of your lower leg. A slender non-weight-bearing and rather fragile bone named the **fibula** lies along the outer or lateral border of the lower leg.

Toe bones have the same name as finger bones—**phalanges**. There are the same number of toe bones as finger bones, a fact that might surprise you because toes are shorter than fingers. Foot bones comparable to the metacarpals and carpals of the hand have slightly different names.

They are called **metatarsals** and **tarsals** in the foot (Figure 6-16). Just as each hand contains five metacarpal bones, each foot contains five metatarsal bones, but the foot has only seven tarsal bones, in contrast to the hand's eight carpals. The largest tarsal bone is the **calcaneus** or heel bone. The bones of the lower extremities are summarized in Table 6-6.

You stand on your feet, so certain features of their structure make them able to support the body's weight. The great toe, for example, is considerably more solid and less mobile than the thumb. The foot bones are held together in such a way as to form springy lengthwise and crosswise arches. These provide great supporting strength and a highly stable base. Strong ligaments and leg muscle tendons normally hold the foot bones firmly in their arched positions. Frequently, however, the foot ligaments and tendons weaken. The arches then flatten, a condition appropriately called *fallen arches* or *flatfeet*.

Two arches extend in a lengthwise direction in a person's foot (Figure 6-17). One lies on the inside part of the foot and is called the **medial longitudinal arch**. The other lies along the outer edge of the foot and is named the **lateral longitu-**

NAME	NUMBER	DESCRIPTION
Os coxae	2	Hipbones; *ilium*—upper flaring part of pelvic bone; *ischium*—lower back part; *pubic bone*—lower front part; *acetabulum*—hip socket; *symphysis pubis*—joint in midline between two pubic bones; *pelvic inlet*—opening into *true pelvis* or pelvic cavity; if pelvic inlet is misshapen or too small, infant skull cannot enter true pelvis for natural birth
Femur	2	Thigh or upper leg bones; *head of femur*—ball-shaped upper end of bone; fits into acetabulum (Muscles are attached to the *greater* and *lesser trochanters* and to the *lateral* and *medial epicondyles*; the *lateral* and *medial condyles* form articulations at the knee.)
Patella	2	Kneecap
Tibia	2	Shinbone; *medial malleolus*—rounded projection at lower end of tibia commonly called *inner anklebone*; muscles are attached to the *tibial tuberosity*
Fibula	2	Long slender bone of lateral side of lower leg; *lateral malleolus*—rounded projection at lower end of fibula commonly called *outer anklebone*
Tarsal bones	14	Form heel and back part of foot; anatomical ankle; largest is the *calcaneus*
Metatarsals	10	Form part of foot to which toes are attached; tarsal and metatarsal bones arranged so that they form three arches in foot; *inner longitudinal arch* and *outer longitudinal arch*, which extend from front to back of foot, and transverse or *metatarsal arch*, which extends across foot
Phalanges	28	Toe bones; three in each toe, two in each great toe

TABLE 6-6 Bones of the Hip Girdle and Lower Extremities

▽ **FIGURE 6-16 Bones of the Right Foot.**
Compare the names and numbers of foot bones
(viewed here from above) with those of the hand bones
shown in Figure 6-14.

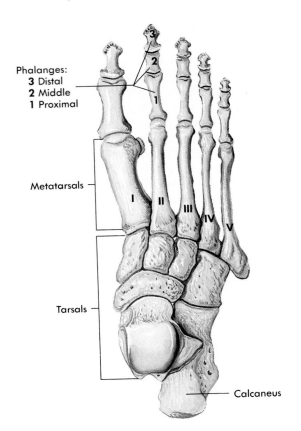

dinal arch. Another arch extends across the ball
of the foot; it is called the **transverse** or **metatarsal
arch.**

▽ ## Differences Between a Man's and a Woman's Skeleton

A man's skeleton and a woman's skeleton differ
in several ways. If you were to examine a male
skeleton and a female skeleton placed side by
side, you would probably notice first the difference in their sizes. Most male skeletons are larger
than most female skeletons, a structural fact that
seems to have no great functional importance.
Structural differences between the male and
female hipbones, however, do have functional
importance. The female pelvis is made so that
the body of a baby can be cradled in it before
birth and can pass through it during birth.
Although the individual male hipbones (os
coxae) are generally larger than the individual
female hipbones, together the male hipbones
form a narrower structure than do the female
hipbones. A man's pelvis is shaped something
like a funnel, but a woman's pelvis has a broader,
shallower shape, more like a basin. (Incidentally,
the word *pelvis* means "basin.") Another difference is that the pelvic inlet or brim is normally much wider in the female than in the male.

▽ **FIGURE 6-17 Arches of the Foot.** Medial
and lateral longitudinal arches are shown.

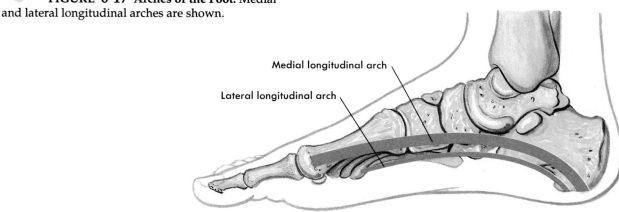

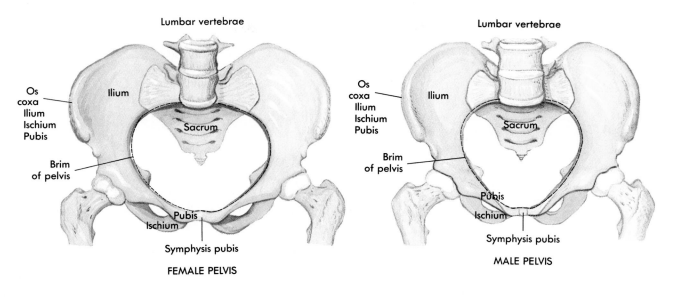

FIGURE 6-18 Comparison of the Male and Female Pelvis. Notice the narrower width of the male pelvis, giving it a more funnel-like shape than the female pelvis.

Figure 6-18 shows this difference clearly. The angle at the front of the female pelvis where the two pubic bones join is wider than it is in the male.

Joints (Articulations)

Every bone in the body, except one, connects to at least one other bone. In other words, every bone but one forms a joint with some other bone. (The exception is the hyoid bone in the neck, to which the tongue anchors.) Most of us probably never think much about our joints unless something goes wrong with them and they do not function properly. Then their tremendous importance becomes painfully clear. Joints hold our bones together securely and at the same time make it possible for movement to occur between the bones—between most of them, that is. Without joints we could not move our arms, legs, or any other of our body parts. Our bodies would, in short, be rigid, immobile hulks. Try, for example, to move your arm at your shoulder joint in as

many directions as you can. Try to do the same thing at your elbow joint. Now examine the shape of the bones at each of these joints on a skeleton or in Figure 6-6. Do you see why you cannot move your arm at your elbow in nearly as many directions as you can at your shoulder?

KINDS OF JOINTS

One method classifies joints into three types according to the degree of movement that they allow:

1 Synarthroses (no movement)
2 Amphiarthroses (slight movement)
3 Diarthroses (free movement)

Differences in the structure of joints account for differences in the degree of movement they make possible.

Synarthroses

A synarthrosis is a joint in which fibrous connective tissue grows between the articulating (joining) bones holding them close together. The joints between cranial bones are synarthroses, commonly called *sutures* (Figure 6-19, *A*).

FIGURE 6-19 Joints of the Skeleton. A, Synarthrotic joint. **B,** Amphiarthrotic joint.

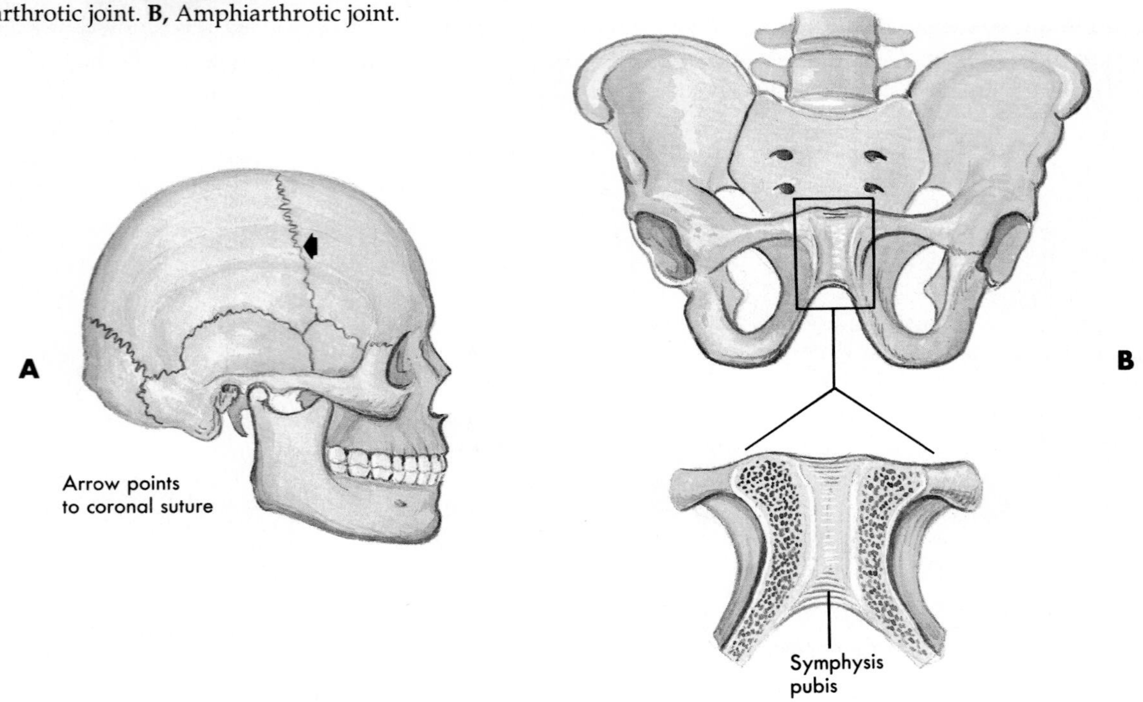

A

Arrow points
to coronal suture

B

Symphysis
pubis

FIGURE 6-20 Herniated Disk. Sagittal section of vertebrae showing normal and herniated disks.

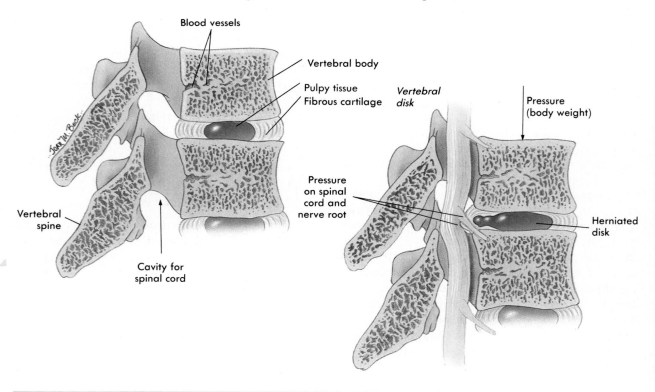

Blood vessels

Vertebral body

Pulpy tissue
Fibrous cartilage

*Vertebral
disk*

Pressure
(body weight)

Pressure
on spinal
cord and
nerve root

Herniated
disk

Vertebral
spine

Cavity for
spinal cord

Amphiarthroses

An amphiarthrosis is a joint in which cartilage connects the articulating bones. The symphysis pubis, the joint between the two pubic bones, is an amphiarthrosis (Figure 6-19, *B*).

Joints between the bodies of the vertebrae are also amphiarthroses. These joints make it possible to flex the trunk forward or sideways and even to circumduct and rotate it. Strong ligaments connect the bodies of the vertebrae, and fibrous disks lie between them. The central core of these intervertebral disks consists of a pulpy, elastic substance that loses some of its resiliency with age. Damage to a disk caused by the pressure of sudden exertion or injury may push its wall into the spinal canal (Figure 6-20). Severe pain may result if the disk presses on the spinal cord. Popularly known as a *slipped disk*, this condition is known to health professionals as a *herniated disk*.

Diarthroses

Fortunately most of our joints by far are diarthroses. Such joints allow considerable movement—sometimes in many directions and sometimes in only one or two directions.

Structure. Diarthroses (freely movable joints) are made alike in certain ways. All have a joint capsule, a joint cavity, and a layer of cartilage over the ends of two joining bones (Figure 6-21). The **joint capsule** is made of the body's strongest and toughest material, fibrous connective tissue, and is lined with a smooth, slippery synovial membrane. The capsule fits over the ends of the two bones somewhat like a sleeve. Because it attaches firmly to the shaft of each bone to form its covering (called the *periosteum; peri* means "around," and *osteon* means "bone"), the joint capsule holds the bones securely together but at the same time permits movement at the joint. The structure of the joint capsule, in other words, helps make possible the joint's function.

Ligaments (cords or bands made of the same strong fibrous connective tissue as the joint capsule) also grow out of the periosteum and lash the two bones together even more firmly.

The layer of **articular cartilage** over the joint ends of bones acts like a rubber heel on a shoe—it absorbs jolts. The **synovial membrane** secretes a lubricating fluid (synovial fluid) that allows easier movement with less friction.

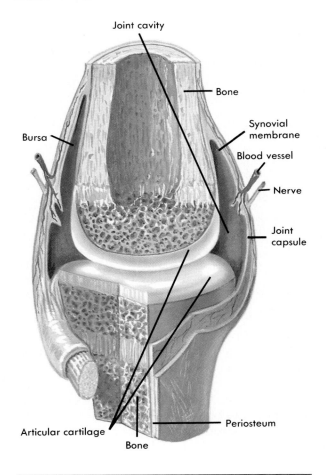

▼ **FIGURE 6-21 Structure of a Diarthrotic Joint.** Each diarthrosis has a joint capsule, a joint cavity, and a layer of cartilage over the ends of the joined bones.

Joint cavity

Bursa

Bone

Synovial membrane

Blood vessel

Nerve

Joint capsule

Periosteum

Articular cartilage

Bone

There are several types of diarthroses, namely, ball-and-socket, hinge, pivot, saddle, gliding, and condyloid (Figure 6-22). Because they differ in structure, they differ also in their possible range of movement. In a ball-and-socket joint, a ball-shaped head of one bone fits into a concave socket of another bone. Shoulder and hip joints, for example, are ball-and-socket joints. Of all the joints in our bodies, these permit the widest range of movements. Think for a moment about how many ways you can move your upper arms. You can move them forward, you can move them backward, you can move them away from the sides of your body, and you can move them back

▼ **FIGURE 6-22 Types of Diarthrotic Joints.** Notice that the structure of each type dictates its function (movement).

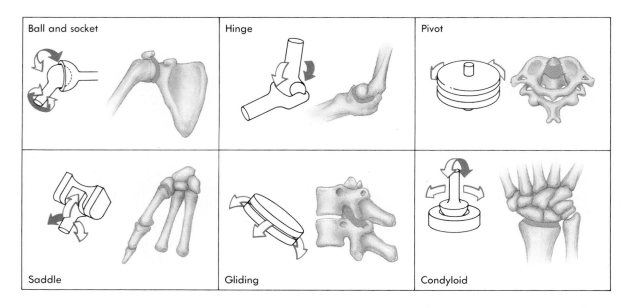

down to your sides. You can also move them around so that you can describe a circle with your hands.

Hinge joints, like the hinges on a door, allow movements in only two directions, namely, flexion and extension. Flexion is bending a part; extension is straightening it out. Elbow and knee joints and the joints in the fingers are hinge joints.

Pivot joints are those in which a small projection of one bone pivots in an arch of another bone. For example, a projection of the axis, the second vertebra in the neck, pivots in an arch of the atlas, the first vertebra in the neck. This rotates the head, which rests on the atlas.

Only one pair of saddle joints exists in the body—between the metacarpal bone of each thumb and a carpal bone of the wrist (the name of this carpal bone is the *trapezium*). Because the articulating surfaces of these bones are saddle-shaped, they make possible the human thumb's great mobility, a mobility no animal's thumb possesses. We can flex, extend, abduct, adduct, and circumduct our thumbs, and most important of

all, we can move our thumbs to touch the tip of any one of our fingers. (This movement is called *opposing the thumb to the fingers.*) Without saddle joints at the base of each thumb, we could not do a simple act such as picking up a pin or grasping a pencil between thumb and forefinger.

Gliding joints are the least movable diarthrotic joints. Their flat articulating surfaces allow limited gliding movements, such as that at the superior and inferior articulating processes between successive vertebrae.

Condyloid joints are those in which a condyle fits into an elliptical socket. An example is the fit of the distal end of the radius into depressions in the carpal bones.

▼ *Skeletal Disorders*

Skeletal disorders include pathological conditions associated with bone, cartilage, ligaments, and joints.

ARTHROSCOPY

Arthroscopy is an imaging technique that allows a physician to examine the internal structure of a joint without the use of extensive surgery. As Figure *A* shows, a narrow tube with lenses and a fiberoptic light source is inserted into the joint space through a small puncture. Isotonic saline (salt) solution is injected through a needle to expand the volume of the synovial space. This spreads the joint structures and makes viewing easier *(B)*.

Although arthroscopy is often used as a diagnostic procedure, it can be used to perform joint surgery. While the surgeon views the internal structure of the joint through the arthroscope or on an attached video monitor, instruments can be inserted through puncture holes and used to repair or remove damaged tissue. Arthroscopic surgery is much less traumatic than previous methods in which the joint cavity was completely opened.

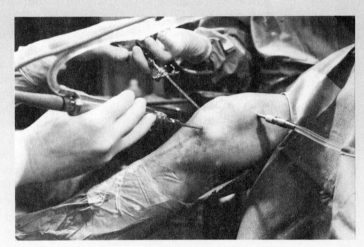

BONE TUMORS AND CANCER

Neoplasms can form benign or malignant tumors of bone tissue (Figure 6-23), cartilage, or other connective tissues. For example, *osteosarcoma* (see Chapter 4) is a common skeletal tumor—and one of the most rapidly fatal. *Chondrosarcoma* (kon-dro-sar-KO-ma) is cancer of skeletal cartilage tissue. *Fibrosarcoma* (see Chapter 4) is a type of malignancy often occurring in bone marrow that causes excessive production of collagenous connective tissue.

METABOLIC BONE DISEASES

Osteoporosis

Osteoporosis (os-tee-o-po-RO-sis) is one of the most common and serious bone diseases. It is characterized by excessive loss of calcified matrix and collagenous fibers from bone. The name *osteoporosis* means "condition of bone pores," referring to the holes or pores formed as bone tissue is lost. Bone porosity results from decreased volume of organic bone matrix and causes the bones to become brittle and easily broken. Osteoporosis occurs most frequently in elderly white women. Although white and black men are also susceptible, black women are seldom affected by it.

Because sex hormones play important roles in stimulating osteoblast activity after puberty, decreasing levels of these hormones in the blood of elderly persons reduces new bone growth and the maintenance of existing bone mass. Therefore

▼ **FIGURE 6-23 Bone Tumor.** This x-ray photograph shows hyperplasia (excessive increase in cell number) in bone tissue, forming a tumor.

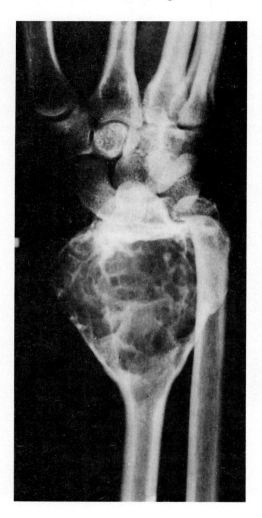

some reabsorption of bone and subsequent loss of bone mass is an accepted consequence of advancing years. However, bone loss in osteoporosis goes far beyond the decrease normally seen in old age. The result is a dangerous pathological condition resulting in bone degeneration, increased susceptibility to "spontaneous fractures"—especially in the neck of the femur and vertebrae—and pathological curvature of the spine. Treatment or preventative measures may include hormone therapy and dietary supplements of calcium and vitamin D to replace deficiencies or to offset intestinal malabsorption.

Osteomalacia

Osteomalacia (os-tee-o-ma-LAY-sha) is another metabolic disorder involving mineral loss in bones. In osteomalacia the volume of bone matrix remains about the same, even though the mass of hard mineral crystals decreases. Recall that in osteoporosis the total volume of organic bone matrix (calcium and collagen) *decreases,* forming holes or pores. The demineralization of bone characteristic of osteomalacia is usually caused by deficiency of vitamin D, although certain drugs may induce it. Vitamin D is required to absorb calcium and phosphate from food. In children, this condition is called **rickets.** Prolonged cases of rickets result in bone deformity. The calcium imbalance caused by rickets may result in death from respiratory muscle spasms.

Paget Disease

Also called *osteitis deformans,* **Paget** (PAJ-et) **disease** is a metabolic disorder affecting older adults. This condition, which is often asymptomatic, is characterized by abnormal bone remodeling. Normal spongy bone is replaced by disorganized, new bone. The abnormal, new bone often grows into marrow spaces and causes thickening of the entire bone.

BONE INFECTION

Osteomyelitis (os-tee-o-my-el-EYE-tis) is the general name for bacterial infections of bone and marrow tissue. *Staphylococcus* bacteria are the most common pathogens in this condition. Besides bacterial infections, bone tissue is also susceptible to damage by viruses, fungi, and other pathogens. Any bone infection is difficult to treat because of the density of the bone and slowness of the healing process compared with other tissues.

BONE FRACTURES

Excessive mechanical stress on bones can result in breaks or fractures (Figure 6-24). Sometimes bone cancer or metabolic bone disorders weaken a bone to the point that it fractures with very little stress. **Open fractures** or *compound fractures,* in

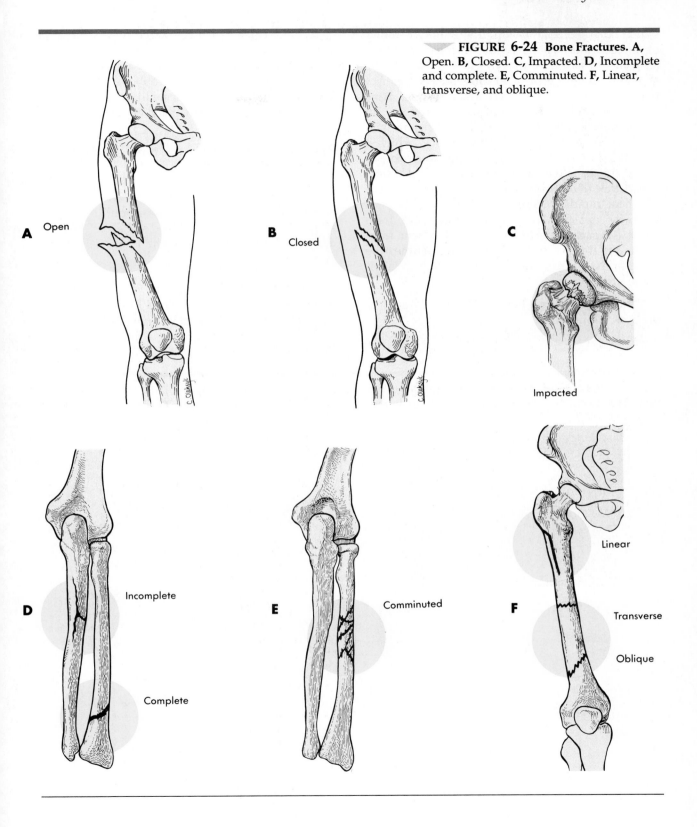

FIGURE 6-24 **Bone Fractures. A,** Open. **B,** Closed. **C,** Impacted. **D,** Incomplete and complete. **E,** Comminuted. **F,** Linear, transverse, and oblique.

A Open

B Closed

C Impacted

D Incomplete

Complete

E Comminuted

F Linear

Transverse

Oblique

which bone pierces the skin, invite the possibility of infection or osteomyelitis. **Closed fractures,** also known as *simple fractures,* do not pierce the skin and so do not pose an immediate danger of bone infection. In **complete fractures** the bone fragments separate completely, whereas in **incomplete fractures** the bone fragments are still partially joined. Incomplete fractures in which a bone is bent but broken only on the outer curve of the bend is often called a *greenstick fracture.* Greenstick fractures, common in children, usually heal rapidly. **Comminuted** (KOM-in-oot-ed) **fractures** are breaks that produce many fragments. **Impacted fractures** occur when bone fragments are driven into each other. Sometimes the angle of the fracture line or crack is used in labeling fracture types:

1 **Linear fracture**—Fracture line is parallel to the bone's long axis.

2 **Transverse fracture**—Fracture line is at a right angle to the bone's long axis.

3 **Oblique fracture**—Fracture line is diagonal to the bone's long axis. If the oblique fracture line seems to spiral around a bone like the stripe on a candy cane, the fracture may be called a *spiral fracture.*

Figure 6-24 summarizes the ways in which fractures can be classified.

After a fracture, a bone usually bleeds, becomes inflamed, and then forms a bony framework called a **callus** (KAL-us) around the injury (Figure 6-25). The callus tissue stabilizes the bone fragments and thus aids in the long healing and remodeling process.

JOINT DISORDERS

Joint disorders can be classified as *noninflammatory joint disease* or *inflammatory joint disease.*

▼ **FIGURE 6-25 Bone Fracture and Repair.** After a fracture **(A)**, there is bleeding and inflammation around the affected area **(B)**. Special tissue forms a bony framework called a callus **(C)** that stabilizes the bone until the repair is complete **(D)**.

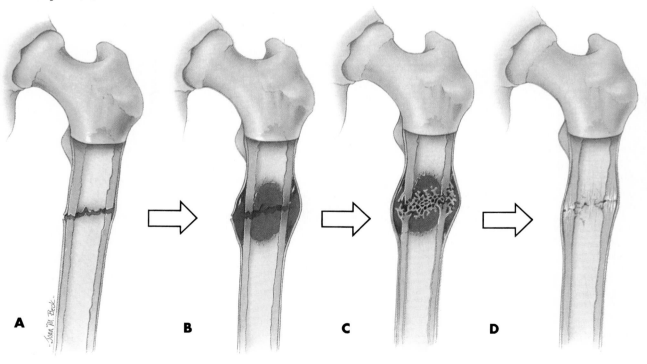

A   B   C   D

Noninflammatory Joint Disease

Noninflammatory joint disease is distinguished from other joint conditions because it does not involve inflammation of the synovial membrane and does not produce systemic signs or symptoms. The most common of these conditions is **osteoarthritis** (os-tee-o-arth-RYE-tis) or *degenerative joint disease (DJD)*. Degeneration of articular cartilage is the characteristic feature of osteoarthritis. The etiology of most cases of DJD is unknown, but advanced age and joint damage or stress are known risk factors.

Inflammatory Joint Disease (Arthritis)

Arthritis (arth-RYE-tis) is a general name for many different inflammatory joint diseases. Arthritis can be caused by a variety of factors, including infection, injury, genetic factors, and autoimmunity. Here is a brief list of major types of arthritis:

1 Rheumatoid (ROO-ma-toyd) **arthritis**—Believed to be a type of autoimmune disease, rheumatoid arthritis involves chronic inflammation of connective tissues. It begins in the synovial membrane and spreads to cartilage and other tissues, often causing severe crippling (Figure 6-26).

2 Gouty arthritis—Gout is a metabolic condition in which uric acid, a nitrogenous waste, increases in the blood. Excess uric acid is deposited as sodium urate crystals in distal joints and other tissues. These crystals trigger the chronic inflammation and tissue damage characteristic of gouty arthritis.

3 Infectious arthritis—A variety of pathogens can infect synovial membrane and other joint tissues. One form of infectious arthritis, *Lyme arthritis* or *Lyme disease* has become a problem throughout most of the United States only in the last couple of decades. Lyme disease was identified in Lyme, Connecticut, in 1975 and has since spread across the continent. Caused by a spirochete bacterium carried by ticks, this condition is characterized by inflammation in the knees or other joints accompanied by a variety of systemic signs and symptoms (see Appendix B).

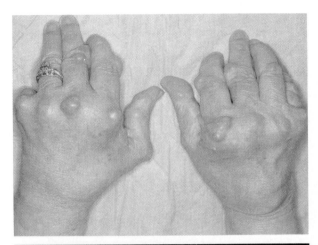
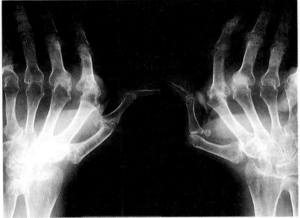

FIGURE 6-26 Rheumatoid Arthritis. The photograph shows a pair of hands whose joints are affected by rheumatoid arthritis. The crippling effects of this condition are even more evident in the x-ray study of the same pair of hands.

OUTLINE SUMMARY

BONE FORMATION AND GROWTH
(Figure 6-1)

A Sequence of development—early cartilage models replaced by calcified bone matrix

B Osteoblasts form new bone, and osteoclasts resorb bone

MICROSCOPIC STRUCTURE OF BONE AND CARTILAGE

A Bone types (Figure 6-2)
  1 Spongy
    a Texture results from needlelike threads of bone called *trabeculae* surrounded by a network of open spaces
    b Found in epiphyses of bones
    c Spaces contain red bone marrow
  2 Compact
    a Structural unit is Haversian system—composed of concentric lamella, lacunae containing osteocytes, and canaliculi, all covered by periosteum

B Cartilage (Figure 6-4)
  1 Cell type called *chondrocyte*
  2 Matrix is gel-like and lacks blood vessels

TYPES OF BONES

A Long—Example: humerus (upper arm)

B Short—Example: carpals (wrist)

C Flat—Example: frontal (skull)

D Irregular—Example: vertebrae (spinal cord)

STRUCTURE OF LONG BONES

A Structural components (Figure 6-5)
  1 Diaphysis or shaft
  2 Medullary cavity containing yellow marrow
  3 Epiphyses or ends of the bone; spongy bone contains red bone marrow
  4 Articular cartilage—covers epiphyses as a cushion
  5 Periosteum—strong membrane covering bone except at joint surfaces
  6 Endosteum—lines medullary cavity

FUNCTIONS OF BONE

A Supports and gives shape to the body

B Protects internal organs

C Helps make movements possible

D Stores calcium

E Hemopoiesis or blood cell formation

DIVISIONS OF SKELETON

Skeleton composed of the following divisions and their subdivisions:

A Axial skeleton
  1 Skull
  2 Spine
  3 Thorax
  4 Hyoid bone

B Appendicular skeleton
  1 Upper extremities, including shoulder girdle
  2 Lower extremities, including hip girdle

C Location and description of bones—see Figures 6-6 to 6-18 and Tables 6-2 to 6-6

DIFFERENCES BETWEEN A MAN'S AND A WOMAN'S SKELETON

A Size—male skeleton generally larger

B Shape of pelvis—male pelvis deep and narrow, female pelvis broad and shallow

C Size of pelvic inlet—female pelvic inlet generally wider, normally large enough for baby's head to pass through it

D Pubic angle—angle between pubic bones of female generally wider

JOINTS (ARTICULATIONS)

A Kinds of joints (Figures 6-19 to 6-22)
  1 Synarthroses (no movement)—fibrous connective tissue grows between articulating bones; for example, sutures of skull
  2 Amphiarthroses (slight movement)—cartilage connects articulating bones; for example, symphysis pubis

3 Diarthroses (free movement)—most joints belong to this class
  a Structures of freely movable joints—joint capsule and ligaments hold adjoining bones together but permit movement at joint
  b Articular cartilage—covers joint ends of bones and absorbs jolts
  c Synovial membrane—lines joint capsule and secretes lubricating fluid
  d Joint cavity—space between joint ends of bones
B Types of freely movable joints—ball-and-socket, hinge, pivot, saddle, gliding, and condyloid

SKELETAL DISORDERS

A Bone tumors and cancer—benign **or m**alignant neoplasms of bone, cartilage, and fibrous tissue
B Metabolic bone diseases
  1 Osteoporosis—excessive loss of bone matrix (mineral and collagen)
  2 Osteomalacia—softening of bone from loss of mineral (but not volume) in bone matrix; called *rickets* in children
  3 Paget disease—osteitis deformans; abnormal bone remodeling in which spongy bone is replaced by disorganized, excessive bone matrix
C Bone infection
  1 Osteomyelitis—general term for bacterial (usually staphylococcal) infection of bone

2 Bone infections may also be caused by viruses, fungi, and other pathogens
D Bone fractures (Figure 6-24)
  1 Open (compound) fractures pierce the skin and closed (simple) fractures do not
  2 Complete fractures involve total separation of bone fragments, and incomplete fractures involve partially separated fragments; comminuted fractures involve many fragments
  3 Fracture lines can be classified by their angle relative to a bone's axis: linear, transverse, and oblique
E Joint disorders
  1 Noninflammatory joint disease does not usually involve inflammation of the synovial membrane
    a Osteoarthritis—degenerative joint disease; degeneration of articular cartilage
  2 Inflammatory joint disease (arthritis)—inflammation of synovial membrane with systemic signs or symptoms
    a Rheumatoid arthritis—autoimmune inflammation of synovial membrane and other structures
    b Gouty arthritis—synovial inflammation caused by gout, a condition in which sodium urate crystals form in joints and other tissues
    c Infectious arthritis—arthritis resulting from infection by a pathogen, as in Lyme arthritis caused by the Lyme disease bacterium

NEW WORDS

amphiarthroses
appendicular skeleton
articular cartilage
articulation
axial skeleton
canaliculi
chondrocytes
compact bone

diaphysis
diarthroses
epiphyses
fontanels
Haversian system
hemopoiesis
lacunae
lamella

medullary cavity
osteoblasts
osteoclasts
osteocytes
pectoral girdle
pelvic girdle
periosteum
red bone marrow

sinus
skull
spine
synarthroses
synovial membrane
thorax
trabeculae
yellow bone marrow

Diseases and Other Clinical Terms

arthritis
callus
chondrosarcoma
closed fracture
comminuted fracture
complete fracture

gout
incomplete fracture
kyphosis
linear fracture
lordosis
mastoiditis

oblique fracture
open fracture
osteoarthritis
osteomalacia
osteomyelitis
osteoporosis

Paget disease
rheumatoid arthritis
rickets
scoliosis
transverse fracture

CHAPTER TEST

1. Spongy bone contains needlelike threads of bone known as _____.
2. The structural units of compact bone are called _____.
3. Bone-forming cells are called _____, whereas bone-resorbing cells are called _____.
4. The hollow shaft of a long bone is also known as the _____.
5. The human skeleton has two main divisions, the _____ skeleton and the _____ skeleton.
6. Spaces or cavities located inside some of the cranial bones are called _____.
7. The supporting framework of the neck is formed by the seven _____ vertebrae.
8. The shoulder blade is also known as the _____, and the collar bone is called the _____.
9. The long bones of the forearm are the _____ and the _____.
10. The large bony process of the ulna that forms the elbow is called the _____ process.
11. There are 8 _____ bones in the wrist and 14 _____ or finger bones in each hand.
12. Each hipbone consists of three separate bones called the _____, _____, and _____, which fuse in the adult to become one bone.
13. The femur of the leg articulates distally with the kneecap or _____ and the shinbone or _____.
14. The metatarsal and tarsal bones are located in the _____.
15. Joints that permit free movement are called _____ joints.
16. Abnormal side-to-side curvature of the vertebral column is called _____.
17. The skeletal disorder characterized by excessive loss of calcified matrix and collagen fibers is called _____.
18. Microbial infection of bone is called _____.
19. A _____ fracture invites the possibility of infection because the skin is pierced.
20. Degenerative joint disease, or _____, involves wearing away of articular cartilage.

Circle the "T" before each true statement and the "F" before each false statement.

T F **21.** Adult bones do not contain living cells.
T F **22.** Red bone marrow functions in hemopoiesis or blood cell formation.
T F **23.** The hollow cylindrical portion of a long bone is called the *epiphysis.*
T F **24.** Haversian systems are components of spongy bone.
T F **25.** Lacunae are found in compact bone and cartilage.
T F **26.** There are more bones in the axial than in the appendicular skeleton.
T F **27.** The term *phalanges* is used to describe the bones of the fingers and the toes.
T F **28.** "Soft spots" in the skull at birth are called *fontanels.*
T F **29.** Synarthroses are freely movable joints.
T F **30.** Synovial membranes are found in diarthrotic joints.

REVIEW QUESTIONS

1 Discuss the mechanism of bone formation and growth.
2 Is it possible to tell whether a child is going to grow taller? If so, how?
3 Compare the structures of compact and spongy bone.
4 Discuss the following cell types: osteocyte, osteoblast, osteoclast, chondrocyte.
5 List the major structural components of a typical long bone, and briefly describe the function of each.
6 List and discuss the generalized functions of the skeletal system.
7 What are the two major subdivisions of the skeleton? What are the major body areas in each subdivision?
8 Give the correct anatomical name for each of the following: collarbone, breastbone, wrist bones, finger bones, forearm bones, thigh bone, hipbone, kneecap, ankle bones, and neck vertebrae.
9 Compare the structures and functions of the arms and legs, pectoral and pelvic girdles, shoulder and hip joints, and hands and feet.
10 Describe one functionally important difference between a male and a female adult skeleton. Why is this difference important?
11 Classify the major types of joints.
12 List and compare the types of freely movable joints.
13 Compare and contrast the causes and changes associated with osteoporosis, osteomalacia, and Paget disease.
14 Describe how a fracture may result in osteomyelitis. Why is it often difficult to treat osteomyelitis?
15 How does osteoarthritis differ from the so-called inflammatory joint diseases?

CLINICAL APPLICATIONS

1 Andrew is a young boy who loves to climb trees. While attempting to climb his favorite oak tree, Andrew fell and fractured his humerus. The radiologist described Andrew's injury as a "greenstick fracture." What does this label tell you about the appearance of the fracture? If given proper medical care, is the injury likely to heal rapidly or slowly?

2 Christine is a young music student at the local college. One of her professors suggested that she analyze her conducting technique by videotaping herself as she conducts the choir. As she replays the tape, Christine notices that her hips and shoulders seem awkwardly bent—even when she is in a formal standing position. What condition might cause this abnormal curve of the trunk? What are some ways of treating this condition?

3 Agnes is an elderly woman with osteoporosis. She recently sustained a severe bone fracture for no apparent reason (she did not fall or otherwise injure herself). The fracture was not treated for some time, and as a result, Agnes developed osteomyelitis. Based on what you know of osteoporosis, how do you explain her mysterious fracture? How can a fracture progress to osteomyelitis?

The Muscular System

C H A P T E R

7

Outline

Objectives

After you have completed this chapter, you should be able to:

1 List, locate in the body, and compare the structure and function of the three major types of muscle tissue.

2 Discuss the microscopic structure of a skeletal muscle sarcomere and motor unit.

3 Discuss how a muscle is stimulated and compare the major types of skeletal muscle contractions.

4 Name, identify on a model or diagram, and give the function of the major muscles of the body discussed in this chapter.

5 List and explain the most common types of movement produced by skeletal muscles.

6 Name and describe the major disorders of skeletal muscles.

Boxed Essays

Muscles Versus Gravity
Occupational Health Problems
Rigor Mortis
Intramuscular Injections

This chapter is devoted to the 40% to 45% of our body weight composed of muscle—a highly specialized tissue that enables the body and its parts to move. Movement occurs because of the ability of muscle cells, also called *muscle fibers*, to contract or shorten. Muscular contractions may cause an eye to blink, move the body as a whole from place to place, cause air to move in and out of the lungs, cause food to move through the digestive tract, or force blood through the body. Few body systems have greater importance for healthy living or for life itself than the muscular system.

All muscle cells shorten or contract by converting chemical energy obtained from food into mechanical energy that is translated into movement. Although we will initially review the three types of muscle tissue introduced earlier (see Chapter 3), the plan for this chapter is to focus on skeletal or voluntary muscle.

Muscle Tissue

If you weigh 120 pounds, about 50 pounds of your weight comes from your muscles, the "red meat" attached to your bones. Under the microscope, these threadlike muscle cells appear in bundles. They are characterized by many crosswise stripes and multiple nuclei (Figure 7-1, *A*). Each fine thread is a muscle cell or, as it is usually called, a *muscle fiber*. This type of muscle tissue has three names: *skeletal muscle*, because it attaches to bone; *striated muscle*, because of its cross stripes or striations; and *voluntary muscle*, because its contractions can be controlled voluntarily.

Besides **skeletal muscle,** the body also contains two other kinds of muscle tissue: cardiac muscle and nonstriated, smooth, or involuntary muscle. **Cardiac muscle,** as its name suggests, composes the bulk of the heart. Cells in this type of muscle appear to branch frequently (Figure 7-1, *B*) and then recombine into a continuous mass of interconnected tissue. Like skeletal muscle cells, they have cross striations. They also have unique dark bands called *intercalated disks* where the plasma membranes of adjacent cardiac fibers come in contact with each other. Cardiac muscle tissue demonstrates the principle that "form follows function." The interconnected nature of cardiac

muscle fibers helps the tissue to contract as a unit and increases the efficiency of the heart muscle in pumping blood.

Nonstriated or **smooth muscle** lacks the cross stripes or striations of skeletal muscle. It has a smooth, even appearance when viewed through a microscope (Figure 7-1, *C*). It is called *involuntary* because we normally do not have control over its contractions. Smooth or involuntary muscle forms an important part of blood vessel walls and of many hollow internal organs (viscera) such as the gut, urethra, and ureters. Because of its location in many visceral structures, it is sometimes called *visceral muscle.* Although we cannot willfully control the action of smooth muscle, its contractions are highly regulated so that, for example, food is passed through the digestive tract or urine is pushed through the ureters into the bladder.

Muscle cells specialize in contraction or shortening. Every movement we make is produced by contractions of skeletal muscle cells. Contractions of cardiac muscle cells keep the blood circulating, and smooth muscle contractions do many things; for instance, they move food into and through the stomach and intestines and contribute to the maintenance of normal blood pressure.

Structure of Skeletal Muscle

A skeletal muscle is an organ composed mainly of striated muscle cells and connective tissue. Most skeletal muscles attach to two bones that have a movable joint between them. In other words, most muscles extend from one bone across a joint to another bone. Also, one of the two bones is usually more stationary in a given movement than the other. The muscle's attachment to this more stationary bone is called its **origin.** Its attachment to the more movable bone is called the muscle's **insertion.** The rest of the muscle (all of it except its two ends) is called the *body* of the muscle (Figure 7-2).

Tendons anchor muscles firmly to bones. Made of dense fibrous connective tissue in the shape of heavy cords, tendons have great strength. They do not tear or pull away from bone easily. Yet any emergency room nurse or physician sees many tendon injuries—severed tendons and tendons torn loose from bones.

▽ **FIGURE 7-1 Muscle Tissue. A,** Skeletal muscle is attached to bone. Skeletal muscle cells are cylindrical in shape, are striated, and have multiple nuclei. **B,** Cardiac muscle is in the heart. Cardiac muscle cells are cyclindrical in shape, are striated, and have a single nucleus. The cells are branched and are connected to each other by intercalated disks. **C,** Smooth muscle is in hollow organs such as the stomach and the intestine. Smooth-muscle cells are tapered at each end and have a single nucleus.

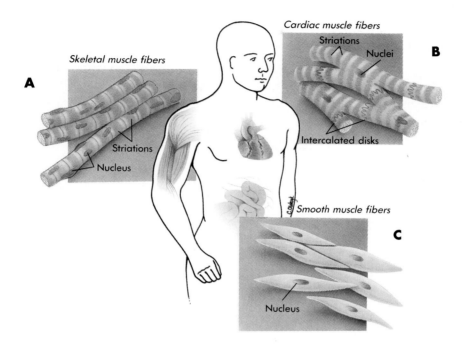

Small fluid-filled sacs called **bursae** lie between some tendons and the bones beneath them (see Figure 6-21). These small sacs are made of connective tissue and are lined with **synovial membrane.** The synovial membrane secretes a slippery lubricating fluid (synovial fluid) that fills the bursa. Like a small, flexible cushion, a bursa makes it easier for a tendon to slide over a bone when the tendon's muscle shortens. **Tendon sheaths** enclose some tendons. Because these tube-shaped structures are also lined with synovial membrane and are moistened with synovial fluid, they, like the bursae, facilitate movement.

MICROSCOPIC STRUCTURE

Muscle tissue consists of specialized contractile cells or **muscle fibers** that are grouped together and arranged in a highly organized way. Each skeletal muscle fiber is itself filled with two kinds

▽ **FIGURE 7-2 Skeletal Muscle Attachments.** A muscle originates at a stable part of the skeleton and inserts at a part moved when the muscle contracts.

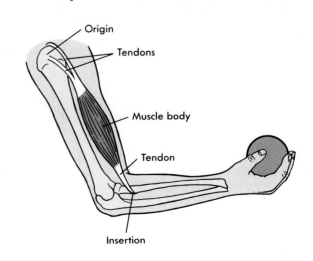

▼ **FIGURE 7-3 Structure of Skeletal Muscle.**
A, Each muscle organ has many muscle fibers, each
containing many bundles of thick and thin filaments.
The diagrams show the overlapping thick and thin
filaments arranged to form adjacent segments called
sarcomeres. During contraction, the thin filaments are
pulled toward the center of each sarcomere, shortening
the whole muscle. **B,** This electron micrograph shows
that the overlapping thick and thin filaments within
each sarcomere creates a pattern of dark striations in
the muscle. The extreme magnification allowed by
electron microscopy has revolutionized our concept of
the structure and function of skeletal muscle and other
tissues.

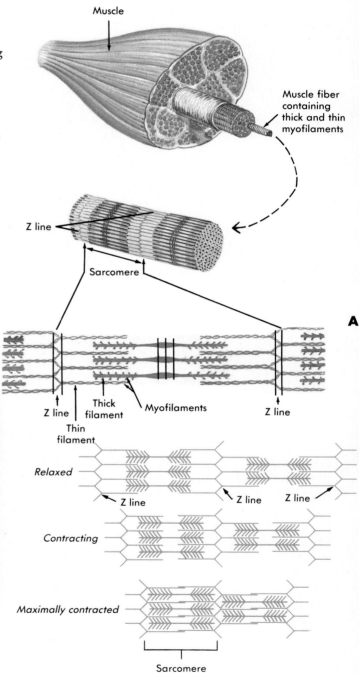

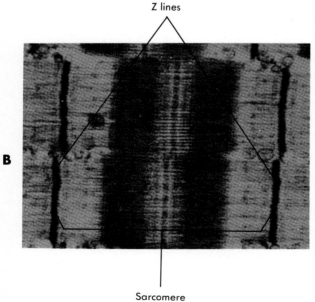

of very fine and threadlike structures that are called **thick** and **thin myofilaments** (my-o-FIL-a-ments). The thick myofilaments are formed from a protein that is called **myosin,** and the thin myofilaments are composed mostly of the protein **actin.** Find the label **sarcomere** (SAR-ko-meer) shown in Figure 7-3. Think of the sarcomere as the basic functional or *contractile unit* of skeletal muscle. Recall that the Haversian system serves as the basic building block of compact bone; the sarcomere serves that function in skeletal muscle. The submicroscopic structure of a sarcomere consists of numerous actin and myosin myofilaments arranged so that, when viewed under a microscope, dark and light stripes or cross striations are seen. The repeating units or sarcomeres are separated from each other by dark bands called *Z lines.*

Although the sarcomeres in the upper portion and in the electron photomicrograph (EM) of Figure 7-3 are in a relaxed state, the thick and thin myofilaments, which are lying parallel to each other, still overlap. Now look at the diagrams in the lower portion of Figure 7-3. Note that contraction of the muscle causes the two types of myofilaments to slide toward each other and shorten the sarcomere and thus the entire muscle. When the muscle relaxes, the sarcomeres can return to resting length, and the filaments resume their resting positions.

An explanation of how a skeletal muscle contracts (that is, how its myofilaments move toward one another to cause shortening of sarcomeres) is provided by the **sliding filament model.** According to this model, during contraction, the thick and thin myofilaments in a muscle fiber first attach to one another by the formation of "bridges" that then act as levers to ratchet or pull the myofilaments past each other. The connecting bridges between the myofilaments can form properly only if calcium is present. During the relaxed state, calcium is within the endoplasmic reticulum (see Chapter 2) in the muscle cell. It is released into the cytoplasm when the muscle is stimulated to contract. In addition to calcium, the shortening of a muscle cell requires energy. This is supplied by the breakdown of adenosine triphosphate (ATP) molecules, the energy storage molecules of the cell.

Functions of Skeletal Muscle

The three primary functions of the muscular system are:

1 Movement
2 Posture or muscle tone
3 Heat production

MOVEMENT

Muscles move bones by pulling on them. Because the length of a skeletal muscle becomes shorter as its fibers contract, the bones to which the muscle attaches move closer together. As a rule, only the insertion bone moves. Look again at Figure 7-2. As the ball is lifted, the shortening of the muscle body pulls the insertion bone toward the origin bone. The origin bone stays put, holding firm, while the insertion bone moves toward it. One tremendously important function of skeletal muscle contractions therefore is to produce movements. Remember this rule: a muscle's insertion bone moves toward its origin bone. It can help you understand muscle actions.

Voluntary muscular movement is normally smooth and free of jerks and tremors because skeletal muscles generally work in coordinated teams, not singly. Several muscles contract while others relax to produce almost any movement you can imagine. Of all the muscles contracting simultaneously, the one mainly responsible for producing a particular movement is called the **prime mover** for that movement. The other muscles that help in producing the movement are called **synergists** (SIN-er-jists). As prime movers and synergist muscles at a joint contract, other muscles, called **antagonists** (an-TAG-o-nists), relax. When those antagonist muscles contract, they produce a movement opposite to that of those prime movers and their synergist muscles.

Locate the biceps brachii, brachialis, and triceps brachii muscles in Figure 7-6. All the muscles in these figures are involved in bending and straightening the forearm at the elbow joint. The biceps brachii is the prime mover during bending, and the brachialis is its helper or synergist muscle. When the biceps brachii and brachialis muscles bend the forearm, the triceps brachii relaxes. Therefore while the forearm bends, the triceps brachii is the antagonistic muscle. While the forearm straightens, these three muscles continue to work as a team. However, during straightening, the triceps brachii becomes the prime mover and the biceps brachii and brachialis become the antagonistic muscles. This combined and coordinated activity is what makes our muscular movements smooth and graceful.

POSTURE

We are able to maintain our body position because of a specialized type of skeletal muscle contraction called **tonic contraction.** Because relatively few of a muscle's fibers shorten at one time in a tonic contraction, the muscle as a whole does not shorten, and no movement occurs. Consequently, tonic contractions do not move any body parts. They do hold muscles in position, however. In other words, muscle tone maintains **posture.** Good posture means that body parts are held in the positions that favor best function. These positions balance the distribution of weight and therefore put the least strain on muscles, tendons, ligaments, and bones. To have good posture in a standing position, for example, you must stand with your head and chest held high, your chin, abdomen, and buttocks pulled in, and your knees bent slightly.

To judge for yourself how important good posture is, consider some of the effects of poor posture. Besides detracting from appearance, poor posture makes a person tire more quickly. It puts an abnormal pull on ligaments, joints, and bones and sometimes leads to deformities. Poor posture crowds the heart, making it harder for it to contract. Poor posture also crowds the lungs, decreasing their breathing capacity.

Skeletal muscle tone maintains posture by counteracting the pull of gravity. Gravity tends to pull the head and trunk down and forward, but the tone in certain back and neck muscles pulls just hard enough in the opposite direction to overcome the force of gravity and hold the head and trunk erect. The tone in thigh and leg muscles puts just enough pull on thigh and leg bones to counteract the pull of gravity on them that would otherwise collapse the hip and knee joints and cause us to fall in a heap.

HEAT PRODUCTION

Healthy survival depends on our ability to maintain a constant body temperature. A fever or elevation in body temperature of only a degree or two above 37° C (98.6° F) is almost always a sign of illness. Just as serious is a fall in body temperature. Any decrease below normal, a condition called **hypothermia** (hye-po-THER-mee-ah), drastically affects cellular activity and normal body function. The contraction of muscle fibers produces most of the heat required to maintain body temperature. Energy required to produce a muscle contraction is obtained from ATP. Most of the energy released during the breakdown of ATP during a muscular contraction is used to shorten the muscle fibers; however, some of the energy is lost as heat during the reaction. This heat helps us to maintain our body temperature at a constant level.

Fatigue

If muscle cells are stimulated repeatedly without adequate periods of rest, the strength of the muscle contraction will decrease, resulting in **fatigue.** If repeated stimulation occurs, the strength of the contraction continues to decrease, and eventually the muscle loses its ability to contract.

During exercise, the stored ATP required for muscle contraction becomes depleted. Formation of more ATP results in rapid consumption of oxygen and nutrients, often outstripping the ability of the muscle's blood supply to replenish them. When oxygen supplies run low, muscle cells switch to a type of energy conversion that does not require oxygen. This process produces lactic acid that may result in muscle soreness after exercise. The term **oxygen debt** describes the continued increased metabolism that must occur in a cell to remove excess lactic acid that accumulates during prolonged exercise. Thus depleted energy reserves are replaced. Labored breathing after the cessation of exercise is required to "pay the debt" of oxygen required for the metabolic effort. This mechanism is a good example of homeostasis at work. The body returns the cells' energy and oxygen reserves to normal, resting levels.

Role of Other Body Systems in Movement

Remember that muscles do not function alone. Other structures such as bones and joints must function with them. Most skeletal muscles cause movements by pulling on bones across movable joints.

The respiratory, circulatory, nervous, muscular, and skeletal systems play essential roles in producing normal movements. This fact has great practical importance. For example, a person might have perfectly normal muscles and still not be able to move normally. He might have a nervous system disorder that shuts off impulses to certain skeletal muscles and thereby results in **paralysis.** Multiple sclerosis acts in this way, but so do some other conditions such as a brain hemorrhage, a brain tumor, or a spinal cord injury.

Skeletal system disorders, especially arthritis, have disabling effects on movement. Muscle functioning, then, depends on the functioning of many other parts of the body. This fact illustrates a principle repeated often in this book. It can be simply stated: Each part of the body is one of many components in a large, interactive system. The normal function of one part depends on the normal function of the other parts.

Motor Unit

Before a skeletal muscle can contract and pull on a bone to move it, the muscle must first be stimulated by nerve impulses. Muscle cells are stimulated by a nerve fiber called a **motor neuron** (Figure 7-4). The point of contact between the nerve ending and the muscle fiber is called a **neu-**

FIGURE 7-4 Motor Neuron. A motor unit consists of one motor neuron and the muscle fibers supplied by its branches.

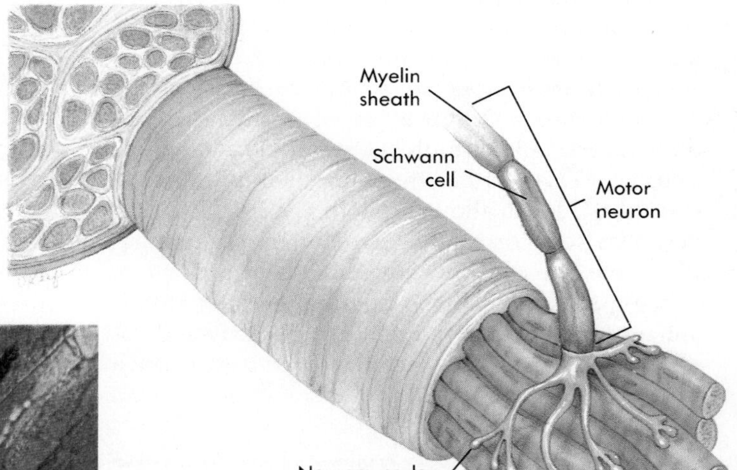

Myelin sheath

Schwann cell

Motor neuron

Neuromuscular junction

Nucleus

Muscle fibers

Myofibrils

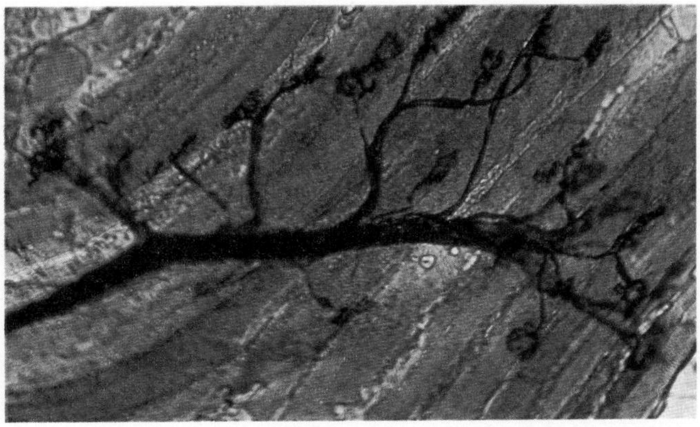

romuscular junction. Specialized chemicals are released by the motor neuron in response to a nervous impulse. These chemicals then generate events within the muscle cell that result in contraction or shortening of the muscle cell. A single motor neuron, with the muscle cells it innervates, is called a **motor unit** (Figure 7-4).

Muscle Stimulus

In a laboratory setting a single muscle fiber can be isolated and subjected to stimuli of varying intensities so that it can be studied. Such experiments show that a muscle fiber does not contract until an applied stimulus reaches a certain level of intensity. The minimal level of stimulation required to cause a fiber to contract is called the **threshold stimulus.**

When a muscle fiber is subjected to a threshold stimulus, it contracts completely. Because of this, muscle cells are said to respond **"all or none."** However, a muscle is composed of many muscle cells that are controlled by different motor units and that have different threshold-stimulus levels. Although each fiber in a muscle such as the biceps brachii responds all or none when subjected to a threshold stimulus, the muscle as a whole does not. This fact has tremendous importance in everyday life. It allows you to pick up a 2-liter bottle of soda or a 20-kg weight by stimulating the same muscle but executing contractions of varying strength. This is possible because different numbers of motor units can be activated for different loads. Once activated, however, each fiber always responds all or none.

Types of Skeletal Muscle Contractions

In addition to the specialized tonic contraction of muscle that maintains a person's muscle tone and posture, other types of contraction also occur. Additional types of muscle contraction include the following:

1 **Twitch contraction**
2 **Tetanic contraction**
3 **Isotonic contraction**
4 **Isometric contraction**

TWITCH AND TETANIC CONTRACTIONS

The term **twitch** describes a quick, jerky response to a single stimulus. Twitch contractions can be easily seen in muscles prepared for research, but they play a minimal role in normal muscle activity. To accomplish the coordinated and fluid muscular movements necessary to accomplish most daily tasks, muscles must contract not in a jerky but in a smooth and sustained way.

A **tetanic contraction** is a more sustained and steady response than a twitch. It is produced by a series of stimuli bombarding the muscle in rapid succession. Individual contractions "melt" together to produce a sustained contraction or *tetanus*. About 30 stimuli per second, for example, evoke a tetanic contraction in certain types of skeletal muscle. Tetanic contraction is not necessarily a maximal contraction in which each muscle fiber responds at the same time. In most cases, only a few areas of the muscle undergo tetanic contractions at any one time.

ISOTONIC CONTRACTION

In most cases, isotonic contraction of muscle produces movement at a joint. With this type of contraction the muscle shortens, and the insertion end moves toward the point of origin (Figure 7-5, *A*). Walking, running, breathing, lifting, and twisting are examples of isotonic contraction.

ISOMETRIC CONTRACTION

Contraction of a skeletal muscle does not always produce movement. Sometimes, it increases the tension within a muscle but does not shorten the muscle. When the muscle does not shorten and no movement results, it is called an *isometric contraction*. The word *isometric* comes from Greek words that mean "equal measure." In other words, a muscle's length during an isometric contraction and during relaxation is about equal. Although muscles do not shorten (and therefore do not produce movement) during isometric contractions, the tension within them increases (Figure 7-5, *B*). Because of this, repeated isometric contractions tend to make muscles grow larger and stronger—hence the popularizing in recent years of isometric exercises as great muscle builders. Pushing against a wall or another immovable object is a good example of isometric exercise. Although no movement occurs and the muscle does not shorten, its internal tension increases dramatically.

▼ **FIGURE 7-5 Types of Muscle Contraction. A,** In isotonic contraction the muscle shortens, producing movement. **B,** In isometric contraction the muscle pulls forcefully against a load but does not shorten.

Isotonic

Isometric

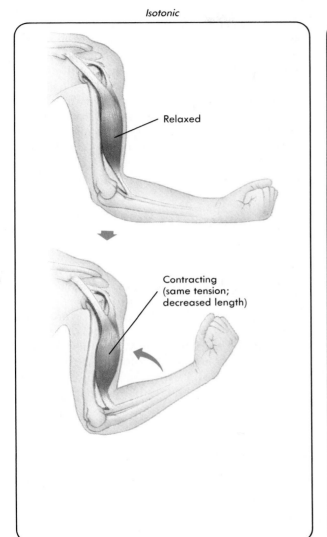

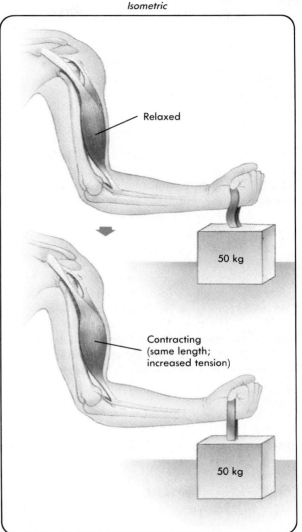

A

B

Effects of Exercise on Skeletal Muscles

Most of us believe that exercise is good for us, even if we have no idea what or how many specific benefits can come from it. Some of the good consequences of regular, properly practiced exercise are greatly improved muscle tone, better pos-ture, more efficient heart and lung function, less fatigue, and looking and feeling better.

Skeletal muscles undergo changes that correspond to the amount of work that they normally do. During prolonged inactivity, muscles usually shrink in mass, a condition called **disuse atrophy.** Exercise, on the other hand, may cause an increase in muscle size called **hypertrophy.**

Muscle hypertrophy can be enhanced by a program of **strength training,** which involves contracting muscles against heavy resistance. Isometric exercises and weight lifting are common strength-training activities. This type of training results in increased numbers of myofilaments in each muscle fiber. Although the actual number of muscle fibers stays the same, the increased number of myofilaments greatly increases the mass of the muscle.

MUSCLES VERSUS GRAVITY

One important function of muscles is to maintain muscle tone that counteracts the force of gravity on our skeleton. Exercise physiologists in the Soviet Union have found that the zero-gravity environment of space promotes a loss of postural muscle tone. Cosmonauts returning from long missions in orbit had to be carried around like invalids for weeks or months until their normal muscle tone returned. To avoid this problem, cosmonauts on later missions imitated the effects of gravity on postural muscles by pulling on resistance machines.

Endurance training, often called **aerobic training,** does not usually result in muscle hypertrophy. Instead, this type of exercise program increases a muscle's ability to sustain moderate exercise over a long period. Aerobic activities such as running, bicycling, or other primarily isotonic movements increase the number of blood vessels in a muscle without significantly increasing its size. The increased blood flow allows a more efficient delivery of oxygen and glucose to muscle fibers during exercise. Aerobic training also causes an increase in the number of mitochondria in the muscle fibers (see pp. 24 to 25). This allows production of more ATP as a rapid energy source.

Skeletal Muscle Groups

In the paragraphs that follow, representative muscles from the most important skeletal muscle groups will be discussed. Table 7-1 identifies and groups muscles according to function and provides information about muscle action and points of origin and insertion. Keep in mind that muscles move bones, and the bones that they move are their insertion bones. Refer to Figure 7-6 often so that you will be able to see a muscle as you read about its placement on the body and its function.

TABLE 7-1   Principal Muscles of the Body

MUSCLE	FUNCTION	INSERTION	ORIGIN
Muscles of the Head and Neck			
Frontal	Raises eyebrow	Skin of eyebrow	Occipital bone
Orbicularis oculi	Closes eye	Maxilla and frontal bone	Maxilla and frontal bone (encircles eye)
Orbicularis oris	Draws lips together	Encircles lips	Encircles lips
Zygomaticus	Elevates corners of mouth and lips	Angle of mouth and upper lip	Zygomatic
Masseter	Closes jaws	Mandible	Zygomatic arch
Temporal	Closes jaw	Mandible	Temporal region of the skull
Sternocleidomastoid	Rotates and extends head	Mastoid process	Sternum and clavicle
Trapezius	Extends head and neck	Scapula	Skull and upper vertebrae

▽ **TABLE 7-1**—cont'd   Principal Muscles of the Body

MUSCLE	FUNCTION	INSERTION	ORIGIN
Muscles that Move the Upper Extremities			
Pectoralis major	Flexes and helps adduct upper arms	Humerus	Sternum, clavicle, and upper rib cartilages
Latissimus dorsi	Extends and helps adduct upper arm	Humerus	Vertebrae and ilium
Deltoid	Abducts upper arm	Humerus	Clavicle and scapula
Biceps brachii	Flexes lower arm	Radius	Clavicle and scapula
Triceps brachii	Extends lower arm	Ulna	Scapula and humerus
Muscles of the Trunk			
External oblique	Compresses abdomen	Midline of abdomen	Lower thoracic cage
Internal oblique	Compresses abdomen	Midline of abdomen	Pelvis
Transversus abdominis	Compresses abdomen	Midline of abdomen	Ribs, vertebrae, and pelvis
Rectus abdominis	Flexes trunk	Lower rib cage	Pubis
Muscles that Move the Lower Extremities			
Iliopsoas	Flexes thigh or trunk	Femur	Ilium and vertebrae
Sartorius	Flexes thigh and rotates lower leg	Tibia	Ilium
Gluteus maximus	Extends thigh	Femur	Ilium, sacrum, and coccyx
Adductor group			
Adductor longus	Adducts thigh	Femur	Pubis
Gracilis	Adducts thigh	Tibia	Pubis
Pectineus	Adducts thigh	Femur	Pubis
Hamstring group			
Semimembranosus	Flexes lower leg	Tibia	Ischium
Semitendinosus	Flexes lower leg	Tibia	Ischium
Biceps femoris	Flexes lower leg	Fibula	Ischium and femur
Quadriceps group			
Rectus femoris	Extends lower leg	Tibia	Ilium
Vastus lateralis, intermedius, and medialus	Extend lower leg	Tibia	Femur
Tibialis anterior	Dorsiflexes foot	Metatarsals (foot)	Tibia
Gastrocnemius	Plantar flexes foot	Calcaneus (heel)	Femur
Soleus	Plantar flexes foot	Calcaneus (heel)	Tibia and fibula
Peroneus group			
Peroneus longus and brevis	Plantar flex foot	Tarsal and metatarsals (ankle and foot)	Tibia and fibula

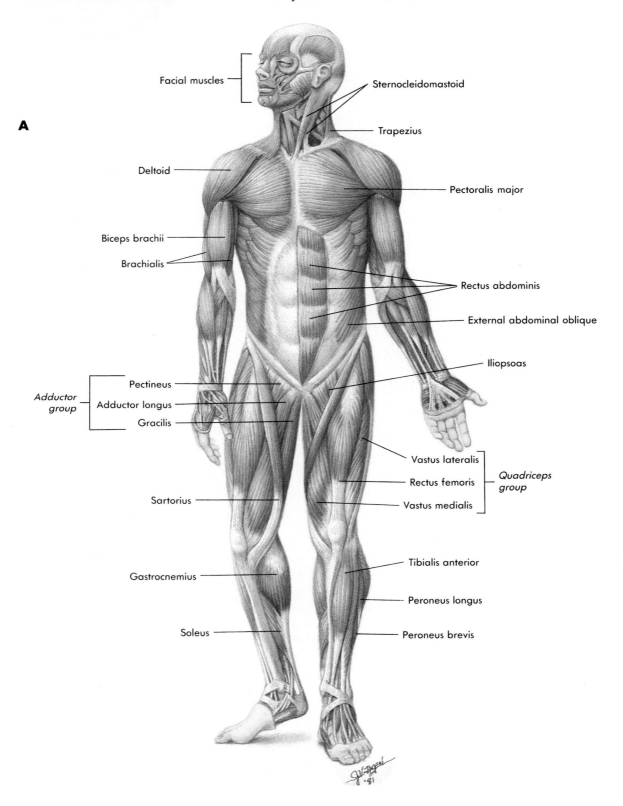

A

Facial muscles

Sternocleidomastoid

Trapezius

Deltoid

Pectoralis major

Biceps brachii

Brachialis

Rectus abdominis

External abdominal oblique

Iliopsoas

Adductor group

Pectineus

Adductor longus

Gracilis

Vastus lateralis

Quadriceps group

Rectus femoris

Vastus medialis

Sartorius

Tibialis anterior

Peroneus longus

Gastrocnemius

Soleus

Peroneus brevis

FIGURE 7-6, cont'd B, Posterior view.

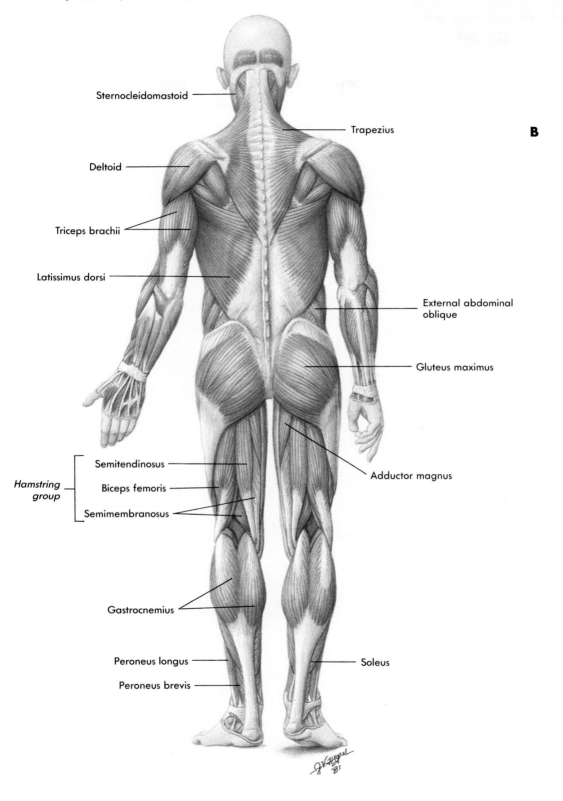

B

OCCUPATIONAL HEALTH PROBLEMS

Some epidemiologists specialize in the field of *occupational health*, the study of health matters related to work or the workplace. Many problems seen by occupational health experts are caused by repetitive motions of the wrists or other joints. Word processors (typists) and meat cutters, for example, are at risk of developing conditions caused by repetitive motion injuries.

One common problem often caused by such repetitive motion is **tenosynovitis** (ten-o-sin-o-VYE-tis)—inflammation of a tendon sheath. Tenosynovitis can be painful, and the swelling characteristic of this condition can limit movement in affected parts of the body. For example, swelling of the tendon sheath around tendons in an area of the wrist known as the *carpal tunnel* can limit movement of the wrist, hand, and fingers. The figure shows the relative positions of the tendon sheath and median nerve within the carpal tunnel. If this swelling, or any other lesion in the carpal tunnel, presses on the *median nerve* a condition called **carpal tunnel syndrome** may result. Because the median nerve connects to the palm and radial side (thumb side) of the hand, carpal tunnel syndrome is characterized by weakness, pain, and tingling in this part of the hand. The pain and tingling may also radiate to the forearm and shoulder. Prolonged or severe cases of carpal tunnel syndrome may be relieved by injection of antiinflammatory agents. A permanent cure is sometimes accomplished by surgical cutting or removal of the swollen tissue pressing on the median nerve.

Repetitive motion and other types of trauma may also cause inflammation of a bursa or **bursitis** (ber-SYE-tis). For example, carpet layers, roofers, and others who work on their knees are prone to bursitis involving the knee joints. Bursitis is most often treated with antiinflammatory agents.

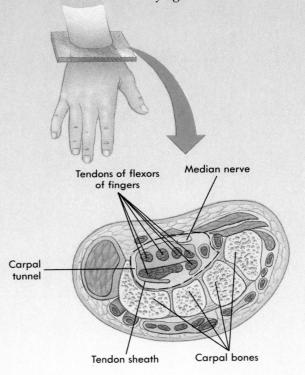

Tendons of flexors of fingers

Median nerve

Carpal tunnel

Tendon sheath

Carpal bones

MUSCLES OF THE HEAD AND NECK

The **muscles of facial expression** (Figure 7-7) allow us to communicate many different emotions nonverbally. Contraction of the **frontal** muscle, for example, allows you to raise your eyebrows in surprise and furrow the skin of your forehead into a frown. The **orbicularis** (or-bik-yoo-LAIR-is) **oris** (OR-is) called the *kissing muscle*, puckers the lips. The **zygomaticus** (zye-go-MAT-ik-us) elevates the corners of the mouth and lips and has been called the *smiling muscle.*

The **muscles of mastication** are responsible for closing the mouth and producing chewing movements. As a group, they are among the strongest muscles in the body. The two largest muscles of the group, identified in Figure 7-7, are the **mas-**

FIGURE **7-7 Muscles of the Head and Neck.** Muscles that produce most facial expressions surround the eyes, nose, and mouth. Large muscles of mastication stretch from the upper skull to the lower jaw. These powerful muscles produce chewing movements. The neck muscles connect the skull to the trunk of the body, rotating the head or bending the neck.

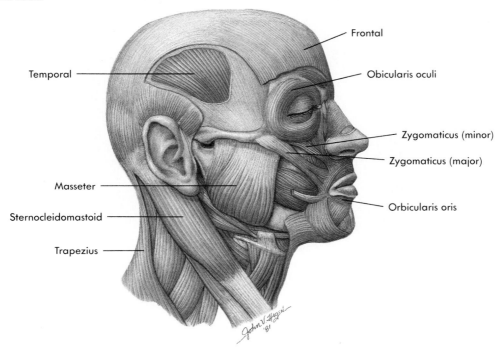

seter (mas-SEE-ter), which elevates the mandible, and the **temporal** (TEM-po-ral), which assists the masseter in closing the jaw.

The **sternocleidomastoid** (stern-o-kli-doe-MAS-toyd) and **trapezius** (tra-PEE-zee-us) muscles are easily identified in Figures 7-6 and 7-7. The two sternocleidomastoid muscles are located on the anterior surface of the neck. They originate on the sternum and then pass up and cross the neck to insert on the mastoid process of the skull on both sides. Working together, they flex the head on the chest. If only one contracts the head is both flexed and tilted to the opposite side. The triangular-shaped trapezius muscles form the line from each shoulder to the neck on its posterior surface. They have a wide line of origin extending from the base of the skull down the spinal column to the last thoracic vertebra. When contracted, the trapezius muscles help elevate the shoulders and extend the head backwards.

MUSCLES THAT MOVE THE UPPER EXTREMITIES

The upper extremity is attached to the thorax by the fan-shaped **pectoralis** (pek-tor-RAL-is) **major**, which covers the upper chest, and by the **latissimus** (la-TIS-i-mus) **dorsi** muscle, which takes its origin from structures over the lower back (Figures 7-6 and 7-8). Both muscles insert on the humerus. The pectoralis major is a flexor, and the latissimus dorsi is an extensor of the upper arm.

The **deltoid** muscle forms the thick, rounded prominence over the shoulder and upper arm (Figure 7-6). The muscle takes its origin from the scapula and clavicle and inserts on the humerus. It is a powerful abductor of the upper arm.

As the name implies, the **biceps brachii** (BRAY-kee-eye) is a two-headed muscle that serves as a primary flexor of the forearm (Figure 7-6). It originates from the bones of the shoulder girdle and inserts on the radius in the forearm.

FIGURE 7-8 Muscles of the Trunk. A, Anterior view showing superficial muscles. **B,** Anterior view showing deeper muscles.

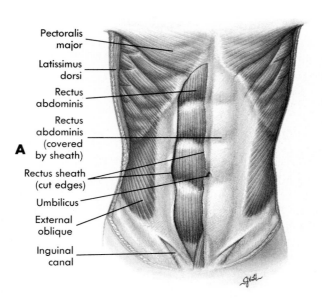
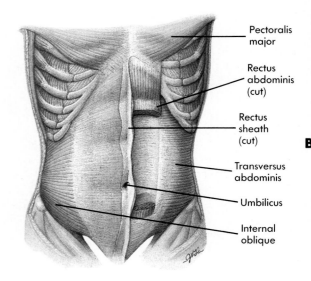

The **triceps brachii** is on the posterior or back surface of the upper arm. It has three heads of origin from the shoulder girdle and inserts into the olecranon process of the ulna. The triceps is an extensor of the elbow and thus performs a straightening function. Because this muscle is responsible for delivering blows during fights, it is often called the *boxer's muscle.*

MUSCLES OF THE TRUNK

The muscles of the anterior or front side of the abdomen are arranged in three layers, with the fibers in each layer running in different directions much like the layers of wood in a sheet of plywood (Figure 7-8). The result is a very strong "girdle" of muscle that covers and supports the abdominal cavity and its internal organs.

The three layers of muscle in the anterolateral (side) abdominal walls are arranged as follows: the outermost layer or **external oblique;** a middle layer or **internal oblique;** and the innermost layer or **transversus abdominis.** In addition to these sheetlike muscles, the band- or strap-shaped **rectus abdominis** muscle runs down the midline of the abdomen from the thorax to the pubis. The rectus abdominis and external oblique muscles can be seen in Figure 7-8. In addition to protecting the abdominal viscera, the rectus abdominis flexes the spinal column.

The *respiratory muscles* will be discussed in Chapter 15. **Intercostal muscles,** located between the ribs, and the sheetlike **diaphragm** separating the thoracic and abdominal cavities change the size and shape of the chest during breathing. As a result, air is moved into or out of the lungs.

MUSCLES THAT MOVE THE LOWER EXTREMITIES

The **iliopsoas** (il-ee-o-SO-us) originates from deep within the pelvis and the lower vertebrae to insert on the lesser trochanter of the femur and capsule of the hip joint. It is generally classified as a flexor of the thigh and is an important postural muscle to stabilize and keep the trunk from falling over backward when you stand. However, if the thigh is fixed so that it cannot move, the iliopsoas flexes the *trunk.* An example would be doing sit-ups.

The **gluteus** (GLOO-tee-us) **maximus** (MAX-i-mus) forms the outer contour and much of the substance of the buttock. It is an important extensor of the thigh (Figure 7-6) and supports the torso in the erect position.

The **adductor muscles** originate on the bony pelvis and insert on the femur. They are located on the inner or medial side of the thighs. These muscles adduct or press the thighs together.

The three **hamstring muscles** are called the *semimembranosus, semitendinosus,* and *biceps femoris.* Acting together, they serve as powerful flexors of the lower leg (Figure 7-6). They originate on the ischium and insert on the tibia or fibula.

The **quadriceps** (KWOD-re-seps) **femoris** muscle group covers the upper thigh. The four thigh muscles—the *rectus femoris* and three *vastus* muscles—extend the lower leg (Figure 7-6 and Table 7-1). One component of the quadriceps group has its origin on the pelvis, and the remaining three originate on the femur; all four insert on the tibia. Only two of the vastus muscles are visible in Figure 7-6. The vastus intermedius is covered by the rectus femoris and is not visible.

The **tibialis** (tib-ee-AL-is) **anterior** muscle (Figure 7-6) is located on the anterior or front surface of the leg. It dorsiflexes the foot. The **gastrocnemius** (gas-trok-NEE-mee-us) is the primary calf muscle. Note in Figure 7-6 that it has two fleshy components arising from both sides of the femur. It inserts through the Achilles tendon into the heel bone or calcaneus. The gastrocnemius is responsible for plantar flexion of the foot; because it is used to stand on tiptoe, it is sometimes called the *toe dancer's muscle.* A group of three muscles called the **peroneus** (pair-o-NEE-us) **group** (Figure 7-6) is found along the sides of the lower leg. As a group, these muscles plantar flex the foot. A long tendon from one component of the group—the *peroneus longus* muscle tendon—forms a support arch for the foot (see Figure 6-17).

▼ *Movements Produced by Skeletal Muscle Contractions*

Types of movement that produce a muscle contraction at any joint depend largely on the shapes of the bones involved and the joint type (see Chapter 6). Muscles acting on some joints produce movement in several directions, whereas only limited movement is possible at other joints. Terms most often used to describe body movements are:

1 **Flexion**
2 **Extension**
3 **Abduction**
4 **Adduction**
5 **Rotation**
6 **Supination** and **pronation**
7 **Dorsiflexion** and **plantar flexion**

Flexion is a movement that makes the angle between two bones at their joint smaller than it was at the beginning of the movement. Most flexions are movements described as bending. If you bend your lower arm at the elbow or your lower leg at the knee, you flex the lower arm and leg. **Extension** movements are the opposite of flexions. They make the angle between two bones at their joint larger than it was at the beginning of the movement. Therefore, extensions are straightening or stretching movements rather than bending movements. Figures 7-9 and 7-10 illustrate flexion and extension of the lower arm and leg.

RIGOR MORTIS

The term **rigor mortis** is a Latin phrase that means "stiffness of death." In a medical context, the term *rigor mortis* refers to the stiffness of skeletal muscles sometimes observed shortly after death. What causes rigor mortis? At the time of death, stimulation of muscle cells ceases. However, some muscle fibers may have been in mid-contraction at the time of death—when the myosin-actin cross-bridges are still intact. ATP is required to release the crossbridges and "energize" them for their next attachment. Because the last of a cell's ATP supply is used up at the time it dies, many crossbridges may be left "stuck" in the contracted position. Thus muscles in a dead body may be stiff because individual muscle fibers ran out of the ATP required to "turn-off" a muscle contraction.

FIGURE 7-9 Flexion and Extension of the Lower Arm. A and B, When the lower arm is flexed at the elbow, the biceps brachii contracts while its antagonist, the triceps brachii, relaxes. **B and C,** When the lower arm is extended, the biceps brachii relaxes while the triceps brachii contracts.

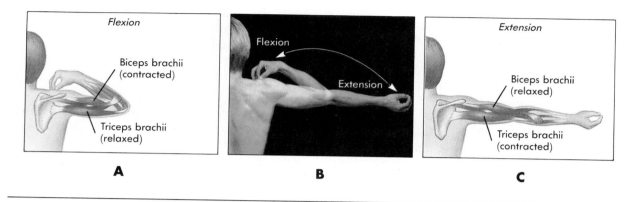

A   **B**   **C**

FIGURE 7-10 Flexion and Extension of the Lower Leg. A and B, When the lower leg flexes at the knee, muscles of the hamstring group contract while their antagonists in the quadriceps femoris group relax. **B and C,** When the lower leg extends, the hamstring muscles relax while the quadriceps femoris muscle contracts.

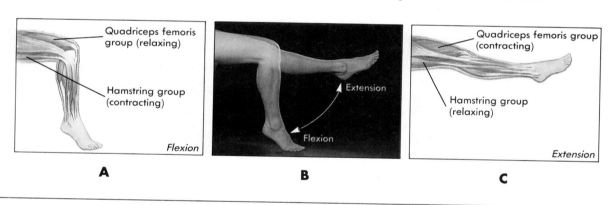

A   **B**   **C**

Abduction means moving a part away from the midline of the body such as moving your arm out to the side. **Adduction** means moving a part toward the midline, such as bringing your arms down to your sides from an elevated position. Figure 7-11, *A*, shows abduction and adduction.

Rotation is movement around a longitudinal axis. You rotate your head by moving your skull from side to side as in shaking your head "no" (Figure 7-11, *B*).

Supination and **pronation** refer to hand positions that result from rotation of the forearm. (The term *prone* refers to the body as a whole lying face down. *Supine* means lying face up.) Supination results in a hand position with the palm turned to the anterior position (as in the anatomical position), and pronation occurs when the palm is turned so that it faces posteriorly (Figure 7-11, *C*).

Dorsiflexion and **plantar flexion** refer to foot movements. In dorsiflexion the dorsum or top of the foot is elevated with the toes pointing upward. In plantar flexion the bottom of the foot is directed downward so that you are in effect standing on your toes (Figure 7-11, *D*).

As you study the illustrations and learn to recognize the muscles discussed in this chapter, you should attempt to group them according to function, as in Table 7-2. You will note, for example,

FIGURE 7-11 Examples of Body Movements.
A, Adduction and abduction. B, Rotation. C, Pronation and supination. D, Dorsiflexion and plantar flexion.

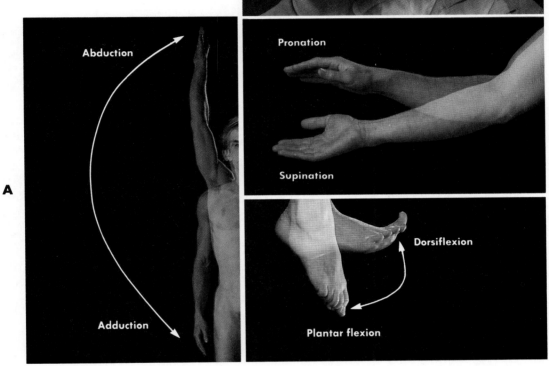

TABLE 7-2 Muscles Grouped According to Function

PART MOVED	FLEXORS	EXTENSORS	ABDUCTORS	ADDUCTORS
Upper arm	Pectoralis major	Latissimus dorsi	Deltoid	Pectoralis major and latissimus dorsi contracting together
Lower arm	Biceps brachii	Triceps brachii	None	None
Thigh	Iliopsoas and sartorius	Gluteus maximus	Gluteus medius	Adductor group
Lower leg	Hamstrings	Quadriceps group	None	None
Foot	Tibialis anterior	Gastrocnemius and soleus	Peroneus longus	Tibialis anterior

INTRAMUSCULAR INJECTIONS

Many drugs are administered by intramuscular injection. If the amount to be injected is 2 ml or less, the deltoid muscle is often selected as the site of injection. Note in Figure *A* that the needle is inserted into the muscle about two-fingers' breadth below the acromion process of the scapula and lateral to the tip of the acromion. If the amount of medication to be injected is 2 to 3 ml, the gluteal area shown in Figure *B* is often used. Injections are made into the gluteus medius muscle near the center of the upper outer quadrant, as shown in the illustration. Another technique of locating the proper injection site is to draw an imaginary diagonal line from a point of reference on the back of the bony pelvis (posterior superior iliac spine) to the greater trochanter of the femur. The injection is given about three-fingers' breadth above and one third of the way down the line. It is important that the sciatic nerve and the superior gluteal blood vessels be avoided during the injection. Proper technique requires a knowledge of the underlying anatomy.

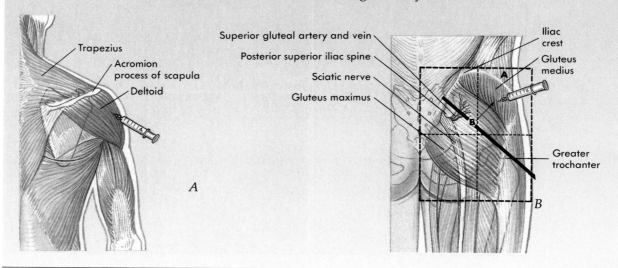

that flexors produce many of the movements used for walking, sitting, swimming, typing, and many other activities. Extensors also function in these activities but perhaps play their most important role in maintaining an upright posture.

Major Muscular Disorders

As you might expect, muscle disorders or **myopathies** (my-OP-ah-theez) generally disrupt the normal movement of the body. In mild cases, these disorders vary from inconvenient to slightly troublesome. Severe muscle disorders, however, can impair the muscles used in breathing—a life-threatening situation.

MUSCLE INJURY

Injuries to skeletal muscles resulting from overexertion or trauma usually result in a muscle **strain** (Figure 7-12). Muscle strains are characterized by muscle pain or **myalgia** (my-AL-jee-ah) and involve overstretching or tearing of muscle fibers. If an injury occurs in the area of a joint and a liga-

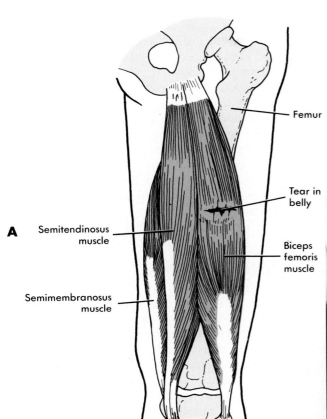

A

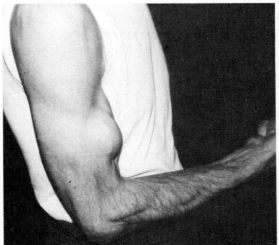

B

FIGURE 7-12 Muscle Strain. A, This illustration shows a muscle strain in the biceps femoris muscle of the hamstring group, in this case a tear in the midportion or "belly" of the muscle. **B,** Severe strain of the biceps brachii muscle. In a severe muscle strain, a muscle may break in two pieces, causing a visible gap in muscle tissue under the skin. Notice how the broken ends of the muscle reflexively contract (spasm) to form a knot of tissue.

ment is damaged, the injury may be called a **sprain.** Any muscle inflammation, including that caused by a muscle strain, is termed **myositis** (my-o-SYE-tis). If tendon inflammation occurs with myositis, as in a *charley horse*, the condition is termed **fibromyositis** (fi-bro-my-o-SYE-tis). Although inflammation may subside in a few hours or days, it usually takes several weeks for damaged muscle fibers to repair themselves. Some damaged muscle cells may be replaced by fibrous tissue, forming scars. Occasionally, hard calcium is deposited in the scar tissue.

Cramps are painful muscle spasms (involuntary twitches). Cramps often result from mild myositis or fibromyositis, but they can be a symp-

tom of any irritation or of an ion and water imbalance.

Minor trauma to the body, especially a limb, may cause a muscle *bruise* or **contusion** (kon-TOO-zhun). Muscle contusions involve local internal bleeding and inflammation. Severe trauma to a skeletal muscle may cause a *crush injury.* Crush injuries not only greatly damage the affected muscle tissue, but the release of muscle fiber contents into the bloodstream can be life threatening. For example, the reddish muscle pigment *myoglobin* can accumulate in the blood and cause kidney failure.

Stress-induced muscle tension can result in myalgia and stiffness in the neck and back and is thought to be one cause of "stress headaches." Headache and back-pain clinics use a variety of strategies to treat stress-induced muscle tension. These treatments include massage, biofeedback, and relaxation training.

MUSCLE INFECTIONS

Several bacteria, viruses, and parasites are known to infect muscle tissue—often producing local or widespread myositis. For example, in trichinosis (see Appendix B), widespread myositis is common. The muscle pain and stiffness that sometimes accompany influenza is another example.

Once a tragically common disease, **poliomyelitis** (po-lee-o-my-el-EYE-tis) is a viral infection of the nerves that control skeletal muscle movement. Although the disease can be asymptomatic, it often causes paralysis that may progress to death. Virtually eliminated in the United States as a result of a comprehensive vaccination program, it still affects millions in other parts of the world.

MUSCULAR DYSTROPHY

Muscular dystrophy (DIS-tro-fee) is not a single disorder but a group of genetic diseases characterized by atrophy (wasting) of skeletal muscle tissues. Some, but not all, forms of muscular dystrophy can be fatal.

The most common form of muscular dystrophy is **Duchenne** (doo-SHEN) **muscular dystrophy (DMD).** This form of the disease is also called *pseudohypertrophy* (meaning "false muscle growth") because the atrophy of muscle is masked by excessive replacement of muscle by fat and fibrous tissue. DMD is characterized by mild leg muscle weakness that progresses rapidly to include the shoulder muscles. The first signs of DMD are apparent at about 3 years of age, and the stricken child is usually severely affected within 5 to 10 years. Death from respiratory or cardiac muscle weakness often occurs by the time the victim is 21 years old.

Many pathophysiologists believe that DMD is caused by a missing fragment in an X chromosome, although other factors may be involved. DMD occurs primarily in boys. Because girls have two X chromosomes and boys only one, genetic diseases involving X chromosome abnormalities are more likely to occur in boys than in girls. This is true because girls with one damaged X chromosome may not exhibit an "X-linked" disease if their other X chromosome is normal (see Chapter 23).

MYASTHENIA GRAVIS

Myasthenia gravis (my-es-THEE-nee-ah GRA-vis) is a chronic disease characterized by muscle weakness, especially in the face and throat. Most forms of this disease begin with mild weakness and chronic muscle fatigue in the face, then progress to wider muscle involvement. When severe muscle weakness causes immobility in all four limbs, a *myasthenic crisis* is said to have occurred. A person in myasthenic crisis is in danger of dying from respiratory failure because of weakness in the respiratory muscles.

Myasthenia gravis is an autoimmune disease in which the immune system attacks muscle cells at the neuromuscular junction (see Figure 7-4). Nerve impulses from motor neurons are then unable to fully stimulate the affected muscle.

OUTLINE SUMMARY

INTRODUCTION

A Muscular tissue enables the body and its parts to move

  1 Movement caused by ability of muscle cells (called *fibers*) to shorten or contract

  2 Muscle cells shorten by converting chemical energy (obtained from food) into mechanical energy, which causes movement

  3 Three types of muscle tissue exist in body (see Chapter 2)

MUSCLE TISSUE

A Types of muscle tissue (Figure 7-1)

  1 Skeletal muscle—also called *striated* or *voluntary muscle*

    a Is 40% to 50% of body weight ("red meat" attached to bones)

    b Microscope reveals crosswise stripes or striations

    c Contractions can be voluntarily controlled

  2 Cardiac muscle—composes bulk of heart

    a Cardiac muscle cells branch frequently

    b Characterized by unique dark bands called *intercalated disks*

    c Interconnected nature of cardiac muscle cells allows heart to contract efficiently as a unit

  3 Nonstriated muscle or involuntary muscle—also called *smooth* or *visceral muscle*

    a Lacks cross stripes or striations when seen under a microscope; appears smooth

    b Found in walls of hollow visceral structures such as digestive tract, blood vessels, and ureters

    c Contractions not under voluntary control; movement caused by contraction is involuntary

B Function—all muscle cells specialize in contraction (shortening)

STRUCTURE OF SKELETAL MUSCLE

A Structure

  1 Each skeletal muscle is an organ composed mainly of skeletal muscle cells and connective tissue

  2 Most skeletal muscles extend from one bone across a joint to another bone

  3 Parts of a skeletal muscle

    a Origin—attachment to the bone that remains relatively stationary or fixed when movement at the joint occurs

    b Insertion—point of attachment to the bone that moves when a muscle contracts

    c Body—main part of the muscle

  4 Muscles attach to bone by tendons—strong cords of fibrous connective tissue; some tendons enclosed in synovial-lined tubes and are lubricated by synovial fluid; tubes called *tendon sheaths*

  5 Bursae—small synovial-lined sacs containing a small amount of synovial fluid; located between some tendons and underlying bones

B Microscopic structure (Figure 7-3)

  1 Contractile cells called *fibers*—grouped into bundles

  2 Fibers contain thick myofilaments (containing protein myosin) and thin myofilaments (composed of actin)

  3 Basic functional (contractile) unit called *sarcomere*; sarcomeres separated from each other by dark bands called *Z lines*

    a Sliding filament model explains mechanism of contraction

      (1) Thick and thin myofilaments slide past each other as a muscle contracts

      (2) Contraction requires calcium and energy-rich ATP molecules

FUNCTIONS OF SKELETAL MUSCLE

A Movement

  1 Muscles produce movement; as a muscle contracts, it pulls the insertion bone closer to the origin bone; movement occurs at the joint between the origin and the insertion

    a Groups of muscles usually contract to produce a single movement

      (1) Prime mover—muscle whose contraction is mainly responsible for producing a given movement

      (2) Synergist—muscle whose contractions help the prime mover produce a given movement

      (3) Antagonist—muscle whose actions oppose the action of a prime mover in any given movement

B Posture

  1 A specialized type of muscle contraction, called *tonic contraction*, enables us to maintain body position

    a In tonic contraction, only a few of a muscle's fibers shorten at one time

    b Tonic contractions produce no movement of body parts

    c Tonic contractions maintain muscle tone called *posture*

(1) Good posture reduces strain on muscles, tendons, ligaments, and bones

(2) Poor posture causes fatigue and may lead to deformity

C Heat production

1 Survival depends on the body's ability to maintain a constant body temperature

a Fever—an elevated body temperature—often a sign of illness

b Hypothermia—a reduced body temperature

2 Contraction of muscle fibers produces most of the heat required to maintain normal body temperature

FATIGUE

A Reduced strength of muscle contraction

B Caused by repeated muscle stimulation without adequate periods of rest

C Repeated muscular contraction depletes cellular ATP stores and outstrips the ability of the blood supply to replenish oxygen and nutrients

D Contraction in the absence of adequate oxygen produces lactic acid, which contributes to muscle soreness

E *Oxygen debt*—term used to describe the metabolic effort required to burn excess lactic acid that may accumulate during prolonged periods of exercise; the body is attempting to return the cells' energy and oxygen reserves to pre-exercise levels

ROLE OF OTHER BODY SYSTEMS IN MOVEMENT

A Muscle functioning depends on the functioning of many other parts of the body

1 Most muscles cause movements by pulling on bones across movable joints

2 Respiratory, circulatory, nervous, muscular, and skeletal systems play essential roles in producing normal movements

3 Multiple sclerosis, brain hemorrhage, and spinal cord injury are examples of how pathological conditions in other body organ systems can dramatically affect movement

MOTOR UNIT (Figure 7-4)

A Stimulation of a muscle by a nerve impulse is required before a muscle can shorten and produce movement

B A motor neuron is the specialized nerve that transmits an impulse to a muscle, causing contraction

C A neuromuscular junction is the specialized point of contact between a nerve ending and the muscle fiber it innervates

D A motor unit is the combination of a motor neuron with the muscle cell or cells it innervates

MUSCLE STIMULUS

A A muscle will contract only if an applied stimulus reaches a certain level of intensity

1 A threshold stimulus is the minimal level of stimulation required to cause a muscle fiber to contract

B Once stimulated by a threshold stimulus, a muscle fiber will contract completely, a response called *all or none*

C Different muscle fibers in a muscle are controlled by different motor units having different threshold-stimulus levels

1 Although individual muscle fibers always respond all or none to a threshold stimulus, the muscle as a whole does not

2 Different motor units responding to different threshold stimuli permit a muscle as a whole to execute contractions of graded force

TYPES OF SKELETAL MUSCLE CONTRACTION

A Twitch and tetanic contractions

1 Twitch contractions are laboratory phenomena and do not play a significant role in normal muscular activity; they are a single contraction of muscle fibers caused by a single threshold stimulus

2 Tetanic contractions are sustained and steady muscular contractions caused by a series of stimuli bombarding a muscle in rapid succession

B Isotonic contractions

1 Contraction of a muscle that produces movement at a joint

2 During isotonic contractions, the muscle shortens, causing the insertion end of the muscle to move toward the point of origin

3 Most types of body movements such as walking and running are caused by isotonic contractions

C Isometric contractions

1 Isometric contractions are muscle contractions that do not produce movement; the muscle as a whole does not shorten

2 Although no movement occurs during isometric contractions, tension within the muscle increases

EFFECTS OF EXERCISE ON SKELETAL MUSCLES

A Exercise, if regular and properly practiced, improves muscle tone and posture, results in more efficient heart and lung functioning, and reduces fatigue

B Effects of exercise on skeletal muscles

1 Muscles undergo changes related to the amount of work they normally do

a Prolonged inactivity causes disuse atrophy

b Regular exercise increases muscle size, called *hypertrophy*

2 Strength training is exercise involving contraction of muscles against heavy resistance
  a Strength training increases the numbers of myofilaments in each muscle fiber, and as a result, the total mass of the muscle increases
  b Strength training does not increase the number of muscle fibers
3 Endurance training—increases a muscle's ability to sustain moderate exercise over a long period; sometimes called *aerobic training*
  a Endurance training allows more efficient delivery of oxygen and nutrients to a muscle via increased blood flow
  b Endurance training does not usually result in muscle hypertrophy

SKELETAL MUSCLE GROUPS (Table 7-1)

A Muscles of the head and neck (Figure 7-7)
  1 Facial muscles
    a Orbicularis oculi
    b Orbicularis oris
    c Zygomaticus
  2 Muscles of mastication
    a Masseter
    b Temporal
  3 Sternocleidomastoid—flexes head
  4 Trapezius—elevate shoulders and extend head
B Muscles that move the upper extremities
  1 Pectoralis major—flexes upper arm
  2 Latissimus dorsi—extends upper arm
  3 Deltoid—abducts upper arm
  4 Biceps brachii—flexes forearm
  5 Triceps brachii—extends forearm
C Muscles of the trunk (Figure 7-8)
  1 Abdominal muscles
    a Rectus abdominis
    b External oblique
    c Internal oblique
    d Transversus abdominis
  2 Respiratory muscles
    a Intercostal muscles
    b Diaphragm
D Muscles that move the lower extremities
  1 Iliopsoas—flexes thigh
  2 Gluteus maximus—extends thigh
  3 Adductor muscles
  4 Hamstring muscles—flex lower leg
    a Semimembranosus
    b Semitendinosus
    c Biceps femoris
  5 Quadriceps femoris group—extend lower leg
    a Rectus femoris
    b Vastus muscles
  6 Tibialis anterior—dorsiflexes foot
  7 Gastrocnemius—plantar flexes foot
  8 Peroneus group—flex foot

TYPES OF MOVEMENTS PRODUCED BY SKELETAL MUSCLE CONTRACTIONS
(Figures 7-9 through 7-11)

A Flexion—movement that decreases the angle between two bones at their joint: bending
B Extension—movement that increases the angle between two bones at their joint: straightening
C Abduction—movement of a part away from the midline of the body
D Adduction—movement of a part toward the midline of the body
E Rotation—movement around a longitudinal axis
F Supination and pronation—hand positions that result from rotation of the forearm; supination results in a hand position with the palm turned to the anterior position; pronation occurs when the palm faces posteriorly
G Dorsiflexion and plantar flexion—foot movements; dorsiflexion results in elevation of the dorsum or top of the foot; during plantar flexion, the bottom of the foot is directed downward

MAJOR MUSCULAR DISORDERS

A Myopathies—muscle disorders; can range from mild to life threatening
B Muscle injury
  1 Strain—injury from overexertion or trauma; involves stretching or tearing of muscle fibers
    a Often accompanied by myalgia (muscle pain)
    b May result in inflammation of muscle (myositis) or of muscle and tendon (fibromyositis)
    c If injury is near a joint and involves ligament damage, it may be called a *sprain*
  2 Cramps are painful muscle spasms (involuntary twitches)
  3 Crush injuries result from severe muscle trauma and may release cell contents that ultimately cause kidney failure
  4 Stress-induced muscle tension can cause headaches and back pain
C Infections
  1 Several bacteria, viruses, and parasites can infect muscles
  2 Poliomyelitis is a viral infection of motor nerves that ranges from mild to life threatening
D Muscular dystrophy
  1 A group of genetic disorders characterized by muscle atrophy
  2 Duchenne (pseudohypertrophic) muscular dystrophy is the most common type
    a Characterized by rapid progression of weakness and atrophy, resulting in death by age 21
    b X-linked inherited disease, affecting mostly boys
E Myasthenia gravis—autoimmune muscle disease characterized by weakness and chronic fatigue

NEW WORDS

abduction	flexion	neuromuscular junction	sarcomere
actin	hypertrophy	origin	sliding filament theory
adduction	insertion	oxygen debt	stimulus
all or none	isometric	plantar flexion	supination
antagonist	isotonic	posture	synergist
bursa	motor neuron	prime mover	tendon
dorsiflexion	motor unit	pronation	tetanic contraction
extension	myofilaments	rotation	tonic contraction
fatigue	myosin		

Diseases and Other Clinical Terms

atrophy	fibromyositis	myalgia	paralysis
carpal tunnel syndrome	hypothermia	myasthenia gravis	poliomyelitis
contusion	muscle strain	myopathy	tenosynovitis
cramps	muscular dystrophy	myositis	

CHAPTER TEST

1. Cardiac muscle:
   a. Cells have no nuclei
   b. Is composed of cells that branch frequently
   c. Lines many hollow internal organs
   d. All of the above are correct
2. The term *origin* refers to:
   a. The attachment of a muscle to a bone that does not move when contraction occurs
   b. Attachment of a muscle to a bone that moves when contraction occurs
   c. The body of a muscle
   d. None of the above are correct
3. A sarcomere:
   a. Is the basic functional or contractile unit of skeletal muscle
   b. Contains only actin myofilaments
   c. Contains only myosin myofilaments
   d. Is found only in smooth or involuntary muscle
4. The muscle mainly responsible for producing a particular movement is called a:
   a. Synergist   c. Antagonist
   b. Prime mover   d. Fixator
5. A motor unit consists of:
   a. Only a motor neuron
   b. A motor neuron and the muscle cells it innervates
   c. Only contracting muscle cells
   d. None of the above are correct

6. Which of the following types of muscle contraction does not produce movement?
   a. Isotonic contraction
   b. Tetanic contraction
   c. Twitch contraction
   d. Isometric contraction
7. Which of the following terms refers to moving a part away from the midline of the body?
   a. Abduction   c. Supination
   b. Adduction   d. Flexion
8. Many hollow internal organs and blood vessel walls contain _____ muscle.
9. Muscles are anchored firmly to bones by _____.
10. Thick myofilaments are formed from a protein called _____.
11. The explanation of muscle contraction resulting from movement of thick and thin myofilaments toward one another is called the _____ theory.
12. The muscles that assist prime movers in producing a particular movement are called _____.
13. Skeletal muscle tone maintains _____ by counteracting the pull of gravity.
14. Depletion of oxygen in a muscle during vigorous and prolonged exercise is called _____ _____.
15. Walking and running are examples of _____ muscle contraction.
16. Standing on your toes is an example of _____ flexion.
17. The term _____ _____ is used to describe shrinkage or decrease in the size of a muscle.
18. The major flexor muscles of the lower leg are the _____.
19. Excessive stretching or tearing of muscle fibers is called _____.
20. Inflammation of muscle and tendon is termed _____.
21. _____ is a viral infection of motor nerves that may progress to life-threatening paralysis of respiratory muscles.
22. _____ is a group of muscle disorders characterized by muscle atrophy and that often progresses to death before age 21.
23. _____ is an autoimmune muscle disease characterized by weakness and chronic fatigue.

Select the most correct answer from Column B for each statement in Column A. (Only one answer is correct.)

COLUMN A

24. _____ Contractile unit of muscle
25. _____ Flexes lower arm
26. _____ Extends lower leg
27. _____ Closes jaw
28. _____ Movable point of attachment
29. _____ Flexes upper arm
30. _____ Extends thigh

COLUMN B

a. Biceps brachii
b. Masseter
c. Sarcomere
d. Quadriceps group
e. Pectoralis major
f. Gluteus maximus
g. Insertion

REVIEW QUESTIONS

1 Compare the three kinds of muscle tissue regarding location, microscopic appearance, and nerve control.
2 Explain flexion, extension, abduction, and adduction; give an example of each.
3 Explain how skeletal muscles, bones, and joints work together to produce movements.
4 Why can a spinal cord injury be followed by muscle paralysis?
5 Can a muscle contract very long if its blood supply is shut off? Give a reason for your answer.
6 What two kinds of muscle contractions do not produce movement?
7 What is the name of the main muscle or muscles that
   a Flex the upper arm?
   b Flex the lower leg?
   c Extend the lower arm?
   d Extend the thigh?
   e Abduct the upper arm?
   f Elevate the mandible?
   g Flex the spinal column?
   h Dorsiflex the foot?

8 Give the approximate location of the biceps brachii, hamstrings, deltoid, pectoralis major, quadriceps femoris group, latissimus dorsi, trapezius, rectus abdominis, and gastrocnemius; tell what movement each produces.
9 Briefly describe changes that gradually take place in bones, joints, and muscles if a person habitually gets too little exercise.
10 What is meant by the term *tetanus*?
11 Discuss the microscopic structure of skeletal muscle tissue.
12 Explain good posture, hypertrophy, sarcomere, muscle tone, isometric contractions, isotonic contractions, tonic contractions, tetanus, threshold, and all or none.
13. What signs and symptoms are likely to accompany a moderate muscle strain?
14. Briefly describe the progression of Duchenne muscular dystrophy.
15 What causes the signs and symptoms of myasthenia gravis?

CLINICAL APPLICATIONS

1 Your nephew Tom has just been diagnosed with *pseudohypertrophic muscular dystrophy*. How did he get this disease? Is his twin sister, Geri, likely to develop the same condition? Are you likely to get this disease? (Hint: see Chapter 23) You know that muscular dystrophy typically causes atrophy or wasting of muscle tissue; yet Tom's leg muscles seem particularly well developed. Tom's physician said that the appearance of Tom's legs is typical for this form of muscular dystrophy. Can you explain this apparent contradiction?
2 Your friend Elena is suffering from a strain of her gastrocnemius muscle. What type of injury is this and where in Elena's body is it located? What symptoms are likely to accompany Elena's injury? What movements should Elena avoid to prevent further injury to the gastrocnemius muscle?
3. Robert has decided to improve his appearance by exercising. He would like to build up his chest and shoulder muscles so that he looks better in the T-shirts he is so fond of wearing. He has decided to play racquetball every day as his primary training program because he knows that he uses his upper body muscles in this sport. After his first game of racquetball, you ask him how he likes his new sport and he can hardly answer you—he seems out of breath. Is Robert's plan likely to help him meet his goal? How do you explain his breathing difficulties?

The Nervous System

C H A P T E R

8

Outline

Objectives

*After you have completed this chapter,
you should be able to:*

1. List the organs and divisions of the nervous system and describe the generalized functions of the system as a whole.

2. Identify the major types of cells in the nervous system and discuss the function of each.

3. Identify the anatomical and functional components of a three-neuron reflex arc. Compare and contrast the propagation of a nerve impulse along a nerve fiber and across a synaptic cleft.

4. Identify the major anatomical components of the brain and spinal cord and briefly comment on the function of each.

5. Identify and discuss the coverings and fluid spaces of the brain and spinal cord.

6. Discuss spinal and cranial nerves.

7. Discuss the anatomical and functional characteristics of the two divisions of the autonomic nervous system.

8. List and describe major disorders of the nervous system.

Boxed Essays

The Blood-Brain Barrier
Suppressing Pain During Exercise
Brain Studies
Lumbar Puncture

The normal body must accomplish a gigantic and enormously complex job—that of keeping itself alive and healthy. Each one of its billions of cells performs some activity that is a part of this function. Control of the body's billions of cells is accomplished mainly by two communication systems: the *nervous system* and the *endocrine system*. Both systems transmit information from one part of the body to another, but they do it in different ways. The nervous system transmits information very rapidly by nerve impulses conducted from one body area to another. The endocrine system transmits information more slowly by chemicals secreted by ductless glands into the bloodstream and circulated from the glands to other parts of the body. Nerve impulses and hormones communicate information to body structures, increasing or decreasing their activities as needed for healthy survival. In other words, the communication systems of the body are also its control and integrating systems. They weld all the body's functions into its one function of keeping itself alive and healthy.

Recall that homeostasis is the balanced and controlled internal environment of the body that is basic to life itself. Homeostasis is possible only if our physiological control and integration systems function properly. Our plan for this chapter is to name the cells, organs, and divisions of the nervous system; discuss the generation of nervous impulses; and then discover how these impulses move between one area of the body and another. We will study not only the major components of the nervous system, such as the brain, spinal cord, and nerves, but also learn about how they function to maintain and regulate homeostasis or respond to disease. In Chapter 9, special senses are discussed.

Organs and Divisions of the Nervous System

The organs of the nervous system as a whole include the brain and spinal cord, the numerous nerves of the body, the specialized sense organs such as the eyes and ears, and the microscopic sense organs such as those found in the skin. The system as a whole consists of two principal divisions called the central nervous system and the peripheral nervous system (Figure 8-1). Because the brain and spinal cord occupy a midline or central location in the body, they are together called the **central nervous system** or **CNS.** Similarly, the usual designation for the nerves of the body is the **peripheral nervous system** or **PNS.** Use of the term *peripheral* is appropriate because nerves extend to outlying or peripheral parts of the body. A subdivision of the peripheral nervous system, called the **autonomic nervous system** or **ANS** consists of structures that regulate the body's automatic or involuntary functions (for example, the heart rate, the contractions of the stomach and intestines, and the secretion of chemical compounds by glands).

Cells of the Nervous System

The two types of cells found in the nervous system are called **neurons** (NOO-rons) or nerve cells and **neuroglia** (noo-ROG-lee-ah), which are specialized connective tissue cells. Neurons conduct impulses, whereas neuroglia support neurons.

NEURONS

Each neuron consists of three parts: a main part called the neuron **cell body,** one or more branching projections called **dendrites** (DEN-drites), and one elongated projection known as an **axon.** Identify each part on the neuron shown in Figure 8-2. Dendrites are the processes or projections that transmit impulses to the neuron cell bodies, and axons are the processes that transmit impulses away from the neuron cell bodies.

The three types of neurons are classified according to the direction in which they transmit impulses: **sensory neurons, motor neurons,** and **interneurons.** Sensory neurons transmit impulses to the spinal cord and brain from all parts of the body. Motor neurons transmit impulses in the opposite direction—away from the brain and spinal cord. They do not conduct impulses to all parts of the body but only to two kinds of tissue—muscle and glandular epithelial tissue. Interneurons conduct impulses from sensory neurons to motor neurons. Sensory neurons are also called *afferent* neurons; motor neurons are called *efferent* neurons, and interneurons are called *central* or *connecting* neurons.

▼ **FIGURE 8-1** Divisions of the Nervous System.

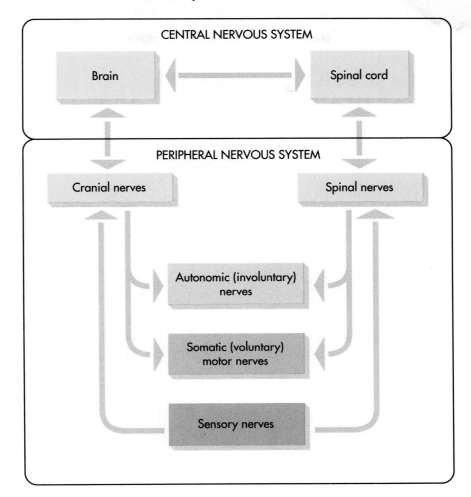

The axon shown in Figure 8-2, *B*, is surrounded by a segmented wrapping of a material called **myelin** (MY-e-lin). Myelin is a white, fatty substance formed by **Schwann cells** that wrap around some axons outside the central nervous system. Such fibers are called **myelinated fibers.** In Figure 8-2, *B*, one such axon has been enlarged to show additional detail. **Nodes of Ranvier** (rahn-vee-AY) are indentations between adjacent Schwann cells.

The outer cell membrane of a Schwann cell is called the **neurilemma** (noo-ri-LEM-mah). The fact that axons in the brain and cord have no neurilemma is clinically significant because it plays an essential part in the regeneration of cut and injured axons. Therefore the potential for regeneration in the brain and spinal cord is far less than it is in the peripheral nervous system.

NEUROGLIA

Neuroglia do not specialize in transmitting impulses. Instead, they are special types of connective tissue cells. Their name is appropriate because it is derived from the Greek word *glia*

FIGURE 8-2 Neuron. A, Diagram of a typical neuron showing dentrites, a cell body, and an axon. **B,** Segment of a myelinated axon cut to show detail of the concentric layers of the Schwann cell filled with myelin. **C,** Photomicrograph of neuron.

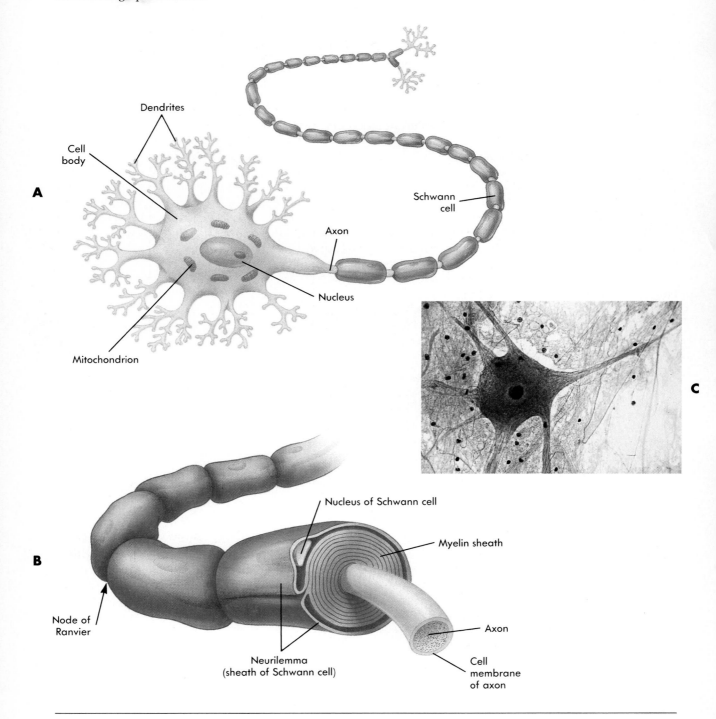

meaning "glue." One function of neuroglia cells is to hold the functioning neurons together and protect them. An important reason for discussing neuroglia is that one of the most common types of brain tumor—called **glioma** (glee-O-mah)—develops from them.

Neuroglia vary in size and shape (Figure 8-3). Some are relatively large cells that look somewhat like stars because of the threadlike extensions that jut out from their surfaces. These neuroglia cells are called **astrocytes** (AS-tro-sites), a word that means "star cells" (Figure 8-3). Their threadlike branches attach to neurons and to small blood vessels, holding these structures close to each other.

Microglia (my-KROG-lee-ah) are smaller than astrocytes. They usually remain stationary, but in inflamed or degenerating brain tissue, they enlarge, move about, and act as microbe-eating scavengers. They surround the microbes, draw them into their cytoplasm, and digest them. Recall from Chapter 2 that phagocytosis is the scientific name for this important cellular process.

The **oligodendroglia** (ol-i-go-den-DROG-lee-ah) help to hold nerve fibers together and also serve another and probably more important function; they produce the fatty myelin sheath that envelops nerve fibers located in the brain and spinal cord.

DISORDERS OF NERVOUS TISSUE
Multiple Sclerosis

A number of diseases are associated with disorders of the oligodendroglia. Because these neuroglial cells are involved in myelin formation, the diseases as a group are called **myelin disorders.** The most common primary disease of the CNS is a myelin disorder called **multiple sclerosis** or **MS.** It is characterized by myelin loss and destruction accompanied by varying degrees of oligodendroglial cell injury and death. The result is demyelination throughout the white matter of the CNS. Hard, plaquelike lesions replace the destroyed myelin, and affected areas are invaded by inflammatory cells. As the myelin around the axons is lost, nerve conduction is impaired, and weakness, incoordination, visual impairment, and speech disturbances occur. Although the disease occurs in both sexes and all age groups, it is most common in women between the ages of 20 and 40 years.

The cause of multiple sclerosis is thought to be related to autoimmunity and to viral infections in some individuals. MS is characteristically relapsing and chronic in nature, but some cases of acute and unremitting disease have been reported. In most instances the disease is prolonged, with remissions and relapses occurring over many years. There is no known cure.

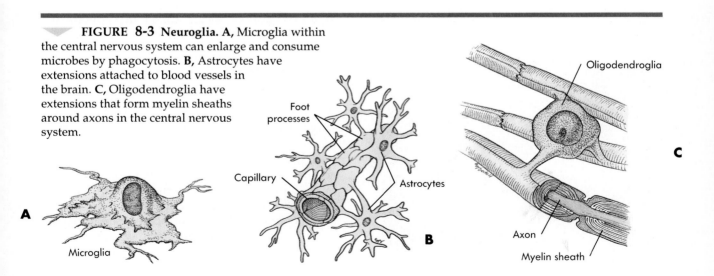

FIGURE 8-3 Neuroglia. A, Microglia within the central nervous system can enlarge and consume microbes by phagocytosis. **B,** Astrocytes have extensions attached to blood vessels in the brain. **C,** Oligodendroglia have extensions that form myelin sheaths around axons in the central nervous system.

Tumors

The general name for tumors arising in nervous system structures is **neuroma** (noo-RO-mah). Tumors do not usually develop directly from neurons but from neuroglia, membrane tissues, and blood vessels. As stated above, a common type of brain tumor—glioma—occurs in neuroglia. Gliomas are usually benign but may still be life threatening. Patients usually show deficits in the area in which the tumor is located (see Figure 8-12, *B*). Because these tumors often develop in deep areas of the brain, they are difficult to treat. Untreated gliomas may grow to a size that disrupts normal brain function, perhaps leading to death.

Multiple neurofibromatosis (noo-ro-fye-bro-mah-TOE-sis) is an inherited disease characterized by numerous fibrous neuromas throughout the body (Figure 8-4). The tumors are benign, appearing first as small nodules in the Schwann cells of cutaneous nerves. In some cases, involvement spreads as large, disfiguring fibrous tumors appear in many areas of the body, including muscles, bones, and internal organs. The disfigurement can be severe, as in the famous case of the "Elephant Man," who suffered crippling deformities in the skull, spinal column, and many other parts of the body.

Most malignant tumors of neuroglia and other nervous tissues do not arise there but are secondary tumors resulting from metastasis of cancer cells from the breast, lung, or other organs.

Nerves

A **nerve** is a group of peripheral nerve fibers (axons) bundled together like the strands of a cable. Because nerve fibers usually have a myelin sheath and myelin is white, nerves are called the **white matter** of the PNS. Bundles of axons in the CNS, called **tracts,** may also be myelinated and thus form this system's white matter. Tissue comprised of cell bodies and unmyelinated axons and dendrites is called **gray matter** because of its characteristic gray appearance.

Figure 8-5 shows that each axon in a nerve is surrounded by a thin wrapping of fibrous connective tissue called the **endoneurium** (en-doe-NOO-ree-um). Groups of these wrapped axons are called **fascicles.** Each fascicle is surrounded by a thin, fibrous **perineurium** (pair-i-NOO-ree-um). A tough, fibrous sheath called the **epineurium** (ep-i-NOO-ree-um) covers the whole nerve.

Reflex Arcs

During every moment of life, nerve impulses speed over neurons to and from the spinal cord and brain. If all impulse conduction ceases, life itself ceases. Only neurons can provide the rapid communication between cells that is necessary for maintaining life. Hormonal messages are the only other kind of communications the body can send, and they travel much more slowly than impulses. They can move from one part of the body to another only via circulating blood. Compared with impulse conduction, circulation is a very slow process.

FIGURE 8-4 Multiple Neurofibromatosis. This photo shows multiple tumors of Schwann cells in nerves of the skin that are characteristic of this inherited condition.

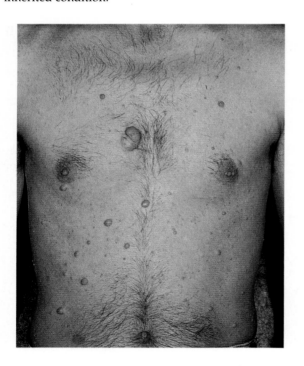

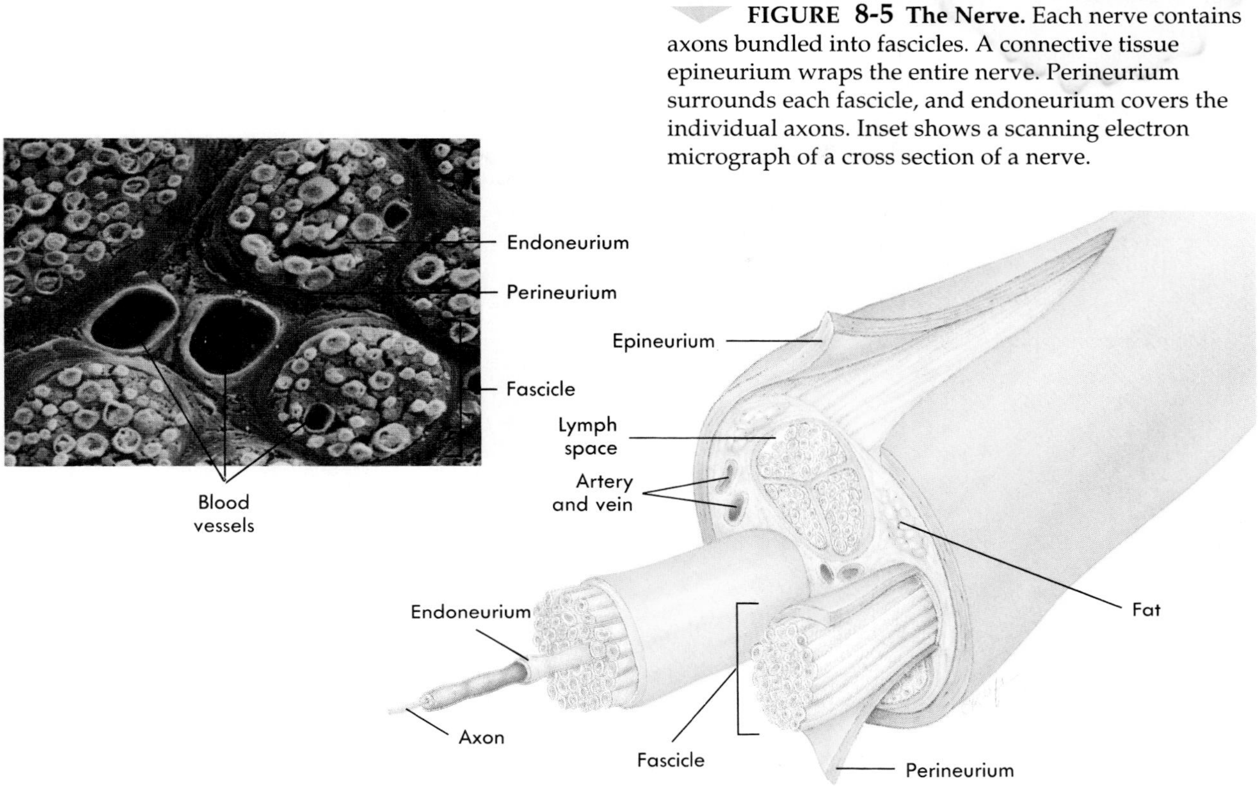

FIGURE 8-5 The Nerve. Each nerve contains axons bundled into fascicles. A connective tissue epineurium wraps the entire nerve. Perineurium surrounds each fascicle, and endoneurium covers the individual axons. Inset shows a scanning electron micrograph of a cross section of a nerve.

Nerve impulses, sometimes called *action potentials,* can travel over trillions of routes—routes made up of neurons because they are the cells that conduct impulses. Hence the routes traveled by nerve impulses are sometimes spoken of as *neuron pathways.* A specialized type of neuron pathway, called a **reflex arc,** is important to nervous system functioning. The simplest kind of reflex arc is a two-neuron arc, so called because it consists of only two types of neurons: sensory neurons and motor neurons. Three-neuron arcs are the next simplest kind. They, of course, consist of all three kinds of neurons: sensory neurons, interneurons, and motor neurons. Reflex arcs are like one-way streets; they allow impulse conduction in only one direction. The next paragraph describes this direction in detail. Look frequently at Figure 8-6 as you read it.

Impulse conduction normally starts in receptors. **Receptors** are the beginnings of dendrites of

sensory neurons. They are often located at some distance from the spinal cord (in tendons, skin, or mucous membranes, for example). In Figure 8-6 the sensory receptors are located in the quadriceps muscle group and the patellar tendon. In the reflex illustrated there, stretch receptors are stimulated as a result of a tap on the patellar tendon from a rubber hammer used by a physician to elicit a reflex during a physical examination. The nerve impulse that is generated, its neurologic pathway, and its ultimate "knee-jerk" effect is an example of the simplest form of a two-neuron reflex arc. In this reflex, only sensory and motor neurons are involved. The nerve impulse generated by stimulation of the stretch receptors travels the length of the sensory neuron's dendrite to its cell body located in the **posterior (dorsal) root ganglion** (GANG-lee-on). A **ganglion** is a group of nerve-cell bodies located in the PNS. This ganglion is located near the spinal cord. Each spinal

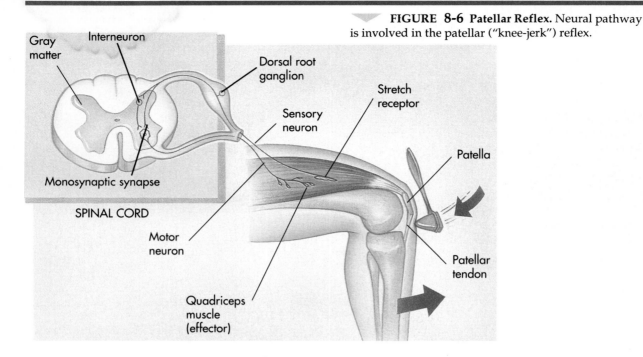

FIGURE 8-6 Patellar Reflex. Neural pathway is involved in the patellar ("knee-jerk") reflex.

ganglion contains not one sensory neuron cell body as shown in Figure 8-6 but hundreds of them. The axon of the sensory neuron travels from the cell body in the dorsal root ganglion and ends near the dendrites of another neuron located in the gray matter of the spinal cord. A microscopic space separates the axon ending of one neuron from the dendrites of another neuron. This space is called a **synapse.** The nerve impulse stops at the synapse, chemical signals are sent across the gap, and the impulse then continues along the dendrites, cell body, and axon of the motor neuron. The motor neuron axon forms a synapse with a structure called an *effector,* an organ that puts nerve signals "into effect."

Effectors are muscles or glands, and muscle contractions and gland secretion are the only kinds of reflexes. The response to impulse conduction over a reflex arc is called a **reflex.** In short, impulse conduction by a reflex arc causes a reflex to occur. In a patellar reflex, the nerve impulses that reach the quadriceps muscle (the effector) result in the classic "knee-jerk" response.

Now turn your attention to the *interneuron* shown in Figure 8-6. Some reflexes involve three rather than two neurons. In these more complex types of responses, an interneuron, in addition to a sensory and motor neuron, is involved. In three-neuron reflexes, the end of the sensory neuron's axon synapses first with an interneuron before chemical signals are sent across a second synapse, resulting in conduction through the motor neuron. For example, application of an irritating stimulus to the skin of the thigh initiates a three-neuron reflex response that causes contraction of muscles to pull the leg away from the irritant—a three-neuron arc reaction called the *withdrawal reflex.* All interneurons lie within the gray matter of the brain or spinal cord. Gray matter forms the H-shaped core of the spinal cord. Because of the presence of an interneuron, three-neuron reflex arcs have two synapses. A two-neuron reflex arc, however, has only a sensory neuron and a motor neuron with one synapse between them.

Identify the motor neuron in Figure 8-6. Observe that its dendrites and cell body, like those of

THE BLOOD-BRAIN BARRIER

Astrocytes have another important function besides supporting neurons and blood vessels. Notice in the figure that the "feet" of the astrocytes form a wall around the outside of blood vessels in the nervous system. This astrocyte wall, with the vessel wall, forms a structure known as the **blood-brain barrier.** The blood-brain barrier allows water, oxygen, carbon dioxide, and a few other substances—such as alcohol—to move between the blood and the tissue of the brain. However, many toxins and pathogens that can enter other tissues through blood vessel walls cannot enter nervous tissue because of this barrier. This adaptation enhances survival because vital brain and nerve tissues are protected from damage. This protective function of the blood-brain barrier has great clinical significance. Drugs used in other parts of the body to treat infections, cancer, and other disorders, often cannot pass through the blood-brain barrier. For example, penicillin and other antibiotics cannot enter the interstitial fluid of brain tissue from the blood. Obviously, this makes development of treatments for brain disorders sometimes very difficult.

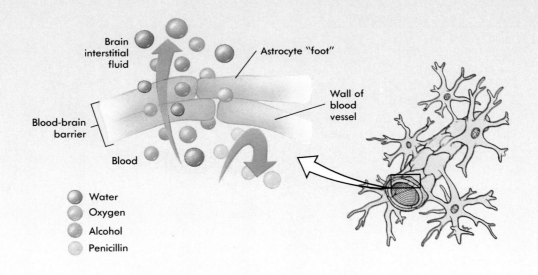

Brain interstitial fluid

Astrocyte "foot"

Wall of blood vessel

Blood-brain barrier

Blood

- Water
- Oxygen
- Alcohol
- Penicillin

an interneuron, are located in the spinal cord's gray matter. The axon of this motor neuron, however, runs through the anterior (ventral) root of the spinal nerve and terminates in a muscle.

Nerve Impulses

What are nerve impulses? Here is one widely accepted definition: a nerve impulse is a self-propagating wave of electrical disturbance that travels along the surface of a neuron's plasma membrane. You might visualize this as a tiny spark sizzling its way along a fuse. Nerve impulses do not continually race along every nerve cell's surface. First they have to be initiated by a stimulus, a change in the neuron's environment. Pressure, temperature, and chemical changes are the usual stimuli. The membrane of each resting neuron has a slight positive charge on the outside

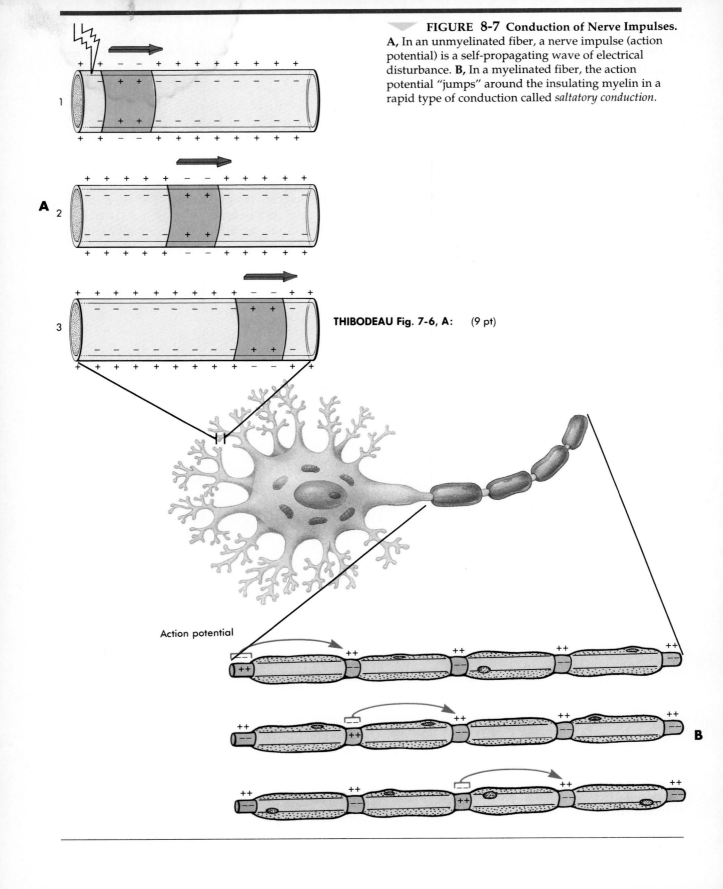

THIBODEAU Fig. 7-6, A: (9 pt)

FIGURE 8-7 Conduction of Nerve Impulses.
A, In an unmyelinated fiber, a nerve impulse (action potential) is a self-propagating wave of electrical disturbance. **B,** In a myelinated fiber, the action potential "jumps" around the insulating myelin in a rapid type of conduction called *saltatory conduction*.

A

1

2

3

B

Action potential

and a negative charge on the inside, as shown in Figure 8-7. This occurs because there is normally an excess of sodium ions (Na$^+$) on the outside of the membrane. When a section of the membrane is stimulated, its permeability to Na$^+$ suddenly increases, and Na$^+$ rushes inward. The inside of the membrane temporarily becomes positive and the outside negative. Although this section of the membrane immediately recovers, the electrical disturbance stimulates a similar change in the next section of membrane. Thus a self-propagating wave of disturbance—a nerve impulse—travels in one direction across the neuron's surface (Figure 8-7, *A*). If the traveling impulse encounters a section of membrane covered with insulating myelin, it simply "jumps" around the myelin. Called **saltatory conduction,** this type of impulse

travel is much faster than is possible in nonmyelinated sections. Saltatory conduction is illustrated in Figure 8-7, *B*.

The Synapse

Transmission of signals from one neuron to the next—across the synapse—is an important part of the nerve conduction process. By definition, a synapse is the place where impulses are transmitted from one neuron, called the **presynaptic neuron,** to another neuron, called the **postsynaptic neuron.** Three structures make up a synapse: a synaptic knob, a synaptic cleft, and the plasma membrane of a postsynaptic neuron. A **synaptic knob** is a tiny bulge at the end of a terminal

FIGURE 8-8 Components of a Synapse. Diagram shows synaptic knob or axon terminal of presynaptic neuron, the plasma membrane of a postsynaptic neuron, and a synaptic cleft. On the arrival of an action potential at a synaptic knob, neurotransmitter molecules are released from vesicles in the knob into the synaptic cleft. The combining of neurotransmitter and receptor molecules in the plasma membrane of the postsynaptic neuron initiates impulse conduction in the postsynaptic neuron.

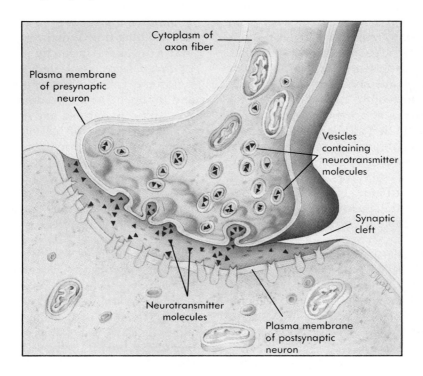

SUPPRESSING PAIN DURING EXERCISE

Research shows that the release of endorphins increases during heavy exercise. Endorphins inhibit pain, so it is no wonder that pain associated with muscle fatigue decreases when endorphins are present. Normally, pain is a warning signal that calls attention to injuries or dangerous circumstances. However, it is better to inhibit severe pain if it stops us from continuing an activity that may be necessary for survival. Athletes and others who exercise heavily have even reported a unique feeling of well-being or euphoria associated with elevated endorphin levels.

branch of a presynaptic neuron's axon (Figure 8-8). Each synaptic knob contains many small sacs or vesicles. Each vesicle contains a very small quantity of a chemical compound called a **neurotransmitter.** When a nerve impulse arrives at the synaptic knob, neurotransmitter molecules are released from the vesicles into the **synaptic cleft.** The synaptic cleft is the space between a synaptic knob and the plasma membrane of a postsynaptic neuron. It is an incredibly narrow space—only about two millionths of a centimeter in width. Identify the synaptic cleft in Figure 8-8. The plasma membrane of a *postsynaptic neuron* has protein molecules embedded in it opposite each synaptic knob. These serve as receptors to which neurotransmitter molecules bind. This binding can initiate an impulse in the postsynaptic neuron by increasing the permeability of that membrane to positive ions.

After the impulse conduction by postsynaptic neurons is initiated, the neurotransmitter activity is rapidly terminated. Either one or both of two mechanisms can cause this. Some neurotransmitter molecules diffuse out of the synaptic cleft back into the synaptic knobs. Other neurotransmitter molecules are metabolized into inactive compounds by specific enzymes.

Neurotransmitters are chemicals by which neurons communicate. As previously noted, at trillions of synapses in the CNS, presynaptic neurons release neurotransmitters that assist, stimulate, or inhibit postsynaptic neurons. At least 30 different compounds have been identified as neurotransmitters. They are not distributed randomly through the spinal cord and brain. Instead, specific neurotransmitters are localized in discrete groups of neurons and released in specific pathways.

For example, the substance named **acetylcholine** (as-e-til-KO-leen) is released at some of the synapses in the spinal cord and at neuromuscular (nerve-muscle) junctions. Other well-known neurotransmitters include **norepinephrine** (nor-ep-i-NEF-rin), **dopamine** (DOE-pa-meen), and **serotonin** (sair-o-TOE-nin). They belong to a group of compounds called **catecholamines** (kat-e-kol-AM-eens), which may play a role in sleep, motor function, mood, and pleasure recognition.

Two morphinelike neurotransmitters called **endorphins** (en-DOR-fins) and **enkephalins** (en-KEF-a-lins) are released at various spinal cord and brain synapses in the pain conduction pathway. These neurotransmitters inhibit conduction of pain impulses. They are natural pain killers.

Parkinson's disease is a chronic nervous disorder resulting from a deficiency of the neurotransmitter dopamine in certain parts of the brain. The group of signs associated with this disorder, a syndrome called *parkinsonism,* includes rigidity and trembling of the head and extremities, a forward tilt of the trunk, and a shuffling manner of walking (Figure 8-9). All of these characteristics result from lack of dopamine leading to misinformation in the parts of the brain that control voluntary muscle movements. Dopamine injection is not an effective treatment because dopamine cannot cross the blood-brain barrier. A breakthrough in treatment of Parkinson's disease came when the drug *levodopa* or L-dopa was found to increase the dopamine levels in afflicted patients. Neurons use L-dopa, which can cross the blood-brain barrier, to make dopamine. For some reason, L-dopa does not always have the desired effects in individual patients, so a number of alternate treatments have been developed. One option that is being explored is surgical grafting of normal dopamine-secreting neurons into the brains of individuals with Parkinson's disease.

FIGURE 8-9 Parkinsonism. Parkinsonism is a syndrome typically found in individuals with Parkinson's disease. The signs include (but are not limited to) rigidity and trembling of the head and extremities, a forward tilt of the trunk, and a shuffling gait with short steps and reduced arm swinging.

FIGURE 8-10 The Nervous System. The brain and spinal cord constitute the *central nervous system* (CNS), and the nerves make up the *peripheral nervous system* (PNS).

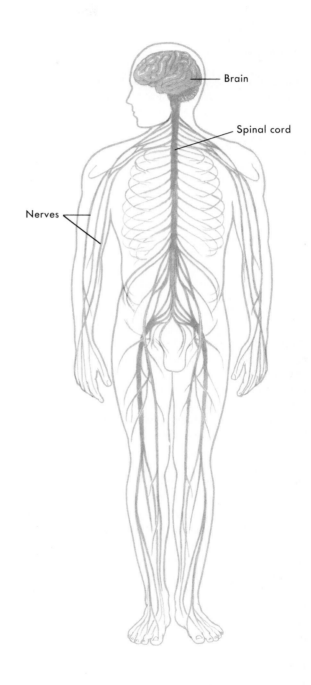

Central Nervous System

The CNS as its name implies, is centrally located. Its two major structures, the brain and spinal cord, are found along the midsagittal plane of the body (Figure 8-10). The brain is protected in the cranial cavity of the skull, and the spinal cord is surrounded in the spinal cavity by the vertebral column. In addition, the brain and spinal cord are covered by protective membranes called **meninges** (me-NIN-jeez), which are discussed in a later section of the chapter.

DIVISIONS OF THE BRAIN

The brain, one of our largest organs, consists of the following major divisions, named in ascending order beginning with most inferior part:

I **Brain stem**
  A **Medulla oblongata**
  B **Pons**
  C **Midbrain**
II **Cerebellum**
III **Diencephalon**
  A **Hypothalamus**
  B **Thalamus**
IV **Cerebrum**

Observe in Figure 8-11 the location and relative sizes of the medulla, pons, cerebellum, and cerebrum. Also identify the midbrain.

Brain Stem

The lowest part of the brain stem is the medulla oblongata. Immediately above the medulla lies the pons and above that the midbrain. Together these three structures are called the *brain stem* (Figure 8-11).

The **medulla oblongata** (ob-long-gah-tah) is an enlarged, upward extension of the spinal cord. It lies just inside the cranial cavity above the large hole in the occipital bone called the *foramen magnum*. Like the spinal cord, the medulla consists of gray and white matter, but their arrangement differs in the two organs. In the medulla, bits of gray matter mix closely and intricately with white matter to form the *reticular formation* (*reticular* means "netlike"). In the spinal cord, gray and white matter do not intermingle; gray matter forms the interior core of the cord, and white matter surrounds it. The **pons** and **midbrain,** like the medulla, consist of white matter and scattered bits of gray matter.

All three parts of the brain stem function as two-way conduction paths. Sensory fibers conduct impulses up from the cord to other parts of the brain, and motor fibers conduct impulses down from the brain to the cord. In addition, many important reflex centers lie in the brain stem. The cardiac, respiratory, and vasomotor centers (collectively called the *vital centers*), for example, are located in the medulla. Impulses from these centers control heartbeat, respirations, and blood vessel diameter (which is important in regulating blood pressure).

Diencephalon

The **diencephalon** (dye-en-SEF-ah-lon) is a small but important part of the brain located between the midbrain below and the cerebrum above. It consists of two major structures: the hypothalamus and the thalamus.

Hypothalamus. The **hypothalamus** (hye-po-THAL-ah-mus), as its name suggests, is located below the thalamus. The posterior pituitary gland, the stalk that attaches it to the undersurface of the brain, and areas of gray matter located in the side walls of a fluid-filled space called the *third ventricle* are extensions of the hypothalamus. Identify the pituitary gland and the hypothalamus in Figure 8-11.

The old adage, "Don't judge by appearances," applies well to appraising the importance of the hypothalamus. Measured by size, it is one of the least significant parts of the brain, but measured by its contribution to healthy survival, it is one of the most important brain structures. Impulses from neurons whose dendrites and cell bodies lie in the hypothalamus are conducted by their axons to neurons located in the spinal cord, and many of these impulses are then relayed to muscles and glands all over the body. Thus the hypothalamus exerts a major control over virtually all internal organs. Among the vital functions that it helps control are the heart beat, constriction and dilation of blood vessels, and contractions of the stomach and intestines.

Some neurons in the hypothalamus function in a surprising way; they make the hormones that the posterior pituitary gland secretes into the blood. Because one of these hormones (called *antidiuretic hormone* or *ADH*) affects the volume of urine excreted, the hypothalamus plays an essential role in maintaining the body's water balance.

Some of the neurons in the hypothalamus function as endocrine (ductless) glands. Their axons secrete chemicals called *releasing hormones* into the blood, which then carries them to the anterior pituitary gland. Releasing hormones, as their name suggests, control the release of certain anterior pituitary hormones. These in turn influence the hormone secretion of other endocrine glands. Thus the hypothalamus indirectly helps control the functioning of every cell in the body.

The hypothalamus is a crucial part of the mechanism for maintaining body temperature. Therefore marked elevations in body temperature in the absence of disease frequently characterizes injuries or other abnormalities of the hypothalamus. In addition, this important center is

FIGURE 8-11 Major Regions of the Central Nervous System. A, Sagittal sections of the brain and spinal cord. **B,** Section of preserved brain.

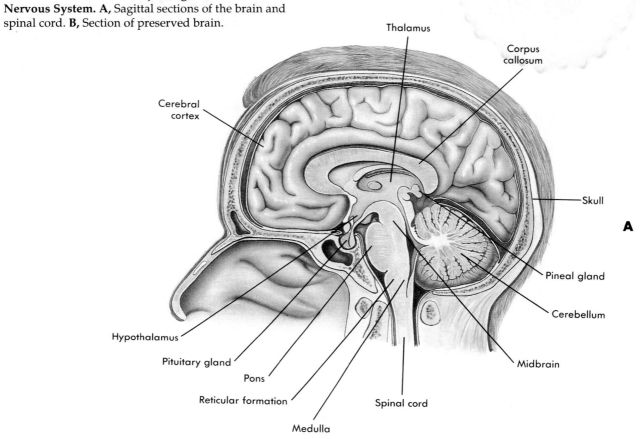

Thalamus

Corpus callosum

Cerebral cortex

Skull

Pineal gland

Cerebellum

Hypothalamus

Pituitary gland

Pons

Reticular formation

Spinal cord

Midbrain

Medulla

A

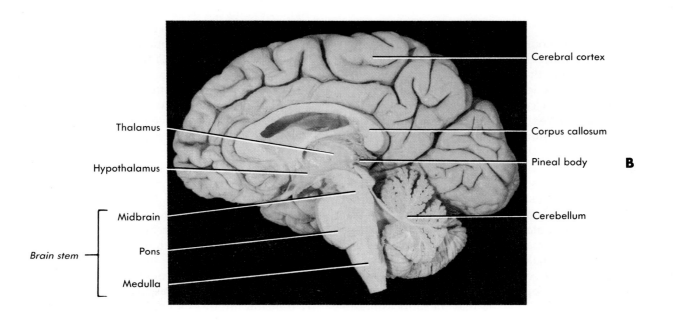

Cerebral cortex

Thalamus

Corpus callosum

Hypothalamus

Pineal body

Midbrain

Brain stem

Pons

Medulla

Cerebellum

B

involved in functions such as the regulation of water balance, sleep cycles, and the control of appetite and many emotions involved in pleasure, fear, anger, sexual arousal, and pain.

Thalamus. Just above the hypothalamus is a dumbbell-shaped section of gray matter called the **thalamus** (THAL-ah-mus). Each enlarged end of the dumbbell lies in a lateral wall of the third ventricle. The thin center section of the thalamus passes from left to right through the third ventricle. The thalamus is composed chiefly of dendrites and cell bodies of neurons that have axons extending up to the sensory areas of the cerebrum. It performs the following functions:

1 It helps produce sensations. Its neurons relay impulses to the cerebral cortex from the sense organs of the body.
2 It associates sensations with emotions. Almost all sensations are accompanied by a feeling of some degree of pleasantness or unpleasantness. The way that these pleasant and unpleasant feelings are produced is unknown except that they seem to be associated with the arrival of sensory impulses in the thalamus.
3 It plays a part in the so-called arousal or alerting mechanism.

Cerebellum

Structure. Look at Figure 8-11 to find the location, appearance, and size of the cerebellum. The cerebellum is the second largest part of the human brain. It lies under the occipital lobe of the cerebrum. In the cerebellum, gray matter composes the outer layer, and white matter composes the bulk of the interior.

Function. Most of our knowledge about cerebellar functions has come from observing patients who have some sort of disease of the cerebellum and from animals who have had the cerebellum removed. From such observations, we know that the cerebellum plays an essential part in the production of normal movements. Perhaps a few examples will make this clear. A patient who has a tumor of the cerebellum frequently loses balance and may topple over and reel like a drunken person when walking. It may be impossible to coordinate muscles normally. Frequent complaints about being clumsy and unable to even

drive a nail or draw a straight line are typical. With the loss of normal cerebellar functioning, the ability to make precise movements is lost. The general functions of the cerebellum, then, are to produce smooth coordinated movements, maintain equilibrium, and sustain normal postures.

Cerebrum

The **cerebrum** (SAIR-e-brum) is the largest and uppermost part of the brain. If you were to look at the outer surface of the cerebrum, the first features you would notice might be its many ridges and grooves. The ridges are called *convolutions* or *gyri* (JYE-rye), and the grooves are called *sulci* (SUL-kye). The deepest sulci are called *fissures;* the longitudinal fissure divides the cerebrum into right and left halves or hemispheres. These halves are almost separate structures except for their lower midportions, which are connected by a structure called the **corpus callosum** (COR-pus kal-LO-sum) (Figure 8-11). Two deep sulci subdivide each cerebral hemisphere into four major lobes and each lobe into numerous convolutions. The lobes are named for the bones that lie over them: the frontal lobe, the parietal lobe, the temporal lobe, and the occipital lobe. Identify these in Figure 8-12, *A.*

A thin layer of gray matter, made up of neuron dendrites and cell bodies, composes the surface of the cerebrum. Its name is the *cerebral cortex.* White matter, made up of bundles of nerve fibers (tracts), composes most of the interior of the cerebrum. Within this white matter, however, are a few islands of gray matter known as the **basal ganglia,** whose functioning is essential for producing automatic movements and postures. Parkinson's disease is a disease of the basal ganglia. Because shaking or tremors are common symptoms of Parkinson's disease, it is also called "shaking palsy."

What functions does the cerebrum perform? This is a hard question to answer briefly because the neurons of the cerebrum do not function alone. They function with many other neurons in many other parts of the brain and in the spinal cord. Neurons of these structures continually bring impulses to cerebral neurons and continually transmit impulses away from them. If all other neurons were functioning normally and only cerebral neurons were not functioning, here are

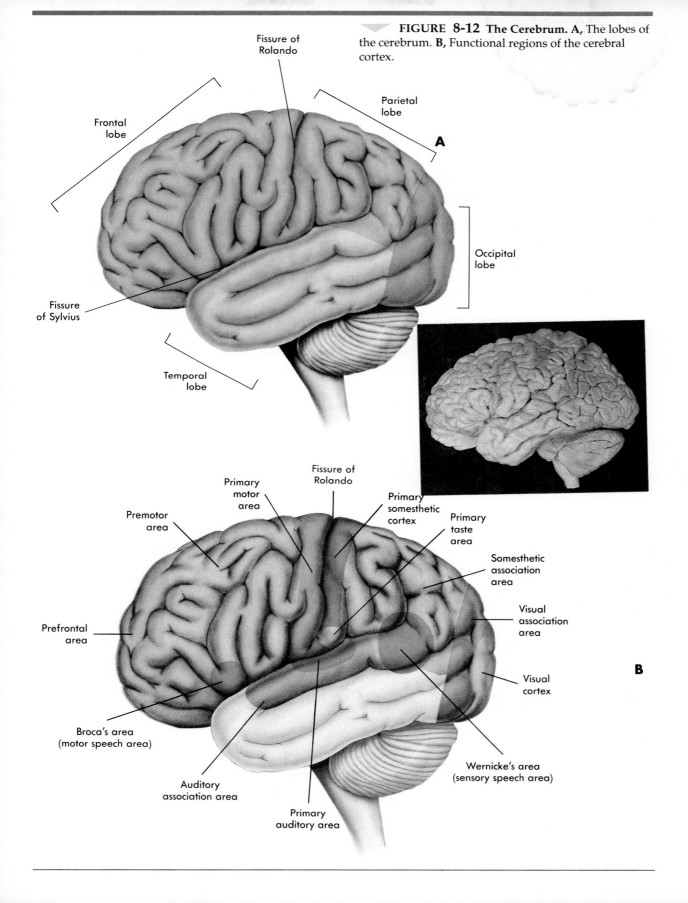

FIGURE 8-12 The Cerebrum. A, The lobes of the cerebrum. **B,** Functional regions of the cerebral cortex.

some of the things that you could not do. You could not think or use your will. You could not remember anything that has ever happened to you. You could not decide to make the smallest movement, nor could you make it. You would not see or hear. You could not experience any of the sensations that make life so rich and varied. Nothing would anger or frighten you, and nothing would bring you joy or sorrow. You would, in short, be unconscious. These terms then, sum up cerebral functions: consciousness, thinking, memory, sensations, emotions, and willed movements. Figure 8-12, *B*, shows the areas of the cerebral cortex essential for willed movements, general sensations, vision, hearing, and normal speech.

It is important to understand that very specific areas of the cortex have very specific functions. For example, the temporal lobe's auditory areas interpret incoming nervous signals from the ear as very specific sounds. The visual area of the cortex in the occipital lobe helps you identify and understand specific images. Localized areas of the cortex are directly related to specific functions, as shown in Figure 8-12, *B*. This explains the very specific symptoms associated with an injury to localized areas of the cerebral cortex after a stroke or traumatic injury to the head. Table 8-1 summarizes the major components of the brain and their main functions.

BRAIN DISORDERS
Destruction of Brain Tissue

Injury or disease can destroy neurons. A common example is the destruction of neurons of the motor area of the cerebrum that results from a **cerebrovascular accident (CVA).** A CVA or *stroke* is a hemorrhage from or cessation of blood flow through cerebral blood vessels. When this happens, the oxygen supply to portions of the brain is disrupted, and neurons cease functioning. If the lack of oxygen is prolonged, the neurons die. If the damage occurs in a motor control area of the brain (see Figure 8-12, *B*), the victim can no longer voluntarily move the parts of the body controlled by the affected areas. Because the paths of motor neurons in the cerebrum cross over in the brain stem, paralysis appears on the side of the body opposite to the side of the brain on which the CVA occurred. The term **hemiplegia** (hem-i-PLEE-jee-ah) refers to paralysis (loss of voluntary muscle control) of one whole side of the body.

One of the most common crippling diseases that appears during childhood, **cerebral palsy,** also results from damage to brain tissue. Cerebral palsy involves permanent, nonprogressive damage to motor control areas of the brain. Such damage is present at birth or occurs shortly after birth and remains throughout life. Possible causes of brain damage include prenatal infections or dis-

TABLE 8-1	Functions of Major Divisions of the Brain	
BRAIN AREA	**FUNCTION**
Brain stem	
Medulla oblongata	Two-way conduction pathway between the spinal cord and higher brain centers; cardiac, respiratory, and vasomotor control center
Pons	Two-way conduction pathway between areas of the brain and other regions of the body; influences respiration
Midbrain	Two-way conduction pathway; relay for visual and auditory impulses
Diencephalon	
Hypothalamus	Regulation of body temperature, water balance, sleep-cycle control, appetite, and sexual arousal
Thalamus	Sensory relay station from various body areas to cerebral cortex; emotions and alerting or arousal mechanisms
Cerebellum	Muscle coordination; maintenance of equilibrium and posture
Cerebrum	Sensory perception, emotions, willed movements, consciousness, and memory

BRAIN STUDIES

Biotechnology has produced many methods for studying the brain without the trauma of extensive exploratory surgery. Some of those methods are listed here:

X-ray photography

Traditional radiography (x-ray photography) of the head sometimes reveals tumors or injuries but does not show the detail of soft tissue necessary to diagnose many brain problems.

Computed tomography (CT)

This imaging technique involves scanning the head with a revolving x-ray generator. X-rays that pass through tissue hit sensors, which send the information to a computer that constructs an image that appears as a "slice of brain" on a video screen (Figure 4-15, *B*). Hemorrhages, tumors, and other lesions can often be detected with CT scanning.

Positron-emission tomography (PET)

PET scanning is a variation of CT scanning in which a radioactive substance is introduced into the blood supply of the brain. The radioactive material shows up as a bright spot on the image. Different substances are taken up by brain cells in different amounts, depending on the level of tissue activity—enabling radiologists to determine the *functional* characteristics of specific parts of the brain (see figure).

Single-photon emission computed tomography (SPECT)

SPECT is similar to PET but uses more stable substances and different detectors. SPECT is used to visualize blood flow patterns in the brain—making it useful in diagnosing CVAs and brain tumors.

Ultrasonography

In this method, high-frequency sound (ultrasound) waves are reflected off anatomical structures to form images—similar to radar. Because it does not use harmful radiation, ultrasonography is often used in diagnosing hydrocephalus or brain tumors in infants.

Magnetic resonance imaging (MRI)

Also called *nuclear magnetic resonance (NMR)* imaging, this scanning method also has the advantage of avoiding the use of harmful radiation. In MRI, a magnetic field surrounding the head induces brain tissues to emit radio waves that can be used by a computer to construct a sectional image. MRI has the added advantage of producing sharper images than CT scanning and ultrasound. This makes it very useful in detecting small brain lesions (Figure 4-15, *C*).

Electroencephalography (EEG)

As discussed in this chapter, electroencephalography is the measurement of the electrical activity of the brain. Typically, changes in voltage are recorded as deflections of a continuous line drawn on graph paper (see Figure 8-13). EEGs are used to detect seizure disorders, sleeping disorders, and other brain abnormalities.

Evoked potential (EP) test

The EP test is similar to EEG, but the brain waves observed are caused (evoked) by specific stimuli such as a flash of light or a sudden sound. Recently, this information has been analyzed by a computer that then produces a color-coded graphic image of the brain generated on a video screen—a *brain electric activity map (BEAM)*. Changes in color represent changes in brain activity evoked by each stimulus given. This technique is useful in diagnosing lesions of the visual or auditory systems because it reveals whether a sensory impulse is reaching the appropriate part of the brain.

EYES CLOSED   EYES OPEN   COMPLEX SCENE

FIGURE 8-13 Electroencephalography. A, Photograph of a person with voltage-sensitive electrodes attached to her skull. Information from these electrodes is used to produce a graphic recording of brain activity—an electroencephalogram (EEG). **B,** An EEG tracing showing activity in four different places in the brain (obtained from four sets of electrodes). Compare the moderate chaotic activity identified as *normal* with the explosive activity that occurs during a seizure.

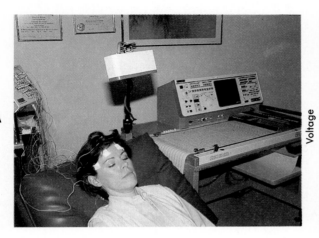

A

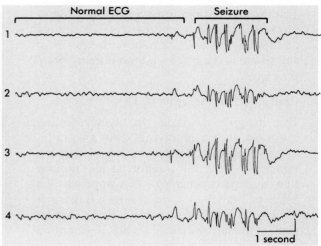

B

eases of the mother; mechanical trauma to the head before, during, or after birth; nerve-damaging poisons; and reduced oxygen supply to the brain. The resulting impairment to voluntary muscle control can manifest itself in a variety of ways. Many people with cerebral palsy exhibit **spastic paralyis,** a type of paralysis characterized by involuntary contractions of affected muscles. In cerebral palsy, spastic paralysis may affect one entire side of the body **(hemiplegia),** both legs **(paraplegia),** both legs and one arm **(triplegia),** or all four extremities **(quadriplegia).**

A variety of degenerative diseases can result in destruction of neurons in the brain. This degeneration can progress to adversely affect memory, attention span, intellectual capacity, personality, and motor control. The general term for this syndrome is **dementia** (de-MEN-shee-ah).

Dementia is characteristic of **Alzheimer's** (ALZ-hye-merz) **disease.** Its characteristic lesions develop in the cortex during the middle to late adult years. Exactly what makes dementia-causing lesions develop in the brains of individuals with Alzheimer's is not known. There is some evidence that this disease has a genetic basis—at least in some families. Other evidence indicates that environmental factors may play a role. Because the exact cause of Alzheimer's disease is still not known, development of an effective treatment has proved difficult. Currently, people diagnosed with Alzheimer's disease are treated by helping them maintain their remaining mental abilities and by looking after their hygiene, nutrition, and other aspects of personal health management.

Huntington's disease (HD) is an inherited disease characterized by *chorea* (involuntary, purposeless movements) that progresses to severe dementia and death. The initial symptoms of this disease first appear between age 30 and 40, with death occurring by age 55. Now that the gene responsible for Huntington's disease has been located, researchers hope that an effective treatment will be found (see Chapter 23).

The human immunodeficiency virus (HIV) that causes *acquired immune deficiency syndrome (AIDS)* can also cause dementia. The immune deficiency characteristic of AIDS results from HIV infection of white blood cells critical to the proper function of the immune system (see Chapter 14). However, the AIDS virus also infects neurons and can cause progressive degeneration of the brain, resulting in dementia.

Seizure Disorders

Some of the most common nervous system abnormalities belong to the group of conditions called *seizure disorders*. These disorders are characterized by **seizures**—sudden bursts of abnormal neuron activity that result in temporary changes in brain function. Seizures may be very mild, causing subtle changes in level of consciousness, motor control, or sensory perception. On the other hand, seizures may be quite severe, resulting in jerky, involuntary muscle contractions called *convulsions* or even unconsciousness.

Recurring or chronic seizure episodes constitute a condition called **epilepsy.** Although some cases of epilepsy can be traced to specific causes such as tumors, trauma, or chemical imbalances, most epilepsy is idiopathic (of unknown cause). Epileptics are often treated with anticonvulsive drugs such as *phenobarbital, phenytoin,* or *valproic acid* that block neurotransmitters in affected areas of the brain. By thus blocking synaptic transmission, such drugs inhibit the explosive bursts of neuron activity associated with seizures. With proper medication, many epileptics lead normal lives without the fear of experiencing uncontrollable seizures.

Diagnosis and evaluation of epilepsy or any seizure disorder often relies on a graphic representation of brain activity called an **electroencephalogram** (el-ek-tro-en-SEF-al-o-gram) **(EEG)** (Figure 8-13). As Figure 8-13, *B* shows, a normal EEG shows the chaotic rise and fall of the electrical activity in different parts of the brain as a series of wavy lines (the so-called brain waves). A seizure manifests itself as an explosive increase in the size and frequency of waves—as seen on the right side of Figure 8-13, *B*. Different classifications of epilepsy are based on the locations in the brain and the duration of these changes in brain activity.

SPINAL CORD

Structure

If you are of average height, your spinal cord is about 42 to 45 cm (17 to 18 inches) long (Figure 8-14). It lies inside the spinal column in the spinal cavity and extends from the occipital bone down to the bottom of the first lumbar vertebra. Place your hands on your hips, and they will line up with your fourth lumbar vertebra. Your spinal cord ends just above this level.

Look now at Figure 8-15. Notice the H-shaped core of the spinal cord. It consists of gray matter and so is composed mainly of dendrites and cell bodies of neurons. Columns of white matter form the outer portion of the spinal cord, and bundles of myelinated nerve fibers—the **spinal tracts** make up the white columns.

Spinal cord tracts provide two-way conduction paths to and from the brain. **Ascending tracts** conduct impulses up the cord to the brain. **Descending tracts** conduct impulses down the cord from the brain. Tracts are functional organizations in that all axons composing one tract serve one general function. For instance, fibers of the spinothalamic tracts serve a sensory function. They transmit impulses that produce sensations of crude touch, pain, and temperature. Other ascending tracts shown in Figure 8-15 include the gracilis and cuneatus tracts, which transmit sensations of touch and pressure up to the brain, and the anterior and posterior spinocerebellar tracts, which transmit information about muscle length to the cerebellum. Descending tracts include the lateral and ventral corticospinal tracts, which transmit impulses controlling many voluntary movements.

Functions

To try to understand spinal cord functions, think about a hotel telephone switchboard. Suppose a guest in Room 108 calls the switchboard operator and asks for Room 520, and in a second or so, someone in that room answers. Very briefly, three events took place: a message traveled into the switchboard, a connection was made in the switchboard, and a message traveled out from the switchboard. The telephone switchboard provided the connection that made possible the completion of the call, or we might say that it transferred the incoming call to an outgoing line. The spinal

FIGURE 8-14 Spinal Cord and Spinal Nerves.

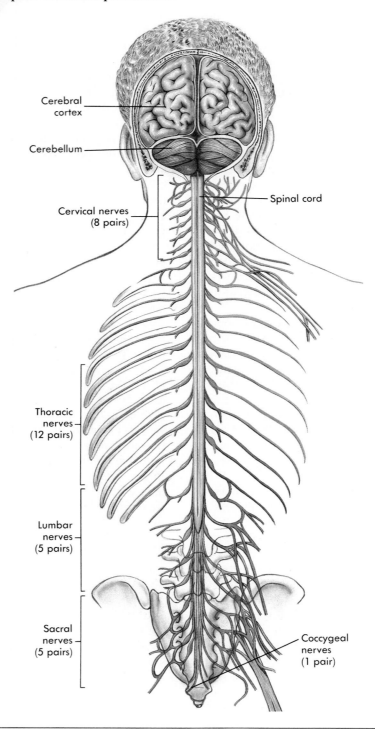

Cerebral cortex

Cerebellum

Cervical nerves (8 pairs)

Spinal cord

Thoracic nerves (12 pairs)

Lumbar nerves (5 pairs)

Sacral nerves (5 pairs)

Coccygeal nerves (1 pair)

FIGURE 8-15 Spinal Cord. Cross section of the spinal cord showing the horns, pathways (nerve tracts), and roots.

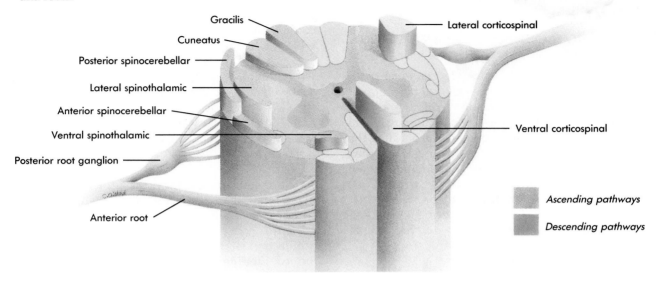

Gracilis

Cuneatus

Posterior spinocerebellar

Lateral spinothalamic

Anterior spinocerebellar

Ventral spinothalamic

Posterior root ganglion

Anterior root

Lateral corticospinal

Ventral corticospinal

Ascending pathways

Descending pathways

cord functions similarly. It contains the centers for thousands and thousands of reflex arcs. Look back at Figure 8-6. The interneuron shown there is an example of a spinal cord reflex center. It switches or transfers incoming sensory impulses to outgoing motor impulses, thereby making it possible for a reflex to occur. Reflexes that result from conduction over arcs whose centers lie in the spinal cord are called *spinal cord reflexes.* Two common kinds of spinal cord reflexes are withdrawal and jerk reflexes. An example of a withdrawal reflex is pulling one's hand away from a hot surface. The familiar knee jerk is an example of a jerk reflex.

In addition to functioning as the primary reflex center of the body, the spinal cord tracts, as previously noted, carry impulses to and from the brain. Sensory impulses travel up to the brain in ascending tracts, and motor impulses travel down from the brain in descending tracts. Therefore if an injury cuts the cord all the way across, impulses can no longer travel to the brain from any part of the body located below the injury, nor can they travel from the brain down to these parts. In short, this kind of spinal cord injury produces a

LUMBAR PUNCTURE

The extension of the meninges beyond the spinal cord makes it possible to perform lumbar punctures without the danger of injuring the spinal cord. A **lumbar puncture** is the withdrawal of some of the CSF from the subarachnoid space in the lumbar region of the spinal cord. The physician inserts a needle just above or below the fourth lumbar vertebra, knowing that the spinal cord ends an inch or more above this level. The fourth lumbar vertebra can be easily located because it lies on a line with the iliac crest. Placing a patient on his side and arching his back by drawing the knees and chest together separates the vertebrae sufficiently to introduce the needle. Lumbar punctures are often performed to withdraw CSF for analysis or to reduce the pressure caused by swelling of the brain or spinal cord after injury or disease.

loss of sensation, which is called **anesthesia** (an-es-THEE-zee-ah), and a loss of the ability to make voluntary movements, which is called **paralysis** (pah-RAL-i-sis).

COVERINGS AND FLUID SPACES OF THE BRAIN AND SPINAL CORD

Nervous tissue is not a sturdy tissue. Even moderate pressure can kill nerve cells, so nature safeguards the chief organs made of this tissue—the spinal cord and the brain—by surrounding them with three fluid-containing membranes called the **meninges** (me-NIN-jeez). The meninges are then surrounded by bone. The spinal meninges form a tubelike covering around the spinal cord and line the bony vertebral foramen of the vertebrae that surround the cord. Look at Figure 8-16, and you can identify the three layers of the spinal

▼ **FIGURE 8-16 Spinal Cord.** The meninges, spinal nerves, and sympathetic trunk are visible in the illustration.

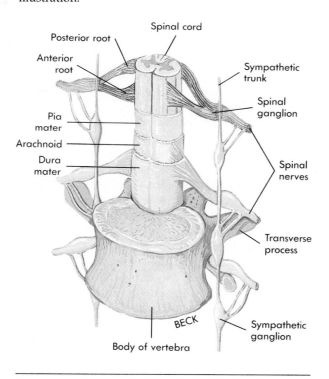

meninges. They are the **dura mater** (DOO-rah MA-ter), which is the tough outer layer that lines the vertebral canal, the **pia** (PEE-ah) **mater,** which is the innermost membrane covering the spinal cord itself, and the **arachnoid** (ah-RAK-noyd), which is the membrane between the dura and the pia mater. The arachnoid resembles a cobweb with fluid in its spaces. The word *arachnoid* means "cobweblike." It comes from *arachne,* which is the Greek word for spider. Arachne is the name of the girl who was changed into a spider by Athena because she boasted of the fineness of her weaving—at least, so an ancient Greek myth tells us.

The meninges that form the protective covering around the spinal cord also extend up and around the brain to enclose it completely (Figure 8-16).

Infection or inflammation of the meninges is termed **meningitis** (men-in-JYE-tis). This condition is most commonly caused by bacteria such as *Neisseria meningitidis* (meningococcus), *Streptococcus pneumoniae,* or *Heamophilus influenzae* (see Appendix B). However, viral infections, mycoses (fungal infections), and tumors may also cause inflammation of the meninges. Patients with meningitis usually complain of severe headaches and neck pain. Those experiencing symptoms should seek immediate attention to get the problem under control. Depending on the primary cause, meningitis may be mild and self-limiting or may progress to a severe, perhaps fatal, condition. If only the spinal meninges are involved, the condition is called *spinal meningitis.*

Fluid fills the subarachnoid spaces between the pia mater and arachnoid in the brain and spinal cord. This fluid is called **cerebrospinal fluid (CSF).** Cerebrospinal (sair-e-bro-SPY-nal) fluid also fills spaces in the brain called cerebral **ventricles.** In Figure 8-17, you can see the irregular shapes of the ventricles of the brain. These illustrations can also help you visualize the location of the ventricles if you remember that these large spaces lie deep inside the brain and that there are two lateral ventricles. One lies inside the right half of the cerebrum (the largest part of the human brain), and the other lies inside the left half of the cerebrum.

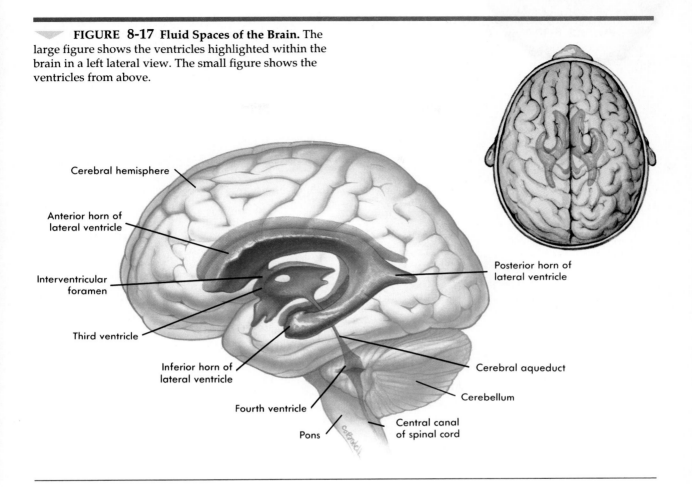

FIGURE 8-17 Fluid Spaces of the Brain. The large figure shows the ventricles highlighted within the brain in a left lateral view. The small figure shows the ventricles from above.

CSF is one of the body's circulating fluids. It forms continually from fluid filtering out of the blood in a network of brain capillaries known as the **choroid plexus** (KO-royd PLEK-sus) and into the ventricles. CSF seeps from the lateral ventricles into the third ventricle and flows down through the cerebral aqueduct (find this in Figure 8-17 and 8-18) into the fourth ventricle. It moves from the fourth ventricle into the small, tubelike central canal of the cord and out into the subarachnoid spaces. Then it moves leisurely down and around the cord and up and around the brain (in the subarachnoid spaces of their meninges) and returns to the blood (in the veins of the brain).

Remembering that this fluid forms continually from blood, circulates, and is resorbed into blood can be useful. It can help you understand certain abnormalities. Suppose a person has a brain tumor that presses on the cerebral aqueduct. This blocks the way for the return of CSF to the blood. Because the fluid continues to form but cannot drain away, it accumulates in the ventricles or in the meninges. Other conditions can cause an accumulation of CSF in ventricles. An example is **hydrocephalus** (hye-dro-SEF-ah-lus) or "water on the brain." One treatment involves surgical placement of a hollow tube or catheter through the blocked channel so that CSF can drain into another location in the body (Figure 8-19).

FIGURE 8-18 Flow of Cerebrospinal Fluid. The fluid produced by filtration of blood by the choroid plexus of each ventricle flows inferiorly through the lateral ventricles, interventricular foramen, third ventricle, cerebral aqueduct, fourth ventricle, and subarachnoid space and to the blood.

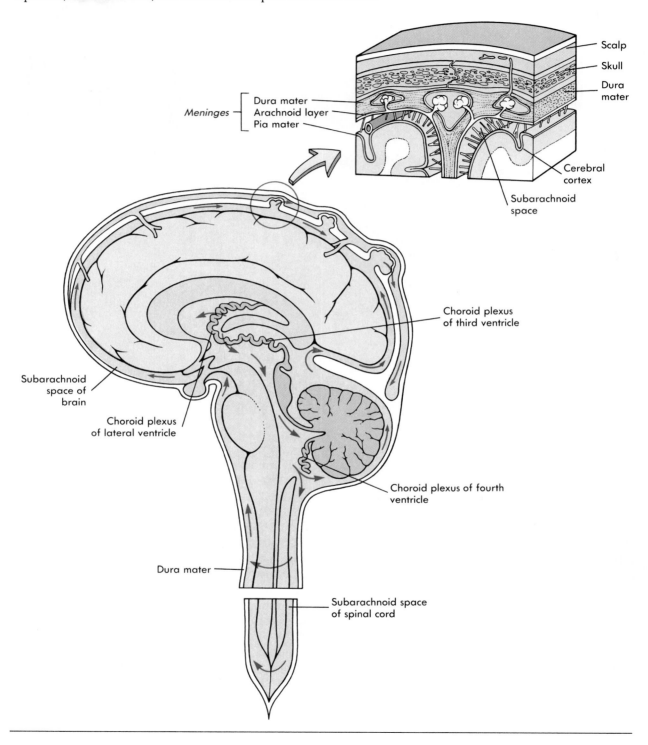

Scalp

Skull

Dura mater

Meninges
- Dura mater
- Arachnoid layer
- Pia mater

Cerebral cortex

Subarachnoid space

Choroid plexus of third ventricle

Subarachnoid space of brain

Choroid plexus of lateral ventricle

Choroid plexus of fourth ventricle

Dura mater

Subarachnoid space of spinal cord

FIGURE 8-19 Hydrocephalus. Hydrocephalus is caused by narrowing or blockage of the pathway for CSF, causing retention of CSF in the ventricles. This condition can be treated by surgical placement of a shunt or tube to drain the excess fluid. Notice in the cross sections of the brain how the ventricles and surrounding tissue return to their normal shape and size after shunt placement.

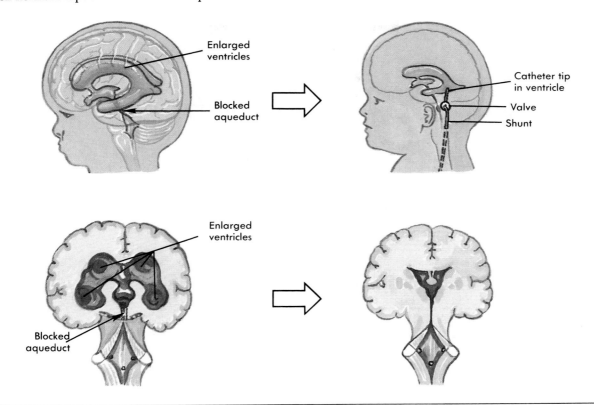

Peripheral Nervous System

The nerves connecting the brain and spinal cord to other parts of the body constitute the **peripheral nervous system** (PNS). This system includes **cranial** and **spinal nerves** that connect the brain and spinal cord, respectively, to peripheral structures such as the skin surface and the skeletal muscles. In addition, other structures in the **autonomic nervous system** or **ANS** are considered part of the PNS. These connect the brain and spinal cord to various glands in the body and to the cardiac and smooth muscle in the thorax and abdomen.

CRANIAL NERVES

Twelve pairs of cranial nerves are attached to the undersurface of the brain, mostly from the brain stem. Figure 8-20 shows the attachments of these nerves. Their fibers conduct impulses between the brain and structures in the head and neck and in the thoracic and abdominal cavities. For instance, the second cranial nerve (optic nerve) conducts impulses from the eye to the brain, where these impulses produce vision. The third cranial nerve (oculomotor nerve) conducts impulses from the brain to muscles in the eye, where they cause contractions that move the eye. The tenth cranial

FIGURE 8-20 Cranial Nerves. View of the undersurface of the brain shows attachments of the cranial nerves (cut here for clarity).

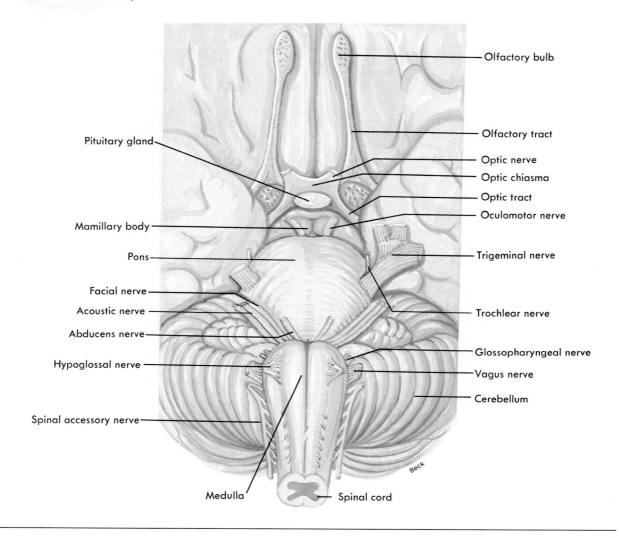

Olfactory bulb

Olfactory tract

Optic nerve

Optic chiasma

Optic tract

Oculomotor nerve

Trigeminal nerve

Trochlear nerve

Glossopharyngeal nerve

Vagus nerve

Cerebellum

Pituitary gland

Mamillary body

Pons

Facial nerve

Acoustic nerve

Abducens nerve

Hypoglossal nerve

Spinal accessory nerve

Medulla

Spinal cord

Beck

nerve (vagus nerve) conducts impulses between the medulla oblongata and structures in the neck and thoracic and abdominal cavities. The names of each cranial nerve and a brief description of their functions are listed in Table 8-2.

SPINAL NERVES

Structure

Thirty-one pairs of nerves are attached to the spinal cord in the following order: eight pairs are attached to the cervical segments, twelve pairs are attached to the thoracic segments, five pairs are attached to the lumbar segments, five pairs are attached to the sacrospinal segments, and one pair is attached to the coccygeal segment (see Figure 8-14). Unlike cranial nerves, spinal nerves have no special names; instead, a letter and number identify each one. C1, for example, indicates the pair of spinal nerves attached to the first segment of the cervical part of

TABLE 8-2	Cranial Nerves	
NERVE*	**CONDUCTS IMPULSES**	**FUNCTIONS**
I Olfactory	From nose to brain	Sense of smell
II Optic	From eye to brain	Vision
III Oculomotor	From brain to eye muscles	Eye movements
IV Trochlear	From brain to external eye muscles	Eye movements
V Trigeminal	From skin and mucous membrane of head and from teeth to brain; also from brain to chewing muscles	Sensations of face, scalp, and teeth, chewing movements
VI Abducens	From brain to external eye muscles	Turning eyes outward
VII Facial	From taste buds of tongue to brain; from brain to face muscles	Sense of taste; contraction of muscles of facial expression
VIII Acoustic	From ear to brain	Hearing; sense of balance
IX Glossopharyngeal	From throat and taste buds of tongue to brain; also from brain to throat muscles and salivary glands	Sensations of throat, taste, swallowing movements, secretion of saliva
X Vagus	From throat, larynx, and organs in thoracic and abdominal cavities to brain; also from brain to muscles of throat and to organs in thoracic and abdominal cavities	Sensations of throat and larynx and of thoracic and abdominal organs; swallowing, voice production, slowing of heartbeat, acceleration of peristalsis (gut movements)
XI Spinal accessory	From brain to certain shoulder and neck muscles	Shoulder movements; turning movements of head
XII Hypoglossal	From brain to muscles of tongue	Tongue movements

*The first letters of the words of the following sentence are the first letters of the names of cranial nerves: "On Old Olympus' Tiny Tops, A Finn And German Viewed Some Hops." Many generations of students have used this or a similar sentence to help them remember the names of cranial nerves.

the cord, and T8 indicates nerves attached to the eighth segment of the thoracic part of the spinal cord. In Figure 8-21 the cervical area of the spine has been dissected to show the emerging spinal nerves in that area. After spinal nerves exit from the spinal cord, they branch to form the many peripheral nerves of the trunk and limbs. Sometimes, nerve fibers from several spinal nerves are reorganized to form a single peripheral nerve. This reorganization can be seen as a network of intersecting branches that is called a **plexus.**

Functions

Spinal nerves conduct impulses between the spinal cord and the parts of the body not supplied by cranial nerves. The spinal nerves shown in Figure 8-21 contain, as do all spinal nerves, sensory and motor fibers. Spinal nerves therefore function to make possible both sensations and movements. A disease or injury that prevents conduction by a spinal nerve thus results in a loss of feeling and a loss of movement in the part supplied by that nerve.

Detailed mapping of the skin surface reveals a close relationship between the source on the cord of each spinal nerve and the part of the body it innervates (Figure 8-22). Knowledge of the segmental arrangement of spinal nerves is useful to physicians. For instance, a neurologist can identify the site of spinal cord or nerve abnormality from the area of the body insensitive to a pinprick. Skin surface areas supplied by a single spinal nerve are called **dermatomes** (DER-mahtomes). A dermatome "map" of the body is shown in Figure 8-22.

FIGURE 8-21 Spinal Nerves. Dissection of the cervical segment of the spinal cord shows emerging cervical nerves. The spinal cord is viewed from behind (posterior aspect).

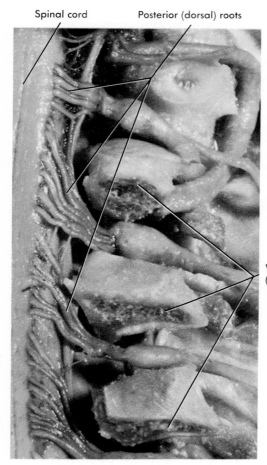

Spinal cord   Posterior (dorsal) roots

Vertebrae (cut)

PERIPHERAL NERVE DISORDERS

Many afflictions of peripheral nerves, or their branches, involve inflammation—or **neuritis** (noo-RYE-tis). You may know someone who suffers from a form of neuritis called **sciatica** (sye-AT-ik-a). This is a painful inflammation of the spinal nerve branch in the thigh called the *sciatic nerve*—the largest nerve in the body. This condition is characterized by nerve pain or **neuralgia** (noor-AL-jee-a). In some cases, it may lead to atrophy of the leg muscles.

Compression or degeneration of the fifth cranial nerve, the trigeminal nerve, may result in a condition called **trigeminal neuralgia** or *tic douloureux* (doo-loo-ROO). This condition is characterized by recurring episodes of stabbing pain radiating from the angle of the jaw along a branch of the trigeminal nerve. Neuralgia of one branch occurs over the forehead and around the eyes. Pain along another branch is felt in the cheek, nose, and upper lip. Neuralgia of the third branch results in stabbing pains in the tongue and lower lip.

Compression, degeneration, or infection of the seventh cranial nerve, the facial nerve, may result in **Bell's palsy.** Bell's palsy is characterized by paralysis of some or all of the facial features innervated by the facial nerve, including the eyelids and mouth. This condition is often temporary but in some cases is irreversible. Plastic surgery is sometimes used to correct permanent disfigurement.

Herpes zoster or **shingles** is a unique viral infection that almost always affects the skin of a single dermatome. It is caused by a varicella zoster virus of chickenpox (see Appendix B). About 3% of the population will suffer from shingles at some time in their lives. In most cases the disease results from reactivation of the varicella virus. The virus probably traveled through a cutaneous nerve and remained dormant in a dorsal root ganglion for years after an episode of the chickenpox. If the body's immunological protective mechanism becomes diminished in an older adult after stress or in an individual undergoing radiation therapy or taking immunosuppressive drugs, the virus may reactivate. If this occurs, the virus travels over the sensory nerve to the skin of a single dermatome. The result is a painful eruption of red, swollen plaques or vesicles that eventually rupture and crust before clearing in 2 to 3 weeks. In severe cases, extensive inflammation, hemorrhagic blisters, and secondary bacterial infection may lead to permanent scarring. In most cases, the eruption of vesicles is preceded by 4 to 5 days of pain, burning, and itching in the affected dermatome. Treatment with antiviral drugs, when started at initial outbreak, helps minimize the effects. Unfortunately, an attack of herpes zoster does not confer lasting immunity. Many individuals suffer three or four episodes in a lifetime.

FIGURE 8-22 Dermatomes. Segmental dermatome distribution of spinal nerves to the front and back of the body. *C,* Cervical segments; *T,* thoracic segments; *L,* lumbar segments; *S,* sacral segments.

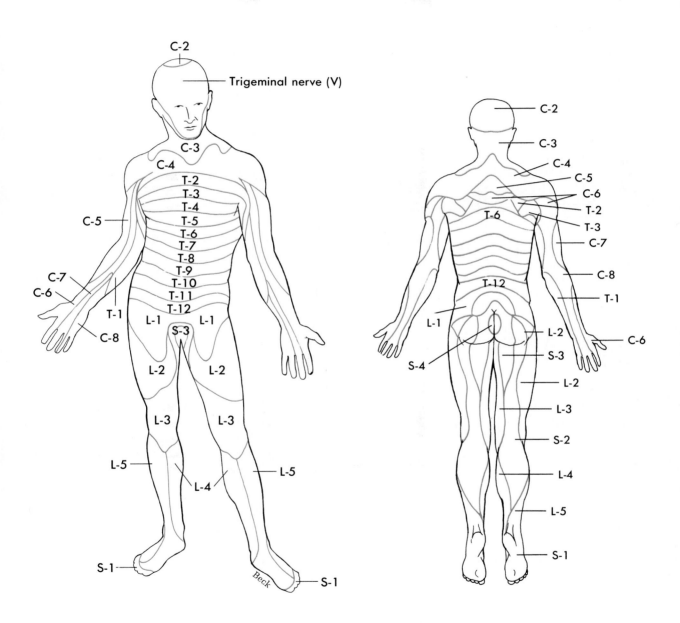

FIGURE 8-23 Innervation of the Major Target Organs by the Autonomic Nervous System. The sympathetic fibers are highlighted with red, and the parasympathetic fibers are highlighted with blue.

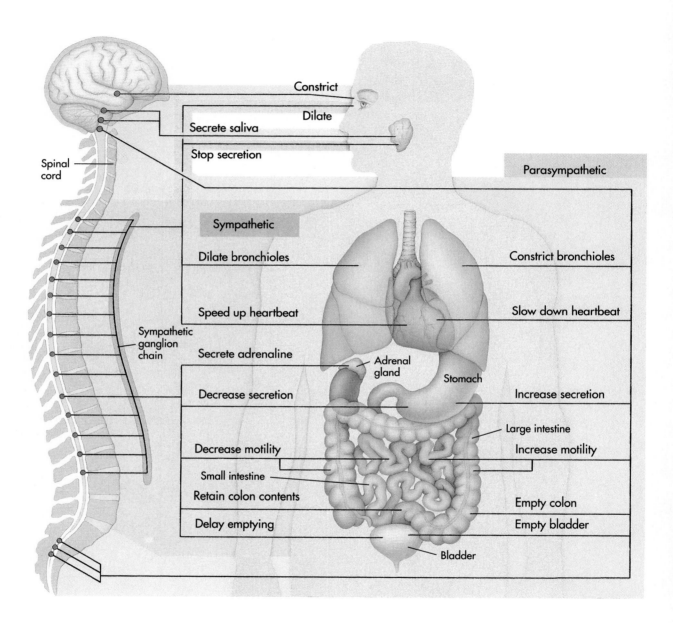

Autonomic Nervous System

The **autonomic nervous system (ANS)** consists of certain motor neurons that conduct impulses from the spinal cord or brain stem to the following kinds of tissues:

1 Cardiac muscle tissue
2 Smooth muscle tissue
3 Glandular epithelial tissue

The ANS consists of the parts of the nervous system that regulate the body's involuntary functions (for example, the heartbeat, contractions of the stomach and intestines, and secretions by glands). On the other hand, the motor nerves that control the voluntary actions of skeletal muscles are sometimes referred to as the *somatic nervous system.*

The autonomic nervous system consists of two divisions called the **sympathetic nervous system** and the **parasympathetic nervous system** (Figure 8-23).

FUNCTIONAL ANATOMY

Autonomic neurons are the motor neurons that make up the ANS. The dendrites and cell bodies of some autonomic neurons are located in the gray matter of the spinal cord or brain stem. Their axons extend from these structures and terminate in ganglia. These autonomic neurons are called **preganglionic neurons** because they conduct impulses between the spinal cord and a ganglion. In the ganglia the axon endings of preganglionic neurons synapse with the dendrites or cell bodies of postganglionic neurons. **Postganglionic neurons,** as their name suggests, conduct impulses from a ganglion to cardiac muscle, smooth muscle, or glandular epithelial tissue.

Autonomic or visceral effectors are the tissues to which autonomic neurons conduct impulses. Specifically, visceral effectors are cardiac muscle that makes up the wall of the heart, smooth muscle that partially makes up the walls of blood vessels and other hollow internal organs, and glandular epithelial tissue that makes up the secreting part of glands.

AUTONOMIC CONDUCTION PATHS

Conduction paths to visceral and somatic effectors from the CNS (spinal cord or brain stem) dif-fer somewhat. Autonomic paths to visceral effectors, as the right side of Figure 8-24 shows, consist of two-neuron relays. Impulses travel over preganglionic neurons from the spinal cord or brain stem to autonomic ganglia. There, they are relayed across synapses to postganglionic neurons, which then conduct the impulses from the ganglia to visceral effectors. Compare the autonomic conduction path with the somatic conduction path illustrated on the left side of Figure 8-25. Somatic motor neurons, like the ones shown here, conduct all the way from the spinal cord or brain stem to somatic effectors with no intervening synapses.

SYMPATHETIC NERVOUS SYSTEM
Structure

Sympathetic preganglionic neurons have dendrites and cell bodies in the gray matter of the thoracic and upper lumbar segments of the spinal cord. The sympathetic system has also been referred to as the *thoracolumbar system.* Look now at the right side of Figure 8-24. Follow the course of the axon of the sympathetic preganglionic neuron shown there. It leaves the spinal cord in the anterior (ventral) root of a spinal nerve. It next enters the spinal nerve but soon leaves it to extend to and through a sympathetic ganglion and terminate in a collateral ganglion. There, it synapses with several postganglionic neurons whose axons extend to terminate in visceral effectors. Also shown in Figure 8-24, branches of the preganglionic axon may ascend or descend to terminate in ganglia above and below their point of origin. All sympathetic preganglionic axons therefore synapse with many postganglionic neurons, and these frequently terminate in widely separated organs. Hence sympathetic responses are usually wide-spread, involving many organs and not just one.

Sympathetic postganglionic neurons have dendrites and cell bodies in sympathetic ganglia. Sympathetic ganglia are located in front of and at each side of the spinal column. Because short fibers extend between the sympathetic ganglia, they look a little like two chains of beads and are often referred to as the *sympathetic chain ganglia.* Axons of sympathetic postganglionic neurons travel in spinal nerves to blood vessels, sweat glands, and arrector hair muscles all over the

FIGURE 8-24 Autonomic Conduction Paths. The left side of the diagram shows that one somatic motor neuron conducts impulses all the way from the spinal cord to a somatic effector. Conduction from the spinal cord to any visceral effector, however, requires a relay of at least two autonomic motor neurons—a preganglionic and a postganglionic neuron, shown on the right side of the diagram.

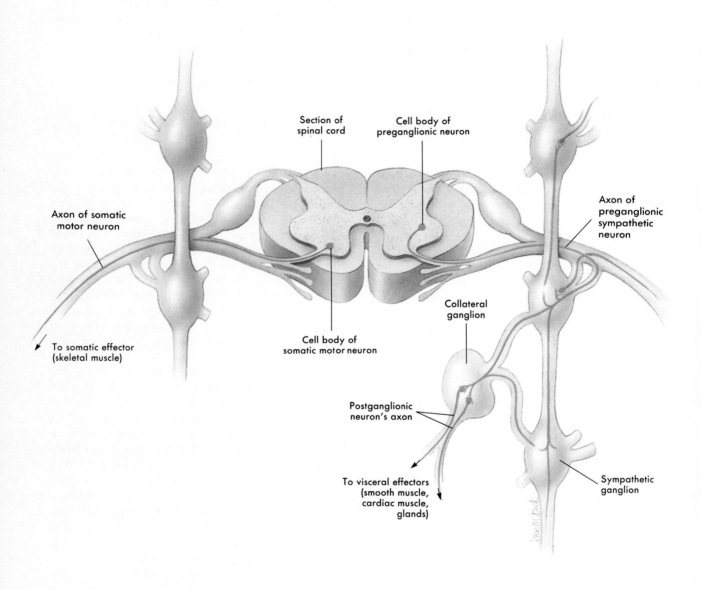

body. Separate autonomic nerves distribute many sympathetic postganglionic axons to various internal organs.

Functions of the Sympathetic Nervous System

The sympathetic nervous system functions as an emergency system. Impulses over sympathetic fibers take control of many internal organs when we exercise strenuously and when strong emotions—anger, fear, hate, anxiety—are elicited. In short, when we must cope with stress of any kind, sympathetic impulses increase to many visceral effectors and rapidly produce widespread changes within our bodies. The middle column of Table 8-3 indicates many sympathetic responses. The heart beats faster. Most blood vessels constrict,

causing blood pressure to increase. Blood vessels in skeletal muscles dilate, supplying the muscles with more blood. Sweat glands and adrenal glands secrete more abundantly. Salivary and other digestive glands secrete sparingly. Digestive tract contractions (peristalsis) become sluggish, hampering digestion. These sympathetic responses make us ready for strenuous muscular work, or they prepare us for *fight or flight*. The group of changes induced by sympathetic control is known as the **fight-or-flight syndrome.**

PARASYMPATHETIC NERVOUS SYSTEM
Structure

The dendrites and cell bodies of parasympathetic preganglionic neurons are located in the gray

TABLE 8-3 Autonomic Functions		
VISCERAL EFFECTORS	**SYMPATHETIC CONTROL**	**PARASYMPATHETIC CONTROL**
Heart muscle	Accelerates heartbeat	Slows heartbeat
Smooth muscle		
Of most blood vessels	Constricts blood vessels	None
Of blood vessels in skeletal muscles	Dilates blood vessels	None
Of the digestive tract	Decreases peristalsis; inhibits defecation	Increases peristalsis
Of the anal sphincter	Stimulates—closes sphincter	Inhibits—opens sphincter for defecation
Of the urinary bladder	Inhibits—relaxes bladder	Stimulates—contracts bladder
Of the urinary sphincters	Stimulates—closes sphincter	Inhibits—opens sphincter for urination
Of the eye		
Iris	Stimulates radial fibers—dilation of pupil	Stimulates circular fibers—constriction of pupil
Ciliary	Inhibits—accommodation for far vision (flattening of lens)	Stimulates—accommodation for near vision (bulging of lens)
Of hairs (pilomotor muscles)	Stimulates—"goose pimples"	No parasympathetic fibers
Glands		
Adrenal medulla	Increases epinephrine secretion	None
Sweat glands	Increases sweat secretion	None
Digestive glands	Decreases secretion of digestive juices	Increases secretion of digestive juices

matter of the brain stem and the sacral segments of the spinal cord. The parasympathetic system has also been referred to as the *craniosacral system.* The preganglionic parasympathetic axons extend some distance before terminating in parasympathetic ganglia in the head and thoracic and abdominal cavities close to the visceral effectors that they control. The dendrites and cell bodies of parasympathetic postganglionic neurons lie in these outlying parasympathetic ganglia, and their short axons extend into nearby structures. Therefore each parasympathetic preganglionic neuron synapses only with postganglionic neurons to a single effector. For this reason, parasympathetic stimulation frequently involves response by only one organ. This is not true of sympathetic responses; sympathetic stimulation usually results in responses by numerous organs.

Functions of the Parasympathetic Nervous System

The parasympathetic system controls visceral effectors under normal conditions. Impulses over parasympathetic fibers, for example, slow heartbeat, increase peristalsis, and increase secretion of digestive juices and insulin (Table 8-3).

AUTONOMIC NEUROTRANSMITTERS

Figure 8-25 reveals information about autonomic neurotransmitters, the chemical compounds released from the axon terminals of autonomic neurons. Observe that three of the axons shown in Figure 8-25—the sympathetic preganglionic axon, the parasympathetic preganglionic axon, and the parasympathetic postganglionic axon—release acetylcholine. These axons are therefore classified as **cholinergic fibers.** Only one type of autonomic

▼ **FIGURE 8-25 Autonomic Neurotransmitters.** Three of the four fiber types are cholinergic, secreting the neurotransmitter acetylcholine *(Ach)* into a synapse. Only the sympathetic postganglionic fiber is adrenergic, secreting norepinephrine *(NE)* into a synapse.

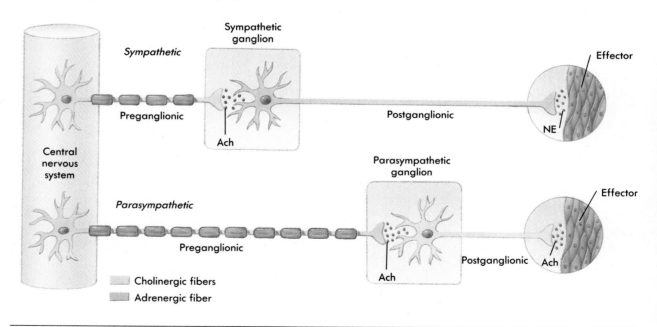

axon releases the neurotransmitter norepinephrine (noradrenaline). This is the axon of a sympathetic postganglionic neuron, and such neurons are classified as **adrenergic fibers.** See p. 204 for other neurons whose axons release acetylcholine and norepinephrine.

AUTONOMIC NERVOUS SYSTEM AS A WHOLE

The function of the **autonomic nervous system** is to regulate the body's automatic, involuntary functions in ways that maintain or quickly restore homeostasis. Many internal organs are doubly innervated by the ANS. In other words, they receive fibers from parasympathetic and sympathetic divisions. Parasympathetic and sympathetic impulses continually bombard them and, as Table 8-3 indicates, influence their function in opposite or antagonistic ways. For example, the heart continually receives sympathetic impulses that make it beat faster and parasympathetic impulses that slow it down. The ratio between these two antagonistic forces determines the actual heart rate.

The name *autonomic nervous system* is something of a misnomer. It seems to imply that this part of the nervous system is independent from other parts. This is not true. Dendrites and cell bodies of preganglionic neurons are located, as observed, in the spinal cord and brain stem. They are continually influenced directly or indirectly by impulses from neurons located above them, notably by some in the hypothalamus and in the parts of the cerebral cortex called the *emotional brain.* Through conduction paths from these areas, emotions can produce widespread changes in the automatic functions of our bodies, in cardiac and smooth muscle contractions, and in secretion by glands. Anger and fear, for example, lead to increased sympathetic activity and the fight-or-flight syndrome. According to some physiologists, the altered state of consciousness known as *meditation* leads to decreased sympathetic activity and a group of changes opposite to those of the fight-or-flight syndrome.

DISORDERS OF THE AUTONOMIC NERVOUS SYSTEM

Stress-Induced Disease

Considering the variety and number of effectors innervated by the autonomic nervous system, it is no wonder that autonomic disorders have varied and far-reaching consequences. This is especially true of stress-induced diseases. Prolonged or excessive physiological response to stress, the fight-or-flight response, can disrupt normal functioning throughout the body. Stress has been cited as an indirect cause or an important risk factor in a number of conditions. Only a few are listed here:

1 Heart disease—Although an extreme episode of stress can precipitate heart failure even in healthy people, chronic stress is known to increase the risk of certain heart disorders. One such condition is stress-induced high blood pressure or *hypertension* that can weaken the heart and blood vessels.

2 Digestive problems—Colitis (colon inflammation) and gastric ulcers, for example, may be precipitated by the changes in digestive secretion and movement that occur during prolonged or repeated stress responses.

3 Reduced resistance to disease—Hormones called *glucocorticoids* released by the adrenal glands during prolonged or repeated stress episodes depress the activity of the immune system. Depressed immune function leads to increased risk of infection and cancer.

Neuroblastoma

Neuroblastoma (noo-ro-blast-O-mah) is a malignant tumor of the sympathetic nervous system. It most often occurs in the developing nervous systems of young children and metastasizes rapidly to other parts of the body. Symptoms often include exaggerated or inappropriate sympathetic effects, including increased heart rate, sweating, and high blood pressure. As with some other forms of cancer, spontaneous remissions may occur.

OUTLINE SUMMARY

ORGANS AND DIVISIONS OF THE NERVOUS SYSTEM (Figure 8-1)

A Central nervous system (CNS)—brain and spinal cord
B Peripheral nervous system (PNS)—all nerves
C Autonomic nervous system (ANS)

CELLS OF THE NERVOUS SYSTEM

A Neurons
  1 Consist of three main parts—dendrites: conduct impulses to cell body of neuron; cell body of neuron; and axon: conducts impulses away from cell body of neuron (Figure 8-2)
  2 Neurons classified according to function—sensory: conduct impulses to the spinal cord and brain; motor: conduct impulses away from brain and spinal cord to muscles and glands; and interneurons: conduct impulses from sensory neurons to motor neurons
B Neuroglia—three main types of connective tissue cells of the CNS (Figure 8-3)
  1 Astrocytes—star-shaped cells that anchor small blood vessels to neurons
  2 Microglia—small cells that move in inflamed brain tissue carrying on phagocytosis
  3 Oligodendroglia—form myelin sheaths on axons in the CNS
C Disorders of nervous tissue
  1 Multiple sclerosis—characterized by myelin loss in central nerve fibers and resulting conduction impairments
  2 Tumors
    a General name for nervous system tumors is *neuroma*
    b Most neuromas are gliomas, neuroglial tumors
    c Multiple neurofibromatosis—characterized by numerous benign tumors that can progress to disfiguring, crippling soft tissue tumors

NERVES (Figure 8-5)

A Nerve—bundle of peripheral axons
  1 Tract—bundle of central axons
  2 White matter—tissue composed primarily of myelinated axons (nerves or tracts)
  3 Gray matter—tissue composed primarily of cell bodies and unmyelinated fibers
B Nerve coverings—fibrous connective tissue
  1 Endoneurium—surrounds individual fibers within a nerve

  2 Perineurium—surrounds a group (fascicle) of nerve fibers
  3 Epineurium—surrounds the entire nerve

REFLEX ARCS

A Nerve impulses are conducted from receptors to effectors over neuron pathways or reflex arcs; conduction by a reflex arc results in a reflex (that is, contraction by a muscle or secretion by a gland)
B Simplest reflex arcs are two-neuron arcs—consist of sensory neurons synapsing in the spinal cord with motor neurons; three-neuron arcs consist of sensory neurons synapsing in the spinal cord with interneurons that synapse with motor neurons (Figure 8-6)

NERVE IMPULSES

A Definition—self-propagating wave of electrical disturbance that travels along the surface of a neuron membrane
B Mechanism
  1 A stimulus increases the permeability of the neuron membrane to positive sodium ions
  2 Inward movement of positive sodium ions leaves a slight excess of negative ions outside at a stimulated point; marks the beginning of a nerve impulse

THE SYNAPSE

A Definition—chemical compounds released from axon terminals (of a presynaptic neuron) into a synaptic cleft
B Neurotransmitters bind to specific receptor molecules in the membrane of a postsynaptic neuron, thereby stimulating an impulse conduction by the membrane
C Names of neurotransmitters—acetylcholine, catecholamines (norepinephrine, dopamine, and serotonin), and other compounds
D Parkinson's disease—characterized by abnormally low levels of dopamine in motor control areas of the brain; patients usually exhibit involuntary trembling and muscle rigidity

CENTRAL NERVOUS SYSTEM

A Divisions of the brain (Figure 8-11 and Table 8-1)
  1 Brain stem
    a Consists of three parts of brain; named in ascending order, they are the medulla oblongata, pons, and midbrain

b Structure—white matter with bits of gray matter scattered through it

c Function—gray matter in the brain stem functions as reflex centers (for example, for heartbeat, respirations, and blood vessel diameter); sensory tracts in the brain stem conduct impulses to the higher parts of the brain; motor tracts conduct from the higher parts of the brain to the spinal cord

2 Diencephalon

a Structure and function of the hypothalamus

   (1) Consists mainly of the posterior pituitary gland, pituitary stalk, and gray matter

   (2) Acts as the major center for controlling the ANS; therefore helps control the functioning of most internal organs

   (3) Controls hormone secretion by anterior and posterior pituitary glands; therefore indirectly helps control hormone secretion by most other endocrine glands

   (4) Contains centers for controlling appetite, wakefulness, pleasure, etc.

b Structure and function of the thalamus

   (1) Dumbbell-shaped mass of gray matter in each cerebral hemisphere

   (2) Relays sensory impulses to cerebral cortex sensory areas

   (3) In some way produces the emotions of pleasantness or unpleasantness associated with sensations

3 Cerebellum

a Second largest part of the human brain

b Helps control muscle contractions to produce coordinated movements so that we can maintain balance, move smoothly, and sustain normal postures

4 Cerebrum

a Largest part of the human brain

b Outer layer of gray matter is the cerebral cortex; made up of lobes; composed mainly of dendrites and cell bodies of neurons

c Interior of the cerebrum composed mainly of white matter (that is nerve fibers arranged in bundles called *tracts*)

d Functions of the cerebrum—mental processes of all types, including sensations, consciousness, memory, and voluntary control of movements

5 Brain disorders

a Destruction of brain tissue

   (1) Cerebrovascular accident (CVA)—hemorrhage from or cessation of blood flow through cerebral blood vessels; a "stroke"

   (2) Cerebral palsy—condition in which damage to motor control areas of the brain before, during, or shortly after birth causes paralysis (usually spastic) of one or more limbs

   (3) Dementia—syndrome that includes progressive loss of memory, shortened attention span, personality changes, reduced intellectual capacity, and motor control deficit

     (a) Alzheimer's disease—brain disorder of the middle and late adult years characterized by dementia

     (b) Huntington's disease—inherited disorder characterized by chorea (purposeless movement) progressing to severe dementia

     (c) HIV virus (that also causes AIDS) can infect neurons and thus cause dementia

b Seizure disorders

   (1) Seizure—sudden burst of abnormal neuron activity that results in temporary changes in brain function

   (2) Epilepsy—many forms, all characterized by recurring seizures

   (3) Electroencephalogram—graphic representation of voltage changes in the brain used to evaluate brain activity

B Spinal cord (Figure 8-14)

1 Outer part is composed of white matter made up of many bundles of axons called *tracts;* interior composed of gray matter made up mainly of neuron dendrites and cell bodies

2 Functions as the center for all spinal cord reflexes; sensory tracts conduct impulses to the brain, and motor tracts conduct impulses from the brain

C Coverings and fluid spaces of the brain and spinal cord

1 Coverings

a Cranial bones and vertebrae

b Cerebral and spinal meninges—the dura mater, the pia mater, and the arachnoid (Figure 8-16)

c Meningitis—inflammation of meninges, often resulting from infection

2 Fluid spaces—subarachnoid spaces of meninges, central canal inside cord, and ventricles in brain (Figure 8-17)

PERIPHERAL NERVOUS SYSTEM

A Cranial nerves (Figure 8-20 and Table 8-2)

1 Twelve pairs—attached to undersurface of the brain

2 Connect brain with the neck and structures in the thorax and abdomen

B Spinal nerves

1 Structure—contain dendrites of sensory neurons and axons of motor neurons

2 Functions—conduct impulses necessary for sensations and voluntary movements

C Peripheral nerve disorders

1 Neuritis—general term referring to nerve inflammation

a Sciatica is inflammation of the sciatic nerve that innervates the legs

b Neuralgia, or muscle pain, often accompanies neuritis

2 Trigeminal neuralgia—recurring episodes of stabbing pain along one or more branches of the trigeminal (fifth cranial) nerve in the head

3 Bell's palsy—paralysis of facial features resulting from damage to the facial (seventh cranial) nerve

4 Herpes zoster or shingles

a Viral infection caused by chickenpox virus that has invaded the dorsal root ganglion and remained dormant until an episode of shingles

b Usually affects a single dermatome, producing characteristic painful plaques or vesicles

AUTONOMIC NERVOUS SYSTEM

A Autonomic nervous system—motor neurons that conduct impulses from the central nervous system to cardiac muscle, smooth muscle, and glandular epithelial tissue; regulates the body's automatic or involuntary functions (Figure 8-24)

B Autonomic neurons—preganglionic autonomic neurons conduct from spinal cord or brain stem to an autonomic ganglion; postganglionic neurons conduct from autonomic ganglia to cardiac muscle, smooth muscle, and glandular epithelial tissue

C Autonomic or visceral effectors—tissues to which autonomic neurons conduct impulses (that is, cardiac and smooth muscle and glandular epithelial tissue)

D Composed of two divisions—the sympathetic system and the parasympathetic system

E Autonomic conduction paths

1 Consist of two-neuron relays (that is, preganglionic neurons from the central nervous system to autonomic ganglia, synapses, postganglionic neurons from ganglia to visceral effectors)

2 In contrast, somatic motor neurons conduct all the way from the CNS to somatic effectors with no intervening synapses

F Sympathetic nervous system

1 Structure

a Dendrites and cell bodies of sympathetic preganglionic neurons are located in the gray matter of the thoracic and upper lumbar segments of the spiral cord

b Axons leave the spinal cord in the anterior roots of spinal nerves, extend to sympathetic or collateral ganglia, and synapse with several postganglionic neurons whose axons extend to spinal or autonomic nerves to terminate in visceral effectors

c A chain of sympathetic ganglia is in front of and at each side of the spinal column

2 Functions

a Serves as the emergency or stress system, controlling visceral effectors during strenuous exercise and strong emotions (anger, fear, hate, or anxiety)

b Group of changes induced by sympathetic control is called the *fight-or-flight syndrome*

G Parasympathetic nervous system

1 Structure

a Parasympathetic preganglionic neurons have dendrites and cell bodies in the gray matter of the brain stem and of the sacral segments of the spinal cord

b Parasympathetic preganglionic neurons terminate in parasympathetic ganglia located in the head and the thoracic and abdominal cavities close to visceral effectors

c Each parasympathetic preganglionic neuron synapses with postganglionic neurons to only one effector

2 Function—dominates control of many visceral effectors under normal, everyday conditions

H Autonomic neurotransmitters

1 Cholinergic fibers—preganglionic axons of parasympathetic and sympathetic systems and parasympathetic postganglionic axons release acetylcholine

2 Adrenergic fibers—axons of sympathetic postganglionic neurons release norepinephrine (noradrenaline)

I Autonomic nervous system as a whole

1 Regulates the body's automatic functions in ways that maintain or quickly restore homeostasis

2 Many visceral effectors are doubly innervated (That is, they receive fibers from parasympathetic and sympathetic divisions and are influenced in opposite ways by the two divisions)

J Disorders of the autonomic nervous system

1 Stress-induced disease

a Prolonged or excessive response to stress can disrupt normal functioning throughout the body

b Examples of stress-induced conditions

(1) Heart disease

(2) Digestive problems

(3) Reduced resistance to disease

2 Neuroblastoma—highly malignant tumor of the sympathetic nervous system, primarily affecting young children

NEW WORDS

acetylcholine
arachnoid
astrocytes
axon
blood-brain barrier
catecholamines
dendrite
dopamine
dura mater
effectors
endorphins

enkephalins
fight-or-flight
   syndrome
ganglion
interneuron
meninges
microglia
motor neuron
multiple sclerosis
myelin
neuroglia

neurons
neurotransmitter
node of Ranvier
norepinephrine
oligodendroglia
parasympathetic
   system
pia mater
postganglionic neurons
postsynaptic neuron
preganglionic neuron

presynaptic neuron
receptors
reflex arc
saltatory conduction
sensory neuron
serotonin
sympathetic system
synapse
synaptic cleft
tract

Diseases and Other Clinical Terms

Alzheimer's disease
anesthesia
Bell's palsy
cerebral palsy
cerebrovascular
   accident (CVA)
dementia
electroencephalogram
   (EEG)

epilepsy
glioma
hemiplegia
herpes zoster (shingles)
Huntington's disease
   (HD)
hydrocephalus
meningitis

multiple sclerosis (MS)
multiple
   neurofibromatosis
neuralgia
neuritis
neuroblastoma
neuroma
paraplegia

Parkinson's disease
quadriplegia
sciatica
seizure
spastic paralysis
trigeminal neuralgia
triplegia

CHAPTER TEST

1. The nervous system as a whole is divided into two principal divisions: the _____ and _____ nervous systems.

2. The tough, fluid-containing membrane surrounding the brain and spinal cord is called the _____.

3. The medical name describing an accumulation of CSF in the ventricles of the brain is _____.

4. The two principal types of specialized cells found in the nervous system are _____ and _____.

5. Nervous system cells that form the myelin sheaths around nerve fibers in the brain and spinal cord are called _____.

6. Nervous impulses travel _____ from a nerve cell body in a single process called the _____.

7. The microscopic space that separates the axon endings of one neuron from the dendrites of another neuron is called a _____.

8. Chemical compounds released from axon terminals into synaptic clefts are called _____.

9. An area of skin supplied by a single spinal nerve is called a _____.

10. Two natural, morphinelike painkillers produced in the brain are called _____ and _____.

11. The two major structures that make up the diencephalon are called the _____ and _____.

12. There are _____ pairs of cranial nerves and _____ pairs of spinal nerves.

13. Motor neurons in the ANS conduct impulses to three kinds of tissues: _____ _____, _____ _____, and _____ _____.

14. The ANS consists of two divisions called the _____ and _____ divisions.

15. Strong emotions and strenuous exercise activate the _____ division of the ANS.

Select the most correct answer from Column B for each statement in Column A. (Only one answer is correct.)

COLUMN A

16. _____ Meninges
17. _____ Transmit impulses to a neuron
18. _____ Capable of phagocytosis
19. _____ Produce myelin
20. _____ Natural pain killer
21. _____ Brain tumor
22. _____ Nerve cell bodies
23. _____ Largest part of the brain
24. _____ Fight-or-flight syndrome

COLUMN B

a. Microglia
b. Schwann cells
c. Pia mater
d. Response to stress
e. Endorphin
f. Gray matter
g. Cerebrum
h. Dendrites
i. Glioma

Circle the T before each true statement and the F before each false statement.

T F 25. Most neuromas are gliomas.
T F 26. Parkinson's disease is treated with dopamine, which crosses the blood-brain barrier easily.
T F 27. Cerebrovascular accidents are commonly known as *strokes.*
T F 28. Alzheimer's disease, HIV infection, and Huntington's disease may all result in dementia.
T F 29. All forms of epilepsy are inherited.
T F 30. Shingles is caused by the same pathogen that causes chickenpox.

REVIEW QUESTIONS

1 What general function does the nervous system perform?
2 What other system performs the same general function as the nervous system?
3 What general functions does the spinal cord perform?
4 What does *CNS* mean? *PNS?*
5 What are the meninges?
6 Why is the medulla considered the most vital part of the brain?
7 What general functions does the cerebellum perform?
8 What general functions does the cerebrum perform?
9 Explain how MS impairs nerve function.
10 Describe Parkinson's disease. Explain how L-dopa helps this condition.
11 List three conditions that can produce severe dementia. How are they alike? How are they different?
12 What general functions do spinal nerves perform?
13 Define briefly each of the terms listed under "New Words."
14 What is a seizure? How is brain activity measured by clinicians and researchers?
15 Identify interneuron, motor neuron, reflex center, sensory neuron, somatic motor neuron, and synapse.
16 Explain the fight-or-flight syndrome.
17 How can stress cause disease?
18 Compare parasympathetic and sympathetic functions.
19 Contrast visceral and somatic effectors.

CLINICAL APPLICATIONS

1 Tony's teachers describe him as a daydreamer. The teachers often find him staring off into space when they are trying to get his attention. When Tony's parents mentioned this to their family physician, the possibility of epilepsy was brought up by the physician. Could Tony's daydreaming be a sign of epilepsy? What test could help confirm such a diagnosis? What signs would one look for in such a test if epilepsy is present?

2 Over the last few years, your friend Angela has developed fibrous nodules in many areas of her skin. She recently confided that she has an inherited disorder of the nervous system that causes these bumps. What disease might Angela have? How can a nervous disorder cause skin lesions?

The Senses

CHAPTER

9

Objectives

After you have completed this chapter,
you should be able to:

1 Classify sense organs as special or general and explain the basic differences between the two groups.

2 Discuss how a stimulus is converted into a sensation.

3 List the major senses.

4 Describe the structure of the eye and the functions of its components.

5 Name and describe the major visual disorders.

6 Discuss the anatomy of the ear and its sensory function in hearing and equilibrium.

7 Name and describe the major forms of hearing impairment.

8 Discuss the chemical receptors and their functions.

9 Discuss the general sense organs and their functions.

*I*f you were asked to name the sense organs, what organs would you name? Can you think of any besides the eyes, ears, nose, and taste buds? Actually there are millions of other sense organs throughout the body in our skin, internal organs, and muscles. They constitute the many sensory receptors that allow us to respond to stimuli such as touch, pressure, temperature, and pain. These microscopic receptors, you will recall, are located at the tips of dendrites of sensory neurons.

Our ability to "sense" changes in our external and internal environments is a requirement for maintaining homeostasis and for survival itself. We can initiate protective reflexes important to homeostasis only if we can sense a change or danger. The problem or danger may approach the body from a distance, such as a falling tree. External dangers may be detected by sight or hearing. If the danger is internal, such as overstretching a muscle, detecting an increase in body temperature (fever), or sensing the pain caused by an ulcer, other receptors make us aware of the problem and permit us to take appropriate action to maintain homeostasis.

Classification of Sense Organs

The sense organs are often classified as: **special** sense organs and **general** sense organs. Special sense organs, such as the eye, are characterized by large and complex organs or by localized groupings of specialized receptors in areas such as the nasal mucosa or tongue. The general sense organs for detecting stimuli such as pain and touch are microscopic receptors widely distributed throughout the body. Other general sense organs include receptors that indicate the tension on our muscles and tendons so that we can maintain balance and muscle tone and be aware of the positions of our body parts. Table 9-1 classifies the special sense organs. Table 9-2 on p. 252 classifies the general sense organs.

Converting a Stimulus into a Sensation

All sense organs, regardless of size, type, or location, have in common some important functional characteristics. First, they must be able to sense or detect a stimulus in their environment. Of course, different sense organs detect and respond to different types of stimuli in different ways. Whether it is light, sound, temperature change, mechanical pressure, or the presence of chemicals identified as taste or smell, the stimulus must be changed into an electrical signal or nerve impulse. This signal is then transmitted over a nervous system "pathway" to the brain, where the sensation is perceived.

Special Sense Organs

THE EYE

When you look at a person's eye, you see only a small part of the whole eye. Three layers of tissue form the eyeball: the **sclera** (SKLE-rah), the **choroid** (KO-royd), and the **retina** (RET-i-nah) (Figure 9-1). The outer layer of sclera consists of tough fibrous tissue. The "white" of the eye is part of the front surface of the sclera. The other part of the front surface of the sclera is called the *cornea* and is sometimes spoken of as the window of the eye because of its transparency. At a casual glance, however, it does not look transparent but appears blue, brown, gray, or green because it lies over the **iris,** the colored part of the eye. A mucous membrane known as the **conjunctiva** (kon-junk-TEE-vah) lines the eyelids and covers the sclera in front. The conjunctiva is kept moist by tears formed in the **lacrimal gland** located in the upper lateral portion of the orbit.

The middle layer of the eyeball, the *choroid,* contains a dark pigment to prevent the scattering of incoming light rays. Two involuntary muscles make up the front part of the choroid. One is the *iris,* the colored structure seen through the cornea, and the other is the *ciliary muscle* (Figure 9-1). The black center of the iris is really a hole in this doughnut-shaped muscle; it is the **pupil** of the eye. Some of the fibers of the iris are arranged like spokes in a wheel. When they contract the pupils dilate, letting in more light rays. Other fibers are circular. When they contract, the pupils constrict, letting in fewer light rays. Normally, the pupils constrict in bright light and dilate in dim light. When we look at distant objects, the ciliary muscle is relaxed, and the lens has only a slightly curved shape. To focus on near objects, however,

▽ **FIGURE 9-1 Horizontal Section Through the Left Eyeball.** The eye is viewed from above.

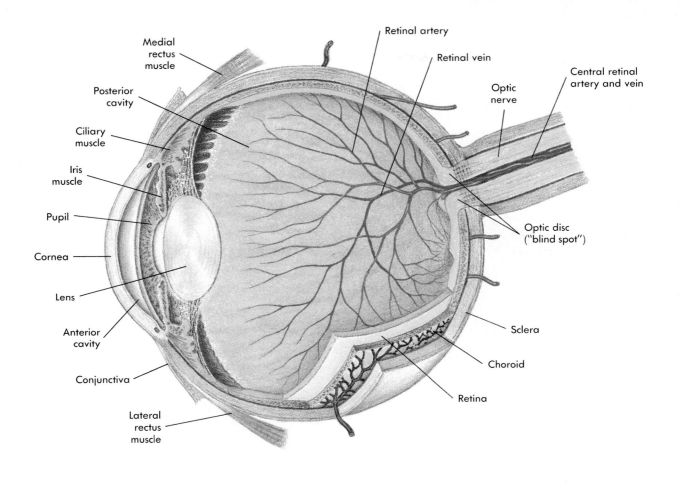

▽ **TABLE 9-1** Special Sense Organs

SENSE ORGAN	SPECIFIC RECEPTOR	TYPE OF RECEPTOR	SENSE
Eye	Rods and cones	Photoreceptor	Vision
Ear	Organ of Corti	Mechanoreceptor	Hearing
	Cristae ampullares	Mechanoreceptor	Balance
Nose	Olfactory cells	Chemoreceptor	Smell
Taste buds	Gustatory cells	Chemoreceptor	Taste

FIGURE 9-2 Cells of the Retina. Photoreceptors called *rods* and *cones* (notice their shapes) detect changes in light and relay the information to bipolar neurons. The bipolar cells, in turn, conduct the information to ganglion cells. The information eventually leaves the eye by way of the optic nerve.

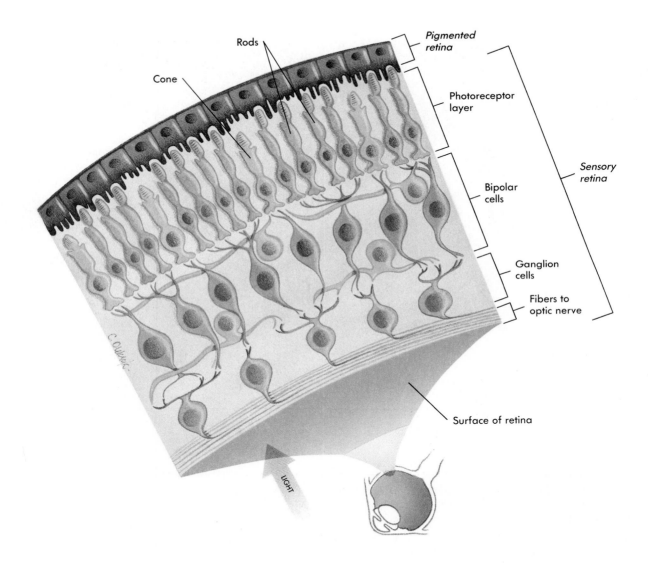

the ciliary muscle must contract. As it contracts, it pulls the choroid coat forward toward the lens, thus causing the lens to bulge and curve even more. Most of us become more farsighted as we grow older and lose the ability to focus on close objects because our lenses lose their elasticity and can no longer bulge enough to bring near objects into focus. **Presbyopia** or "oldsightedness" is the name for this condition.

The *retina* or innermost layer of the eyeball contains microscopic receptor cells, called *rods* and *cones* because of their shapes. Dim light can stimulate the rods, but fairly bright light is necessary to stimulate the cones. In other words, **rods** are the receptors for night vision and **cones** for daytime vision. There are three kinds of cones; each is sensitive to a different color: red, green, or blue. Scattered throughout the central portion of the retina, these three types of cones allow us to distinguish between different colors.

Fluids fill the hollow inside of the eyeball. They maintain the normal shape of the eyeball and help refract light rays; that is, the fluids bend light rays to bring them to focus on the retina. **Aqueous humor** is the name of the watery fluid in front of the lens (in the anterior cavity of the eye), and **vitreous humor** is the name of the jellylike fluid behind the lens (in the posterior cavity). Aqueous humor is constantly being formed, drained, and replaced in the anterior cavity. If drainage is blocked for any reason, the internal pressure within the eye will increase, and damage that could lead to blindness will occur. This condition is called **glaucoma** (glaw-KO-mah).

The **lens** of the eye lies directly behind the pupil. It is held in place by a ligament attached to the ciliary muscle. In most young people, the lens is transparent and somewhat elastic so that it is capable of changing shape. Exposure to ultraviolet (UV) radiation in sunlight causes the lens to become hard and milky. Over the years, repeated exposure to sunlight may cause **cataracts** or milky spots on the lens. Large or numerous cataracts may cause blindness. Cataracts can be removed surgically and replaced with artificial lenses. Many scientists believe that wearing sunglasses that filter UV radiation from an early age will help prevent the formation of cataracts.

Visual Pathway

Light is the stimulus that results in vision (that is, our ability to see objects as they exist in our environment). Light enters the eye through the pupil and is *refracted* or bent so that it is focused on the retina. Refraction occurs as light passes through the cornea, the aqueous humor, the lens, and the vitreous humor on its way to the retina.

The innermost layer of the retina contains the rods and cones, which are the *photoreceptor* cells of the eye (Figure 9-2). They respond to a light stimulus by producing a nervous impulse. The rod and cone photoreceptor cells synapse with neurons in the bipolar and ganglionic layers of the retina. Nervous signals eventually leave the retina and exit the eye through the optic nerve on the posterior surface of the eyeball. No rods or cones are present in the area of the retina where the optic nerve fibers exit. The result is a "blind spot" known as the *optic disc* (Figure 9-1).

After leaving the eye, the optic nerves enter the brain and travel to the visual cortex of the occipital lobe. In this area of the brain, *visual interpretation* of the nervous impulses that were generated by light stimuli in the rods and cones of the retina result in "seeing."

VISUAL DISORDERS

Healthy vision requires three basic processes: formation of an image on the retina (refraction), stimulation of rods and cones, and conduction of nerve impulses to the brain. Malfunction of any of these processes can disrupt this chain of processes, producing a visual disorder.

Refraction Disorders

Focusing a clear image on the retina is essential for good vision. In the normal eye, light rays enter the eye and are focused into a clear, upside-down image on the retina (Figure 9-3, *A*). The brain can easily right the upside-down image in our conscious perception but cannot correct an image that is not sharply focused. If our eyes are elongated the image focuses in front of the retina rather than on it. The retina receives only a fuzzy image. This condition, called **myopia** or *nearsightedness*, can be corrected by using contact lenses or glasses (Figure 9-3, *B* and *C*). If our eyes are shorter than normal, the image focuses behind the reti-

FIGURE 9-3 Refraction. The upper figure shows how light is refracted in the normal eye to form a well-focused image **(A).** The lower figures show the abnormal and corrected refraction observed in myopia **(B** and **C)** and hyperopia **(D** and **E).**

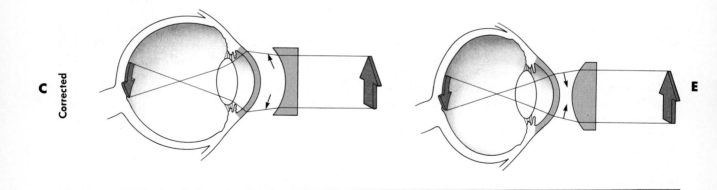

na, also producing a fuzzy image. This condition, called **hyperopia** or *farsightedness,* can also be corrected by lenses (Figure 9-3, *D* and *E*).

A variety of other conditions can prevent the formation of a clear image on the retina. For example, the inability to focus the lens properly as we age, or presbyopia, has already been discussed. Older individuals can compensate for presbyopia by using "reading glasses" when near vision is needed. An irregularity in the cornea or lens, a condition called **astigmatism** (a-STIG-ma-tizm), can also be corrected with glasses or contact lenses. *Cataracts,* cloudy spots in the the eye's lens that develop as we age, may also interfere with focusing. Cataracts are especially troublesome in dim light because weak beams of light cannot pass through the cloudy spots the way some brighter light can. This fact accounts for the trouble many older adults have with their "night vision." Although the tendency to develop cataracts is inherited, the disorder is relatively simple to treat with lens implant surgery.

Infections of the eye and its associated structures also have the potential to impair vision, sometimes permanently. Most eye infections start out in the conjunctiva, producing an inflammation response known as "pink-eye" or **conjunctivitis** (kon-junk-ti-VYE-tis). You may recall from Chapter 4 that a variety of different pathogens can cause conjunctivitis. For example, the bacterium *Chlamydia trachomatis* that commonly infects the reproductive tract can cause a chronic infection called *chlamydial conjunctivitis* or **trachoma**. Because chlamydia and other pathogens often inhabit the birth canal, antibiotics are routinely applied to the eyes of newborns to prevent conjunctivitis. Highly contagious *acute bacterial conjunctivitis,* characterized by drainage of a mucous pus (Figure 9-4), is most commonly caused by bacteria such as *Staphylococcus* and *Haemophilus*. Conjunctivitis may produce lesions on the inside of the eyelid that can damage the cornea and thus impair vision. Occasionally infections of the conjunctiva spread to the tissues of the eye proper and cause permanent injury—even total blindness. Besides infection, conjunctivitis may also be caused by allergies. The red, itchy, watery eyes commonly associated with allergic reactions to pollen and other substances results from an allergic inflammation response of the conjunctiva. Allergy and hypersensitivity reactions are discussed further in Chapter 14.

Yet another manner in which refraction can be disrupted occurs when one eye does not focus on the same object as the other eye. Normally, we use *binocular* ("two-eyed") vision in which both eyes aim toward the same object at the same time. Because the eyes are separated by a short distance, the images formed in each eye do not match exactly—a fact that allows *depth perception.* If the positioning of the eyes cannot be coordinat-

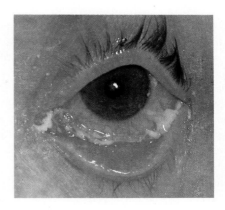

FIGURE 9-4 Acute Bacterial Conjunctivitis. Notice the discharge of mucous pus characteristic of this highly contagious infection of the conjunctiva.

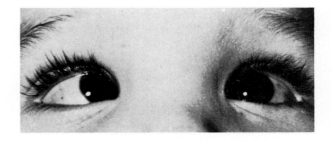

FIGURE 9-5 Strabismus. This child exhibits convergence of the eyes or "cross-eyes"—a common form of strabismus.

ed, a condition called **strabismus** (strah-BIS-mus) results. Strabismus or "cross-eyes" is a common condition in which the eyes are so far off in their center of focus that the brain cannot mesh the two resulting images into a single picture. In some cases, the eyes seem to diverge or face outward to the side, and in others the eyes converge or cross. The photo in Figure 9-5 shows a child with convergent strabismus. Usually the brain compensates for missing or unusual elements in the visual field, but in strabismus the abnormality is too severe for the brain to compensate. If this condi-

VISUAL ACUITY

Visual *acuity* is the clearness or sharpness of visual perception. Acuity is affected by our focusing ability, the efficiency of the retina, and the proper function of the visual pathway and processing centers in the brain.

One common way to measure visual acuity is to use the familiar test chart on which letters or other objects of various sizes and shapes are printed. The subject is asked to identify the smallest object that he or she can see from a distance of 20 feet (6.1 m). The resulting determination of visual acuity is expressed as a double number such as "20-20." The first number represents the distance (in feet) between the subject and the test chart—the standard being 20. The second number represents the number of feet a person with normal acuity would have to stand to see the same objects clearly. Thus a finding of 20-20 is normal because the subject can see at 20 feet what a person with normal acuity can see at 20 feet. A person with 20-100 vision can see objects at 20 feet that a person with normal vision can see at 100 feet.

People whose acuity is worse than 20-200 after correction are considered to be *legally blind*. Legal blindness is the designation used to identify the severity of a wide variety of visual disorders so that laws that involve visual acuity can be enforced. For example, laws that govern the awarding of driving licenses require that drivers have a minimum level of visual acuity.

tion is not corrected early, visual centers in the brain will learn to ignore information from one eye—causing permanent blindness in the affected eye. Strabismus is usually caused by paralysis, weakness, or other abnormality affecting the external muscles of the eye (see Figure 9-1). Whatever the cause, strabismus can often be corrected by exercises that train the eyes to focus together, by corrective lenses, or by corrective surgery.

Disorders of the Retina

Damage to the retina impairs vision because even a well-focused image cannot be perceived if some or all of the light receptors do not function properly. For example, in a condition called *retinal detachment,* part of the retina falls away from the tissue supporting it. This condition often results from normal aging, eye tumors, or from sudden blows to the head—as in a sporting injury. Common warning signs include the sudden appearance of floating spots that may decrease over a period of weeks and odd "flashes of light" that appear when the eye moves. If left untreated, the retina may detach completely and cause total blindness in the affected eye. A number of treatments are available for correcting retinal detachment. A traditional approach is laser therapy. A newer approach involves placing a tight collar around the eyeball to increase pressure within the eye. The high pressure of the vitreous humor holds the retina in place against the rear of the eyeball.

Diabetes mellitus, a disorder involving the hormone insulin, may cause a condition known as **diabetic retinopathy** (ret-in-OP-ah-thee). In this disorder, diabetes causes small hemorrhages in retinal blood vessels that disrupts the oxygen supply to the photoreceptors. The eye responds by building new but abnormal vessels that block vision and may cause detachment of the retina. Diabetic retinopathy is considered one of the leading causes of blindness in the United States. Fortunately, treatments developed over the last two decades have improved the outlook in this regard. For example, a type of laser therapy in which laser beams are used to seal off hemorrhaging retinal vessels has been used successfully in many cases.

Another common condition that can damage the retina is *glaucoma*. Recall that glaucoma is excessive *intraocular* (in-trah-AHK-yoo-lar) *pressure* caused by abnormal accumulation of aqueous humor. As fluid pressure against the retina increases above normal, blood flow through the retina slows. Reduced blood flow causes degeneration of the retina and thus a loss of vision. Although acute forms of glaucoma can occur, most cases develop slowly over a period of years. This chronic form may not produce symptoms, especially in its early stages. For this reason, routine eye examinations typically include a screening test for glaucoma. As chronic glaucoma progresses, damage first appears at the edges of the retina—causing a gradual loss of peripheral vision commonly known as "tunnel vision." Blurred vision and headaches may also occur. As the damage becomes more extensive, "halos" are seen around bright lights. If untreated, glaucoma eventually produces total, permanent blindness.

Degeneration of the retina can cause difficulty seeing at night or in dim light. This condition, called **nyctalopia** (nik-ta-LO-pee-ah) or "night blindness," can also be caused by a deficiency of vitamin A. Vitamin A is needed to make *photopigment* in rod cells. Photopigment is a light-sensitive chemical that triggers stimulation of the visual nerve pathway. A lack of vitamin A may result in a lack of photopigment in rods, a condition that impairs dim-light vision.

The leading cause of permanent blindness in the elderly is progressive degeneration of the central part of the retina. Called *age-related macular degeneration,* this condition affects the part of the retina that is most essential to good vision—the central region called the *macula.* The exact cause of the degeneration is unknown, but the risk for developing this condition increases with age after reaching 50. Other known risk factors include cigarette smoking and a family history of the disorder.

Retinal disorders are sometimes inherited. Most forms of **color blindness** are caused by genes on the X chromosome that produce abnormal photopigments in the cones. Each of three photopigments in cones is sensitive to one of the primary colors of light: green, blue, and red. In many color-blind individuals, the green-sensitive photopigment is missing or deficient; at other times, the red-sensitive photopigment is abnormal. (Deficiency of the blue-sensitive photopigment is very rare.) Color-blind individuals see colors, but they cannot distinguish between them normally. Because color blindness is an X-linked genetic trait, more men than women have this condition (see Chapter 23). Although color blindness is an abnormality, it is not usually considered a clinical disease.

Colored figures are often used to screen individuals for color blindness (Figure 9-6). A person with red-green color blindness cannot see the number *74* in Figure 9-6, *A*, whereas a person with normal color vision can. To determine which photopigment is deficient, red or green, a color blind person may try a figure similar to Figure 9-6, *B*. Persons with a deficiency of red-sensitive photopigment can distinguish only the number 2; those deficient in green-sensitive photopigment can only see the number 4.

Disorders of the Visual Pathway

Damage or degeneration in the optic nerve, the brain, or any part of the visual pathway between

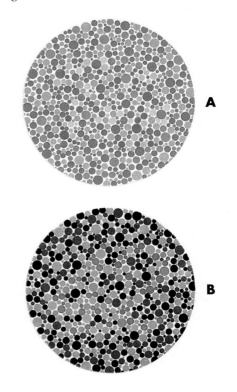

FIGURE 9-6 Color Vision Screening Figures. **A,** People with normal color vision can see 74 in this mosaic; people with red-green color blindness cannot. **B,** This mosaic is used to classify the type of red-green color blindness a patient has. If the patient sees only the 2, the red-sensitive cones are abnormal. If only the 4 is seen, the green-sensitive cones are abnormal.

LASER THERAPY

Advances in laser technology have produced a number of applications in the medical field, especially in treating eye problems. For some time now, it has been common practice to use the intense light from lasers to repair detached retinas. An **ophthalmologist** (eye physician) directs the laser beam at different points in the retina and makes tiny burns. Each burned area eventually forms a small, fibrous scar that holds the retina in place.

Laser therapy may soon replace procedures such as *radial keratotomy,* in which cuts made in the cornea cause it to change shape and thus correct refraction errors. Recently, researchers have been testing a new laser procedure for correcting refraction problems. Working without heat, a special laser that shoots "bursts" of light is used to microscopically shave the cornea, giving it a new shape. Although this therapy needs further testing, it promises to eliminate the need for eyeglasses or contact lenses in correcting many vision problems.

---

them can impair vision. For example, the pressure associated with glaucoma can also damage the optic nerve. Diabetes, already cited as a cause of retina damage, can also cause degeneration of the optic nerve.

Damage to the visual pathway does not always result in total loss of sight. Depending on where the damage occurs, only a part of the visual field may be affected. For example, a certain form of neuritis often associated with multiple sclerosis can cause loss of only the center of the visual field—a condition called **scotoma** (sko-TOE-mah).

A cerebrovascular accident (CVA) or stroke can cause vision impairment when the resulting tissue damage occurs in one of the regions of the brain that process visual information (see Figure 8-12). For example, damage to an area that processes information about colors may result in a rare condition called *acquired cortical color blind-*

ness. This condition is characterized by difficulty in distinguishing any color—not just one or two colors as in the more common inherited forms of color blindness.

THE EAR

In addition to its role in hearing, the ear also functions as the sense organ of equilibrium and balance. As we shall later see, the stimulation or "trigger" that activates receptors involved with hearing and equilibrium is mechanical, and the receptors themselves are called *mechanoreceptors* (mek-an-o-ree-SEP-tors). Physical forces that involve sound vibrations and fluid movements are responsible for initiating nervous impulses eventually perceived as sound and balance.

The ear is much more than a mere appendage on the side of the head. A large part of the ear—and by far its most important part—lies hidden from view deep inside the temporal bone. It is divided into the following anatomical areas (Figure 9-7):

1 **External ear**
2 **Middle ear**
3 **Inner (internal) ear**

External Ear

The external ear has two parts: the **auricle** (AW-ri-kul) or pinna and the **external auditory canal.** The auricle is the appendage on the side of the head surrounding the opening of the external auditory canal. The canal itself is a curving tube about 2.5 cm (1 inch) in length. It extends into the temporal bone and ends at the **tympanic** (tim-PAN-ik) **membrane** or **eardrum,** which is a partition between the external and middle ear. The skin of the auditory canal, especially in its outer one third, contains many short hairs and **ceruminous** (se-ROO-mi-nus) **glands** that produce a waxy substance called *cerumen* that may collect in the canal and impair hearing by absorbing or blocking the passage of sound waves. Sound waves traveling through the external auditory canal strike the tympanic membrane and cause it to vibrate.

Middle Ear

The middle ear is a tiny and very thin epithelium-lined cavity hollowed out of the temporal bone. It

FIGURE 9-7 The Ear. External, middle, and inner ear.

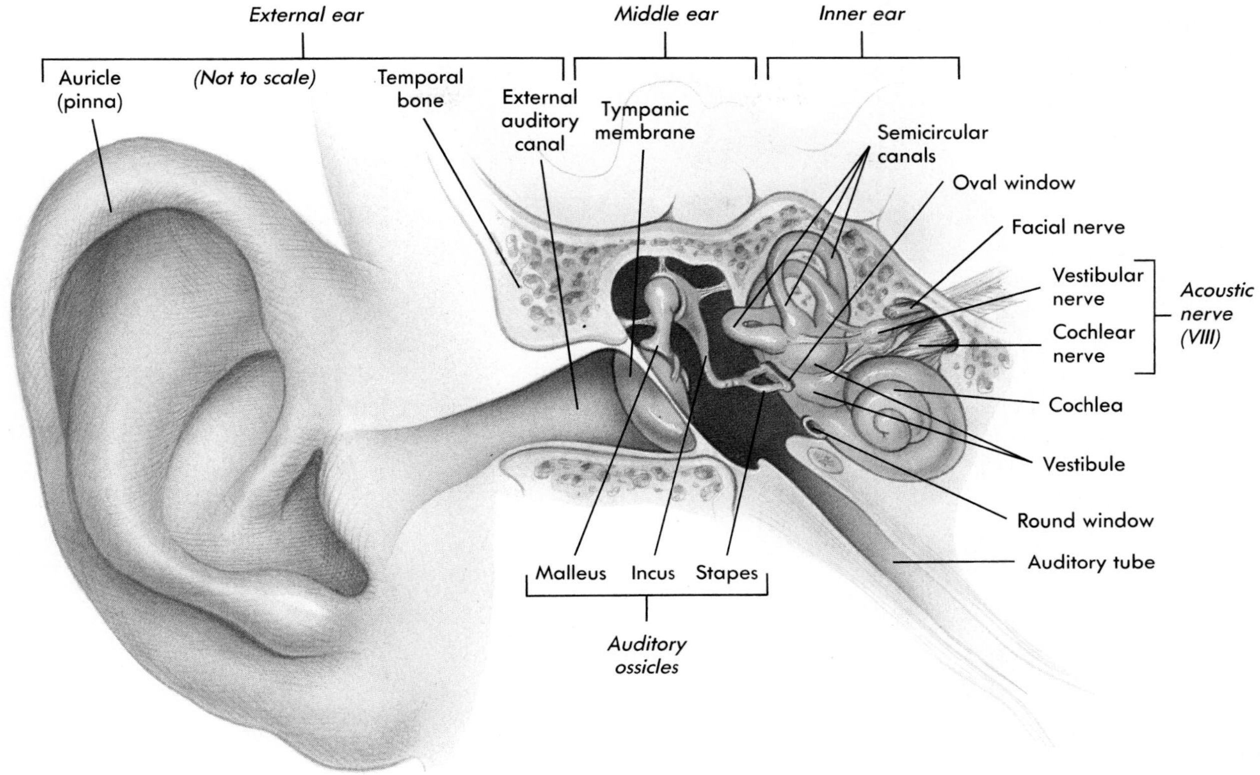

houses three very small bones. The names of these ear bones, called **ossicles** (OS-si-kuls), describe their shapes—**malleus** (hammer), **incus** (anvil), and **stapes** (stirrup). The "handle" of the malleus attaches to the inside of the tympanic membrane, and the "head" attaches to the incus. The incus attaches to the stapes, and the stapes presses against a membrane that covers a small opening, the *oval window.* The oval window separates the middle ear from the inner ear. When sound waves cause the eardrum to vibrate, that movement is transmitted and amplified by the ear ossicles as it passes through the middle ear. Movement of the stapes against the oval window causes movement of fluid in the inner ear.

A point worth mentioning, because it explains the frequent spread of infection from the throat to the ear, is the fact that a tube—the **auditory** or **eustachian** (yoo-STAY-shen) **tube**—connects the throat with the middle ear. The epithelial lining of the middle ears, auditory tubes, and throat are extensions of one continuous membrane. Consequently a sore throat may spread to produce a middle ear infection called *otitis* (o-TIE-tis) *media* (ME-dee-ah).

Inner Ear

The activation of specialized mechanoreceptors in the inner ear generates nervous impulses that result in hearing and equilibrium. Anatomically, the inner ear consists of three spaces in the temporal bone, assembled in a complex maze called the **bony labyrinth** (LAB-i-rinth). This odd-shaped bony space is filled with a watery fluid called **perilymph** (PAIR-i-limf) and is divided into the following parts: **vestibule** (VES-ti-byool),

FIGURE 9-8 The Inner Ear. The bony labyrinth (orange) is the hard outer wall of the entire inner ear, and includes semicircular canals, vestibule, and cochlea. Within the bony labyrinth is the membranous labyrinth (purple), which is surrounded by perilymph and filled with endolymph. Each ampulla in the vestibule contains a crista ampullaris that detects changes in head position and sends sensory impulses through the vestibular nerve to the brain. The inset shows a section of the membranous cochlea. Hair cells in the organ of Corti detect sound and send the information through the cochlear nerve. The vestibular and cochlear nerves join to form the eighth cranial nerve.

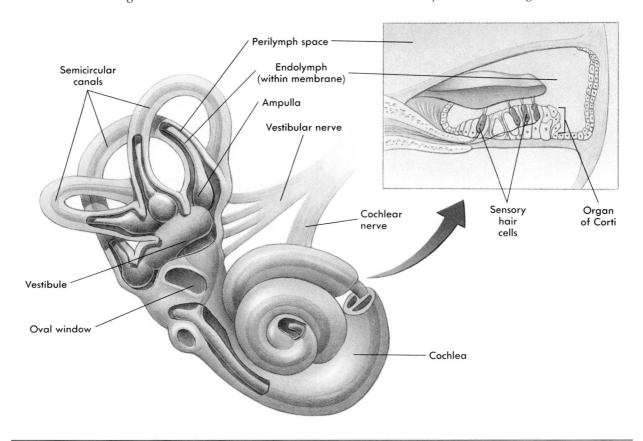

SWIMMER'S EAR

External otitis or *swimmer's ear* is a common infection of the external ear in athletes. It can be bacterial or fungal in origin and is usually associated with prolonged exposure to water. The infection generally involves, at least to some extent, the auditory canal and auricle. The ear as a whole is tender, red, and swollen. Treatment of swimmer's ear usually involves antibiotic therapy and prescription analgesics.

semicircular canals, and **cochlea** (KOK-lee-ah). The vestibule is adjacent to the oval window between the semicircular canals and the cochlea (Figure 9-8). Note in Figure 9-8 that a balloonlike membranous sac is suspended in the perilymph and follows the shape of the bony labyrinth much like a "tube within a tube." This is the **membranous labyrinth,** and it is filled with a thicker fluid called **endolymph** (EN-doe-limf).

The specialized mechanoreceptors for balance and equilibrium are located in the three semicircular canals and the vestibule. The three half-circle semicircular canals are oriented at right angles to one another (Figure 9-8). Within each canal is a

▼ **FIGURE 9-9 Effect of Sound Waves on Cochlear Structures.** Sound waves strike the tympanic membrane and cause it to vibrate. This vibration causes the membrane of the oval window to vibrate. This vibration causes the perilymph in the bony labyrinth of the cochlea to move, which causes the endolymph in the membranous labyrinth of the cochlea or cochlear duct to move. This movement of endolymph stimulates hair cells on the organ of Corti to generate a nerve impulse. The nerve impulse travels over the cochlear nerve, which becomes a part of the eighth cranial nerve. Eventually, nerve impulses reach the auditory cortex and are interpreted as sound.

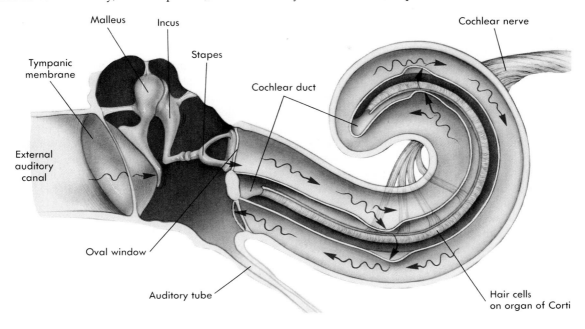

specialized receptor called a **crista** (KRIS-tah) **ampullaris** (am-pyoo-LAIR-is), which generates a nerve impulse when you move your head. The sensory cells in the cristae ampullares have hair-like extensions that are suspended in the endolymph. The sensory cells are stimulated when movement of the head causes the endolymph to move, thus causing the hairs to bend. Nerves from other receptors in the vestibule join those from the semicircular canals to form the **vestibular nerve,** a division of the acoustic nerve or cranial nerve VIII (Figure 9-8). Eventually, nervous impulses passing through this nerve reach the cerebellum and medulla. Other connections from these areas result in impulses reaching the cerebral cortex.

The organ of hearing, which lies in the snail-shaped cochlea, is the **organ of Corti** (KOR-tie). It is surrounded by endolymph filling the membranous cochlea or **cochlear duct,** which is the membranous tube within the bony cochlea. Specialized hair cells on the organ of Corti generate nerve impulses when they are bent by the movement of endolymph set in motion by sound waves (Figures 9-8 and 9-9).

HEARING DISORDERS

Hearing problems can be divided into two basic categories: *conduction impairment* and *nerve impairment.* Conduction impairment refers to the blocking of sound waves as they travel through the external and middle ear to the sensory receptors of the inner ear (the conduction pathway). Nerve impairment results in insensitivity to sound because of inherited or acquired nerve damage.

The most obvious cause of conduction impairment is blockage of the external auditory canal. Waxy buildup of cerumen commonly blocks conduction of sound toward the tympanic membrane. Foreign objects, tumors, and other matter can block conduction in the external or middle ear. An inherited bone disorder called **otosclero-**

sis (o-toe-skle-RO-sis) impairs conduction by causing structural irregularities in the stapes. Otosclerosis usually first appears during childhood or early adulthood as **tinnitus** (tin-EYE-tus) or "ringing in the ear."

Temporary conduction impairment often results from ear infection or **otitis**. As stated earlier, the structure of the auditory tube makes the middle ear prone to bacterial or viral *otitis media*. Otitis media often produces swelling and pus formation that blocks the conduction of sound through the middle ear. Permanent damage to structures of the middle ear occasionally occurs in severe cases. Untreated otitis media can lead to mastoiditis.

Hearing loss due to nerve impairment is common in the elderly. Called **presbycusis** (pres-be-KYOO-sis), this progressive hearing loss associated with aging results from degeneration of nerve tissue in the ear and vestibulocochlear nerve. A similar type of hearing loss occurs after chronic exposure to loud noises that damages receptors in the organ of Corti. Because different sound *frequencies* (tones) stimulate different regions of the organ of Corti, hearing impairment is limited to only frequencies associated with the damaged portion of the organ of Corti. For example, the portion of the organ of Corti that degenerates first in presbycusis is normally stimulated by high-frequency sounds. Thus the inability to hear high-pitched sounds is common among the elderly.

Nerve damage can also occur in **Meniere's** (may-nee-ERZ) **disease,** a chronic inner ear disease of unknown cause. Meniere's disease is characterized by tinnitus, progressive nerve deafness, and *vertigo* (sensation of spinning).

We have seen how the rods and cones of the eye, called *photoreceptors*, are stimulated by light and how specialized mechanoreceptors, which respond to movement, are involved with hearing and equilibrium. Another category of receptors, the *chemoreceptors* (kee-mo-ree-SEP-tors), respond to chemicals. This type of receptor is responsible for our sense of taste and smell.

THE TASTE RECEPTORS

The chemical receptors that generate nervous impulses resulting in the sense of taste are called

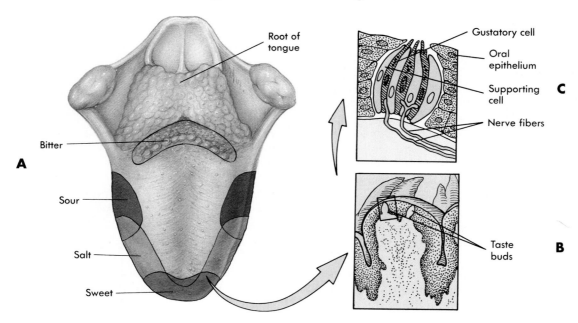

▼ **FIGURE 9-10 The Tongue. A,** Dorsal surface and regions sensitive to various tastes. **B,** Section through a papilla with taste buds on the side. **C,** Enlarged view of a section through a taste bud.

taste buds. About 10,000 of these microscopic receptors are found on the sides of much larger structures on the tongue called **papillae** (pah-PIL-ee) and also as portions of other tissues in the mouth and throat. Nervous impulses are generated by specialized cells in taste buds, called **gustatory** (GUS-tah-toe-ree) **cells.** They respond to dissolved chemicals in the saliva that bathe the tongue and mouth tissues (Figure 9-10).

Only four kinds of taste sensations—sweet, sour, bitter, and salty—result from stimulation of taste buds. All other flavors result from a combination of taste bud and olfactory receptor stimulation. In other words, the myriads of tastes recognized are not tastes alone but tastes plus odors. For this reason a cold that interferes with the stimulation of the olfactory receptors by odors from foods in the mouth markedly dulls taste sensations. Nervous impulses generated by stimulation of taste buds travel primarily through two cranial nerves (VII and IX) to end in the specialized taste area of the cerebral cortex.

THE SMELL RECEPTORS

The chemical receptors responsible for the sense of smell are located in a small area of epithelial tissue in the upper part of the nasal cavity (Figure 9-11). The location of the **olfactory receptors** is somewhat hidden, and we are often forced to forcefully sniff air to smell delicate odors. Each olfactory cell has a number of specialized cilia that sense different chemicals and cause the cell to respond by generating a nervous impulse. To be detected by olfactory receptors, chemicals must be dissolved in the watery mucus that lines the nasal cavity.

Although the olfactory receptors are extremely sensitive (that is, stimulated by even very slight odors), they are also easily fatigued—a fact that explains why odors that are at first very noticeable are not sensed at all after a short time. After the olfactory cells are stimulated by odor-causing chemicals, the resulting nerve impulse travels through the olfactory nerves in the olfactory bulb and tract and then enters the thalamic and olfac-

FIGURE 9-11 Olfactory Structures. Gas molecules stimulate olfactory cells in the nasal epithelium. Sensory information is then conducted along nerves in the olfactory bulb and olfactory tract to sensory processing centers in the brain.

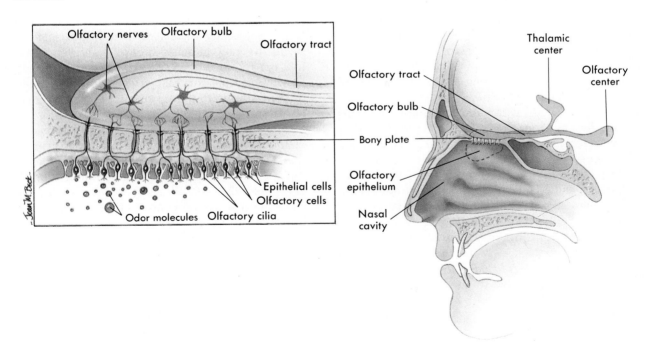

COCHLEAR IMPLANTS

Recent advances in electronic circuitry are now being used to correct some forms of nerve deafness. If the hairs on the organ of Corti are damaged, nerve deafness results—even if the vestibulocochlear nerve is healthy. A new surgically implanted device can improve this form of hearing loss by eliminating the need for the sensory hairs. As you can see in the figure, a transmitter just outside the scalp sends external sound information to a receiver under the scalp (behind the auricle). The receiver translates the information into an electrical code that is relayed down an electrode to the cochlea. The electrode, wired to the organ of Corti, stimulates the vestibulocochlear nerve endings directly. Thus even though the cochlear hair cells are damaged, sound can still be perceived.

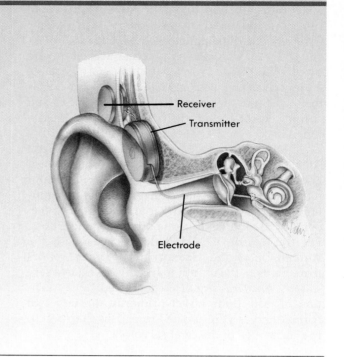

tory centers of the brain, where the nervous impulses are interpreted as specific odors. The pathways taken by olfactory nerve impulses and the areas where these impulses are interpreted are closely associated with areas of the brain important in memory and emotion. For this reason, we may retain vivid and long-lasting memories of particular smells and odors.

Temporary reduction of sensitivity to smells often results from colds and other nasal infections. Progressive reduction of the sense of smell is often seen in smokers because of the damaging effects of the pollutants in tobacco smoke. In olfaction, as with all the special senses, advancing age often brings a structural degeneration that results in reduced function. It is no wonder that many older adults become isolated and depressed when their contact with the outside world, the special senses, is gradually lost. Caring health professionals recognize these signs of aging and provide assistance needed by their aged patients to enjoy life.

General Sense Organs

Groups of highly specialized and localized receptors are typically associated with the special senses. In the general sense organs, however, receptors are found in almost every part of the body. To demonstrate this fact, try touching any point of your skin with the tip of a toothpick. You can hardly miss stimulating at least one receptor and almost instantaneously experiencing a sensation of touch. Stimulation of some receptors leads to the sensation of heat; stimulation of others gives the sensation of cold, and stimulation of still others gives the sensation of pain or pressure. General sense receptors are listed in Table 9-2 and illustrated in Figure 9-12. When special receptors in the muscles and joints are stimulated, you sense the position of the different parts of the body and know whether they are moving and in which direction they are moving without even looking at them. Perhaps you have never realized that you have this sense of position and

FIGURE 9-12 General Sense Receptors. A, Meissner's corpuscle. **B,** Pacinian corpuscle. **C,** Free nerve ending. **D,** Ruffini's corpuscle. **E,** Krause's end-bulb.

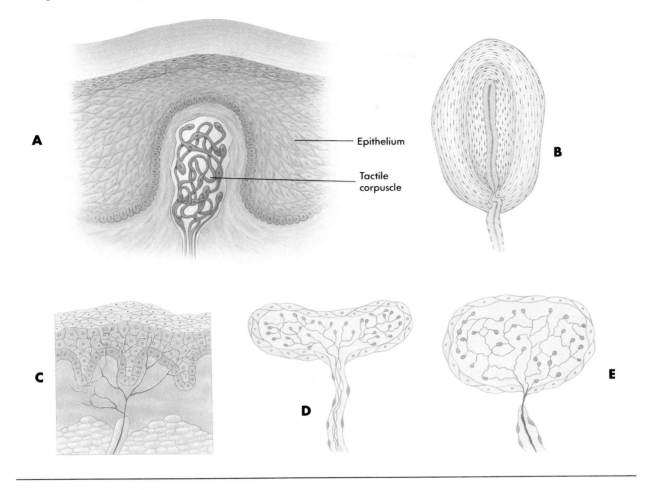

movement—a sense called **proprioception** (pro-pree-oh-SEP-shun) or **kinesthesia** (kin-es-THEE-zee-ah).

Disruption of general sense organs can occur by a variety of mechanisms. For example, third-degree burns can completely destroy general sense receptors throughout the affected area. Temporary impairment of general sense receptors occurs when the blood flow to them is slowed.

This commonly occurs when you put your legs in a position that presses your legs in a way that reduces blood flow. When you try to stand up, you cannot feel your legs because the general sense organs are temporarily impaired. You may not even be able to walk because you cannot tell where your legs are without looking at them. As blood flow returns, reactivation of the sense organs may produce a tingling sensation.

▽   **TABLE 9-2**   General Sense Organs

TYPE	MAIN LOCATIONS	GENERAL SENSES
Free nerve endings (naked nerve endings)	Skin and mucosa (epithelial layers)	Pain, crude touch, and possibly temperature
Encapsulated nerve endings		
Meissner's corpuscles	Skin (in papillae of dermis) and fingertips and lips (numerous)	Fine touch and vibration
Ruffini's corpuscles	Skin (dermal layer) and subcutaneous tissue of fingers	Touch and pressure
Pacinian corpuscles	Subcutaneous, submucous, and subserous tissues; around joints; in mammary glands and external genitals of both sexes	Pressure and vibration
Krause's end-bulbs	Skin (dermal layer), subcutaneous tissue, mucosa of lips and eyelids, and external genitals	Touch and possibly cold
Golgi tendon receptors	Near junction of tendons and muscles	Proprioception
Muscle spindles	Skeletal muscles	Proprioception

OUTLINE SUMMARY

CLASSIFICATION OF SENSE ORGANS

A Special sense organs (Table 9-1)
  1 Large and complex organs
  2 Localized grouping of specialized receptors
B General sense organs (Table 9-2)
  1 Often exist as individual cells or receptor units
  2 Widely distributed throughout the body

CONVERTING A STIMULUS INTO A SENSATION

A All sense organs have common functional characteristics
  1 All are able to detect a particular stimulus
  2 A stimulus is converted into a nerve impulse
  3 A nerve impulse is perceived as a sensation in the CNS

SPECIAL SENSE ORGANS

A The eye (Figure 9-1)
  1 Layers of eyeball
    a Sclera—tough outer coat; "white" of eye; cornea is transparent part of sclera over iris

    b Choroid—pigmented vascular layer prevents scattering of light; front part of this layer made of ciliary muscle and iris, the colored part of the eye; the pupil is the hole in the center of the iris; contraction of iris muscle dilates or constricts pupil
    c Retina (Figure 9-2)—innermost layer of the eye; contains rods (receptors for night vision) and cones (receptors for day vision and color vision)
  2 Conjunctiva—mucous membrane covering the front surface of the sclera and lining the eyelid
  3 Lens—transparent body behind the pupil; focuses light rays on the retina
  4 Eye fluids
    a Aqueous humor—in the anterior cavity in front of the lens
    b Vitreous humor—in the posterior cavity behind the lens
  5 Visual pathway
    a Innermost layer of retina contains rods and cones
    b Impulse travels from the rods and cones through the bipolar and ganglionic layers of retina (Figure 9-2)

c Nerve impulse leaves the eye through the optic nerve; the point of exit is free of receptors and is therefore called a *blind spot*

d Visual interpretation occurs in the visual cortex of the cerebrum

B Visual disorders

  1 Refraction disorders (Figure 9-3)

    a Myopia (nearsightedness) is often caused by elongation of the eyeball

    b Hyperopia (farsightedness) is often caused by a shortened eyeball

    c Astigmatism is distortion due to an irregularity of the cornea or lens

    d Conjunctivitis (inflammation of the conjunctiva) can interfere with refraction

      (1) Trachoma—chronic chlamydial infection

      (2) Acute bacterial conjunctivitis—highly contagious infection that produces a discharge of mucous pus (Figure 9-4)

      (3) Conjunctivitis can be caused by allergies

    e Strabismus—improper alignment of eyes (Figure 9-5)

      (1) Eyes can converge (cross) or diverge

      (2) If not corrected, can cause blindness

  2 Disorders of the retina

    a Retinal detachment can be a complication of aging, eye tumors, or head trauma

    b Diabetic retinopathy—damage to retina from hemorrhages and growth of abnormal vessels associated with diabetes mellitus

    c Glaucoma—intraocular pressure that decreases blood flow in retina and thus causes retinal degeneration

    d Nyctalopia (night blindness) or the inability to see in dim light is caused by retinal degeneration or lack of vitamin A

    e Macular degeneration—progressive degeneration of central part of retina; leading cause of permanent blindness in elderly

    f Red-green color blindness is an X-linked genetic condition in which certain colors are not seen properly; it is caused by an abnormality in the cones' photopigments

  3 Disorders of the visual pathway

    a Degeneration of the optic nerve resulting from diabetes, glaucoma, and other causes can impair vision

    b Scotoma is the loss of only the central visual field when only certain nerve pathways are damaged

    c Cerebrovascular accidents (CVAs) can damage visual processing centers; example is acquired cortical color blindness

C The ear

  1 The ear functions in hearing and in equilibrium and balance

    a Receptors for hearing and equilibrium are mechanoreceptors

  2 Divisions of the ear (Figure 9-7)

    a External ear

      (1) Auricle (pinna)

      (2) External auditory canal

        (a) Curving canal 2.5 cm (1 inch) in length

        (b) Contains ceruminous glands

        (c) Ends at the tympanic membrane

    b Middle ear

      (1) Houses ear ossicles—malleus, incus, and stapes

      (2) Ends in the oval window

      (3) The auditory (eustachian) tube connects the middle ear to the throat

      (4) Inflammation called *otitis media*

    c Inner ear (Figure 9-8)

      (1) Bony labyrinth filled with perilymph

      (2) Subdivided into the vestibule, semicircular canals, and cochlea

      (3) Membranous labyrinth filled with endolymph

      (4) The receptors for balance in the semicircular canals are called *cristae ampulares*

      (5) Specialized hair cells on the organ of Corti respond when bent by the movement of surrounding endolymph set in motion by sound waves (Figure 9-9)

D Hearing disorders

  1 Conduction impairments

    a Can be caused by blockage of the external or middle ear (e.g., cerumen and tumors)

    b Otosclerosis—inherited bone disorder involving irregularity of the stapes; it first appears as tinnitus (ringing) then progresses to hearing loss

    c Otitis—ear inflammation caused by infection; can produce swelling and fluids that block sound conduction

  2 Nerve impairment

    a Presbycusis—progressive nerve deafness associated with aging

    b Progressive nerve deafness can also result from chronic exposure to loud noise

c Meniere's disease—chronic inner ear disorder characterized by tinnitus, nerve deafness, and vertigo

E The taste receptors (Figure 9-10)
  1 Receptors are chemoreceptors called *taste buds*
  2 Cranial nerves VII and IX carry gustatory impulses
  3 Only four kinds of taste sensations—sweet, sour, bitter, salty
  4 Gustatory and olfactory senses work together

F The smell receptors (Figure 9-11)
  1 Receptors for fibers of olfactory or cranial nerve I lie in olfactory mucosa of nasal cavity
  2 Olfactory receptors are extremely sensitive but easily fatigued

3 Odor-causing chemicals initiate a nervous signal that is interpreted as a specific odor by the brain

GENERAL SENSE ORGANS (Table 9-2)

A Distribution is widespread; single-cell receptors are common

B Examples (Figure 9-12)
  1 Free nerve endings—pain and crude touch
  2 Meissner's corpuscles—fine touch and vibration
  3 Ruffini's corpuscles—touch and pressure
  4 Pacinian corpuscles—pressure and vibration
  5 Krause's end-bulbs—touch
  6 Golgi tendon receptors—proprioception
  7 Muscle spindles—proprioception

$\mathcal{N}$EW WORDS

aqueous humor	endolymph	organ of Corti	pupil
auricle	gustatory cells	ossicles	retina
chemoreceptor	kinesthesia	papillae	sclera
choroid	lacrimal gland	perilymph	semicircular canals
cochlea	lens	photoreceptor	tympanic membrane
conjunctiva	mechanoreceptor	proprioception	vitreous humor
crista ampullaris			

Diseases and Other Clinical Terms

astigmatism	hyperopia	otosclerosis	strabismus
cataracts	Meniere's disease	presbycusis	tinnitus
color blindness	myopia	presbyopia	trachoma
conjunctivitis	nyctalopia	scotoma	vertigo
glaucoma	otitis		

$\mathcal{C}$HAPTER TEST

1. _____ sense organs are widely distributed throughout the body.
2. The receptors involved in vision are called _____ and _____.
3. The three layers of tissue that form the eyeball are the _____, the _____, and the _____.
4. The colored structure seen through the cornea of the eye is called the _____.
5. The fluid found in the anterior cavity of the eye is called _____ _____.
6. The loss of lens transparency is a condition called _____.
7. The area of the retina that helps to reduce light scattering is called the _____.
8. The type of receptors involved with hearing and equilibrium are called _____.
9. The major anatomical areas of the ear are called: _____ ear; _____ ear; and _____ ear.
10. The ear ossicle attached to the tympanic membrane is called the _____.
11. The bony labyrinth is filled with fluid called _____.
12. The hearing sense organ is called the organ of _____.
13. Sense receptors of the type used in taste and smell are called _____.
14. Receptor cells in the taste buds are called _____ cells.

Select the most correct answer from Column B for each statement in Column A. (Only one answer is correct.)

COLUMN A

15. _____ White portion of eyeball
16. _____ Receptors for night vision
17. _____ Kinesthesia
18. _____ Blind spot
19. _____ External ear
20. _____ Ear drum
21. _____ Middle ear infection
22. _____ Receptor for equilibrium
23. _____ Receptor for smell

COLUMN B

a. Rods
b. Auricle
c. Optic disc
d. Otitis media
e. Olfactory cell
f. Crista ampullaris
g. Sclera
h. Tympanic membrane
i. Proprioception

Circle the T before each true statement and the F before each false statement.

T   F   **24.** Nearsightedness can also be called *hyperopia*.
T   F   **25.** Conjunctivitis can be caused by bacterial or viral infection.
T   F   **26.** Complications of diabetes mellitus can cause vision impairment by destroying the retina and/or the optic nerve.
T   F   **27.** Red-green color blindness results from lack of vitamin A.
T   F   **28.** Conduction deafness involves damage to the organ of Corti or vestibulocochlear nerve.
T   F   **29.** Otosclerosis causes deafness by changing the structure of the ear.
T   F   **30.** The hearing loss associated with aging results from nerve impairment.

REVIEW QUESTIONS

1 List, compare, and contrast special and general sense organs.
2 What are the functional characteristics required for a sense organ to convert a stimulus into a sensation?
3 How does a cataract cause a problem with visual quality?
4 How do rods and cones differ in terms of their ability to serve as receptors for vision?
5 Give examples of visual disorders that disrupt refraction, affect stimulation of the retina, and involve nerve transmission.
6 Which pair or pairs of cranial nerves would you nickname "seeing nerves," "hearing nerves," "smelling nerves," and "tasting nerves"?
7 Explain the anatomical relationship between the bony and the membranous labyrinths.
8 Trace and discuss the path of a sound wave as it influences the structures involved in hearing.

9 Compare and contrast conduction deafness and nerve deafness. Give examples of each.
10 Explain how photoreceptors, mechanoreceptors, and chemoreceptors differ in function.
11 Briefly explain why most elderly people have difficulty in seeing objects close to their eyes.
12 What is the function of the ear ossicles?
13 Discuss the sense of balance or equilibrium, including the name, location, and mechanism of action of the specialized receptors involved.
14 Discuss the organ of Corti. Where is it located? How is it stimulated?
15 Explain the difference between a tongue papilla, a taste bud, and a gustatory cell.
16 Discuss how odors are detected and perceived.
17 Identify and give the functions of five general sense organs.
18 How can sensory impairments lead to psychological or social problems?

CLINICAL APPLICATIONS

1 Roger is legally blind. His vision impairment is a complication of diabetes mellitus. Can you describe what structural changes in Roger's eyes have caused his blindness? As you were helping him cross the street, a fellow pedestrian suddenly stumbled into your path. Without any signal from you, Roger jumped back to avoid hitting the other person. If Roger is blind, how could he have reacted this way?

2 As a child, Mrs. Stark was tested for color blindness and told that she had normal color vision. She never had any problems distinguishing colors until shortly after her retirement, when suddenly she lost her sense of color. She describes her perception of the world as being "like a black and white movie." She cannot distinguish yellows, oranges, blues, greens, reds, or any other colors. What might have caused Mrs. Stark's problems?

3 You have just been diagnosed with *otitis media*. Describe what has happened to your body to produce this condition. If left untreated, what are the possible outcomes of this condition?

The Endocrine System

C H A P T E R

10

Objectives

*After you have completed this chapter,
you should be able to:*

1 Distinguish between endocrine and exocrine glands and define the terms *hormone* and *prostaglandin*.

2 Identify and locate the primary endocrine glands and list the major hormones produced by each gland.

3 Describe the mechanisms of steroid and protein hormone action.

4 Explain how negative and positive feedback mechanisms regulate the secretion of endocrine hormones.

5 Explain the primary mechanisms of endocrine disorders.

6 Identify the principal functions of each major endocrine hormone and describe the conditions that may result from hyposecretion or hypersecretion.

7 Define *diabetes insipidus, diabetes mellitus, gigantism, goiter, cretinism, glycosuria.*

*H*ave you ever seen a giant or a dwarf? Have you ever known anyone who had "sugar diabetes" or a goiter? If so, you have had visible proof of the importance of the endocrine system for normal development and health.

The **endocrine system** performs the same general functions as the nervous system: communication and control. The nervous system provides rapid, brief control by fast-traveling nerve impulses. The endocrine system provides slower but longer-lasting control by **hormones** (chemicals) secreted into and circulated by the blood.

The organs of the endocrine system are located in widely separated parts of the body—in the neck; the cranial, thoracic, abdominal, and pelvic cavities; and outside of the body cavities. Note the names and locations of the endocrine glands shown in Figure 10-1.

All organs of the endocrine system are glands, but not all glands are organs of the endocrine system. Of the two types of glands in the body—**exocrine glands** and **endocrine glands**—only endocrine glands belong to this system. Exocrine glands secrete their products into ducts that empty onto a surface or into a cavity. For example, sweat glands produce a watery secretion that empties onto the surface of the skin. Salivary glands are also exocrine glands, secreting saliva that flows into the mouth. Endocrine glands are ductless glands. They secrete chemicals known as **hormones** into intercellular spaces. From there, the hormones diffuse directly into the blood and are carried throughout the body. Each hormone molecule may then bind to a cell that has specific receptors for that hormone, triggering a reaction in the cell. Such a cell is called a **target organ** cell. The list of endocrine glands and their target organs continues to grow. The names, locations, and functions of the well-known endocrine glands are given in Figure 10-1 and Table 10-1.

In this chapter you will read about the functions of the main endocrine glands and discover why their importance is almost impossible to exaggerate. Hormones are the main regulators of metabolism, growth and development, reproduction, and many other body activities. They play important roles in maintaining homeostasis—fluid and electrolyte, acid-base, and energy balances, for example. Hormones make the difference between normalcy and many kinds of abnormalities such as dwarfism, gigantism, and sterility. They are important not only for the healthy survival of each one of us but also for the survival of the human species.

Mechanisms of Hormone Action

A hormone causes its target cells to respond in particular ways; this has been the subject of intense interest and research. The two major classes of hormones—**protein hormones** and **steroid hormones**—differ in the mechanisms by which they influence target organ cells.

The most widely accepted theory of protein hormone action is called the **second messenger hypothesis.** A *hypothesis* (hye-POTH-e-sis) is a proposed explanation that explains observed phenomena. The second messenger hypothesis attempts to explain why protein hormones cause specific effects in target organs but do not "recognize" or act on other organs of the body. According to this concept a protein hormone acts as a "first messenger" (that is, it delivers its chemical message from the cells of an endocrine gland to highly specific membrane receptor sites on the cells of a target organ). This interaction between a hormone and its specific receptor site on the cell membrane of a target organ cell is often compared to the fitting of a unique key into a lock. After the hormone is attached to its specific receptor site, a number of chemical reactions occur. These reactions activate molecules within the cell called *second messengers.* One example of this mechanism occurs when the hormone-receptor interaction changes energy-rich ATP molecules inside the cell into **cyclic AMP** (adenosine monophosphate). Cyclic AMP serves as the second messenger, delivering information inside the cell that regulates the cell's activity. For example, cyclic AMP causes thyroid cells to respond by secreting thyroid hormones. Cyclic AMP is only one of several second messengers that have been discovered.

In summary, protein hormones serve as first messengers, providing communication between endocrine glands and target organs. Another molecule, such as cyclic AMP, then acts as the second messenger, providing communication within

Text continued on p. 263.

FIGURE 10-1 Location of the Endocrine Glands. Thymus gland is shown at maximum size at puberty.

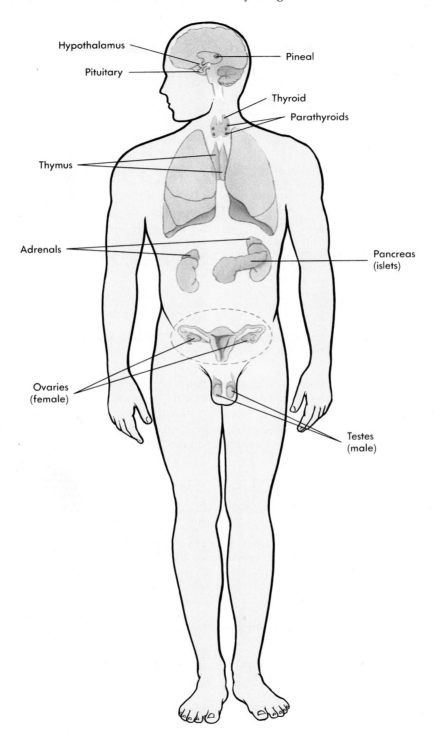

TABLE 10-1	Endocrine Glands, Hormones, and Their Functions	
GLAND/HORMONE	**FUNCTION**	**DYSFUNCTION**[*]
Anterior Pituitary		
Thyroid-stimulating hormone (TSH)	Tropic hormone Stimulates secretion of thyroid hormones	*Hypersecretion:* overstimulation of thyroid *Hyposecretion:* understimulation of thyroid
Adrenocorticotropic hormone (ACTH)	Tropic hormone Stimulates secretion of adrenal cortex hormones	*Hypersecretion:* overstimulation of adrenal cortex *Hyposecretion:* understimulation of adrenal cortex
Follicle-stimulating hormone (FSH)	Tropic hormone *Female:* stimulates development of ovarian follicles and secretion of estrogens *Male:* stimulates seminiferous tubules of testes to grow and produce sperm	*Hyposecretion:* lack of sexual development and sterility
Luteinizing hormone (LH)	Tropic hormone *Female:* stimulates maturation of ovarian follicle and ovum; stimulates secretion of estrogen; triggers ovulation; stimulates development of corpus luteum (luteinization) *Male:* stimulates interstitial cells of the testes to secrete testosterone	*Hyposecretion:* lack of sexual development and sterility
Melanocyte-stimulating hormone (MSH)	Stimulates synthesis and dispersion of melanin pigment in the skin	*Hypersecretion:* darkening of skin
Growth hormone (GH)	Stimulates growth in all organs; mobilizes food molecules, causing an increase in blood glucose concentration	*Hypersecretion:* gigantism (pre-adult); acromegaly (mature adult) *Hyposecretion:* dwarfism (preadult)
Prolactin (lactogenic hormone)	Stimulates breast development during pregnancy and milk secretion after pregnancy	*Hypersecretion:* inappropriate lactation in men or nonnursing women *Hyposecretion:* insufficient lactation in nursing women
Posterior Pituitary[†]		
Antidiuretic hormone (ADH)	Stimulates retention of water by the kidneys	*Hypersecretion:* abnormal water retention *Hyposecretion:* diabetes insipidus
Oxytocin	Stimulates uterine contractions at the end of pregnancy; stimulates the release of milk into the breast ducts	*Hypersecretion:* inappropriate ejection of milk in lactating women *Hyposecretion:* prolonged or difficult labor and delivery (uncertain)
Hypothalamus		
Releasing hormones (several)	Stimulate the anterior pituitary to release hormones	*Hypersecretion:* hypersecretion by anterior pituitary *Hyposecretion:* hyposecretion by anterior pituitary
Inhibiting hormones (several)	Inhibit the anterior pituitary's secretion of hormones	*Hypersecretion:* hyposecretion by anterior pituitary *Hyposecretion:* hypersecretion by anterior pituitary

[*]In some cases, signs of hyposecretion result from target cell abnormality rather than from actual hyposecretion of a hormone.
[†]Posterior pituitary hormones are synthesized in the hypothalamus but released from axon terminals in the posterior pituitary.

▽　**TABLE 10-1**—cont'd　Endocrine Glands, Hormones, and Their Functions

GLAND/HORMONE	FUNCTION	DYSFUNCTION*
Thyroid		
Thyroxine (T_4) and triiodothyronine (T_3)	Stimulate the energy metabolism of all cells	*Hypersecretion:* hyperthyroidism, Graves' disease *Hyposecretion:* hypothyroidism; cretinism (pre-adult); myxedema (adult); goiter
Calcitonin	Inhibits the breakdown of bone; causes a decrease in blood calcium concentration	*Hypersecretion:* possible hypocalcemia *Hyposecretion:* possible hypercalcemia
Parathyroid		
Parathyroid hormone (PTH)	Stimulates the breakdown of bone; causes an increase in blood calcium concentration	*Hypersecretion:* possible hypercalcemia *Hyposecretion:* possible hypocalcemia
Adrenal Cortex		
Mineralocorticoids: aldosterone	Regulate electrolyte and fluid homeostasis	*Hypersecretion:* increased water retention *Hyposecretion:* abnormal water loss (dehydration)
Glucocorticoids: cortisol (hydrocortisone)	Stimulate gluconeogenesis, causing an increase in blood glucose concentration; also have antiinflammatory and antiimmunity, antiallergy effects	*Hypersecretion:* Cushing's syndrome *Hyposecretion:* Addison's disease
Sex hormones (androgens)	Stimulate sexual drive in the female but have negligible effects in the male	*Hypersecretion:* premature sexual development; masculinization of female *Hyposecretion:* no significant effect
Adrenal Medulla		
Epinephrine (adrenaline) and norepinephrine	Prolong and intensify the sympathetic nervous response during stress	*Hypersecretion:* stress effects *Hyposecretion:* no significant effect
Pancreatic Islets		
Glucagon	Stimulates liver glycogenolysis, causing an increase in blood glucose concentration	(uncertain)
Insulin	Promotes glucose entry into all cells, causing a decrease in blood glucose concentration	*Hypersecretion:* severe hypoglycemia (insulin shock) *Hyposecretion:* diabetes mellitus
Ovary		
Estrogens	Promotes development and maintenance of female sexual characteristics (see Chapter 19)	*Hypersecretion:* premature sexual development (female) and infertility *Hyposecretion:* lack of sexual development (female), infertility, and osteoporosis
Progesterone	Promotes conditions required for pregnancy (see Chapter 19)	*Hyposecretion:* sterility
Testis		
Testosterone	Promotes development and maintenance of male sexual characteristics (see Chapter 19)	*Hypersecretion:* premature sexual development (male); muscle hypertrophy *Hyposecretion:* lack of sexual development (male)

Continued.

TABLE 10-1—cont'd   Endocrine Glands, Hormones, and Their Functions

GLAND/HORMONE	FUNCTION	DYSFUNCTION*
Thymus Thymosin	Promotes development of immune system cells	*Hyposecretion:* depression of immune system functions
Placenta Chorionic gonadotropin, estrogens, progesterone	Promote conditions required during early pregnancy	*Hyposecretion:* spontaneous abortion (miscarriage)
Pineal Melatonin	Inhibits tropic hormones that affect the ovaries; may be involved in the body's internal clock	*Hypersecretion:* winter depression and other possible effects
Heart (atria) Atrial natriuretic hormone (ANH)	Regulates fluid and electrolyte homeostasis	(uncertain)

FIGURE 10-2 Mechanism of Protein Hormone Action. The hormone acts as "first messenger," delivering its message via the bloodstream to a membrane receptor in the target organ cell much like a key fits into a lock. The "second messenger" causes the cell to respond and perform its specialized function.

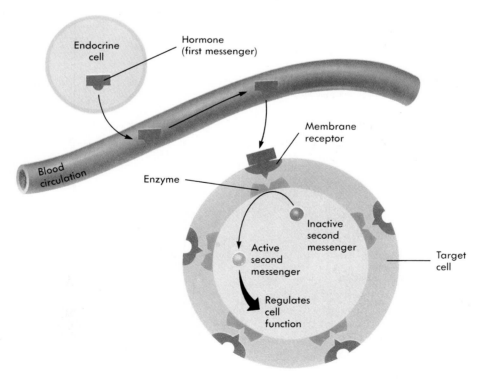

FIGURE 10-3 Negative Feedback. The secretion of most hormones is regulated by negative feedback mechanisms that tend to reverse any deviations from normal. In this example, an increase in blood glucose triggers secretion of insulin. Since insulin promotes glucose uptake by cells, the blood glucose level is restored to its lower, normal level.

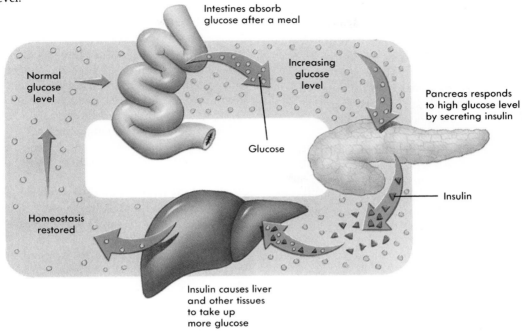

a hormone's target cells. Figure 10-2 summarizes the mechanism of hormone action as explained by the second messenger hypothesis.

The action of small, lipid-soluble steroid hormones such as estrogen is not explained by the second messenger hypothesis. Because they are lipid soluble, steroid hormones can pass intact directly through the cell membrane of the target organ cell. Once inside the cell they combine with specific receptors and then enter the nucleus to influence cell activity by acting on specific genes.

Regulation of Hormone Secretion

The regulation of hormone levels in the blood depends on a highly specialized homeostatic mechanism called **negative feedback.** The principle of negative feedback can be illustrated by using the hormone insulin as an example. When released from endocrine cells in the pancreas, insulin lowers blood sugar levels. Normally, elevated blood sugar levels occur after a meal, after the absorption of sugars from the digestive tract takes place. The elevated blood sugar stimulates the release of insulin from the pancreas. Insulin then assists in the transfer of sugar from the blood into cells, and blood sugar levels drop. Low blood sugar levels then cause endocrine cells in the pancreas to cease the production and release of insulin. These responses are *negative*. Therefore the homeostatic mechanism is called a **negative feedback** control mechanism because they reverse the change in blood sugar level (Figure 10-3).

Positive feedback mechanisms, which are uncommon, amplify changes rather than reverse them. Usually, such amplification threatens homeostasis, but in some situations, it can help the body maintain its stability. For example, during labor, the muscle contractions that push the

STEROID ABUSE

Some steroid hormones are called **anabolic steroids** because they stimulate the building of large molecules (anabolism). Specifically, they stimulate the building of proteins in muscle and bone. Steroids such as testosterone and its synthetic derivatives are often abused by athletes and others who want to increase their performance. The anabolic effects of the hormones increase the mass and strength of skeletal muscles.

Unfortunately, steroid abuse has other consequences. It disrupts the normal negative feedback control of hormones throughout the body and may result in tissue damage, sterility, mental imbalance, and many life-threatening metabolic problems.

baby through the birth canal become stronger by means of a positive feedback mechanism that regulates secretion of the hormone oxytocin.

Mechanisms of Endocrine Disease

Diseases of the endocrine system are numerous, varied, and sometimes spectacular. Tumors or other abnormalities frequently cause the glands to secrete too much or too little of their hormones. Production of too much hormone by a diseased gland is called **hypersecretion.** If too little hormone is produced, the condition is called **hyposecretion.**

A variety of endocrine disorders that appear to result from hyposecretion are actually caused by a problem in the target cells. If the usual target cells of a particular hormone have damaged receptors, too few receptors, or some other abnormality, they will not respond to that hormone properly. In other words, lack of target cell response could be a sign of hyposecretion or a sign of target cell insensitivity. *Diabetes mellitus*, for example, can result from insulin hyposecretion or from the target cells' insensitivity to insulin.

Endocrinologists, scientists who specialize in endocrine function, have developed a variety of strategies for treating endocrine disorders. Surgical or chemical treatment of tumors or damaged tissue is useful in some cases. Another common strategy is the use of pharmacological preparations of hormones. For example, insulin injections are used in treating some forms of diabetes mellitus. The availability of synthetic hormones produced with genetic engineering technology has revolutionized the treatment of many endocrine disorders. Synthetic hormones are cheaper and more widely available than natural human hormones, and they do not carry the same risk of contamination with viruses or other dangerous substances.

Table 10-1 summarizes some of the major disorders of the endocrine system. Refer to this table often as you study the individual glands and hormones of the endocrine system. You may also want to refer to Appendix B, which also contains a table listing major endocrine disorders.

Prostaglandins

Prostaglandins (PGs) or tissue hormones are important and extremely powerful substances found in a wide variety of tissues. They play an important role in communication and the control of many body functions but do not meet the definition of a typical hormone. The term *tissue hormone* is appropriate because in many instances a prostaglandin is produced in a tissue and diffuses only a short distance to act on cells within that tissue. Typical hormones influence and control activities of widely separated organs; typical prostaglandins influence activities of neighboring cells.

The prostaglandins in the body can be divided into several groups. Three classes of prostaglandins—prostaglandin A (PGA), prostaglandin E (PGE), and prostaglandin F (PGF)—are among the best known. Prostaglandins have profound effects on many body functions. They influence respiration, blood pressure, gastrointestinal secretions, and the reproductive system. Researchers believe that most prostaglandins regulate cells by influencing the production of cyclic AMP.

Pituitary Gland

The **pituitary** (pi-TOO-i-tair-ee) **gland** is a small but mighty structure. Although no larger than a pea, it is really two endocrine glands. One is called the **anterior pituitary gland** or *adenohypophysis* (ad-e-no-hye-POF-i-sis), and the other is called the **posterior pituitary gland** or *neurohypophysis* (noo-ro-hye-POF-i-sis). Differences between the two glands are suggested by their names—*adeno* means "gland," and *neuro* means "nervous." The adenohypophysis has the structure of an endocrine gland, whereas the neurohypophysis has the structure of nervous tissue. Hormones secreted by the adenohypophysis serve very different functions from those released from the neurohypophysis.

The protected location of this dual gland suggests its importance. The pituitary gland lies buried deep in the cranial cavity, in the small depression of the sphenoid bone that is shaped like a saddle and called the *sella turcica* (Turkish saddle). A stemlike structure, the pituitary stalk, attaches the gland to the undersurface of the brain. More specifically, the stalk attaches the pituitary body to the hypothalamus.

ANTERIOR PITUITARY GLAND HORMONES

The anterior pituitary gland secretes several major hormones. Each of the four hormones, listed as a **tropic** (TRO-pik) **hormone** in Table 10-1 stimulates another endocrine gland to grow and secrete its hormones. Because the anterior pituitary gland exerts this control over the structure and function of the thyroid gland, the adrenal cortex, the ovarian follicles, and the corpus luteum, it is sometimes referred to as the *master gland.*

Thyroid-stimulating hormone (TSH) acts on the thyroid gland. As its name suggests, it stimulates the thyroid gland to increase secretion of thyroid hormone.

The **adrenocorticotropic** (ad-re-no-kor-ti-ko-TRO-pik) **hormone (ACTH)** acts on the adrenal cortex. It stimulates the adrenal cortex to increase in size and to secrete larger amounts of its hormones, especially of cortisol (hydrocortisone).

Follicle-stimulating hormone (FSH) stimulates the primary ovarian follicles in an ovary to start growing and to continue developing to maturity (that is, to the point of ovulation). FSH also stimulates follicle cells to secrete estrogens. In the male, FSH stimulates the seminiferous tubules to grow and form sperm.

Luteinizing (LOO-te-nye-zing) **hormone (LH)** acts with FSH to perform several functions. It stimulates a follicle and ovum to complete their growth to maturity, it stimulates follicle cells to secrete estrogens, and it causes ovulation (rupturing of the mature follicle with expulsion of its ripe ovum). Because of this function, LH is sometimes called the *ovulating hormone.* Finally, LH stimulates the formation of a golden body, the corpus luteum, in the ruptured follicle; the process is called *luteinization.* This function, of course, is the one that earned LH its title of *luteinizing hormone.* The male pituitary gland also secretes LH; it was formerly called *interstitial cell-stimulating hormone (ICSH)* because it stimulates interstitial cells in the testes to develop and secrete testosterone, the male sex hormone.

The **melanocyte-stimulating hormone (MSH)** causes a rapid increase in the synthesis and dispersion of melanin (pigment) granules in specialized skin cells.

Another important hormone secreted by the anterior pituitary gland is **growth hormone.** Growth hormone (GH) speeds up the movement of digested proteins (amino acids) out of the blood and into the cells, and this accelerates the cells' anabolism (build up) of amino acids to form tissue proteins; hence this action promotes normal growth. Growth hormone also affects fat and carbohydrate metabolism; it accelerates fat catabolism (break-down) but slows glucose catabolism. This means that less glucose leaves the blood to enter cells, and therefore the amount of glucose in the blood increases. Thus growth hormone and insulin have opposite effects on blood glucose. Insulin decreases blood glucose, and growth hormone increases it. Too much insulin in the blood produces **hypoglycemia** (hye-po-glye-SEE-me-ah) (lower than normal blood glucose concentration). Too much growth hormone produces **hyperglycemia** (higher than normal blood glucose concentration). This type of hyperglycemia is called *pituitary diabetes.*

Hypersecretion of growth hormone during the early years of life produces a condition called **gi-**

A

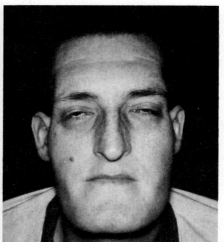

B

FIGURE **10-4** **Growth Hormone Abnormalities. A,** Generations of students have viewed this photo of growth hormone abnormalities. The man to the far left exhibits gigantism caused by hypersecretion of growth hormone. The man on the far right exhibits pituitary dwarfism caused by hyposecretion of growth hormone. **B,** Acromegaly. Notice the large head, exaggerated projection of the lower jaw and protrusion of the frontal bone.

gantism (jye-GAN-tizm). The name suggests the obvious characteristics of this condition. The child grows to a giant size. If the anterior pituitary gland secretes too much growth hormone after the normal growth years, the disease called **acromegaly** (ak-row-MEG-ah-lee) develops. Characteristics of this disease are enlargment of the bones of the hands, feet, jaws, and cheeks. The facial appearance that is typical of acromegaly results from the combination of bone and soft tissue overgrowth. A prominent forehead and large nose are characteristic. In addition, patients with acromegaly may have enlarged skin pores and an overgrown mandible. Figure 10-4 illustrates the major characteristics of gigantism and acromegaly.

Hyposecretion of growth hormone during the growth years often produces pituitary **dwarfism.** Pituitary dwarfs usually have a body frame of normal proportions but are much smaller in overall size. Dwarfism caused by other conditions may produce an oddly proportioned body frame. Figure 10-4, *A,* illustrates both gigantism and pituitary dwarfism.

The anterior pituitary gland also secretes **prolactin** (pro-LAK-tin) or lactogenic hormone. During pregnancy, prolactin stimulates the breast development necessary for eventual lactation (milk secretion). Also, soon after delivery of a baby, prolactin stimulates the breasts to start secreting milk, a function suggested by prolactin's other name, lactogenic hormone.

▽ **FIGURE 10-5 Pituitary Hormones.** Principal anterior and posterior pituitary hormones and their target organs.

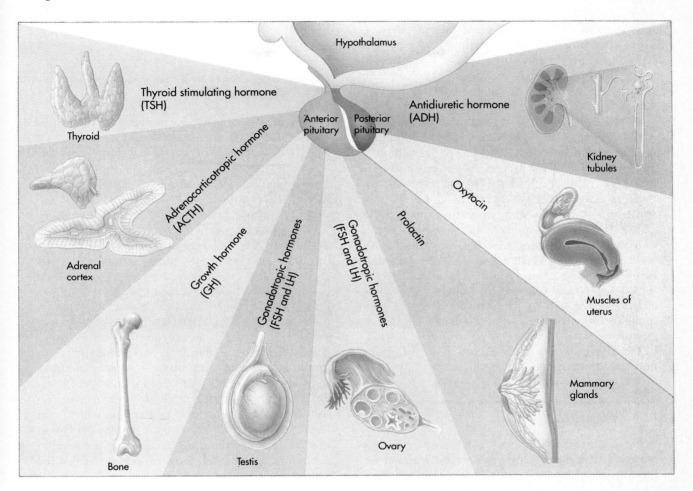

For a brief summary of anterior pituitary hormone target organs and functions, see Figure 10-5.

POSTERIOR PITUITARY GLAND HORMONES

The posterior pituitary gland releases two hormones—**antidiuretic (an-tie-dye-yoo-RET-ik) hormone (ADH)** and **oxytocin** (ok-see-TOE-sin). ADH accelerates the reabsorption of water from urine in kidney tubules back into the blood. With more water moving out of the tubules into the blood, less water remains in the tubules, and therefore less urine leaves the body. The name *antidiuretic hormone* is appropriate because *anti-* means "against" and *diuretic* means "increasing the volume of urine excreted." Therefore antidiuretic means "acting against an increase in urine volume"; in other words, ADH acts to decrease urine volume.

The posterior pituitary hormone oxytocin is secreted by a woman's body before and after she has a baby. Oxytocin stimulates contraction of the smooth muscle of the pregnant uterus and is believed to initiate and maintain labor. This is why physicians sometimes prescribe oxytocin injections to induce or increase labor. Oxytocin also performs a function important to a newborn baby. It causes the glandular cells of the breast to release milk into ducts from which a baby can obtain it by sucking. The right side of Figure 10-5 summarizes posterior pituitary functions.

Diabetes insipidus (dye-ah-BEE-tes in-SIP-i-dus) is the name of the disease caused by hyposecretion of antidiuretic hormone. This condition is marked by the elimination of extremely large volumes of urine each day. In untreated cases, patients may excrete as much as 25 to 30 L of urine in a 24-hour period. In addition to voiding an abnormally large quantity of urine, these individuals suffer from great thirst, dehydration, and serious electrolyte imbalances. Treatment of diabetes insipidus requires the administration of replacement quantities of antidiuretic hormone, either by injection or by absorption of the hormone through the mucosa of the nose when given as a nasal spray.

Disorders of the anterior and posterior pituitary are summarized in Table 10-1.

Hypothalamus

In discussing the hormones, ADH and oxytocin, we noted that these hormones were *released* from the posterior lobe of the pituitary. Actual production of these two hormones occurs in the hypothalamus of the brain. Two groups of specialized neurons in the hypothalamus synthesize the posterior pituitary hormones, which then pass down along axons into the pituitary gland. Release of ADH and oxytocin into the blood is controlled by nervous stimulation.

In addition to oxytocin and ADH, the hypothalamus also produces substances called **releasing** and **inhibiting hormones.** These substances are produced in the hypothalamus and then travel directly through a specialized blood capillary system to the anterior pituitary gland, where they cause the release of anterior pituitary hormones or, in a number of instances, inhibit their production and their release into the general circulation.

The combined nervous and endocrine functions of the hypothalamus allow it to play a dominant role in the regulation of many body functions related to homeostasis. Examples include the regulation of body temperature, appetite, and thirst.

Thyroid Gland

Earlier in this chapter, we mentioned that some endocrine glands are not located in a body cavity. The thyroid is one of these. It lies in the neck just below the larynx (Figure 10-6).

The thyroid gland secretes two thyroid hormones, **thyroxine** (thye-ROK-sin) or T_4 and **triiodothyronine** (try-eye-o-doe-THY-ro-neen) or T_3. It also secretes the hormone **calcitonin** (kal-si-TOE-nin). Of the two thyroid hormones, T_4 is the more important and more abundant. One molecule of T_4 contains four atoms of iodine, and one molecule of T_3 as its name suggests, contains three iodine atoms. For T_4 to be produced in adequate amounts, the diet must contain sufficient iodine.

Most endocrine glands do not store their hormones but secrete them directly into the blood as they are produced. The thyroid gland is different in that it stores considerable amounts of the thyroid hormones in the form of a colloid compound seen in Figure 10-7. The colloid material is stored in the follicles of the gland, and when the thyroid hormones are needed, they are released from the colloid and secreted into the blood.

T_4 and T_3 influence every one of the trillions of cells in our bodies. They make them speed up their release of energy from foods. In other words, these thyroid hormones stimulate cellular metabolism. This has far-reaching effects. Because all body functions depend on a normal supply of energy, they all depend on normal thyroid secretion. Even normal mental and physical growth and development depend on normal thyroid functioning.

Calcitonin decreases the concentration of calcium in the blood by first acting on bone to inhibit its breakdown. With less bone being resorbed, less calcium moves out of bone into blood, and, as a result, the concentration of calcium in blood decreases. An increase in calcitonin secretion

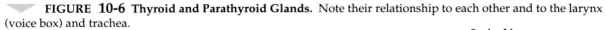

FIGURE 10-6 Thyroid and Parathyroid Glands. Note their relationship to each other and to the larynx (voice box) and trachea.

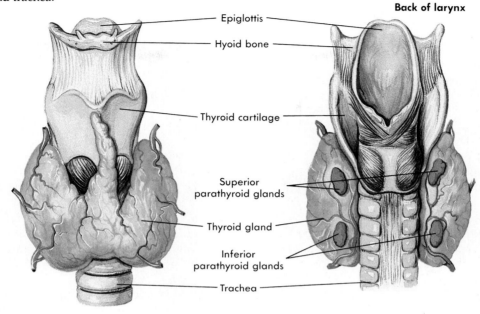

Back of larynx

Epiglottis

Hyoid bone

Thyroid cartilage

Superior parathyroid glands

Thyroid gland

Inferior parathyroid glands

Trachea

FIGURE 10-7 Thyroid Gland Tissue. Note that each of the follicles is filled with colloid. The colloid serves as a storage medium for the thyroid hormones.

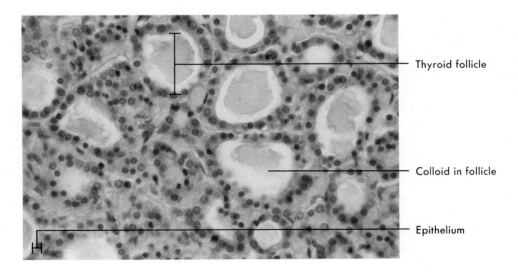

Thyroid follicle

Colloid in follicle

Epithelium

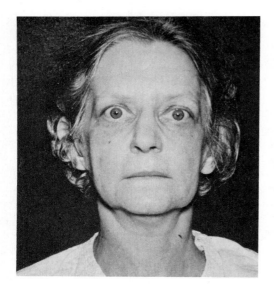

FIGURE 10-8 Hyperthyroidism. Note the prominent, protruding eyes (exophthalmos) of this woman with Graves' disease.

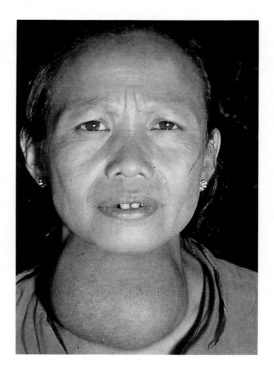

FIGURE 10-9 Simple Goiter. The enlarged thyroid gland appears as a swelling of the neck. This condition results from a low dietary intake of iodine.

PROSTAGLANDIN THERAPY

Although much research is yet to be done, prostaglandins are already playing an important role in treatment of diverse conditions such as high blood pressure, asthma, and ulcers. Because some prostaglandins have local muscle-relaxing effects, they can relax muscles in the walls of blood vessels to reduce blood pressure. In asthma, prostaglandins administered in a nebulizer (mist applicator) relax the muscles that constrict air flow during an asthma attack. Some gastric ulcers can be treated with prostaglandins that decrease stomach acid secretion.

Pharmacologists, scientists who study drug actions, have discovered that prostaglandins may also be involved in some traditional therapies. For example, evidence indicates that aspirin and its derivatives *(salicylates)* produce some of their effects by blocking prostaglandins involved in the inflammation response.

FIGURE 10-10 Myxedema. This condition results from hyposecretion of the thyroid gland during the adult years. Note the edema around the eyes and the facial puffiness.

quickly follows any increase in blood calcium concentration, even if it is a slight one. This causes blood calcium concentration to decrease to its normal level. Calcitonin thus helps maintain homeostasis of blood calcium. It prevents a harmful excess of calcium in the blood, a condition called **hypercalcemia** (hye-per-kal-SEE-me-ah), from developing.

Hyperthyroidism (hye-per-THYE-royd-izm) or oversecretion of the thyroid hormones dramatically increases the metabolic rate. Food material is burned by the cells at an excessive rate, and individuals who suffer from this condition lose weight, have an increased appetite, and show signs of nervous irritability. They appear restless, jumpy, and excessively active. Many patients with hyperthyroidism also have very prominent, almost protruding eyes—a condition called *exophthalmos* (Figure 10-8). Hyperthyroidism with exophthalmos is characteristic of **Graves' disease,** an inherited condition that occurs 5 times more frequently in women than in men.

Hypothyroidism (hye-po-THYE-royd-izm) or undersecretion of thyroid hormones can be caused by and result in a number of different conditions. Low dietary intake of iodine causes a painless enlargement of the thyroid gland called **simple goiter** (GOY-ter), shown in Figure 10-9. This condition was once common in areas of the United States where the iodine content of the soil and water was inadequate. The use of iodized salt has dramatically reduced the incidence of simple goiter caused by low iodine intake. In simple goiter the gland enlarges in an attempt to compensate for the lack of iodine in the diet necessary for the synthesis of thyroid hormones.

Hyposecretion of thyroid hormones during the formative years leads to a condition called **cretinism** (KREE-tin-izm). It is characterized by a low metabolic rate, retarded growth and sexual development, and, frequently, mental retardation. Later in life, deficient thyroid hormone secretion produces the disease called **myxedema** (mik-se-DEE-mah). The low metabolic rate that characterizes myxedema leads to lessened mental and physical vigor, weight gain, loss of hair, and an accumulation of mucous fluid in the subcutaneous tissue that is often most noticeable around the eyes (Figure 10-10).

Disorders of thyroid secretion are summarized in Table 10-1.

Parathyroid Glands

The **parathyroid glands** are small glands. There are usually four of them, and they are found on the back of the thyroid gland (Figure 10-6). The parathyroid glands secrete **parathyroid hormone (PTH).**

Parathyroid hormone increases the concentration of calcium in the blood—the opposite effect of the thyroid gland's calcitonin. Whereas calcitonin acts to decrease the amount of calcium being resorbed from bone, parathyroid hormone acts to increase it. Parathyroid hormone stimulates bone-resorbing cells or osteoclasts to increase their breakdown of bone's hard matrix, a process that frees the calcium stored in the matrix. The released calcium then moves out of bone into blood, and this in turn increases the blood's calci-

▼ **FIGURE 10-11 Regulation of Blood Calcium Levels.** Calcitonin and parathyroid hormones have antagonistic (opposite) effects on calcium concentration in the blood.

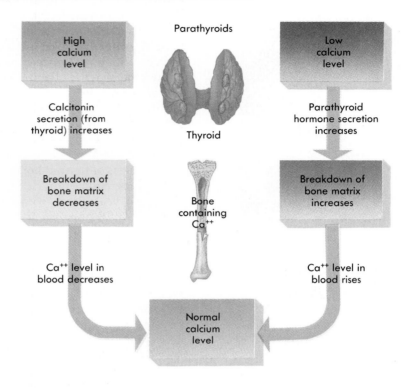

| High calcium level |
| Calcitonin secretion (from thyroid) increases |
| Breakdown of bone matrix decreases |

Parathyroids

Thyroid

Bone containing Ca^{++}

| Low calcium level |
| Parathyroid hormone secretion increases |
| Breakdown of bone matrix increases |

Ca^{++} level in blood decreases

Ca^{++} level in blood rises

Normal calcium level

um concentration. For a summary of the antagonistic effects of calcitonin and parathyroid hormone, see Figure 10-11. This is a matter of life-and-death importance because our cells are extremely sensitive to changing amounts of blood calcium. They cannot function normally with too much or too little calcium. For example, with too much blood calcium, brain cells and heart cells soon do not function normally; a person becomes mentally disturbed, and the heart may stop. However, with too little blood calcium, nerve cells become overactive, sometimes to such an extreme degree that they bombard muscles with so many impulses that the muscles go into spasms.

Disorders of parathyroid secretion are summarized in Table 10-1.

▼ *Adrenal Glands*

As you can see in Figures 10-1 and 10-12, an adrenal gland curves over the top of each kidney. From the surface an adrenal gland appears to be only one organ, but it is actually two separate endocrine glands: the **adrenal cortex** and the **adrenal medulla.** Does this two-glands-in-one structure remind you of another endocrine organ? (See p. 265). The cortex is the outer part of an adrenal gland, and the medulla is its inner part. Adrenal cortex hormones have different names and actions from adrenal medulla hormones.

ADRENAL CORTEX

Three different zones or layers of cells make up the **adrenal cortex** (Figure 10-12). Starting with

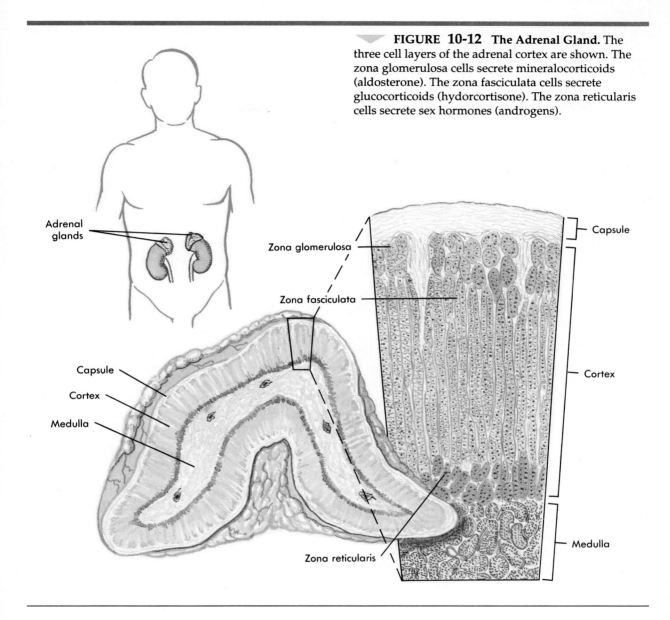

FIGURE 10-12 The Adrenal Gland. The three cell layers of the adrenal cortex are shown. The zona glomerulosa cells secrete mineralocorticoids (aldosterone). The zona fasciculata cells secrete glucocorticoids (hydorcortisone). The zona reticularis cells secrete sex hormones (androgens).

the zone or layer directly under the outer capsule of the gland and proceeding to the center (that is, going from superficial to deep), their names are:

1 Zona (ZO-nah) glomerulosa (glo-mare-yoo-LO-sah)
2 Zona (ZO-nah) fasciculata (fas-ik-yoo-LAH-tah)
3 Zona (ZO-nah) reticularis (re-tic-yoo-LAIR-is)

Hormones secreted by the three cell layers or zona of the adrenal cortex are called **corticoids** (KOR-ti-koyds). The outer zone of adrenal cortex cells or zona glomerulosa secretes hormones called **mineralocorticoids** (min-er-al-o-KOR-ti-koyds) or **MCs** for short. The main mineralocorticoid is the hormone **aldosterone** (al-DOS-ste-rone). The middle zone or zona fasciculata secretes **glucocorticoids** (gloo-ko-KOR-ti-koyds) or **GCs. Cortisol** or **hydrocortisone** is the chief glucocorticoid. The innermost or deepest zone of the cortex or zona reticularis secretes small amounts of **sex hormones.** Sex hormones secreted by the adrenal cortex resemble testosterone. We shall now discuss briefly the functions of these three kinds of adrenal cortical hormones.

As their name suggests, **mineralocorticoids** help control the amount of certain mineral salts (mainly sodium chloride) in the blood. Aldosterone is the chief mineralocorticoid. Remember its main functions—to increase the amount of sodium and decrease the amount of potassium in the blood. These changes, in turn, lead to other profound changes. Aldosterone increases blood sodium and decreases blood potassium by influencing the kidney tubules. It causes them to speed up their reabsorption of sodium back into the blood so that less of it will be lost in the urine. At the same time, aldosterone causes the tubules to increase their secretion of potassium so that more of this mineral will be lost in the urine. The effects of aldosterone speed up kidney reabsorption of water.

One of the important functions of glucocorticoids is to help maintain normal blood glucose concentration. Glucocorticoids increase **gluconeogenesis** (gloo-ko-nee-o-JEN-e-sis), a process that converts amino acids or fatty acids to glucose. It is performed mainly by liver cells. Glucocorticoids act in several ways to increase gluconeogenesis. They promote the breakdown of tissue proteins to amino acids, especially in muscle cells. Amino acids thus formed move out of the tissue cells into blood and circulate to the liver. Liver cells then change them to glucose by the process of gluconeogenesis. The newly formed glucose leaves the liver cells and enters the blood. This of course increases blood glucose concentration.

In addition to performing these functions that are necessary for maintaining normal blood glucose concentration, glucocorticoids also play an essential part in maintaining normal blood pressure. They act in a complicated way to make it possible for two other hormones secreted by the adrenal medulla to partially constrict blood vessels, a condition necessary for maintaining normal blood pressure. Also, glucocorticoids act with these hormones from the adrenal medulla to pro-

duce an antiinflammatory effect. They bring about a normal recovery from inflammations produced by many kinds of agents. The use of hydrocortisone to relieve skin rashes, for example, is based on the antiinflammatory effect of glucocorticoids.

Another effect produced by glucocorticoids is called their *antiimmunity, antiallergy effect.* Glucocorticoids bring about a decrease in the number of certain cells that produce antibodies, substances that make us immune to some factors and allergic to others.

When extreme stimuli act on the body, they produce an internal state or condition known as *stress.* Surgery, hemorrhage, infections, severe burns, and intense emotions are examples of extreme stimuli that bring on stress. The normal adrenal cortex responds to the condition of stress by quickly increasing its secretion of glucocorticoids. This fact is well established. What is still not known, however, is whether the increased amount of glucocorticoids helps the body cope successfully with stress. Increased glucocorticoid secretion is only one of many ways in which the body responds to stress, but it is one of the first stress responses, and it brings about many of the other stress responses. Examine Figure 10-13 to discover what stress responses are produced by a high concentration of glucocorticoids in the blood.

The sex hormones secreted by the zona reticularis are weak male hormones **(androgens)** similar to testosterone. These hormones are secreted in small amounts in males and females. In females, these androgens stimulate the female sexual drive. In males, so much testosterone is secreted by the testes that adrenal androgens are physiologically insignificant.

ADRENAL MEDULLA

The **adrenal medulla,** or inner portion of the adrenal gland shown in Figure 10-12, secretes the hormones, **epinephrine** (ep-i-NEF-rin) and **norepinephrine** (nor-ep-i-NEF-rin).

Our bodies have many ways to defend themselves against enemies that threaten their well-being. A physiologist might say that the body resists stress by making many stress responses. We have just discussed increased glucocorticoid secretion. An even faster-acting stress response is

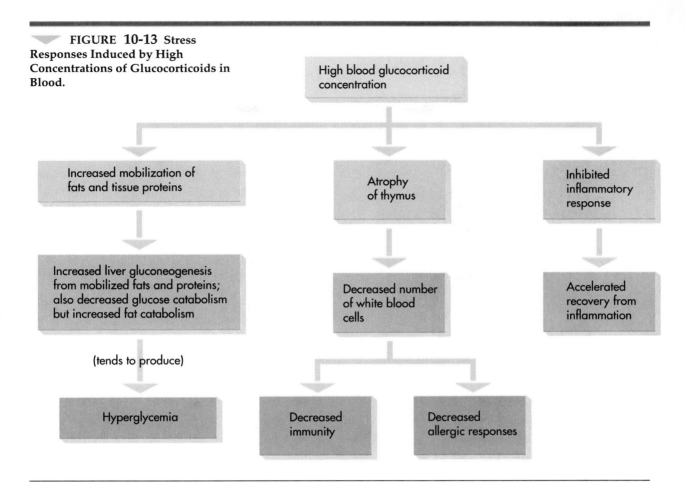

FIGURE 10-13 Stress Responses Induced by High Concentrations of Glucocorticoids in Blood.

increased secretion of hormones by the adrenal medulla. This occurs very rapidly because nerve impulses conducted by sympathetic nerve fibers stimulate the adrenal medulla. When stimulated, it literally squirts epinephrine and norepinephrine into the blood. Like glucocorticoids, these hormones may help the body resist stress. Unlike glucocorticoids, they are not essential for maintaining life. On the other hand, glucocorticoids, the hormones from the adrenal cortex, may help the body resist stress and are essential for life.

Suppose you suddenly faced some threatening situation. Imagine that a gunman threatened to kill you or that your doctor told you that you had to have a dangerous operation. Almost instantaneously, the medullas of your two adrenal glands would be galvanized into feverish activity. They would quickly secrete large amounts of epinephrine (adrenaline) into your blood. Many of your body functions would seem to be supercharged. Your heart would beat faster; your blood pressure would rise; more blood would be pumped to your skeletal muscles; your blood would contain more glucose for more energy, and so on. In short, you would be geared for strenuous activity, for "fight or flight." Epinephrine prolongs and intensifies changes in body function brought about by the stimulation of the sympathetic subdivision of the autonomic nervous system. Recall from Chapter 8 that sympathetic or adrenergic nerve fibers release epinephrine and norepinephrine as neurotransmitter substances.

The close functional relationship between the nervous and the endocrine systems is perhaps

▼ **FIGURE 10-14 Cushing's Syndrome.** This condition results from hypersecretion of glucocorticoid hormone by a tumor of the zona fasciculata of the adrenal cortex. **A,** First diagnosed with Cushing's syndrome. **B,** Four months later.

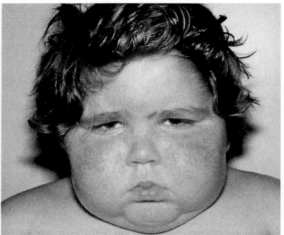

most noticeable in the body's response to stress. The term **general-adaptation syndrome** or **GAS** is often used to describe how the body mobilizes different defense mechanisms when threatened by harmful stimuli. In generalized stress conditions the hypothalamus acts on the anterior pituitary gland to cause the release of ACTH, which stimulates the adrenal cortex to secrete glucocorticoids. In addition, the sympathetic subdivision of the autonomic nervous system is stimulated with the adrenal medulla, so the release of epinephrine and norepinephrine occurs to assist the body in responding to the stressful stimulus. Unfortunately, during periods of prolonged stress, glucocorticosteroids may have harmful side effects because they are antiinflammatory and cause blood vessels to constrict. For example, decreased immune activity in the body may promote the spread of infections; in addition, prolonged blood vessel constriction may lead to increased blood pressure.

Injury, disease states, or malfunction of the adrenal glands can result in hypersecretion or hyposecretion of several different hormones.

Tumors of the adrenal cortex located in the zona fasciculata often result in the production of abnormally large amounts of glucocorticoids. The medical name for the collection of symptoms that characterize hypersecretion of glucocorticoids is **Cushing's syndrome.** For some reason more women than men develop Cushing's syndrome. Its most noticeable features are the so-called moon face (Figure 10-14) and the buffalo hump on the upper back that develop because of the redistribution of body fat. Individuals with Cushing's syndrome also have elevated blood sugar levels and suffer frequent infections. Surgical removal of a glucocorticoid-producing tumor may result in dramatic improvement of the moon-face symptom within only 6 months.

Tumors that affect the zona reticularis often produce testosterone-like sex hormones called

FIGURE 10-15 Results of a Virilizing Tumor. This young girl has a virilizing tumor of the zona reticularis of the adrenal cortex. The tumor secretes androgens, thereby producing masculinizing effects that resemble the secondary sex characteristics of men.

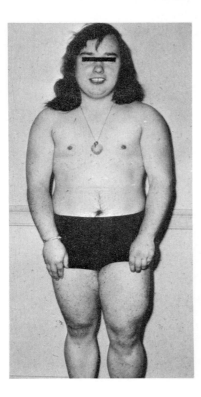

androgens. As a result, the symptoms of hypersecetion often resemble the male secondary sexual characteristics such as beard growth, development of body hair, and increased muscle mass. If these masculinizing symptoms appear in a woman (Figure 10-15), the cause is frequently a **virilizing** (VEER-il-eye-zing) **tumor** of the adrenal cortex. The term *virile* is from the Latin word *virilis* meaning "male" or "masculine."

Deficiency or hyposecretion of adrenal cortex hormones results in a condition called **Addison's disease.** Reduced cortical hormone levels result in muscle weakness, reduced blood sugar, nausea, loss of appetite, and weight loss.

Disorders of adrenal secretion are summarized in Table 10-1.

Pancreatic Islets

All the endocrine glands discussed so far are big enough to be seen without a magnifying glass. The **pancreatic islets** or **islets of Langerhans,** in contrast, are too tiny to be seen without a microscope. These glands are merely little clumps of cells scattered like islands in a sea among the pancreatic cells that secrete the pancreatic digestive juice (Figure 10-16).

There are two kinds of cells in the pancreatic islets: the *alpha cells* and *beta cells.* Alpha cells secrete a hormone called **glucagon,** whereas beta cells secrete one of the most famous of all hormones, **insulin.**

Glucagon accelerates a process called **liver glycogenolysis** (glye-ko-jen-OL-i-sis). Glycogenolysis is a chemical process by which the glucose stored in the liver cells in the form of glycogen is converted to glucose. This glucose then leaves the liver cells and enters the blood. Glucagon therefore increases blood glucose concentration.

Insulin and glucagon are antagonists. In other words, insulin decreases blood glucose concentration; glucagon increases it. Insulin is the only hormone that can decrease blood glucose concentration. Other hormones, however, increase its concentration. We have already named three of these: glucocorticoids, growth hormone, and glucagon. Insulin decreases blood glucose by accelerating its movement out of the blood, through cell membranes, and into cells. As glucose enters the cells at a faster rate, the cells increase their metabolism of glucose. Briefly then, insulin decreases blood glucose and increases glucose metabolism.

If the pancreatic islets secrete a normal amount of insulin, a normal amount of glucose enters the cells, and a normal amount of glucose stays behind in the blood. ("Normal" blood glucose is about 80 to 120 mg of glucose in every 100 ml of blood during fasting.) If the pancreatic islets secrete too much insulin, as they sometimes do when a person has a tumor of the pancreas, more glucose than usual leaves the blood to enter the

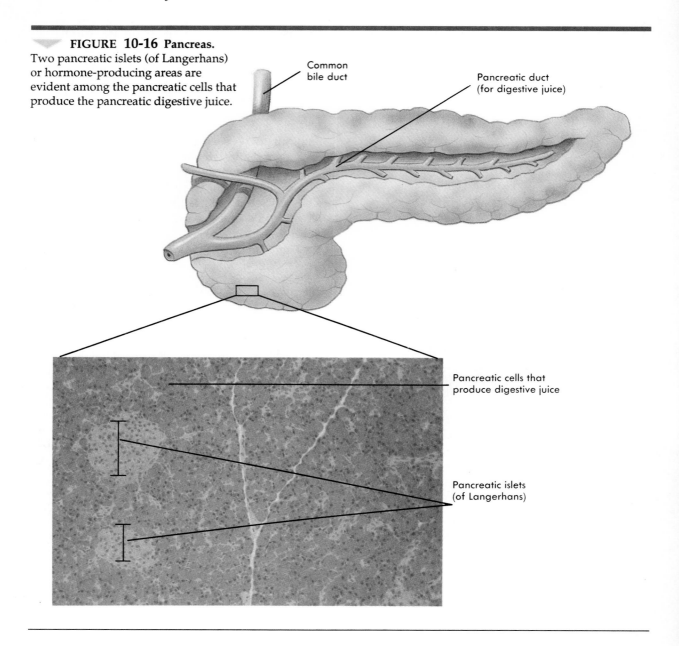

FIGURE 10-16 Pancreas.
Two pancreatic islets (of Langerhans) or hormone-producing areas are evident among the pancreatic cells that produce the pancreatic digestive juice.

Common bile duct

Pancreatic duct (for digestive juice)

Pancreatic cells that produce digestive juice

Pancreatic islets (of Langerhans)

cells, and blood glucose decreases. If the pancreatic islets secrete too little insulin, as they do in type I (insulin-dependent) **diabetes** (dye-ah-BEE-tes) **mellitus** (mell-EYE-tus), less glucose leaves the blood to enter the cells, so the blood glucose increases, sometimes to even three or more times the normal amount. Most cases of type II (non-insulin-dependent) diabetes mellitus result from biochemical problems with target cells, preventing the normal effects of insulin on its target cells and thus also raising blood glucose levels.

Screening tests for all types of diabetes mellitus rely on the fact that the blood glucose level is elevated in this condition. Today, most screening is done with a simple test with a drop of blood. Subjects with a high blood glucose level are suspected of having diabetes mellitus. Testing for sugar in the urine is another common screening procedure. In diabetes mellitus, excess glucose is filtered out of the blood by the kidneys and lost in the urine, producing the condition **glycosuria** (glye-ko-SOO-ree-ah).

Disorders of pancreatic islet secretion are summarized in Table 10-1.

Female Sex Glands

A woman's primary sex glands are her two ovaries. Each ovary contains two different kinds of glandular structures: the ovarian follicles and the corpus luteum. **Ovarian follicles** secrete estrogen, the "feminizing hormone." The **corpus luteum** chiefly secretes progesterone but also some estrogen. We shall save our discussion of the structure of these endocrine glands and the functions of their hormones for Chapter 21.

Male Sex Glands

Some of the cells of the testes produce the male sex cells called **sperm.** Other cells in the testes, male reproductive ducts, and glands produce the liquid portion of the male reproductive fluid called *semen.* The interstitial cells in the testes secrete the male sex hormone called **testosterone** directly into the blood. These cells of the testes are therefore the male endocrine glands. Testosterone is the "masculinizing hormone." Chapter 21 contains more information about the structure of the testes and the functions of testosterone.

Thymus

The thymus is located in the mediastinum (Figure 10-1), and in infants it may extend up into the neck as far as the lower edge of the thyroid gland. Like the adrenal gland, the thymus has a cortex and medulla. Both portions are composed largely of lymphocytes (white blood cells). As part of the body's immune system, the endocrine function of

the thymus is not only important but essential. This small structure (it weighs more than 28 grams [about an ounce]) plays a critical part in the body's defenses against infections—in its vital immunity mechanism.

The hormone **thymosin** (THYE-mo-sin) has been isolated from thymus tissue and is considered responsible for its endocrine activity. Thymosin is actually a group of several hormones that together play an important role in the development and function of the body's immune system.

Suppression of the immune system sometimes occurs in certain disease states and in patients who are undergoing massive chemotherapy or radiotherapy for the treatment of cancer. Such individuals are said to be "immunosuppressed" and are extremely susceptible to infections. Thymosin may prove useful as an activator of the immune system in such patients.

Placenta

The placenta functions as a temporary endocrine gland. During pregnancy, it produces **chorionic** (KO-ree-on-ik) **gonadotropins** (gon-ah-doe-TRO-pins), so called because they are tropic hormones secreted by cells of the **chorion** (KO-ree-on), the outermost membrane that surrounds the baby during development in the uterus. In addition to chorionic gonadotropins, the placenta also produces estrogen and progesterone. During pregnancy, the kidneys excrete large amounts of chorionic gonadotropins in the urine. This fact, discovered more than a half century ago, led to the development of the now-familiar pregnancy tests.

Pineal Gland

The pineal gland is a small, pine-cone-shaped gland near the roof of the third ventricle of the brain (see Figure 8-11). It is easily located in a child but becomes fibrous and encrusted with calcium deposits as a person ages. The pineal gland produces a number of hormones in very small quantities, with **melatonin** being the most significant. Melatonin inhibits the tropic hormones that

EXERCISE AND DIABETES MELLITUS

Type I (insulin-dependent) diabetes mellitus is characterized by high blood glucose concentration because the lack of sufficient insulin prevents glucose from entering cells. However, exercise physiologists have found that aerobic training increases the number of insulin receptors in target cells and the insulin affinity (attraction) of the receptors. This condition allows a small amount of insulin to have a greater effect than it would have otherwise had. Thus exercise reduces the severity of the diabetic condition.

All forms of diabetes benefit from properly planned exercise therapy. Not only is this form of treatment natural and cost effective, but it also helps reduce or prevent other problems such as obesity and heart disease.

affect the ovaries, and it is thought to be involved in regulating the onset of puberty and the menstrual cycle in women. Because the pineal gland receives and responds to sensory information from the optic nerves, it is sometimes called the *third eye.* The pineal gland uses information regarding changing light levels to adjust its output of melatonin; melatonin levels increase during the night and decrease during the day. This cyclic variation is thought to be an important timekeeping mechanism for the body's internal clock.

Abnormal secretion of or sensitivity to melatonin has recently been implicated in a number of disorders. One dramatic example is so-called *winter depression*. Patients with this condition exhibit signs of clinical depression only during the winter months, when nights are long. Apparently, unusually high melatonin levels associated with long winter nights causes psychological effects in these patients. A treatment that has been successful in some cases involves the use of bright lights for a few hours after sundown. The pineal gland seems to be tricked into responding as if the patient is experiencing a long summer day, thus secreting less melatonin (see Table 10-1).

Other Endocrine Structures

Continuing research into the endocrine system has shown that nearly every organ and system has an endocrine function. Tissues in the kidneys, stomach, intestines, and other organs secrete hormones that regulate a variety of essential human functions. One of the most recently discovered hormones is **atrial natriuretic hormone (ANH).** Secreted by cells in the wall of the heart's atria (upper chambers), ANH is an important regulator of fluid and electrolyte homeostasis. ANH is an antagonist to aldosterone. Aldosterone stimulates the kidney to retain sodium ions and water, whereas ANH stimulates loss of sodium ions and water.

OUTLINE SUMMARY

MECHANISMS OF HORMONE ACTION

A Endocrine glands secrete chemicals (hormones) into the blood (Figure 10-1)

B Hormones perform general functions of communication and control but a slower, longer-lasting type of control than that provided by nerve impulses

C Cells acted on by hormones are called *target organ cells*

D Protein hormones (first messengers) bind to receptors on the target cell membrane, triggering second messengers to affect the cell's activities (Figure 10-2)

E Steroid hormones bind to receptors within the target cell and influence cell activity by acting on specific genes

REGULATION OF HORMONE SECRETION

A Hormone secretion is controlled by homeostatic feedback

B Negative feedback—mechanisms that reverse the direction of a change in a physiological system (Figure 10-3)

C Positive feedback—(uncommon) mechanisms that amplify physiological changes

MECHANISMS OF ENDOCRINE DISEASE
(Table 10-1)

A Hypersecretion—secretion of an excess of hormone

B Hyposecretion—insufficent hormone secretion

C Target cell insensitivity produces results similar to hyposecretion

D Endocrinologists have developed many different strategies for treatment (for example, surgery and hormone therapy)

PROSTAGLANDINS

A Prostaglandins (PGs) are powerful substances found in a wide variety of body tissues

B PGs are often produced in a tissue and diffuse only a short distance to act on cells in that tissue

C Several classes of PGs include prostaglandin A (PGA), prostaglandin E (PGE), and prostaglandin F (PGF)

D PGs influence many body functions, including respiration, blood pressure, gastrointestinal secretions, and reproduction

PITUITARY GLAND (Figure 10-5)

A Anterior pituitary gland (adenohypophysis)
  1 Names of major hormones
    a Thyroid-stimulating hormone (TSH)
    b Adrenocorticotropic hormone (ACTH)
    c Follicle-stimulating hormone (FSH)
    d Luteinizing hormone (LH)
    e Melanocyte-stimulating hormone (MSH)
    f Growth hormone (GH)
    g Prolactin (lactogenic hormone)
  2 Functions of major hormones
    a TSH—stimulates growth of the thyroid gland; also stimulates it to secrete thyroid hormone
    b ACTH—stimulates growth of the adrenal cortex and stimulates it to secrete glucocorticoids (mainly cortisol)
    c FSH—initiates growth of ovarian follicles each month in the ovary and stimulates one or more follicles to develop to the stage of maturity and ovulation; FSH also stimulates estrogen secretion by developing follicles; stimulates sperm production in the male
    d LH—acts with FSH to stimulate estrogen secretion and follicle growth to maturity; causes ovulation; causes luteinization of the ruptured follicle and stimulates progesterone secretion by corpus luteum; causes interstitial cells in the testes to secrete testosterone in the male
    e MSH—causes a rapid increase in the synthesis and spread of melanin (pigment) in the skin
    f GH—stimulates growth by accelerating protein anabolism; also accelerates fat catabolism and slows glucose catabolism; by slowing glucose catabolism, tends to increase blood glucose to higher than normal level (hyperglycemia)
      (1) Hypersecretion during childhood results in gigantism and during adulthood results in acromegaly
      (2) Hyposecretion during childhood results in pituitary dwarfism
    g Prolactin or lactogenic hormone—stimulates breast development during pregnancy and secretion of milk after the delivery of the baby

B Posterior pituitary gland (neurohypophysis)
  1 Names of hormones
    a Antidiuretic hormone (ADH)
      (1) Hyposecretion causes diabetes insipidus, characterized by excessive volumes of urine
    b Oxytocin

2 Functions of hormones
  a ADH—accelerates water reabsorption from urine in the kidney tubules into the blood, thereby decreasing urine secretion
  b Oxytocin—stimulates the pregnant uterus to contract; may initiate labor; causes glandular cells of the breast to release milk into ducts

HYPOTHALAMUS

A Actual production of ADH and oxytocin occurs in the hypothalamus
B After production in the hypothalamus, hormones pass along axons into the pituitary gland
C The secretion and release of posterior pituitary hormones is controlled by nervous stimulation
D The hypothalamus controls many body functions related to homeostasis (temperature, appetite, and thirst)

THYROID GLAND (Figure 10-6)

A Names of hormones
  1 Thyroid hormone—thyroxine (T_4) and triiodothyronine (T_3)
  2 Calcitonin
B Functions of hormones
  1 Thyroid hormone—accelerates catabolism (increases the body's metabolic rate)
  2 Calcitonin—decreases the blood calcium concentration by inhibiting breakdown of bone, which would release calcium into the blood
C Hyperthyroidism (hypersecretion of thyroid hormones) increases metabolic rate
  1 Characterized by restlessness and exophthalmos (protuding eyes)
  2 Graves' disease is an inherited form of hyperthyroidism
D Hypothyroidism (hyposecretion of thyroid hormones)
  1 May result from different conditions
  2 Simple goiter—painless enlargement of thyroid due to dietary deficiency of iodine
  3 Hyposecretion during early development may result in cretinism (retardation) and during adulthood in myxedema (characterized by edema and sluggishness)

PARATHYROID GLANDS (Figure 10-6)

A Name of hormone—parathyroid hormone (PTH)
B Function of hormone—increases blood calcium concentration by increasing the breakdown of bone with the release of calcium into the blood

ADRENAL GLANDS (Figure 10-12)

A Adrenal cortex
  1 Names of hormones (corticoids)
    a Glucocorticoids (GCs)—chiefly cortisol (hydrocortisone)
    b Mineralocorticoids (MCs)—chiefly aldosterone
    c Sex hormones—small amounts of male hormones (androgens) secreted by adrenal cortex of both sexes
  2 Cell layers (zonae)
    a Zona glomerulosa—outermost layer, secretes mineralocorticoids
    b Zona fasciculata—middle layer, secretes glucocorticoids
    c Zona reticularis—deepest or innermost layer, secretes sex hormones
  3 Mineralocorticoids—increase blood sodium and decrease body potassium concentrations by accelerating kidney tubule reabsorption of sodium and excretion of potassium
  4 Functions of glucocorticoids
    a Help maintain normal blood glucose concentration by increasing gluconeogenesis—the formation of "new" glucose from amino acids produced by the breakdown of proteins, mainly those in muscle tissue cells; also the conversion to glucose of fatty acids produced by the breakdown of fats stored in adipose tissue cells
    b Play an essential part in maintaining normal blood pressure—make it possible for epinephrine and norepinephrine to maintain a normal degree of vasoconstriction, a condition necessary for maintaining normal blood pressure
    c Act with epinephrine and norepinephrine to produce an antiinflammatory effect, to bring about normal recovery from inflammations of various kinds
    d Produce antiimmunity, antiallergy effect; bring about a decrease in the number of lymphocytes and plasma cells and therefore a decrease in the amount of antibodies formed
    e Secretion of glucocorticoid quickly increases when the body is thrown into a condition of stress; high blood concentration of glucocorticoids, in turn, brings about many other stress responses (Figure 10-13)
B Adrenal medulla
  1 Names of hormones—epinephrine (adrenaline) and norepinephrine
  2 Functions of hormones—help the body resist stress by intensifying and prolonging the effects of sympathetic stimulation; increased epinephrine secretion is the first endocrine response to stress

C Adrenal abnormalities
1 Hypersecretion of glucocorticoids causes Cushing's syndrome: moon face, hump on back, elevated blood sugar levels, frequent infections
2 Hypersecretion of adrenal androgens may result from a virilizing tumor and cause masculinization of affected women
3 Hyposecretion of cortical hormones may result in Addison's disease: muscle weakness, reduced blood sugar, nausea, loss of appetite, and weight loss

PANCREATIC ISLETS (Figure 10-16)

A Names of hormones
1 Glucagon—secreted by alpha cells
2 Insulin—secreted by beta cells
B Functions of hormones
1 Glucagon increases the blood glucose level by accelerating liver glycogenolysis (conversion of glycogen to glucose)
2 Insulin decreases the blood glucose by accelerating the movement of glucose out of the blood into cells, which increases glucose metabolism by cells
C Diabetes mellitus
1 Type I (insulin dependent) results from hyposecretion of insulin
2 Type II (non–insulin dependent) results from target cell insensitivity to insulin
3 Glucose cannot enter cells and thus blood glucose levels rise, producing glycosuria (glucose in the urine)

FEMALE SEX GLANDS

The ovaries contain two structures that secrete hormones—the ovarian follicles and the corpus luteum; see Chapter 21

MALE SEX GLANDS

The interstitial cells of testes secrete the male hormone testosterone: see Chapter 21

THYMUS

A Name of hormone—thymosin
B Function of hormone—plays an important role in the development and function of the body's immune system

PLACENTA

A Name of hormones—chorionic gonadotropins, estrogens, and progesterone
B Functions of hormones—maintain the corpus luteum during pregnancy

PINEAL GLAND

A A cone-shaped gland near the roof of the third ventricle of the brain
1 Glandular tissue predominates in children and young adults
2 Becomes fibrous and calcified with age
B Called *third eye* because of its influence on secretory activity is related to the amount of light entering the eyes
C Secretes melatonin, which
1 Inhibits ovarian activity
2 Regulates the body's internal clock
D Abnormal secretion of (or sensitivity to) melatonin may produce winter depression, a form of depression that occurs when exposure to sunlight is low and melatonin levels are high

OTHER ENDOCRINE STRUCTURES

A Many organs (for example, the stomach, intestines, and kidney) produce endocrine hormones
B The atrial wall of the heart secretes atrial natriuretic hormone (ANH), which stimulates sodium loss from the kidneys

NEW WORDS

corticoids	glucocorticoids	luteinization	second messenger
diuresis	gluconeogenesis	mineralocorticoids	steroids
endocrine	glycogenolysis	negative feedback	stress
exocrine	hormone	prostaglandins	target organ cell

Diseases and Other Clinical Terms

acromegaly	diabetes insipidus	goiter	hyperglycemia
Addison's disease	diabetes mellitus	Graves' disease	hypoglycemia
cretinism	exophthalmos	hypercalcemia	myxedema
Cushing's syndrome	gigantism		

CHAPTER TEST

1. Chemicals secreted directly into the blood by endocrine glands are called _____.
2. Glands that discharge their secretions into ducts are called _____ glands.
3. Cyclic AMP is said to serve as a _____ _____ providing communication within the cells acted on by protein hormones.
4. The regulation of hormone levels in the blood depends on a homeostatic mechanism that reverses physiological changes, called _____ feedback.
5. Another name for the group of powerful "tissue hormones" found in many body tissues is _____.
6. Production of too much hormone by a diseased gland is called _____.
7. The hormone that acts on the thyroid gland and is produced by the pituitary gland is called _____ _____ hormone.
8. Growth hormone and prolactin are both secreted by the _____ _____ gland.
9. The letters ADH stand for _____ hormone.
10. One of the primary thyroid hormones is called T_4 or _____.
11. Low dietary intake of iodine may cause an enlargement of the thyroid gland, called simple _____.
12. The layer of the adrenal cortex that secretes mineralocorticoids is called the zona _____.
13. Epinephrine and norepinephrine are secreted by the adrenal _____.
14. The alpha cells of the pancreatic islets secrete _____.
15. Inadequate insulin secretion is associated with the disease _____ _____.
16. The corpus luteum of the ovary secretes chiefly _____.
17. The male sex hormone is called _____.
18. The hormone of the thymus gland is called _____.
19. The hormone melatonin is secreted by the _____ gland.
20. In the female the ovarian follicles secrete the hormone _____.

Circle the T before each true statement and the F before each false statement.

T F 21. Endocrine and exocrine glands discharge their secretions directly into the blood.
T F 22. Hormones act on target organ cells.
T F 23. The placenta is considered an endocrine gland.
T F 24. The second messenger hypothesis is used to explain the action of steroid hormones.
T F 25. Prostaglandins are also called *tissue hormones.*
T F 26. The term *adenohypophysis* is used to describe the posterior pituitary gland.
T F 27. Hyposecretion of antidiuretic hormone causes diabetes insipidus.
T F 28. Oxytocin and ADH are produced in the anterior pituitary gland.
T F 29. Parathyroid hormone stimulates breakdown of bone with release of calcium into the blood.
T F 30. The zona fasciculata of the adrenal cortex secretes sex hormones.
T F 31. High blood glucocorticoid concentration inhibits the inflammatory response.
T F 32. Hyposecretion of glucocorticoids results in Addison's disease.
T F 33. Beta cells of the pancreatic islets secrete insulin.
T F 34. Glycogenolysis converts glycogen to glucose.
T F 35. The thymus does not function as an endocrine gland.

REVIEW QUESTIONS

1 What endocrine glands are located in the abdominal cavity, cranial cavity, mediastinum, neck, and pelvic cavity?

2 How are the prostaglandins similar or dissimilar to regular hormones?

3 What endocrine gland secretes ACTH, aldosterone, calcitonin, chorionic gonadotropins, epinephrine, growth hormone, insulin, oxytocin, and progesterone?

4 Many changes occur in the body when it is in a condition of stress (for example, after a person has had major surgery). Name two endocrine glands that greatly increase their secretion of hormones in times of stress. Name the hormones they secrete.

5 Metabolism changes when the body is in a condition of stress. How does the metabolism of proteins, of fats, and of carbohydrates change, and what hormones cause the changes?

6 What hormones, with a high concentration present in the blood, make us less immune to infectious diseases?

7 What hormone is called the *water-retaining hormone* because it decreases the amount of urine formed?

8 What hormone causes potassium loss from the body through the urine?

9 What hormone is called the *salt-retaining hormone* because it makes the kidneys resorb sodium into the blood more rapidly so that less sodium is lost in the urine?

10 Name two or more hormones that increase blood glucose.

11 What hormone speeds up the rate of catabolism (that is, makes you burn up your foods faster)?

12 What is the main function of ACTH, ADH, calcitonin, epinephrine, insulin, parathyroid hormone, thyroid hormone, and testosterone?

13 Which hormone stimulates ovulation?

14 Cretinism results from abnormal secretion of what hormone?

15 Describe diabetes mellitus.

16 Does diabetes insipidus result from abnormal secretion of ADH or insulin?

17 What endocrine disorder produces gigantism?

18 What hormone decreases blood calcium concentration? What effect does this have on bone breakdown?

19 What hormone or hormones prepare the body for strenuous activity—for "fight or flight," in other words?

20 What hormone is important in the development and function of the immune system?

21 Name the hormone found at high levels only in the urine of pregnant women.

CLINICAL APPLICATIONS

1 George, the chief executive officer of a major institution, was jogging around his summer home when he became distressed as what seemed to be an irregularity of his heart rhythm. His assistants immediately rushed George to a hospital where he was diagnosed as having atrial fibrillation (uncoordinated contractions of the upper heart chambers). George was even more distressed to hear he had a specific heart condition, fearing it might disrupt his very active lifestyle. His physicians informed him that the overactivity of his heart—and perhaps other organs—was due to hyperthyroidism. Explain how hyperthyroidism could cause George's problems. What strategies might his physicians have available in treating him?

2 In George's case (see item *1*), the attending physicians chose to surgically remove part of the thyroid in an attempt to control George's hyperthyroidism. What precautions ought George's surgeons take in removing this tissue? (Hint: what anatomical structures in the thyroid area should they avoid cutting or removing?)

3 Your friend Lynn has type I diabetes mellitus. What therapy is likely to help her regain control of her metabolism and thus avoid possible tissue or organ damage? Lynn has told you that her condition, if untreated, results in "starvation" of cells in her body. You know that this condition is characterized by hyperglycemia (elevated blood glucose), so you wonder how the cells could starve if they have an excess of nutrients available. How is this possible?

Blood

Outline

Objectives

*After you have completed this chapter,
you should be able to:*

1 Describe the primary functions of blood.

2 List the formed elements of blood and identify the most important function of each.

3 Discuss anemia in terms of red blood cell numbers and hemoglobin content.

4 Explain the steps involved in blood clotting.

5 Describe ABO and Rh blood typing.

6 Define the following medical terms associated with blood: *hematocrit, leukocytosis, leukopenia, polycythemia, sickle cell, phagocytosis, acidosis, thrombosis, erythroblastosis fetalis, serum, fibrinogen, Rh factor, anemia, hemophilia, thrombocytopenia.*

7 Name two common disorders associated with each type of blood cell.

The next few chapters deal with **transportation** and **protection,** two of the body's most important functions. Have you ever thought of what would happen if the transportation ceased in your city or town? Or what would happen if the police, fire fighters, and members of the armed services stopped doing their jobs? Food would become scarce, garbage would pile up, and no one would protect you or your property. Stretch your imagination just a little, and you can imagine many disastrous results. Similarly, lack of transportation and protection for the cells—the "individuals" of the body—threatens the homeostasis of the body. The systems that provide these vital services for the body are the **circulatory system** and **lymphatic system.** In this chapter, we will discuss the primary transportation fluid—blood. Blood not only performs vital pickup and delivery services, but it also provides much of the protection necessary to withstand foreign "invaders." Blood vessels and the heart are discussed later, in Chapters 12 and 13. Lymphatic vessels, lymph nodes, and immunity are discussed in Chapter 14.

▼ *Blood Composition*

FORMED ELEMENTS

Blood is a fluid tissue that has many kinds of chemicals dissolved in it and millions upon millions of cells floating in it. The liquid (extracellular) part is called **plasma.** Suspended in the plasma are many different types of cells and cell fragments, which make up the **formed elements** of blood. There are three main types and several subtypes of formed elements:

1 Red blood cells (RBCs) or **erythrocytes** (e-RITH-ro-sites)
2 White blood cells (WBCs) or **leukocytes** (LOO-ko-sites)
   a Granular leukocytes (have granules in their cytoplasm)
      (1) Neutrophils
      (2) Eosinophils
      (3) Basophils
   b Nongranular **leukocytes** (do not have granules in their cytoplasm)
      (1) Lymphocytes
      (2) Monocytes
3 Platelets or **thrombocytes** (THROM-bo-sites)

Table 11-1 lists the functions of these different kinds of blood cells and shows what each looks like under the microscope.

It is difficult to believe how many blood cells and cell fragments there are in the body. For instance, 5,000,000 RBCs, 7500 WBCs, and 300,000 platelets in 1 cubic millimeter (mm^3) of blood (approximately 1 drop) are considered normal RBC, WBC, and platelet counts. Because RBCs, WBCs, and platelets are always being destroyed, the body must continuously make new ones at a staggering rate to take their place; a few million RBCs are manufactured *each second!*

Two kinds of connective tissue—**myeloid tissue** and **lymphatic tissue**—make blood cells for the body. Recall that formation of new blood cells is called *hemopoiesis.* Myeloid tissue is better known as *red bone marrow.* In the adult, it is chiefly in the sternum, ribs, and hipbones. A few other bones such as the vertebrae, clavicles, and cranial bones also contain small amounts of this valuable substance. Red bone marrow forms all types of blood cells except some lymphocytes and monocytes. Most of these others are formed by lymphatic tissue, which is located chiefly in the lymph nodes, thymus, and spleen.

As blood cells mature, they move into the circulatory vessels. Erythrocytes circulate up to 4 months before they break apart and their components are removed from the bloodstream by the liver. Granular leukocytes often have a lifespan of only a few days, but nongranular leukocytes may live over 6 months.

MECHANISMS OF BLOOD DISEASE

Most blood diseases are disorders of the formed elements. Thus it is not surprising that the basic mechanism of many blood diseases is the failure of the blood-producing myeloid and lymphatic tissues to form blood cells properly. In many cases, this failure is the result of damage by toxic chemicals or radiation. In other cases, it results from an inherited defect, viral infection, or even cancer.

If bone marrow failure is the suspected cause of a particular blood disorder, a sample of myeloid tissue may be drawn into a syringe from inside the pelvic bone (iliac crest) or the sternum. This procedure, called *aspiration biopsy cytology (ABC)* allows examination of the tissue that may help confirm or reject a tentative diagnosis. If the

▼ **FIGURE 11-1 RBCs.** Color-enhanced scanning electron micrograph shows the detailed structure of normal RBCs.

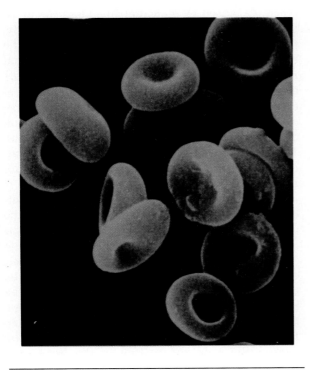

▼ **TABLE 11-1**   Classes of Blood Cells

BLOOD CELL	FUNCTION
Erythrocyte	Oxygen and carbon dioxide transport
Neutrophil	Immune defenses (phagocytosis)
Eosinophil	Defense against parasites and allergic responses
Basophil	Inflammatory response
B-lymphocyte	Antibody production (precursor of plasma cells)
T-lymphocyte	Cellular immune response
Monocyte	Immune defenses (phagocytosis)
Platelet	Blood clotting

bone marrow is severely damaged, the choice of a **bone marrow transplant** may be offered to the patient. In this procedure, myeloid tissue from a compatible donor is introduced into the recipient intravenously. If the recipient's immune system does not reject the new tissue, which is always a danger in tissue transplants, the donor cells may establish a colony of new, healthy tissue in the bone marrow.

RED BLOOD CELLS

As you can see in Figure 11-1, RBCs have an unusual shape. The cell is "caved in" on both sides so that each one has a thin center and thicker edges. Notice also that mature RBCs have no nucleus. Figure 11-1 shows RBCs photographed with a scanning electron microscope. With this instrument, extremely small objects can be enlarged far more than is possible with a standard light microscope, and, as you can see in the illustration, objects appear more three-dimensional. Because of the large numbers of RBCs and their unique shape, their total surface area is enormous. It provides an area larger than a football field for the exchange of oxygen and carbon dioxide between the blood and the body's cells.

RBCs have several functions. They transport oxygen to other cells in the body. A red pigment

▼ **FIGURE 11-2 Hematocrit Tubes Showing Normal Blood, Anemia, and Polycythemia.** Note the buffy coat located between the packed RBCs and the plasma. **A,** A normal percent of RBCs. **B,** Anemia (a low percent of RBCs). **C,** Polycythemia (a high percent of RBCs).

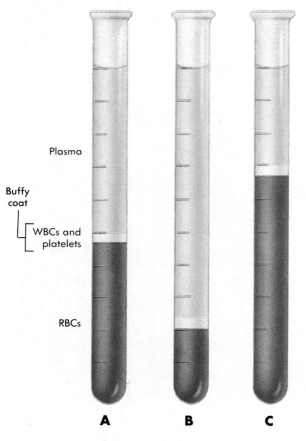

Plasma

Buffy coat

WBCs and platelets

RBCs

A   **B**   **C**

occur if the hemoglobin in RBCs is inadequate, even if adequate numbers of RBCs are present.

Changes in RBC Number

Anemias caused by an actual change in the number of RBCs can occur if blood is lost by hemorrhage as with accidents or bleeding ulcers or if the blood-forming tissues cannot maintain normal numbers of blood cells. Such failures occur because of cancer, radiation (x-ray) damage, and certain types of infections. If bone marrow produces an *excess* of RBCs, the result is a condition called **polycythemia** (pol-ee-sye-THEE-me-ah). The blood in individuals suffering from this condition may contain so many RBCs that it may become too thick to flow properly.

A common laboratory test called the **hematocrit** can tell a physician a great deal about the volume of RBCs in a blood sample. If whole blood is placed in a special hematocrit tube and then "spun down" in a centrifuge, the heavier formed elements will quickly settle to the bottom of the tube. During the procedure, RBCs are forced to the bottom of the tube first. The WBCs and platelets then settle out in a layer called the **buffy coat.** In Figure 11-2 the buffy coat can be seen between the packed RBCs on the bottom of the hematocrit tube and the liquid layer of plasma above. Normally about 45% of the blood volume consists of RBCs. For a patient with anemia, the percentage of RBCs drops, and for a patient with polycythemia, it increases dramatically (Figure 11-2).

Another laboratory test can give a more accurate report of the number of RBCs per unit of blood volume. Simply called an **RBC count,** this test requires counting the number of red blood cells in a blood sample of known volume. Originally RBC counts were done with a **hemocytometer,** a microscope slide with a counting grid etched on it. The current practice is to use a faster, more accurate automated blood cell counter. Normal RBC counts range from 4.5 to 5.5 million cells per cubic millimeter (mm^3) of blood.

One type of anemia characterized by an abnormally low number of red blood cells is **aplastic** (ay-PLAS-tik) **anemia.** Although idiopathic forms of this disease occur, most cases result from destruction of bone marrow by toxic chemicals, drugs or radiation. Less commonly, aplastic anemia results from bone marrow destruction by cancer. Because tissues that produce other formed ele-

called **hemoglobin** (hee-mo-GLO-bin) in RBCs unites with oxygen to form **oxyhemoglobin** (ok-see-hee-mo-GLO-bin). This oxygen-hemoglobin complex makes possible the efficient transport of large quantities of oxygen to body cells.

RED BLOOD CELL DISORDERS

The term **anemia** (ah-NEE-me-ah) is used to describe a number of different disease conditions caused by an inability of the blood to carry sufficient oxygen to the body cells. Anemias can result from inadequate numbers of RBCs or a deficiency of oxygen-carrying hemoglobin. Thus anemia can

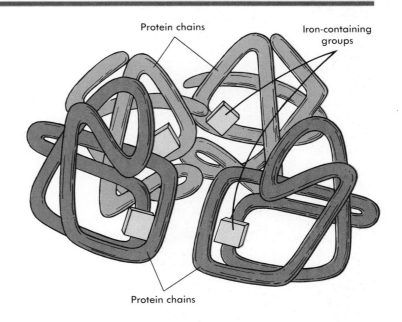

FIGURE 11-3 The Hemoglobin Molecule. This large molecule is composed of four protein subunits or chains that each hold an iron-containing chemical group at its core. The iron gives hemoglobin its oxygen-carrying capacity.

Protein chains

Iron-containing groups

Protein chains

ments are also affected, aplastic anemia is usually accompanied by a decreased number of white blood cells and platelets. Bone marrow transplants have been successful in treating some cases.

Pernicious (per-NISH-us) **anemia** is another disorder characterized by a low number of red blood cells. Pernicious anemia sometimes results from a dietary deficiency of vitamin B_{12}. Vitamin B_{12} is used in the formation of new red blood cells in the bone marrow. In many cases, pernicious anemia results from the failure of the stomach lining to produce *intrinsic factor*—the substance that allows vitamin B_{12} to be absorbed. Pernicious anemia can be fatal if not successfully treated. One method of treatment involves intramuscular injections of vitamin B_{12}.

Folate-deficiency anemia is similar to pernicious anemia because it also causes a decrease in the RBC count resulting from a vitamin deficiency. In this condition, *folic acid* (vitamin B_9) is deficient. Folic acid deficiencies are common among alcoholics and other malnourished individuals. Treatment involves taking vitamin supplements until a balanced diet can be restored.

Of course, a significant reduction in the number of red blood cells can occur as a result of blood loss. *Acute blood-loss anemia* often occurs after hemorrhages associated with trauma, extensive surgeries, or other situations involving a sudden loss of blood. *Chronic blood-loss anemia* is also called *anemia of chronic disease* because it often results from frequent or long-term episodes of blood loss associated with chronic diseases such as cancer or slow-bleeding ulcers.

Changes in Hemoglobin

The amount and quality of hemoglobin within red blood cells is just as important as the number of RBCs. In hemoglobin disorders, RBCs are sometimes classified as *hyperchromic* (abnormally high hemoglobin content) or *hypochromic* (abnormally low hemoglobin content). Total hemoglobin content is usually measured by breaking all the red blood cells in a sample of blood, producing a solution of free hemoglobin. The hemoglobin solution is then placed in an optical device that measures the color density of the hemoglobin in the sample. Normal hemoglobin ranges from 12 to 14 grams per 100 ml (g/100 ml) for adult females and from 14 to 16 g/100 ml for adult males. A hemoglobin value less than 10 g/100 ml indicates anemia.

Iron is a critical component of the hemoglobin molecule (Figure 11-3). Without adequate iron in

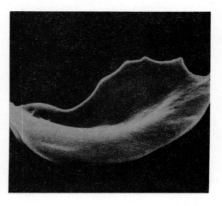

▽ **FIGURE 11-4 Sickle Cell.** A sickle-shaped red blood cell typical of sickle cell anemia.

BLOOD DOPING

A number of athletes have reportedly improved their performance by a practice called **blood doping.** A few weeks before an important event, an athlete draws some blood. The RBCs are separated and frozen. Just before competition, the RBCs are thawed and injected into the athlete. The increased hematocrit that results slightly improves the oxygen-carrying capacity of the blood, which theoretically improves performance. This method is judged to be an unfair and unwise practice in athletics but has recently been investigated for use by the military. U.S. armed services have reported that blood doping of soldiers just before combat could improve their endurance for up to 10 days.

the diet, the body cannot manufacture enough hemoglobin. The result is **iron deficiency anemia**—a worldwide medical problem. If hemoglobin falls below the normal level, as it does in this type of anemia, it starts an unhealthy chain reaction: less hemoglobin, less oxygen transported to cells, slower breakdown and use of nutrients by cells, less energy produced by cells, decreased cellular functions. If you understand this relationship between hemoglobin and energy, you can correctly guess that an anemic person's chief complaint will probably be that he or she feels "so tired all the time." RBCs perform another essential function; besides transporting oxygen, they also transport carbon dioxide.

The term **hemolytic** (hee-mo-LIT-ik) **anemia** applies to any of a variety of inherited blood disorders characterized by abnormal types of hemoglobin. The term *hemolytic* means "pertaining to blood breakage" and emphasizes the fact that abnormal hemoglobin often causes red blood cells to become distorted and easily broken.

Abnormal hemoglobin can be separated from normal hemoglobin by **electrophoresis** (e-lek-tro-foe-REE-sis). In this process, an electric current is passed through a hemoglobin solution—causing the hemoglobin molecules to migrate to different parts of the solution according to the weight of each molecule. Once separated, the abnormal hemoglobin is measured and analyzed.

An example of a hemolytic anemia is **sickle cell anemia.** Sickle cell anemia is a severe, sometimes fatal, hereditary disease. A person who inherits only one defective gene develops a form of the disease called *sickle cell trait.* In sickle cell trait, red blood cells contain a small proportion of a type of hemoglobin that is less soluble than normal. It forms solid crystals when the blood oxygen level is low, causing distortion and fragility of the red blood cell. If two defective genes are inherited (one from each parent), then more of the defective hemoglobin is produced, and the distortion of red blood cells becomes severe. Figure 11-4 shows the characteristic sickle shape of many of the cells.

The frequency of occurrence of the abnormal sickle-cell gene is an example of an interesting epidemiological phenomenon. Because sickle cell trait provides resistance to the parasite that causes malaria (see Appendix B), sickle cell disorders persist in areas of the world in which malaria is still prevalent (Figure 11-5). This situation results from the fact that people without sickle cell trait more often die of malaria before producing offspring than those with the malaria-resistant sickle cell trait. Thus the "bad" sickle cell gene is more

FIGURE **11-5 Relationship Between the Frequency of Sickle Cell Trait and the Distribution of Malaria.** The distribution of the most deadly form of malaria in Africa **(A)** correlates closely with the frequency of occurrence of the sickle cell gene **(B).** The sickle cell trait provides resistance to malaria, and thus carriers are more likely than noncarriers to survive and reproduce—spreading the abnormal gene further in the population.

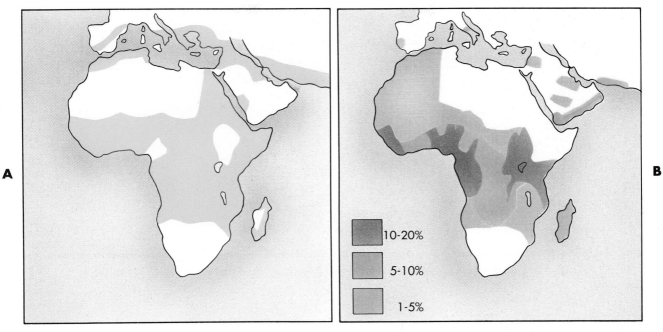

A

B

10-20%

5-10%

1-5%

Distribution of *falciparum* malaria   Frequency of sickle cell gene

likely to be transmitted to the next generation than the "good" genes for normal hemoglobin.

Another type of hemolytic anemia is **thalassemia** (thal-as-SEE-me-ah). Like sickle cell anemia, thalassemia is an inherited disorder with a mild and a severe form. *Thalassemia major*, the severe form, occurs when two defective genes are inherited. *Thalassemia minor*, which occurs when only one defective gene is inherited, may be so mild that it is asymptomatic. The defective genes associated with thalassemia adversely affect the production of one of two types of protein chains in the hemoglobin molecule (Figure 11-3). This causes much of the hemoglobin to be abnormal and decreases the overall amount of hemoglobin. Children born with thalassemia major often fail to thrive, but those born with thalassemia minor sometimes live relatively long, healthy lives.

Table 11-2 summarizes the laboratory results that are typically found in the major types of anemia. As you look at this table, compare and contrast the characteristics of each anemic condition listed.

WHITE BLOOD CELLS

WBCs have a function that is perhaps slightly less vital than that of RBCs but that nevertheless often saves our lives. They defend the body from microorganisms that have succeeded in invading the tissues or bloodstream. For example, **neutrophils** (NOO-tro-fils) (Figure 11-6, *A*) and **monocytes** (MON-o-sites) (Figure 11-6, *B*) engulf microbes. They actually take them into their own cell bodies and digest them in the process of **phagocytosis** (see p. 29). The cells that carry on this process are called **phagocytes** (FAG-o-sites)

▼ **TABLE 11-2** Laboratory Results for Types of Anemia

ANEMIA	FOLATE CONTENT	HEMATOCRIT	HEMOGLOBIN CONTENT	IRON CONTENT	RBC SIZE (VOLUME)	VITAMIN B$_{12}$ CONTENT
Aplastic anemia	Normal	Low to normal	Low to normal	High	Normal to slightly high	Normal
Pernicious anemia	Normal	Low	Low	High	High	Low
Folate-deficiency anemia	Low	Low	Low	High	High	Normal
Acute blood-loss anemia	Normal	Low to normal	Low to normal	Normal	Slightly low	Normal
Chronic blood-loss anemia	Normal	Low	Low	Low	Low to normal	Normal
Iron-deficiency anemia	Normal	Low	Low	Low	Low	Normal
Hemolytic anemia (sickle-cell anemia and thalassemia)	Normal	Low	Low	Normal to high	Normal to high	Normal

▼ **FIGURE 11-6 Leukocytes in Human Blood Smears. A,** Neutrophil. **B,** Monocyte. **C,** Lymphocyte. **D,** Basophil.

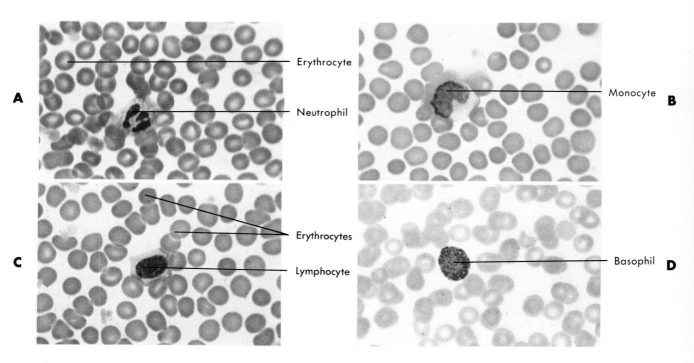

FIGURE 11-7 Phagocytosis. This colorized transmission electron micrograph shows a phagocytic cell consuming a foreign particle in a manner similar to that used by human WBCs. Extensions of the plasma membrane literally reach out and grab a particle and then digest the particle within an intracellular vesicle.

Foreign particle
about to be consumed

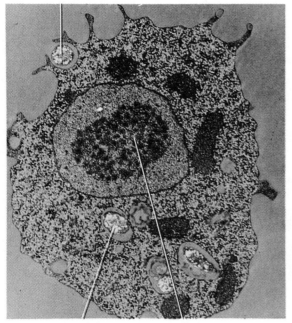

Foreign particle being
digested by the cell

Nucleus

(Figure 11-7). The neutrophils are the most numerous of the phagocytes.

WBCs of the type called **lymphocytes** (LIM-fo-sites) (Figure 11-6, *C*) also help protect us against infections, but they do it by a process different from phagocytosis. Lymphocytes function in the immune mechanism, the complex process that makes us immune to infectious diseases. The immune mechanism starts to operate, for example, when microbes invade the body. In some way, their presence stimulates lymphocytes to start multiplying and become active immune cells. Lymphocytes called *B-lymphocytes* begin to actively produce specific antibodies that inhibit

the microbes. Other lymphocytes, called *T-lymphocytes*, may also become involved, directly attacking the microbes or aiding in the function of B-lymphocytes. Details of the immune system are discussed in Chapter 14.

Eosinophils (ee-o-SIN-o-fils) are granulocytic WBCs that help protect the body from parasites and the numerous irritants that cause allergies. They are also capable of phagocytosis. **Basophils** (BAY-so-fils) (Figure 11-6, *D*) also function in allergic reactions. These leukocytes, which are less abundant than other types, also secrete a number of important substances. For example, they secrete the potent chemical **heparin,** which helps prevent the clotting of blood as it flows through the blood vessels of the body.

WHITE BLOOD CELL DISORDERS

The term **leukopenia** (loo-ko-PEE-nee-ah) refers to an abnormally low WBC count (under 5000 WBCs/mm^3 of blood). A number of disease conditions may affect the immune system and decrease the amount of circulating WBCs. Acquired immune deficiency syndrome or AIDS, which will be discussed in Chapter 14, is one example of a disease characterized by marked leukopenia.

Leukocytosis (loo-ko-sye-TOE-sis) refers to an abnormally high WBC count (that is, over 10,000 WBCs/mm^3 of blood). It is a much more common problem than leukopenia and almost always accompanies infections. There is also a malignant disease, **leukemia** (loo-KEE-mee-ah), in which the number of WBCs increases tremendously. The buffy coat is thicker and more noticeable in the hematocrit of blood from patients with leukemia because of the elevated WBC counts. You may have heard of this disease as "blood cancer." As in all cancers, the extra cells do not function properly.

A special type of white blood cell count called a **differential WBC count** reveals more information than a regular WBC count. In a differential WBC count, the proportions of each type of white blood cell are reported as percentages of the total WBC count. Because all disorders do not affect each WBC type the same way, the differential WBC count is a valuable diagnostic tool. For example, parasite infestations often cause an increase in the proportion of eosinophils in the

FIGURE 11-8 Blood Clotting. The extremely complex clotting mechanism can be distilled into three basic steps: release of platelet factors at the injury site, formation of thrombin, and trapping of red blood cells in fibrin to form a clot.

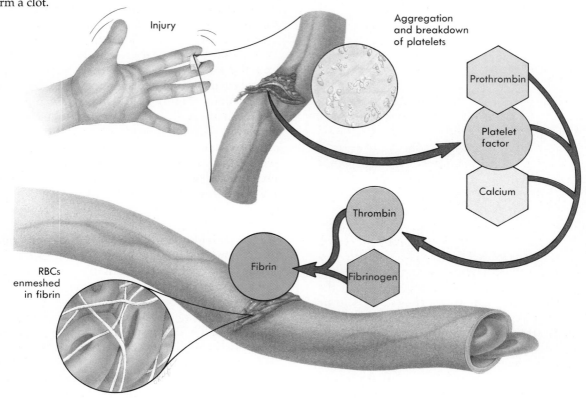

blood because this type of WBC specializes in defending against parasites (Table 11-1).

PLATELETS AND BLOOD CLOTTING

Platelets, the third main type of formed element, play an essential part in blood clotting. Your life might someday be saved just because your blood can clot. A clot plugs up torn or cut vessels and stops bleeding that otherwise might prove fatal.

The story of how blood clots is the story of a chain of rapid-fire reactions. The first step in the chain is some kind of an injury to a blood vessel that makes a rough spot in its lining. (Normally the lining of blood vessels is extremely smooth.) Almost immediately, some blood platelets break up as they flow over the rough spot in the vessel's lining and release a substance into the blood that leads to the formation of other substances called **platelet factors.** Platelets become "sticky" at the point of injury and soon accumulate near the opening in the broken blood vessel. As platelet numbers increase, so does the release of platelet factors. Soon the next step in the blood-clotting mechanism is triggered. Platelet factors combine with **prothrombin** (pro-THROM-bin) (a protein in normal blood), calcium, and other substances to form **thrombin** (THROM-bin). In the last step, thrombin reacts with **fibrinogen** (fi-BRIN-o-jen) (another protein in normal blood) to change it to a fibrous gel that is called **fibrin.** Under the microscope, fibrin looks like a tangle of fine threads with RBCs caught in the tangle. Figure 11-8 illustrates the steps in the blood-clotting mechanism.

The clotting mechanism contains clues for ways to stop bleeding by speeding up blood clotting. For example, you might simply apply gauze to a bleeding surface. Its slight roughness would cause more platelets to break down and release more platelet factors. These additional factors would then make the blood clot more quickly.

Physicians sometimes prescribe vitamin K before surgery. Why? The reason is to make sure that the patient's blood will clot fast enough to prevent hemorrhage. Vitamin K stimulates liver cells to increase the synthesis of prothrombin. More prothrombin in blood allows faster production of thrombin during clotting and thus faster clot formation.

CLOTTING DISORDERS

Unfortunately, clots sometimes form in unbroken blood vessels of the heart, brain, lungs, or some other organ—a dreaded thing because clots may produce sudden death by shutting off the blood supply to a vital organ. When a clot stays in the place where it formed, it is called a **thrombus** (THROM-bus) and the condition is spoken of as **thrombosis** (throm-BO-sus). If part of the clot dislodges and circulates through the bloodstream, the dislodged part is then called an **embolus** (EM-bo-lus), and the condition is called an **embolism** (EM-bo-lizm). Suppose that your doctor told you that you had a clot in one of your coronary arteries. Which diagnosis would he make—coronary thrombosis or coronary embolism—if he thought that the clot had formed originally in the coronary artery as a result of the accumulation of fatty material in the vessel wall? Physicians now have some drugs that they can use to help prevent thrombosis and embolism. Heparin, for example, can be used to prevent excessive clotting. Heparin inhibits the conversion of prothrombin to thrombin, preventing formation of a thrombus. Dicumarol, an oral *anticoagulant*, is also frequently used to prevent excessive clotting. Dicumarol blocks the stimulating effect of vitamin K on the liver, and consequently the liver cells make less prothrombin. The prothrombin content soon falls low enough to prevent abnormal clotting.

A type of X-linked inherited disorder called **hemophilia** (he-mo-FI-lee-ah) results from a failure to form blood clotting factor VIII, IX, or XI. These clotting factors are necessary to complete the clotting process illustrated in Figure 11-8. Thus hemophilia is characterized by a relative inability to form blood clots. Because minor blood vessel injuries are common in ordinary life, hemophilia can be a life-threatening condition. Although rare, hemophilia has become well known for two unfortunate reasons. The first is the fact that the royal families of Europe have been affected by a particularly severe form of this disease. The second is the announcement that persons with hemophilia are at a higher than average risk of contracting blood-borne diseases such as viral hepatitis and AIDS. Because hemophilia is often treated with clotting factors derived from donated whole blood, they risk infection from virus-contaminated blood. Blood banks are constantly working to improve screening procedures. Unfortunately, some tests for viruses do not return positive until the level of virus in the blood is high. Persons with hemophilia may have this risk removed in the near future with the advent of factor production through genetic engineering. This would remove risk of contamination.

A more common type of clotting disorder results from a decrease in the platelet count—a condition called **thrombocytopenia** (throm-bo-si-toe-PEE-nee-ah). This condition is characterized by bleeding from many small blood vessels throughout the body, most visibly in the skin and mucous membranes. Although a number of different mechanisms can result in thrombocytopenia, the usual cause is bone marrow destruction by chemicals, radiation, or cancer. Occasionally, drugs cause thrombocytopenia as a side effect. In such cases, stopping use of the drug usually solves the problem.

BLOOD PLASMA

Blood plasma is the liquid part of the blood or blood minus its formed elements. It consists of water with many substances dissolved in it. All of the chemicals needed by cells to stay alive—food, oxygen, and salts, for example—have to be brought to them by the blood. Food and salts are dissolved in plasma; so, too, is a small amount of oxygen. (Most of the oxygen in the blood is carried in the RBCs as oxyhemoglobin.) Wastes that cells must get rid of are dissolved in plasma and transported to the excretory organs. And, finally, the hormones that help control cells' activities and the antibodies that help protect us against microorganisms are dissolved in plasma.

Blood **serum** is plasma minus its clotting factors. Serum is obtained from whole blood by allowing it to clot in the bottom of a tube and then pouring off the liquid serum. Serum still contains antibodies, so it can be used to treat patients that have a need for specific antibodies.

Many people seem curious about how much blood they have. The amount depends on how big they are and whether they are male or female. A big person has more blood than a small person, and a man has more blood than a woman. But as a general rule, most adults probably have between 4 and 6 L of blood. It normally accounts for about 7% to 9% of the total body weight.

The volume of the plasma part of blood is usually a little more than half the volume of whole blood. Examples of normal volumes follow: plasma volume—2.6 L; blood cell volume—2.4 L; total blood volume—5 L.

Blood is alkaline with a pH between 7.35 and 7.45; it rarely reaches even the neutral point (see Appendix A). If the alkalinity of your blood decreases toward neutral, you are a very sick person; in fact, you have **acidosis.** However, even in this condition, blood almost never becomes the least bit acid; it just becomes less alkaline than normal.

COMPLETE BLOOD COUNT

One of the most useful and frequently performed clinical blood tests is called the *complete blood count* or simply the *CBC.* The CBC is a collection of tests whose results, when interpreted as a whole, can yield an enormous amount of information regarding a person's health. Standard RBC, WBC, and thrombocyte counts, the differential WBC count, hematocrit, hemoglobin content, and other characteristics of the formed elements are usually included in this battery of tests. Normal ranges for blood values included in most CBC tests are found in Appendix D at the back of this book.

Laboratory analysis of plasma often reveals a great deal about the body because it reflects the content of the entire internal environment. Most tests rely on automated procedures such as *sequential multiple analysis (SMA)* that test for many different plasma components. Such tests are often designated SMA-12 or SMA-18, depending on the number of plasma components tested.

One common use of plasma tests is in determining the concentration of food molecules in the blood. In people with diabetes, for example, blood glucose levels must be checked frequently to make sure that treatments are working. Many persons with diabetes have learned to do these blood tests at home. Because we now realize the role of high blood cholesterol levels in increasing the risk of several diseases, many people have submitted to blood screening tests. If a screening test shows a high level of cholesterol, a full *lipid panel* may be suggested. The lipid panel is a series of tests for different types of cholesterol and other lipids. The number and proportions of lipids (see Appendices A and D) help determine the best course of treatment.

The presence of certain enzymes in the plasma often indicate specific pathological events. For example, *transaminase* (tranz-AM-in-ace) is an enzyme found in high concentration just after massive tissue damage—as in a heart attack. Other enzymes, such as *alkaline phosphatase* (FOS-fat-ace), are found in high levels in liver and bone diseases. *Acid phosphatase* levels are commonly elevated in prostate cancer and after severe trauma. Elevated transaminase levels accompanied by elevated *lactic dehydrogenase (LDH)* levels are characteristic of Duchenne muscular dystrophy (DMD) and heart muscle damage. Normal values of these plasma enzymes are given in Appendix D.

Blood Types

ABO SYSTEM

Blood types are identified by certain antigens on RBCs. An antigen (AN-ti-jen) is a substance that can stimulate the body to make antibodies. Almost all substances that act as antigens are foreign proteins. In other words, they are not the body's own natural proteins but are proteins that have entered the body from the outside—by infection, transfusion, or some other method.

FIGURE 11-9 Results of Different Combinations of Donor and Recipient Blood. The left columns show the recipient's blood characteristics and the top row shows the donor's blood type.

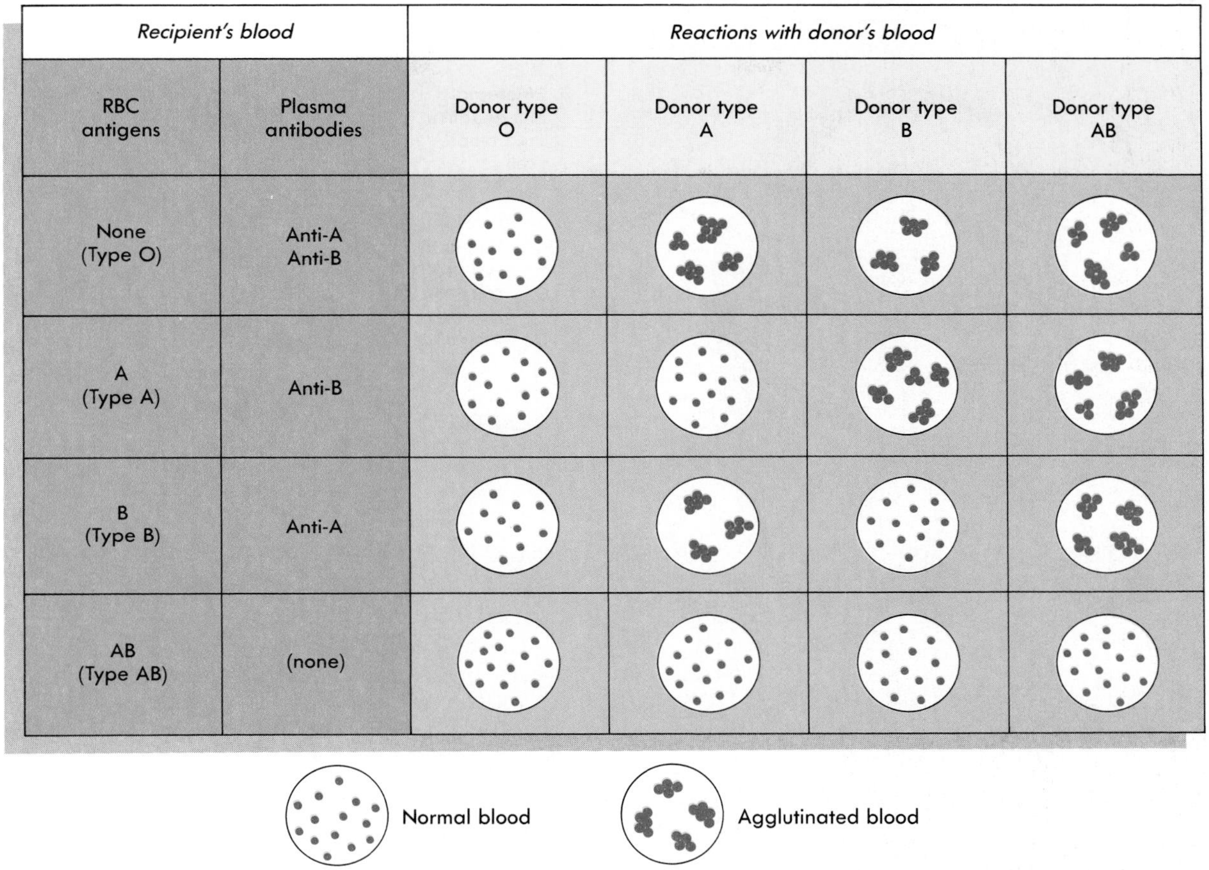

Recipient's blood		Reactions with donor's blood			
RBC antigens	Plasma antibodies	Donor type O	Donor type A	Donor type B	Donor type AB
None (Type O)	Anti-A Anti-B				
A (Type A)	Anti-B				
B (Type B)	Anti-A				
AB (Type AB)	(none)				

 Normal blood    Agglutinated blood

The word *antibody* can be defined in terms of what causes its formation or in terms of how it functions. Defined the first way, an **antibody** (an-ti-BOD-ee) is a substance made by the body in response to stimulation by an antigen. Defined according to its functions, an antibody is a substance that reacts with the antigen that stimulated its formation. Many antibodies react with their antigens to clump or agglutinate (ah-GLOO-tin-ate) them. In other words, they cause their antigens to stick together in little clusters.

Every person's blood is one of the following four blood types in the ABO system of typing blood (Figure 11-9):

1 Type A
2 Type B
3 Type AB
4 Type O

Suppose that you have type A blood (as do about 41% of Americans). The letter *A* stands for a certain type of antigen (a protein) in the plasma membrane of your RBCs since birth. Because you were born with type A antigen, your body does not form antibodies to react with it. In other words, your blood plasma contains no anti-A antibodies. It does, however, contain anti-B antibodies. For some unknown reason, these antibod-

FIGURE 11-10 Erythroblastosis Fetalis. A, Rh-positive blood cells enter the mother's blood stream during delivery of an Rh-positive baby. If not treated, the mother's body will produce anti-Rh antibodies. **B,** A later pregnancy involving an Rh-negative baby is normal because there are no Rh antigens in the baby's blood. **C,** A later pregnancy involving an Rh-positive baby may result in erythroblastosis fetalis. Anti-Rh antibodies enter the baby's blood supply and cause agglutination of RBCs with the Rh antigen.

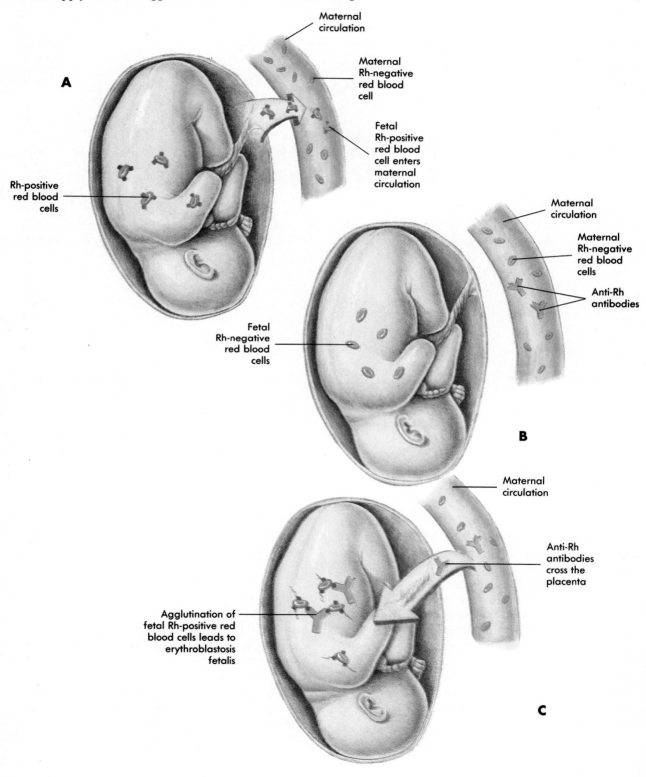

ies are present naturally in type A blood plasma. The body did not form them in response to the presence of the B antigen. In summary, then, in type A blood the RBCs contain type A antigen and the plasma contains anti-B antibodies.

Similarly, in type B blood, the RBCs contain type B antigen, and the plasma contains anti-A antibodies. In type AB blood, as its name indicates, the RBCs contain both type A and type B antigens, and the plasma contains neither anti-A nor anti-B antibodies. The opposite is true of type O blood; its RBCs contain neither type A nor type B antigens, and its plasma contains both anti-A and anti-B antibodies.

Harmful effects or even death can result from a blood transfusion if the donor's RBCs become agglutinated by antibodies in the recipient's plasma. If a donor's RBCs do not contain any A or B antigen, they of course cannot be clumped by anti-A or anti-B antibodies. For this reason the type of blood that contains neither A nor B antigens—namely, type O blood—can be used in an emergency as donor blood without the danger of anti-A or anti-B antibodies clumping its RBCs. Type O blood has therefore been called **universal donor** blood. Similarly, blood type AB has been called **universal recipient** blood because it contains neither anti-A nor anti-B antibodies in its plasma. Therefore it does not clump any donor's RBCs containing A or B antigens. In a normal clinical setting, however, all blood intended for transfusion is matched carefully to the blood of the recipient for a variety of factors.

Figure 11-9 shows the results of different combinations of donor and recipient blood.

Rh SYSTEM

You may be familiar with the term **Rh-positive** blood. It means that the RBCs of this type of blood contain an antigen called the Rh factor. If, for example, a person has type AB, Rh-positive blood, his red blood cells contain type A antigen, type B antigen, and the Rh factor antigen. The term *Rh* is used because this important blood cell antigen was first discovered in the blood of Rhesus monkeys.

In **Rh-negative** blood the RBCs do not contain the Rh factor. Plasma never naturally contains anti-Rh antibodies. But if Rh-positive blood cells are introduced into an Rh-negative person's body, anti-Rh antibodies soon appear in the blood plasma. In this fact lies the danger for a baby born to an Rh-negative mother and an Rh-positive father. If the baby inherits the Rh-positive trait from his father, the Rh factor on his RBCs may stimulate the mother's body to form anti-Rh antibodies. Then, if she later carries another Rh-positive fetus, it may develop a disease called **erythroblastosis** (e-rith-ro-blas-TOE-sis) **fetalis** (fe-TAL-is), caused by the mother's Rh antibodies reacting with the baby's Rh-positive cells (Figure 11-10).

Some Rh-negative mothers who carry an Rh-positive baby are treated with a protein marketed as RhoGAM. RhoGAM stops the mother's body from forming anti-Rh antibodies and thus prevents the possibility of harm to the next Rh-positive baby.

OUTLINE SUMMARY

BLOOD COMPOSITION (Table 11-1)

A Formed elements
  1 Kinds
    a RBCs (erythrocytes)
    b WBCs (leukocytes)
      (1) Granular leukocytes—neutrophils, eosinophils, and basophils
      (2) Nongranular leukocytes—lymphocytes, and monocytes
    c Platelets or thrombocytes
  2 Numbers
    a RBCs—4½ to 5 million per mm^3 of blood
    b WBCs—5000 to 10,000 per mm^3 of blood
    c Platelets—300,000 per mm^3 of blood
  3 Formation—red bone marrow (myeloid tissue) forms all blood cells except some lymphocytes and monocytes, which are formed by lymphatic tissue in the lymph nodes, thymus, and spleen
B Mechanisms of blood disease
  1 Most blood disorders result from failure of blood-producing tissues
  2 Diseased bone marrow can sometimes be replaced by tissue transplanted from a donor
C RBCs
  1 Structure—disk-shaped, without nuclei
  2 Functions—transport oxygen and carbon dioxide
D Red blood cell disorders
  1 Anemia—inability of blood to carry adequate oxygen to tissues, characterized by:
    a Abnormal red blood cell numbers
    b Deficiency of hemoglobin
  2 Changes in RBC number
    a Hematocrit—test in which a centrifuge is used to separate whole blood into formed elements and liquid fraction (normal RBC volume is about 45% of total)
    b RBC count—hemocytometer or automatic counter is used to calculate actual number of RBCs (4.5 to 5.5 million/mm^3 is normal)
    c Aplastic anemia—low RBC number due to bone marrow destruction
    d Pernicious anemia—low RBC number due to lack of available vitamin B_{12}
    e Folate-deficiency anemia—low RBC number due to lack of folic acid (vitamin B_9)
  3 Changes in hemoglobin
    a Both amount and quality of hemoglobin is important

    b Hemoglobin concentration is often measured with optical devices; normal ranges:
      (1) Female: 12-14 g/100 ml
      (2) Male: 14-16 g/100 ml
    c Iron deficiency anemia—low hemoglobin due to lack of iron
    d Hemolytic anemia—abnormal hemoglobin that causes deformation and fragility of red blood cells
    e Abnormal types of hemoglobin can be separated with electrophoresis for further study
    f Sickle cell anemia—inherited disorder in which abnormal hemoglobin causes characteristic RBC deformities
    g Thalassemia—inherited disorder in which a small amount of abnormal hemoglobin is produced
E WBCs
  1 General function—defense
  2 Neutrophils and monocytes carry out phagocytosis
  3 Lymphocytes produce antibodies (B-lymphocytes) or directly attack foreign cells (T-lymphocytes)
  4 Eosinophils protect against irritants that cause allergies
  5 Basophils produce heparin, which inhibits clotting
F White blood cell disorders
  1 Leukopenia—abnormally low WBC count
  2 Leukocytosis—abnormally high WBC count
  3 Leukemia—cancer: elevated WBC count; cells do not function properly
G Platelets and blood clotting (Figure 11-8)
  1 Platelets play an essential role in blood clotting
  2 Blot clot formation
    a Release of platelet factors at the injury site
    b Formation of thrombin
    c Trapping of red blood cells in fibrin to form a clot
H Clotting disorders
  1 Thrombosis—formation of an inappropriate stationary clot (thrombus)
  2 Embolism—dislodged blood clot (embolus)
  3 Hemophilia—X-linked inheritance of inability to form essential clotting factors
  4 Thrombocytopenia—abnormally small number of platelets

I Blood plasma
  1 Definition—blood minus its cells
  2 Composition—water containing many dissolved substances (for example, foods, salts, and hormones)
  3 Amount of blood—varies with size and sex; 4 to 6 L about average; about 7% to 9% of body weight
  4 Slightly alkaline
  5 Plasma is often analyzed to determine the concentration of food molecules, enzymes, and other clinically important plasma components

BLOOD TYPES

A ABO system (Figure 11-10)
  1 Type A blood—type A antigens in RBCs; anti-B type antibodies in plasma
  2 Type B blood—type B antigens in RBCs; anti-A type antibodies in plasma
  3 Type AB blood—type A and type B antigens in RBCs; no anti-A or anti-B antibodies in plasma
  4 Type O blood—no type A or type B antigens in RBCs; both anti-A and anti-B antibodies in plasma
B Rh system
  1 Rh-positive blood—Rh factor antigen present in RBCs
  2 Rh-negative blood—no Rh factor present in RBCs; no anti-Rh antibodies present naturally in plasma; anti-Rh antibodies, however, appear in the plasma of Rh-negative persons if Rh-positive RBCs have been introduced into their bodies
  3 Erythroblastosis fetalis—caused by the mother's Rh antibodies reacting with the baby's Rh-positive RBCs

NEW WORDS

antibodies	fibrin	lymphocyte	plasma
antigens	fibrinogen	monocyte	prothrombin
basophil	hemoglobin	neutrophil	serum
buffy coat	heparin	oxyhemoglobin	thrombocyte
eosinophil	leukocyte	phagocyte	thrombin
erythrocyte			

Diseases and Other Clinical Terms

acidosis	embolus	hemophilia	sickle cell anemia
anemia	erythroblastosis fetalis	leukemia	thalassemia
aplastic anemia	folate-deficiency	leukocytosis	thrombocytopenia
differential WBC count	anemia	leukopenia	thrombosis
electrophoresis	hematocrit	pernicious anemia	thrombus
embolism	hemolytic anemia	polycythemia	

CHAPTER TEST

1. Erythrocytes is another name for:
   a. WBCs
   b. RBCs
   c. Platelets
   d. Thrombocytes
2. Which of the following is classified as a granular leukocyte?
   a. Monocyte
   b. Neutrophil
   c. Lymphocyte
   d. None of the above are correct
3. Which of the following terms refers to an abnormally low WBC count?
   a. Anemia
   b. Leukemia
   c. Leukocytosis
   d. Leukopenia
4. The most numerous of the phagocytes are the:
   a. Neutrophils
   b. RBCs
   c. Basophils
   d. Lymphocytes
5. Heparin secretion is a function of:
   a. Monocytes
   b. Neutrophils
   c. Basophils
   d. Eosinophils
6. Vitamin K:
   a. Increases prothrombin levels in blood
   b. Is necessary for normal blood clotting
   c. Stimulates liver cells
   d. All of the above are correct
7. A thrombus:
   a. Is a stationary blood clot
   b. Is a moving blood clot
   c. Is seldom serious
   d. Is a normal component of plasma
8. Erythroblastosis fetalis is caused by:
   a. Anti-Rh antibodies in the father's blood
   b. Anti-Rh antibodies in the mother's blood
   c. Both of the above are correct
   d. None of the above are correct

9. The liquid part of blood that has not clotted is called _____.
10. A better-known name for myeloid tissue is _____ _____ _____.
11. An excess of RBCs is called _____.
12. Sickle cell anemia and thalassemia are caused by abnormalities in the production of _____.
13. Proportions of white blood cell types can be determined by performing a _____ WBC count.
14. _____ is an X-linked inherited disorder characterized by a failure to produce essential blood-clotting factors.
15. A moving blood clot is called an _____.
16. A substance that can stimulate the body to produce antibodies is called an _____.
17. Type _____ blood contains both A and B antigens and neither anti-A nor anti-B antibodies.
18. The blood cell antigen first discovered in the blood of Rhesus monkeys is the _____ factor.

REVIEW QUESTIONS

1 What is the normal number of RBCs per mm^3 of blood? WBCs? Platelets?
2 Name the granular and nongranular leukocytes.
3 What two kinds of connective tissue make blood cells for the body?
4 Suppose that your doctor told you that your "red count was 3 million." What does "red count" mean? Might the doctor say that you had any of the following conditions—anemia, leukocytosis, leukopenia, polycythemia—with a RBC count of this amount? If so, which one?
5 What does the hematocrit blood test measure? What is the normal value for this test?
6 If you had appendicitis or some other acute infection, would your WBC count be more likely to be 2000, 7000, or 15,000 per mm^3? Give a reason for your answer.
7 Your circulatory system is the transportation system of your body. Mention some of the substances it transports and tell whether each is carried in blood cells or in the blood plasma.

8 Briefly describe three ways in which blood cells defend the body against foreign cells.
9 Briefly explain what happens when blood clots, including what makes it start to clot.
10 You hear that a friend has a "coronary thrombosis." What does this mean to you?
11 Define *antibody* and *antigen* as they apply to blood typing.
12 Identify the four blood types.
13 What is the difference between ABO blood type and the Rh blood type?
14 Why might a physician prescribe vitamin K before surgery for a patient with a history of bleeding problems?
15 Name and describe some blood disorders caused by abnormalities of the bone marrow.
16 Name and describe some inherited blood disorders.

CLINICAL APPLICATIONS

1 Angela's physician suspects that Angela has just suffered a *myocardial infarction,* or heart attack. She tells Angela that she is going to take a blood sample so that the hospital lab can perform a test to confirm her diagnosis. What information can Angela's blood yield to help the physician?

2 Yvonne has just been told that she has a condition called *pernicious anemia.* She looked up the definition of this disease in a dictionary and learned that it is caused by a decreased availability of vitamin B_{12} needed for manufacturing RBCs. Yvonne promptly went to the local pharmacy to buy some vitamin B_{12} tablets to help her overcome this condition. Is this a wise course of action?

3 Your brother has just been diagnosed as having anemia, but your mother cannot seem to remember the specific type. She shows you a copy of your young brother's CBC results, but no diagnosis is stated. Based on the results given below, what would you guess is your brother's condition? (Hint: see Table 11-2.) Are you likely to develop this condition?

Folate content: normal
Hematocrit: low
Hemoglobin content: low
Iron content: slightly high
RBC size (mean corpuscular volume): high
Vitamin B_{12} content: normal

The Heart and Heart Disease

CHAPTER

12

Objectives

After you have completed this chapter, you should be able to:

1 Discuss the location, size, and position of the heart in the thoracic cavity and identify the heart chambers, sounds, and valves.

2 Describe the major types of cardiac valve disorders.

3 Trace blood through the heart and compare the functions of the heart chambers on the right and left sides.

4 Explain how a myocardial infarction might occur.

5 List the anatomical components of the heart conduction system and discuss the features of the normal electrocardiogram.

6 Describe the major types of cardiac arrhythmia.

7 List and describe the possible causes of heart failure.

Differing amounts of nutrients and waste products enter and leave the fluid surrounding each body cell continually. In addition, requirements for hormones, body salts, water, and other critical substances constantly change. However, homeostasis or constancy of the body fluid contents surrounding the billions of cells that make up our bodies is required for survival. The system that supplies our cells' transportation needs is the **circulatory system.** The levels of dozens of substances in the blood can remain constant even though the absolute amounts that are needed or produced may change because we have this extremely effective system that transports these substances to or from each cell as circumstances change.

In this chapter, we will discuss the heart—the pump that keeps blood moving around a closed circuit of blood vessels. As appropriate, we will include discussions of **heart disease,** a group of disorders that together constitute the leading cause of death in the United States. Chapter 13 continues the study of the circulatory system by discussing the blood vessels in health and disease.

Location, Size, and Position of the Heart

No one needs to be told where the heart is or what it does. Everyone knows that the heart is in the chest, that it beats night and day to keep the blood flowing, and that if it stops, life stops.

Most of us probably think of the heart as located on the left side of the body. As you can see in Figure 12-1, the heart is located between the lungs in the lower portion of the mediastinum. Draw an imaginary line through the middle of the trachea in Figure 12-1 and continue the line down through the thoracic cavity to divide it into right and left halves. Note that about two thirds of the mass of the heart is to the left of this line and one third to the right.

The heart is often described as a triangular organ, shaped and sized roughly like a closed fist. In Figure 12-1 you can see that the **apex** or blunt point of the lower edge of the heart lies on the diaphragm, pointing toward the left. Doctors and nurses often listen to the heart sounds by placing a stethoscope on the chest wall directly over the apex of the heart. Sounds of the so-called apical beat are easily heard in this area (that is, in the space between the fifth and sixth ribs on a line even with the midpoint of the left clavicle).

The heart is positioned in the thoracic cavity between the sternum in front and the bodies of the thoracic vertebrae behind. Because of this placement, it can be compressed or squeezed by application of pressure to the lower portion of the body of the sternum using the heel of the hand. Rhythmic compression of the heart in this way can maintain blood flow in cases of cardiac arrest and, if combined with effective artificial respiration, the resulting procedure, called **cardiopulmonary resuscitation (CPR),** can be lifesaving.

Anatomy

HEART CHAMBERS

If you cut open a heart, you can see many of its main structural features (Figure 12-2). This organ is hollow, not solid. A partition divides it into right and left sides. The heart contains four cavities or hollow chambers. The two upper chambers are called **atria** (AY-tree-ah), and the two lower chambers are called **ventricles** (VEN-tri-kuls). The atria are smaller than the ventricles, and their walls are thinner and less muscular. Atria are often called *receiving chambers* because blood enters the heart through veins that open into these upper cavities. Eventually, blood is pumped from the heart into arteries that exit from the ventricles; therefore the ventricles are sometimes referred to as the *discharging chambers* of the heart. Each heart chamber is named according to its location. Thus there is a right and left atrial chamber above and a right and left ventricular chamber below. The wall of each heart chamber is composed of cardiac muscle tissue usually referred to as the **myocardium** (my-o-KAR-dee-um). The septum between the atrial chambers is called the *interatrial septum;* the *interventricular septum* separates the ventricles.

Each chamber of the heart is lined by a thin layer of very smooth tissue called the **endocardium** (en-doe-KAR-dee-um) (Figure 12-2). Inflammation of this lining is referred to as **endocarditis** (en-doe-kar-DYE-tis). If inflamed, the endocardial lining can become rough and abrasive to RBCs passing over its surface. Blood flowing over a

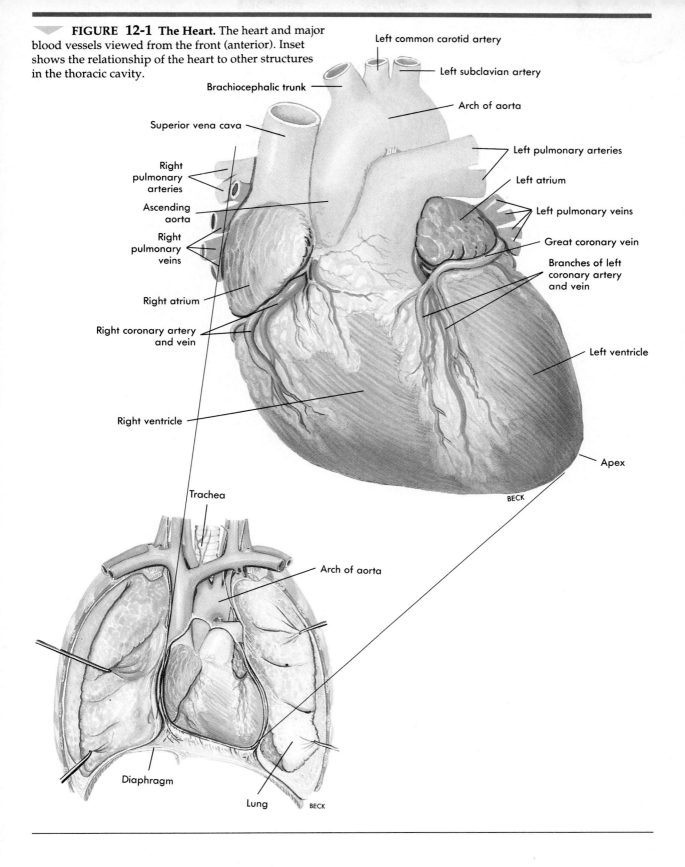

FIGURE 12-1 The Heart. The heart and major blood vessels viewed from the front (anterior). Inset shows the relationship of the heart to other structures in the thoracic cavity.

Left common carotid artery

Left subclavian artery

Brachiocephalic trunk

Arch of aorta

Superior vena cava

Left pulmonary arteries

Right pulmonary arteries

Left atrium

Ascending aorta

Left pulmonary veins

Right pulmonary veins

Great coronary vein

Branches of left coronary artery and vein

Right atrium

Right coronary artery and vein

Left ventricle

Right ventricle

Apex

BECK

Trachea

Arch of aorta

Diaphragm

Lung

BECK

FIGURE 12-2 An Internal View of the Heart.

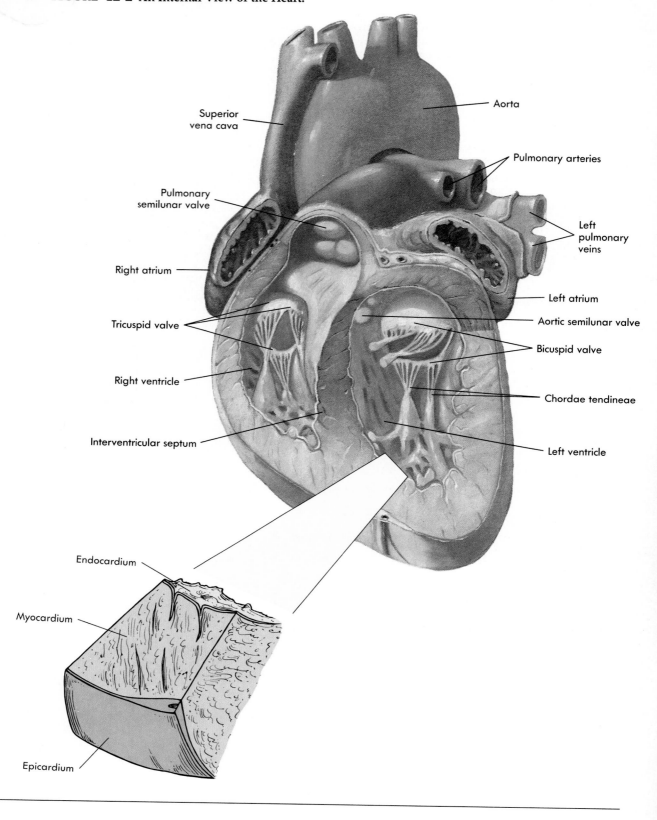

Superior
vena cava

Aorta

Pulmonary arteries

Pulmonary
semilunar valve

Left
pulmonary
veins

Right atrium

Left atrium

Tricuspid valve

Aortic semilunar valve

Bicuspid valve

Right ventricle

Chordae tendineae

Interventricular septum

Left ventricle

Endocardium

Myocardium

Epicardium

rough surface is subject to clotting, and a **thrombosis** (throm-BO-sis) or clot may form (see Chapter 11). Unfortunately, rough spots caused by endocarditis or injuries to blood vessel walls often cause the release of platelet factors. The result is often the formation of a fatal blood clot.

THE PERICARDIUM AND PERICARDITIS

The heart has a covering and a lining. Its covering, called the **pericardium** (pair-i-KAR-dee-um), consists of two layers of fibrous tissue with a small space in between. The inner layer of the pericardium is called the **visceral pericardium** or **epicardium** (ep-i-KAR-dee-um). It covers the heart the way an apple skin covers an apple. The outer layer of pericardium is called the **parietal pericardium.** It fits around the heart like a loose-fitting sack, allowing enough room for the heart to beat. It is easy to remember the difference between the *endocardium*, which lines the heart chambers, and the *epicardium*, which covers the surface of the heart (Figure 12-2), if you understand the meaning of the prefixes *endo-* and *epi-*. *Endo-* comes from the Greek word meaning "inside" or "within" and *epi-* from the Greek word meaning "upon" or "on" (see Appendix C).

The two pericardial layers slip against each other without friction when the heart beats because these are serous membranes with moist, not dry, surfaces. A thin film of pericardial fluid furnishes the lubricating moistness between the heart and its enveloping pericardial sac.

If the pericardium becomes inflamed, a condition called **pericarditis** (pair-i-kar-DYE-tis) results. Pericarditis may be caused by a variety of factors: trauma, viral or bacterial infection, tumors, and other factors. The pericardial edema that characterizes this condition often causes the visceral and parietal pericardium to rub together—causing severe chest pain. Pericardial fluid, pus, or blood (in the case of an injury) may accumulate in the space between the two pericardial layers and impair the pumping action of the heart. This is termed *pericardial effusion* and may develop into a serious compression of the heart called **cardiac tamponade** (tam-po-NAHD).

HEART ACTION

The heart serves as a muscular pumping device for distributing blood to all parts of the body. Contraction of the heart is called **systole** (SIS-toe-lee), and relaxation is called **diastole** (dye-AS-toe-lee). When the heart beats (that is, when it contracts), the atria contract first (atrial systole), forcing blood into the ventricles. Once filled, the two ventricles contract (ventricular systole) and force blood out of the heart (Figure 12-3). For the heart to be efficient in its pumping action, more than just the rhythmic contraction of its muscular fibers is required. The direction of the blood flow must be directed and controlled. This is accomplished by four sets of valves that are located at the entrance and near the exit of the ventricles.

HEART VALVES AND VALVE DISORDERS

The two valves that separate the atrial chambers above from the ventricles below are called **AV** or **atrioventricular** (ay-tree-o-ven-TRIK-yoo-lar) **valves.** The two AV valves are called the **bicuspid** or **mitral** (MY-tral) **valve,** located between the left atrium and ventricle, and the **tricuspid valve,** located between the right atrium and ventricle. The AV valves prevent backflow of blood into the atria when the ventricles contract. Locate the AV valves in Figures 12-2 and 12-3. Note that a number of stringlike structures called **chordae tendineae** (KOR-dee ten-DIN-ee) attach the AV valves to the wall of the heart.

The **semilunar** (sem-i-LOO-nar) **valves** are located between the two ventricular chambers and the large arteries that carry blood away from the heart when contraction occurs (Figure 12-3). The ventricles, like the atria, contract together. Therefore the two semilunar valves open and close at the same time. The **pulmonary semilunar valve** is located at the beginning of the pulmonary artery and allows blood going to the lungs to flow out of the right ventricle but prevents it from flowing back into the ventricle. The **aortic semilunar valve** is located at the beginning of the aorta and allows blood to flow out of the left ventricle up into the aorta but prevents backflow into this ventricle.

Disorders of the cardiac valves can have several effects. For example, a congenital defect in valve structure can result in mild to severe pumping inefficiency. **Incompetent valves** leak, allowing some blood to flow back into the chamber from which it came. **Stenosed valves** are valves that are narrower than normal, slowing blood flow from a heart chamber.

FIGURE 12-3 Heart Action. A, During atrial systole (contraction) cardiac muscle in the atrial wall contracts, forcing blood through the AV valves and into the ventricles. **B,** During ventricular systole that follows, the AV valves close, and blood is forced out of the ventricles through the semilunar valves and into the arteries. **C,** The photograph shows the pulmonary semilunar valve as seen from above (superior view).

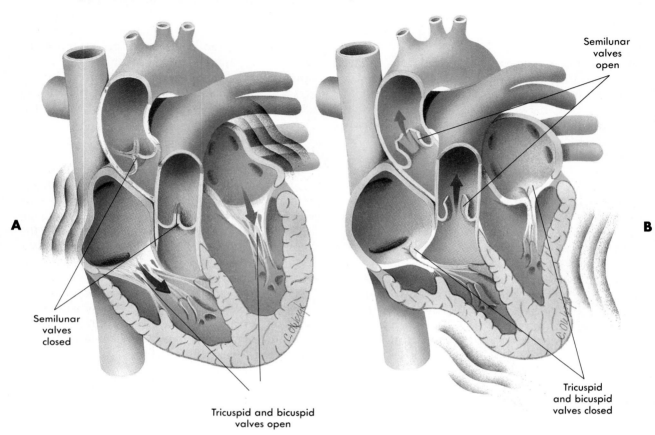

A

B

Semilunar
valves
open

Semilunar
valves
closed

Tricuspid and bicuspid
valves open

Tricuspid
and bicuspid
valves closed

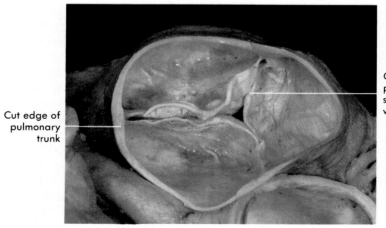

C Cut edge of
pulmonary
trunk

Cusp of
pulmonary
semilunar
valve

Rheumatic heart disease is cardiac damage resulting from a delayed inflammatory response to streptococcal infection that occurs most often in children. A few weeks after an improperly treated streptococcal infection, the cardiac valves and other tissues in the body may become inflamed—a condition called *rheumatic fever*. If severe, the inflammation can result in stenosis or other deformities of the valves, chordae tendineae, or myocardium.

Mitral valve prolapse (MVP), a condition affecting the bicuspid or mitral valve, has a genetic basis in some cases but can result from rheumatic fever or other factors. A prolapsed mitral valve is one whose flaps extend back into the left atrium, causing incompetence (leaking) of the valve (Figure 12-4). Although this condition is common, occurring in up to 1 in every 20 people, most cases are asymptomatic. In severe cases, patients suffer chest pain and fatigue.

FIGURE 12-4 Mitral Valve Prolapse. The normal mitral valve *(left)* prevents backflow of blood from the left ventricle into the left atrium during ventricular systole (contraction). The prolapsed mitral valve *(inset)* permits leakage because the valve flaps billow backward, parting slightly.

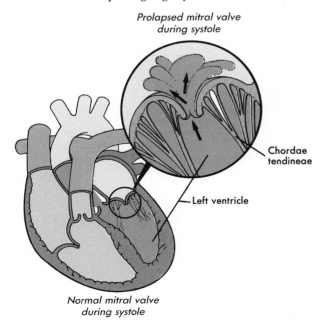

Prolapsed mitral valve during systole

Chordae tendineae

Left ventricle

Normal mitral valve during systole

Damaged or defective cardiac valves can often be replaced surgically. Animal valves and artificial valves made from synthetic materials are frequently used in these valve replacement procedures.

Heart Sounds

When a stethoscope is placed on the anterior chest wall, two distinct sounds can be heard. They are rhythmical and repetitive sounds that are often described as **lub dup.** Disorders of the cardiac valves are often diagnosed by changes in these normal valve sounds of the heart.

The first or *lub* sound is caused by the vibration and abrupt closure of the AV valves as the ventricles contract (Figure 12-3). Closure of the AV valves prevents blood from rushing back up into the atria during contraction of the ventricles. This first sound is of longer duration and lower pitch than the second. The pause between this first sound and the *dup* or second sound is shorter than that after the second sound and the *lub dup* of the next systole (contraction). The second heart sound is caused by the closing of both the semilunar valves when the ventricles undergo diastole (relax) (Figure 12-3).

Abnormal heart sounds called **heart murmurs** are often caused by disorders of the valves. For example, incompetent valves may cause a swishing sound as a "lub" or "dup" ends. Stenosed valves, on the other hand, often cause swishing sounds just before a "lub" or "dup."

Blood Flow Through the Heart

The heart acts as two separate pumps. The right atrium and the right ventricle perform a task quite different from the left atrium and the left ventricle. When the heart "beats," first the atria contract simultaneously. This is atrial systole. Then the ventricles fill with blood, and they, too, contract together during ventricular systole. Although the atria contract as a unit followed by the ventricles below, the right and left sides of the heart act as separate pumps. As we study the blood flow through the heart, the separate functions of the two pumps will become clearer.

FIGURE 12-5 Blood Flow Through the Circulatory System. In the pulmonary circulatory system, blood is pumped from the right side of the heart to the gas-exchange tissues of the lungs. In the systemic circulation, blood is pumped from the left side of the heart to all other tissues of the body.

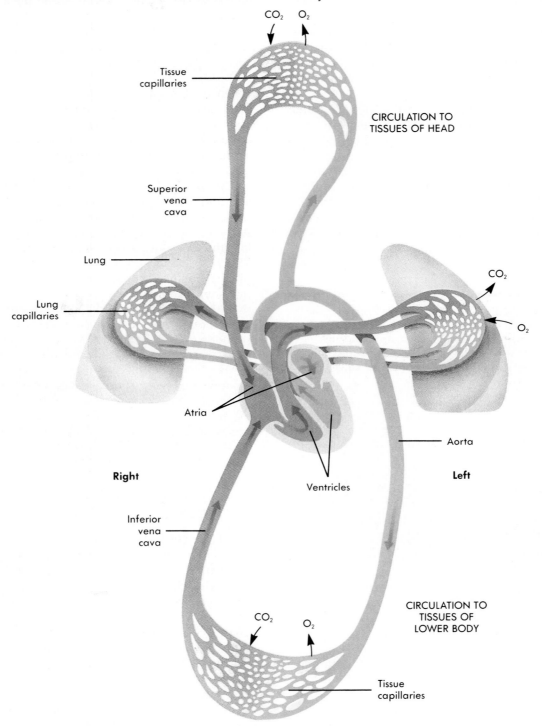

Note in Figure 12-5 that blood enters the right atrium through two large veins called the **superior vena** (VEE-nah) **cava** (KAY-vah) and **inferior vena cava.** The right heart pump receives oxygen-poor blood from the veins. After entering the right atrium, it is pumped through the right AV or tricuspid valve and enters the right ventricle. When the ventricles contract, blood in the right ventricle is pumped through the pulmonary semilunar valve into the **pulmonary artery** and eventually to the lungs, where oxygen is added and carbon dioxide is lost.

As you can see in Figure 12-5, blood rich in oxygen returns to the left atrium of the heart through four **pulmonary veins.** It then passes through the left AV or bicuspid valve into the left ventricle. When the left ventricle contracts, blood is forced through the aortic semilunar valve into the **aorta** (ay-OR-tah) and is distributed to the body as a whole.

As you can tell from Figure 12-5, the two sides of the heart actually pump blood through two separate "circulations" and function as two separate pumps. The **pulmonary circulation** involves movement of blood from the right ventricle to the lungs, and the **systemic circulation** involves movement of blood from the left ventricle through the body. The pulmonary and systemic circulations are discussed in Chapter 13.

Coronary Circulation and Coronary Heart Disease

To sustain life, the heart must pump blood throughout the body on a regular and ongoing basis. As a result, the heart muscle or myocardium requires a constant supply of blood containing nutrients and oxygen to function effectively. The delivery of oxygen and nutrient-rich arterial blood to cardiac muscle tissue and the return of oxygen-poor blood from this active tissue to the venous system is called the **coronary circulation** (Figure 12-6).

FIGURE 12-6 Coronary Circulation. A, Arteries. **B,** Veins. Both are anterior views of the heart. Vessels near the anterior surface are more darkly colored than vessels of the posterior surface seen through the heart.

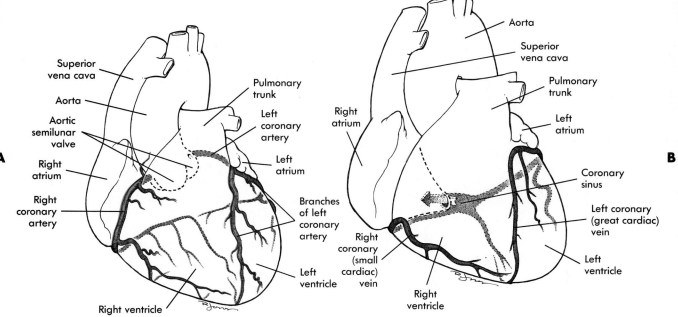

FIGURE 12-7 Coronary Bypass. In coronary bypass surgery, blood vessels are "harvested" from other parts of the body and used to construct detours around blocked coronary arteries. Artificial vessels can also be used.

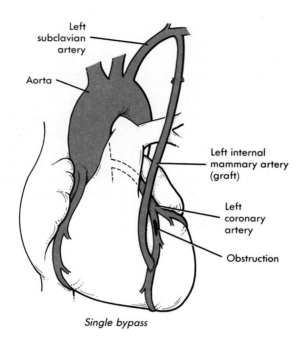

Single bypass

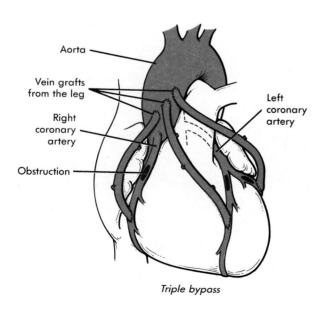

Triple bypass

Blood flows into the heart muscle by way of two small vessels that are surely the most famous of all the blood vessels—the **right** and **left coronary arteries**—famous because coronary heart disease kills many thousands of people every year. The coronary arteries are the aorta's first branches. The openings into these small vessels lie behind the flaps of the aortic semilunar valves. In both coronary thrombosis and coronary embolism (EM-bo-lizm), a blood clot occludes or plugs up some part of a coronary artery. Blood cannot pass through the occluded vessel and so cannot reach the heart muscle cells it normally supplies. Deprived of oxygen, these cells soon die or are damaged. In medical terms, a **myocardial** (my-o-KAR-dee-al) **infarction** (in-FARK-shun) or tissue death occurs. A myocardial infarction or "heart attack" is a common cause of death during middle and late adulthood. Recovery from a myocardial infarction is possible if the amount of heart tissue damaged was small enough so that the remaining undamaged heart muscle can pump blood effectively enough to supply the needs of the rest of the heart, as well as the body.

Coronary arteries may also become blocked as a result of **atherosclerosis,** a type of "hardening of the arteries" in which lipids and other substances build up on the inside wall of blood vessels. Mechanisms of atherosclerosis are discussed further in Chapter 13. Coronary atherosclerosis has increased dramatically over the last few decades to become the leading cause of death in western countries. Many pathophysiologists believe this increase results from a change in lifestyle. They cite several important risk factors associated with coronary atherosclerosis: cigarette smoking, high-fat and high-cholesterol diets, and hypertension (high blood pressure).

The term **angina** (an-JYE-nah) **pectoris** (PEK-tor-is) is used to describe the severe chest pain that occurs when the myocardium is deprived of adequate oxygen. It is often a warning that the coronary arteries are no longer able to supply enough blood and oxygen to the heart muscle. **Coronary bypass surgery** is a frequent treatment for those who suffer from severely restricted coronary artery blood flow. In this procedure, veins are "harvested" or removed from other areas of the body and used to bypass partial blockages in coronary arteries (Figure 12-7).

After blood has passed through the capillary beds in the myocardium, it flows into **coronary veins,** which empty into the **coronary sinus** and finally into the right atrium.

Cardiac Cycle

The beating of the heart is a regular and rhythmic process. Each complete heart beat is called a **cardiac cycle** and includes the contraction (systole) and relaxation (diastole) of atria and ventricles. Each cycle takes about 0.8 seconds to complete if the heart is beating at an average rate of about 72 beats per minute. The term **stroke volume** refers to the volume of blood ejected from the ventricles during each beat. **Cardiac output** or the volume of blood pumped by one ventricle per minute averages about 5 L in a normal, resting adult.

Valve disorders, coronary artery blockage, or myocardial infarction can all decrease stroke volume and thus may decrease cardiac output. Decreased cardiac output can result in fatigue or even death.

Conduction System of the Heart

Cardiac muscle fibers can contract rhythmically on their own. However, they must be coordinated by electrical signals (impulses) if the heart is to pump effectively. Although the rate of the cardiac muscle's rhythm is controlled by autonomic nerve signals, the heart has its own built-in conduction system (Figure 12-8) for coordinating contractions during the cardiac cycle. Four specialized structures embedded in the wall of the heart generate

FIGURE 12-8 Conduction System of the Heart. Specialized cardiac muscle cells in the wall of the heart rapidly conduct an electrical impulse throughout the myocardium. The signal is initiated by the SA node (pacemaker) and spreads to the rest of atrial myocardium and to the AV node. The AV node then initiates a signal that is conducted through the ventricular myocardium by way of the AV bundle (of His) and Purkinje fibers.

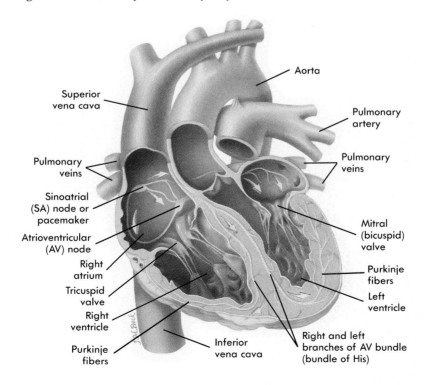

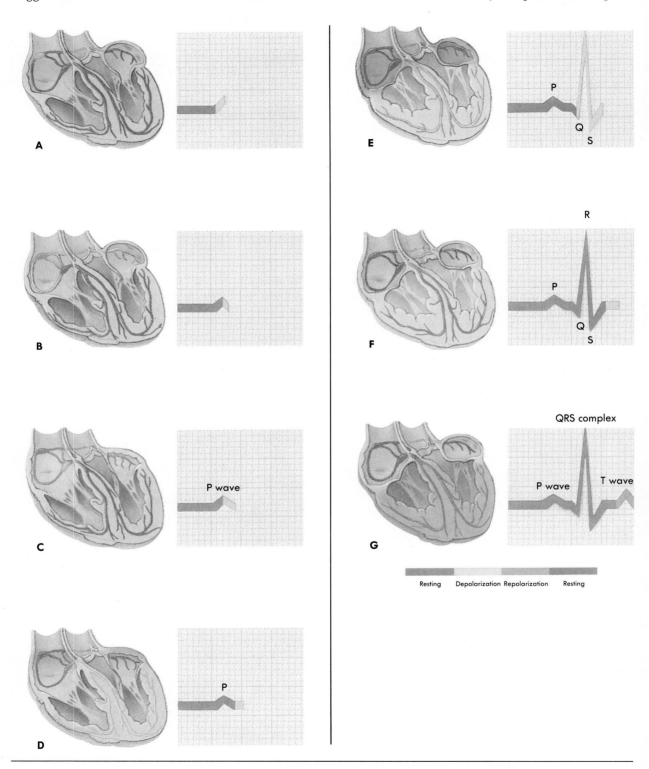

FIGURE 12-9 Events Represented by the Electrocardiogram (ECG). A through **C,** The P wave represents the depolarization of cardiac muscle tissue in the SA node and atrial walls. **C** and **D,** Before the QRS complex is observed, the AV node and AV bundle depolarize. **E** and **F,** The QRS complex occurs as the atrial walls repolarize and the ventricular walls depolarize. **G,** The T wave is observed as the ventricular walls repolarize. Depolarization triggers contraction in the affected muscle tissue. Thus cardiac muscle contraction occurs *after* depolarization begins.

and then conduct impulses through heart muscle to cause first the atria and then the ventricles to contract. The names of the structures that make up this conduction system follow:

1 **Sinoatrial** (sye-no-AY-tree-al) **node,** which is sometimes called the *SA node* or the *pacemaker*
2 **Atrioventricular** (ay-tree-o-ven-TRIK-yoo-lar) **node** or **AV node**
3 **AV bundle** or **bundle of HIS**
4 **Purkinje** (pur-KIN-jee) **fibers**

Impulse conduction normally starts in the heart's pacemaker, namely, the SA node. From there, it spreads, as you can see in Figure 12-8 in all directions through the atria. This causes the atria to contract. When impulses reach the AV node, it relays them by way of the bundle of His and Purkinje fibers to the ventricles, causing them to contract. Normally, therefore, a ventricular beat follows each atrial beat.

ELECTROCARDIOGRAPHY

The specialized structures of the heart's conduction system generate tiny electrical currents that spread through surrounding tissues to the surface of the body. This fact is of great clinical significance because these electrical signals can be picked up from the body surface and transformed into visible tracings by an instrument called an **electrocardiograph** (e-lek-tro-KAR-dee-o-graf).

The **electrocardiogram** (e-lek-tro-KAR-dee-o-gram) or **ECG** is the graphic record of the heart's electrical activity. Skilled interpretation of these ECG records may sometimes make the difference between life and death. A normal ECG tracing is shown in Figure 12-9.

A normal ECG tracing has three very characteristic deflections or waves called the **P wave,** the **QRS complex,** and the **T wave.** These deflections represent the electrical activity that regulates the contraction or relaxation of the atria or ventricles. The term *depolarization* describes the electrical activity that triggers contraction of the heart muscle. *Repolarization* begins just before the relaxation phase of cardiac muscle activity. In the normal ECG shown in Figure 12-9, *G,* the small P wave occurs with depolarization of the atria. The QRS complex occurs as a result of depolarization of the ventricles, and the T wave results from electrical activity generated by repolarization of the ventricles. You may wonder why no visible record of atrial repolarization is noted in a normal ECG. The reason is simply that the deflection is very small and is hidden by the large QRS complex that occurs at the same time.

Damage to cardiac muscle tissue that is caused by a myocardial infarction or disease affecting the heart's conduction system results in distinctive changes in the ECG. Therefore ECG tracings are extremely valuable in the diagnosis and treatment of heart disease.

CARDIAC ARRHYTHMIA

Various conditions such as endocarditis or myocardial infarction can damage the heart's conduction system and thereby disturb the rhythmical beating of the heart. The term **arrhythmia** (ay-RITH-me-ah) refers to an abnormality of heart rhythm.

One kind of arrhythmia is called a **heart block.** In *AV node block,* impulses are blocked from getting through to the ventricular myocardium, resulting in the ventricles contracting at a much slower rate than normal. On an ECG, there may be a large distance between the P wave and the R peak of the QRS complex. *Complete heart block* occurs when the P waves do not match up with

ECHOCARDIOGRAPHY

Although the stethoscope is still the basic tool of the **cardiologist,** or heart specialist, more sophisticated methods for detecting abnormalities in heart valve function are available. One widely used technique is **echocardiography** (ek-o-kar-dee-OG-ra-fee). Ultrasound (extremely high-pitched sound) directed toward the heart is reflected back (echoed) by the tissues. Like an airport's radar, a detector picks up the echoed ultrasound and produces an image showing different regions of blood and heart tissues. As the valves and other structures move during a series of heart beats, the image changes. A cardiologist can examine a continuous recording called an *echocardiogram* and determine the nature of a valve problem or other heart disorder.

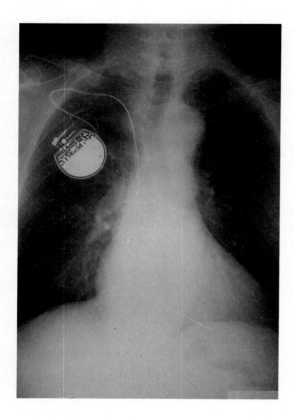

▼ **FIGURE 12-10 Artificial Pacemaker.** This x-ray photograph shows the stimulus generator in the subcutaneous tissue of the chest wall. Thin, flexible wires extend through veins to the heart, where timed electrical impulses stimulate the myocardium.

the QRS complexes at all—as in an ECG that shows 2 or more P waves for every QRS complex. A physician may treat heart block by implanting in the heart an **artificial pacemaker,** a battery-operated device implanted under the skin and connected by thin wires to the myocardium (Figure 12-10). This device stimulates the myocardium with timed electrical impulses that cause ventricular contractions at a rate fast enough to maintain an adequate circulation of blood.

Bradycardia (bray-dee-KAR-dee-ah) is a slow heart rhythm—below 50 beats per minute. Slight bradycardia is normal during sleep and in con-ditioned athletes while they are awake (but at rest). Abnormal bradycardia can result from improper autonomic nervous control of the heart or from a damaged SA node. If the problem is severe, artificial pacemakers can be used to increase the heart rate by taking the place of the SA node. For example, *demand pacemakers* take over SA node function only when the heart rate falls below a level programmed into the pacemaker by the physician.

Tachycardia (tak-ee-KAR-dee-ah) is a rapid heart rhythm—over 100 beats per minute. Tachycardia is normal during and after exercise and during the stress response. Abnormal tachycardia can result from improper autonomic control of the heart, blood loss or shock, the action of drugs and toxins, fever, and other factors.

Sinus arrhythmia is a variation in heart rate during the breathing cycle. Typically, the rate increases during inspiration and decreases during expiration. The causes of sinus arrhythmia are not clear. This phenomenon is common in young people and does not require treatment.

Premature contractions or *extrasystoles* (eks-trah-SIS-tol-ees) are contractions that occur before the next expected contraction in a series of cardiac cycles. For example, *premature atrial contractions (PACs)* may occur shortly after the ventricles contract—an early P wave on the ECG. Premature contractions often occur with lack of sleep, too much caffeine or nicotine, alcoholism, or heart damage. Frequent premature contractions can lead to **fibrillation** (fib-ril-AY-shun), a condition in which cardiac muscle fibers contract out of step with each other. This event can be seen in an ECG as the absence of regular P waves or abnormal QRS and T waves. In fibrillation, the affected heart chambers do not effectively pump blood. *Atrial fibrillation* occurs commonly in mitral stenosis, rheumatic heart disease, and infarction of the atrial myocardium. This condition can be treated with drugs such as digoxin (a *digitalis* preparation) or by *defibrillation*—application of an electric shock to force cardiac muscle fibers to contract in rhythm. *Ventricular fibrillation* is an immediately life-threatening condition in which the lack of ventricular pumping suddenly stops the flow of blood to vital tissues. Unless ventricular fibrillation is corrected immediately by defibrillation or some other method, death may occur within minutes.

HEART MEDICATIONS

Although numerous drugs are used in the treatment of heart disease, a few have proven themselves as the basic tools of the cardiologist.

Anticoagulants

These are drugs, including aspirin, that prevent clot formation in patients with valve damage or who have experienced a myocardial infarction. Coumadin and dicumarol are examples of commonly used oral anticoagulants.

Beta-adrenergic blockers

These are drugs that block norepinephrine receptors in cardiac muscle and thus reduce the rate and strength of the heart beat. Such drugs can help correct certain arrhythmias, as well as reduce the amount of oxygen required by the myocardium. Propranolol and related drugs are beta-adrenergic blockers.

Calcium-channel blockers

These are drugs that block the flow of calcium into cardiac muscle cells, thus reducing heart contractions. Calcium-channel blockers may be used in treating certain arrhythmias and coronary heart disease. Some examples include diltiazem, nicardipine, and nifedipine.

Digitalis

This is a drug that slows and increases the strength of cardiac contractions. This drug is important in the treatment of congestive heart failure and certain arrhythmias. Digoxin is one of several commonly used digitalis preparations.

Nitroglycerin

This drug dilates (widens) coronary blood vessels, thus increasing the flow of oxygenated blood to the myocardium. Nitroglycerin is often used to prevent or relieve angina pectoris.

Tissue plasminogen activator (TPA)

Usually a synthetic version of a naturally occurring substance from the walls of blood vessels, TPA activates a substance in the blood called *plasminogen*, which dissolves clots that may be blocking coronary arteries. Another preparation called *streptokinase*, an enzyme produced by *Streptococcus* bacteria, has similar effects.

Heart Failure

Heart failure is the inability of the heart to pump enough blood to sustain life. Heart failure can be the result of many different heart diseases. Valve disorders can reduce the pumping efficiency of the heart enough to cause heart failure. **Cardiomyopathy** (kar-dee-o-my-OP-ath-ee) or disease of the myocardial tissue may reduce pumping effectiveness. A specific event, such as myocardial infarction, can result in myocardial damage that causes heart failure. Arrhythmias, such as complete heart block or ventricular fibrillation, can also impair the pumping effectiveness of the heart and thus cause heart failure.

Failure of the right side of the heart or *right-sided heart failure* accounts for about one fourth of all cases of heart failure. Right-sided heart failure often results from the progression of disease that begins in the left side of the heart. Failure of the left side of the heart results in reduced pumping of blood returning from the lungs. Blood backs up into the pulmonary circulation, then into the right heart—causing an increase in pressure that the right side of the heart simply cannot overcome. Right-sided heart failure can also be caused by lung disorders that obstruct normal pulmonary blood flow and thus overload the right side of the heart—a condition called **cor pulmonale** (kor pul-mon-AHL-ee).

▼ **FIGURE 12-11 Human Heart Prepared for Transplantation into a Patient.**

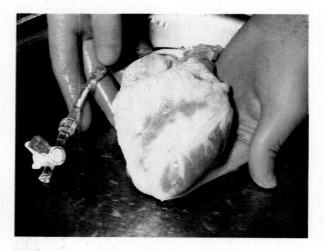

Congestive heart failure (CHF) or *simply left-sided heart failure* is the inability of the left ventricle to pump blood effectively. Most often, such failure results from myocardial infarction caused by coronary artery disease. It is called *congestive heart failure* because it decreases pumping pressure in the systemic circulation, which in turn causes the body to retain fluids. Portions of the systemic circulation thus become congested with extra fluid. As stated above, left-sided heart failure also causes congestion of blood in the pulmonary circulation, termed *pulmonary edema—*possibly leading to right heart failure.

Patients in danger of death because of heart failure may be candidates for heart *transplants* or heart *implants*. Heart transplants are surgical procedures in which healthy hearts from recently deceased donors replace the hearts of patients with heart disease (Figure 12-11). Unfortunately, a continuing problem with this procedure is the tendency of the body's immune system to reject the new heart as a foreign tissue. More details about the rejection of transplanted tissues are found in Chapter 14. Heart implants are artificial hearts that are made of biologically inert synthetic materials. The publicity surrounding the implantation of the Jarvik-7 artificial heart in patient Barney Clark in 1982 led many people to believe that the era of artificial hearts had arrived. However, the Jarvik-7 and most other artificial hearts require cumbersome external pumps and other devices that limit the recipient's mobility severely. Even more important, they are not efficient enough to serve as more than a temporary solution until a transplant or other medical treatment is available. Although technology in this area is improving rapidly, the promise of a practical and permanent heart replacement is still unfulfilled.

OUTLINE SUMMARY

LOCATION, SIZE, AND POSITION OF THE HEART

A Triangular organ located in mediastinum with two thirds of the mass to the left of the body midline and one third to the right; the apex on the diaphragm; shape and size of a closed fist (the Figure 12-1)

B Cardiopulmonary resuscitation (CPR)—heart lies between the sternum in front and the bodies of the thoracic vertebrae behind; rhythmic compression of the heart between the sternum and vertebrae can maintain blood flow during cardiac arrest; if combined with artificial respiration procedure, it can be life saving

ANATOMY OF THE HEART

A Heart chambers (Figure 12-2)
1 Two upper chambers are called *atria* (receiving chambers)—right and left atria
2 Two lower chambers called *ventricles* (discharging chambers)—right and left ventricles
3 Wall of each heart chamber is composed of cardiac muscle tissue called *myocardium*
4 Endocardium—smooth lining of heart chambers—inflammation of endocardium called *endocarditis*

B The pericardium and pericarditis
1 Pericardium is a two-layered fibrous sac with a lubricated space between the two layers
2 Inner layer is called *visceral pericardium* or *epicardium*
3 Outer layer called *parietal pericardium*
4 Pericarditis—inflammation of the pericardium
5 Cardiac tamponade—compression of the heart caused by fluid building up between the visceral pericardium and parietal pericardium

C Heart action
Contraction of the heart is called *systole*; relaxation is called *diastole.*

D Heart valves and valve disorders (Figure 12-3)
Four valves keep blood flowing through the heart; prevent backflow (two atrioventricular or AV and two semilunar valves)
1 Tricuspid—at the opening of the right atrium into the ventricle
2 Bicuspid (mitral)—at the opening of the left atrium into the ventricle
3 Pulmonary semilunar—at the beginning of the pulmonary artery
4 Aortic semilunar—at the beginning of the aorta
5 Incompetent valves "leak," allowing some blood back into the chamber from which it came

6 Stenosed valves are narrower that normal, reducing blood flow
7 Rheumatic heart disease—cardiac damage resulting from a delayed inflammatory response to strep infection
8 Mitral valve prolapse (MVP)—incompetence of mitral valve because its edges extend into the left atrium when the left ventricle contracts

HEART SOUNDS

A Two distinct heart sounds in every heartbeat or cycle—"lub-dup"
B First (lub) sound is caused by the vibration and closure of AV valves during contraction of the ventricles
C Second (dup) sound is caused by the closure of the semilunar valves during relaxation of the ventricles
D Heart murmurs—abnormal heart sounds often caused by abnormal valves

BLOOD FLOW THROUGH THE HEART
(Figure 12-5)

A Heart acts as two separate pumps—the right atrium and ventricle performing different functions from the left atrium and ventricle
B Sequence of blood flow: venous blood enters the right atrium through the superior and inferior venae cavae—passes from the right atrium through the tricuspid valve to the right ventricle; from the right ventricle through the pulmonary semilunar valve to the pulmonary artery to the lungs—blood from the lungs to the left atrium, passing through the bicuspid (mitral) valve to left ventricle; Blood in the left ventricle is pumped through the aortic semilunar valve into the aorta and is distributed to the body as a whole

CORONARY CIRCULATION AND CORONARY HEART DISEASE

A Blood, which supplies oxygen and nutrients to the myocardium of the heart, flows through the right and left coronary arteries
B Blockage of blood flow through the coronary arteries can cause myocardial infarction (heart attack)
C Atherosclerosis (type of "hardening of arteries" in which lipids build up on the inside wall of blood vessels) can partially or totally block coronary blood flow
D Angina pectoris—chest pain caused by inadequate oxygen to the heart

CARDIAC CYCLE

A Heart beat is regular and rhythmic—each complete beat called a *cardiac cycle*—average is about 72 beats per minute

B Each cycle, about 0.8 seconds long, subdivided into systole (contraction phase) and diastole (relaxation phase)

C Stroke volume is the volume of blood ejected from one ventricle with each beat

D Cardiac output is amount of blood that one ventricle can pump each minute—average is about 5 L per minute at rest

CONDUCTION SYSTEM OF THE HEART
(Figure 12-8)

A Normal structure and function
1 SA (sinoatrial) node, the pacemaker—located in the wall of the right atrium near the opening of the superior vena cava
2 AV (atrioventricular) node—located in the right atrium along the lower part of the interatrial septum
3 AV bundle (bundle of His)—located in the septum of the ventricle
4 Purkinje fibers—located in the walls of the ventricles

B Electrocardiography (Figure 12-9)
1 Specialized conduction system structures generate and transmit the electrical impulses that result in contraction of the heart
2 These tiny electrical impulses can be picked up on the surface of the body and transformed into visible tracings by a machine called an *electrocardiograph*
3 The visible tracing of these electrical signals is called an *electrocardiogram* or *ECG*
4 The normal ECG has three deflections or waves called the *P wave*, the *QRS complex*, and the *T wave*
a P wave—associated with depolarization of the atria
b QRS complex—associated with depolarization of the ventricles
c T wave—associated with repolarization of the ventricles

C Cardiac arrhythmia—abnormality of heart rhythm
1 Heart block—conduction of impulses is blocked
a Complete heart block—impaired AV node conduction, producing complete dissociation of P waves from QRS complexes
b Can be treated by implanting an artificial pacemaker
2 Bradycardia—slow heart rate (under 60 beats/min)
3 Tachycardia—rapid heart rate (over 100 beats/min)
4 Sinus arrhythmia—variation in heart rate during breathing cycle
5 Premature contraction (extrasystole)—contraction that occurs sooner than expected in a normal rhythm
6 Fibrillation—condition in which cardiac muscle fibers are "out of step," producing no effective pumping action

HEART FAILURE

A Heart failure—inability to pump enough returned blood to sustain life; it can be caused by many different heart diseases

B Right-sided heart failure—failure of the right side of the heart to pump blood, usually because the left side of the heart is not pumping effectively

C Left-sided heart failure (congestive heart failure)—inability of the left ventricle to pump effectively, resulting in congestion of the systemic and pulmonary circulations

D Diseased hearts can be replaced by donated living hearts (transplants) or by artificial hearts (implants), although both procedures have yet to be perfected

NEW WORDS

atrioventricular (AV) valve
atrium
bicuspid valve
cardiac output
chordae tendineae
coronary circulation

coronary sinus
diastole
endocardium
epicardium
mitral valve
myocardium
P wave

pacemaker
pericardium
pulmonary circulation
Purkinje fibers
QRS complex
semilunar valve

sinoatrial node
systemic circulation
systole
T wave
tricuspid valve
ventricle

Diseases and Other Clinical Terms

angina pectoris
arrhythmia
atherosclerosis
bradycardia
cardiac tamponade
cardiologist
cardiomyopathy
cardiopulmonary
  resuscitation (CPR)

congestive heart failure
  (CHF)
cor pulmonale
echocardiography
electrocardiogram
  (ECG)
endocarditis

fibrillation
heart disease
heart failure
incompetent valves
mitral valve prolapse
  (MVP)
myocardial infarction

pericarditis
premature contractions
rheumatic heart disease
sinus arrhythmia
stenosed valves
tachycardia

CHAPTER TEST

1. The two upper chambers of the heart are called _____, and the two lower chambers are called _____.
2. Cardiac muscle tissue, which forms the wall of each heart chamber, is called the _____.
3. The covering sac around the heart is called the _____.
4. The two valves that separate the upper from the lower chambers of the heart are called _____ valves.
5. The valve located between the right atrium and ventricle is called the _____ valve.
6. Blood enters the right atrium through two large veins called the superior and inferior _____.
7. Blood passing through the pulmonary semilunar valve enters the _____ artery.
8. Severe chest pain caused by inadequate blood flow to the myocardium is called _____.
9. The sinoatrial or SA node is often called the _____ of the heart.
10. A graphic recording of the heart's electrical activity is called an _____.
11. In a normal electrocardiogram the QRS complex occurs as a result of depolarization of the _____.

Select the most correct answer from Column B for each statement in Column A. (Only one answer is correct.)

COLUMN A	COLUMN B
12. _____ Receiving chambers of heart	**a.** Pericarditis
13. _____ Inflamed heart covering	**b.** P wave
14. _____ Attach valves to heart wall	**c.** Coronary arteries
15. _____ Supply blood to heart muscle	**d.** Heart attack
16. _____ Myocardial infarction	**e.** Chordae tendineae
17. _____ Depolarization of atria	**f.** Systole
18. _____ Heart contraction	**g.** Atria

Circle the T before each true statement and the F before each false statement.

T F   19. Heart disease is the leading cause of death in the United States.
T F   20. Stenosed heart valves can result from rheumatic heart disease.
T F   21. The only possible cause of coronary artery blockage is atherosclerosis, a type of "hardening of the arteries."
T F   22. Bradycardia is a rapid heart rate (above 100 beats/min).
T F   23. Premature contractions of the heart (extrasystoles) can lead to cardiac fibrillation.
T F   24. Disease of the myocardium is also known as *cardiomyopathy.*
T F   25. Congestive heart failure can cause right-sided heart failure.

REVIEW QUESTIONS

1 Describe the position of the heart in the mediastinum.

2 What is the lining of the heart called? The muscle layer of the wall? The covering of the heart?

3 What is cardiopulmonary resuscitation (CPR)?

4 What is meant by the term *cardiac cycle?*

5 A patient is told she has mitral valve prolapse (MVP). Where is the mitral valve and what is MVP? Explain how a severe prolapse could lead to congestive heart failure.

6 What is the difference between a stenosed valve and an incompetent valve? How can they be distinguished using a stethoscope?

7 You hear that a friend has blocked coronary arteries. What are the possible causes? What could result if the condition is not treated? How can a coronary blockage be treated?

8 What two hearts sounds are heard with a stethoscope? What causes these sounds?

9 Trace the flow of blood through the heart.

10 Name the three major features of a normal ECG. What does each represent?

11 Define each of these types of cardiac arrhythmia: *heart block, bradycardia, tachycardia, sinus arrhythmia, premature contraction.*

12 What is meant the by term *heart failure?*

CLINICAL APPLICATIONS

1 You are visiting a friend in the hospital. Beside her bed is a video monitor that displays your friend's ECG. She asks you what the large spikes represent—can you tell her? Suddenly the ECG line becomes completely disorganized, with no discernible P, QRS, or T waves. What may have happened? What, if anything, should be done for her?

2 Your classmate, Vivian, told you during lunchtime today that she has been diagnosed as having mitral valve prolapse. Can you describe this structural abnormality and describe its possible effects on heart function?

3 Uncle John is about to undergo coronary bypass surgery. His surgeon carefully explained Uncle John's condition and the surgical procedure to correct it, but your uncle was too upset to pay close attention. Now that he has calmed a bit, he realizes that he has very little idea of what "his triple-bypass surgery" is all about. Can you describe to your uncle the probable condition of his heart and explain the concept of the planned surgery?

Circulation of the Blood

C H A P T E R

13

Objectives

After you have completed this chapter,
you should be able to:

1 Describe the structure and function of each major type of blood vessel: artery, vein, and capillary.

2 List the major disorders of blood vessels and explain how they develop.

3 Trace the path of blood through the systemic, portal, and fetal circulations.

4 Identify and discuss the factors involved in the generation of blood pressure and how they relate to each other.

5 Define pulse and locate the major pulse points on the body.

6 Explain what is meant by the term *circulatory shock* and describe the major types.

*I*n the previous chapter, we discussed the basic structure and function of the circulatory system's pump: the heart. In this chapter, we will continue our explanation of how blood circulates through the internal environment of the body. First, the structure and function of blood vessels will be discussed in some detail. We will then explain how the vessels fit together into routes for the conduction of blood. The last part of this chapter will deal with blood pressure, the driving force of blood circulation. We will, of course, also discuss major circulatory disorders as appropriate throughout the chapter.

Blood Vessels

KINDS

Arterial blood is pumped from the heart through a series of large distribution vessels—the **arteries.** The largest artery in the body is the aorta. Arteries subdivide into vessels that become progressively smaller and finally become tiny **arterioles** (ar-TEER-ee-ols) that control the flow into microscopic exchange vessels called **capillaries** (KAP-i-lair-ees). In the so-called **capillary beds,** the exchange of nutrients and respiratory gases occurs between the blood and tissue fluid around the cells. Blood exits or is drained from the capillary beds when it enters the small **venules** (VEN-yools), which join with other venules and increase in size, becoming **veins.** The largest veins are the superior vena cava and the inferior vena cava.

As noted in Chapter 12, the arteries carry blood away from the heart toward capillaries. The veins carry blood toward the heart away from capillaries, and the capillaries carry blood from the tiny arterioles into tiny venules. The aorta carries blood out of the left ventricle of the heart, and the venae cavae return blood to the right atrium after the blood has circulated throughout the body.

In 1628, William Harvey, an English physician, first described and demonstrated how blood circulated. A clever set of experiments enabled him to work out a correct scheme of blood circulation, even though he could not see the microscopic capillaries that connect the arterial and venous flow.

STRUCTURE

Arteries, veins, and capillaries differ in structure. Three coats or layers are found in both arteries and veins (Figure 13-1). The outermost layer is called the **tunica externa.** Note that smooth muscle tissue is found in the middle layer or **tunica media** of arteries and veins. However, the muscle layer is much thicker in arteries than in veins. Why is this important? Because the thicker muscle layer in the artery wall is able to resist great pressures generated by ventricular systole. In arteries, the tunica media plays a critical role in maintaining blood pressure and controlling blood distribution in the body. This is a smooth muscle, so it is controlled by the autonomic nervous system.

A thin layer of elastic and white fibrous tissue covers an inner layer of endothelial cells called the **tunica interna** in arteries and veins. The tunica interna is actually a single layer of squamous epithelial cells called **endothelium** (en-doe-THEE-lee-um) that lines the inner surface of the entire circulatory system.

As you can see in Figure 13-1, veins have a unique structural feature not present in arteries. They are equipped with one-way valves that prevent the backflow of blood. When a surgeon cuts into the body, only arteries, arterioles, veins, and venules can be seen. Capillaries cannot be seen because they are microscopic. The most important structural feature of capillaries is their extreme thinness—only one layer of flat, endothelial cells composes the capillary membrane. Instead of three layers or coats, the capillary wall is composed of only one—the tunica interna. Substances such as glucose, oxygen, and wastes can quickly pass through it on their way to or from the cells. Smooth muscle cells that are called **precapillary sphincters** guard the entrance to the capillary and determine into which capillary blood will flow.

FUNCTIONS

Arteries, veins, and capillaries have different functions. Arteries and arterioles distribute blood from the heart to capillaries in all parts of the body. In addition, by constricting or dilating, arterioles help maintain arterial blood pressure at a normal level. Venules and veins collect blood

FIGURE 13-1 Artery and Vein. Schematic drawings of an artery and a vein show comparative thicknesses of the three layers: the outer layer or tunica externa, the muscle layer or tunica media, and the tunica interna lined with endothelium. Note that the muscle and outer layer are much thinner in veins than in arteries and that veins have valves.

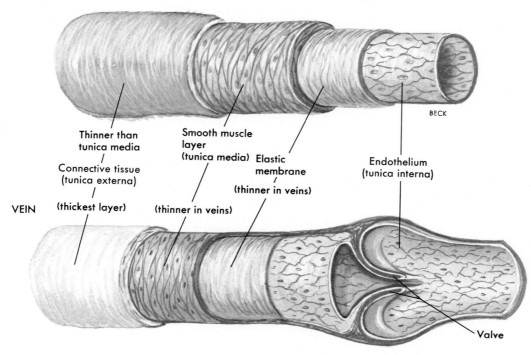

from capillaries and return it to the heart. They also serve as blood reservoirs because they can expand to hold a larger volume of blood or constrict to hold a much smaller amount. Capillaries function as exchange vessels. For example, glucose and oxygen move out of the blood in capillaries into interstitial fluid and on into cells. Carbon dioxide and other substances move in the opposite direction (that is, into the capillary blood from the cells). Fluid is also exchanged between capillary blood and interstitial fluid (see Chapter 19).

Study Figure 13-2 and Table 13-1 on p. 332 to learn the names of the main arteries of the body and Figure 13-3 and Table 13-2 on p. 332 for the names of the main veins.

CHANGES IN BLOOD FLOW DURING EXERCISE

Not only does the overall rate of blood flow increase during exercise, but the relative blood flow through the different organs of the body also changes. During exercise, blood is routed away from the kidneys and digestive organs and toward the skeletal muscles, cardiac muscle, and skin.

FIGURE 13-2 Principal Arteries of the Body.

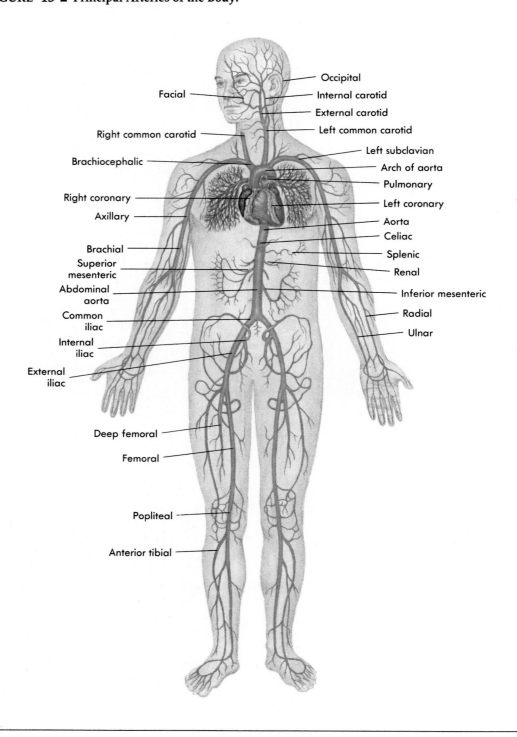

FIGURE 13-3 Principal Veins of the Body.

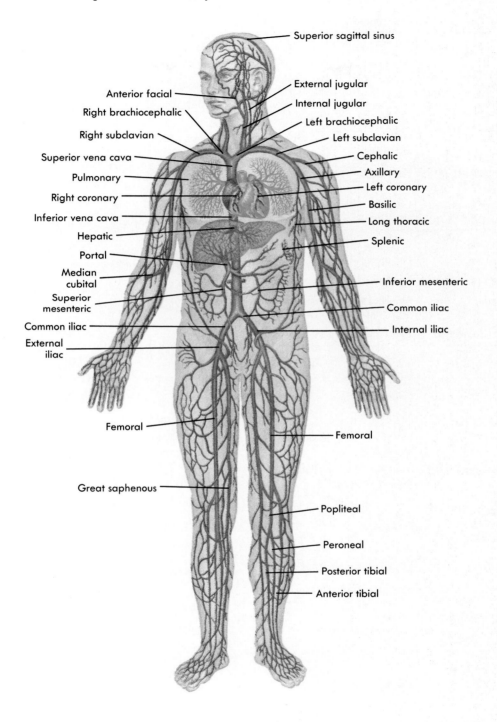

- Superior sagittal sinus
- Anterior facial
- Right brachiocephalic
- Right subclavian
- Superior vena cava
- Pulmonary
- Right coronary
- Inferior vena cava
- Hepatic
- Portal
- Median cubital
- Superior mesenteric
- Common iliac
- External iliac
- Femoral
- Great saphenous

- External jugular
- Internal jugular
- Left brachiocephalic
- Left subclavian
- Cephalic
- Axillary
- Left coronary
- Basilic
- Long thoracic
- Splenic
- Inferior mesenteric
- Common iliac
- Internal iliac
- Femoral
- Popliteal
- Peroneal
- Posterior tibial
- Anterior tibial

TABLE 13-1 The Major Arteries

ARTERY	TISSUES SUPPLIED
Head and Neck	
Occipital	Posterior head and neck
Facial	Mouth, pharynx, and face
Internal carotid	Anterior brain and meninges
External carotid	Superficial neck, face, eyes, and larynx
Right common carotid	Right side of the head and neck
Left common carotid	Left side of the head and neck
Thorax	
Left subclavian	Left upper extremity
Brachiocephalic	Head and arm
Arch of aorta	Branches to head, neck, and upper extremities
Right and left coronary	Heart muscle
Abdomen	
Celiac	Stomach, spleen, and liver
Splenic	Spleen
Renal	Kidneys
Superior mesenteric	Lower half of the large intestine
Inferior mesenteric	Small intestine; first half of the large intestine
Upper Extremity	
Axillary	Axilla (armpit)
Brachial	Arm
Radial	Lateral side of the hand
Ulnar	Medial side of the hand
Lower Extremity	
Internal iliac	Pelvic viscera and rectum
External iliac	Genitalia and lower trunk muscles
Deep femoral	Deep thigh muscles
Femoral	Thigh
Popliteal	Leg and foot
Anterior tibial	Leg

TABLE 13-2 The Major Veins

VEIN	TISSUES DRAINED
Head and Neck	
Superior sagittal sinus	Brain
Anterior facial	Anterior and superficial face
External jugular	Superficial tissues of the head and neck
Internal jugular	Sinuses of the brain
Thorax	
Brachiocephalic	Viscera of the thorax
Subclavian	Upper extremities
Superior vena cava	Head, neck, upper extremities
Pulmonary	Lungs
Right and left coronary	Heart
Inferior vena cava	Lower body
Abdomen	
Hepatic	Liver
Long thoracic	Abdominal and thoracic muscles
Hepatic portal	Liver and gallbladder
Splenic	Spleen
Superior mesenteric	Small intestine, most of colon
Inferior mesenteric	Descending colon and rectum
Upper Extremity	
Cephalic	Lateral arm
Axillary	Axilla and arm
Basilic	Medial arm
Median cubital	Cephalic vein (to basilic vein)
Lower Extremity	
External iliac	Lower limb
Internal iliac	Pelvic viscera
Femoral	Thigh
Great saphenous	Leg
Popliteal	Lower leg
Peroneal	Foot
Anterior tibial	Deep anterior leg, dorsal foot
Posterior tibial	Deep posterior leg and plantar aspect of foot

Disorders of Blood Vessels

DISORDERS OF ARTERIES

As you may have gathered from the previous discussion, arteries contain blood that is maintained at a relatively high pressure. This means the arterial walls must be able to withstand a great deal of force, or they will burst. The arteries must also stay free of obstruction; otherwise they cannot deliver their blood to the capillary beds (and thus the tissues they serve).

A common type of vascular disease that occludes (blocks) arteries and weakens arterial walls is called **arteriosclerosis** (ar-tee-ree-o-skle-RO-sis) or *hardening of the arteries.* Arteriosclerosis is characterized by thickening of arterial walls that progresses to hardening as calcium deposits form. The thickening and calcification reduce the flow of blood to the tissues. If the blood flow slows down too much, **ischemia** (is-KEE-me-ah) results. Ischemia or decreased blood supply to a tissue involves the gradual death of cells and may lead to complete tissue death—a condition called **necrosis** (ne-KRO-sis). If a large section of tissue becomes necrotic, it may begin to decay. Necrosis that has progressed this far is called **gangrene** (GANG-green). Because of the tissue damage involved, arteriosclerosis is not only painful—it is life-threatening. For example, ischemia of heart muscle can lead to *myocardial infarction* (see Chapter 12).

There are several types of arteriosclerosis, but perhaps the most well known is *atherosclerosis,* described in Chapter 12 as the blockage of arteries by lipids and other matter (Figure 13-4). Eventually, the fatty deposits in the arterial walls become fibrous and perhaps calcified—resulting in sclerosis (hardening). High blood levels of triglycerides and cholesterol, which may be caused by a high-fat, high-cholesterol diet, smoking, and a genetic predisposition, are associated with atherosclerosis. (See the Chemistry Appendix for a discussion of triglycerides and cholesterol.)

In general, arteriosclerosis develops with advanced age, diabetes, high-fat and high-cholesterol diets, hypertension (high blood pressure), and smoking. Arteriosclerosis can be treated by

FIGURE 13-4 Partial Blockage of an Artery in Atherosclerosis. Atherosclerotic plaque develops from the deposition of fats and other substances in the wall of the artery.

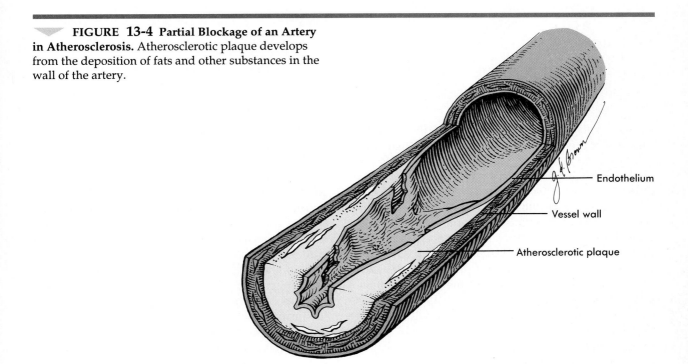

Endothelium

Vessel wall

Atherosclerotic plaque

FIGURE 13-5 Balloon Angioplasty. A, A catheter is inserted into vessel until it reaches the affected region. **B,** A probe with a metal tip is pushed out the end of the catheter into the blocked region of the vessel. **C,** The balloon is inflated, pushing the walls of the vessel outward. Sometimes metal coils or tubes are inserted to keep the vessel open.

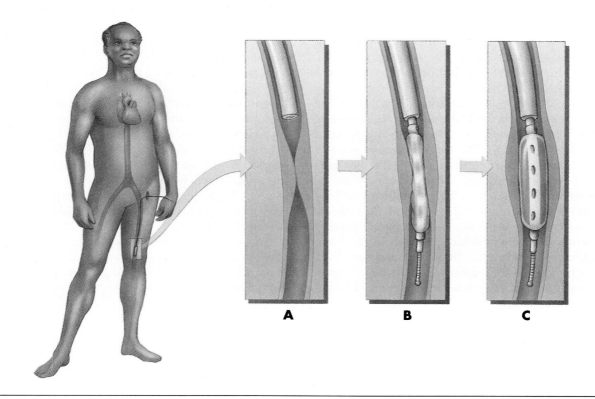

A   **B**   **C**

drugs called *vasodilators* that trigger the smooth muscles of the arterial walls to relax, thus causing the arteries to dilate (widen). Some cases of atherosclerosis are treated by mechanically opening the affected area of an artery, a type of procedure called **angioplasty** (AN-jee-o-plas-tee). In one procedure, a deflated balloon attached to a long tube called a *catheter* is inserted into a partially blocked artery and inflated (Figure 13-5). As the balloon inflates, the *plaque* (fatty deposits and tissue) is pushed outward, and the artery widens to allow near-normal blood flow. In a similar procedure, metal springs or mesh tubes called *stents* are inserted in affected arteries and hold them open. Other types of angioplasty use lasers, drills, or spinning loops of wire to clear the way for nor-

mal blood flow. Severely affected arteries can also be surgically bypassed or replaced, as discussed in Chapter 12.

Damage to arterial walls caused by arteriosclerosis or other factors may lead to the formation of an **aneurysm** (AN-yoor-izm). An aneurysm is a section of an artery that has become abnormally widened because of a weakening of the arterial wall. Aneurysms sometimes form a saclike extension of the arterial wall. One reason aneurysms are dangerous is because they, like atherosclerotic plaques, promote the formation of thrombi (abnormal clots). A thrombus may cause an embolism (blockage) in the heart or some other vital tissue. Another reason aneurysms are dangerous is their tendency to burst, causing severe hemorrhaging that may result in death. A brain aneurysm may lead to a *stroke* or **cerebrovascular accident (CVA).** A stroke results from ischemia of brain tissue caused by an embolism or ruptured aneurysm. Depending on the amount of tissue affected and the place in the brain the CVA occurs, effects of a stroke may range from hardly noticeable to crippling to fatal.

DISORDERS OF VEINS

Varicose veins are veins in which blood tends to pool rather than continue on toward the heart. Varicosities, also called **varices** (VAIR-i-seez) (singular, *varix*), most commonly occur in *superficial veins* near the surface of the body (Figure 13-6). The *great saphenous vein*, the largest superficial vein of the leg (Figure 13-3), often becomes varicose in people that stand for long periods. The force of gravity slows the return of venous blood to the heart in such cases, causing blood-engorged veins to dilate. As the veins dilate, the distance between the flaps of venous valves widen—eventually making them incompetent (leaky). Incompetence of valves causes even more pooling in affected veins—a positive-feedback phenomenon.

Hemorrhoids (HEM-or-oyds) or *piles* are varicose veins in the rectum. Excessive straining during defecation can create pressures that cause hemorrhoids. The unusual pressures of carrying a child during pregnancy predisposes expectant mothers to hemorrhoids and other varicosities.

Varicose veins can be treated by supporting the dilated veins from the outside. For instance, support stockings can reduce blood pooling in the great saphenous vein. Surgical removal of varicose veins can be performed in severe cases. Advanced cases of hemorrhoids are often treated this way. Symptoms of milder cases can be relieved by removing the pressure that caused the condition.

A number of factors can cause **phlebitis** (fle-BYE-tis) or vein inflammation. Irritation by an intravenous catheter, for example, is a common cause of vein inflammation. **Thrombophlebitis** (throm-bo-fle-BYE-tis) is acute phlebitis caused by clot (thrombus) formation. Veins are more likely sites of thrombus formation than arteries because venous blood moves more slowly and is under less pressure. Thrombophlebitis is characterized

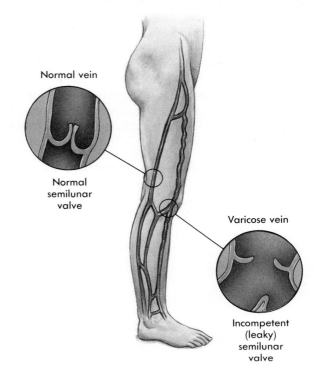

FIGURE 13-6 Varicose Veins. Veins near the surface of the body—especially in the legs—may bulge and cause venous valves to leak. The resulting enlarged veins are called *varicose veins* or *varices.*

Normal vein

Normal semilunar valve

Varicose vein

Incompetent (leaky) semilunar valve

by pain and discoloration of the surrounding tissue. If a piece of a clot breaks free, it may cause an embolism when it blocks a blood vessel. **Pulmonary embolism,** for example, could result when an embolus lodges in the circulation of the lung. Pulmonary embolism can lead to death quickly if too much blood flow is blocked.

Circulation of Blood

SYSTEMIC AND PULMONARY CIRCULATION

The term **circulation of blood** is self-explanatory, meaning that blood flows through vessels that are arranged to form a circuit or circular pattern. Blood flow from the left ventricle of the heart through blood vessels to all parts of the body and back to the right atrium of the heart was described in Chapter 12 as the **systemic circulation.** The left ventricle pumps blood into the aorta. From there, it flows into arteries that carry it into the tissues and organs of the body. As indicated in Figure 13-7, within each structure, blood moves from arteries to arterioles to capillaries. There, the vital two-way exchange of substances occurs between blood and cells. Next, blood flows out of each organ by way of its venules and then its veins to drain eventually into the inferior or superior vena cava. These two great veins return venous blood to the right atrium of the heart to complete the systemic circulation. But the blood has not quite come full circle back to its starting point in the left ventricle. To do this and start on its way again, it must first flow through another circuit, referred to in Chapter 12 as the **pulmonary circulation.** Observe in Figure 13-7 that venous blood moves from the right atrium to the right ventricle to the pulmonary artery to lung arterioles and capillaries. There, the exchange of gases between the blood and air takes place, converting the deep crimson typical of venous blood to the scarlet of arterial blood. This oxygenated blood then flows on through lung venules into

FIGURE 13-7 Diagram of Blood Flow in the Circulatory System. Blood leaves the heart through arteries, then travels through arterioles, capillaries, venules, and veins before returning to the opposite side of the heart. Compare this figure with Figure 11-4.

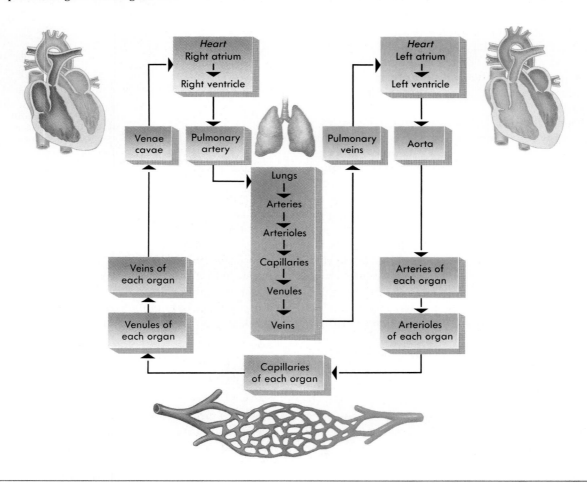

four pulmonary veins and returns to the left atrium of the heart. From the left atrium, it enters the left ventricle to be pumped again through the systemic circulation.

HEPATIC PORTAL CIRCULATION

The term **hepatic portal circulation** refers to the route of blood flow through the liver. Veins from the spleen, stomach, pancreas, gallbladder, and intestines do not pour their blood directly into the inferior vena cava as do the veins from other abdominal organs. Instead, they send their blood to the liver by means of the hepatic portal vein (Figure 13-8). It then must pass through the liver before it reenters the regular venous return to the heart. Blood leaves the liver by way of the hepatic veins, which drain into the inferior vena cava. As noted in Figure 13-7, blood normally flows from arteries to arterioles to capillaries to venules to veins and back to the heart. Blood flow in the hepatic portal circulation, however, does not follow this typical route. Venous blood, which would ordinarily return directly to the heart, is sent instead through a second capillary bed in the liver. The hepatic portal vein shown in Figure 13-8 is located between two capillary beds—one set

FIGURE 13-8 Hepatic Portal Circulation. In this very unusual circulation, a vein is located between two capillary beds. The hepatic portal vein collects blood from capillaries in visceral structures located in the abdomen and empties it into the liver. Hepatic veins return blood to the inferior vena cava. (Organs are not drawn to scale.)

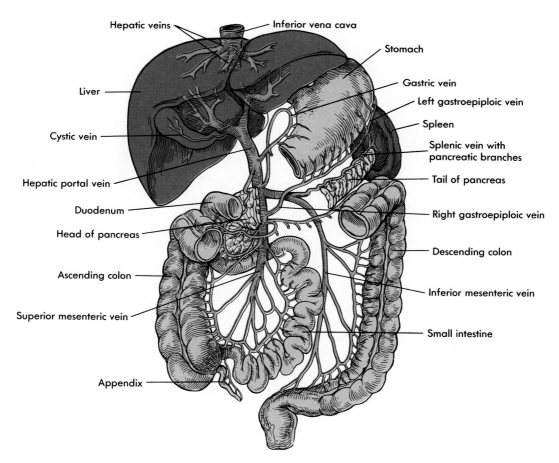

FIGURE 13-9 The Fetal Circulation.

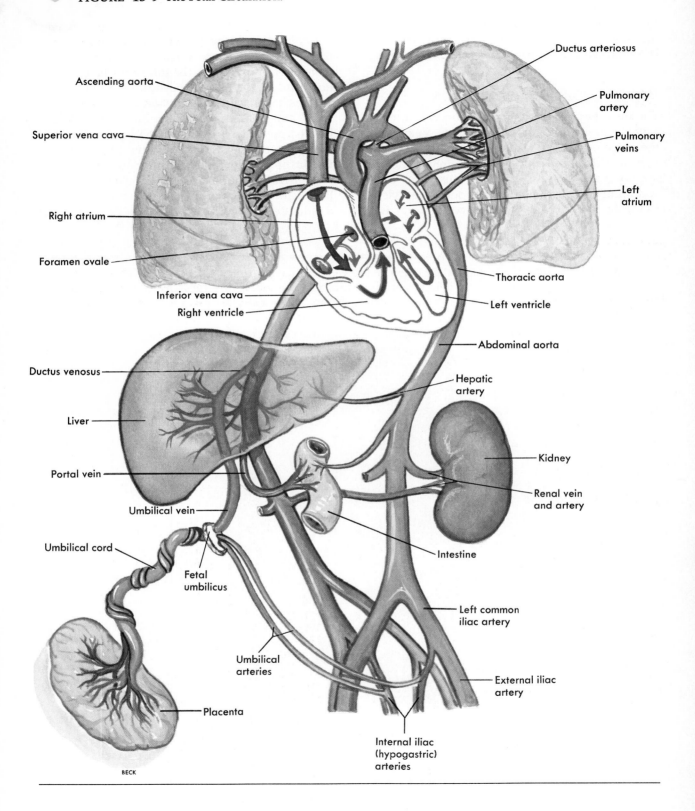

Ascending aorta

Superior vena cava

Right atrium

Foramen ovale

Inferior vena cava

Right ventricle

Ductus venosus

Liver

Portal vein

Umbilical vein

Umbilical cord

Fetal umbilicus

Umbilical arteries

Placenta

Ductus arteriosus

Pulmonary artery

Pulmonary veins

Left atrium

Thoracic aorta

Left ventricle

Abdominal aorta

Hepatic artery

Kidney

Renal vein and artery

Intestine

Left common iliac artery

External iliac artery

Internal iliac (hypogastric) arteries

BECK

in the digestive organs and the other in the liver. From the liver capillary beds, the path of blood returns to its normal route.

The detour of venous blood through a second capillary bed in the liver before its return to the heart serves some valuable purposes. For example, when a meal is being absorbed, the blood in the portal vein contains a higher-than-normal concentration of glucose. Liver cells remove the excess glucose and store it as glycogen; therefore blood leaving the liver usually has a normal blood glucose concentration. Liver cells also remove and detoxify various poisonous substances that may be present in the blood. The hepatic portal system is an excellent example of how "structure follows function" in helping the body maintain homeostasis.

FETAL CIRCULATION

Circulation in the body before birth differs from circulation after birth because the fetus must secure oxygen and food from maternal blood instead of from its own lungs and digestive organs. For the exchange of nutrients and oxygen to occur between fetal and maternal blood, specialized blood vessels must carry the fetal blood to the **placenta** (plah-SEN-tah), where the exchange occurs, and then return it to the fetal body. Three vessels (shown in Figure 13-9 as part of the **umbilical cord**) accomplish this purpose. They are the two small **umbilical arteries** and a single, much larger **umbilical vein.**

The movement of blood in the umbilical vessels may seem unusual at first in that the umbilical vein carries oxygenated blood, and the umbilical artery carries oxygen-poor blood. Remember that arteries are vessels that carry blood away from the heart, whereas veins carry blood toward the heart, regardless of the oxygen supply they may happen to have.

Another structure unique to fetal circulation is called the **ductus venosus** (DUK-tus ve-NO-sus). As you can see in Figure 13-9, it is actually a continuation of the umbilical vein. It serves as a shunt, allowing most of the blood returning from the placenta to bypass the immature liver of the developing baby and empty directly into the inferior vena cava. Two other structures in the developing fetus allow most of the blood to bypass the developing lungs, which remain collapsed until

birth. The **foramen ovale** (fo-RAY-men o-VAL-ee) shunts blood from the right atrium directly into the left atrium, and the **ductus arteriosus** (DUK-tus ar-teer-ee-O-sus) connects the aorta and the pulmonary artery.

At birth, the specialized fetal blood vessels and shunts must be rendered nonfunctional. When the newborn infant takes its first deep breaths, the circulatory system is subjected to increased pressure. The result is closure of the foramen ovale and rapid collapse of the umbilical blood vessels, the ductus venosus, and ductus arteriosus.

Several congenital disorders result from the failure of the circulatory system to shift from the fetal route of blood flow at the time of birth. The ductus arteriosus may fail to close, for example, and allow deoxygenated blood to bypass the lungs. Similarly, the foramen ovale may fail to close and remain as a so-called hole in the heart that allows blood to bypass the pulmonary circulation. In such cases, a light-skinned baby may appear bluish because of the lack of oxygen in the systemic arterial blood. This condition of bluish tissue coloration is called **cyanosis** (sye-a-NO-sis).

Blood Pressure

A good way to understand blood pressure might be to try to answer a few questions about it. What is blood pressure? Just what the words say— blood pressure is the pressure or push of blood.

Where does blood pressure exist? It exists in all blood vessels, but it is highest in the arteries and lowest in the veins. In fact, if we list blood vessels in order according to the amount of blood pressure in them and draw a graph, as in Figure 13-10, the graph looks like a hill, with aortic blood pressure at the top and vena caval pressure at the bottom. This blood pressure "hill" is spoken of as the *blood pressure gradient.* More precisely, the blood pressure gradient is the difference between two blood pressures. The blood pressure gradient for the entire systemic circulation is the difference between the average or mean blood pressure in the aorta and the blood pressure at the termination of the venae cavae where they join the right atrium of the heart. The mean blood pressure in the aorta, given in Figure 13-10, is 100 mm of mercury (mm Hg), and the pressure at the termina-

FIGURE 13-10 Pressure Gradients in Blood Flow. Blood flows down a "blood pressure hill" from arteries, where blood pressure is highest, into arterioles, where it is somewhat lower, into capillaries, where it is still lower, and so on. All numbers on the graph indicate blood pressure measured in millimeters of mercury. The top figure, 100 mm, represents the average pressure in the aorta.

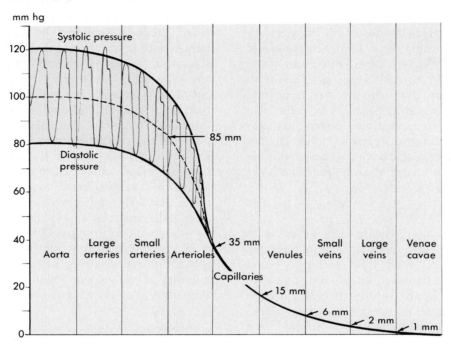

tion of the venae cavae is 0. Therefore, with these typical normal figures, the systemic blood pressure gradient is 100 mm Hg (100 minus 0).

Why is it important to understand blood pressure? What is its function? The blood pressure gradient is vitally involved in keeping the blood flowing. When a blood pressure gradient is present, blood circulates; conversely, when a blood pressure gradient is not present, blood does not circulate. For example, suppose that the blood pressure in the arteries were to decrease so that it became equal to the average pressure in arterioles. There would no longer be a blood pressure gradient between arteries and arterioles, and therefore there would no longer be a force to move blood out of arteries into arterioles. Circulation would stop, in other words, and very soon life itself would cease. This is why when arterial blood pressure is observed to be falling

rapidly, whether in surgery or elsewhere, emergency measures must be quickly started to try to reverse this fatal trend.

What we have just said in the preceding paragraph may start you wondering about why high blood pressure (meaning, of course, high arterial blood pressure) and low blood pressure are bad for circulation. High blood pressure or **hypertension** is bad for several reasons. For one thing, if it becomes too high, it may cause the rupture of one or more blood vessels (for example, in the brain, as happens in a stroke). But low blood pressure also can be dangerous. If arterial pressure falls low enough, circulation and life cease. Massive hemorrhage, which dramatically reduces blood pressure, kills in this way.

What causes blood pressure, and what makes blood pressure change from time to time? The direct cause of blood pressure is the volume of blood in the vessels. The larger the volume of blood in the arteries, for example, the more pressure the blood exerts on the walls of the arteries, or the higher the arterial blood pressure.

Conversely, the less blood in the arteries, the lower the blood pressure tends to be. Hemorrhage demonstrates this relation between blood volume and blood pressure. In hemorrhage a pronounced loss of blood occurs, and this decrease in the volume of blood causes blood pressure to drop. In fact, the major sign of hemorrhage is a rapidly falling blood pressure.

The volume of blood in the arteries is determined by how much blood the heart pumps into the arteries and how much blood the arterioles drain out of them. The diameter of the arterioles plays an important role in determining how much blood drains out of arteries into arterioles.

The strength and the rate of the heartbeat affect cardiac output and therefore blood pressure. Each time the left ventricle contracts, it squeezes a certain volume of blood (the stroke volume) into the aorta and on into other arteries. The stronger that each contraction is, the more blood it pumps into the aorta and arteries. Conversely, the weaker that each contraction is, the less blood it pumps. Suppose that one contraction of the left ventricle pumps 70 ml of blood into the aorta, and suppose that the heart beats 70 times a minute; 70 ml times 70 equals 4900 ml. Almost 5 L of blood would enter the aorta and arteries every minute (the cardiac output). Now suppose that the heartbeat were to become weaker and that each contraction of the left ventricle pumps only 50 instead of 70 ml of blood into the aorta. If the heart still contracts 70 times a minute, it will obviously pump much less blood into the aorta—only 3500 ml instead of the more normal 4900 ml per minute. This decrease in the heart's output decreases the volume of blood in the arteries, and the decreased arterial blood volume decreases arterial blood pressure. In summary, the strength of the heartbeat affects blood pressure in this way; a stronger heartbeat increases blood pressure, and a weaker beat decreases it.

The rate of the heartbeat may also affect arterial blood pressure. You might reason that, when the heart beats faster, more blood enters the aorta, and therefore the arterial blood volume and blood pressure increase. This is true only if the stroke volume does not decrease sharply when the heart rate increases. Often, however, when the heart beats faster, each contraction of the left ventricle takes place so rapidly that it has little time to fill, and it squeezes out much less blood than usual into the aorta. For example, suppose that the heart rate speeded up from 70 to 100 times a minute and that at the same time its stroke volume decreased from 70 to 40 ml. Instead of a cardiac output of 70 × 70 or 4900 ml per minute, the cardiac output would have changed to 100 × 40 or 4000 ml per minute. Arterial blood volume decreases under these conditions, and therefore blood pressure also decreases, even though the heart rate has increased. What generalization can we make? We can only say that an increase in the rate of the heartbeat increases blood pressure, and a decrease in the rate decreases blood pressure. But whether a change in the heart rate actually produces a similar change in blood pressure depends on whether the stroke volume also changes and in which direction.

Another factor that we ought to mention in connection with blood pressure is the viscosity of blood, or in plainer language, its thickness. If blood becomes less viscous than normal, blood pressure decreases. For example, if a person suffers a hemorrhage, fluid moves into the blood from the interstitial fluid. This dilutes the blood and decreases its viscosity, and blood pressure then falls because of the decreased viscosity. After hemorrhage, whole blood or plasma is preferred to saline solution for transfusions. The reason is that saline solution is not a viscous liquid and so cannot keep blood pressure at a normal level.

No one's blood pressure stays the same all the time. It fluctuates, even in a perfectly healthy individual. For example, it goes up when a person exercises strenuously. Not only is this normal, but the increased blood pressure serves a good purpose. It increases circulation to bring more blood to muscles each minute and thus supplies them with more oxygen and food for more energy.

A normal average arterial blood pressure is about 120/80, or 120 mm Hg systolic pressure (as the ventricles contract) and 80 mm Hg diastolic pressure (as the ventricles relax). Remember, however, that what is "normal" varies somewhat among individuals and also with age.

BLOOD PRESSURE READINGS

A device called a **sphygmomanometer** (sfig-mo-ma-NAH-me-ter) has been used to measure blood pressures in clinical situations. The traditional sphygmomanometer is an inverted tube of mercury (Hg) with a balloonlike air cuff attached via an air hose. The air cuff is placed around a limb, usually the subject's upper left arm as shown in the figure. A stethoscope sensor is placed over a major artery (the *brachial artery* in the figure) to listen for the arterial pulse. A hand-operated pump fills the air cuff, increasing the air pressure and pushing the column of mercury higher. While listening through the stethoscope, the operator opens the air cuff's outlet valve and slowly reduces the air pressure around the limb. Loud, tapping *Korotkoff sounds* suddenly begin when the cuff pressure measured by the mercury column equals the systolic pressure—usually about 120 mm. As the air pressure surrounding the arm continues to decrease, the Korotkoff sounds disappear. The pressure measurement at which the sounds disappear is equal to the diastolic pressure—usually about 80 mm. The subject's blood pressure is then expressed as systolic pressure (the maximum arterial pressure during each cardiac cycle) over the diastolic pressure (the minimum arterial pressure), such as 120/80. Mercury sphygmomanometers have been replaced in many clinical settings by nonmercury devices that similarly measure the maximum and minimum arterial blood pressures.

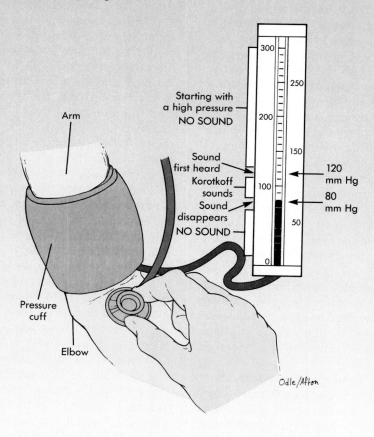

Arm

Starting with a high pressure
NO SOUND

Sound first heard
Korotkoff sounds
Sound disappears
NO SOUND

300
250
200
150
100
50
0

120 mm Hg
80 mm Hg

Pressure cuff

Elbow

Odle/Afton

The venous blood pressure, as you can see in Figure 13-10, is very low in the large veins and falls almost to 0 by the time blood leaves the venae cavae and enters the right atrium. The venous blood pressure within the right atrium is called the **central venous pressure.** This pressure level is important because it influences the pressure that exists in the large peripheral veins. If the heart beats strongly, the central venous pressure is low as blood enters—and leaves—the heart chambers efficiently. However, if the heart is weakened, central venous pressure increases, and the flow of blood into the right atrium is slowed. As a result, a person suffering heart failure, who is sitting at rest in a chair, often has distended external jugular veins as blood "backs up" in the venous network.

Five mechanisms help to keep venous blood moving back through the circulatory system and into the right atrium. They include a strongly beating heart, an adequate arterial blood pressure, valves in veins, the "milking action" of skeletal muscles as they contract, and changing pressures in the chest cavity caused by breathing.

Pulse

What you feel when you take a pulse is an artery expanding and then recoiling alternately. To feel a pulse, you must place your fingertips over an artery that lies near the surface of the body and over a bone or other firm base. The pulse is a valuable clinical sign. It can provide information, for example, about the rate, strength, and rhythmicity of the heart beat. It is also easily determined with little or no danger or discomfort. There are nine major "pulse points" named after the arteries felt. Locate each pulse point on Figure 13-11 and on your own body.

The following pulse points are located on each side of the head and neck: (1) over the superficial temporal artery in front of the ear, (2) the common carotid artery in the neck along the front edge of the sternocleidomastoid muscle, and (3) over the facial artery at the lower margin of the mandible at a point below the corner of the mouth.

A pulse is also detected at three points in the upper limb: (1) in the axilla over the axillary artery; (2) over the brachial artery at the bend of the elbow along the inner or medial margin of the biceps brachii muscle, and (3) at the radial artery at the wrist. The so-called radial pulse is the most frequently monitored and easily accessible.

The pulse can also be felt at three locations in the lower extremity: (1) over the femoral artery in the groin, (2) at the popliteal artery behind and just proximal to the knee, and (3) at the dorsalis pedis artery on the front surface of the foot, just below the bend of the ankle joint.

Knowledge of pulse points can be as valuable in treatment as it is in diagnosis. Minor bleeding

FIGURE 13-11 Pulse Points. Each pulse point is named after the artery with which it is associated.

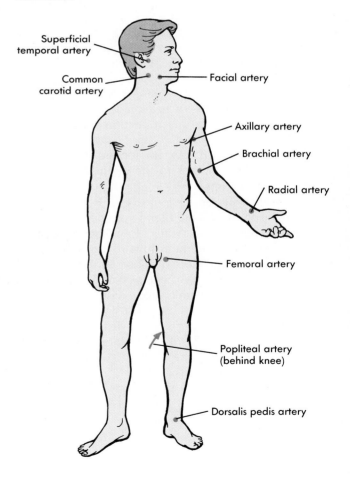

or *hemorrhage* usually stops within minutes by the normal clotting process, but severe hemorrhaging may require intervention to prevent hypovolemic shock. Because arteries at pulse points are easily palpated (felt) and have a firm base behind them, they are logical points to apply pressure to slow massive bleeding. For example, bleeding from the femoral artery or one of its branches in the leg can be slowed by pressing hard at the femoral pressure point to squeeze the artery shut.

Circulatory Shock

The term **circulatory shock** refers to the failure of the circulatory system to adequately deliver oxygen to the tissues, resulting in the impairment of cell function throughout the body. If left untreated, circulatory shock may lead to death. Circulatory failure has a variety of causes, all of which somehow reduce the flow of blood through the blood vessels of the body. Because of the variety of causes, circulatory shock is often classified into the following types:

1 Cardiogenic (kar-dee-o-JEN-ik) **shock** results from any type of heart failure, such as that after severe myocardial infarction (heart attack), heart infections, and other heart conditions. Because the heart can no longer pump blood effectively, blood flow to the tissues of the body decreases or stops.

2 Hypovolemic (hye-po-vo-LEE-mik) **shock** results from the loss of blood volume in the blood vessels (*hypovolemia* means "low blood volume"). Reduced blood volume results in low blood pressure and reduced flow of blood to tissues. Hemorrhage is a common cause of blood volume loss leading to hypovolemic shock. Hypovolemia can also be caused by loss of interstitial fluid, causing the drain of blood plasma out of the vessels and into the tissue spaces. Loss of interstitial fluid is common in chronic diarrhea or vomiting, dehydration, intestinal blockage, severe or extensive burns, and other conditions.

3 Neurogenic (noo-ro-JEN-ik) **shock** results from widespread dilation of blood vessels caused by an imbalance in autonomic stimulation of smooth muscles in vessel walls. You may recall from Chapter 8 that autonomic effectors such as smooth muscle tissues are controlled by a balance of stimulation from the sympathetic and parasympathetic divisions of the autonomic nervous system. Normally, sympathetic stimulation maintains the muscle tone that keeps blood vessels at their usual diameter. If sympathetic stimulation is disrupted by an injury to the spinal cord or medulla, depressive drugs, emotional stress, or some other factor, blood vessels dilate significantly. Widespread vasodilation reduces blood pressure, thus reducing blood flow.

4 Anaphylactic (an-a-fi-LAK-tik) **shock** results from an acute type of allergic reaction called *anaphylaxis*. Anaphylaxis causes the same kind of blood vessel dilation characteristic of neurogenic shock.

5 Septic shock results from complications of *septicemia*, a condition in which infectious agents release toxins into the blood. The toxins often dilate blood vessels, causing shock. The situation is usually made worse by the damaging effects of the toxins on tissues combined with the increased cell activity caused by the accompanying fever. One type of septic shock is *toxic shock syndrome (TSS)*, which usually results from stapholococcal infections that begin in the vagina of menstruating women and spread to the blood (see Appendix B).

The body has a number of mechanisms that compensate for the changes that occur during shock. However, these mechanisms may fail to compensate for changes that occur in severe cases, perhaps resulting in death.

OUTLINE SUMMARY

BLOOD VESSELS

A Kinds
  1 Arteries—carry blood away from the heart
  2 Veins—carry blood toward the heart
  3 Capillaries—carry blood from the arterioles to the venules
B Structure (Figure 13-1)
  1 Arteries
    a Tunica interna—inner layer of endothelial cells
    b Tunica media—smooth muscle, thick in arteries; important in blood pressure regulation
    c Tunica externa—thin outer layer of elastic tissue
  2 Capillaries—microscopic vessels
    a Only layer is the tunica interna
  3 Veins
    a Tunica interna—inner layer; valves prevent retrograde movement of blood
    b Tunica media—smooth muscle; thin in veins
    c Tunica externa—heavy layer in many veins
C Functions
  1 Arteries—distribution of nutrients, gases, etc., with movement of blood under high pressure; assist in maintaining the arterial blood pressure
  2 Capillaries—serve as exchange vessels for nutrients, wastes, and fluids
  3 Veins—collect blood for return to the heart; low pressure vessels
D Names of main arteries—see Figure 13-2 and Table 13-1
E Names of main veins—see Figure 13-3 and Table 13-2

DISORDERS OF BLOOD VESSELS

A Disorders of arteries—arteries must withstand high pressure and remain free of blockage
  1 Arteriosclerosis—hardening of arteries
    a Reduces flow of blood, possibly causing ischemia that may progress to necrosis (or gangrene)
    b Atherosclerosis—disorder in which lipids and other matter blocks arteries
    c May be corrected by vasodilators (vessel-relaxing drugs) or angioplasty (mechanical widening of vessels) or surgical replacement
  2 Aneurysm—abnormal widening of arterial wall
    a Aneurysms promote formation of thrombi that may obstruct vital tissues
    b Aneurysms may burst, resulting in life-threatening hemorrhaging

  c Cerebrovascular accident (CVA) or stroke—ischemia of brain tissue caused by embolism or hemorrhage
B Disorders of veins—veins are low-pressure vessels
  1 Varicose veins (varices)—enlarged veins in which blood pools
    a Hemorrhoids are varicose veins in the rectum
    b Treatments include supporting affected veins and surgical removal
  2 Thrombophlebitis—vein inflammation (phlebitis) accompanied by clot (thrombus) formation; may result in fatal pulmonary embolism

CIRCULATION OF BLOOD

A Plan of circulation—refers to the blood flow through the vessels arranged to form a circuit or circular pattern (Figure 13-7)
B Types of circulation
  1 Systemic circulation
    a Carries blood throughout the body
    b Path goes from left ventricle through aorta, smaller arteries, arterioles, capillaries, venules, venae cavae, to right atrium
  2 Pulmonary circulation
    a Carries blood to and from the lungs; arteries deliver deoxygenated blood to the lungs for gas exchange
    b Path goes from right ventricle through pulmonary arteries, lungs, pulmonary veins, to left atrium
  3 Hepatic portal circulation (Figure 13-8)
    a Unique blood route through the liver
    b Vein (hepatic portal vein) exists between two capillary beds
    c Assists with homeostasis of blood glucose levels
  4 Fetal circulation (Figure 13-9)
    a Refers to circulation before birth
    b Modifications required for fetus to efficiently secure oxygen and nutrients from the maternal blood
    c Unique structures include the placenta, umbilical arteries and vein, ductus venosus, ductus arteriosus, and foramen ovale

BLOOD PRESSURE

A Blood pressure is push or force of blood in the blood vessels
B Highest in arteries, lowest in veins (Figure 13-10)

C Blood pressure gradient causes blood to circulate—liquids can flow only from the area where pressure is higher to where it is lower

D Blood volume, heartbeat, and blood viscosity are main factors that produce blood pressure

E Blood pressure varies within normal range from time to time

PULSE

A Definition—alternate expansion and recoil of the blood vessel wall

B Places where you can count the pulse easily (Figure 13-11)

CIRCULATORY SHOCK

A Circulatory shock—failure of the circulatory system to deliver oxygen to the tissues adequately, resulting in cell impairment

B When the cause is known, shock can be classified by this scheme:

1 Cardiogenic shock—caused by heart failure

2 Hypovolemic shock—caused by a drop in blood volume that causes blood pressure (and blood flow) to drop

3 Neurogenic shock—caused by nerve condition that relaxes (dilates) blood vessels and thus reduces blood flow

4 Anaphylactic shock—caused by a type of severe allergic reaction characterized by blood vessel dilation

5 Septic shock—results from complications of septicemia (toxins in blood resulting from infection)

NEW WORDS

arteriole	diastolic pressure	hepatic portal	umbilical
artery	ductus arteriosus	circulation	vein
capillary	ductus venosus	pulse	venule
central venous pressure	foramen ovale		

Diseases and Other Clinical Terms

anaphylactic shock	circulatory shock	hypovolemic shock	sphygmomanometer
aneurysm	cyanosis	ischemia	thrombophlebitis
angioplasty	gangrene	necrosis	varices
arteriosclerosis	hemorrhoid	neurogenic shock	varicose vein
cardiogenic shock	hypertension	septic shock	
cerebrovascular accident (CVA)			

CHAPTER TEST

1. Blood is carried toward the heart by large vessels called _____.
2. The muscular middle layer of arteries and veins is called the _____ _____.
3. Blood vessels that serve as exchange vessels for nutrients, wastes, and fluids are called _____.
4. The large high-pressure vessels of the circulatory system are called _____, and the large low-pressure vessels are called _____.
5. Blood flow from the left ventricle of the heart to all parts of the body and back to the right atrium of the heart is called the _____ circulation.
6. In portal circulation a portal vein is located between two _____ beds.
7. In fetal circulation blood is shunted from the right atrium directly into the left atrium by passing through the _____ _____ of the heart.
8. A condition in which an artery bulges and forms a saclike extension can be called the _____.
9. Circulatory _____ is the failure of the circulatory system to deliver oxygen to the tissues adequately, resulting in cell impairment.
10. Arterial blood pressure is often expressed as _____ pressure (in mm Hg) over the _____ pressure (in mm Hg).

Select the most correct answer from Column B for each statement in Column A. (Only one answer is correct.)

COLUMN A

11. _____ Cerebrovascular accident
12. _____ Hardening of arteries
13. _____ Hemorrhoids
14. _____ Vein inflammation associated with clot formation
15. _____ Delivers blood to the liver
16. _____ High blood pressure
17. _____ Alternate expansion and recoil of a vessel wall
18. _____ Circulatory failure due to heart failure
19. _____ Toxic shock syndrome

COLUMN B

a. Arteriosclerosis
b. Cardiogenic shock
c. Hepatic portal circulation
d. Hypertension
e. Pulse
f. Septic shock
g. Stroke
h. Thrombophlebitis
i. Varicose veins

REVIEW QUESTIONS

1 What are the major differences among arteries, veins, and capillaries? How are they alike?
2 How do arteriosclerosis and aneurysms disrupt the normal function of arteries?
3 How do varicose veins develop? What complications can arise from varicose veins?
4 Describe the route taken by blood as it moves from the right atrium to the liver and back to the right atrium. As it travels, does blood pass through the systemic circulation? The pulmonary circulation? A portal circulation?
5 What major anatomical features are present in the fetal circulatory system that are not present in the normal adult circulatory system? What is the function of each? What circulatory changes must occur at the time of birth?
6 How does cardiac output affect blood pressure? What happens if the blood pressure rises or falls from its normal value?
7 Explain what causes a pulse in an artery. Locate the major pulse points on your body.
8 What is shock? Describe the major types of shock.

CLINICAL APPLICATIONS

1 Kevin has just learned that he has hypercholesterolemia (high blood cholesterol). On the advice of his physician, he is starting a regular exercise program. How might this affect Kevin's cholesterol problem? (Hint: see Appendix A.) What vascular disorder might Kevin develop if he is not able to correct his hypercholesterolemia?

2 Leo is a middle-age man who has recently been experiencing pain in his legs, especially when he walks for even moderate distances. His physician tells him that he has atherosclerosis in a major artery in the affected leg. Why does this cause pain when Leo walks? What treatments might Leo's physician recommend to correct this problem? Explain how each will improve Leo's condition.

3 If balloon angioplasty is used to correct mitral valve stenosis (see Chapter 12), what route must the catheter travel if it enters at the femoral artery?

The Lymphatic System and Immunity

C H A P T E R

14

Objectives

*After you have completed this chapter,
you should be able to:*

1 Describe the generalized functions of the lymphatic system and list the primary lymphatic structures.

2 Define and compare nonspecific and specific immunity, inherited and acquired immunity, and active and passive immunity.

3 Name the major disorders associated with the lymphatic system.

4 Discuss the major types of immune system molecules and indicate how antibodies and complements function.

5 Discuss and contrast the development and functions of B and T cells.

6 Compare and contrast humoral and cell-mediated immunity.

7 Describe the mechanisms of allergy, autoimmunity, and isoimmunity.

8 List the major types of immune deficiencies and explain their causes.

*A*ll of us live in a hostile and dangerous environment. Each day we are faced with potentially harmful toxins, disease-causing bacteria, viruses, and even cells from our own bodies that have been transformed into cancerous invaders (see Chapter 4). Fortunately, we are protected from this staggering variety of differing biological enemies by a remarkable set of defense mechanisms. We refer to this protective "safety net" as the **immune system.**

This chapter deals with the immune system. This system is characterized by structural components, many of them lymphatic organs, and by a functional group of specialized cells and molecules that protect us from infection and disease. This chapter begins with an overview of the lymphatic system, discussing vessels that help maintain fluid balance and lymphoid tissues that help protect the internal environment. We will then discuss the concept of immunity and the ways that highly specialized cells and molecules provide us with effective and very specific resistance to disease.

The Lymphatic System

LYMPH AND LYMPH VESSELS

Maintaining the constancy of the fluid around each body cell is possible only if numerous homeostatic mechanisms function effectively together in a controlled and integrated response to changing conditions. We know from Chapters 11 through 13 that the circulatory system provides a key role in bringing many needed substances to cells and then removing the waste products that accumulate as a result of metabolism. This exchange of substances between blood and tissue fluid occurs in capillary beds. Many additional substances that cannot enter or return through the capillary walls, including excess fluid and protein molecules, are returned to the blood as **lymph.** Lymph is a specialized fluid formed in the tissue spaces that is transported by way of specialized **lymphatic vessels** to eventually reenter the circulatory system. In addition to lymph and the lymphatic vessels, the lymphatic system includes lymph nodes and specialized lymphatic organs such as the thymus and spleen (Figure 14-1).

Lymph forms in this way: blood plasma filters out of the capillaries into the microscopic spaces between tissue cells because of the pressure generated by the pumping action of the heart. There, the liquid is called **interstitial fluid** or tissue fluid. Much of the interstitial fluid goes back into the blood by the same route it came out (that is, through the capillary membrane). The remainder of the interstitial fluid enters the lymphatic system before it returns to the blood. The fluid, now called *lymph*, enters a network of tiny blind-ended tubes distributed in the tissue spaces. These tiny vessels, called *lymphatic capillaries*, permit excess tissue fluid and some other substances such as dissolved protein molecules to leave the tissue spaces.

Lymphatic and blood capillaries are similar in many ways. Both types of vessels are microscopic and both are formed from sheets consisting of a cell layer of simple squamous epithelium called *endothelium* (en-doe-THEE-lee-um). The flattened endothelial cells that form blood capillaries, however, fit tightly together so that large molecules cannot enter or exit from the vessel. The "fit" between endothelial cells forming the lymphatic capillaries is not as tight. As a result, they are more porous and allow larger molecules, including proteins and other substances, as well as the fluid itself, to enter the vessel and eventually return to the general circulation. The movement of lymph in the lymphatic vessels is one way. Unlike blood, lymph does not flow over and over again through vessels that form a circular route.

Lymph flowing through the lymphatic capillaries next moves into successively larger and larger vessels called *lymphatic venules* and *veins* and eventually empties into two terminal vessels called the **right lymphatic duct** and the **thoracic duct,** which empty their lymph into the blood in veins in the neck region. Lymph from about three fourths of the body eventually drains into the thoracic duct, which is the largest lymphatic vessel in the body. Lymph from the right upper extremity and from the right side of the head, neck, and upper torso flows into the right lymphatic duct (Figure 14-2). The lymphatic vessels have a "beaded" appearance caused by the presence of valves that assist in maintaining a one-way flow of lymph. Note in Figure 14-1 that the thoracic duct in the abdomen has an enlarged pouchlike structure called the **cisterna chyli** (sis-TER-nah KI-li) that serves as a storage area for lymph mov-

▼ **FIGURE 14-1 Principal Organs of the Lymphatic System.**

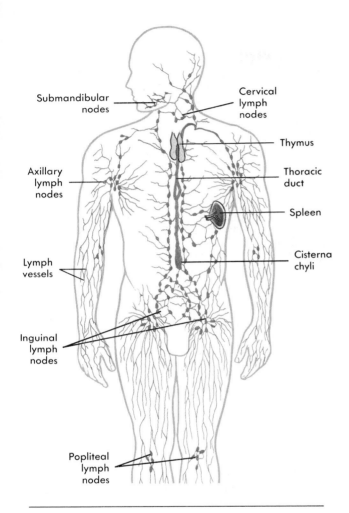

- Submandibular nodes
- Cervical lymph nodes
- Thymus
- Axillary lymph nodes
- Thoracic duct
- Spleen
- Lymph vessels
- Cisterna chyli
- Inguinal lymph nodes
- Popliteal lymph nodes

▼ **FIGURE 14-2 Lymph Drainage.** The right lymphatic duct drains lymph from the upper right quarter of the body into the right subclavian vein. The thoracic duct drains lymph from the rest of the body into the left subclavian vein.

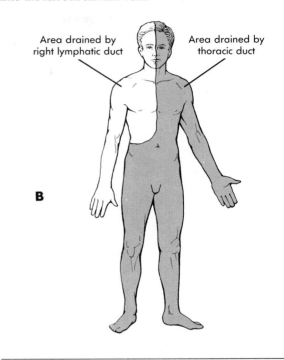

- Area drained by right lymphatic duct
- Area drained by thoracic duct

B

ing toward its point of entry into the venous system.

Lymphatic capillaries in the wall of the small intestine are given the special name of **lacteals** (LAK-tee-als). They transport fats obtained from food to the bloodstream and are discussed in Chapter 16.

LYMPHEDEMA

Lymphedema (lim-fe-DEE-mah) is an abnormal condition in which tissues exhibit swelling (edema) because of the accumulation of lymph.

Lymph may accumulate in tissue when the lymph vessels are partially blocked. This may result from a congenital abnormality or from **lymphangitis** (lim-fan-JYE-tis), that is, lymph vessel inflammation. Lymphangitis is characterized by thin, red streaks extending from an infected region. The infectious agent that causes lymphangitis may eventually spread to the blood stream, causing septicemia ("blood poisoning") and possibly death from septic shock.

Rarely, lymphedema may be caused by small parasitic worms that infest the lymph vessels. When such infestation blocks the flow of lymph, edema of the tissues drained by the affected vessels occurs. In severe cases, the tissues swell so much that the limbs look as if they belonged to an elephant! For this reason, the condition is often called **elephantiasis**—literally "condition of being like an elephant."

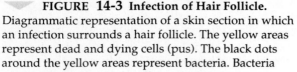

FIGURE 14-3 Infection of Hair Follicle.
Diagrammatic representation of a skin section in which an infection surrounds a hair follicle. The yellow areas represent dead and dying cells (pus). The black dots around the yellow areas represent bacteria. Bacteria entering the node via the afferent lymphatics are filtered out.

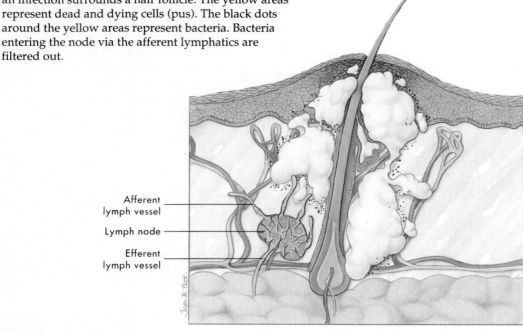

Afferent lymph vessel

Lymph node

Efferent lymph vessel

LYMPH NODES

As lymph moves from its origin in the tissue spaces toward the thoracic or right lymphatic ducts and then into the venous blood, it is filtered by moving through **lymph nodes,** which are located in clusters along the pathway of lymphatic vessels. Some of these nodes may be as small as a pinhead, and others, may be as large as a lima bean. With the exception of a comparatively few single nodes, most lymph nodes occur in groups or clusters in certain areas. Figure 14-1 shows the locations of the clusters of greatest clinical importance. You may have experienced swelling of the submandibular nodes during a "sore throat" or as a result of an infected tooth. The structure of the lymph nodes makes it possible for them to perform two important functions: defense and white blood cell formation.

Defense Function: Biological Filtration

Figure 14-3 shows the structure of a typical lymph node. In this example, a small node located next to an infected hair follicle is shown filtering bacteria from lymph. Lymph nodes perform biological filtration, a process in which cells (phagocytic cells in this case) alter the contents of the filtered fluid. Biological filtration of bacteria and other abnormal cells by phagocytosis prevents local infections from spreading. Note in Figure 14-3 that lymph enters the node through four **afferent** (from the Latin "to carry toward") **lymph vessels.** These vessels deliver lymph to the node. When passing through the node, lymph is filtered so that injurious particles such as bacteria, soot, and cancer cells are removed and prevented from entering the blood and circulating through the body. Lymph exits the node through an **efferent** (from the Latin "to carry away from") **lymph vessel.**

Clusters of lymph nodes allow a very effective biological filtration of lymph flowing from specific body areas. A knowledge of lymph node location and function is important in clinical medicine. For example, a school nurse monitoring the progress of a child with an infected finger will watch the elbow and axillary regions for swelling and tenderness of the lymph nodes, a condition called **lymphadenitis** (limf-ad-en-EYE-tis). These nodes filter lymph returning from the hand and may become infected by the bacteria they trap. A surgeon uses knowledge of lymph node function when removing lymph nodes under the arms (axillary nodes) and in other areas during an operation for breast cancer. These nodes may contain cancer cells filtered out of the lymph drained from the breast. Cancer of the breast is one of the most common forms of this disease in women. Unfortunately, cancer cells from a single tumorous growth in the breast often spread to other areas of the body through the lymphatic system. Figure 14-4 shows how lymph from the breast drains into many different and widely placed nodes.

LYMPHOMA

As you may recall from Chapter 4, *lymphoma* is a term that refers to lymphatic tumors. Lymphomas are most often malignant but in rare cases can be benign. The two principal categories of lymphoma are **Hodgkin's disease** and **non-Hodgkin's lymphoma.** All types of lymphoma characteristically cause painless enlargements of the lymph nodes in the neck and other regions. This sign is followed by anemia, weight loss, weakness, fever, and spread to other lymphatic tissues.

In later stages, the lymphoma spreads to many other areas of the body. When discovered early, lymphoma can be successfully treated with intensive radiation and chemotherapy. Lymphoma occurs more often in men than in women.

THYMUS

The **thymus** (THYE-mus) (Figure 14-1) is a small lymphoid tissue organ located in the mediastinum, extending upward in the midline of the neck. It is composed of lymphocytes in a unique epithelial tissue framework. The thymus is largest at puberty and even then weighs only more than 28 g—over an ounce. Although small in size the thymus plays a central and critical role in the body's vital immunity mechanism. First, it is a source of lymphocytes before birth and is then especially important in the "maturation" or development of specialized lymphocytes that then leave the thymus and circulate to the spleen, tonsils, lymph nodes, and other lymphatic tissues. These **T-lymphocytes** or T cells are critical to the functioning of the immune system and are discussed later. They develop under the influence of

FIGURE 14-4 Lymphatic Drainage of the Breast. Note the extensive network of nodes that receive lymph from the breast.

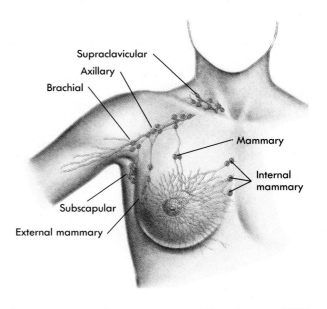

Supraclavicular
Axillary
Brachial
Mammary
Internal mammary
Subscapular
External mammary

LYMPHEDEMA AFTER BREAST SURGERY

Surgical procedures called *mastectomies,* in which some or all of the breast tissues are removed, are sometimes done to treat breast cancer. Because cancer cells can spread so easily through the extensive network of lymphatic vessels associated with the breast (see Figure 14-4), the lymphatic vessels and their nodes are sometimes also removed. Occasionally, such procedures interfere with the normal flow of lymph fluid from the arm. When this happens, tissue fluid may accumulate in the arm—resulting in lymphedema.

a hormone secreted by the thymus, **thymosin** (THYE-mo-sin). The thymus essentially completes its work early in childhood and is then gradually replaced by fat and connective tissue, a process called *involution.*

TONSILS

Masses of lymphoid tissue called **tonsils** are located in a protective ring under the mucous membranes in the mouth and back of the throat (Figure 14-5). They help protect us against bacteria that may invade tissues in the area around the openings between the nasal and oral cavities. The **palatine tonsils** are located on each side of the throat. The **pharyngeal tonsils,** known as **adenoids** (AD-e-noyds) when they become swollen, are near the posterior opening of the nasal cavity. A third type of tonsil, the **lingual tonsils,** are near the base of the tongue. The tonsils serve as the first line of defense from the exterior and as such are subject to chronic infection or **tonsillitis** (tahn-sil-LYE-tis). They may have to be removed surgically if antibiotic therapy is not successful or if swelling impairs breathing.

FIGURE **14-5 Location of the Tonsils.** Small segments of the roof and floor of the mouth have been removed to show the protective ring of tonsils (lymphoid tissue) around the internal opening of the nose and throat.

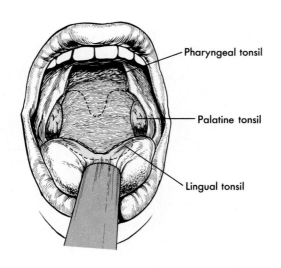

Pharyngeal tonsil

Palatine tonsil

Lingual tonsil

SPLEEN

The spleen is the largest lymphoid organ in the body. It is located high in the upper left quadrant of the abdomen lateral to the stomach (Figure 14-1). Although the spleen is protected by the lower ribs, it can be injured by abdominal trauma. The spleen has a very rich blood supply and may contain over 1 pint of blood. If damaged and bleeding, surgical removal, called a **splenectomy** (splen-NEK-toe-mee), may be required to stop the loss of blood.

After entering the spleen, blood flows through dense, pulplike accumulations of lymphocytes. As blood flows through the pulp, the spleen removes by filtration and phagocytosis many bacteria and other foreign substances, destroys worn out RBCs and salvages the iron found in hemoglobin for future use, and serves as a reservoir for blood that can be returned to the circulatory system when needed.

Splenomegaly (sple-no-MEG-ah-lee), or abnormal spleen enlargement, is observed in a variety of disorders. For example, infectious conditions such as scarlet fever, syphilis, and typhoid fever are characterized by splenomegaly. Spleen enlargement sometimes accompanies hypertension. Splenomegaly also accompanies some forms of hemolytic anemia in which red blood cells appear to be broken apart at an abnormally fast rate. Surgical removal of the spleen often cures such cases.

The Immune System

The body's defense mechanisms protect us from disease-causing microorganisms that invade our bodies, foreign tissue cells that may have been transplanted into our bodies, and our own cells that have turned malignant or cancerous. The body's specific defense system is called the **immune system.** The immune system makes us immune (that is, able to resist these enemies). Unlike other systems of the body, which are made up of groups of organs, the immune system is made up of billions of cells and trillions of molecules.

Nonspecific immunity is maintained by mechanisms that attack any irritant or abnormal substance that threatens the internal environment. In

TABLE 14-1   Specific Immunity

TYPE	EXAMPLE
Inborn immunity	Immunity to certain diseases (for example, canine distemper) is inherited.
Acquired immunity	
Natural immunity	Exposure to the causative agent is not deliberate.
Active (exposure)	A child develops measles and acquires an immunity to a subsequent infection.
Passive (exposure)	A fetus receives protection from the mother through the placenta, or an infant receives protection via the mother's milk.
Artificial immunity	Exposure to the causative agent is deliberate.
Active (exposure)	Injection of the causative agent, such as a vaccination against polio, confers immunity.
Passive (exposure)	Injection of protective material (antibodies) that was developed by another individual's immune system is given.

other words, nonspecific immunity confers general protection rather than protection from certain kinds of invading cells or chemicals. The skin and mucous membranes, for example, are mechanical barriers to prevent entry into the body of bacteria and many other substances such as toxins and harmful chemicals. Tears and mucus also contribute to nonspecific immunity. Tears wash harmful substances from the eyes, and mucus traps foreign material that may enter the respiratory tract. Phagocytosis of bacteria by WBCs is a nonspecific form of immunity.

Specific immunity includes protective mechanisms that confer very specific protection against certain types of invading bacteria or other toxic materials. Specific immunity involves memory and the ability to recognize and respond to certain particular harmful substances or bacteria. For example, when the body is first attacked by particular bacteria or viruses, disease symptoms may occur as the body fights to destroy the invading organism. However, if the body is exposed a second time, no symptoms occur because the organism is destroyed quickly—the person is said to be **immune.** Immunity to one type of disease-causing bacteria or virus does not protect the body against others. Immunity can be very selective.

Immunity to disease is classified as **inherited** or **acquired.** If inherited, it is described as *inborn immunity* (Table 14-1). Humans are immune from birth to certain diseases that affect other animals. Distemper, for example, is an often fatal viral disease in dogs. The virus does not produce symptoms in humans; we have an inborn or inherited immunity to the disease.

Acquired immunity may be further classified as "natural" or "artificial" depending on how the body is exposed to the harmful agent. Natural exposure is not deliberate and occurs in the course of everyday living. We are naturally exposed to many disease-causing agents on a regular basis. Artificial exposure is called *immunization* and is the deliberate exposure of the body to a potentially harmful agent.

Natural and artificial immunity may be "active" or "passive." Active immunity occurs when an individual's own immune system responds to a harmful agent, regardless of whether that agent was naturally or artificially encountered. Passive immunity results when immunity to a disease that has developed in another individual or animal is transferred to an individual who was not previously immune. For example, antibodies in a mother's milk confer passive immunity to her nursing infant. Active immunity generally lasts longer than passive immunity. Passive immunity, although temporary, provides immediate protection. Table 14-1 lists the various forms of specific immunity and gives examples of each.

Immune System Molecules

The immune system functions because of adequate amounts of highly specialized protein molecules and unique cells. The protein molecules critical to immune system functioning are called **antibodies** (AN-ti-bod-ees) and **complements** (KOM-ple-ments). Antibody molecules outnumber the cells of the immune system by about 100 million to one.

ANTIBODIES
Definition

Antibodies are protein compounds that are normally present in a person's body. A defining characteristic of an antibody molecule is the uniquely shaped concave regions, called *combining sites,* on its surface. Another defining characteristic is the ability of an antibody molecule to combine with a specific compound that is called an **antigen** (AN-ti-jen). All antigens are compounds

FIGURE 14-6 Antibody Function. Antibodies produce humoral immunity by binding to specific antigens to form antigen-antibody complexes, which produce changes that inactivate or kill invading cells.

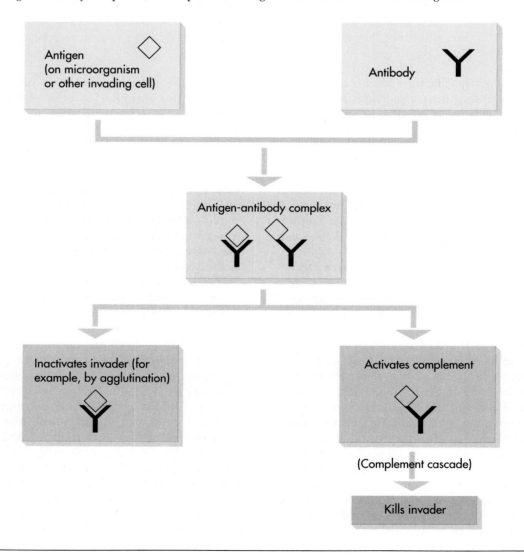

whose molecules have small regions on their surfaces that are uniquely shaped to fit into the combining sites of a specific antibody molecule as precisely as a key fits into a specific lock. Antigens are usually foreign proteins, most often the molecules in the surface membranes of invading or diseased cells such as microorganisms or cancer cells.

Functions

In general, antibodies produce **humoral** or **antibody-mediated immunity** by changing antigens so that they cannot harm a person's body (Figure 14-6). To do this an antibody must first bind to its specific antigen. This forms an antigen-antibody complex. The antigen-antibody complex then acts in one or more ways to make the antigen, or the cell on which it is present, harmless.

For example, if the antigen is a toxin, a substance poisonous to body cells, the toxin is neutralized, or made nonpoisonous, by becoming part of an antigen-antibody complex. Or if antigens are molecules in the surface membranes of invading cells, when antibodies combine with them, the resulting antigen-antibody complexes may agglutinate the enemy cells (that is, make them stick together in many clumps). Then macrophages or other phagocytes can rapidly destroy them by ingesting and digesting large numbers of them at one time.

Another important function of antibodies is promotion and enhancement of phagocytosis. Certain antibody fractions help promote the attachment of phagocytic cells to the object they will engulf. As a result, the contact between the phagocytic cell and its victim is enhanced, and the

MONOCLONAL ANTIBODIES

Techniques that have permitted biologists to produce large quantities of pure and very specific antibodies have resulted in dramatic advances in medicine. As a new medical technology, the development of **monoclonal antibodies** has been compared in importance with advances in recombinant DNA or genetic engineering.

Monoclonal antibodies are specific antibodies produced or derived from a population or culture of identical, or **monoclonal**, cells. In the past, antibodies produced by the immune system against a specific antigen had to be "harvested" from serum containing literally hundreds of other antibodies. The total amount of a specific antibody that could be recovered was very limited, so the cost of recovery was high. Monoclonal antibody techniques are based on the ability of immune system cells to produce individual antibodies that bind to and react with very specific antigens. We know, for example, that if the body is exposed to the varicella virus of chickenpox, WBCs will produce an antibody that will react very specifically with that virus and no other. With mono-clonal antibody techniques, lymphocytes that are produced by the body after the injection of a specific antigen are "harvested" and then "fused" with other cells that have been transformed to grow and divide indefinitely in a tissue culture medium. These fused or hybrid cells, called *hybridomas* (HYE-brid-o-mahs) continue to produce the same antibody produced by the original lymphocyte. The result is a rapidly growing population of identical or monoclonal cells that produce large quantities of a very specific antibody. Monoclonal antibodies have now been produced against a wide array of different antigens, including disease-producing organisms and various types of cancer cells.

The availability of very pure antibodies against specific disease-producing agents is the first step in the commercial preparation of diagnostic tests that can be used to identify viruses, bacteria, and even specific cancer cells in the blood or other body fluids. The use of monoclonal antibodies may serve as the basis for specific treatment of many human diseases.

FIGURE 14-7 Complement Fixation. A, Complement molecules activated by antibodies form doughnut-shaped complexes in a bacterium's plasma membrane. **B,** Holes in the complement complex allow sodium (Na^+) and then water (H_2O) to diffuse into the bacterium. **C,** After enough water has entered, the swollen bacterium bursts.

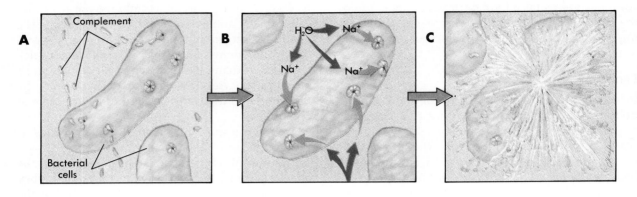

object is more easily ingested. This process contributes to the efficiency of immune system phagocytic cells, which is described on pp. 358-360.

Probably the most important way in which antibodies act is one we will consider last. It is a process called **complement fixation.** In many instances, when antigens that are molecules on an antigenic or foreign cell's surface combine with antibody molecules, they change the shape of the antibody molecule slightly but just enough to expose two previously hidden regions. These are called **complement-binding sites.** Their exposure initiates a series of events that kill the cell on whose surface they take place. The next section describes these events.

COMPLEMENT PROTEINS

Complement is the name used to describe a group of 14 proteins normally present in an inactive state in blood. These proteins are activated by exposure of complement-binding sites. The result is formation of highly specialized antigen-antibody complexes that target foreign cells for destruction. The process is a rapid-fire cascade or sequence of events collectively called **complement fixation.** In this process a doughnut-shaped assemblage—complete with a hole in the middle—is formed when antibodies, antigens in the invading cell's plasma membrane, and complement molecules combine.

Complement fixation kills invading cells of various types. How? In effect, by drilling a hole in their plasma membranes! The tiny holes allow sodium to rapidly diffuse into the cell; then water follows through osmosis. The cell literally bursts as the internal osmotic pressure increases (Figure 14-7).

Immune System Cells

The primary cells of the immune system include:

1 Phagocytes
   a Neutrophils
   b Monocytes
   c Macrophages
2 Lymphocytes
   a T-lymphocytes
   b B-lymphocytes

PHAGOCYTES

Phagocytic WBCs are an important part of the immune system. In Chapter 11, phagocytes were described as cells derived from the bone marrow that carry on phagocytosis or ingestion and digestion of foreign cells or particles (Figure 14-8). The most important phagocytes are neutrophils and monocytes. These blood phagocytes migrate out of the blood and into the tissues in response to an infection. The neutrophils are functional but short

INTERFERON

Interferon (in-ter-FEER-on) is a small protein compound that plays a very significant role in producing immunity from viral infections. It is produced by T cells within hours after they have been infected by a virus. The interferon released from the T cells protects other cells by interfering with the ability of the virus to reproduce as it moves from cell to cell. In the past, thousands of pints of blood had to be processed to harvest tiny quantities of leukocyte (T cell) interferon for study. Synthetic human interferon is now being "manufactured" in bacteria as a result of gene-splicing techniques and is available in quantities sufficient for clinical use. Synthetic interferon decreases the severity of many virus-related diseases including chickenpox and measles. Interferon also shows promise as an anti-cancer agent. It has been shown to be effective in treating breast, skin, and other forms of cancer.

FIGURE 14-8 Phagocytosis. This series of scanning electron micrographs shows the progressive steps in phagocytosis of damaged RBCs by a macrophage. **A,** RBCs *(R)* attach to the macrophage *(M)*. **B,** Plasma membrane of the macrophage begins to enclose the RBC. **C,** The RBCs are almost totally ingested by the macrophage.

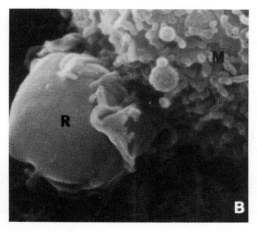

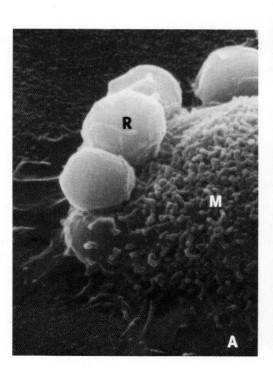

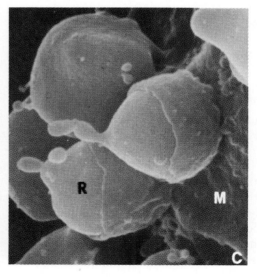

lived in the tissues. Once in the tissues, monocytes develop into phagocytic cells called **macrophages** (MAK-ro-fay-jes). Some macrophages "wander" through the tissues to engulf bacteria wherever they find them. Other macrophages become permanent residents of other organs. Macrophages found in spaces between liver cells, for example, are called *Kupffer's cells*, whereas those that ingest particulate matter in the small air sacs of the lungs are called *dust cells*. Macrophages can also be found in the spleen and lymph nodes and on the lining membranes of the abdominal and thoracic cavities. Specialized antibodies that bind to and coat certain foreign particles help macrophages function effectively. They serve as "flags" that alert the macrophage to the presence of foreign material, infectious bacteria, or cellular debris. They also help bind the phagocyte to the foreign material so that it can be engulfed more effectively (Figure 14-8).

LYMPHOCYTES

The most numerous cells of the immune system are the lymphocytes; they are ultimately responsible for antibody production. Several million strong, lymphocytes continually patrol the body, searching out any enemy cells that may have entered. Lymphocytes circulate in the body's fluids. Huge numbers of them wander vigilantly through most of its tissues. Lymphocytes densely populate the body's widely scattered lymph nodes and its other lymphatic tissues, especially the thymus gland in the chest and the spleen and liver in the abdomen.

There are two major types of lymphocytes, designated as *B-* and *T-lymphocytes* but usually called **B cells** and **T cells.**

Development of B Cells

All lymphocytes that circulate in the tissues arise from primitive cells in the bone marrow called *stem cells* and go through two stages of development. The first stage of B-cell development, transformation of stem cells into immature B cells, occurs in the liver and bone marrow before birth but only in the bone marrow in adults. Because this process was first discovered in a bird organ called the *bursa,* these cells were named B cells.

Immature B cells are small lymphocytes that have synthesized and inserted into their cytoplas-

mic membranes numerous molecules of one specific kind of antibody (Figure 14-9). These antibody-bearing immature B cells leave the tissue where they were formed, enter the blood, and are transported to their new place of residence, chiefly the lymph nodes. There they act as seed cells. Each immature B cell undergoes repeated mitosis (cell division) and forms a clone of immature B cells. A **clone** is a family of many identical cells all descended from one cell. Because all the cells in a clone of immature B cells have descended from one immature B cell, all of them bear the same surface antibody molecules as did their single ancestor cell.

The second stage of the development of B cells changes an immature B cell into an activated B cell. Not all immature B cells undergo this change. They do so only if an immature B cell comes in contact with certain protein molecules—antigens—whose shape fits the shape of the immature B cell's surface antibody molecules. If this happens, the antigens lock onto the antibodies and by so doing change the immature B cell into an activated B cell. Then the activated B cell, by dividing rapidly and repeatedly, develops into clones of two kinds of cells, **plasma cells** and **memory cells** (Figure 14-9). Plasma cells secrete copious amounts of antibody into the blood—reportedly, 2000 antibody molecules per second by each plasma cell for every second of the few days that it lives. Antibodies circulating in the blood constitute an enormous, mobile, ever-on-duty army.

Memory cells can secrete antibodies but do not immediately do so. They remain in reserve in the lymph nodes until they are contacted by the same antigen that led to their formation. Then, very quickly, the memory cells develop into plasma cells and secrete large amounts of antibody. Memory cells, in effect, seem to remember their ancestor-activated B cell's encounter with its appropriate antigen. They stand ready, at a moment's notice, to produce antibody that will combine with this antigen.

Function of B Cells

B cells function indirectly to produce humoral immunity. Recall that humoral immunity is resistance to disease organisms produced by the actions of antibodies binding to specific antigens

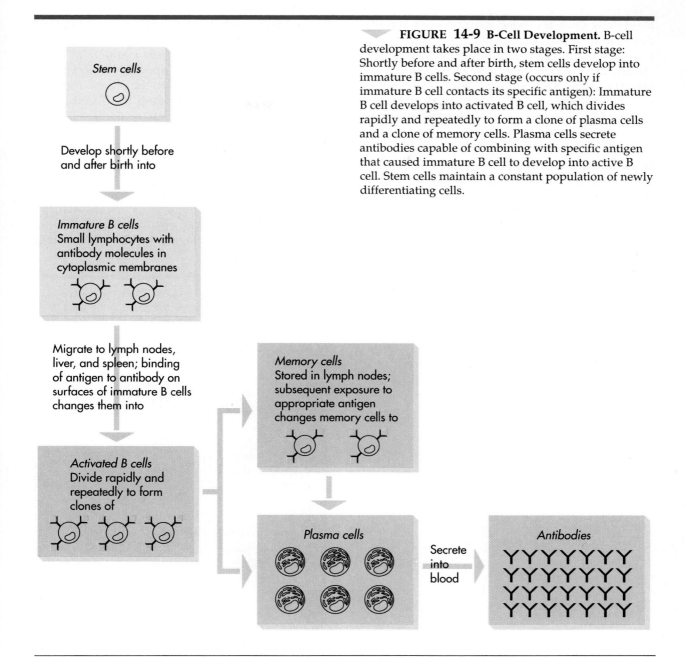

FIGURE 14-9 B-Cell Development. B-cell development takes place in two stages. First stage: Shortly before and after birth, stem cells develop into immature B cells. Second stage (occurs only if immature B cell contacts its specific antigen): Immature B cell develops into activated B cell, which divides rapidly and repeatedly to form a clone of plasma cells and a clone of memory cells. Plasma cells secrete antibodies capable of combining with specific antigen that caused immature B cell to develop into active B cell. Stem cells maintain a constant population of newly differentiating cells.

while circulating in body fluids. Activated B cells develop into plasma cells. Plasma cells secrete antibodies into the blood; they are the "antibody factories" of the body. These antibodies, like other proteins manufactured for extracellular use, are formed on the endoplasmic reticulum of the cell. The B cell's responsibility in building these antibodies is revealed by the extensive development of the endoplasmic reticulum in these cells.

Development of T Cells

T cells are lymphocytes that have undergone their first stage of development in the thymus gland.

Stem cells from the bone marrow seed the thymus, and shortly before and after birth, they develop into T cells. The newly formed T cells stream out of the thymus into the blood and migrate chiefly to the lymph nodes, where they take up residence. Embedded in each T cell's cytoplasmic membrane are protein molecules shaped so that they can fit only one specific kind of antigen molecule. The second stage of T-cell development takes place when and if a T cell comes into contact with its specific antigen. If this happens, the antigen binds to the protein on the T cell's surface, thereby changing the T cell into a sensitized T cell (Figure 14-10).

Functions of T Cells

Sensitized T cells (Figure 14-11) produce cell-mediated immunity. As the name suggests, **cell-mediated immunity** is resistance to disease organisms resulting from the actions of cells—chiefly sensitized T cells. Some sensitized T cells kill invading cells directly (Figure 14-12). When bound

▼ **FIGURE 14-10 T-Cell Development.** The first stage takes place in the thymus gland shortly before and after birth. Stem cells maintain a constant population of newly differentiating cells as they are needed. The second stage occurs only if a T cell contacts antigen, which combines with certain proteins on the T cell's surface.

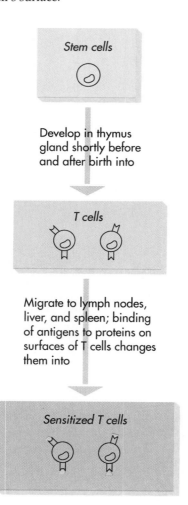

▼ **FIGURE 14-11 T Cells.** The blue spheres seen in this scanning electron microscope view are T cells attacking a much larger cancer cell. The cells are a significant part of our defense against cancer and other types of foreign cells.

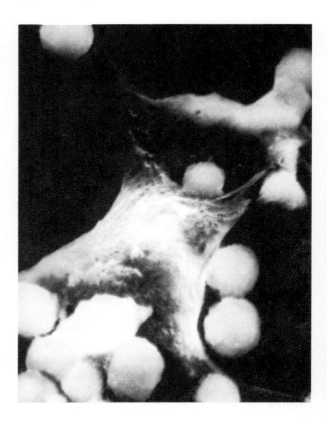

to antigens on an invading cell's surface, they release a substance that acts as a specific and lethal poison against the bound cell. Many sensitized cells produce their deadly effects indirectly by means of compounds that they release into the area around enemy cells. Among these are a substance that attracts macrophages into the neighborhood of the enemy cells. The assembled macrophages then destroy the cells by phagocytosing (ingesting and digesting) them (Figure 14-11).

Hypersensitivity of the Immune System

Hypersensitivity (hye-per-sen-si-TIV-i-tee) is an inappropriate or excessive response of the immune system. There are three types: allergy, autoimmunity, and isoimmunity.

ALLERGY

The term **allergy** is used to describe hypersensitivity of the immune system to relatively harmless

FIGURE 14-12 T-Cell Function. Sensitized T cells produce cell-mediated immunity by releasing various compounds in the vicinity of invading cells. Some act directly, and some act indirectly to kill invading cells.

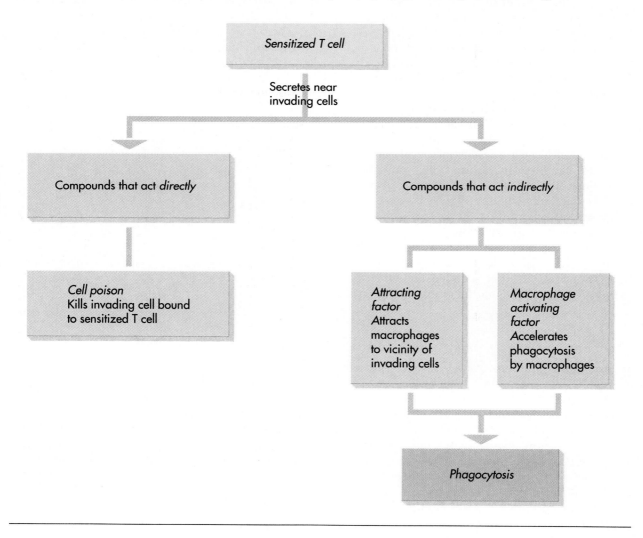

IMMUNIZATION

Active immunity can be established artificially by using a technique called *vaccination*. The original vaccine was a live *cowpox* virus that was injected into healthy people to cause a mild cowpox infection. The term vaccine literally means "cow substance." Because the cowpox virus is similar to the deadly *smallpox* virus, vaccinated individuals developed antibodies that imparted immunity against both cowpox and smallpox viruses.

Modern vaccines work on a similar principle; substances that trigger the formation of antibodies against specific pathogens are introduced orally or by injection. Some of these vaccines are killed pathogens or live, attenuated (weakened) pathogens. Such pathogens still have their specific antigens intact, so they can trigger formation of the proper antibodies, but they are no longer *virulent* (able to cause disease). Although it is rare, these vaccines sometimes backfire and actually cause an infection. Many of the newer vaccines get around this potential problem by using only the part of the pathogen that contains antigens. Because the disease-causing portion is missing, such vaccines cannot cause infection.

The amount of antibodies in a person's blood produced in response to vaccination or an actual infection is called the *antibody titer*. As you can see in the graph, the initial injection of vaccine triggers a rise in the antibody titer that gradually diminishes. Often, a *booster shot* or second injection is given to keep the antibody titer high or to raise it to a level that is more likely to prevent infection. The secondary response is more intense than the primary response because memory B cells are standing ready to produce a large number of antibodies at a moment's notice. A later accidental exposure to the pathogen will trigger an even more intense response—thus preventing infection.

Toxoids are similar to vaccines but use an altered form of a bacterial toxin (poisonous chemical) to stimulate production of antibodies. Injection of toxoids imparts protection against toxins, whereas administration of vaccines imparts protection against pathogenic organisms and viruses.

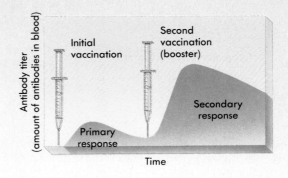

environmental antigens. Antigens that trigger an allergic response are often called **allergens** (AL-ler-jenz). One in six Americans has a genetic predisposition to exhibiting an allergy of some kind.

Immediate allergic responses involve antigen-antibody reactions. Before such a reaction occurs, a susceptible person must be exposed to an allergen repeatedly—triggering the production of antibodies. After a person is thus *sensitized,* exposure to an allergen causes antigen-antibody reactions that trigger the release of histamine, kinins, and other inflammatory substances. These responses usually cause typical allergy symptoms such as runny nose, conjunctivitis, and *urticaria* (hives). In some cases, however, these substances may cause constriction of the airways, relaxation of blood vessels, and irregular heart rhythms that can progress to a life-threatening condition called *anaphylactic shock* (see Chapter 13). Drugs called *antihistamines* are sometimes used to relieve the symptoms of this type of allergy.

Delayed allergic responses, on the other hand, involve cell-mediated immunity. In *contact dermatitis,* for example, T cells trigger events that lead to local skin inflammation a few hours or days after initial exposure to an antigen. Exposure to poison ivy, soaps, and certain cosmetics may cause contact dermatitis in this manner (Figure 14-13). Hypersensitive individuals may use *hypoallergenic* products (products without common allergens) to avoid such allergic reactions.

AUTOIMMUNITY

Autoimmunity (aw-toe-im-YOO-ni-tee) is an inappropriate and excessive response to self-antigens. Disorders that result from autoimmune responses are called **autoimmune diseases.** Examples of autoimmune diseases are given in Appendix B. Self-antigens are molecules that are native to a person's body and that are used by the immune system to identify components of "self." In autoimmunity, the immune system inappropriately attacks these antigens.

A common autoimmune disease is **systemic lupus erythematosus (SLE),** or simply *lupus*. Lupus is a chronic inflammatory disease that affects many tissues in the body: joints, blood vessels, kidney, nervous system, and skin. The name *lupus erythematosus* refers to the red rash that often develops on the face of those afflicted with SLE. The "systemic" part of the name comes from the fact that the disease affects many systems throughout the body. The systemic nature of SLE results from the production of antibodies against many different self-antigens.

ISOIMMUNITY

Isoimmunity (eye-so-im-YOO-ni-tee) is excessive reaction of the immune system to antigens from a different individual of the same species. It is important in pregnancy and tissue transplants.

During pregnancy, antigens from the fetus may enter the mother's blood supply and sensitize her immune system. Antibodies that are formed as a result of this sensitization may enter the fetal circulation and cause an inappropriate immune reaction. One example, erythroblastosis fetalis, was discussed in Chapter 11. Other pathological conditions may also be caused by damage to developing fetal tissues resulting from attack by the mother's immune system. Examples include congenital heart defects, Graves' disease, and myasthenia gravis.

Tissue or organ **transplants** are medical procedures in which tissue from a donor is surgically

FIGURE 14-13 Contact Dermatitis. Dermatitis or skin inflammation can result from contact with allergens—substances that trigger allergic responses in hypersensitive individuals.

grafted into the body. For example, skin grafts are often done to repair damage caused by burns. Donated whole blood tissue is often transfused into a recipient after massive hemorrhaging. A kidney is sometimes removed from a living donor and grafted into a person suffering from kidney failure. Unfortunately, the immune system sometimes reacts against foreign antigens present in the grafted tissue, causing what is often called a *rejection syndrome*. The antigens most commonly involved in transplant rejection are called **human lymphocyte antigens (HLAs).**

Rejection of grafted tissues can occur in two ways. One is called *host versus graft rejection* because the recipient's immune system recognizes foreign HLAs and attacks them, destroying the donated tissue. The other is *graft versus host rejection* because the donated tissue (for example, bone marrow) attacks the recipient's HLAs, destroying tissue throughout the recipient's body. Graft versus host rejection may lead to death.

There are two ways to prevent rejection syndrome. One strategy is called *tissue typing* in which HLAs and other antigens of a potential donor and recipient are identified. If they match, tissue rejection is unlikely to occur. Another strategy is the use of **immunosuppressive** (im-yoo-no-su-PRES-iv) **drugs** in the recipient. Immunosuppressive drugs such as *cyclosporine* and *prednisone* suppress the immune system's ability to attack the foreign antigens in the donated tissue.

Immune System Deficiency

Immune deficiency or *immunodeficiency* is the failure of immune system mechanisms in defending against pathogens. Immune system failure usually results from disruption of lymphocyte (B or T cell) function. The chief characteristic of immune deficiency is the development of unusual or recurring severe infections or cancer. Although immune deficiency by itself does not cause death, the resulting infections or cancer can. There are two broad categories of immune deficiencies, based on the mechanism of lymphocyte dysfunction: *congenital* and *acquired*.

CONGENITAL IMMUNE DEFICIENCY

Congenital immune deficiency, which is rare, results from improper lymphocyte development before birth. Depending on the stage of development of stem cells (B or T cells) during which the defect occurs, different diseases can result. For example, improper B-cell development can cause insufficiency or absence of antibodies in the blood. If stem cells are disrupted, a condition called **severe combined immune deficiency (SCID)** results. In most forms of SCID, humoral immunity and cell-mediated immunity are defective. Temporary immunity can be imparted to children with SCID by injecting them with a preparation of antibodies (gamma globulin). Bone marrow transplants, which replace the defective stem cells with healthy donor cells, have proved effective in treating some cases of SCID.

ACQUIRED IMMUNE DEFICIENCY

Acquired immune deficiency develops after birth (and is not related to genetic defects). A number of factors can contribute to acquired immune deficiency: nutritional deficiencies, immunosuppressive drugs or other medical treatments, trauma, stress, and viral infection.

One of the best known examples of acquired immune deficiency is **acquired immune deficiency syndrome (AIDS)**. AIDS was first recognized as a new disease by the Centers for Disease Control in 1981. This syndrome is caused by the *human immunodeficiency virus* or *HIV*. HIV, a retrovirus, contains RNA that produces its own DNA inside infected cells. The viral DNA often becomes part of the cell's DNA. When the viral DNA is activated, it directs the cell to synthesize viral RNA and viral proteins—producing new retroviruses. The HIV virus thus "steals" raw materials from the cell. When this occurs in T cells, the cell is destroyed and immunity is impaired. As the T cell dies, it releases new retroviruses that can spread the HIV infection.

Although HIV can invade several types of human cells, it has its most obvious effects in T cells. When T-cell function is impaired, infectious organisms and cancer cells can grow and spread much more easily than normal. Unusual conditions, such as *pneumocystosis* (a protozoal condition) and *Kaposi's sarcoma* (a type of skin cancer) may also appear. Because their immune system is deficient, AIDS patients usually die from one of these infections or cancers.

After they are infected with HIV, T cells may not show signs of AIDS for months or years. AIDS does not appear until the viral DNA is activated and begins to destroy many T cells. A milder form of the disease that generally precedes full-blown development is called *AIDS-related complex (ARC)*. ARC produces fever, diarrhea, weight loss, and swollen lymph nodes.

There are several strategies for controlling AIDS and related conditions. Many agencies are trying to slow the spread of AIDS by educating people about how to avoid contact with the HIV retrovirus. HIV is spread by direct contact of body fluids, so preventing such contact reduces HIV transmission. Sexual relations, blood transfusions, and intravenous use of contaminated needles are the usual modes of HIV transmission. Research teams are working on vaccines to impart immunity to people who are not yet infected by HIV. Like many viruses, such as those that cause the common cold, HIV changes rapidly enough to make development of a vaccine difficult.

Another way to halt the disease is by means of chemicals such as *azidothymidine (AZT)* that block HIV's ability to reproduce within infected cells. At least 80 such compounds are currently being evaluated for use in halting the progress of HIV infections.

OUTLINE SUMMARY

THE LYMPHATIC SYSTEM (Figure 14-1)

A Lymph—fluid in the tissue spaces that carries protein molecules and other substances back to the blood

B Lymphatic vessels—permit only one-way movement of lymph
1 Lymphatic capillaries—tiny blind-ended tubes distributed in tissue spaces
   a Microscopic in size
   b Sheets consisting of one cell layer of simple squamous epithelium
   c Poor "fit" between adjacent cells results in porous walls
   d Called *lacteals* in the intestinal wall (for fat transportation)
2 Right lymphatic duct
   a Drains lymph from the right upper extremity and right side of head, neck, and upper torso
3 Thoracic duct
   a Largest lymphatic vessel
   b Has an enlarged pouch along its course, called *cisterna chyli*
   c Drains lymph from about three fourths of the body (Figure 14-2)

C Lymphedema—swelling (edema) of tissues due to blockage of lymph vessels
1 Lymphangitis—inflammation of lymph vessels, may progress to septicemia (blood infection)
2 Elephantiasis—severe lymphedema of limbs resulting from parasite infestation of lymph vessels

D Lymph nodes
1 Filter lymph (Figure 14-3)
2 Located in clusters along the pathway of lymphatic vessels
3 Functions include defense and WBC formation
4 Flow of lymph: to node via several afferent lymph vessels and drained from node by a single efferent lymph vessel

E Lymphoma—malignant tumor of lymph nodes
1 Two types: Hodgkin's disease and non-Hodgkin's lymphoma

F Thymus
1 Lymphoid tissue organ located in mediastinum
2 Total weight of about 28 g—over an ounce
3 Plays a vital and central role in immunity
4 Produces T-lymphocytes or T cells

5 Secretes hormone called *thymosin*
6 Lymphoid tissue is replaced by fat in the process called *involution*

G Tonsils (Figure 14-5)
1 Composed of three masses of lymphoid tissue around the openings of the mouth and throat
   a Palatine tonsils ("the tonsils")
   b Pharyngeal tonsils (adenoids)
   c Lingual tonsils
2 Subject to chronic infection
3 Enlargement of pharyngeal tonsils may impair breathing

H Spleen
1 Largest lymphoid organ in body
2 Located in upper left quadrant of abdomen
3 Often injured by trauma to abdomen
4 Surgical removal called *splenectomy*
5 Functions include phagocytosis of bacteria and old RBCs; acts as a blood reservoir
6 Splenomegaly—enlargement of the spleen

THE IMMUNE SYSTEM (Table 14-1)

A Protects body from pathological bacteria, foreign tissue cells, and cancerous cells

B Made up of specialized cells and molecules

C Nonspecific immunity
1 Skin—mechanical barrier to bacteria and other harmful agents
2 Tears and mucus—wash eyes and trap and kill bacteria

D Specific immunity—ability of body to recognize, respond to, and remember harmful substances or bacteria

E Inherited or inborn immunity—inherited immunity to certain diseases from birth

F Acquired immunity
1 Natural immunity—exposure to causative agent is not deliberate
   a Active—active disease produces immunity
   b Passive—immunity passes from mother to fetus through placenta or from mother to child through mother's milk
2 Artificial immunity—exposure to causative agent is deliberate
   a Active—vaccination results in immunity
   b Passive—protective material developed in another individual's immune system and given to previously nonimmune individual

IMMUNE SYSTEM MOLECULES

A Antibodies
  1 Protein compounds with specific combining sites
  2 Combining sites attach antibodies to specific antigens (foreign proteins), forming an antigen-antibody complex—called *humoral* or *antibody-mediated immunity*
  3 Antigen-antibody complexes may:
    a Neutralize toxins
    b Clump or agglutinate enemy cells
    c Promote phagocytosis
B Complement proteins
  1 Group of 14 proteins normally present in blood in inactive state
  2 Complement fixation
    a Important mechanism of action for antibodies
    b Causes cell lysis by permitting entry of water through a defect created in the plasma membrane (Figure 14-7)

IMMUNE SYSTEM CELLS

A Phagocytes—ingest and destroy foreign cells or other harmful substances via phagocytosis
  1 Types
    a Neutrophils
    b Monocytes
    c Macrophages (Figure 14-8)
      (1) Kupffer's cells (liver)
      (2) Dust cells (lung)
B Lymphocytes
  1 Most numerous of immune system cells
  2 Development of B cells—primitive stem cells migrate from bone marrow and go through two stages of development (Figure 14-9)
    a First stage—stem cells develop into immature B cells; takes place in the liver and bone marrow before birth and in the bone marrow only in adults; immature B cells are small lymphocytes with antibody molecules (which they have synthesized) in their plasma membranes; migrate chiefly to lymph nodes
    b Second stage—immature B cell develops into activated B cell; initiated by immature B cell's contact with antigens, which bind to its surface antibodies; activated B cell, by dividing repeatedly, forms two clones of cells; plasma cells and memory cells; plasma cells secrete antibodies into blood; memory cells stored in lymph nodes; if subsequent exposure to antigen that activated B cell occurs, memory cells become plasma cells and secrete antibodies

  3 Function of B cells—indirectly, B cells produce humoral immunity; activated B cells develop into plasma cells; plasma cells secrete antibodies into the blood; circulating antibodies produce humoral immunity (Figure 14-9)
  4 Development of T cells—stem cells from bone marrow migrate to thymus gland (Figure 14-10)
    a Stage 1—stem cells develop into T cells; occurs in thymus during few months before and after birth; T cells migrate chiefly to lymph nodes
    b Stage 2—T cells develop into sensitized T cells; occurs when, and if, antigen binds to T cell's surface proteins
  5 Functions of T cells—produce cell-mediated immunity; kill invading cells by releasing a substance that poisons cells and also by releasing chemicals that attract and activate macrophages to kill cells by phagocytosis (Figure 14-11)

HYPERSENSITIVITY OF THE IMMUNE SYSTEM

A Inappropriate or excessive immune response
B Allergy—hypersensitivity to harmless environmental antigens (allergens)
  1 Immediate allergic responses usually involve humoral immunity
  2 Delayed allergic responses usually involve cell-mediated immunity
C Autoimmunity—inappropriate, excessive response to self-antigens
  1 Causes autoimmune diseases
  2 Systemic lupus erythematosus (SLE)—chronic inflammatory disease caused by numerous antibodies attacking a variety of tissues
D Isoimmunity—excessive reaction to antigens from another human
  1 May occur between mother and fetus during pregnancy
  2 May occur in tissue transplants (causing rejection syndrome)

IMMUNE SYSTEM DEFICIENCY

A Congenital immune deficiency or immunodeficiency (rare)
  1 Results from improper lymphocyte development before birth
  2 Severe combined immune deficiency (SCID)—caused by disruption of stem cell development
B Acquired immune deficiency
  1 Develops after birth
  2 Acquired immune deficiency syndrome (AIDS)—caused by HIV infection of T cells

NEW WORDS

afferent
antibodies
antigen
B cells (lymphocytes)
cell-mediated
    immunity

clone
combining sites
complement
complement fixation

efferent
humoral immunity
interferon
lymph

macrophage
memory cells
plasma cells
T cells (lymphocytes)

Diseases and Other Clinical Terms

acquired immune
    deficiency syndrome
    (AIDS)
allergen
allergy
autoimmunity
elephantiasis
Hodgkin's disease

human lymphocyte
    antigen (HLA)
hypersensitivity
immune deficiency
immunosuppressive
    drugs
isoimmunity
lymphangitis

lymphedema
monoclonal antibodies
non-Hodgkin's
    lymphoma
severe combined
    immune deficiency
    (SCID)

splenectomy
splenomegaly
systemic lupus
    erythematosus (SLE)
tonsillitis
transplant

CHAPTER TEST

1. The largest lymphatic vessel in the body is called the _____ _____.
2. Phagocytic cells located in the _____ _____ perform biological filtration.
3. Deliberate exposure to a causative agent produces _____ immunity.
4. The two protein molecules most critical to immune system functioning are called _____ and _____.
5. Antibodies are characterized by their ability to combine with very specific compounds called _____.
6. Antibodies make antigens unable to harm the body—a process called _____ immunity.
7. Plasma cells produce _____.
8. The most numerous cells of the immune system are the _____.
9. The most important phagocytes of the immune system are the _____ and _____.
10. B cells and T cells are the two major types of _____ in the immune system.
11. B cells indirectly produce _____ immunity in the body.
12. Sensitized T cells produce _____ immunity in the body.

Circle the T before each true statement and the F before each false statement.

T F **13.** Lymphatic vessels permit only one-way movement of lymph.
T F **14.** The right lymphatic duct is the largest lymph vessel in the body.
T F **15.** Vaccination is an example of active artificial immunity.
T F **16.** Cells of the immune system outnumber antibody molecules by about 100 million to one.
T F **17.** Antibodies function to produce cell-mediated immunity.
T F **18.** After leaving the blood and entering the tissue space, monocytes are called *macrophages*.
T F **19.** Macrophages are capable of phagocytosis.
T F **20.** Complement fixation is an important way in which antibodies act.
T F **21.** Antigen-antibody complexes may neutralize toxins.
T F **22.** Activated B cells develop into Kupffer's cells.
T F **23.** Memory cells develop from sensitized T cells.
T F **24.** T cells begin as stem cells that develop in the thymus gland.

Select the most correct answer from Column B for each statement in Column A. (Only one answer is correct.)

COLUMN A

24. _____ Tumor of lymph nodes
25. _____ Spleen enlargement
26. _____ Hypersensitivity to harmless environmental antigens
27. _____ Excessive reaction to antigens from another human
28. _____ Inappropriate response to self-antigens
29. _____ Congenital failure of the immune system
30. _____ Acquired failure of the immune system

COLUMN B

a. AIDS
b. Allergy
c. Autoimmunity
d. Isoimmunity
e. Lymphoma
f. SCID
g. Splenomegaly

REVIEW QUESTIONS

1 Sometimes a woman's arm becomes very swollen for a while after removal of a breast and the nearby lymph nodes and lymphatic vessels, including some of those in the upper arm. Can you think of any reason why swelling occurs?

2 Discuss two functions of the lymph nodes.

3 How does the "circulation" of lymph differ from that of blood?

4 Explain nonspecific immunity, specific immunity, inherited immunity, acquired immunity, natural immunity, artificial immunity, active immunity, and passive immunity.

5 Explain the terms *antigen* and *antibody*.

6 What are combining sites? Complement binding sites?

7 Name four kinds of cells that constitute part of the immune system. Which are most numerous? How numerous are they estimated to be?

8 Name the two major immune system molecules.

9 What are the two major types of lymphocytes called? What makes these names appropriate?

10 Describe briefly the two stages of development that B cells undergo.

11 Describe briefly the two stages of development that T cells undergo.

12 What cells secrete antibodies?

13 Explain the function of memory cells.

14 Describe the function of activated B cells.

15 Describe the function of sensitized T cells.

16 Explain each of these types of immune hypersensitivity and give examples of each: allergy, autoimmunity, isoimmunity.

17 What are the possible causes of immune deficiency? Give examples.

18 How does HIV infection cause AIDS? How does HIV spread from person to person?

CLINICAL APPLICATIONS

1 After an infection in the groin, a young boy complains of painful swelling in his leg. On examination, you see that the entire leg is swollen, but the opposite leg is fine. The attending physician explains that this is a complication of the recent groin infection, which involved the lymph nodes in that area. Can you explain how this may have caused the boy's leg to swell? Why isn't the other leg affected?

2 Keith was sledding in the snow with his friends when he accidentally hit a tree. After examining Keith, the emergency room physician concluded that he rup-tured his spleen in the accident. How might a ruptured spleen be treated? What might happen if Keith's family or physician delays this treatment? Does Keith need his spleen to survive?

3 A number of years ago there was a famous case of a boy born with severe combined immune deficiency (SCID). His physicians placed him in a pathogen-free chamber that resembled a giant glass bubble. What purpose was served by doing this? What treatments are available today that might have helped this boy?

The Respiratory System

CHAPTER

15

Outline

Objectives

After you have completed this chapter, you should be able to:

1 Discuss the generalized functions of the respiratory system.
2 List the major organs of the respiratory system and describe the function of each.
3 Compare, contrast, and explain the mechanism responsible for the exchange of gases that occurs during internal and external respiration.
4 List and discuss the volumes of air exchanged during pulmonary ventilation.
5 Identify and discuss the mechanisms that regulate respiration.
6 Identify and describe the major disorders of the upper respiratory tract.
7 Identify and describe the major disorders of the lower respiratory tract.

Boxed Essays

Hiccup
Oxygen Therapy
Heimlich Maneuver

No one needs to be told how important the **respiratory system** is. The respiratory system serves the body much as a lifeline to an oxygen tank serves a deep-sea diver. Think how panicked you would feel if suddenly your lifeline became blocked—if you could not breathe for a few seconds! Of all the substances that cells and therefore the body as a whole must have to survive, oxygen is by far the most crucial. A person can live a few weeks without food, a few days without water, but only a few minutes without oxygen. Constant removal of carbon dioxide from the body is just as important for survival as a constant supply of oxygen.

The organs of the respiratory system are designed to perform two basic functions; they serve as an **air distributor** and as a **gas exchanger** for the body. The respiratory system ensures that oxygen is supplied to and carbon dioxide is removed from the body's cells. The process of respiration therefore is an important **homeostatic mechanism.** By constantly supplying adequate oxygen and by removing carbon dioxide as it forms, the respiratory system helps maintain a constant environment that enables our body cells to function effectively.

In addition to air distribution and gas exchange, the respiratory system effectively **filters, warms,** and **humidifies** the air we breathe. Respiratory organs or organs closely associated with the respiratory system, such as the **sinuses,** also influence speech or sound production and make possible the sense of smell or **olfaction** (ol-FAK-shun). This chapter will present information on both the normal and pathological structure and function of the respiratory system. Discussion will also include facts about gas exchange and the nervous system's control of respiration.

sacs called **alveoli** (al-VEE-o-li). Figure 15-1 shows the extensive branching of the "respiratory tree" in both lungs. Think of this air distribution system as an "upside-down tree." The trachea or windpipe then becomes the trunk and the bronchial tubes the branches. This idea will be developed when the types of bronchi and the alveoli are studied in more detail later in the chapter. A network of capillaries fits like a hairnet around each microscopic alveolus. Incidentally, this is a good place for us to think again about a principle already mentioned several times, namely, that structure and function are intimately related. The function of alveoli—in fact, the function of the entire respiratory system—is to distribute air close enough to blood for a gas exchange to take place between air and blood. The passive transport process of **diffusion,** which was described in Chapter 2, is responsible for the exchange of gases that occurs in the respiratory system. You may want to review the discussion of diffusion on pp. 27 to 28 before you study the mechanism of gas exchange that occurs in the lungs and body tissues.

Two characteristics about the structure of alveoli assist in diffusion and make them able to perform this function admirably. First, the wall of each alveolus is made up of a single layer of cells and so are the walls of the capillaries around it. This means that, between the blood in the capillaries and the air in the alveolus, there is a barrier probably less than 1 micron thick! This extremely thin barrier is called the **respiratory membrane** (Figure 15-2). Second, there are millions of alveoli. This means that together they make an enormous surface (approximately 100 square meters, an area many times larger than the surface of the entire body) where larger amounts of oxygen and carbon dioxide can rapidly be exchanged.

Structural Plan

Respiratory organs include the **nose, pharynx** (FAIR-inks), **larynx** (LAIR-inks), **trachea** (TRAY-kee-ah), **bronchi** (BRONG-ki), and **lungs.**

The basic structural design of this organ system is that of a tube with many branches ending in millions of extremely tiny, very thin-walled

Respiratory Tracts

The respiratory system is often divided into upper and lower tracts or divisions to assist in the description of symptoms associated with common respiratory problems such as a cold. The organs of the upper respiratory tract are located outside of the thorax or chest cavity, whereas

FIGURE **15-1** **Structural Plan of the Respiratory Organs Showing the Pharynx, Trachea, Bronchi, and Lungs.** The inset shows the alveolar sacs where the interchange of oxygen and carbon dioxide takes place through the walls of the grapelike alveoli. Capillaries surround the alveoli.

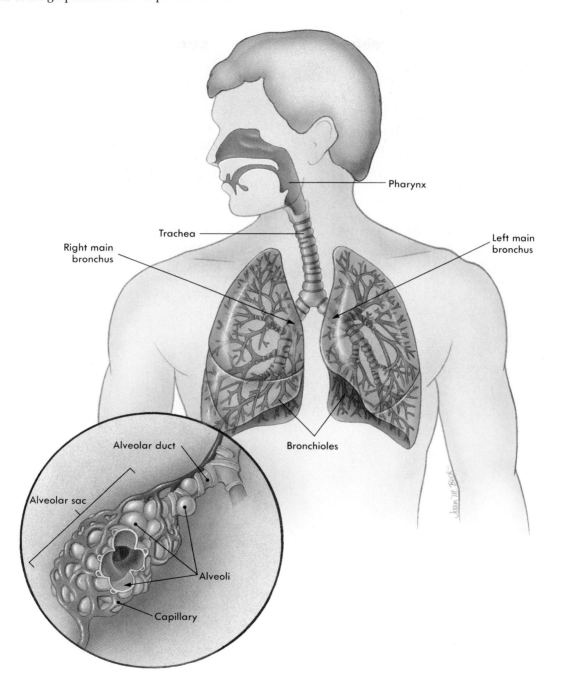

FIGURE 15-2 The Gas-Exchange Structures of the Lung. Each alveolus is continually ventilated with fresh air. The inset shows a magnified view of the respiratory membrane composed of the alveolar wall (surfactant, epithelial cells, and basement membrane), interstitial fluid, and the wall of a pulmonary capillary (basement membrane and endothelial cells). The gases, CO_2 (carbon dioxide) and O_2 (oxygen), diffuse across the respiratory membrane.

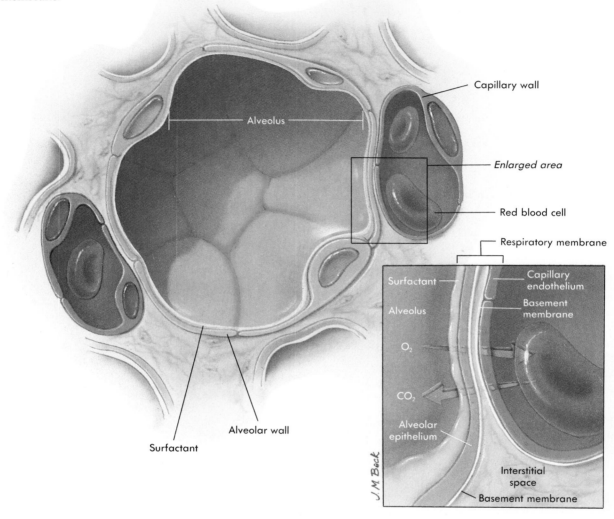

those in the lower tract or division are located almost entirely within it. The **upper respiratory tract** is composed of the nose, pharynx, and larynx. The **lower respiratory tract** or division consists of the trachea, all segments of the bronchial tree, and the lungs.

Respiratory Mucosa

Before beginning the study of individual organs in the respiratory system, it is important to review the histology or microscopic anatomy of the **respiratory mucosa**—the membrane that lines

FIGURE **15-3** **Respiratory Mucosa Lining the Trachea.** A layer of mucus covers the hairlike cilia.

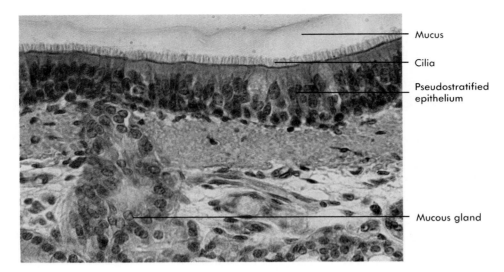

Mucus

Cilia

Pseudostratified epithelium

Mucous gland

most of the air distribution tubes in the system. Do not confuse the respiratory membrane with the respiratory mucosa! The **respiratory membrane** (Figure 15-2) separates the air in the alveoli from the blood in surrounding capillaries. The **respiratory mucosa** (Figure 15-3) is covered with mucus and lines the tubes of the respiratory tree.

Recall that in addition to serving as air distribution passageways or gas exchange surfaces, the anatomical components of the respiratory tract and lungs cleanse, warm, and humidify inspired air. Air entering the nose is generally contaminated with one or more common irritants; examples include insects, dust, pollen, and bacterial organisms. A remarkably effective air purification mechanism removes almost every form of contaminant before inspired air reaches the alveoli or terminal air sacs in the lungs.

The layer of protective mucus that covers a large portion of the membrane that lines the respiratory tree serves as the most important air purification mechanism. Over 125 ml of respiratory mucus is produced daily. It forms a continuous sheet called a *mucus blanket* that covers the lining of the air distribution tubes in the respiratory tree.

This layer of cleansing mucus moves upward to the pharynx from the lower portions of the bronchial tree on millions of hairlike cilia that cover the epithelial cells in the respiratory mucosa (Figure 15-3). The microscopic cilia that cover epithelial cells in the respiratory mucosa beat or move only in one direction. The result is movement of mucus toward the pharynx. Cigarette smoke paralyzes these cilia and results in accumulations of mucus and the typical smoker's cough, an effort to clear the secretions.

Nose

Air enters the respiratory tract through the **external nares** (NA-rees) or nostrils. It then flows into the right and left **nasal cavities,** which are lined by respiratory mucosa. A partition called the *nasal septum* separates these two cavities.

The surface of the nasal cavities is moist from mucus and warm from blood flowing just under it. Nerve endings responsible for the sense of smell (olfactory receptors) are located in the nasal mucosa. Four **paranasal sinuses**—frontal, maxil-

FIGURE 15-4 The Paranasal Sinuses. The anterior view shows the anatomical relationship of the paranasal sinuses to each other and to the nasal cavity. The inset is a lateral view of the position of the sinuses.

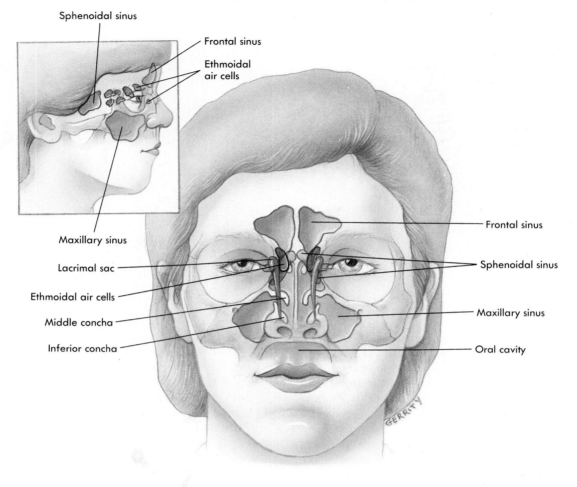

lary, sphenoidal, and ethmoidal—drain into the nasal cavities (Figure 15-4). Because the mucosa that lines the sinuses is continuous with the mucosa that lines the nose, sinus infections, called **sinusitis** (sye-nyoo-SYE-tis), often develop from colds in which the nasal mucosa is inflamed. The paranasal sinuses are lined with a mucous membrane that assists in the production of mucus for the respiratory tract. In addition, these hollow spaces help lighten the skull bones and serve as resonant chambers for the production of sound.

Note in Figure 15-5 that three shelflike structures called **conchae** (KONG-kee) protrude into the nasal cavity on each side. The mucosa-covered conchae greatly increase the surface over which air must flow as it passes through the naval cavity. As air moves over the conchae and through the nasal cavities, it is warmed and humidified. This helps explain why breathing through the nose is more effective in humidifying inspired air than is breathing through the mouth. If an individual who is ill requires supplemental oxygen, it is first bubbled through water to

▼ **FIGURE 15-5 Sagittal Section of the Head and Neck.** The nasal septum has been removed, exposing the right lateral wall of the nasal cavity so that the nasal conchae can be seen. Note also the divisions of the pharynx and the position of the tonsils.

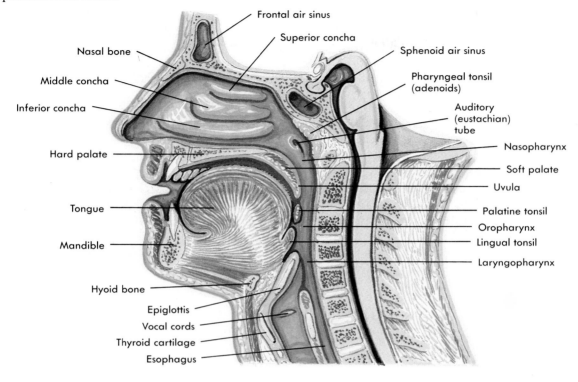

reduce the amount of moisture that would otherwise have to be removed from the lining of the respiratory tree to humidify it. Administration of "dry" oxygen pulls water from the mucosa and results in respiratory discomfort and irritation.

▼ *Pharynx*

The **pharynx** is the structure that many of us call the throat. It is about 12.5 cm (5 inches) long and can be divided into three portions (Figure 15-5). The uppermost part of the tube just behind the nasal cavities is called the **nasopharynx** (nay-zo-FAIR-inks). The portion behind the mouth is called the **oropharynx** (o-ro-FAIR-inks). The last or lowest segment is called the **laryngopharynx** (lah-ring-go-FAIR-inks). The pharynx as a whole

serves the same purpose for the respiratory and digestive tracts as a hallway serves for a house. Air and food pass through the pharynx on their way to the lungs and the stomach respectively. Air enters the pharynx from the two nasal cavities and leaves it by the larynx; food enters it from the mouth and leaves it by the esophagus. The right and left **auditory** or *eustachian* (yoo-STAY-she-an) **tubes** open into the nasopharynx; they connect the middle ears with the nasopharynx (Figure 15-5). This connection permits equalization of air pressure between the middle and the exterior ear. The lining of the auditory tubes is continuous with the lining of the nasopharynx and middle ear. Therefore just as sinus infections can develop from colds in which the nasal mucosa is inflamed, middle ear infections can develop as a result of inflammation of the nasopharynx.

Masses of lymphatic tissue called **tonsils** are embedded in the mucous membrane of the pharynx (see p. 354). The **pharyngeal** (fa-RIN-jee-al) **tonsils** or **adenoids** (AD-e-noyds) are in the nasopharynx. The **palatine tonsils** are located in the oropharynx (Figure 15-5). Both tonsils are generally removed in a **tonsillectomy** (ton-si-LEK-toe-mee). Although this surgical procedure is still rather common, the number of tonsillectomies reported each year continues to decrease as new and more effective antibiotics become available. Physicians now recognize the value of lymphatic tissue in the body defense mechanism and delay removal of the tonsils—even in cases of inflammation or **tonsillitis**—unless antibiotic treatment is ineffective. When the pharyngeal tonsils become swollen, they are referred to as adenoids. Such swelling caused by infections may make it difficult or impossible for air to travel from the nose into the throat. In these cases the individual is forced to breathe through the mouth.

Larynx

The **larynx** or voice box is located just below the pharynx. It is composed of several pieces of cartilage. You know the largest of these pieces (the

FIGURE 15-6 The Larynx. A, Sagittal section of the larynx. **B,** Superior view of the larynx. **C,** Photograph of the larynx taken with an endoscope (optical device) inserted through the mouth and pharynx to the epiglottis.

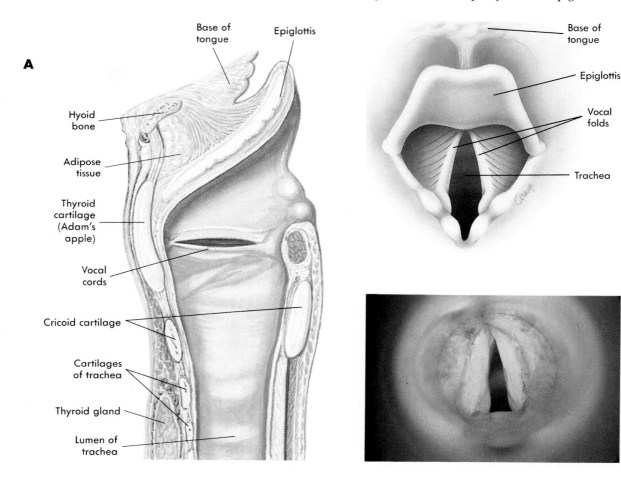

thyroid cartilage) as the "Adam's apple" (Figure 15-6).

Two short fibrous bands called the **vocal cords** stretch across the interior of the larynx. Muscles that attach to the cartilages of the larynx can pull on these cords in such a way that they become either tense or relaxed. When they are tense, the voice sounds high pitched; when they are relaxed, it sounds low pitched. The space between the vocal cords is called the **glottis.**

Another cartilage, the **epiglottis** (ep-i-GLOT-is) partially covers the opening into the larynx (Figure 15-6). The epiglottis acts somewhat like a trapdoor, closing off the larynx during swallowing and preventing food from entering the trachea.

Disorders of the Upper Respiratory Tract

UPPER RESPIRATORY INFECTION

Any infection localized in the mucosa of the upper respiratory tract (nose, pharynx, and larynx) can be called an *upper respiratory infection (URI)*. Although the general designation URI is often used, such infections are sometimes named for the specific structure involved:

1 Rhinitis (rye-NYE-tis) or *coryza* (koe-RYE-za), is inflammation of the nasal mucosa and is often caused by nasal infections. In some cases, it is caused by nasal irritants or an allergic reaction to airborne allergens. Rhinitis usually involves swelling of the nasal mucosa (often obstructing breathing) and excessive discharge of mucus (that is, a runny nose). Rhinitis may be caused by common cold viruses or by *influenza* (the "flu"). See Chapter 4 and Appendix B for more details on these infections.

2 Pharyngitis (fair-in-JYE-tis) or *sore throat*, is inflammation or infection of the pharynx (throat). Pain, redness, and difficulty in swallowing are characteristic of pharyngitis. Pharyngitis may be caused by any of several pathogens, including the streptococcal bacteria that cause "strep throat" (see Appendix B).

3 Laryngitis (lair-in-JYE-tis) is inflammation of the mucous lining of the larynx. It is characterized by edema of the vocal cords, which results in hoarseness or loss of voice. Although laryngitis often results from an infection, it can be caused by overuse of the voice, smoking, sudden temperature changes, and other factors. In children under 5 years of age, acute laryngitis may cause difficulty in breathing, a condition often called "croup".

URIs are rather common, occurring several times a year in most individuals, because the upper respiratory tract is easily accessible to common airborne pathogens. Because the upper respiratory mucosa is continuous with the mucous lining of the sinuses, the eustachian tube and middle ear, and lower respiratory tract, URIs have the unfortunate tendency to spread. It is not unusual therefore to see a common cold progress to become sinusitis or *otitis media* (middle ear infection).

ANATOMICAL DISORDERS

Deviated septum is a condition in which the nasal septum strays from the midline of the nasal cavity. Nobody's nasal septum is *exactly* on the midsagittal plane, but most are fairly close. Some people, however, are born with a congenital defect of the septum that results in blockage to one or both sides of the nasal cavity. Others acquire a deviated septum after birth as a result of damage from an injury or infection. In either case, surgical correction of the anatomical abnormality often results in normal breathing through the nose.

Injury to the nose occurs relatively frequently because the nose projects some distance from the front of the head. Usually, common bumps and other injuries cause little if any serious damage. Occasionally, **epistaxis** (ep-i-STAKS-is) or nosebleed, occurs. The most common cause of nosebleed is a strong bump or blow, but it can result from severe inflammation or rubbing (as in rhinitis), hypertension, or even brain injury. Because of the rich blood supply close to the inside surface of the nasal cavity, even minor nosebleeds can produce a great deal of blood—causing them to appear more serious than they are.

Trachea

The **trachea** or windpipe is a tube about 11 cm (4½ inches) long that extends from the larynx in the neck to the bronchi in the chest cavity (Figures 15-1 and 15-7). The trachea performs a simple but

FIGURE 15-7 Cross Section of the Trachea.
Inset shows where the section was cut.

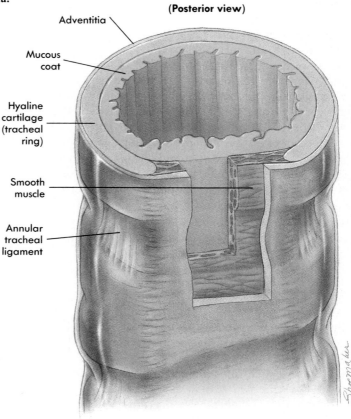

(Posterior view)

Adventitia

Mucous coat

Hyaline cartilage (tracheal ring)

Smooth muscle

Annular tracheal ligament

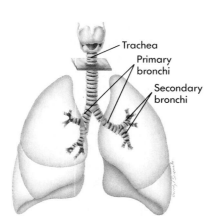

Trachea

Primary bronchi

Secondary bronchi

vital function; it furnishes part of the open passageway through which air can reach the lungs from the outside.

By pushing with your fingers against your throat about an inch above the sternum, you can feel the shape of the trachea or windpipe. Only if you use considerable force can you squeeze it closed. Nature has taken precautions to keep this lifeline open. Its framework is made of an almost noncollapsible material—15 or 20 C-shaped rings of cartilage placed one above the other with only a little soft tissue between them (Figure 15-7). The trachea is lined by the typical respiratory mucosa. Glands below the ciliated epithelium help produce the blanket of mucus that continually moves upward toward the pharynx.

Despite the structural safeguard of cartilage rings, closing of the trachea sometimes occurs. A tumor or an infection may enlarge the lymph nodes of the neck so much that they squeeze the trachea shut, or a person may aspirate (breathe in) a piece of food or something else that blocks the windpipe. Because air has no other way to get to the lungs, complete tracheal obstruction causes death in a matter of minutes. Choking on food and other substances caught in the trachea kills over 4000 people each year and is the fifth major cause of accidental deaths in the United States. A lifesaving technique developed by Dr. Henry Heimlich (see p. 393), is now widely used to free the trachea of ingested food or other foreign objects.

If the air pathway to the trachea is blocked and cannot be cleared by the Heimlich maneuver or other first aid procedures, medical personnel may perform an emergency *tracheostomy*. The term *tracheostomy* literally means "making a mouth in the trachea"—an apt description for this procedure. First, a *tracheotomy* ("cut in the trachea") is made through the skin of the neck just below the larynx and between the cartilage rings in the tracheal wall. Next, a hollow breathing tube is inserted into the incision to form an artificial "mouth" for breathing. Because the blocked upper respiratory tract is bypassed, the subject can begin breathing.

Bronchi, Bronchioles, And Alveoli

STRUCTURE AND FUNCTION

Recall that one way to picture the thousands of air tubes that make up the lungs is to think of an upside-down tree. The trachea is the main trunk of this tree; the right bronchus (the tube leading into the right lung) and the left bronchus (the tube leading into the left lung) are the trachea's first branches or **primary bronchi.** In each lung, they branch into smaller or **secondary bronchi** whose walls, like those of the trachea and bronchi, are kept open by rings of cartilage for air passage. These bronchi divide into smaller and smaller tubes, ultimately branching into tiny tubes whose walls contain only smooth muscle. These very small passageways are called **bronchioles.** The bronchioles subdivide into microscopic tubes called **alveolar ducts,** which resemble the main stem of a bunch of grapes (Figure 15-8). Each alveolar duct ends in several **alveolar sacs,** each of which resembles a cluster of grapes, and the wall of each alveolar sac is made up of numerous **alveoli,** each of which resembles a grape.

Alveoli are very effective in gas exchange, mainly because they are extremely thin walled; each alveolus lies in contact with a blood capillary, and there are millions of alveoli in each lung. The surface of the respiratory membrane inside the alveolus is covered by a substance called **surfactant** (sur-FAK-tant). This important substance helps reduce surface tension in the alveoli and keep them from collapsing as air moves in and out during respiration.

RESPIRATORY DISTRESS

Respiratory distress results from the body's relative inability to inflate the alveoli of the lungs normally. *Respiratory distress syndrome* is a condition most often caused by absence or impairment of the surfactant in the fluid that lines the alveoli.

Infant respiratory distress syndrome or **IRDS** is a very serious, life-threatening condition that often affects prematurely born infants of less than 37 weeks' gestation or those who weigh less than 2.2 kg (5 lbs) at birth. IRDS is the leading cause of

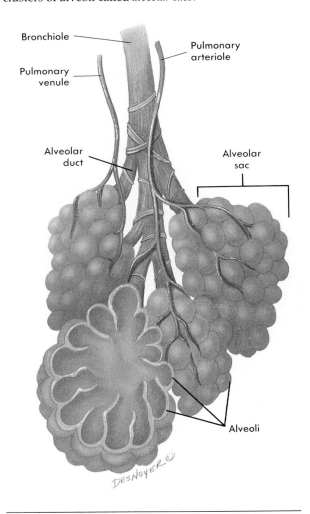

FIGURE 15-8 Alveoli. Bronchioles subdivide to form tiny tubes called *alveolar ducts,* which end in clusters of alveoli called *alveolar sacs.*

Bronchiole

Pulmonary arteriole

Pulmonary venule

Alveolar duct

Alveolar sac

Alveoli

DESNOYER

FIGURE 15-9 Lungs. The trachea is an airway that branches to form a treelike formation of bronchi and bronchioles. Note that the right lung has three lobes and that the left lung two lobes.

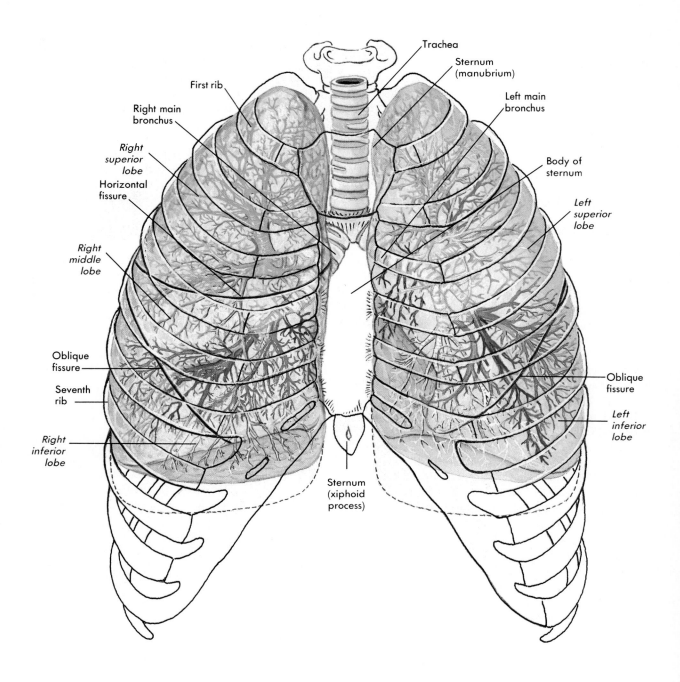

death among premature infants in the United States, claiming over 5000 premature babies each year. The disease, characterized by a lack of **surfactant** in the alveolar air sacs, affects 50,000 babies annually.

Surfactant is manufactured by specialized cells in the walls of the alveoli. Surfactant reduces the surface tension of the fluid on the free surface of the alveolar walls and permits easy movement of air into and out of the lungs. The ability of the body to manufacture this important substance is not fully developed until shortly before birth—normally about 40 weeks after conception.

In newborn infants who are unable to manufacture surfactant, many air sacs collapse during expiration because of the increased surface tension. The effort required to reinflate these collapsed alveoli is much greater than that needed to reinflate normal alveoli with adequate surfactant. The baby soon develops labored breathing, and symptoms of respiratory distress appear shortly after birth.

In the past, treatment of IRDS was limited to keeping the alveoli open so that delivery and exchange of oxygen and carbon dioxide could occur. To accomplish this, a tube is inserted into the respiratory tract, and oxygen-rich air delivered under sufficient pressure to keep the alveoli from collapsing at the end of expiration. A newer treatment involves delivering air under pressure and applying prepared surfactant directly into the baby's airways by means of a tube.

Adult respiratory distress syndrome (ARDS) is caused by impairment or removal of surfactant in the alveoli. For example, accidental inhalation of foreign substances such as water, vomit, smoke, or chemical fumes can cause ARDS. Edema of the alveolar tissue can impair surfactant and reduce the alveoli's ability to stretch, causing respiratory distress.

Lungs and Pleura

The **lungs** are fairly large organs. The right lung has three lobes and the left lung has two. Figure 15-9 shows the relationship of the lungs to the rib cage at the end of a normal expiration. The normal chest x-ray film in Figure 15-10 was taken at

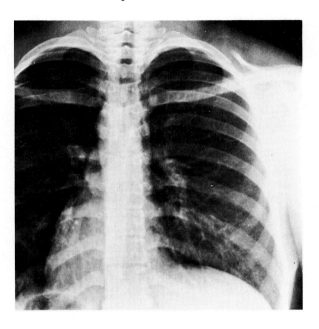

FIGURE 15-10 Normal Chest X-Ray Film. This is a posterior view of the thoracic organs at the end of a maximal inspiration.

the end of a maximal inspiration. Note that the lungs in the chest x-ray film fill the entire chest cavity (all but the space in the center occupied mainly by the heart and large blood vessels). The narrow, superior position of each lung, under the collarbone, is the apex; the broad, inferior portion resting on the diaphragm, is the base.

The **pleura** covers the outer surface of the lungs and lines the inner surface of the rib cage. The pleura resembles other serous membranes in structure and function. Like the peritoneum or pericardium, the pleura is a serous membrane (see Chapter 5)—an extensive, thin, moist, slippery layer of epithelial cells. It lines a large, closed cavity of the body and covers the organs located within it. The parietal pleura lines the walls of the thoracic cavity; the visceral pleura covers the lungs, and the intrapleural space lies between the two pleural membranes (Figure 15-11).

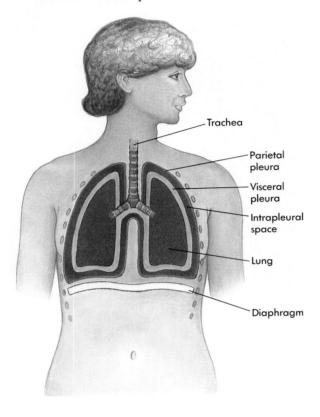

▽ **FIGURE 15-11 Lungs and Pleura.** A serous membrane lines the thoracic wall (parietal pleura) and then folds inward near the bronchi to cover the lung (visceral pleura). The intrapleural space contains a small amount of serous pleural fluid.

Trachea

Parietal pleura

Visceral pleura

Intrapleural space

Lung

Diaphragm

HICCUP

The term **hiccup** (HIK-up) is used to describe an involuntary, spasmodic contraction of the diaphragm. When such a contraction occurs, generally at the beginning of an inspiration, the glottis suddenly closes producing the characteristic sound.

Hiccups lasting for extended periods of time can be disabling. They may be produced by irritation of the phrenic nerve or the sensory nerves in the stomach or by direct injury or pressure on certain areas of the brain. Fortunately, most cases of hiccups last only a few minutes and are harmless.

Pleurisy (PLOO-ri-see) is an inflammation of the parietal pleura, characterized by difficulty in breathing and stabbing pain. The discomfort and restriction of normal breathing associated with pleurisy are caused by the constant rubbing back and forth of the visceral and parietal pleurae during breathing. Pleurisy can be caused by tumors, infections (such as pneumonia and tuberculosis), and other factors.

Normally the pleural space contains just enough fluid to make both portions of the pleura moist and slippery and able to glide easily against each other as the lungs expand and deflate with each breath. However, the pleural space some-

times fills with other substances and increases the pressure on the lung from the outside—causing it to collapse. Collapse of the lung for any reason is called **atelectasis** (at-el-EK-tas-is). While collapsed, the lung cannot be ventilated, making the affected lung useless in breathing. For example, a puncture wound to the chest wall or a rupture of the visceral pleura may cause **pneumothorax** (noo-mo-THOR-aks) (Figure 15-12). Pneumothorax (literally "air in the thorax") is the presence of air in the pleural space on one side of the chest. An injury or disease may also cause **hemothorax** (hee-mo-THOR-aks), the presence of blood in the pleural space. Both conditions are potentially life threatening unless medical treatment is received.

▽ *Respiration*

Respiration means exchange of gases (oxygen and carbon dioxide) between a living organism and its environment. If the organism consists of only one cell, gases can move directly between it and the environment. If, however, the organism consists of billions of cells, as do our bodies, most of its cells are too far from the air for a direct exchange of gases. To overcome this difficulty, a pair of organs—the lungs—provides a place where air and a circulating fluid (blood) can come close enough to each other for oxygen to move out of

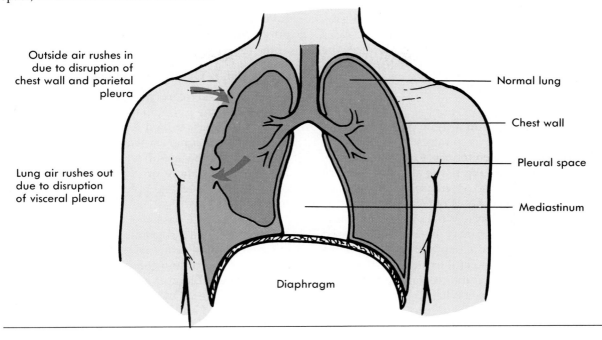

▼ **FIGURE 15-12 Pneumothorax.** Air in the pleural space may accumulate when the visceral pleura ruptures and air from the lung rushes out or when atmospheric air rushes in through a wound in the chest wall and parietal pleura. In either case, the lung collapses and normal respiration is impaired. If blood accumulates in the pleural space, the condition is called *hemothorax.*

Outside air rushes in due to disruption of chest wall and parietal pleura

Lung air rushes out due to disruption of visceral pleura

Normal lung

Chest wall

Pleural space

Mediastinum

Diaphragm

the air into blood while carbon dioxide moves out of the blood into air. Breathing or **pulmonary ventilation** is the process that moves air into and out of the lungs. It makes possible the exchange of gases between air in the lungs and in the blood. This exchange is often called **external respiration.** In addition, exchange of gases occurs between the blood and the cells of the body—a process called **internal respiration.** *Cellular respiration* refers to the actual use of oxygen by cells in the process of metabolism, which is discussed in Chapter 17.

MECHANICS OF BREATHING

Pulmonary ventilation or breathing has two phases. **Inspiration** or inhalation moves air into the lungs, and **expiration** or exhalation moves air out of the lungs. The lungs are enclosed within the thoracic cavity. Thus changes in the shape and size of the thoracic cavity result in changes in the air pressure within that cavity and in the lungs. This difference in air pressure causes the movement of air into and out of the lungs. Air moves from an area where pressure is high to an area where pressure is lower. Respiratory muscles are responsible for the changes in the shape of the thoracic cavity that cause the air movements involved in breathing.

Inspiration

Inspiration occurs when the chest cavity enlarges. As the thorax enlarges, the lungs expand along with it, and air rushes into them and down into the alveoli. Muscles of respiration that are classified as **inspiratory muscles** include the **diaphragm** (DYE-a-fram) and the external intercostals. The diaphragm is the dome-shaped muscle separating the abdominal cavity from the thoracic cavity. The diaphragm flattens out when it contracts during inspiration. Instead of protruding up into the chest cavity, it moves down toward the abdominal cavity. Thus the contraction or flattening of the diaphragm makes the chest cavity longer from top to bottom. The diaphragm is the most important muscle of inspiration. Nerve impulses passing through the *phrenic nerve* stimulate the diaphragm to contract. The external intercostal muscles are located between the ribs. When they contract, they enlarge the thorax by increasing the size of the cavity from front to back and from side to side. Contraction of the inspiratory muscles increases the volume of the thoracic cavity and reduces the air pressure within it, drawing air into the lungs (Figure 15-13).

Expiration

Quiet expiration is ordinarily a passive process that begins when the inspiratory muscles relax. The thoracic cavity then returns to its smaller size. The elastic nature of lung tissue also causes these organs to "recoil" and decrease in size as air leaves the alveoli and flows outward through the respiratory passageways. When we speak, sing, or do heavy work, we may need more forceful expiration to increase the rate and depth of ventilation. During more forceful expiration, the **expiratory muscles** (internal intercostals and abdominal muscles) contract. When contracted, the internal intercostal muscles depress the rib cage and decrease the front-to-back size of the thorax.

Contraction of the abdominal muscles pushes the abdominal organs against the underside of the diaphragm, thus elevating it and making it more "domeshaped." The result is to further shorten or decrease the top to bottom size of the thoracic cavity. As the thoracic cavity decreases in size, the air pressure within it increases and air flows out of the lungs (Figure 15-13).

EXCHANGE OF GASES IN LUNGS

Blood pumped from the right ventricle of the heart enters the pulmonary artery and eventually enters the lungs. It then flows through the thousands of tiny lung capillaries that are in close proximity to the air-filled alveoli (Figure 15-1). External respiration or the exchange of gases between the blood and alveolar air occurs by diffusion.

Diffusion is a passive process that results in movement down a concentration gradient; that is, substances move from an area of high concentration to an area of low concentration of the diffusing substance. Blood flowing into lung capillaries is low in oxygen. Oxygen is continually removed from the blood and used by the cells of the body. By the time it enters the lung capillaries, it is low

FIGURE 15-13 Mechanics of Breathing. During *inspiration*, the diaphragm contracts, increasing the volume of the thoracic cavity. This increase in volume results in a decrease in pressure, which causes air to rush into the lungs. During *expiration*, the diaphragm returns to an upward position, reducing the volume in the thoracic cavity. Air pressure increases then, forcing air out of the lungs. The insets show the classic model in which a jar represents the rib cage, a rubber sheet represents the diaphragm, and a balloon represents the lungs.

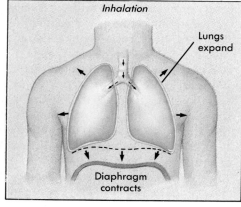

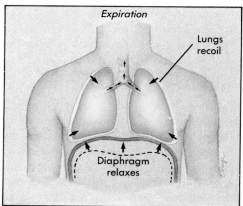

in oxygen content. Because alveolar air is rich in oxygen, diffusion causes movement of oxygen from the area of high concentration (alveolar air) to the area of low concentration (capillary blood). Note in Figure 15-14 that most of the oxygen (O_2) entering the blood combines with hemoglobin (Hb) in the RBCs to form **oxyhemoglobin** (ok-see-HEE-mo-glo-bin) (HbO_2) so that the oxygen can be carried to the tissues and used by the body cells.

Diffusion of carbon dioxide (CO_2) also occurs between blood in lung capillaries and alveolar air.

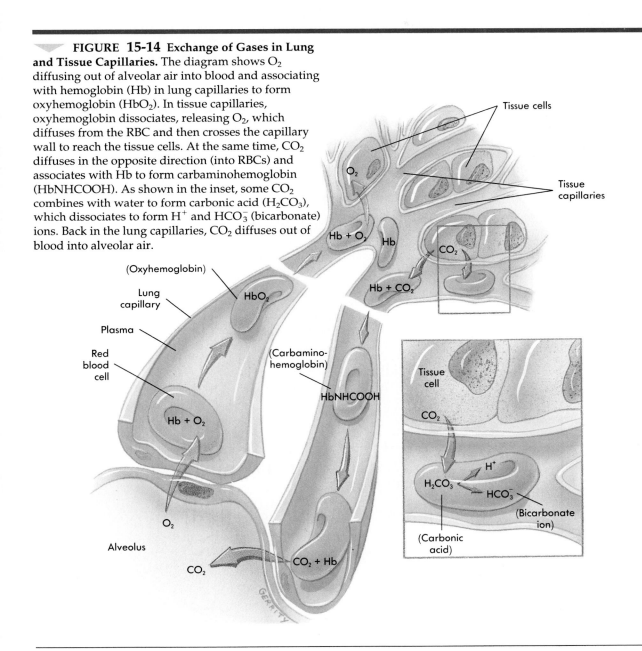

FIGURE 15-14 Exchange of Gases in Lung and Tissue Capillaries. The diagram shows O_2 diffusing out of alveolar air into blood and associating with hemoglobin (Hb) in lung capillaries to form oxyhemoglobin (HbO_2). In tissue capillaries, oxyhemoglobin dissociates, releasing O_2, which diffuses from the RBC and then crosses the capillary wall to reach the tissue cells. At the same time, CO_2 diffuses in the opposite direction (into RBCs) and associates with Hb to form carbaminohemoglobin (HbNHCOOH). As shown in the inset, some CO_2 combines with water to form carbonic acid (H_2CO_3), which dissociates to form H^+ and HCO_3^- (bicarbonate) ions. Back in the lung capillaries, CO_2 diffuses out of blood into alveolar air.

Blood flowing through the lung capillaries is high in carbon dioxide. Most carbon dioxide is carried as bicarbonate ion (HCO_3^-) in the blood. Some, as noted in Figure 15-14, combines with the hemoglobin in RBCs to form **carbaminohemoglobin** (kar-bam-i-no-HEE-mo-glo-bin) (HbNHCOOH). As body cells remove oxygen from circulating blood, they add the waste product carbon dioxide to it. As a result, the blood in pulmonary capillaries eventually becomes low in oxygen and high in carbon dioxide. Diffusion of carbon dioxide results in its movement from an area of high concentration in the pulmonary capillaries to an area of low concentration in alveolar air. Then from the alveoli, carbon dioxide leaves the body in the expired air.

EXCHANGE OF GASES IN TISSUES

The exchange of gases that occurs between blood in tissue capillaries and the body cells is called *internal respiration*. As you would expect, the direction of movement of oxygen and carbon dioxide during internal respiration is just the opposite of that noted in the exchange that occurs

FIGURE 15-15 Pulmonary Ventilation Volumes. The chart shows a training like that produced with a spirometer. During normal, quiet breathing, about 500 ml of air is moved into and out of the respiratory tract, an amount called the *tidal volume*. During forceful breathing (like that during and after heavy exercise), an extra 3300 ml can be inspired (the inspiratory reserve volume), and an extra 1000 ml or so can be expired (the expiratory reserve volume). The largest volume of air that can be moved in and out during ventilation is called the *vital capacity*. Air that remains in the respiratory tract after a forceful expiration is called the *residual volume*.

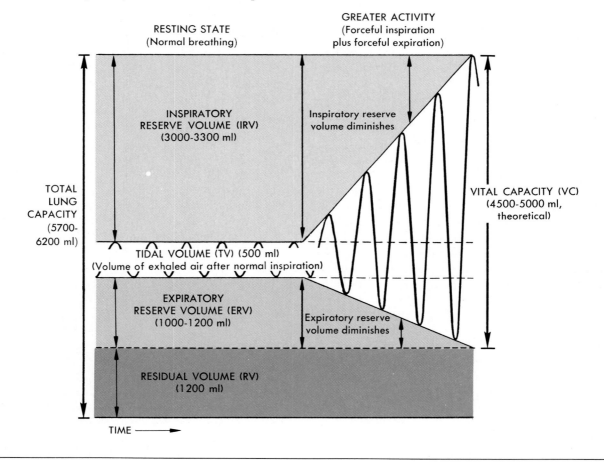

during external respiration when, gases are exchanged between the blood in the lung capillaries and the air in alveoli. As shown in Figure 15-14, oxyhemoglobin breaks down into oxygen and hemoglobin in the tissue capillaries. Oxygen molecules move rapidly out of the blood through the tissue capillary membrane into the interstitial fluid and on into the cells that compose the tissues. The oxygen is used by the cells in their metabolic activities. Diffusion results in the movement of oxygen from an area of high concentration to an area of low concentration in the cells where it is needed. While this is happening, carbon dioxide molecules leave the cells, entering the tissue capillaries where bicarbonate ions are formed and where hemoglobin molecules unite with carbon dioxide to form carbaminohemoglobin. Once again, diffusion is responsible for the movement of carbon dioxide from an area of high concentration in the cells to an area of lower concentration in the capillary blood. In other words, oxygenated blood enters tissue capillaries and is changed into deoxygenated blood as it flows through them. In the process of losing oxygen, the waste product carbon dioxide is picked up and transported to the lungs for removal from the body.

VOLUMES OF AIR EXCHANGED IN PULMONARY VENTILATION

A special device called a **spirometer** is used to measure the amount of air exchanged in breathing. Figure 15-15 illustrates the various pulmonary volumes, which can be measured as a subject breathes into a spirometer. We take 500 ml (about a pint) of air into our lungs with each normal inspiration and expel it with each normal expiration. Because this amount comes and goes regularly like the tides of the sea, it is referred to as the **tidal volume (TV)**. The largest amount of air that we can breathe out in one expiration is known as the **vital capacity (VC)**. In normal young men, this is about 4800 ml. Tidal volume and vital capacity are frequently measured in patients with lung or heart disease, conditions that often lead to abnormal volumes of air being moved in and out of the lungs.

Observe the area in Figure 15-15 that represents the **expiratory reserve volume (ERV)**. This is the amount of air that can be forcibly exhaled after expiring the tidal volume. Compare this with the area in Figure 15-15 that represents the **inspiratory reserve volume (IRV)**. The IRV is the amount of air that can be forcibly inspired over and above a normal inspiration. As the tidal volume increases, the ERV and IRV decrease. Note in Figure 15-15 that vital capacity (VC) is the total of tidal volume, inspiratory reserve volume, and expiratory reserve volume—or expressed in another way: VC = TV + IRV + ERV. **Residual volume (RV)** is simply the air that remains in the lungs after the most forceful expiration.

Regulation of Respiration

We know that the body uses oxygen to obtain energy for the work it has to do. The more work the body does, the more oxygen that must be delivered to its millions of cells. One way this is accomplished is by increasing the rate and depth of respirations. Although we may take only about 12 to 18 breaths a minute when we are not moving about, we take considerably more than this when we are exercising. Not only do we take more breaths, but our tidal volume also increases.

To help supply cells with more oxygen when they are doing more work, automatic adjustments occur not only in respirations but also in circulation. Most notably, the heart beats faster and harder and therefore pumps more blood through the body each minute. This means that the millions of RBCs make more round trips between the lungs and tissues each minute and so deliver more oxygen per minute to tissue cells.

Working cells not only require more oxygen, but they also produce more waste products such as carbon dioxide and certain metabolic acids. The increase in respirations during exercise shows us how the body automatically regulates its vital functions. By increasing the rate and depth of respiration, we can adjust to the varying demands for increased oxygen while increasing the elimination of metabolic waste products in expired air.

Normal respiration depends on proper functioning of the muscles of respiration. These muscles are stimulated by nervous impulses that originate in **respiratory control centers** located in the medulla and pons of the brain. These centers are in turn regulated by a number of inputs from

OXYGEN THERAPY

Oxygen therapy is the administration of oxygen, often by a registered *respiratory therapist*, to patients suffering from **hypoxia** (hi-POK-see-a)—an insufficient oxygen supply to the tissues. Respiratory problems that involve a decrease in ventilation or a lack of efficient gas exchange in the lungs often require treatment with oxygen therapy.

Oxygen (O_2) gas is commonly stored in and dispensed from small, green storage tanks that hold the gas under high pressure until it is used. Because the oxygen dispensed from such tanks is cold and very dry, it must first be warmed and moistened to prevent damage to the respiratory tract. This is usually done by simply bubbling the oxygen through warm water as it leaves the tank. The oxygen gas may then pass through a mask or through tubes that lead into the nasal passage (nasal prongs).

Oxygen therapy is used in a variety of situations in which hypoxia may occur. One example is when breathing stops during *respiratory arrest*. Emergency medical personnel may ventilate the lungs artificially using **CPR (cardiopulmonary resuscitation).** After the subject is breathing again, oxygen therapy may be used to quickly restore normal oxygen levels. The hypoxia that often accompanies chronic pulmonary disease and heart problems such as severe arrhythmia or myocardial infarctions may also be treated with oxygen therapy. Oxygen therapy is used to relieve hypoxia during general anesthesia and postoperative recovery.

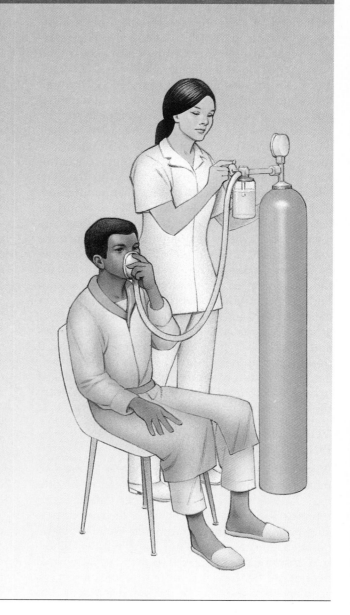

receptors located in varying areas of the body. These receptors can sense the need for changing the rate or depth of respirations to maintain homeostasis. Certain receptors sense carbon dioxide or oxygen levels, whereas others sense blood acid levels or the amount of stretch in lung tissues. The two most important control centers are in the medulla and are called the **inspiratory cen**ter and the **expiratory center.** Centers in the pons have a modifying function. Under rest conditions, neurons in the inspiratory and expiratory centers "fire" at a rate that will produce a normal breathing rate of about 12 to 18 breaths a minute.

The depth and rate of respiration can be influenced by many "inputs" to the respiratory control centers from other areas of the brain or from spe-

FIGURE **15-16** **Regulation of Respiration.** Respiratory control centers in the brain stem control the basic rate and depth of breathing. The brain stem also receives input from other parts of the body; information from chemoreceptors and stretch receptors can alter the basic breathing pattern, as can emotional and sensory input. Despite these controls, the cerebral cortex can override the "automatic" control of breathing to some extent to accomplish activities such as singing or blowing up a balloon.

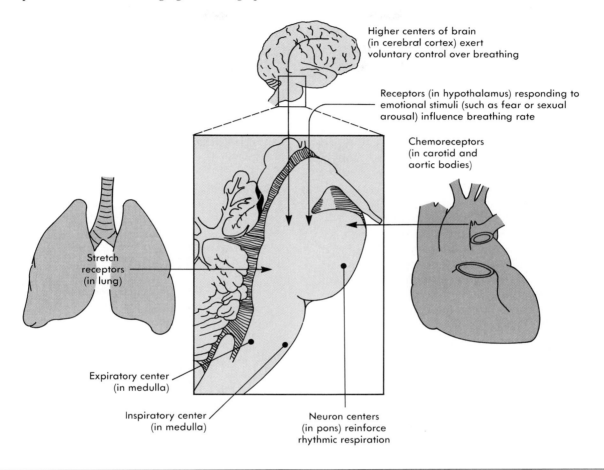

Higher centers of brain (in cerebral cortex) exert voluntary control over breathing

Receptors (in hypothalamus) responding to emotional stimuli (such as fear or sexual arousal) influence breathing rate

Chemoreceptors (in carotid and aortic bodies)

Stretch receptors (in lung)

Expiratory center (in medulla)

Inspiratory center (in medulla)

Neuron centers (in pons) reinforce rhythmic respiration

cialized receptors located outside of the central nervous system (Figure 15-16).

CEREBRAL CORTEX

The cerebral cortex can influence respiration by modifying the rate at which neurons "fire" in the inspiratory and expiratory centers of the medulla. In other words, an individual may voluntarily speed up or slow down the breathing rate or greatly change the pattern of respiration during activities. This ability permits us to change respi-

ratory patterns and even to hold our breath for short periods to accommodate activities such as underwater swimming, speaking, or eating. This voluntary control of respiration, however, has limits. As indicated in a later section, other factors such as blood carbon dioxide levels are much more powerful in controlling respiration than conscious control. Regardless of cerebral intent to the contrary, we resume breathing when our bodies sense the need for more oxygen or if carbon dioxide levels increase to certain levels.

RECEPTORS INFLUENCING RESPIRATION

Chemoreceptors

Chemoreceptors (KEE-mo-ree-SEP-tors) located in the carotid and aortic bodies are specialized receptors that are sensitive to increases in blood carbon dioxide level and decreases in blood oxygen level. They also can sense and respond to increasing blood acid levels. The carotid body receptors are found at the point where the common carotid arteries divide, and the aortic bodies are small clusters of chemosensitive cells that lie adjacent to the aortic arch near the heart (Figure 15-16). When stimulated by increasing levels of blood carbon dioxide, decreasing oxygen levels, or increasing blood acidity, these receptors send nerve impulses to the respiratory regulatory centers that in turn modify respiratory rates.

Pulmonary Stretch Receptors

Specialized stretch receptors in the lungs are located throughout the pulmonary airways and in the alveoli (Figure 15-16). Nervous impulses generated by these receptors influence the normal pattern of breathing and protect the respiratory system from excess stretching caused by harmful overinflation. When the tidal volume of air has been inspired, the lungs are expanded enough to stimulate stretch receptors that then send inhibitory impulses to the inspiratory center. Relaxation of inspiratory muscles occurs, and expiration follows. After expiration, the lungs are sufficiently deflated to inhibit the stretch receptors, and inspiration is then allowed to start again.

▽ *Types of Breathing*

A number of terms are used to describe breathing patterns. Eupnea (YOOP-nee-ah), for example, refers to a normal respiratory rate. During eupnea, the need for oxygen and carbon dioxide exchange is being met, and the individual is usually not aware of the breathing pattern. The terms hyperventilation and hypoventilation describe very rapid and deep or slow and shallow respirations, respectively. Hyperventilation sometimes results from a conscious voluntary effort preceding exertion or from psychological factors—"hysterical hyperventilation." Dyspnea (DISP-nee-ah) refers to labored or difficult breathing and is often associated with hypoventilation. Dyspnea that is relieved by moving into an upright or sitting position is called orthopnea (orth-OP-nee-ah). If breathing stops completely for a brief period, regardless of cause, it is called apnea (AP-nee-ah). A series of cycles of alternating apnea and hyperventilation is called Cheyne-Stokes (chain-stokes) respiration (CSR). CSR occurs in critical diseases such as congestive heart failure, brain injuries, or brain tumors. CSR may also occur in a drug overdose. Failure to resume breathing after a period of apnea is called respiratory arrest. Examples of breathing patterns are summarized in Table 15-1.

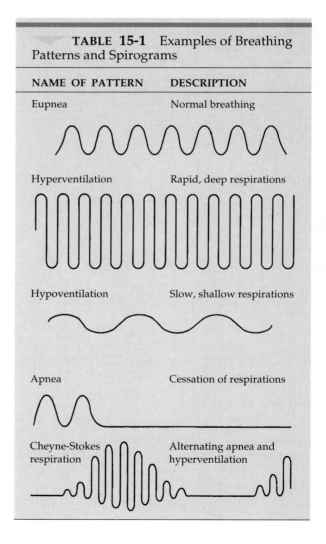

TABLE 15-1 Examples of Breathing Patterns and Spirograms

NAME OF PATTERN	DESCRIPTION
Eupnea	Normal breathing
Hyperventilation	Rapid, deep respirations
Hypoventilation	Slow, shallow respirations
Apnea	Cessation of respirations
Cheyne-Stokes respiration	Alternating apnea and hyperventilation

HEIMLICH MANEUVER

The **Heimlich maneuver** is an effective and often lifesaving technique that can be used to open a windpipe that is suddenly obstructed. The maneuver (see figures) uses air already present in the lungs to expel the object obstructing the trachea. Most accidental airway obstructions result from pieces of food aspirated (breathed in) during a meal; the condition is sometimes referred to as a *café coronary*. Other objects such as chewing gum or balloons are frequently the cause of obstructions in children.

Individuals trained in emergency procedures must be able to tell the difference between airway obstruction and other conditions such as heart attacks that produce similar symptoms. The key question they must ask the person who appears to be choking is, "Are you choking?" A person with an obstructed airway will not be able to speak, even while conscious. The Heimlich maneuver, if the victim is standing, consists of the rescuer's grasping the victim with both arms around the victim's waist just below the rib cage and above the navel. The rescuer makes a fist with one hand, grasps it with the other, and then delivers an upward thrust against the diaphragm just below the xiphoid process of the sternum. Air trapped in the lungs is compressed, forcing the object that is choking the victim out of the airway.

Technique if victim can be lifted (see A)

1 The rescuer stands behind the victim and wraps both arms around the victim's chest slightly below the rib cage and above the navel. The victim is allowed to fall forward with the head, arms, and chest over the rescuer's arms.
2 The rescuer makes a fist with one hand and grasps it with the other hand, pressing the thumb side of the fist against the victim's abdomen just below the end of the xiphoid process and above the navel.
3 The hands only are used to deliver the upward subdiaphragmatic thrusts. It is performed with sharp flexion of the elbows, in an upward rather than inward direction, and is usually repeated four times. It is very important *not* to compress the rib cage or actually press on the sternum during the Heimlich maneuver.

Technique if victim has collapsed or cannot be lifted (see B)

1 Rescuer places victim on floor face up.
2 Facing victim, rescuer straddles the victim's hips.
3 Rescuer places one hand on top of the other, with the bottom hand on the victim's abdomen slightly above the navel and below the rib cage.
4 Rescuer performs a forceful upward thrust with the heel of the bottom hand, repeating several times if necessary.

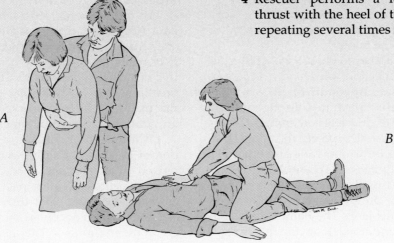

A

B

Disorders of the Lower Respiratory Tract

LOWER RESPIRATORY INFECTION

Acute bronchitis (brahn-KYE-tis) is a common condition characterized by acute inflammation of the bronchi most commonly caused by infection. Because the trachea is often also involved, the condition may be called *tracheobronchitis*. This condition is often preceded by an URI that seems to move down into the trachea and bronchi after several days. Acute bronchitis often starts with a nonproductive cough that progresses to a deep cough that produces *sputum* (SPYOO-tum) containing mucus and pus.

Pneumonia (noo-MO-nee-ah) is an acute inflammation of the lungs in which the alveoli and bronchi become plugged with thick fluid (exudate). The vast majority of pneumonia cases result from infection by *Streptococcus pneumoniae* bacteria, but can be caused by several other bacteria, viruses, and fungi (see Appendix B). Pneumonia pathogens are often opportunistic. In other words, these microorganisms may be present in the respiratory tract of otherwise healthy people, kept in check by harmless microorganisms. During times of stress (infection or other disease, malnutrition, or other weakening condition), the pneumonia pathogens may gain the upper hand and cause a full-blown infection. Pneumonia is characterized by a high fever, severe chills, headache, cough, and chest pain. Types include *lobar pneumonia* (affecting entire lobes of the lung), *bronchopneumonia* (affecting scattered portions of the bronchial tree), and *aspiration pneumonia* (noninfectious pneumonia caused by inhalation of vomit or other foreign matter).

Tuberculosis (too-ber-kyoo-LO-sis) **(TB)** is a chronic bacillus infection caused by *Mycobacterium tuberculosis* (see Appendix B). TB is a highly contagious disease transmitted through inhalation or swallowing of droplets contaminated with the TB bacillus. It usually affects the lungs and surrounding tissues but can invade any other tissue or organ. Early stages of TB are characterized by fatigue, chest pain, pleurisy, weight loss, and fever. As the disease progresses, lung hemorrhage and dyspnea may develop. The name *tuberculosis* literally means "condition of having tubercles," which describes the protective capsules the body forms around colonies of TB bacilli. Successful treatment requires a combination of drugs and other therapies for an extended period—usually longer than one year. TB is still a major cause of death in many poor, densely populated regions of the world. It has recently reemerged as an important health problem in some major U.S. cities.

RESTRICTIVE PULMONARY DISORDERS

Restrictive pulmonary disorders involve restriction (reduced stretch) of the alveoli, as the name implies. Because they inhibit inspiration, restrictive disorders reduce pulmonary volumes and capacities such as inspiratory reserve volume and vital capacity. Some restrictive disorders arise in connective tissue of the lung itself. For example, inflammation or fibrosis (scarring) of lung tissue caused by exposure to asbestos, coal, or silicon dust can restrict alveoli. Restriction of breathing can also be caused by the pain that accompanies pleurisy or mechanical injuries.

OBSTRUCTIVE PULMONARY DISORDERS

A number of different conditions may cause obstruction of the airways. For example, exposure to cigarette smoke and other common air pollutants can trigger a reflexive constriction of bronchial airways. Obstructive disorders may obstruct *inspiration* and *expiration*, whereas restrictive disorders mainly restrict *inspiration*. In obstructive disorders, the total lung capacity may be normal, or even high, but the time it takes to inhale or exhale maximally is significantly increased.

The term **chronic obstructive pulmonary disease (COPD)** is often used to describe the progressive, irreversible obstruction of air flow that may result from a variety of preexisting obstructive disorders. People with COPD have difficulty breathing, especially during expiration. Some of the major disorders observed in COPD victims are summarized here and in Figure 15-17.

1 Chronic bronchitis is a chronic inflammation of the bronchi and bronchioles. It is characterized by edema and excessive mucus production, which block air passages. People with chronic bronchitis have difficulty with exhaling and often cough deeply as they try to dislodge the accumulating mucus. The major cause of chronic

FIGURE 15-17 Disorders Commonly Observed in Chronic Obstructive Pulmonary Disease (COPD).

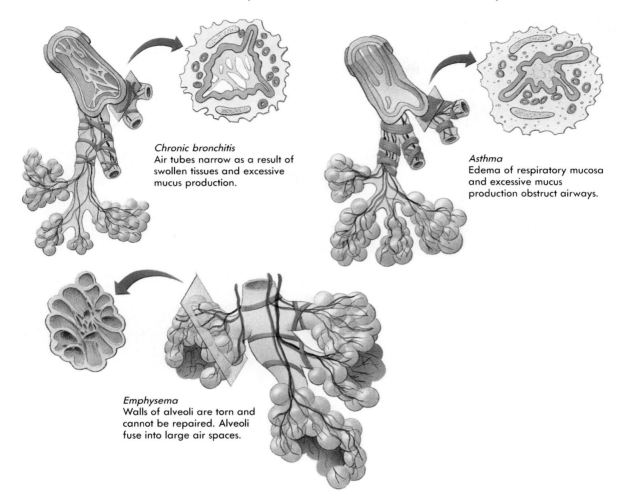

Chronic bronchitis
Air tubes narrow as a result of swollen tissues and excessive mucus production.

Asthma
Edema of respiratory mucosa and excessive mucus production obstruct airways.

Emphysema
Walls of alveoli are torn and cannot be repaired. Alveoli fuse into large air spaces.

bronchitis is cigarette smoking or exposure to cigarette smoke. Exposure to other air pollutants may also cause chronic bronchitis.

2 Emphysema (em-fi-SEE-ma) may result from the progression of chronic bronchitis or other conditions as air becomes trapped within alveoli and causes them to enlarge. As the alveoli enlarge, their walls rupture and then fuse into large irregular spaces (Figure 15-18). The rupture of alveoli reduces the total surface area of the lung, making breathing difficult. Emphysema vic-tims often develop *hypoxia* or oxygen deficiency in the internal environment.

3 Asthma (AZ-mah) is an obstructive disorder characterized by recurring spasms of the smooth muscle in the walls of the bronchial air passages. The muscle contractions narrow airways, making breathing difficult. Inflammation (edema and excessive mucus production) usually accompa-nies the spasms, further obstructing the airways. Asthma can be triggered by stress, heavy exercise, infection, or inhaling allergens or other irritants.

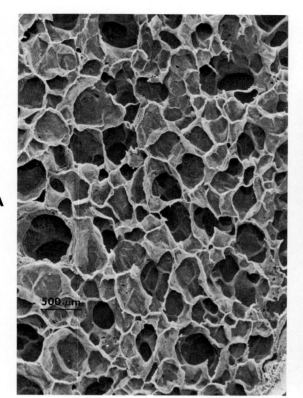

A

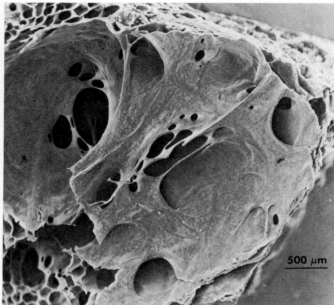

B

FIGURE 15-18 Emphysema. The effects of emphysema can be seen in these scanning electron micrographs of lung tissue. **A,** Normal lung with many small alveoli. **B,** Lung tissue affected by emphysema. Notice that the alveoli have merged into large air spaces, reducing the surface area for gas exchange.

LUNG CANCER

Lung cancer is a malignancy of pulmonary tissue that not only destroys the vital gas-exchange tissues of the lungs but like other cancers may also invade other parts of the body (metastasis). Lung cancer most often develops in damaged or diseased lungs. The most common predisposing condition associated with lung cancer is cigarette smoking (accounting for about 75% of lung cancer cases). Other factors thought to cause lung cancer include exposure to "second-hand" cigarette smoke, asbestos, chromium, coal products, petroleum products, rust, and ionizing radiation (as in radon gas).

Lung cancer may be arrested if detected early in routine chest x-ray films or other diagnostic procedures. Depending on the exact type of malignancy involved and the extent of lung involvement, a number of strategies are available for treatment. Chemotherapy can cause a cure or remission in selected cases, as can radiation therapies. Surgery is the most effective treatment known, but only half of the persons diagnosed as having lung cancer are good candidates for surgery because the damage is too extensive. In a **lobectomy** (lo-BEK-toe-me) only the affected lobe of a lung is removed. **Pneumonectomy** (noo-mo-NEK-toe-me) is the surgical removal of an entire lung.

OUTLINE SUMMARY

STRUCTURAL PLAN (Figure 15-1)

Basic plan of respiratory system would be similar to an inverted tree if it were hollow; leaves of the tree would be comparable to alveoli, with the microscopic sacs enclosed by networks of capillaries

RESPIRATORY TRACTS

A Upper respiratory tract—nose, pharynx, and larynx
B Lower respiratory tract—trachea, bronchial tree, and lungs

RESPIRATORY MUCOSA (Figure 15-3)

A Specialized membrane that lines the air distribution tubes in the respiratory tree
B Over 125 ml of mucus produced each day forms a "mucus blanket" over much of the respiratory mucosa
C Mucus serves as an air purification mechanism by trapping inspired irritants such as dust and pollen
D Cilia on mucosal cells beat in only one direction, moving mucus upward to pharynx for removal

NOSE

A Structure
  1 Nasal septum separates interior of nose into two cavities
  2 Mucous membrane lines nose
  3 Frontal, maxillary, sphenoidal, and ethmoidal sinuses drain into nose (Figure 15-4)
B Functions
  1 Warms and moistens inhaled air
  2 Contains sense organs of smell

PHARYNX

A Structure (Figure 15-5)
  1 Pharynx (throat) about 12.5 cm (5 inches) long
  2 Divided into nasopharynx, oropharynx, and laryngopharynx
  3 Two nasal cavities, mouth, esophagus, larynx, and auditory tubes all have openings into pharynx
  4 Pharyngeal tonsils and openings of auditory tubes open into nasopharynx; tonsils found in oropharynx
  5 Mucous membrane lines pharynx
B Functions
  1 Passageway for food and liquids
  2 Air distribution; passageway for air

LARYNX

A Structure (Figure 15-6)
  1 Several pieces of cartilage form framework
    a Thyroid cartilage (Adam's apple) is largest
    b Epiglottis partially covers opening into larynx
  2 Mucous lining
  3 Vocal cords stretch across interior of larynx
B Functions
  1 Air distribution; passageway for air to move to and from lungs
  2 Voice production

DISORDERS OF THE UPPER RESPIRATORY TRACT

A Upper respiratory infection (URI)
  1 Rhinitis (coryza)—nasal inflammation, as in a cold, influenza, or allergy
  2 Pharyngitis (sore throat)—inflammation or infection of the pharynx
  3 Laryngitis—inflammation of the larynx resulting from infection or irritation
B Anatomical disorders
  1 Deviated septum—septum that is abnormally far from the midsagittal plane (congenital or acquired)
  2 Epistaxis (bloody nose) can result from mechanical injuries to the nose, hypertension, or other factors

TRACHEA (WINDPIPE)

A Structure (Figure 15-7)
  1 Tube about 11 cm ($4\frac{1}{2}$ inches) long that extends from larynx into the thoracic cavity
  2 Mucous lining
  3 C-shaped rings of cartilage hold trachea open
B Function—passageway for air to move to and from lungs
C Obstruction
  1 Blockage of trachea occludes the airway and if complete causes death in minutes
  2 Tracheal obstruction causes over 4000 deaths annually in the United States
  3 Heimlich maneuver (p. 393) is a lifesaving technique used to free the trachea of obstructions
  4 Tracheostomy—surgical procedure in which a tube is inserted into an incision in the trachea so that a person with a blocked airway can breathe

BRONCHI, BRONCHIOLES, AND ALVEOLI

A Structure
  1 Trachea branches into right and left bronchi
  2 Each bronchus branches into smaller and smaller tubes eventually leading to bronchioles
  3 Bronchioles end in clusters of microscopic alveolar sacs, the walls of which are made up of alveoli (Figure 15-8)
B Function
  1 Bronchi and bronchioles—air distribution; passageway for air to move to and from alveoli
  2 Alveoli—exchange of gases between air and blood
C Respiratory distress—relative inability to inflate the alveoli
  1 Infant respiratory distress syndrome (IRDS)—leading cause of death in premature infants resulting from lack of surfactant production in alveoli
  2 Adult respiratory distress syndrome (ARDS)—impairment of surfactant by inhalation of foreign substances or other conditions

LUNGS AND PLEURA

A Structure (Figure 15-9)
  1 Size—large enough to fill the chest cavity, except for middle space occupied by heart and large blood vessels
  2 Apex—narrow upper part of each lung, under collarbone
  3 Base—broad lower part of each lung; rests on diaphragm
  4 Pleura—moist, smooth, slippery membrane that lines chest cavity and covers outer surface of lungs; reduces friction between the lungs and chest wall during breathing (Figure 15-11)
B Function—breathing (pulmonary ventilation)
C Pleurisy—inflammation of the pleura
D Atelectasis—collapse of the lung (alveoli); can be caused by:
  1 Pneumothorax—presence of air in the pleural space
  2 Hemothorax—presence of blood in the pleural space

RESPIRATION

A Mechanics of breathing (Figure 15-13)
  1 Pulmonary ventilation includes two phases called *inspiration* (movement of air into lungs) and *expiration* (movement of air out of lungs)
  2 Changes in size and shape of thorax cause changes in air pressure within that cavity and in the lungs
  3 Air pressure differences actually cause air to move into and out of the lungs

B Inspiration
  1 Active process—air moves into lungs
  2 Inspiratory muscles include diaphragm and external intercostals
    a Diaphragm flattens during inspiration—increases top to bottom length of thorax
    b External intercostals—contraction elevates the ribs and increases the size of the thorax from the front to the back and from side to side
  3 The increase in the size of the chest cavity reduces pressure within it, and air enters the lungs
C Expiration
  1 Quiet expiration is ordinarily a passive process
  2 During expiration, thorax returns to its resting size and shape
  3 Elastic recoil of lung tissues aids in expiration
  4 Expiratory muscles used in forceful expiration are internal intercostals and abdominal muscles
    a Internal intercostals—contraction depresses the rib cage and decreases the size of the thorax from the front to back
    b Contraction of abdominal muscles elevates the diaphragm, thus decreasing size of the thoracic cavity from the top to bottom
  5 Reduction in the size of the thoracic cavity increases its pressure and air leaves the lungs
D Exchange of gases in lungs (Figure 15-14)
  1 Carbaminohemoglobin breaks down into carbon dioxide and hemoglobin
  2 Carbon dioxide moves out of lung capillary blood into alveolar air and out of body in expired air
  3 Oxygen moves from alveoli into lung capillaries
  4 Hemoglobin combines with oxygen, producing oxyhemoglobin
E Exchange of gases in tissues
  1 Oxyhemoglobin breaks down into oxygen and hemoglobin
  2 Oxygen moves out of tissue capillary blood into tissue cells
  3 Carbon dioxide moves from tissue cells into tissue capillary blood
  4 Hemoglobin combines with carbon dioxide, forming carbaminohemoglobin
F Volumes of air exchanged in pulmonary ventilation (Figure 15-15)
  1 Volumes of air exchanged in breathing can be measured with a spirometer
  2 Tidal volume (TV) amount normally breathed in or out with each breath
  3 Vital capacity (VC)—largest amount of air that one can breathe out in one expiration
  4 Expiratory reserve volume (ERV)—amount of air that can be forcibly exhaled after expiring the tidal volume

5 Inspiratory reserve volume (IRV)—amount of air that can be forcibly inhaled after a normal inspiration

6 Residual volume (RV)—air that remains in the lungs after the most forceful expiration

7 Rate—usually about 12 to 18 breaths a minute; much faster during exercise

G Regulation of respiration (Figure 15-16)

 1 Regulation of respiration permits the body to adjust to varying demands for oxygen supply and carbon dioxide removal

 2 Most important central regulatory centers in medulla are called *respiratory control centers* (inspiratory and expiratory centers)

  a Under resting conditions nervous activity in the respiratory control centers produces a normal rate and depth of respirations (12 to 18 per minute)

 3 Respiratory control centers in the medulla are influenced by "inputs" from receptors located in other body areas:

  a Cerebral cortex—voluntary (but limited) control of respiratory activity

  b Chemoreceptors—respond to changes in carbon dioxide, oxygen, and blood acid levels—located in carotid and aortic bodies

  c Pulmonary stretch receptors—respond to the stretch in lungs, thus protecting respiratory organs from overinflation

TYPES OF BREATHING

A Eupnea—normal breathing

B Hyperventilation—rapid and deep respirations

C Hypoventilation—slow and shallow respirations

D Dyspnea—labored or difficult respirations

E Orthopnea—dyspnea relieved by moving into an upright or sitting position

F Apnea—stopped respiration

G Cheyne-Stokes respiration (CSR)—cycles of alternating apnea and hyperventilation associated with critical conditions

H Respiratory arrest—failure to resume breathing after a period of apnea

DISORDERS OF THE LOWER RESPIRATORY TRACT

A Lower respiratory infection

 1 Acute bronchitis or tracheobronchitis—inflammation of the bronchi or bronchi and trachea caused by infection (usually resulting from the spread of a URI)

 2 Pneumonia—acute inflammation (infection) in which lung airways become blocked with thick exudate

 3 Tuberculosis (TB)—chronic, highly contagious lung infection characterized by tubercles in the lung; can progress to involve tissues outside the lungs and plurae

B Restrictive pulmonary disorders reduce the ability of lung tissues to stretch (as during inspiration)

 1 Factors inside the lungs, such as fibrosis (scarring) or inflammation, may restrict breathing

 2 Factors outside the lungs, such as pain of injury or pleurisy, may restrict breathing

C Obstructive pulmonary disorders

 1 Obstruct breathing

 2 Chronic obstructive pulmonary disease (COPD) can develop from preexisting obstructive conditions

  a Chronic bronchitis—chronic inflammation of the bronchial tree

  b Emphysema—reduced surface area of lungs due to rupture or other damage to alveoli

  c Asthma—recurring spasms of the airways accompanied by edema and mucus production

D Lung cancer—malignant tumor of the lungs, occasionally treatable with surgery, chemotherapy, or radiation

New Words

alveoli	expiratory reserve volume (ERV)	paranasal sinuses	spirometer
aortic body		pharynx	surfactant
bronchi	inspiratory reserve volume (IRV)	pulmonary ventilation	tidal volume (TV)
carbaminohemoglobin		residual volume (RV)	trachea
carotid body	larynx	respiration	vital capacity (VC)
conchae	oxyhemoglobin	respiratory membrane	

Diseases and Other Clinical Terms

apnea	deviated septum	hypoventilation	pneumothorax
asthma	dyspnea	laryngitis	respiratory arrest
atelectasis	emphysema	lobectomy	respiratory distress syndrome
bronchitis	epistaxis	orthopnea	
Cheyne-Stokes respiration (CSR)	eupnea	pharyngitis	rhinitis
	Heimlich maneuver	pleurisy	sinusitis
chronic obstructive pulmonary disease (COPD)	hemothorax	pneumonectomy	tonsillectomy
	hyperventilation	pneumonia	tuberculosis

CHAPTER TEST

1. The exchange of gases in the respiratory system depends on the passive transport process of _____.
2. Surface tension in the alveoli is reduced by a substance called _____.
3. The trachea is located in the _____ respiratory tract.
4. The portion of the pharynx located behind the mouth is called the _____.
5. The adenoids or pharyngeal tonsils are located in the _____.
6. The voice box is also known as the _____.
7. During swallowing, food is prevented from entering the trachea by the _____.
8. The presence of air in the pleural space is called _____.
9. The exchange of gases between air in the lungs and the blood is called _____ respiration.
10. Most oxygen entering the blood combines with hemoglobin in the RBCs to form _____.
11. The largest amount of air we can breathe out in one expiration is known as the _____ _____.
12. The most important respiratory control centers are located in the _____.
13. Gas exchange in the lung takes places across the thin moist membranes of the _____, which are surrounded by a dense network of capillaries.
14. Pneumothorax or hemothorax may result in the collapse of a lung, a condition called _____.
15. Lack of surfactant in a premature infant results in a condition called _____ _____ _____ _____.
16. A URI localized in the nasal cavity is called *coryza* or _____.
17. An acute infection of the lungs in which alveoli and bronchioles become plugged with thick exudate is called _____.
18. An obstructive pulmonary disorder characterized by recurring episodes of muscle spasms in the bronchi is called _____.

Select the most correct answer from Column B for each statement in Column A. (Only one answer is correct.)

COLUMN A

19. _____ Maximum expiration
20. _____ "Adam's apple"
21. _____ Lines thoracic cavity
22. _____ Pulmonary ventilation
23. _____ Labored breathing
24. _____ Carotid body
25. _____ Major muscle of inspiration

COLUMN B

a. Diaphragm
b. Thyroid cartilage
c. Chemoreceptor
d. Breathing
e. Vital capacity
f. Dyspnea
g. Parietal pleura

REVIEW QUESTIONS

1 Discuss the location, microscopic structure, and functions of the respiratory mucosa.
2 List the paranasal air sinuses.
3 Discuss the functions of the nose in respiration.
4 What and where are the pharynx and larynx?
5 What structures open into the pharynx?
6 What are the anatomical subdivisions of the pharynx?
7 Where are the tonsils and adenoids located?
8 Why does sinusitis or middle ear infection occur so frequently after a common cold?
9 What function do the C-shaped rings of cartilage serve in the trachea?
10 What and where is the "Adam's apple"?
11 Define parietal pleura, pleural space, pleurisy, atelectaxis, hemothorax, pneumothorax, and visceral pleura.
12 How does the lack of adequate surfactant cause respiratory distress? What causes a lack of surfactant?
13 Do breathing and respiration mean the same thing? Define each term.

14 Briefly explain how oxygen and carbon dioxide can move between alveolar air, blood, and tissue cells.
15 Explain the following equation:

$$VC = TV + IRV + ERV$$

16 What does *residual volume* mean?
17 How can a brain hemorrhage affect the respiratory system?
18 Discuss the anatomy of the "respiratory membrane."
19 Compare the structure, location, and functions of the respiratory membrane and the respiratory mucosa.
20 Discuss the Heimlich maneuver.
21 Discuss the respiratory control mechanisms.
22 What is pneumonia? Does it only occur as a result of infection or can it occur under other circumstances?
23 What is meant by the term COPD? How does bronchitis obstruct breathing? Asthma?
24 Name some respiratory disorders caused by cigarette smoking.

CLINICAL APPLICATIONS

1 Curtis was having fun alongside a neighborhood swimming pool when he was accidentally pushed into the pool. Although he is a good swimmer, the suddenness of the fall caught him off-guard and he inhaled some water before he was able to gain control of the situation. Luckily, a nearby swimmer assisted Curtis to the edge of the pool, but Curtis continued to have great difficulty in breathing. Can you name the syndrome that Curtis must be exhibiting? Explain what has happened to Curtis's lungs to cause his breathing difficulty.

2 Walter has *aspergillosis* in his lungs. This disease has caused a partial blockage of both of his bronchi. Does Walter have a restrictive or obstructive condition? What signs would you look for to confirm your diagnosis if Walter uses a spirometer to test his breathing? What type of pathogenic organism caused Walter's problem? (Hint: see Appendix B.)

3 While chatting with your friend at an expensive restaurant, she suddenly stops in midswallow and looks panicked. When you ask what is wrong, she indicates that she can't speak and runs toward the restroom. What should you do? Assume that your first aid does not work—what procedure might emergency medical personnel use to help your friend?

The Digestive System

Outline

Objectives

*After you have completed this chapter,
you should be able to:*

1 List in sequence each of the component parts or segments of the alimentary canal from the mouth to the anus and identify the accessory organs of digestion.

2 List and describe the four layers of the wall of the alimentary canal. Compare the lining layer in the esophagus, stomach, small intestine, and large intestine.

3 List and describe the major disorders of the digestive organs.

4 Discuss the basics of protein, fat, and carbohydrate digestion and give the end products of each process.

5 Define and contrast mechanical and chemical digestion.

6 Define: *peristalsis, bolus, chyme, jaundice, ulcer,* and *diarrhea.*

Boxed Essays

Mumps
Exercise and Fluid Uptake
Oral Rehydration Therapy

The principal structure of the **digestive system** is an irregular tube, open at both ends, called the **alimentary** (al-i-MEN-tar-ee) **canal** or the **gastrointestinal** (gas-tro-in-TES-ti-nal) **(GI) tract.** In the adult, this hollow tube is about 9 m (29 feet) long. Although this may seem strange, food or other material that enters the digestive tube is not really inside the body. Most parents of young children quickly learn that a button or pebble swallowed by their child will almost always pass unchanged and with little difficulty through the tract. Think of the tube as a passageway that extends through the body like a hallway through a building. Food must be broken down or **digested** and then absorbed through the walls of the digestive tube before it can actually enter the body and be used by cells. The breakdown or digestion of food material is both mechanical and chemical in nature. The teeth are used to physically break down food material before it is swallowed. The churning of food in the stomach then continues the mechanical breakdown process. Chemical breakdown results from the action of digestive enzymes and other chemicals acting on food as it passes through the GI tract. In chemical digestion, large food molecules are reduced to smaller molecules that can be absorbed through the lining of the intestinal wall and then distributed to body cells for use. This process of altering the chemical and physical composition of food so that it can be absorbed and used by body cells is known as *digestion,* and it is the function of the digestive system. Part of the digestive system, the large intestine, serves also as an organ of elimination, ridding the body of the waste material or **feces** resulting from the digestive process. Table 16-1 names both main and accessory digestive organs. Note that the accessory organs include the teeth, tongue, gallbladder, and appendix, as well as a number of glands that secrete their products into the digestive tube.

Foods undergo three kinds of processing in the body: **digestion, absorption,** and **metabolism.** Digestion and absorption are performed by the organs of the digestive system (Figure 16-1). Metabolism is performed by all body cells. In this chapter, we shall discuss the normal structure and function of the digestive system organs, associated disease states, and the processes of digestion and absorption. In Chapter 17, we will discuss the metabolism of food after it has been absorbed.

Wall of the Digestive Tract

The digestive tract has been described as a tube that extends from the mouth to the anus. The wall of this digestive tube is fashioned of four layers of tissue (Figure 16-2). The inside or hollow space within the tube is called the **lumen.** The four layers, named from the inside coat to the outside of the tube, follow:

1 Mucosa or mucous membrane
2 Submucosa
3 Muscularis
4 Serosa

Although the same four tissue coats form the organs of the alimentary tract, their structures vary in different organs. The **mucosa** of the esophagus, for example, is composed of tough and stratified abrasion-resistant epithelium. The mucosa of the remainder of the tract is a delicate layer of simple columnar epithelium designed for absorption and secretion. The mucus produced by either type of epithelium coats the lining of the alimentary canal.

The **submucosa,** as the name implies, is a connective tissue layer that lies just below the mucosa. It contains many blood vessels and nerves. The two layers of muscle tissue called the **muscularis** have an important function in the digestive process. By a wavelike, rhythmic contraction of the muscular coat, called **peristalsis** (pair-i-STAL-sis), food material is moved through the digestive tube. In addition, the contraction of the muscularis also assists in the mixing of food with digestive juice and in the further mechanical breakdown of larger food particles.

The **serosa** is the outermost covering or coat of the digestive tube. In the abdominal cavity it is composed of the parietal peritoneum. The loops of the digestive tract are anchored to the posterior wall of the abdominal cavity by a large double fold of peritoneal tissue called the **mesentery** (MES-en-tair-ee).

Mouth

The **mouth** or **oral cavity** is a hollow chamber with a roof, a floor, and walls. Food enters or is ingested into the digestive tract through the mouth, and the process of digestion begins imme-

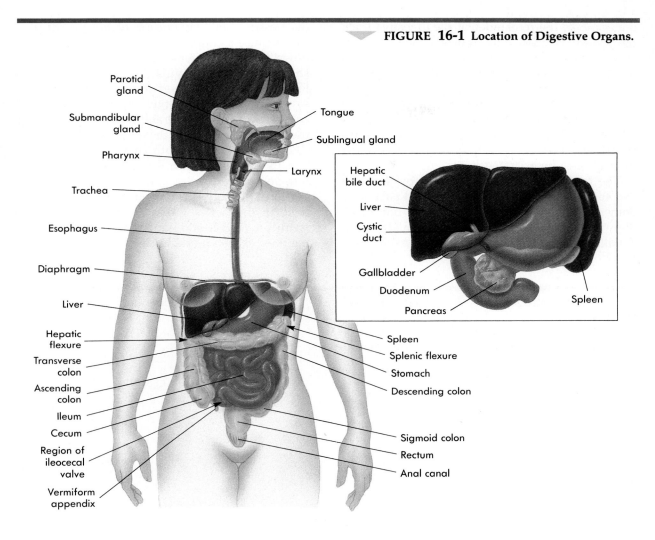

FIGURE 16-1 Location of Digestive Organs.

MAIN ORGAN	ACCESSORY ORGAN	MAIN ORGAN	ACCESSORY ORGAN
Mouth	Teeth and tongue	Large intestine	Vermiform appendix
Pharynx (throat)	Salivary glands	Cecum	
Esophagus (foodpipe)	Parotid	Colon	
	Submandibular	Ascending colon	
Stomach	Sublingual	Transverse colon	
Small intestine		Descending colon	
Duodenum	Liver	Sigmoid colon	
Jejunum	Gallbladder	Rectum	
Ileum	Pancreas	Anal canal	

TABLE 16-1 Organs of the Digestive System

diately. Like the remainder of the digestive tract, the mouth is lined with mucous membrane. It may be helpful if you review the structure and function of mucous membranes in Chapter 5. Typically, mucous membranes line hollow organs, such as the digestive tube, that open to the exterior of the body. Mucus produced by the lining of the GI tract protects the epithelium from digestive juices and lubricates food passing through the lumen.

The roof of the mouth is formed by the **hard** and **soft palates** (Figure 16-3). The hard palate is a bony structure in the anterior or front portion of the mouth, formed by parts of the palatine and maxillary bones. The soft palate is located above the posterior or rear portion of the mouth. It is soft because it consists chiefly of muscle. Hanging down from the center of the soft palate is a cone-shaped process, the **uvula** (YOO-vyoo-lah). If you look in the mirror, open your mouth wide, and say "Ah," you can see the uvula. The uvula and soft palate prevent food and liquid from entering the nasal cavities above the mouth.

The floor of the mouth consists of the tongue and its muscles. The tongue is made of skeletal muscle covered with mucous membrane. It is anchored to bones in the skull and to the hyoid bone in the neck. A thin membrane called the **frenulum** (FREN-yoo-lum) attaches the tongue to the floor of the mouth. Occasionally the frenulum

FIGURE 16-2 Section of the Small Intestine. The four layers typical of walls of the gastrointestinal tract are shown. Circular folds of mucous membrane called *plicae* increase the surface area of the lining coat.

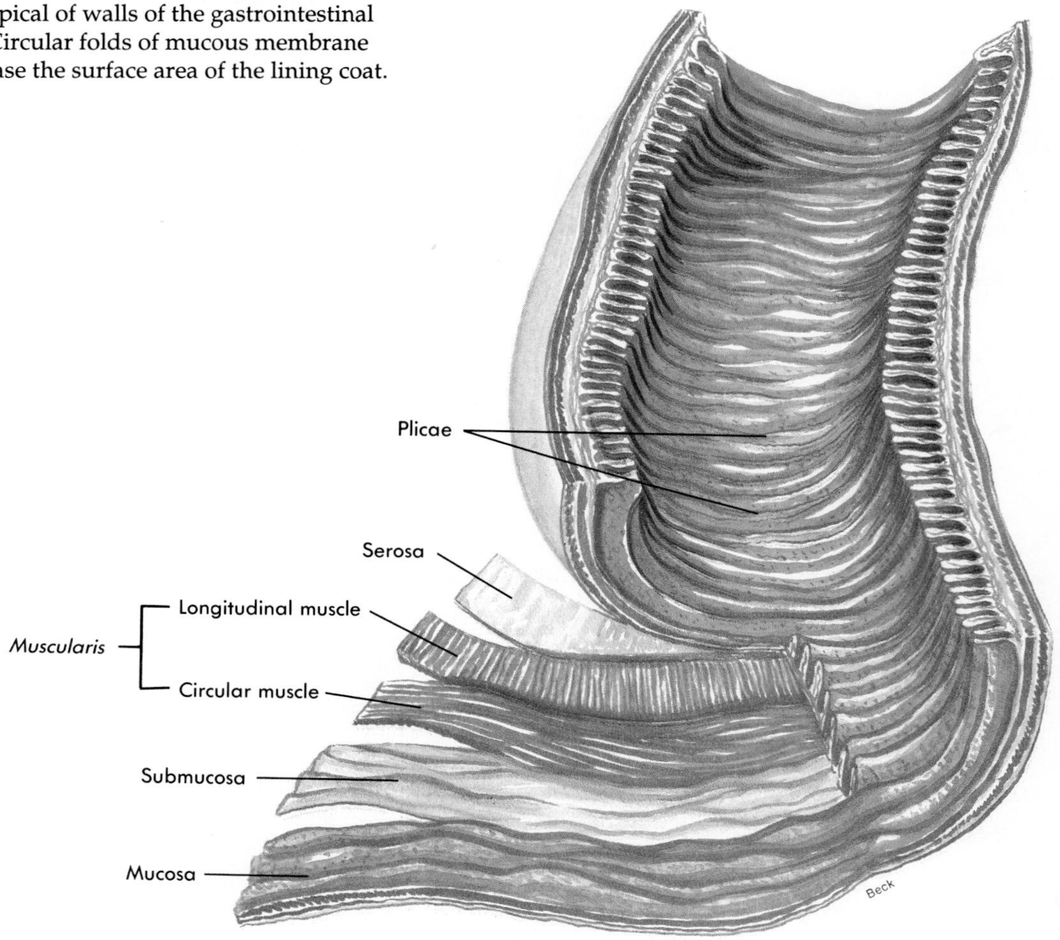

Plicae

Serosa

Muscularis
- Longitudinal muscle
- Circular muscle

Submucosa

Mucosa

Beck

is too short to allow free movements of the tongue. Individuals with this condition cannot enunciate words normally and are said to be "tongue-tied." Note in Figure 16-4 that the tongue can be divided into a blunt rear portion called the *root*, a pointed *tip*, and a central *body*.

Have you ever noticed the many small elevations on the surface of your tongue? They are **papillae**. The largest are the **vallate** type, which form an inverted V-shaped row of about 10 to 12 mushroomlike elevations. The taste buds, which contain sensory receptors for salty, sour, sweet, and bitter compounds, are located on the papillae (see Chapter 8).

Teeth

TYPES OF TEETH

The shape and placement of the teeth assist in their functions. The types of teeth follow:

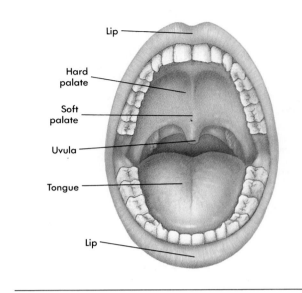

FIGURE 16-3 The Mouth Cavity.

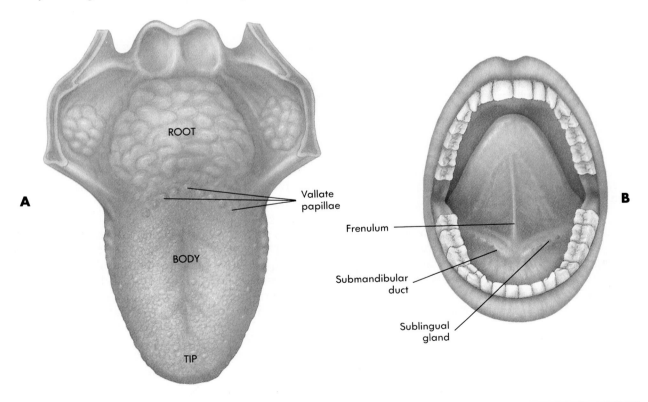

FIGURE 16-4 The Tongue. A, Surface. **B,** Mouth cavity showing the undersurface of the tongue.

FIGURE 16-5 The Deciduous (Baby) Teeth and Adult Teeth. In the deciduous set, there are no premolars and only two pairs of molars in each jaw. Generally the lower teeth erupt before the corresponding upper teeth.

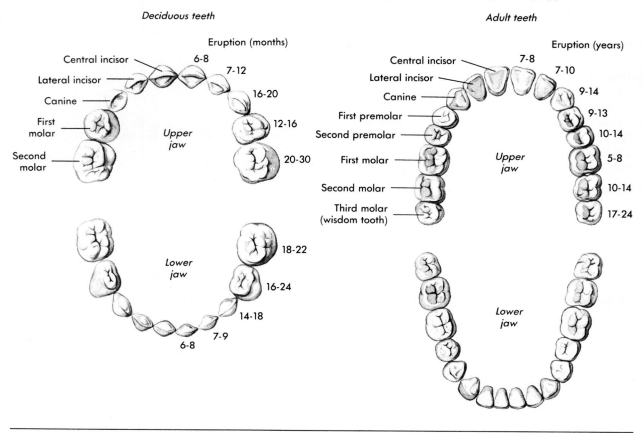

Deciduous teeth

Eruption (months)

Central incisor — 6-8
Lateral incisor — 7-12
Canine — 16-20
First molar —
Second molar — 12-16

Upper jaw

20-30

Lower jaw

18-22
16-24
14-18
7-9
6-8

Adult teeth

Eruption (years)

Central incisor — 7-8
Lateral incisor — 7-10
Canine — 9-14
First premolar — 9-13
Second premolar — 10-14
First molar — 5-8
Second molar — 10-14
Third molar (wisdom tooth) — 17-24

Upper jaw

Lower jaw

1 Incisors
2 Canines
3 Premolars
4 Molars

Note in Figure 16-5 that the incisors have a sharp cutting edge. They have a cutting function during **mastication** (mas-ti-KAY-shun) or chewing of food. The canine teeth are sometimes called **cuspids.** They pierce or tear the food being eaten. This tooth type is particularly apparent in meat-eating mammals such as dogs. Premolars or **bicuspids** and molars or **tricuspids** have large, flat surfaces with two or three grinding or crushing "cusps" on their surface. They provide breakdown of food in the mouth. After food has been chewed,

it is formed into a small, rounded mass called a **bolus** (BO-lus) so that it can be swallowed.

By the time a baby is 2 years old, the child probably has his full set of 20 baby teeth. When a young adult is somewhere between 17 and 24 years old, a full set of 32 permanent teeth is generally present. The average age for cutting the first tooth is about 6 months, and the average age for losing the first baby tooth and starting to cut the permanent teeth is about 6 years. Figure 16-5 gives the names of the teeth and shows which ones are lacking in the deciduous or baby set.

TYPICAL TOOTH

A typical tooth can be divided into three main parts: crown, neck, and root. The **crown** is the

FIGURE 16-6 Longitudinal Section of a Tooth. A molar is sectioned to show its bony socket and details of its three main parts: crown, neck, and root. Enamel (over the crown) and cementum (over the neck and root) surround the dentin layer. The pulp contains nerves and blood vessels.

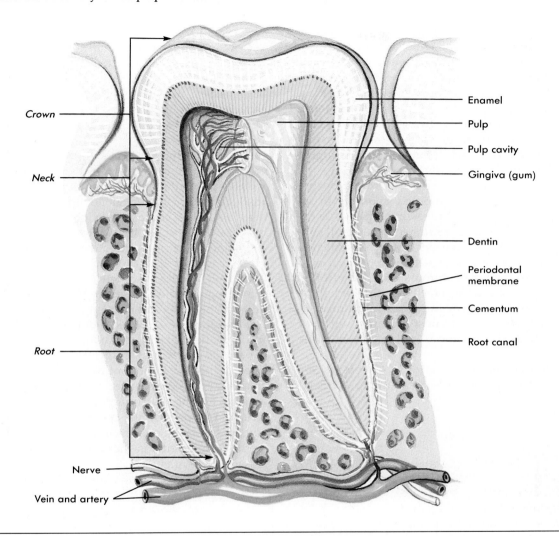

portion that is exposed and visible in the mouth. It is covered by enamel—the hardest tissue in the body. Enamel is ideally suited to withstand the grinding that occurs during the chewing of hard and brittle foods. In addition to enamel, the outer shell of each tooth is covered by dentin and cementum (Figure 16-6). Dentin makes up the greatest portion of the tooth shell. It is covered by enamel in the crown and by cementum in the neck and root areas. The center of the tooth contains a pulp cavity consisting of connective tissue, blood and lymphatic vessels, and sensory nerves.

The **neck** of a tooth is the narrow portion, shown in Figure 16-6, surrounded by the pink gingiva or gum tissue. It joins the crown of the tooth to the root. The **root** fits into the socket of the upper or lower jaw. A fibrous **periodontal membrane** lines each tooth socket.

Disorders of the Mouth and Teeth

Infections, cancer, and other disorders of the mouth and teeth can result in a variety of serious complications. Such conditions may cause pain or damage to the mouth and teeth that makes chewing and swallowing difficult—perhaps causing a person to reduce the intake of food. Reduced food intake may result in life-threatening malnutrition. Even if malnutrition does not occur, mouth infections or cancer can easily travel to nearby tissues: the nasal cavity (then on to the sinuses, middle ear, and brain) or pharynx (and on to the esophagus, larynx, and thoracic organs).

Cancer of the mouth may result from exposure to carcinogens found in tobacco smoke or in so-called smokeless tobacco (chewing tobacco), especially in combination with heavy alcohol consumption. Smokers often develop white patches, or **leukoplakia** (loo-ko-PLAK-ee-ah), which may develop into malignant tumors. Lip cancer often results from the carcinogenic effects of sunlight—which can be avoided by the use of lip balms containing sunscreen. The most common form of mouth cancer is squamous cell carcinoma (see Chapter 4).

Tooth decay or dental **caries** (KAIR-ees) is one of the most common diseases in the civilized world. It is a disease of the enamel, dentin, and cementum of teeth that results in the formation of a permanent defect called a *cavity*. Most people living in the United States, Canada, and Europe are significantly affected by the disease. Decay occurs on tooth surfaces where food debris, acid-secreting bacteria, and plaque accumulate.

If the disease is untreated, tooth decay results in infection, loss of teeth, and inflammation of the soft tissues in the mouth. Bacteria may also invade the paranasal sinuses or extend to the surface of the face and neck, causing serious complications.

Gingivitis (jin-ji-VYE-tis) is the general term for inflammation or infection of the gums. Most cases of gingivitis result from poor oral hygiene—inadequate brushing and no flossing. Gingivitis may also be a complication of other conditions such as diabetes mellitus, vitamin deficiency, or pregnancy. **Vincent's infection** is a highly contagious bacterial infection that causes gingivitis. Also called *Vincent's angina* or *trench mouth*, it may progress to cause severe bleeding and ulceration of the gums and mucosa of the mouth.

Periodontitis (pair-ee-o-don-TIE-tis) is the inflammation of the periodontal membrane or *periodontal ligament* that anchors the tooth to the bone of the jaw. Periodontitis is often a complication of advanced or untreated gingivitis, and may spread to the surrounding bony tissue. Destruction of periodontal membrane and bone results in loosening and eventually complete loss of teeth. Periodontitis is the leading cause of tooth loss among adults.

MUMPS

Mumps is an acute viral disease characterized by swelling of the parotid salivary glands. Most of us think of mumps as a so-called childhood disease because it most often affects children between the ages of 5 and 15 years of age. However, it can occur in adults—often producing a more severe infection. The mumps infection can affect other tissues in addition to the parotid gland, including the joints, pancreas, myocardium, and kidneys. In about 25% of infected men, mumps causes inflammation of the testes or *orchitis*. Orchitis resulting from mumps only very rarely causes enough damage to render a man sterile.

Salivary Glands

Three pairs of salivary glands—the parotids, submandibulars, and sublinguals—secrete most (about 1 L) of the saliva produced each day in the adult. The salivary glands (Figure 16-7) are typical of the accessory glands associated with the digestive system. They are located outside of the digestive tube itself and must convey their secretions by way of ducts into the tract.

The **parotid glands,** largest of the salivary glands, lie just below and in front of each ear at the angle of the jaw—an interesting anatomical position because it explains why people who have mumps (an infection of the parotid gland) often

FIGURE 16-7 Location of the Salivary Glands.

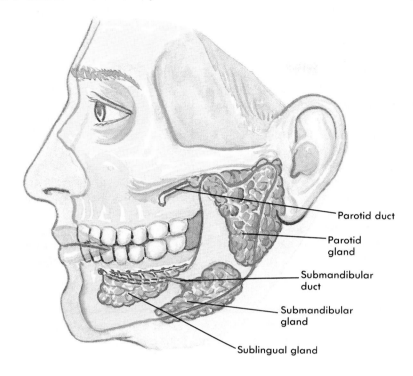

Parotid duct
Parotid gland
Submandibular duct
Submandibular gland
Sublingual gland

complain that it hurts when they open their mouths or chew; these movements squeeze the tender, inflamed gland. To see the openings of the parotid ducts, look in a mirror at the insides of your cheeks opposite the second molar tooth on either side of the upper jaw.

The ducts of the **submandibular glands** open into the mouth on either side of the lingual frenulum (Figure 16-4). The ducts of the **sublingual glands** open into the floor of the mouth.

Saliva contains mucus and a digestive enzyme that is called **salivary amylase** (AM-i-lase). Mucus moistens the food and allows it to pass with less friction through the esophagus and into the stomach. Salivary amylase begins the chemical digestion of carbohydrates.

Pharynx

The **pharynx** is a tubelike structure made of muscle and lined with mucous membrane. Observe its location in Figure 16-1. Because of its location behind the nasal cavities and mouth, it functions as part of the respiratory and digestive systems. Air must pass through the pharynx on its way to the lungs, and food must pass through it on its way to the stomach. The pharynx as a whole is subdivided into three anatomical components, as described in Chapter 15.

Esophagus

The **esophagus** (e-SOF-ah-gus) or foodpipe is the muscular, mucus-lined tube that connects the pharynx with the stomach. It is about 25 centimeters (10 inches) long. The esophagus serves as a dynamic passageway for food, pushing the food toward the stomach.

Heartburn is often described as a burning sensation characterized by pain and a feeling of fullness beneath the sternum. It is a common problem caused by irritation of esophageal mucosa by acid stomach contents that reenter the esophagus. Even very small quantities of this very acidic material can cause discomfort and even inflammation. Heartburn is a common symptom of **hiatal hernia** (hye-AY-tal HER-nee-ah). A *hernia* results from an organ being pushed through a wall that normally acts as a barrier. In hiatal hernia, the stomach pushes through the gap or hiatus in the diaphragm that allows the esophagus to pass through it. Often, the lower esophagus becomes enlarged, allowing some food to bypass a valvelike sphincter muscle and flow upward into the esophagus. Antacids, such as calcium carbonate, which are taken by mouth, relieve the symptoms of heartburn quickly by neutralizing the acid causing the discomfort. Antacids also prevent any permanent damage to the esophageal mucosa.

Stomach

The **stomach** (Figure 16-8) lies in the upper part of the abdominal cavity just under the diaphragm. It serves as a pouch that food enters after it has been chewed, swallowed, and passed through the esophagus. The stomach looks small after it is emptied, not much bigger than a large sausage, but it expands considerably after a large meal. Have you ever felt so uncomfortably full after eating that you could not take a deep breath? If so, it probably meant that your stomach was so full of food that it occupied more space than usual and pushed up against the diaphragm. This made it hard for the diaphragm to contract and move downward as much as necessary for a deep breath.

After food has entered the stomach by passing through the muscular **cardiac sphincter** (SFINGK-

FIGURE 16-8 Stomach. A portion of the anterior wall has been cut away to reveal the muscle layers of the stomach wall. Notice that the mucosa lining the stomach forms folds called *rugae*.

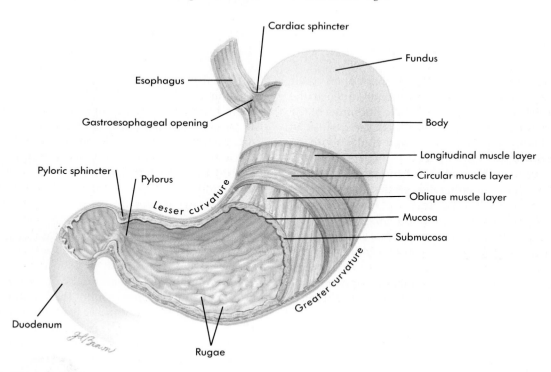

ter) at the end of the esophagus, digestion continues. Sphincters are rings of muscle tissue. The cardiac sphincter keeps food from reentering the esophagus when the stomach contracts.

Contraction of the stomach's muscular walls mixes the food thoroughly with the gastric juice and breaks it down into a semisolid mixture called **chyme** (KIME). Gastric juice contains hydrochloric acid and enzymes that function in the digestive process. Chyme formation is a continuation of the mechanical digestive process that begins in the mouth.

Note in Figure 16-8 that three layers of smooth muscle are in the stomach wall. The muscle fibers that run lengthwise, around, and obliquely make the stomach one of the strongest internal organs—well able to break up food into tiny particles and to mix them thoroughly with gastric juice to form chyme. Stomach muscle contractions result in **peristalsis,** which propels food down the digestive tract. Mucous membrane lines the stomach; it contains thousands of microscopic **gastric glands** that secrete gastric juice and hydrochloric acid into the stomach. When· the stomach is empty, its lining lies in folds called **rugae.**

The three divisions of the stomach shown in Figure 16-8 are the **fundus, body,** and **pylorus** (pie-LOR-us). The fundus is the enlarged portion to the left of and above the opening of the esophagus into the stomach. The body is the central part of the stomach, and the pylorus is its lower narrow section, which joins the first part of the small intestine. Partial digestion occurs after food is held in the stomach by the **pyloric** (pi-LOR-ik) **sphincter** muscle. The smooth muscle fibers of the sphincter stay contracted most of the time and thereby close off the opening of the pylorus into the small intestine. Note also in Figure 16-8 that the upper right border of the stomach is known as the **lesser curvature,** and the lower left border is called the **greater curvature.** After food has been in the stomach for about 3 hours, chyme passes through the pyloric sphincter into the first part of the small intestine.

DISORDERS OF THE STOMACH

Gastroenterology (gas-tro-en-ter-AHL-o-jee) is the study of the stomach (*gastro-*) and intestines

(*entero-*) and their diseases. The stomach is the potential site of numerous diseases and conditions, some of which are briefly described in this section. Many of these disorders are characterized by one or more of these signs and symptoms:

1 **gastritis** (gas-TRY-tis)—stomach inflammation
2 **anorexia** (an-or-EKS-ee-ah)—chronic loss of appetite
3 **nausea** (NAW-zee-ah)—unpleasant feeling that often leads to vomiting
4 **emesis** (EM-e-sis)—vomiting (Figure 16-9)

The pyloric sphincter is of clinical importance because **pylorospasm** (pie-LO-ro-spasm) is a fairly common condition in infants. The pyloric muscle fibers do not relax normally in this condition. As a result, food is not allowed to leave the stomach, and the infant vomits food instead of digesting and absorbing it. The condition is relieved by the administration of a drug that relaxes smooth muscles. Another abnormality of the pyloric sphincter is called **pyloric stenosis** (pie-LO-rik ste-NO-sis), an obstructive narrowing of its opening.

An *ulcer* is an open wound or sore in an area of the digestive system that is acted on by acid gastric juice. The two most common sites for ulcers are the stomach (gastric ulcers) and the upper part of the small intestine or duodenum (duodenal ulcers). Although most people think of ulcers as occurring in the stomach, most are duodenal. Ulcers cause disintegration, loss, and death of tissue as they erode the layers in the wall of the stomach or duodenum. Left untreated, ulcers cause persistent pain and may perforate the wall of the digestive tube, causing massive hemorrhage and widespread inflammation of the abdominal cavity and its contents. Usually perforation does not occur, but small, repeated hemorrhages over long periods cause anemia. There is an old saying in medicine, "No acid, no ulcer." Most experts now agree that too much gastric acid secretion (that is, prolonged hyperacidity) is one of the most important factors in ulcer formation. If the protective layer of mucus is insufficient or if there is inadequate dilution and buffering of acid gastric juices by swallowed food and the alkaline juices of the small intestine, ulcers

▽ **FIGURE 16-9 Emesis.** This illustration shows the key events of the vomiting reflex—a common sign of gastrointestinal disorders.

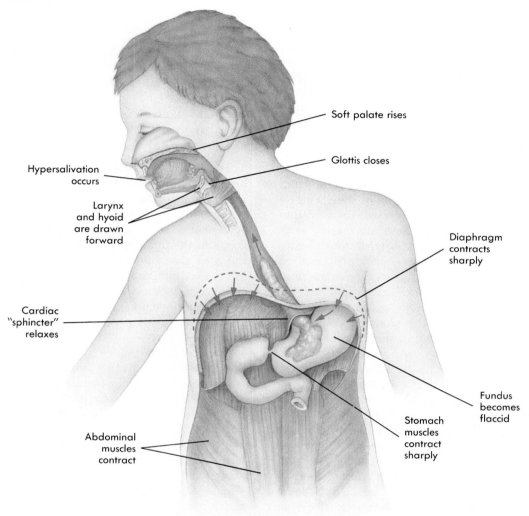

Soft palate rises

Glottis closes

Hypersalivation occurs

Larynx and hyoid are drawn forward

Diaphragm contracts sharply

Cardiac "sphincter" relaxes

Fundus becomes flaccid

Stomach muscles contract sharply

Abdominal muscles contract

may form. Hyperacidity is influenced by nervous system factors and by anxiety, other emotional states, and stress.

The drugs ranitidine (Zantac), cimetidine (Tagamet), and other medications that reduce hydrochloric acid formation in the stomach are widely prescribed in the treatment of ulcers. In addition to excess acid, a bacterium called *Campylobacter pylori*, found in many ulcer patients, may also be a cause of both ulcers and chronic indigestion. This bacterium, which was first discovered in 1982 and linked to gastrointestinal inflammation in 1987, now joins hyperacidity as a potential cause of ulcers.

Stomach cancer has been linked to excessive alcohol consumption, use of chewing tobacco, and eating smoked or heavily preserved food. Unfortunately, there is no practical way to screen the general population for stomach cancer in its earliest stages. Most stomach cancers, usually *adenocarcinomas*, have already metastasized before they are found because patients treat themselves for the early warning signs of heartburn, belching, and nausea. Later warning signs of stomach cancer include chronic indigestion, vomiting, anorexia, stomach pain, and blood in the feces. Surgical removal of the malignant tumors has been the most successful method of treating stomach cancer.

Small Intestine

The **small intestine** seems to be misnamed if you look at its length—it is roughly 7 meters (20 feet) long. However, it is noticeably smaller in diameter, than the large intestine, so in this respect its name is appropriate. Different names identify different sections of the small intestine. In the order in which food passes through them, they are the **duodenum** (doo-o-DEE-num), **jejunum,** (je-JOO-num), and **ileum** (IL-ee-um).

The mucous lining of the small intestine, like that of the stomach, contains thousands of microscopic glands. These **intestinal glands** secrete the intestinal digestive juice. Another structural feature of the lining of the small intestine makes it especially well suited to absorption of food and water; it is not perfectly smooth, as it appears to the naked eye. Instead, the intestinal lining is thrown into multiple circular folds called **plicae** (PLYE-kee) (Figure 16-10). These folds are themselves covered with thousands of tiny "fingers" called **villi** (VILL-eye). Under the microscope, the villi can be seen projecting into the hollow interior of the intestine. Inside each villus lies a rich network of blood capillaries that absorb the products of carbohydrate and protein digestion (sugars and amino acids). Millions and millions of villi jut inward from the mucous lining. Imagine the lining as perfectly smooth without any villi; think how much less surface area there would be for contact between capillaries and intestinal lining. Consider what an advantage a large contact area offers for faster absorption of food from the intestine into the blood and lymph—one more illustration that structure and function are intimately related.

Note also in Figure 16-10 that each villus in the intestine contains a lymphatic vessel or **lacteal** that absorbs lipid or fat materials from the chyme passing through the small intestine. In addition to the thousands of villi that increase surface area in the small intestine, each villus is itself covered by epithelial cells, which have a brushlike border composed of **microvilli.** The microvilli further increase the surface area of each villus for absorption of nutrients.

Most of the chemical digestion occurs in the first subdivision of the small intestine or duodenum. The duodenum is C shaped (Figure 16-11) and curves around the head of the pancreas. The acid chyme enters the duodenum from the stomach. This area is the site of frequent ulceration (duodenal ulcers.) The middle third of the duodenum contains the openings of ducts that empty pancreatic digestive juice and bile from the liver into the small intestine. As you can see in Figure 16-11, the two openings are called the **minor** and **major duodenal papillae.** Occasionally a gallstone blocks the major duodenal papilla, causing symptoms such as severe pain, jaundice, and digestive problems. Smooth muscle in the wall of the small intestine contracts to produce peristalsis, the wavelike contraction that moves food through the tract.

DISORDERS OF THE SMALL INTESTINE

Many disorders of the small intestine involve inflammation, a condition termed **enteritis** (en-ter-EYE-tis). If the stomach is also inflamed, the condition is termed **gastroenteritis** (gas-tro-en-ter-EYE-tis). Bacterial toxins or other irritants in the chyme, including stomach acid, can cause enteritis. Irritation or inflammation in the duodenum can produce a feeling of nausea that leads to emesis (vomiting). Because the duodenum may be emptied with the stomach during vomiting, it is common to observe yellowish or brownish bile in the vomit.

Malabsorption syndrome is a general term referring to a group of symptoms resulting from the failure of the small intestine to absorb nutrients properly. These symptoms include anorexia, abdominal bloating, cramps, anemia, and fatigue. A number of underlying conditions can cause malabsorption syndrome. For example, certain enzyme deficiencies can result in an absorption failure because there are no digested nutrients to absorb. Cystic fibrosis and other genetic conditions can also cause malabsorption syndrome.

FIGURE 16-10 The Small Intestine. Note that the folds of mucosa are covered with villi and that each villus is covered with epithelium, which increases the surface area for absorption of food.

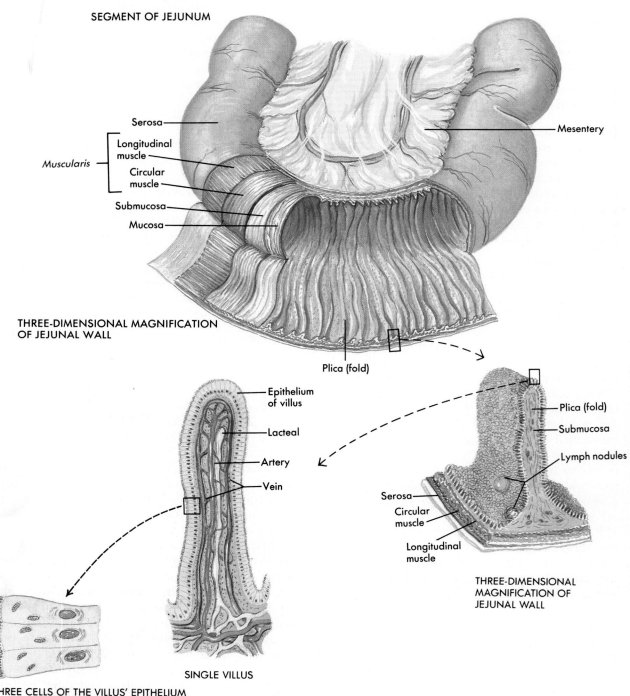

SEGMENT OF JEJUNUM

Serosa

Mesentery

Muscularis
Longitudinal muscle
Circular muscle

Submucosa

Mucosa

THREE-DIMENSIONAL MAGNIFICATION OF JEJUNAL WALL

Plica (fold)

Epithelium of villus

Lacteal

Artery

Vein

Plica (fold)

Submucosa

Lymph nodules

Serosa

Circular muscle

Longitudinal muscle

THREE-DIMENSIONAL MAGNIFICATION OF JEJUNAL WALL

SINGLE VILLUS

THREE CELLS OF THE VILLUS' EPITHELIUM SHOWING BRUSH BORDER (MICROVILLI)

▼ **FIGURE 16-11 The Gallbladder and Bile Ducts.** Obstruction of the hepatic or common bile duct by stone or spasm blocks the exit of bile from the liver, where it is formed, and prevents bile from being ejected into the duodenum.

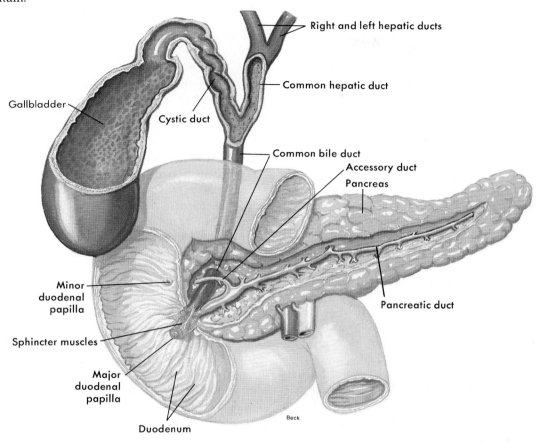

Right and left hepatic ducts

Common hepatic duct

Gallbladder

Cystic duct

Common bile duct

Accessory duct

Pancreas

Minor duodenal papilla

Pancreatic duct

Sphincter muscles

Major duodenal papilla

Duodenum

Beck

Liver and Gallbladder

The liver is so large that it fills the entire upper right section of the abdominal cavity and even extends partway into the left side. Because its cells secrete a substance called **bile** into ducts, the liver is classified as an exocrine gland; in fact, it is the largest gland in the body.

Look again at Figure 16-11. First, identify the hepatic ducts. They drain bile out of the liver, a fact suggested by the name "hepatic," which comes from the Greek word for liver (*hepar*). Next, notice the duct that drains bile into the small intestine (duodenum), the common bile duct. It is formed by the union of the common hepatic duct with the cystic duct.

Because fats form large globules, they must be broken down into smaller particles to increase the surface area for digestion. This is the function of bile. It mechanically breaks up or **emulsifies** (e-MUL-se-fye) fats. When chyme containing lipid or fat enters the duodenum, it initiates a mechanism that contracts the gallbladder and forces bile into the small intestine. Fats in chyme stimulate or "trigger" the secretion of the hormone **cholecystokinin** (ko-le-sis-toe-KYE-nin) or **CCK** from the intestinal mucosa of the duodenum. This hormone then stimulates the contraction of the gallbladder, and bile flows into the duodenum. Between meals, a lot of the bile moves up the cystic duct into the gallbladder, which is located on the undersurface of the liver. The gallbladder thus concentrates and stores bile produced in the liver.

DISORDERS OF THE LIVER AND GALLBLADDER

Gallstones are stones or **calculi** (KAL-kyoo-lee) made of crystallized bile pigments and calcium salts (Figure 16-12). **Cholelithiasis** (ko-le-li-THIGH-ah-sis) literally means "condition of having bile (gall) stones" and often occurs in the presence of gallbladder inflammation or **cholecystitis** (ko-le-sis-TIE-tis).

Visualize a gallstone blocking the common bile duct (choledocholithiasis) shown in Figure 16-11. Bile could not then drain into the duodenum. Feces would then appear gray-white because it is the pigments from bile that gives feces its characteristic color. Furthermore, excessive amounts of bile would be absorbed into the blood. A yellowish skin discoloration called **jaundice** (JAWN-dis) would result. Often, pain accompanies this condition. The pain is called *biliary colic.* Obstruction of the common hepatic duct also leads to jaundice. Because bile cannot then drain out of the liver, excessive amounts of it are absorbed. Because bile is not resorbed from the gallbladder, no jaundice occurs if only the cystic duct is blocked.

Hepatitis (hep-a-TIE-tis) is a general term referring to inflammation of the liver. Hepatitis is characterized by jaundice, liver enlargement, anorexia, abdominal discomfort, gray-white feces, and dark urine. A number of different conditions can produce hepatitis. Alcohol, drugs, or other toxins may cause hepatitis. It may also be a complication of bacterial or viral infection or parasite infestation. *Hepatitis A,* for example, results from infection by the hepatitis A virus. Contaminated food is often a source of infection. Hepatitis A occurs commonly in young people and ranges in severity from mild to life-threatening. Another viral hepatitis, *hepatitis B,* is usually more severe. It is also called *serum hepatitis* because it is often transmitted by contaminated blood serum.

Hepatitis, chronic alcohol abuse, malnutrition, or infection may lead to a degenerative liver condition known as **cirrhosis** (si-ROE-sis). The liver's ability to regenerate damaged tissue is well known, but it has its limits. For example, when the toxic effects of alcohol accumulate faster than the liver can regenerate itself, damaged tissue is replaced with fibrous or fatty tissue instead of normal tissue. *Cirrhosis* is the name given to such degeneration. No matter what the cause of liver cirrhosis, the symptoms are the same: nausea, anorexia, gray-white stools, weakness, and pain. If the cause of cirrhosis is removed and high-protein foods eaten, the liver may be able to repair itself—given enough time. If the damage

FIGURE 16-12 Gallstones. This gallbladder, which has been removed from a patient and cut open, contains numerous calculi or gallstones.

is extensive, a liver transplant may be the only hope of saving someone with cirrhosis of the liver.

Acute or chronic liver disorders such as hepatitis or cirrhosis can block the flow of blood through the liver, thus causing it to back up into the *hepatic portal circulation* (Figure 13-8). As a result, the blood pressure in the hepatic circulation increases abnormally—a condition called *portal hypertension*. To relieve the pressure, new veins that connect to the systemic veins are formed. This often causes the systemic veins lining the esophagus, stomach, and other organs to widen and become varicose. If these varicosities rupture after erosion by stomach acid, vomiting of blood occurs. This massive bleeding may result in death.

Pancreas

The pancreas lies behind the stomach in the concavity produced by the C shape of the duodenum. It is an exocrine gland that secretes pancreatic juice into ducts and an endocrine gland that secretes hormones into the blood. Pancreatic juice is the most important digestive juice. It contains enzymes that digest all three major kinds of foods. It also contains sodium bicarbonate, an alkaline substance that neutralizes the hydrochloric acid in the gastric juice that enters the intestines. Pancreatic juice enters the duodenum of the small intestine at the same place that bile enters. As you can see in Figure 16-11, the common bile and pancreatic ducts open into the duodenum at the major duodenal papilla.

FIGURE 16-13 Horizontal (Transverse) Section of the Abdomen. The photograph of a cadaver section shows the relative position of some of the major digestive organs of the abdomen. Such a view is typical in newer imaging methods such as computed tomography (CT) scanning and magnetic resonance imaging (MRI).

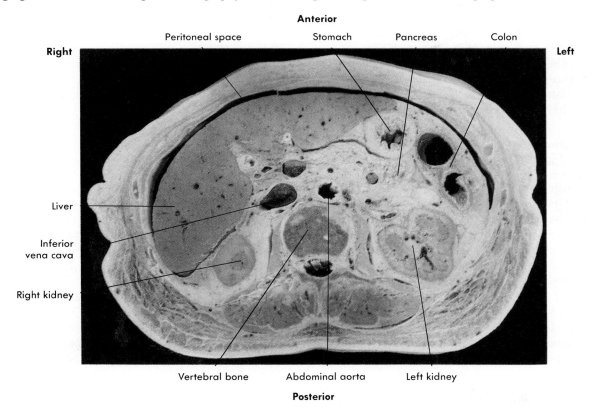

Between the cells that secrete pancreatic juice into ducts lie clusters of cells that have no contact with any ducts. These are the pancreatic islets (of Langerhans), which secrete the hormones of the pancreas described in Chapter 10.

Locate the pancreas and nearby structures in Figure 16-13, which shows a transverse section of a human cadaver.

Besides the endocrine disorders discussed in Chapter 10, such as diabetes mellitus, the pancreas may be involved in a number of other diseases. For example, **pancreatitis** (pan-kree-a-TIE-tis) or inflammation of the pancreas can be caused by a variety of factors. *Acute pancreatitis* usually results from blockage of the common bile duct. The blockage causes pancreatic enzymes to "back up" into the pancreas and digest it. This is a very serious condition; about half of all cases prove to be fatal. Another condition that blocks the flow of pancreatic enzymes is *cystic fibrosis (CF)*. You may recall from Chapter 3 that this inherited disorder disrupts cell transport and causes exocrine glands to produce excessively thick secretions. Thick pancreatic secretions may build up and block pancreatic ducts, disrupting the flow of pancreatic enzymes and damaging the pancreas.

Another serious pancreatic disorder is pancreatic cancer. Usually a form of *adenocarcinoma*, pancreatic cancer claims the lives of nearly all its victims within 5 years after diagnosis.

◤ *Large Intestine*

The **large intestine** is only about 1.5 meters (5 feet) in length. As the name implies, it has a much larger diameter than the small intestine. It forms the lower or terminal portion of the digestive tract. Undigested and unabsorbed food material enters the large intestine after passing through a sphincterlike structure (Figure 16-14) called the **ileocecal** (il-ee-ó-SEE-kal) **valve.** The word *chyme* is no longer appropriate in describing the contents of the large intestine. Chyme, which has the consistency of soup and is found in the small intestine, changes to the consistency of fecal matter as water and salts are reabsorbed during its passage through the small intestine. During its movement through the large intestine, material that escaped digestion in the small intestine is

acted on by bacteria. As a result of this bacterial action, additional nutrients may be released from cellulose and other fibers and absorbed. In addition to their digestive role, bacteria in the large intestine have other important functions. They are responsible for the synthesis of vitamin K needed for normal blood clotting and for the production of some of the B-complex vitamins. Once formed, these vitamins are absorbed from the large intestine and enter the blood.

Although some absorption of water, salts, and vitamins occurs in the large intestine, this segment of the digestive tube is not as well suited for absorption as is the small intestine. Salts, especially sodium, are absorbed by active transport, and water is moved into the blood by osmosis. No villi are present in the mucosa of the large intestine. As a result, much less surface area is available for absorption, and the efficiency and speed of movement of substances through the wall of the large intestine is lower than in the small intestine. Normal passage of material through the large intestine takes about 3 to 5 days.

The subdivisions of the large intestine are listed below in the order in which food passes through them.

1 Cecum
2 Ascending colon
3 Transverse colon
4 Descending colon
5 Sigmoid colon
6 Rectum
7 Anal canal

These areas can be studied and identified by tracing the passage of material from its point of entry into the large intestine at the ileocecal valve to its elimination from the body through the external opening called the **anus.**

Note in Figure 16-14 that the ileocecal valve opens into a pouchlike area called the **cecum** (SEE-kum). The opening itself is about 5 or 6 cm (2 inches) above the beginning of the large intestine. Food residue in the cecum flows upward on the right side of the body in the **ascending colon.** The **hepatic** or **right colic flexure** is the bend between the ascending colon and the **transverse colon,** which extends across the front of the abdomen from right to left. The **splenic** or **left colic flexure** marks the point where the **descend-**

FIGURE 16-14 Divisions of the Large Intestine.

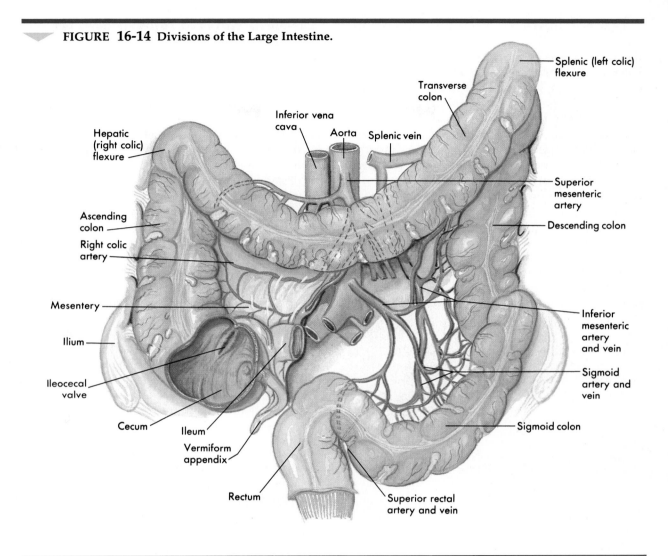

Splenic (left colic) flexure

Transverse colon

Inferior vena cava

Aorta

Splenic vein

Hepatic (right colic) flexure

Superior mesenteric artery

Descending colon

Ascending colon

Right colic artery

Inferior mesenteric artery and vein

Mesentery

Sigmoid artery and vein

Ilium

Ileocecal valve

Sigmoid colon

Cecum

Ileum

Vermiform appendix

Rectum

Superior rectal artery and vein

---

ing colon turns downward on the left side of the abdomen. The **sigmoid colon** is the S-shaped segment that terminates in the **rectum.** The terminal portion of the rectum is called the **anal canal,** which ends at the external opening or anus.

Two sphincter muscles stay contracted to keep the anus closed except during defecation. Smooth or involuntary muscle composes the **inner anal sphincter,** but striated, or voluntary, muscle composes the outer one. This anatomical fact sometimes becomes highly important from a practical standpoint. For example, often after a person has had a stroke, the voluntary anal sphincter at first becomes paralyzed, and the individual has no control at this time over bowel movements.

DISORDERS OF THE LARGE INTESTINE

Many of the more common disorders of the large intestine relate to *motility* or rate of movement of the intestinal contents. Abnormally rapid motility through the large intestine may result in *diarrhea,* and abnormally slow motility may result in *constipation.* These and other conditions are briefly described in this section.

Diarrhea (dye-ah-REE-ah) usually occurs when the intestinal contents move so quickly that the resulting feces are more fluid than normal. Diarrhea is characterized by frequent passing of watery feces. In inflammatory conditions such as *dysentery,* the watery feces also may contain mucus, blood, or pus (see Appendix B). Diarrhea may also be accompanied by abdominal cramps—a symptom caused by excessive contractions of the intestinal muscles. The increased intestinal motility that causes diarrhea often results from the presence of bacterial toxins, parasites, or other irritants. The intestines reflexively "speed up," a mechanism that quickly disposes of the irritant. Because of the water loss involved, untreated diarrhea may quickly lead to dehydration—and possibly convulsions or death.

Constipation results from decreased intestinal motility. If passage of feces through the large intestine is prolonged beyond 5 days, the feces lose volume and become more solid because of excessive water reabsorption. This reduces stimulation of the bowel-emptying reflex, resulting in retention of feces—a positive-feedback effect that makes the condition even worse.

Acute constipation often results from intestinal blockage, tumors, or **diverticulitis** (dye-ver-ti-kyoo-LIE-tis). Diverticulitis is an inflammation of abnormal saclike outpouchings of the intestinal wall called *diverticula.* Diverticula often develop in adults over 50 years of age who eat low-fiber foods. Treatment of acute constipation usually involves treatment of the underlying cause. The use of laxatives and enemas to induce defecation should be avoided.

Colitis (ko-LIE-tis) refers to any inflammatory condition of the large intestine. Symptoms of colitis include diarrhea and abdominal cramps or constipation. Some forms of colitis may also produce bleeding and intestinal ulcers. Colitis may be a result of emotional stress, as in *irritable bowel syndrome.* It may also result from an autoimmune disease, as in *ulcerative colitis.* Another type of autoimmune colitis is *Crohn's disease,* which often also affects the small intestine. If more conservative treatments fail, colitis may be corrected by surgical removal of the affected portions of the colon.

Flatulence is the presence of air or other gases (*flatus*) in the gastrointestinal tract. In the stomach, air swallowed with saliva or food often accumulates in the stomach. In the large intestine, gases such as methane (CH_4) produced by intestinal bacteria may accumulate. In either case, the resulting distension (expansion) of the digestive wall may cause "gas pains" until the gas is expelled.

Colorectal (ko-lo-REK-tal) **cancer** is a malignancy, usually *adenocarcinoma,* of the colon or rectum. Colorectal cancer occurs most frequently after the age of 50, and a low-fiber, high fat diet and genetic predisposition are known risk factors. Early warning signs of this common type of cancer include changes in bowel habits, fecal blood, rectal bleeding, abdominal pain, unexplained anemia or weight loss, and fatigue. Surgical removal or isolation of affected portions of the colon may require a **colostomy** (ko-LAH-sto-me) (Figure 16-15). Colostomy is a surgical procedure in which

EXERCISE AND FLUID UPTAKE

Replacement of fluids lost during exercise, primarily through sweating, is essential for maintaining homeostasis. Nearly everyone increases his or her intake of fluids during and after exercise. The main limitation to efficient fluid replacement is how quickly fluid can be absorbed, rather than how much a person drinks. Very little water is absorbed until it reaches the intestines, where it is absorbed almost immediately. Thus, the rate of **gastric emptying** into the intestine is critical.

Large volumes of fluid leave the stomach and enter the intestines more rapidly than small volumes. However, large volumes may be uncomfortable during exercise. Cool fluids (8° to 13° C) empty more quickly than warm fluids. Fluids with a high solute concentration empty slowly and may cause nausea or stomach cramps. Thus, large amounts of cool, dilute, or isotonic fluids are best for replacing fluids quickly during exercise.

FIGURE 16-15 Colostomy. Feces are usually collected in a removable bag attached to the stoma or opening.

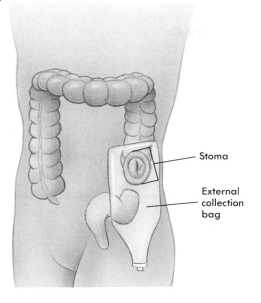

Stoma

External collection bag

FIGURE 16-16 The Large Intestine. A special x-ray technique produces a clear image of the large intestine and its position relative to the skeleton.

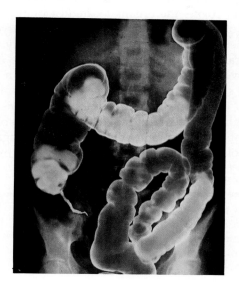

an artificial anus is created on the abdominal wall by cutting the colon and bringing the cut end or ends out to the surface to form an opening called a *stoma*. Immediately after the surgery, a great deal of care must then be taken to irrigate the altered colon with an isotonic solution—because the colon can no longer be evacuated in the normal way.

Appendix and Appendicitis

The **vermiform appendix** (Latin *vermiformis* from *vermis*—"worm" and *forma*—"shape") is, as the name implies, a wormlike, tubular structure. Although it serves no important digestive function in humans, it contains lymphatic tissue and may play a minor role in the immunologic defense mechanisms of the body described in Chapter 14. Note in Figure 16-14 that the appendix is directly attached to the cecum. The appendix contains a blind, tubelike interior lumen that communicates with the lumen of the large intestine 3 cm (1 inch) below the opening of the

ileocecal valve into the cecum. If the mucous lining of the appendix becomes inflamed, the resulting condition is the well-known affliction, **appendicitis.** As you can see in Figures 16-14 and 16-16, the appendix is very close to the rectal wall. For patients with suspected appendicitis, a physician often evaluates the appendix by a digital rectal examination.

The opening between the lumen of the appendix and the cecum is quite large in children and young adults—a fact of great clinical significance because food or fecal material trapped in the appendix will irritate and inflame its mucous lining, causing appendicitis. The opening between the appendix and the cecum is often completely obliterated in elderly persons, which explains the low incidence of appendicitis in this population.

If infectious material becomes trapped in an inflammed appendix, the appendix may rupture and release the material into the abdominal cavity. Infection of the peritoneum and other abdominal organs may result—with sometimes tragic consequences.

Peritoneum

The **peritoneum** is a large, moist, slippery sheet of serous membrane that lines the abdominal cavity and covers the organs located in it, including most of the digestive organs. The parietal layer of the peritoneum lines the abdominal cavity. The visceral layer of the peritoneum forms the outer or covering layer of each abdominal organ. The small space between the parietal and visceral layers is called the *peritoneal space*. It contains just enough peritoneal fluid to keep both layers of the peritoneum moist and able to slide freely against each other during breathing and digestive move-

ments (Figure 16-17). Organs outside of the peritoneum are said to be retroperitoneal.

EXTENSIONS

The two most prominent extensions of the peritoneum are the mesentery and the greater omentum. The **mesentery,** an extension between the parietal and visceral layers of the peritoneum, is shaped like a giant, pleated fan. Its smaller edge attaches to the lumbar region of the posterior abdominal wall, and its long, loose outer edge encloses most of the small intestine, anchoring it to the posterior abdominal wall. The **greater omentum** is a pouchlike extension of the visceral

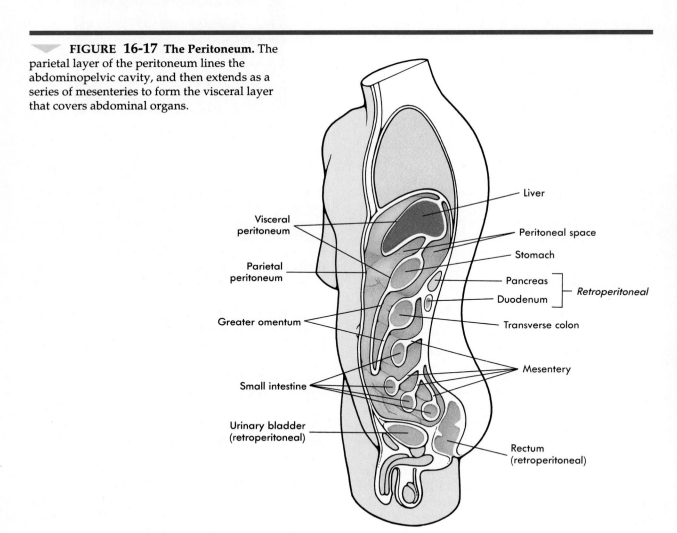

▽ **FIGURE 16-17 The Peritoneum.** The parietal layer of the peritoneum lines the abdominopelvic cavity, and then extends as a series of mesenteries to form the visceral layer that covers abdominal organs.

Visceral peritoneum
Parietal peritoneum
Greater omentum
Small intestine
Urinary bladder (retroperitoneal)

Liver
Peritoneal space
Stomach
Pancreas
Duodenum — *Retroperitoneal*
Transverse colon
Mesentery
Rectum (retroperitoneal)

peritoneum from the lower edge of the stomach, part of the duodenum, and the transverse colon. Shaped like a large apron, it hangs down over the intestines, and because spotty deposits of fat give it a lacy appearance, it has been nicknamed the *lace apron*. It usually envelops a badly inflamed appendix, walling it off from the rest of the abdominal organs.

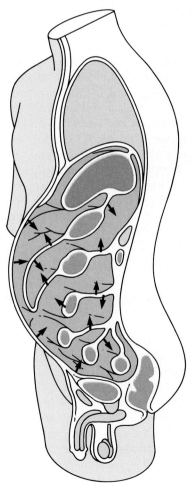

FIGURE 16-18 Ascites. Ascites results from an accumulation of fluid in the peritoneal space. The arrows indicate water filtering out of the peritoneal blood vessels resulting from hypertension or diffusing out of the vessels because of an osmotic imbalance in the blood.

PERITONITIS

Peritonitis (pair-i-toe-NYE-tis) is the inflammation of the peritoneum resulting from a bacterial infection or another irritating condition. Peritonitis most commonly results from an infection that occurs after the rupture of the appendix or other abdominopelvic organ. It is characterized by abdominal distension, pain, nausea, vomiting, tachycardia (rapid heart rate), fever, dehydration, and other signs and symptoms. Circulatory shock progressing to heart failure may result.

ASCITES

Ascites (ah-SEE-teez) is the abnormal accumulation of fluid in the peritoneal space (Figure 16-18). Fluid enters the peritoneal space from the blood because of local hypertension (high blood pressure) or an osmotic imbalance in the plasma (low plasma protein). This condition may be accompanied by abdominal swelling and decreased urinary output. It commonly occurs as a complication of cirrhosis, congestive heart failure, kidney disease, peritonitis, cancer, or malnutrition.

 ORAL REHYDRATION THERAPY

Simple diarrhea can result in life-threatening dehydration if significant water loss occurs during a short period of time. Diarrhea is one of the leading causes of infant mortality in developing countries. New attempts to educate mothers about the dangers of diarrhea and to provide them with a simple, easily prepared remedy have saved hundreds of thousands of lives in recent years. This treatment is called *oral rehydration therapy*. It involves liberal doses of an easily prepared solution containing sugar and salt. This salt-sugar solution replaces nutrients and electrolytes lost in diarrheal fluid. Because the replacement fluid can be prepared from readily available and inexpensive ingredients, it is particularly valuable in the treatment of infant diarrhea in third-world countries.

Digestion

Digestion, a complex process that occurs in the alimentary canal, consists of physical and chemical changes that prepare food for absorption. **Mechanical digestion** breaks food into tiny particles, mixes them with digestive juices, moves them along the alimentary canal, and finally eliminates the digestive wastes from the body. Chewing or mastication, swallowing or deglutition (deg-loo-TISH-un), peristalsis, and defecation are the main processes of mechanical digestion. **Chemical digestion** breaks down large, nonabsorbable food molecules into smaller, absorbable molecules—molecules that are able to pass through the intestinal mucosa into blood and lymph. Chemical digestion consists of numerous chemical reactions catalyzed by enzymes in saliva, gastric juice, pancreatic juice, and intestinal juice.

CARBOHYDRATE DIGESTION

Very little digestion of carbohydrates (starches and sugars) occurs before food reaches the small intestine. Salivary amylase usually has little time to do its work because so many of us swallow our food so fast. Gastric juice contains no carbohydrate-digesting enzymes. But after the food reaches the small intestine, pancreatic and intestinal juice enzymes digest the starches and sugars. A pancreatic enzyme (amylase) starts the process by changing starches into a double sugar, namely, maltose. Three intestinal enzymes—maltase, sucrase, and lactase—digest double sugars by changing them into simple sugars, chiefly glucose (dextrose). Maltase digests maltose (malt sugar), sucrase digests sucrose (ordinary cane sugar), and lactase digests lactose (milk sugar). The end products of carbohydrate digestion are the so-called simple sugars; the most abundant is glucose.

PROTEIN DIGESTION

Protein digestion starts in the stomach. Two enzymes (rennin and pepsin) in the gastric juice cause the giant protein molecules to break up into somewhat simpler compounds. Pepsinogen, a component of gastric juice, is converted into active pepsin enzyme by hydrochloric acid (also in gastric juice). In the intestine, other enzymes (trypsin in the pancreatic juice and peptidases in the intestinal juice) finish the job of protein digestion. Every protein molecule is made up of many amino acids joined together. When enzymes have split up the large protein molecule into its separate amino acids, protein digestion is completed. Hence the end product of protein digestion is amino acids. For obvious reasons, the amino acids are also referred to as *protein building blocks.*

FAT DIGESTION

Very little carbohydrate and fat digestion occurs before food reaches the small intestine. Most fats are undigested until after emulsification by bile in the duodenum (that is, fat droplets are broken into very small droplets). After this takes place, pancreatic lipase splits up the fat molecules into fatty acids and glycerol (glycerin). The end products of fat digestion, then, are fatty acids and glycerol.

• • •

Table 16-2 summarizes the main facts about chemical digestion. Enzyme names indicate the type of food digested by the enzyme. For example, the name *amylase* indicates that the enzyme digests carbohydrates (starches and sugars), *protease* indicates a protein-digesting enzyme, and *lipase* means a fat-digesting enzyme. When carbohydrate digestion has been completed, starches (polysaccharides) and double sugars (disaccharides) have been changed mainly to glucose, a simple sugar (monosaccharide). The end products of protein digestion, on the other hand, are amino acids. Fatty acid and glycerol are the end products of fat digestion.

Absorption

After food is digested, it is absorbed; that is, it moves through the mucous membrane lining of the small intestine into the blood and lymph. In other words, food absorption is the process by which molecules of amino acids, glucose, fatty acids, and glycerol go from the inside of the intestines into the circulating fluids of the body. Absorption of foods is just as essential as digestion of foods. The reason is fairly obvious. As long as food stays in the intestines, it cannot nour-

TABLE 16-2 Chemical Digestion

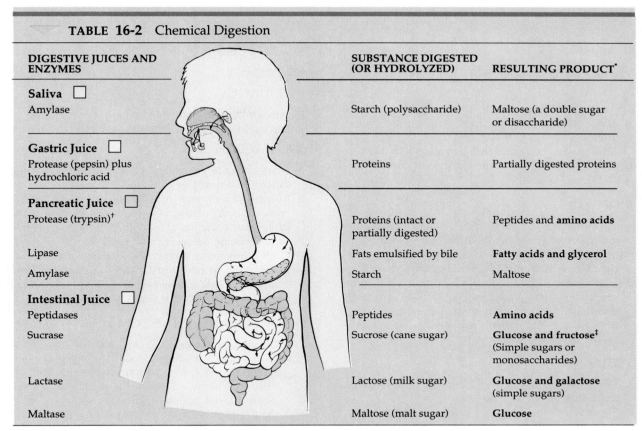

DIGESTIVE JUICES AND ENZYMES	SUBSTANCE DIGESTED (OR HYDROLYZED)	RESULTING PRODUCT[*]
Saliva ☐		
Amylase	Starch (polysaccharide)	Maltose (a double sugar or disaccharide)
Gastric Juice ☐		
Protease (pepsin) plus hydrochloric acid	Proteins	Partially digested proteins
Pancreatic Juice ☐		
Protease (trypsin)[†]	Proteins (intact or partially digested)	Peptides and **amino acids**
Lipase	Fats emulsified by bile	**Fatty acids and glycerol**
Amylase	Starch	Maltose
Intestinal Juice ☐		
Peptidases	Peptides	**Amino acids**
Sucrase	Sucrose (cane sugar)	**Glucose and fructose**[‡] (Simple sugars or monosaccharides)
Lactase	Lactose (milk sugar)	**Glucose and galactose** (simple sugars)
Maltase	Maltose (malt sugar)	**Glucose**

[*]Substances in boldface type are end products of digestion (that is, completely digested foods ready for absorption).
[†]Secreted in inactive form (trypsinogen); activated by enterokinase, an enzyme in the intestinal juice.
[‡]Glucose is also called *dextrose;* fructose is called *levulose.*

ish the millions of cells that compose all other parts of the body. Their lives depend on the absorption of digested food and its transportation to them by the circulating blood.

Structural adaptations of the digestive tube, including folds in the lining mucosa, villi, and microvilli, increase the absorptive surface and the efficiency and speed of transfer of materials from the intestinal lumen to body fluids. Many salts such as sodium are actively transported through the intestinal mucosa. Water follows by osmosis. Other nutrients are also actively transported into the blood of capillaries in the intestinal villi. Fats enter the lymphatic vessels or *lacteals* found in intestinal villi.

OUTLINE SUMMARY

WALL OF THE DIGESTIVE TRACT (Figure 16-2)

The wall of the digestive tube is formed by four layers:
A Mucosa—mucous epithelium
B Submucosa—connective tissue
C Muscularis—two or three layers of smooth muscle
D Serosa—serous membrane that covers the outside of abdominal organs

MOUTH

A Roof—formed by hard palate (parts of maxillary and palatine bones) and soft palate, an arch-shaped muscle separating mouth from pharynx; uvula, a downward projection of soft palate (Figure 16-3)
B Floor—formed by tongue and its muscles; papillae, small elevations on mucosa of tongue; taste buds, found in many papillae; lingual frenulum, fold of mucous membrane that helps anchor tongue to floor of mouth (Figure 16-4)

TEETH

A Names of teeth—incisors, cuspids, bicuspids, and tricuspids
B Twenty teeth in temporary set; average age for cutting first tooth about 6 months; set complete at about 2 years of age
C Thirty-two teeth in permanent set; 6 years about average age for starting to cut first permanent tooth; set complete usually between ages of 17 and 24 years (Figure 16-5)
D Structures of a typical tooth—crown, neck, and root (Figure 16-6)

DISORDERS OF THE MOUTH AND TEETH

A Mouth and tooth disorders can spread to other tissues
B Mouth cancer usually results from tobacco use; lip cancer from exposure to sunlight
C Caries—tooth decay; may lead to other problems
D Gingivitis—gum inflammation or infection
  1 Usually results from poor oral hygiene
  2 Vincent's infection—contagious bacterial infection of gums
E Periodontitis—inflammation of periodontal membrane; leading cause of tooth loss among adults

SALIVARY GLANDS (Figure 16-7)

A Parotid glands
B Submandibular glands
C Sublingual glands

PHARYNX

ESOPHAGUS

Heartburn—esophageal pain caused by backflow of stomach acid; may be a complication of hiatal hernia

STOMACH (Figure 16-8)

A Size—expands after large meal; about size of large sausage when empty
B Pylorus—lower part of stomach; pyloric sphincter muscle closes opening of pylorus into duodenum
C Wall—many smooth muscle fibers; contractions produce churning movements (peristalsis)
D Lining—mucous membrane; many microscopic glands that secrete gastric juice and hydrochloric acid into stomach; mucous membrane lies in folds (rugae) when stomach is empty

DISORDERS OF THE STOMACH

1 Gastroenterology—study of stomach and intestines and their diseases; gastric diseases often exhibit these signs or symptoms: gastritis (inflammation), anorexia (appetite loss), nausea (upset stomach), and emesis (vomiting)
2 Pylorospasm—abnormal spasms of the pyloric sphincter; pyloric stenosis is similar, also narrowing the pyloric opening
3 Ulcers—open wounds caused by the acid in gastric juice
  a Often occur in duodenum or stomach
  b May be associated with the bacterium *Campylobacter pylori*
4 Stomach cancer is associated with consumption of alcohol or preserved food and use of chewing tobacco

SMALL INTESTINE (Figure 16-10)

A Size—about 7 meters (20 feet) long but only 2 cm or so in diameter
B Divisions
  1 Duodenum
  2 Jejunum
  3 Ileum
C Wall—contains smooth muscle fibers that contract to produce peristalsis
D Lining—mucous membrane; many microscopic glands (intestinal glands) secrete intestinal juice; villi (microscopic finger-shaped projections from surface of mucosa into intestinal cavity) contain blood and lymph capillaries

E Disorders of the small intestine
  1 Enteritis—intestinal inflammation; gastroenteritis—inflammation of stomach and intestines
  2 Malabsorption syndrome—group of symptoms resulting from failure to absorb nutrients properly (for example, anorexia, abdominal bloating, cramps, anemia, and fatigue)

LIVER AND GALLBLADDER

A Size and location—liver is largest gland; fills upper right section of abdominal cavity and extends over into left side
B Liver secretes bile
C Ducts (Figure 16-11)
  1 Hepatic—drains bile from liver
  2 Cystic—duct by which bile enters and leaves gallbladder
  3 Common bile—formed by union of hepatic and cystic ducts; drains bile from hepatic or cystic ducts into duodenum
D Gallbladder
  1 Location—undersurface of the liver
  2 Function—concentrates and stores bile produced in the liver
E Disorders of the liver and gallbladder
  1 Gallstones—calculi (stones) made of crystallized bile pigments and calcium salts
    a Cholelithiasis—condition of having gallstones
    b Cholecystitis—inflammation of the gallbladder; may accompany cholelithiasis
    c Can obstruct bile canals, causing jaundice
  2 Hepatitis—liver inflammation
    a Characterized by liver enlargement, jaundice, anorexia, discomfort, gray-white feces, and dark urine
    b Caused by a variety of factors: toxins, bacteria, viruses, and parasites
  3 Cirrhosis—degeneration of liver tissue involving replacement of normal (but damaged) tissue with fibrous and fatty tissue
  4 Portal hypertension—high blood pressure in the hepatic portal veins caused by obstruction of blood flow in a diseased liver; may cause varicosities of surrounding systemic veins

PANCREAS

A Location—behind stomach
B Functions
  1 Pancreatic cells secrete pancreatic juice into pancreatic ducts; main duct empties into duodenum
  2 Pancreatic islets (of Langerhans)—cells not connected with pancreatic ducts; secrete hormones glucagon and insulin into the blood

C Pancreatic disorders
  1 Pancreatitis—inflammation of pancreas; acute pancreatitis results from blocked ducts that force pancreatic juice to backflow, digesting the gland
  2 Cystic fibrosis—thick secretions block flow of pancreatic juice
  3 Pancreatic cancer is very serious—fatal in the majority of cases

LARGE INTESTINE (Figure 16-14)

A Divisions
  1 Cecum
  2 Colon—ascending, transverse, descending, and sigmoid
  3 Rectum
B Opening to exterior—anus
C Wall—contains smooth muscle fibers that contract to produce churning, peristalsis, and defecation
D Lining—mucous membrane
E Disorders of the large intestine often relate to abnormal motility (rate of movement of contents)
  1 Diarrhea results from abnormally increased intestinal motility; may result in dehydration or convulsions
  2 Constipation results from decreased intestinal motility
  3 Diverticulitis (inflammation of abnormal outpouchings called *diverticula*) may cause constipation
  4 Colitis is the general name for any inflammatory condition of the large intestine
  5 Flatulence is the presence of gas in the gastrointestinal tract
  6 Colorectal cancer is a common malignancy of the colon and rectum associated with advanced age; low-fiber, high-fat diets; and genetic predisposition

APPENDIX AND APPENDICITIS

A Blind tube off cecum; no important digestive function in humans
B Appendicitis—inflammation or infection of appendix; if appendix ruptures, infectious material may spread to other organs

PERITONEUM (Figure 16-17)

A Definitions—peritoneum, serous membrane lining abdominal cavity and covering abdominal organs; parietal layer of peritoneum lines abdominal cavity; visceral layer of peritoneum covers abdominal organs; peritoneal space lies between parietal and visceral layers

B Extensions—largest ones are the mesentery and greater omentum; mesentery is extension of parietal peritoneum, which attaches most of small intestine to posterior abdominal wall; greater omentum, or "lace apron," hangs down from lower edge of stomach and transverse colon over intestines

C Peritonitis—inflammation of peritoneum resulting from infection or other irritant; often a complication of ruptured appendix

D Ascites—abnormal accumulaton of fluid in peritoneal space, often causing bloating of abdomen

DIGESTION (Table 16-2)

Meaning—changing foods so that they can be absorbed and used by cells

A Mechanical digestion—chewing, swallowing, and peristalsis break food into tiny particles, mix them well with digestive juices, and move them along the digestive tract

B Chemical digestion—breaks up large food molecules into compounds having smaller molecules; brought about by digestive enzymes

C Carbohydrate digestion—mainly in small intestine
  1 Pancreatic amylase—changes starches to maltose
  2 Intestinal juice enzymes
    a Maltase changes maltose to glucose
    b Sucrase changes sucrose to glucose
    c Lactase changes lactose to glucose

D Protein digestion—starts in stomach; completed in small intestine
  1 Gastric juice enzymes, rennin and pepsin, partially digest proteins
  2 Pancreatic enzyme, trypsin, completes digestion of proteins to amino acids
  3 Intestinal enzymes, peptidases, complete digestion of partially digested proteins to amino acids

E Fat digestion
  1 Bile contains no enzymes but emulsifies fats (breaks fat droplets into very small droplets)
  2 Pancreatic lipase changes emulsified fats to fatty acids and glycerol in small intestine

ABSORPTION

A Meaning—digested food moves from intestine into blood or lymph

B Where absorption occurs—foods and most water from small intestine; some water also absorbed from large intestine

$\mathcal{N}$EW WORDS

absorption	feces	mesentery	plica
alimentary canal	frenulum	papilla	rugae
bolus	lumen	peristalsis	uvula
chyme	mastication	peritoneum	villus
digestion			

Diseases and Other Clinical Terms

anorexia	colostomy	gastroenteritis	nausea
appendicitis	constipation	gastroenterology	pancreatitis
ascites	cystic fibrosis	gingivitis	periodontitis
calculi	diarrhea	heartburn	peritonitis
caries	diverticulitis	hepatitis	pyloric stenosis
cholecystitis	dysentery	hiatal hernia	pylorospasm
cholelithiasis	emesis	jaundice	ulcer
cirrhosis	enteritis	leukoplakia	Vincent's infection
colitis	flatulence	malabsorption	
colorectal cancer	gastritis	syndrome	

CHAPTER TEST

1. In the digestive system the breakdown of food material is _____ and _____ in nature.
2. Undigested waste material resulting from the digestive process is called _____.
3. The study of the stomach, the intestines, and their diseases is called _____.
4. Inflammation of the small intestine is called _____.
5. The liver and gallbladder are classified as _____ organs of digestion.
6. The hollow space within the digestive tube is called the _____.
7. The inside or lining coat of the digestive tube is called the _____.
8. The roof of the mouth is formed by the hard and soft _____.
9. The portion of a tooth that is exposed and visible is called the _____.
10. The teeth serve cutting and grinding functions during the chewing of food—a process called _____.
11. Saliva contains the enzyme called salivary _____.
12. Food moves from the pharynx to the stomach by passing through the _____.
13. The semisolid mixture of food and gastric juice in the stomach is called _____.
14. The movement of food through the digestive tract results from contractions called _____.
15. The middle segment of the small intestine is called the _____.

Select the most correct answer from Column B for each statement in Column A. (Only one answer is correct.)

COLUMN A

16. _____ Waste material
17. _____ Accessory organ of digestion
18. _____ Double fold of peritoneum
19. _____ Hardest tissue in body
20. _____ Semisolid mixture
21. _____ Division of stomach
22. _____ Yellowish skin condition
23. _____ Digestive enzyme
24. _____ Inflammation of the stomach
25. _____ Vomiting

COLUMN B

a. Liver
b. Enamel
c. Fundus
d. Feces
e. Jaundice
f. Mesentery
g. Lipase
h. Chyme
i. Emesis
j. Gastritis

Circle the T before each true statement and the F before each false statement.

T F 26. The salivary glands are considered accessory organs of digestion.
T F 27. The ileum is the portion of the small intestine found between the duodenum and jejunum.
T F 28. The large intestine is classified as an accessory organ of digestion.
T F 29. The serosa is the outermost coat of the digestive tube.
T F 30. The uvula is attached to the soft palate.
T F 31. In humans there are 32 deciduous (baby) teeth and 20 permanent teeth.
T F 32. No chemical digestion can occur in the mouth.
T F 33. The pyloric sphincter is located between the esophagus and stomach.
T F 34. Plica, villi, and microvilli all increase the surface area of the small intestine.
T F 35. Dental caries is one of the most common diseases in the civilized world.
T F 36. Most experts agree that excess stomach acid plays an important role in ulcer formation.

REVIEW QUESTIONS

1 What organs form the gastrointestinal tract?
2 Identify the jejunum, cecum, colon, duodenum, and ileum.
3 If you inserted 9 inches of an enema tube through the anus, the tip of the tube would probably be in what structure?
4 How many teeth should an adult have?
5 How many teeth should a child 2¹/₂ years old have? Would he have some of each of the following teeth: incisors, canines, premolars, and molars? If not, which ones would he not have?
6 How does dental caries develop? Gingivitis? Periodontitis? What are possible complications of these conditions?
7 Identify the pancreatic islets, parotid glands, pylorus, rugae, and villi.
8 How do gastric ulcers form? Duodenal ulcers?

9 Describe diarrhea and constipation in terms of intestinal motility. What factors might cause each of these conditions?
10 In what organ does the digestion of starches begin?
11 What digestive juice contains no enzymes?
12 Only one digestive juice contains enzymes for digesting all three kinds of food. Which juice is this? In what organ does it do its work?
13 What kinds of food are not digested in the stomach?
14 Which digestive juice emulsifies fats?
15 What three digestive juices act on foods in the small intestine?
16 What juices digest carbohydrates? Proteins? Fats?
17 Where are simple sugars and amino acids absorbed into blood capillaries? Where are lipids absorbed into lacteals?
18 Where is most of the water absorbed from the lumen of the digestive tract?

CLINICAL APPLICATIONS

1 Give the sequence of events in which a failure to brush and floss the teeth leads to loss of teeth.

2 You have been diagnosed as having a duodenal ulcer. Your physician prescribed the drug *famotidine (Pepcid)*. This drug has an action similar to cimetidine, but it is much more potent. How will famotidine help your ulcer? Your brother also has an ulcer, but his physician has prescribed *sucralfate (Carafate)*, a medication that has an antipepsin action and tends to adhere to membrane injuries. Explain how sucralfate might help your brother's ulcer.

3 Fred has been experiencing recurring episodes of heartburn, especially when he bends over to lace his shoes or when he lies down in bed. His physician believes Fred may have a hiatal hernia. To confirm this diagnosis, she has scheduled Fred for a "barium swallow." In this test, a barium-containing fluid that blocks x-rays is swallowed so that the stomach appears as a bright mass in radiograph. If Fred has a hiatal hernia, what should the radiologist see on the radiograph? Can you explain what caused the symptoms that brought Fred in to see his physician?

Nutrition and Metabolism

C H A P T E R

17

Outline

Objectives

*After you have completed this chapter,
you should be able to:*

1 Define and contrast *catabolism* and *anabolism*.

2 Describe the metabolic roles of carbohydrates, fats, proteins, vitamins, and minerals.

3 Define basal metabolic rate and list some factors that affect it.

4 Describe three disorders associated with eating or metabolism.

5 Discuss the physiological mechanisms that regulate body temperature.

*N*utrition and *metabolism* are words that are often used together—but what do they mean? *Nutrition* is a term that refers to the food (nutrients) that we eat. Proper nutrition requires a balance of the three basic food types, *carbohydrates, fats,* and *proteins,* plus essential *vitamins* and *minerals.* Malnutrition is a deficiency or imbalance in the consumption of food, vitamins, and minerals.

A good phrase to remember in connection with the word metabolism is "use of foods" because basically this is what metabolism is—the use the body makes of foods after they have been digested, absorbed, and circulated to cells. It uses them in two ways: as an energy source and as building blocks for making complex chemical compounds. Before they can be used in these two ways, foods have to be *assimilated.* Assimilation occurs when food molecules enter cells and undergo many chemical changes there. All the chemical reactions that release energy from food molecules make up the process of *catabolism,* a vital process because it is the only way that the body has of supplying itself with energy for doing any work. The many chemical reactions that build food molecules into more complex chemical compounds constitute the process of *anabolism.* Catabolism and anabolism make up the process of metabolism.

This chapter explores nutrient metabolism, the importance of vitamins and minerals, metabolic and eating disorders, and regulation of body temperature.

Role of the Liver

As we discussed in Chapter 16, the liver plays an important role in the mechanical digestion of lipids because it secretes *bile.* As you recall, bile breaks large fat globules into smaller droplets of fat that are more easily broken down. In addition, liver cells perform other functions necessary for healthy survival. They play a major role in the metabolism of all three kinds of foods. They help maintain a normal blood glucose concentration by carrying on complex and essential chemical reactions. Liver cells also carry on the first steps of protein and fat metabolism and synthesize several kinds of protein compounds. They release them into the blood, where they are called the *blood proteins* or *plasma proteins.* Prothrombin and fibrino-gen, two of the plasma proteins formed by liver cells, play essential parts in blood clotting (see p. 296). Another protein made by liver cells, albumin, helps maintain normal blood volume. Liver cells detoxify various poisonous substances such as bacterial products and certain drugs. Liver cells store several substances, notably iron and vitamins A, D, E, and K.

The liver is assisted by an interesting structural feature of the blood vessels that supply it. As you may recall from Chapter 13, the hepatic portal vein delivers blood directly from the gastrointestinal tract to the liver (see Figure 13-8). This arrangement allows blood that has just absorbed nutrients and other substances to be processed by the liver before being distributed in the body. Thus excess nutrients and vitamins can be stored, and toxins can be removed from the bloodstream.

Nutrient Metabolism

CARBOHYDRATE METABOLISM

Carbohydrates are the preferred energy food of the body. They are composed of smaller "building blocks"—primarily *glucose.* (See Appendix A). Human cells catabolize (break down) glucose rather than other substances as long as enough glucose enters them to supply their energy needs. Two series of chemical reactions, occurring in a precise sequence, make up the process of glucose catabolism. **Glycolysis** (glye-KOL-i-sis) is the name given the first series of reactions; **citric acid cycle** is the name of the second series. Glycolysis, as Figure 17-1 shows, changes glucose to pyruvic acid, and then the citric acid cycle changes the pyruvic acid to carbon dioxide. Glycolysis takes place in the cytoplasm of a cell, whereas the citric acid cycle goes on in the mitochondria, the cell's miniature power plants. Glycolysis uses no oxygen; it is an **anaerobic** (an-er-O-bik) process. The citric acid cycle, in contrast, is an oxygen-using or **aerobic** (aer-O-bik) process.

While the chemical reactions of glycolysis and the citric acid cycle occur, energy stored in the glucose molecule is being released. Almost instantaneously, however, more than half of it is put back into storage, not in glucose molecules but in the molecules of another compound, adenosine triphosphate (ATP). The rest of the energy originally stored in the glucose molecule

FIGURE 17-1 Catabolism of Glucose. Glycolysis splits one molecule of glucose (six carbon atoms) into two molecules of pyruvic acid (three carbon atoms each). The citric acid cycle converts each pyruvic acid molecule into three carbon dioxide molecules (one carbon atom each).

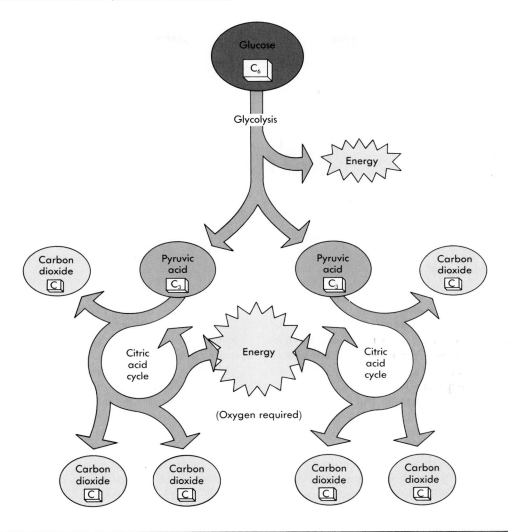

is released as heat. ATP serves as the direct source of energy for doing cellular work in all kinds of living organisms from one-cell plants to billion-cell animals, including man. Among biological compounds, therefore, ATP ranks as one of the most important. The energy stored in ATP molecules differs in two ways from the energy stored in food molecules; the energy in ATP molecules can be released almost instantaneously, and it can be used directly to do cellular work.

Release of energy from food molecules occurs much more slowly because it accompanies the long series of chemical reactions that make up the process of catabolism. For some reason, energy released from food molecules cannot be used directly for doing cellular work. It must first be transferred to ATP molecules and be released explosively from them.

As Figure 17-2 shows, ATP comprises an adenosine group and three phosphate groups.

FIGURE **17-2** **ATP. A,** The structure of ATP. A single adenosine group *(A)* has three attached phosphate groups *(P)*. The high-energy bonds between the phosphate groups can release chemical energy to do cellular work. **B,** ATP energy cycle. ATP stores energy in its last high-energy phosphate bond. When that bond is later broken, energy is released to do cellular work. The ADP and phosphate groups that result can be resynthesized into ATP capturing additional energy from nutrient catabolism.

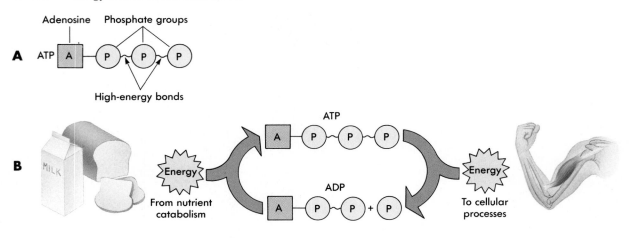

The capacity of ATP to store large amounts of energy is found in the high-energy bonds that hold the phosphate groups together, illustrated as curvy lines. When a phosphate group breaks off the molecule, an adenosine diphosphate (ADP) molecule and free phosphate group result. Energy that had been holding the phosphate bond together is freed to do cellular work (muscle fiber contractions, for example). As you can see in Figure 17-2, the ADP and phosphate are reunited by the energy produced by carbohydrate catabolism, making ATP a reusable energy-storage molecule. Only enough ATP for immediate cellular requirements is made at any one time. Glucose that is not needed is anabolized into larger molecules that are stored for later use.

Glucose anabolism is a process called **glycogenesis** (glye-ko-JEN-e-sis). Carried on chiefly by liver and muscle cells, glycogenesis consists of a series of reactions that join glucose molecules together, like many beads in a necklace, to form *glycogen,* a compound sometimes called *animal starch.*

Something worth noticing is that the amount of nutrients in the blood normally does not change very much, not even when we go without food for many hours, when we exercise and use a lot of food for energy, or when we sleep and use little food for energy. The amount of glucose in our blood, for example, usually stays at about 80 to 120 mg in 100 ml of blood.

Several hormones help regulate carbohydrate metabolism to keep blood glucose normal. **Insulin** is one of the most important of these. It acts in some way not yet definitely known to make glucose leave the blood and enter the cells at a more rapid rate. As insulin secretion increases, more glucose leaves the blood and enters the cells. The amount of glucose in the blood therefore decreases as the rate of glucose metabolism in cells increases (see p. 277). Too little insulin secretion, such as that which occurs with diabetes mellitus, produces the opposite effects. Less glucose leaves the blood and enters the cells; more glucose therefore remains in the blood, and less glucose is metabolized by cells. In other words, high blood glucose (hyperglycemia) and a low rate of glucose metabolism characterize insulin deficiency. Insulin is the only hormone that lowers the blood glucose level. Several other hormones, on the other hand, increase it. Growth hormone secreted by the anterior pituitary gland, hydrocortisone secreted by the adrenal cortex, epinephrine secreted by the adrenal medulla and

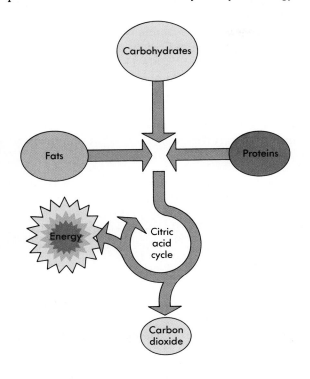

FIGURE 17-3 Catabolism of Nutrients. Fats, carbohydrates, and proteins can be converted to products that enter the citric acid cycle to yield energy.

| TABLE 17-1 Amino Acids | |
ESSENTIAL	NONESSENTIAL
Histidine*	Alanine
Isoleucine	Arginine
Leucine	Asparagine
Lysine	Aspartic acid
Methionine	Cysteine
Phenylalanine	Glutamic acid
Threonine	Glutamine
Tryptophan	Glycine
Tyrosine†	Proline
Valine	Serine

*Essential in infants only.
†Can be synthesized from phenylalanine.

PROTEIN METABOLISM

In a healthy person, proteins are catabolized to release energy to a very small extent. When fat reserves are low, as they are in the starvation that accompanies certain eating disorders such as anorexia nervosa, the body can start to use its protein molecules as an energy source. After this shift occurs, death may quickly follow because vital proteins in the muscles and nerves are catabolized (Figure 17-3).

A more common situation in normal bodies is protein anabolism, the process by which the body builds amino acids into complex protein compounds (for example, enzymes and proteins that form the structural part of the cell). Proteins are assembled from a pool of 20 different kinds of amino acids. If any one type of amino acid is deficient, several vital proteins cannot be synthesized—a serious health threat. One way your body maintains a constant supply of amino acids is by synthesizing them from other compounds already present in the body. Only about half of the required 20 types of amino acid can be made by the body, however. The remaining types of amino acids must be supplied in the diet. Nutritionists often refer to the amino acids that must be in the diet as *essential* amino acids. Table 17-1 lists amino acids according to whether they are considered essential in the diet or *nonessential* (synthesized by the body).

glucagon secreted by the pancreatic islets are four of the most important hormones that increase blood glucose. More information about these hormones and others that help control metabolism appears in Chapter 10.

FAT METABOLISM

Fats, like carbohydrates, are primarily energy foods. If cells have inadequate amounts of glucose to catabolize, they immediately shift to the catabolism of fats for their energy supply. This happens normally when a person goes without food for many hours. It happens abnormally in diabetic individuals. Because of an insulin deficiency, too little glucose enters the cells of a diabetic person to supply all energy needs. Result? The cells catabolize fats to make up the difference (Figure 17-3). In all persons, fats not needed for catabolism are anabolized and stored in adipose tissue.

Vitamins and Minerals

One glance at the label of any packaged food product reveals the importance we place on vitamins and minerals. We know that carbohydrates, fats, and proteins are used by our bodies to build important molecules and to provide energy. So why do we need vitamins and minerals?

First, let's discuss the importance of vitamins. Vitamins are organic molecules needed in small quantities for normal metabolism throughout the body. Vitamin molecules attach to enzymes and help them work properly. Many enzymes are totally useless without the appropriate vitamins to activate them. Most vitamins cannot be made by the body, so we must eat them in our food. The body can store fat-soluble vitamins—A, D, E, and K—in the liver for later use. Because the body cannot store water-soluble vitamins such as B vitamins and vitamin C, they must be continually supplied in the diet. Table 17-2 lists some of the more well-known vitamins, their sources, functions, and symptoms of deficiency.

Vitamin deficiency or **avitaminosis** (a-vye-tah-min-OS-is) can lead to severe metabolic problems. For example, *avitaminosis C* (vitamin C deficiency)

TABLE 17-2 Major Vitamins

VITAMIN	DIETARY SOURCE	FUNCTIONS	SYMPTOMS OF DEFICIENCY
Vitamin A	Green and yellow vegetables, dairy products, and liver	Maintains epithelial tissue and produces visual pigments	Night blindness and flaking skin
B-complex vitamins			
B_1 (thiamine)	Grains, meat, and legumes	Helps enzymes in the citric acid cycle	Nerve problems (beriberi), heart muscle weakness, and edema
B_2 (riboflavin)	Green vegetables, organ meats, eggs, and dairy products	Aids enzymes in the citric acid cycle	Inflammation of skin and eyes
B_3 (niacin)	Meat and grains	Helps enzymes in the citric acid cycle	Pellagra (scaly dermatitis and mental disturbances) and nervous disorders
B_5 (pantothenic acid)	Organ meat, eggs, and liver	Aids enzymes that connect fat and carbohydrate metabolism	Loss of coordination (rare)
B_6 (pyridoxine)	Vegetables, meats, and grains	Helps enzymes that catabolize amino acids	Convulsions, irritability, and anemia
B_{12} (cyanocobalamin)	Meat and dairy products	Involved in blood production and other processes	Pernicious anemia
Biotin	Vegetables, meat, and eggs	Helps enzymes in amino acid catabolism and fat and glycogen synthesis	Mental and muscle problems (rare)
Folic acid	Vegetables	Aids enzymes in amino acid catabolism and blood production	Digestive disorders and anemia
Vitamin C (ascorbic acid)	Fruits and green vegetables	Helps in manufacture of collagen fibers	Scurvy and degeneration of skin, bone, and blood vessels
Vitamin D (calciferol)	Dairy products and fish liver oil	Aids in calcium absorption	Rickets and skeletal deformity
Vitamin E (tocopherol)	Green vegetable and seeds	Protects cell membranes from being catabolized	Muscle and reproductive disorders (rare)

can lead to **scurvy** (SKER-vee) (Figure 17-4). Scurvy results from the inability of the body to manufacture and maintain collagen fibers. As you may have gathered from your studies thus far, collagen fibers compose the connective tissues that hold most of the body together. In scurvy, the body literally falls apart in the same way that a neglected house eventually falls apart. More details about scurvy and other types of avitaminosis are given in Appendix B.

Some forms of **hypervitaminosis** (hye-per-vye-tah-min-OS-is)—or vitamin excess—can be just as serious as a deficiency of vitamins. For example, chronic *hypervitaminosis A* can occur if large amounts of vitamin A (over 10 times the U.S. Recommended Daily Allowance) are consumed daily over a period of 3 months or more. This condition first manifests itself with dry skin, hair loss, anorexia (appetite loss), and vomiting—but may progress to severe headaches and mental disturbances, liver enlargement, and occasionally cirrhosis. Acute hypervitaminosis A, characterized by vomiting, abdominal pain, and headache, can occur if a massive overdose is ingested. Excesses of the fat-soluble vitamins (A, D, E, and K) are generally more serious than excesses of the water-soluble vitamins (B complex and C).

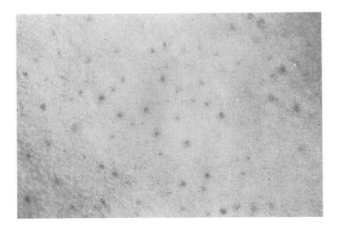

▼ **FIGURE 17-4 Scurvy.** Scurvy impairs the normal maintainance of collagen-containing connective tissues, causing bleeding and ulceration of the skin, gums, and other tissues.

Minerals are just as important as vitamins. Minerals are inorganic elements or salts found naturally in the earth. Like vitamins, mineral ions can attach to enzymes and help them work. Minerals also function in a variety of other vital chemical reactions. For example, sodium, calcium, and other minerals are required for nerve conduction and for contraction in muscle fibers. Without these minerals, the brain, heart, and respiratory tract would cease to function. Information about some of the more important minerals is summarized in Table 17-3.

Like vitamins, minerals are beneficial only when taken in the proper amounts. Many of the minerals listed in Table 17-3 are required in trace amounts. Any intake of such minerals beyond the

CHOLESTEROL

Cholesterol is a type of lipid that has many uses in the body (see Appendix A). The body derives steroid hormones from cholesterol (see Chapter 10) and uses cholesterol to stabilize the phospholipid bilayer that forms the plasma membrane and membranous organelles of the cells. So why does such a useful substance have such a bad reputation? The reason lies in the fact that an *excess* of cholesterol in the blood, a condition called **hypercholesterolemia** (hye-per-koles-ter-ol-EE-me-ah), increases the risk of developing *atherosclerosis*. You may recall from Chapter 13 that atherosclerosis develops into a type of arteriosclerosis or "hardening of the arteries" that can lead to heart disease, stroke, and other problems. Hypercholesterolemia occurs most often in people with a genetic predisposition but is certainly affected by other factors such as diet and exercise. People with hypercholesterolemia are encouraged to switch to diets low in cholesterol and saturated fats and to participate in aerobic exercise, both of which tend to lower blood cholesterol levels. Appendix A discusses different types of cholesterol and their roles in health and disease.

▽ **TABLE 17-3** Major Minerals

MINERAL	DIETARY SOURCE	FUNCTIONS	SYMPTOMS OF DEFICIENCY
Calcium (Ca)	Dairy products, legumes, and vegetables	Helps blood clotting, bone formation, and nerve and muscle function	Bone degeneration and nerve and muscle malfunction
Chlorine (Cl)	Salty foods	Aids in stomach acid production and acid-base balance	Acid-base imbalance
Cobalt (Co)	Meat	Helps vitamin B_{12} in blood cell production	Pernicious anemia
Copper (Cu)	Seafood, organ meats, and legumes	Involved in extracting energy from the citric acid cycle and in blood production	Fatigue and anemia
Iodine (I)	Seafood and iodized salt	Aids in thyroid hormone synthesis	Goiter (thyroid enlargement) and decrease of metabolic rate
Iron (Fe)	Meat, eggs, vegetables, and legumes	Involved in extracting energy from the citric acid cycle and in blood production	Fatigue and anemia
Magnesium (Mg)	Vegetables and grains	Helps many enzymes	Nerve disorders, blood vessel dilation, and heart rhythm problems
Manganese (Mn)	Vegetables, legumes, and grains	Helps many enzymes	Muscle and nerve disorders
Phosphorus (P)	Dairy products and meat	Aids in bone formation and is used to make ATP, DNA, RNA, and phospholipids	Bone degeneration and metabolic problems
Potassium (K)	Seafood, milk, fruit, and meats	Helps muscle and nerve function	Muscle weakness, heart problems, and nerve problems
Sodium (Na)	Salty foods	Aids in muscle and nerve function and fluid balance	Weakness and digestive upset
Zinc (Zn)	Many foods	Helps many enzymes	Metabolic problems

recommended trace amount may become toxic—perhaps even life-threatening.

▽ *Metabolic Rates*

The **basal metabolic rate (BMR)** is the rate at which food is catabolized under basal conditions (that is, when the individual is resting but awake, is not digesting food, and is not adjusting to a cold external temperature). Or, stated differently, the BMR is the number of calories of heat that must be produced per hour by catabolism just to keep the body alive, awake, and comfortably warm. For the determination of the BMR, the amount of oxygen the individual inhales in a specific length of time is measured, and this figure is used to calculate the BMR, stated in kilocalories per hour (kcal/hr).

The BMR is an indirect measure of thyroid gland functioning. So, too, is the amount of protein-bound iodine (PBI) in venous blood. A higher than normal concentration of PBI indicates a higher than normal secretion of thyroid hormone and a higher than normal metabolic rate. Thyroid hormone tests have largely replaced BMR or PBI

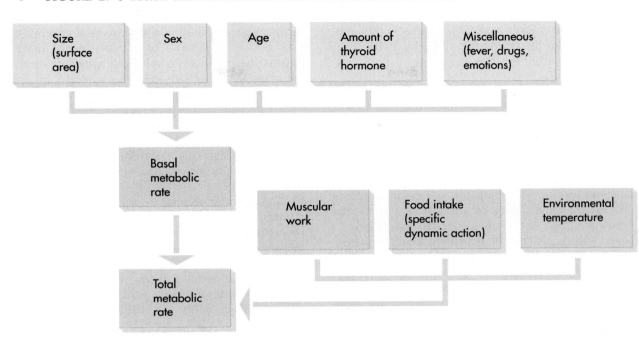

FIGURE **17-5** Factors that Determine the Basal and Total Metabolic Rates.

determination in clinical use because thyroid hormone tests are direct measurements of gland function and hence are more accurate.

BMR represents the amount of food that a person's body must catabolize each day for him or her to stay alive and awake in a comfortably warm environment. To provide energy for muscular work and digestion and absorption of food, an additional amount of food must be catabolized. The amount of additional food depends mainly on how much work the individual does. The more active he or she is, the more food that the body must catabolize and the higher the total metabolic rate will be. The **total metabolic rate (TMR)** is the total amount of energy used by the body per day (Figure 17-5).

When the number of calories in your food intake equals your TMR, your weight remains constant (except for possible variations resulting from water retention or water loss). When your food intake provides more calories than your TMR, you gain weight; when your food intake

VITAMIN SUPPLEMENTS FOR ATHLETES

Because a deficiency of vitamins *(avitaminosis)* can cause poor athletic performance, many athletes regularly consume vitamin supplements. However, research suggests that vitamin supplementation has little or no effect on a person's athletic performance. A reasonably well-balanced diet supplies more than enough vitamins for even the elite athlete. The use of vitamin supplements therefore has fueled somewhat of a controversy among exercise experts. Opponents of vitamin supplements cite the cost and the possibility of liver damage associated with some forms of *hypervitaminosis*, whereas supporters cite the benefit of protecting against vitamin deficiency.

provides fewer calories than your TMR, you lose weight. These weight control principles rarely fail to operate. Nature does not forget to count calories. Reducing diets make use of this knowledge. They contain fewer calories than the TMR of the individual eating the diet.

Metabolic and Eating Disorders

Disorders characterized by a disruption or imbalance of normal metabolism can be caused by several different factors. For example, *inborn errors of metabolism* are a group of genetic conditions involving a deficiency or absence of a particular enzyme. Specific enzymes are required by cells to carry out each step of every metabolic reaction. Although an abnormal genetic code may affect the production of only a single enzyme, the resulting abnormal metabolism may have widespread effects. Specific diseases resulting from inborn errors of metabolism, such as *phenylketonuria,* are discussed in Chapter 23.

A number of metabolic disorders are complications of other conditions. For example, you may recall from Chapter 10 that both hyperthyroidism and hypothyroidism have profound effects on the basal metabolic rate (BMR). Diabetes mellitus affects metabolism throughout the body when an insulin deficiency limits the amount of glucose available for use by the cells.

Some metabolic disorders result from normal mechanisms in the body that maintain homeostasis. For example, the body has several mechanisms that maintain a relatively constant level of glucose in the blood—glucose required by cells for life-sustaining catabolism. As mentioned earlier in this chapter, during *starvation* or in certain *eating disorders,* these mechanisms are taken to the extreme as they attempt to maintain blood glucose homeostasis. A few of the more well-known eating and nutrition disorders are briefly described here:

1 Anorexia nervosa is a behavioral disorder characterized by chronic refusal to eat, often because of an abnormal fear of becoming obese. This condition is most commonly seen in teenage girls and young adult women and is often linked to emotional stress. Treatment is usually directed at solving the resulting nutritional deficit first, then dealing with the underlying behavioral problem.

2 Bulimia (bu-LEE-me-ah) is a behavioral disorder characterized by insatiable craving for food alternating with periods of self-deprivation. The self-deprivation that follows a "food binge" is often accompanied by depression. People with a form of this disorder called *bulimarexia* (bu-lee-mah-REK-see-ah) purposely induce the vomiting reflex to purge themselves of the food they just ate. Excessive vomiting in this way can have a variety of consequences, including damage to the esophagus, pharynx, mouth, and teeth by stomach acid.

3 Obesity is not an eating disorder itself but may be a symptom of chronic overeating behavior. Like anorexia nervosa and bulimia, eating disorders characterized by chronic overeating usually have an underlying emotional cause. Obesity is defined as an abnormal increase in the proportion of fat in the body. Most of the excess fat is stored in the subcutaneous tissue and around the viscera. Obesity is a risk factor for a variety of life-threatening diseases, including many forms of cancer and heart disease.

4 Protein-calorie malnutrition (PCM) is an abnormal condition resulting from a deficiency of calories in general and protein in particular. PCM is likely to result from reduced intake of food but

MEASURING ENERGY

Physiologists studying metabolism must be able to express a quantity of energy in mathematical terms. The unit of energy measurement most often used is the **calorie (cal).** A calorie is the amount of energy needed to raise the temperature of 1 g of water 1° C. Because physiologists often deal with very large amounts of energy, the larger unit, *kilocalorie (kcal)* or *Calorie* (notice the upper-case C), is used. There are 1000 cal in 1 kcal or Calorie. Nutritionists prefer to use *Calorie* when they express the amount of energy stored in a food.

TABLE 17-4   Some Causes of Protein-Calorie Malnutrition

CONDITION	IMPACT ON NUTRIENTS
Conditions that Reduce Nutrient Intake	
Anorexia	Absence of appetite; reduced motivation to eat
Dysphagia	Difficulty in swallowing; inhibition of normal eating
Gastrointestinal obstruction	Inability of food to be digested or absorbed
Nausea	Upset stomach; discomfort, which inhibits appetite
Pain	Discomfort, which discourages eating
Poverty	Inability to acquire proper nutrients
Social isolation	Absence of social cues or motivation for eating
Substance abuse	Reduction or replacement of the motivation to eat
Tooth problems	Difficulty in chewing, which discourages or prevents eating
Conditions that Increase Loss of Nutrients	
Diarrhea	Increased intestinal motility, which reduces absorption of nutrients
Glycosuria	Loss of glucose in the urine
Hemorrhage	Loss of blood and the nutrients it contains
Malabsorption	Failure to properly absorb nutrients, which causes nutrients to pass unabsorbed
Conditions that Increase the Use of Nutrients by the Body	
Burns	Loss of nutrients from damaged tissues
Fever	Increased temperature and metabolic rate, which increase rate of nutrient catabolism
Infection	Increased immune activity and tissue repair, which increase the rate of nutrient use
Trauma and surgery	Increased immune activity, tissue repair, and homeostatic-compensating mechanisms, which increase the rate of nutrient use
Tumors	Increased tissue growth, which increases the rate of nutrient use

may also be caused by increased nutrient loss or increased use of nutrients by the body. Table 17-4 summarizes a few of the many conditions that may lead to PCM. Mild cases occur frequently in illness; as many as one in five patients admitted to the hospital are significantly malnourished. More severe cases of PCM are likely to occur in parts of the world where food, especially protein-rich food, is relatively unavailable. There are two forms of advanced PCM: *marasmus* and *kwashiorkor* (kwah-shee-OR-kor). Marasmus results from an overall lack of calories and proteins, such as when sufficient quantities of food are not available. Marasmus is characterized by progressive wasting of muscle and subcutaneous tissue

accompanied by fluid and electrolyte imbalances. Kwashiorkor results from a protein deficiency in the presence of sufficient calories, as when a child is weaned from milk to low-protein foods. Kwashiorkor also causes wasting of tissues, but unlike marasmus, it also causes pronounced ascites (abdominal bloating) and flaking dermatitis. The ascites results from a deficiency of plasma proteins, which changes the osmotic balance of the blood and thus promotes osmosis of water from the blood into the peritoneal space (see Figure 16-21).

Nutrition disorders, including many specific deficiency diseases, are summarized in Appendix B.

Body Temperature

Considering the fact that over 60% of the energy released from food molecules during catabolism is converted to heat rather than being transferred to ATP, it is no wonder that maintaining a constant body temperature is a challenge. Maintaining homeostasis of body temperature or thermoregulation is the function of the hypothalamus. The hypothalamus operates a variety of negative-feedback mechanisms that keep body temperature in its normal range (36.2° to 37.6° C or 97° to 100° F).

The skin is often involved in negative-feedback loops that maintain body temperature. When the body is overheated, blood flow to the skin increases (Figure 17-6). Warm blood from the body's core can then be cooled by the skin, which acts as a radiator. At the skin, heat can be lost from blood by the following mechanisms:

1 Radiation—flow of heat waves away from the blood

FIGURE 17-6 The Skin as a Thermoregulatory Organ. When homeostasis requires that the body conserve heat, blood flow in the warm organs of the body's core increases *(left)*. When heat must be lost to maintain the stability of the internal environment, flow of warm blood to the skin increases *(right)*. Heat can be lost from the blood and skin by means of radiation, conduction, convection, and evaporation.

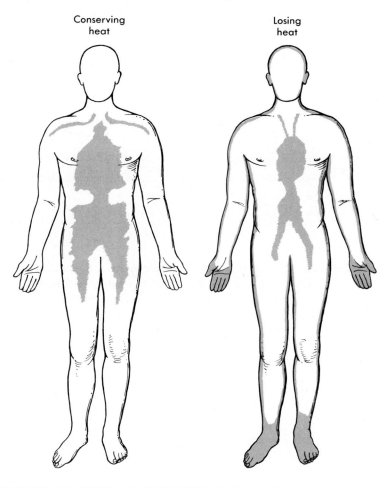

Conserving heat

Losing heat

2 Conduction—transfer of heat energy to the skin and then the external environment

3 Convection—transfer of heat energy to air that is continually flowing away from the skin

4 Evaporation—absorption of heat by water (sweat) vaporization

When necessary, heat can be conserved by reducing blood flow in the skin (Figure 17-6).

A number of other mechanisms can be called on to help maintain the homeostasis of body temperature. Heat-generating muscle activity such as shivering and secretion of metabolism-regulating hormones are two of the body's processes that can be altered to adjust the body's temperature. The concept of using feedback control loops in homeostatic mechanisms was introduced in Chapter 1.

ABNORMAL BODY TEMPERATURE

Maintenance of a body temperature within a narrow range is necessary for normal functioning of the body. As Figure 17-7 shows, straying too far out of the normal range of body temperatures can have very serious physiological consequences. A few important conditions related to body temperature follow:

1 Fever—As explained in Chapter 4, a fever or *febrile* state is an unusually high body temperature associated with a systemic inflammation response. In the case of infections, chemicals called *pyrogens* (literally "fire-makers") cause the thermostatic control centers of the hypothalamus

FIGURE 17-7 Body Temperature. This diagram, modeled after a thermometer, shows some of the physiological consequences of abnormal body temperature. The normal range of body temperature under a variety of conditions is shown in the inset.

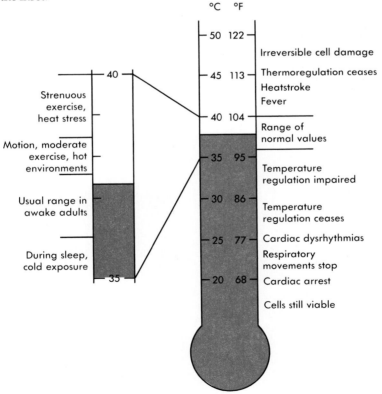

to produce a fever. Because the body's "thermostat" is reset to a higher setting, a person feels a need to warm up to this new temperature and often experiences "chills" as the febrile state begins. The high body temperature associated with infectious fever is thought to enhance the body's immune responses, eliminating the pathogen. Strategies aimed at reducing the temperature of a febrile person are normally counteracted by the body's heat-generating mechanisms and have the effect of further weakening the infected person. Under ordinary circumstances, it is best to let the fever "break" on its own after the pathogen is destroyed.

2 Malignant hyperthermia (MH) is an inherited condition characterized by abnormally increased body temperature (hyperthermia) and muscle rigidity when exposed to certain anasthetics (succinylcholine, for example). The drug *dantrolene (Dantrium)*, which inhibits heat-producing muscle contractions, has been used to prevent or relieve effects of this condition.

3 Heat exhaustion occurs when the body loses a large amount of fluid resulting from heat-loss mechanisms. This usually happens when environmental temperatures are high. Although a normal body temperature is maintained, the loss of water and electrolytes can cause weakness, vertigo, nausea, and possibly loss of consciousness. Heat exhaustion may also be accompanied by skeletal muscle cramps that are often called *heat cramps*. Heat exhaustion is treated with rest (in a cool environment) accompanied by fluid replacement.

4 Heat stroke or *sunstroke* is a severe, sometimes fatal, condition resulting from the inability of the body to maintain a normal temperature in an extremely warm environment. Such thermoregulatory failure may result from factors such as old age, disease, drugs that impair thermoregulation, or simply overwhelming elevated environmental temperatures. Heat stroke is characterized by body temperatures of 41° C (105° F) or higher, tachycardia, headache, and hot, dry skin. Confusion, convulsions, or loss of consciousness may occur. Unless the body is cooled and body fluids replaced immediately, death may result.

5 Hypothermia (hye-po-THER-me-ah) is the inability to maintain a normal body temperature in extremely cold environments. Hypothermia is characterized by body temperatures lower than 35° C (95° F), shallow and slow respirations, and a faint, slow pulse. Hypothermia is usually treated by slowly warming the affected person's body.

6 Frostbite is local damage to tissues caused by extremely low temperatures. Damage to tissues results from formation of ice crystals accompanied by a reduction in local blood flow. Necrosis (tissue death) and even gangrene (decay of dead tissue) can result from frostbite.

OUTLINE SUMMARY

DEFINITIONS

A Nutrition—food, vitamins, and minerals that are ingested and assimilated into the body

B Metabolism—process of using food molecules as energy sources and as building blocks for our own molecules

C Catabolism—breaks food molecules down, releasing their stored energy; oxygen used in catabolism

D Anabolism—builds food molecules into complex substances

ROLE OF THE LIVER

A Processes blood immediately after it leaves the gastrointestinal tract

  1 Helps maintain normal blood glucose level

  2 Site of protein and fat metabolism

  3 Removes toxins from the blood

NUTRIENT METABOLISM

A Carbohydrates are primarily catabolized for energy (Figure 17-1), but small amounts are anabolized by glycogenesis (a series of chemical reactions that changes glucose to glycogen—occurs mainly in liver cells where glycogen is stored)

B Blood glucose (imprecisely, blood sugar)—normally stays between about 80 and 120 mg per 100 ml of blood; insulin accelerates the movement of glucose out of the blood into cells, therefore decreases blood glucose and increases glucose catabolism

C Adenosine triphosphate (ATP)—molecule in which energy obtained from breakdown of foods is stored; serves as a direct source of energy for cellular work (Figure 17-2)

D Fats catabolized to yield energy and anabolized to form adipose tissue (Figure 17-3)

E Proteins primarily anabolized and secondarily catabolized; protein anabolism requires essential amino acids (those required in the diet) and nonessential amino acids (those that can be made by the body)

VITAMINS AND MINERALS

A Vitamins—organic molecules that are needed in small amounts for normal metabolism (Table 17-2)

  1 Avitaminosis—deficiency of a vitamin, such as avitaminosis C (vitamin C deficiency)

  2 Hypervitaminosis—excess of a vitamin, such as hypervitaminosis A (excess of vitamin A)

B Minerals—inorganic molecules required by the body for normal function (Table 17-3)

METABOLIC RATES

A Basal metabolic rate (BMR)—rate of metabolism when a person is lying down but awake and not digesting food and when the environment is comfortably warm

B Total metabolic rate (TMR)—the total amounts of energy, expressed in calories, used by the body per day (Figure 17-5)

C Protein-bound iodine (PBI)—one indirect measure of thyroid secretion and of metabolic rate

D Thyroid hormone tests—direct measurements of thyroid secretion; now often used clinically

METABOLIC AND EATING DISORDERS

A Inborn errors of metabolism—genetic conditions involving deficient or abnormal metabolic enzymes

B Hormonal imbalances may cause metabolic problems

C Eating disorders usually have an underlying emotional cause

D Examples of eating and nutritional disorders

  1 Anorexia nervosa is characterized by chronic refusal to eat

  2 Bulimia involves an alternating pattern of craving of food followed by a period of self-denial; in bulimarexia, the self-denial triggers self-induced vomiting

  3 Obesity—abnormally high proportion of fat in the body; may be a symptom of an eating disorder

  4 Protein-calorie malnutrition (PCM)—results from a deficiency of calories in general and proteins in particular

    a May be a complication of a preexisting condition (Table 17-4)

    b Marasmus—type of advanced PCM caused by an overall lack of calories and protein, characterized by tissue wasting and fluid and electrolyte imbalances

    c Kwashiorkor—type of advanced PCM caused by a lack of protein in the presence of sufficient calories; similar to marasmus but distinguished by ascites and flaking dermatitis

BODY TEMPERATURE

A Hypothalamus—regulates the homeostasis of body temperature through a variety of processes

B Skin—can cool the body by losing heat from the blood through four processes: radiation, conduction, convection, evaporation (Figure 17-6)

C Abnormal body temperature can have serious physiological consequences

1 Fever (febrile state)—unusually high body temperature associated with systemic imflammation response

2 Malignant hyperthermia—inherited condition that causes increased body temperature (hyperthermia) and muscle rigidity when exposed to certain anesthetics

3 Heat exhaustion—results from loss of fluid as the body tries to cool itself; may be accompanied by heat cramps

4 Heat stroke (sunstroke)—overheating of body resulting from failure of thermoregulatory mechanisms in a warm environment

5 Hypothermia—reduced body temperature resulting from failure of thermoregulatory mechanisms in a cold environment

6 Frostbite—local tissue damage caused by extreme cold; may result in necrosis or gangrene

NEW WORDS

anabolism
basal metabolic rate (BMR)
calorie

catabolism
citric acid cycle
glycogenesis

glycolysis
kilocalorie
thermoregulation

total metabolic rate (TMR)
vitamin

Diseases and Other Clinical Terms

anorexia nervosa
avitaminosis
bulimarexia
bulimia
frostbite

heat exhaustion
heat stroke
hypercholesterolemia
hypervitaminosis

hypothermia
malignant hyperthermia
obesity

phenylketonuria (PKU)
protein-calorie malnutrition (PCM)
scurvy

CHAPTER TEST

1. The type of metabolism that involves the breakdown of food molecules is called _____.
2. The type of metabolism that involves the synthesis of large molecules is called _____.
3. *Nutrition* refers to the food that we eat, whereas _____ refers to the use of foods after they have entered the cells.
4. The _____ vein delivers blood from the gastrointestinal tract directly to the liver.
5. The chemical process that changes glucose into pyruvic acid, producing energy, is called _____.
6. The series of reactions in liver cells that joins glucose molecules together to form glycogen is called _____.
7. When the body runs low on carbohydrates several hours after a meal, it begins catabolizing _____ instead.
8. _____ are organic molecules needed in small quantities for normal metabolism throughout the body.
9. The molecule in which energy from catabolism is stored until it is used by the cell is called _____.
10. The rate at which food is catabolized in a resting individual who is not digesting a meal is called the _____.
11. Thermoregulation is controlled mainly by the _____ in the brain.
12. Heat exhaustion may progress to become _____ _____ if the body's thermoregulatory mechanisms fail.
13. _____ occurs when tissue in local areas is damaged by extreme or prolonged cold temperatures.

Circle the T before each true statement and the F before each false statement.

T F 14. The citric acid cycle changes glucose into pyruvic acid, releasing energy for the cell's use.
T F 15. Insulin promotes the entry of glucose into body cells.
T F 16. Anorexia nervosa can result in death.
T F 17. Minerals can also be called vitamins.
T F 18. Sodium is a mineral that is found in only a few types of food.
T F 19. Measuring oxygen consumption or blood concentration of protein-bound iodine can be used to determine a person's rate of metabolism.
T F 20. The skin is an important organ for maintaining the body's temperature homeostasis.

REVIEW QUESTIONS

1 Briefly and clearly explain anabolism, catabolism, metabolism, and nutrition.
2 What adaptive advantage is gained by detouring blood from the gastrointestinal tract to the liver before returning it to the heart?
3 In words or in a sketched diagram, describe the metabolic pathway taken by a glucose molecule when energy is extracted from it.
4 Liver cells perform a process that prevents the blood glucose level from getting too high just after a large meal. What is it?
5 How does the body get energy during fasting?
6 How are fats used by the body? Proteins?
7 What is the difference between essential and non-essential amino acids?
8 Explain the metabolic roles of vitamins and minerals.
9 Explain what is meant by the term *metabolic rate*. What is the difference between basal metabolic rate and total metabolic rate.
10 How can a person's metabolic rate be measured? What units of measurement are used?
11 Explain why you think the following statement is true or false: "If you do not want to gain or lose weight but just stay the same, you must eat just enough food to supply the kilocalories of your BMR. If you eat more than this, you will gain; if you eat less than this, you will lose."
12 What is protein-calorie malnutrition? What factors or conditions may lead to this nutritional disorder?
13 Compare and contrast anorexia nervosa and bulimia.
14 How does the body maintain the body temperature within a normal range?
15 Name the four mechanisms by which the skin removes heat from the body.
16 Briefly explain the mechanisms for each of these abnormalities: fever, malignant hyperthermia, heat exhaustion, heat stroke, hypothermia, frostbite.

CLINICAL APPLICATIONS

1 A friend of yours is helping you chop firewood on a hot day. She complains of muscle cramps and nausea but has a normal body temperature. What has happened to her? How would you help your friend?
2 While looking through an old family album, you can't help but notice that your great-great-great-grandfather's smile reveals that he has no teeth. When asked why this ancestor lost his teeth at an early age, your grandmother replies that he suffered from scurvy as a merchant marine and lost all his teeth as a result. Is this possible? Can you explain how scurvy can cause the loss of teeth?
3 Andrea is planning to adopt a totally vegetarian diet—a diet that includes no meats or animal products. Her friends have voiced some concern that her new diet may not contain certain essential amino acids. What is an essential amino acid? Why must her diet contain these nutrients?

The Urinary System

C H A P T E R

18

Objectives

*After you have completed this chapter,
you should be able to:*

1 Identify the major organs of the urinary system and give the generalized function of each.

2 Name the parts of a nephron and describe the role each component plays in the formation of urine.

3 Explain the importance of filtration, tubular reabsorption, and tubular secretion in urine formation.

4 Discuss the mechanisms that control urine volume.

5 Explain how the kidneys act as vital organs in maintaining homeostasis.

6 List the major renal and urinary disorders and explain the mechanism of each.

s you might guess from its name, the urinary system performs the functions of producing and excreting urine from the body. What you might not guess so easily is how essential these functions are for the maintenance of homeostasis and healthy survival. The constancy of body fluid volumes and the levels of many important chemicals depend on normal urinary system function. Unless the urinary system operates normally, the normal composition of blood cannot be maintained long, and serious consequences soon follow. The kidneys "clear" or clean the blood of the many waste products continually produced as a result of metabolism of foodstuffs in the body cells. As nutrients are burned for energy, the waste products produced must be removed from the blood, or they quickly accumulate to toxic levels—a condition called **uremia** (yoo-REE-mee-ah) or **uremic poisoning.** The kidneys also play a vital role in maintaining electrolyte, water, and acid-base balances in the body. In this chapter, we will discuss the structure and function of each organ of the urinary system. There are two kidneys, two ureters, one bladder, and one urethra (Figure 18-1). We will also discuss disease conditions produced by abnormal functioning of the urinary system.

Kidneys

LOCATION

To locate the kidneys on your own body, stand erect and put your hands on your hips with your thumbs meeting over your backbone. When you are in this position, your kidneys lie just above your thumb; in short, the kidneys lie just above the waistline. Usually the right kidney is a little lower than the left. They are located under the muscles of the back and behind the parietal peritoneum (the membrane that lines the abdominal cavity). Because of this retroperitoneal location, a surgeon can operate on a kidney without cutting through the peritoneum. A heavy cushion of fat normally encases each kidney and helps hold it in place.

Note the relatively large diameter of the renal arteries in Figure 18-1. Normally a little over 20% of the total blood pumped by the heart each minute enters the kidneys. The rate of blood flow

through this organ is among the highest in the body. This is understandable because one of the main functions of the kidney is to remove waste products from the blood. Maintenance of a high rate of blood flow and normal blood pressure in the kidney is essential for the formation of urine.

INTERNAL STRUCTURE

If you were to slice through a kidney from side to side and open it like the pages of a book, you would see the structures shown in Figure 18-2. Identify each of the following parts:

1 **Cortex** (KOR-teks)—the outer part of the kidney (The word *cortex* comes from the Latin word for "bark" or "rind," so the cortex of an organ is its outer layer; each kidney and adrenal gland, as well as the brain, has a cortex.)
2 **Medulla** (me-DUL-ah)—the inner portion of the kidney
3 **Pyramids** (PIR-ah-mids)—the triangular divisions of the medulla of the kidney
4 **Papilla** (pah-PIL-ah) (pl. *papillae*)—narrow, innermost end of a pyramid
5 **Pelvis**—(the kidney or renal pelvis) an expansion of the upper end of a ureter (the tube that drains urine into the bladder)
6 **Calyx** (KAY-liks) (pl. *calyces*)—a division of the renal pelvis (The papilla of a pyramid opens into each calyx.)

MICROSCOPIC STRUCTURE

More than a million microscopic units called **nephrons** (NEF-rons) make up each kidney's interior. The shape of a nephron is unique, unmistakable, and admirably suited to its function of producing urine. It looks a little like a tiny funnel with a very long stem, but it is an unusual stem in that it is highly convoluted (that is, it has many bends in it). The nephron is composed of two principle parts: the **renal corpuscle** located in the cortex of the kidney and the **renal tubule.** Locate the following parts of a nephron in Figures 18-3 to 18-4.

1 **Renal corpuscle**
   a **Bowman's capsule**—the cup-shaped top of a nephron (The sacklike Bowman's capsule surrounds the glomerulus.)

Text continued on p. 457.

FIGURE 18-1 Urinary System. A, Location of urinary system organs. **B,** X-ray film of the urinary organs.

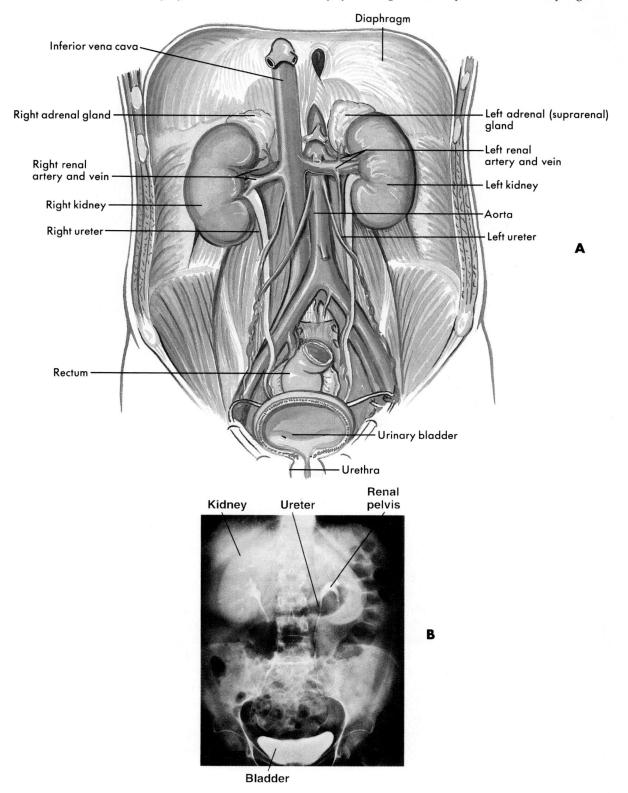

Diaphragm

Inferior vena cava

Right adrenal gland

Left adrenal (suprarenal) gland

Left renal artery and vein

Right renal artery and vein

Left kidney

Right kidney

Aorta

Right ureter

Left ureter

A

Rectum

Urinary bladder

Urethra

Kidney

Ureter

Renal pelvis

B

Bladder

FIGURE 18-2 Coronal Section Through the Right Kidney.

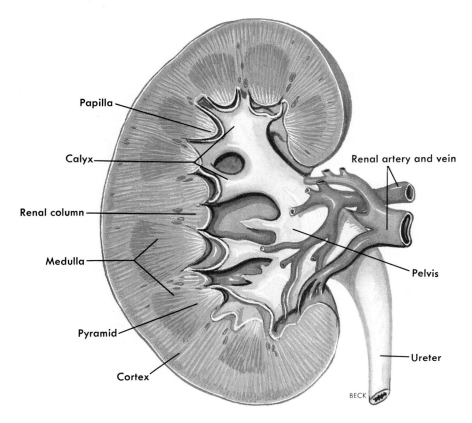

Papilla

Calyx

Renal column

Medulla

Pyramid

Cortex

Renal artery and vein

Pelvis

Ureter

BECK

URINARY CATHETERIZATION

Urinary catheterization is the passage or insertion of a hollow tube or catheter through the urethra into the bladder for the withdrawal of urine. It is a medical procedure commonly performed on many patients who undergo prolonged surgical or diagnostic procedures or who experience problems with urinary retention.

Correct catheterization procedures require aseptic techniques to prevent the introduction of infectious bacteria into the urinary system. Clinical studies have proved that improper catheterization techniques cause bladder infections (cystitis) in hospitalized patients. One landmark study confirmed the high percentage of catheterized patients who develop cystitis (almost 8%) and found that, of those who developed such infections, a significant number died. The results of the study are sobering and point out the need for extensive training of health professionals in this area.

▼ **FIGURE 18-3 Magnified Wedge Cut from a Renal Pyramid.** Note that the renal corpuscles (glomeruli surrounded by Bowman's capsule) and both proximal and distal convoluted tubules are located in the cortex of the kidney. The medulla contains the loop of Henle and collecting tubules. Urine from the collecting tubules exits from the pyramid thorough the papilla and enters the calyx and renal pelvis before flowing into the ureter. The inset shows a scanning electron micrograph of several glomeruli and their associated blood vessels. Note that the blood vessel that brings blood to the glomerulus (afferent arteriole) has a larger diameter than the blood vessel that drains blood from it (efferent arteriole).

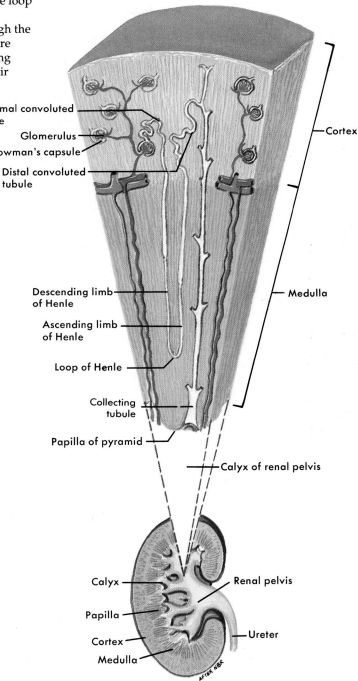

Proximal convoluted tubule

Renal corpuscle { Glomerulus
Bowman's capsule

Distal convoluted tubule

Descending limb of Henle

Ascending limb of Henle

Loop of Henle

Collecting tubule

Papilla of pyramid

Cortex

Medulla

Calyx of renal pelvis

Calyx

Papilla

Cortex

Medulla

Renal pelvis

Ureter

AFTER HBR

Glomerular capillaries

Efferent arteriole   Afferent arteriole

FIGURE 18-4 The Nephron Unit. Cross sections from the four segments of the renal tubule are shown. The differences in appearance in tubular cells seen in the cross sections reflect the differing functions of each nephron segment.

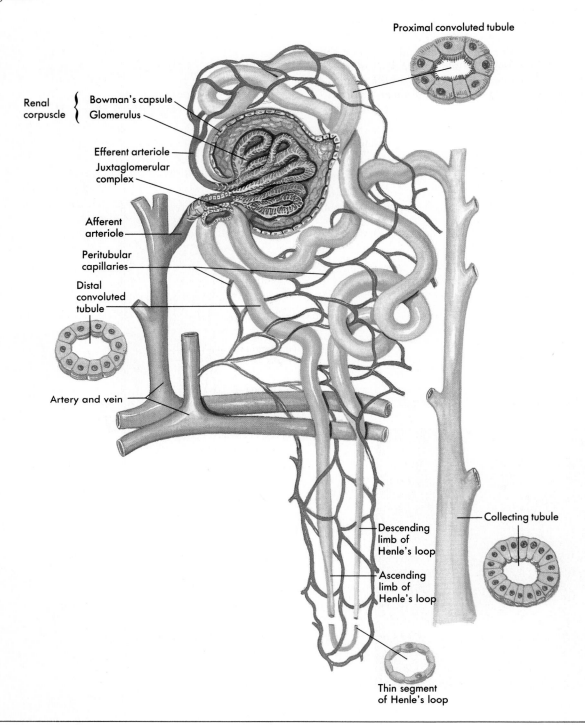

b **Glomerulus** (glo-MAIR-yoo-lus) (pl. *glomeruli*)—a network of blood capillaries tucked into Bowman's capsule (Note in Figures 18-3 and 18-4 that the small artery [arteriole] that delivers blood to the glomerulus, **[afferent arteriole]** is larger in diameter than the blood vessel that drains blood from it **[efferent arteriole]** and that it is relatively short. This explains the high blood pressure that exists in the glomerular capillaries. This high pressure is required to filter wastes from the blood.)

2 **Renal tubule**

a **Proximal convoluted tubule**—the first segment of a renal tubule (It is called *proximal* because it lies nearest the tubule's origin from Bowman's capsule, and *convoluted* because it has several bends.)

b **Loop of Henle** (HEN-lee)—the extension of the proximal tubule (Observe that the loop of Henle consists of a straight descending limb, a hairpin loop, and a straight ascending limb.)

c **Distal convoluted tubule**—the part of the tubule distal to the ascending limb of the loop of Henle (It is the extension of the ascending limb.)

d **Collecting tubule**—a straight (that is, not convoluted) part of a renal tubule (Distal tubules of several nephrons join to form a single collecting tubule or duct.)

FUNCTION

The kidneys are vital organs. The function they perform, that of forming urine, is essential for homeostasis and maintenance of life. Early in the process of urine formation, fluid, electrolytes, and wastes from metabolism are *filtered* from the blood and enter the nephron. Additional wastes may be *secreted* into the tubules of the nephron as substances useful to the body are reabsorbed into the blood. Normally the kidneys balance the amount of many substances entering and leaving the blood over time so that normal concentrations can be maintained. In short, the kidneys adjust their output to equal the intake of the body. By eliminating wastes and adjusting fluid balance, the kidneys play an essential part in maintaining homeostasis. Homeostasis cannot be maintained—nor can life itself—if the kidneys fail and the condition is not soon corrected. Nitrogenous waste products accumulate as a result of protein breakdown and quickly reach toxic levels if not excreted. If kidney function ceases because of injury or disease, life can be maintained by using an **artificial kidney** to cleanse the blood of wastes.

Excretion of toxins and of waste products containing nitrogen such as urea and ammonia represents only one of the important responsibilities of the kidney. The kidney also plays a key role in regulating the levels of many chemical substances in the blood such as chloride, sodium, potassium, and bicarbonate. The kidneys also regulate the proper balance between body water content and salt by selectively retaining or excreting both substances as requirements demand. In addition, the cells of the *juxtaglomerular complex* (Figure 18-4) function in blood pressure regulation. When blood pressure is low, these cells secrete a hormone that initiates constriction of blood vessels and thus raises blood pressure. It is easy to understand why the kidneys are often considered to be the most important homeostatic organs in the body.

Formation of Urine

The kidney's 2 million or more nephrons form urine by a series of three processes: (1) filtration, (2) reabsorption, and (3) secretion (Figure 18-5). Urine formation begins with the process of **filtration,** which goes on continually in the renal corpuscles (Bowman's capsules plus their encased glomeruli). Blood flowing through the glomeruli exerts pressure, and this glomerular blood pressure is high enough to push water and dissolved substances out of the glomeruli into the Bowman's capsule. Briefly, glomerular blood pressure causes filtration through the glomerular-capsular membrane. If the glomerular blood pressure drops below a certain level, filtration and urine formation cease. Hemorrhage, for example, may cause a precipitous drop in blood pressure followed by kidney failure.

Glomerular filtration normally occurs at the rate of 125 ml per minute. The following simple calculations may help you visualize how enormous this volume is:

FIGURE 18-5 Formation of Urine. Diagram shows the steps in urine formation in successive parts of a nephron: filtration, reabsorption, and secretion.

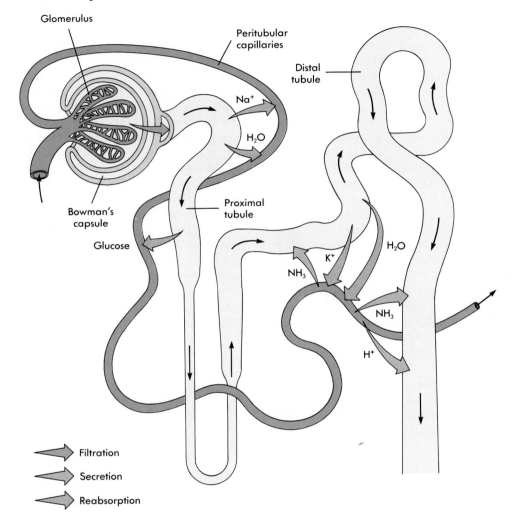

125 ml/min × 60 min/hr = 7500 ml of glomerular
filtrate per hour
7500 ml/hr × 24 hrs/day = 180,000 ml or 180 L
(about 190 quarts) of fluid filtered
out of glomerular blood per day

Obviously no one ever excretes anywhere near 180 L of urine per day. Why? Because most of the fluid that leaves the blood by glomerular filtration, the first process in urine formation, returns to the blood by the second process—reabsorption.

Reabsorption is the movement of substances out of the renal tubules into the blood capillaries located around the tubules (peritubular capillaries). Water, glucose and other nutrients, and sodium and other ions are substances that are reabsorbed. Reabsorption begins in the proximal convoluted tubules and continues in the loop of Henle, distal convoluted tubules, and collecting tubules.

Large amounts of water—approximately 178 L per day—are reabsorbed by osmosis from the

TABLE 18-1 Functions of Parts of Nephron in Urine Formation

PART OF NEPHRON	PROCESS IN URINE FORMATION	SUBSTANCES MOVED
Glomerulus	Filtration	Water and solutes (for example, sodium and other ions, glucose and other nutrients filtering out of glomeruli into Bowman's capsules)
Proximal tubule	Reabsorption	Water and solutes
Loop of Henle	Reabsorption	Sodium and chloride ions
Distal and collecting tubules	Reabsorption	Water, sodium, and chloride ions
	Secretion	Ammonia, potassium ions, hydrogen ions, and some drugs

proximal tubules. In other words, nearly 99% of the 180 L of water that leave the blood each day by glomerular filtration returns to the blood by proximal tubule reabsorption.

The nutrient glucose is entirely reabsorbed from the proximal tubules. It is actively transported out of them into peritubular capillary blood. None of this valuable nutrient is wasted by being lost in the urine. However, exceptions occur. For example, in *diabetes mellitus,* if blood glucose concentration increases above a certain level, the tubular filtrate then contains more glucose than kidney tubule cells can reabsorb. Some of the glucose therefore remains behind in the urine. Glucose in the urine **(glycosuria)** (glye-ko-SOO-ree-ah), is a well-known sign of diabetes mellitus.

Sodium ions and other ions are only partially reabsorbed from renal tubules. For the most part, sodium ions are actively transported back into blood from the tubular urine. The amount of sodium reabsorbed varies from time to time; it depends largely on salt intake. In general the greater the amount of salt intake, the less the amount of salt reabsorption and therefore the greater the amount of salt excreted in the urine. Also, the less the salt intake, the greater the salt reabsorption and the less salt excreted in the urine. By varying the amount of salt reabsorbed, the body usually can maintain homeostasis of the blood's salt concentration. This is an extremely important matter because cells are damaged by either too much or too little salt in the fluid around them (see box about tonicity, p. 28).

Secretion is the process by which substances move into urine in the distal and collecting tubules from blood in the capillaries around these tubules. In this respect, secretion is reabsorption in reverse. Whereas reabsorption moves substances out of the urine into the blood, secretion moves substances out of the blood into the urine. Substances secreted are hydrogen ions, potassium ions, ammonia, and certain drugs. Hydrogen ions, potassium ions, and drugs are secreted by being actively transported out of the blood into tubular urine. Ammonia is secreted by diffusion. Kidney tubule secretion plays a crucial role in maintaining the body's acid-base balance (see Chapter 18).

In summary, the following processes occurring in successive portions of the nephron accomplish the function of urine formation (Table 18-1):

1 **Filtration**—of water and dissolved substances out of the blood in the glomeruli into Bowman's capsule
2 **Reabsorption**—of water and dissolved substances out of the kidney tubules back into the blood (This process prevents substances needed by the body from being lost in the urine. Usually, 97% to 99% of water filtered out of the glomerular blood is retrieved from the tubules.)
3 **Secretion**—of hydrogen ions, potassium ions, and certain drugs

CONTROL OF URINE VOLUME

The body has ways to control the amount and composition of the urine that it excretes. It does this mainly by controlling the amount of water and dissolved substances that are reabsorbed by the convoluted tubules. For example, a hormone (antidiuretic hormone or ADH) from the posterior pituitary gland decreases the amount of urine by making collecting tubules permeable to water. If no ADH is present, the tubules are practically impermeable to water, so little or no water is reabsorbed from them. When ADH is present in the blood, collecting tubules are permeable to water and water is reabsorbed from them. As a result, less water is lost from the body as urine, or more water is retained from the tubules—whichever way you wish to say it. At any rate, for this reason ADH is accurately described as the "water-retaining hormone." You might also think of it as the "urine-decreasing hormone."

The hormone aldosterone, secreted by the adrenal cortex, plays an important part in controlling the kidney tubules' reabsorption of salt. Primarily it stimulates the tubules to reabsorb sodium salts at a faster rate. Secondarily, aldosterone also increases tubular water reabsorption.

The term *salt- and water-retaining hormone* therefore is a descriptive nickname for aldosterone.

Another hormone, atrial natriuretic hormone (ANH) secreted from the heart's atrial wall, has the opposite effect of aldosterone. ANH stimulates kidney tubules to secrete more sodium and thus lose more water. Thus ANH is a *salt- and water-losing hormone.* The body secretes ADH, aldosterone, and ANH in different amounts, depending on the homeostatic balance of body fluids at any particular moment.

Sometimes the kidneys do not excrete normal amounts of urine as a result of kidney disease, cardiovascular disease, or stress. Here are some terms associated with abnormal amounts of urine:

1 **Anuria** (ah-NOO-ree-ah)—absence of urine
2 **Oliguria** (ol-i-GOO-ree-ah)—scanty amounts of urine
3 **Polyuria** (pol-e-YOO-ree-ah)—unusually large amounts of urine

TABLE 18-2　Characteristics of Urine*

NORMAL CHARACTERISTICS	ABNORMAL CHARACTERISTICS
Color Transparent yellow, amber or straw colored	Abnormal colors or cloudiness, which may indicate presence of blood, bile, bacteria, drugs, food pigments, or high-solute concentration
Compounds Mineral ions (for example, Na, Cl, K)	Acetone
Nitrogenous wastes: ammonia, creatinine, urea, uric acid	Albumin
Suspended solids (sediment)†: bacteria, blood cells, casts (solid matter)	Bile
Urine pigments	Glucose
Odor Slight odor	Acetone odor, which is common in diabetes mellitus
pH 4.6-8.0 (freshly voided urine is generally acidic)	High in alkalosis; low in acidosis
Specific Gravity 1.001-1.035	High specific gravity can cause precipitation of solutes and formation of kidney stones

*For more-detailed information, see Appendix D, Table 2.
†Occasional trace amounts.

URINALYSIS

The physical, chemical, and microscopic examination of urine is termed *urinalysis*. Like blood, urine is a fluid that reveals much about the function of the body. Changes in the normal characteristics of urine or the appearance of abnormal urine characteristics may be a sign of disease. Table 18-2 lists the characteristics of urine.

Physical characteristics of urine such as color, turbidity (cloudiness), odor, and specific gravity (density), are general indicators of the composition of urine. For example, an abnormally dark color may indicate the presence of excess bile pigments—possibly a sign of liver disease. On the other hand, a dark color may signify the presence of hemoglobin—a sign of bleeding somewhere in the urinary system. Or, it may simply indicate insufficient fluid intake. In other words, a change in physical characteristics may be a sign that *something* is wrong, but it will not provide the detailed information that a chemical analysis provides.

Chemical analysis of urine usually provides information about pH, urea concentration, or the presence of abnormal chemicals such as glucose, acetone, albumin (protein), or bile. As mentioned previously, the presence of a significant amount of glucose in the urine (glycosuria) is a well-known sign of diabetes mellitus.

Urine specimens are often spun in a *centrifuge* to force suspended particles to the bottom of a test tube. The sediment that forms is then examined with a microscope to detect the presence of abnormal cells or other particles. Blood cells may indicate bleeding or infection somewhere along the urinary tract. An abnormal urine sample may contain numerous *casts*, small particles formed by deposits of minerals or cells on the walls of renal tubules that break off into the urine. A large number of casts may indicate any of several kidney disorders, depending on the composition of the casts.

In clinical and laboratory situations a standard urinalysis is often referred to as a "routine and microscopic" urinalysis, or simply an "R and M." The "routine" portion is a series of physical and chemical tests, whereas the "microscopic" part refers to the study of urine sediment with a microscope. This series of laboratory tests provides the variety of information often necessary for a physician to make a diagnosis. Tests included in typical routine and microscopic urinalysis studies are listed in Appendix D.

Ureters

Urine drains out of the collecting tubules of each kidney into the renal pelvis and down the ureter into the urinary bladder (Figure 18-1). The **renal pelvis** is the basinlike upper end of the ureter located inside the kidney. Ureters are narrow tubes less than 6 mm (¼ inch) wide and 25 to 30 cm (10 to 12 inches) long. Mucous membranes line both ureters and each renal pelvis. Note in Figure 18-6 that the ureter has a thick, muscular wall. Contraction of the muscular coat produces peristaltic-type movements that assist in moving urine down the ureters into the bladder. The lining membrane of the ureters is richly supplied with sensory nerve endings.

FIGURE 18-6 Ureter Cross Section. Note the thick layer of muscle around the tube.

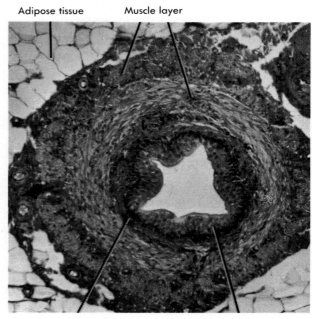

Adipose tissue   Muscle layer

Connective tissue   Transitional epithelium

Urinary Bladder

The empty urinary bladder lies in the pelvis just behind the pubic symphysis. When full of urine, it projects upward into the lower portion of the abdominal cavity.

Elastic fibers and involuntary muscle fibers in the wall of the urinary bladder make it well suit- ed for expanding to hold variable amounts of urine and then contracting to empty itself. Mucous membrane lines the urinary bladder. The lining is loosely attached to the deeper muscular layer so that the bladder is very wrinkled and lies in folds called *rugae* when it is empty. When the bladder is filled, its inner surface is smooth. Note in Figure 18-7 that one triangular area on the back

FIGURE 18-7 The Male Urinary Bladder. This view (with bladder cut to show the interior) shows how the prostate gland surrounds the urethra as it exits from the bladder. The glands associated with the male reproductive system are further discussed in Chapter 19.

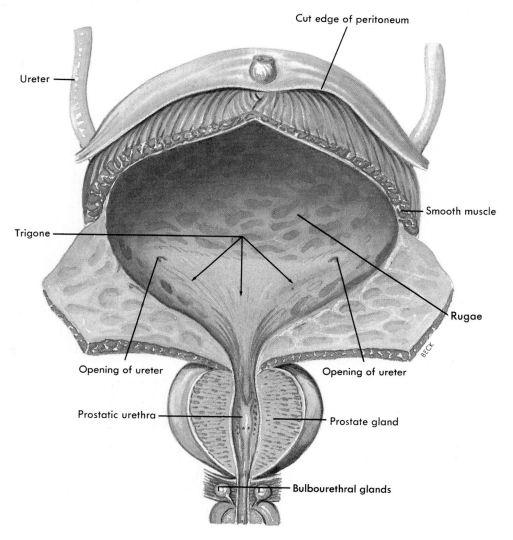

Cut edge of peritoneum
Ureter
Trigone
Smooth muscle
Rugae
Opening of ureter
Opening of ureter
Prostatic urethra
Prostate gland
Bulbourethral glands

or posterior surface of the bladder is free of rugae. This area, called the *trigone,* is always smooth. There, the lining membrane is tightly fixed to the deeper muscle coat. The trigone extends between the openings of the two ureters above and the point of exit of the urethra below.

Urethra

To leave the body, urine passes from the bladder, down the urethra, and out of its external opening, the **urinary meatus.** In other words, the urethra is the lowest part of the urinary tract. The same sheet of mucous membrane that lines each renal pelvis, the ureters, and the bladder extends down into the urethra, too; this is a structural feature worth noting because it accounts for the fact that an infection of the urethra may spread upward through the urinary tract. The urethra is a narrow tube; it is only about 4 cm (1½ inches) long in a woman, and it is about 20 cm (8 inches) long in a man. In a man, the urethra has two functions: (1) it is the terminal portion of the urinary tract, and (2) it is the passageway for movement of the reproductive fluid (semen) from the body. In a woman, the urethra is a part of only the urinary tract.

Micturition

The terms **micturition** (mik-too-RISH-un), **urination** (yoor-i-NAY-shun), and **voiding** refer to the passage of urine from the body or the emptying of the bladder. This is a reflex action in infants or very young children. Although there is considerable variation between individuals, most children between 2 and 3 years of age learn to urinate voluntarily and also to inhibit voiding if the urge comes at an inconvenient time.

Two **sphincters** (SFINGK-ters) or rings of muscle tissue guard the pathway leading from the bladder. The **internal urethral sphincter** is located at the bladder exit, and the **external urethral sphincter** circles the urethra just below the neck of the bladder. When contracted, both sphincters seal off the bladder and allow urine to accumulate without leaking to the exterior. The internal urethral sphincter is involuntary, and the external urethral sphincter is composed of striated muscle and is under voluntary control.

The muscular wall of the bladder permits this organ to accommodate a considerable volume of urine with very little increase in pressure until a volume of 300 to 400 ml is reached. As the volume of urine increases, the need to void may be noticed at volumes of 150 ml, but micturition in adults does not normally occur much below volumes of 350 ml. As the bladder wall stretches, nervous impulses are transmitted to the second, third, and fourth sacral segments of the spinal cord, and an **emptying reflex** is initiated. The reflex causes contraction of the muscle of the bladder wall and relaxation of the internal sphincter. Urine then enters the urethra. If the external sphincter, which is under voluntary control, is relaxed, micturition occurs. Voluntary contraction of the external sphincter suppresses the emptying reflex until the bladder is filled to capacity with urine and loss of control occurs. Contraction of this powerful sphincter also abruptly terminates urination voluntarily.

Higher centers in the brain also function in micturition by integrating bladder contraction and internal and external sphincter relaxation, with the cooperative contraction of pelvic and abdominal muscles. Urinary **retention** is a condition in which no urine is voided. The kidneys produce urine, but the bladder for one reason or another cannot empty itself. In urinary **suppression** the opposite is true. The kidneys do not produce any urine, but the bladder retains the ability to empty itself.

Incontinence (in-KON-ti-nens) is a condition in which the patient voids urine involuntarily. It frequently occurs in patients who have suffered a stroke or spinal cord injury. If the sacral segments of the spinal cord are injured, some loss of bladder function always occurs. Although the voiding reflex may be reestablished to some degree, the bladder does not empty completely. In these individuals the residual urine is often the cause of repeated bladder infections or **cystitis** (sis-TIE-tis). Complete destruction or transection of the sacral cord produces a condition called an *automatic bladder.* Totally cut off from spinal innervation, the bladder musculature acquires some automatic action and periodic but unpredictable voiding occurs.

Renal and Urinary Disorders

You may have experienced the discomfort and pain of a bladder infection or know someone who has. Bladder infection is the most common urinary disorder, but it usually is not serious if promptly treated. A number of renal and urinary disorders are very serious, however. Any disorder that significantly reduces the effectiveness of the kidneys is immediately life-threatening. In this section, we will discuss some life-threatening kidney diseases, as well as a few of the less serious but more common disorders.

OBSTRUCTIVE DISORDERS

Obstructive urinary disorders are abnormalities that interfere with normal urine flow anywhere in the urinary tract. The severity of obstructive disorders depends on the location of the interference and the degree to which the flow of urine is impaired. Obstruction of urine flow usually results in "backing up" of the urine, perhaps all the way to the kidney itself. When urine backs up into the kidney, causing swelling of the renal pelvis and calyces, the condition is called **hydronephrosis** (hye-dro-nef-RO-sis). Some of the more important obstructive conditions are summarized in the following paragraphs.

Renal Calculi

Renal calculi or *kidney stones* are crystallized mineral chunks that develop in the renal pelvis or calyces. Calculi develop when calcium and other minerals, such as uric acid, crystallize on the renal papillae, then break off into the urine. *Staghorn calculi* are large, branched stones that form in the pelvis and branched calyces.

If the stones are small enough, they will simply flow through the ureters and eventually be voided with the urine. Larger stones may obstruct the ureters, causing an intense pain called **renal colic** as rhythmic muscle contractions of the ureter attempt to dislodge it. Hydronephrosis may occur if the stone does not move from its obstructing position.

Drugs that increase urine output may be used in combination with drugs that dissolve stones to encourage voiding of the stones. Lasers, conventional surgery, and high-energy sound waves have also been used to break up large stones, a process called *lithotripsy* (lith-o-TRIP-see).

Neurogenic Bladder

As mentioned in an earlier discussion of micturition, disruption of the nerve input to the bladder results in loss of normal control of voiding. Such paralysis or abnormal activity of the bladder is termed **neurogenic** (noo-ro-JEN-ik) **bladder**. Neurogenic bladder is characterized by involuntary retention of urine, subsequent distention (bulging) of the bladder, and perhaps a burning sensation or fever with chills.

Tumors

Tumors of the urinary system typically obstruct urine flow, possibly causing hydronephrosis in one or both kidneys. Most kidney tumors are malignant neoplasms called *renal cell carcinomas.* They usually occur only in one kidney. Renal cell carcinoma metastasizes most often to the lungs and bone tissue. Bladder cancer occurs about as frequently as renal cancer (each accounts for about two in every hundred cancer cases). Renal and bladder cancer have few symptoms early in their development, other than traces of blood in the urine or **hematuria** (he-mat-YOO-ree-ah). As the cancer develops, pelvic pain and symptoms of urinary obstruction may occur.

URINARY TRACT INFECTIONS

Most *urinary tract infections (UTIs)* are caused by bacteria, most often gram-negative types. UTIs can involve the urethra, bladder, ureter, and kidneys. Common types of urinary tract infections are summarized in the following paragraphs.

Urethritis

Urethritis (yoo-reth-RYE-tis) is inflammation of the urethra that commonly results from bacterial infection, often *gonorrhea* (Appendix B). Nongonococcal urethritis is usually caused by *chlamydial* infection (Appendix B). Males (particularly infants) suffer from urethritis more often than females.

Cystitis

Cystitis (sis-TIE-tis) is a term that refers to an inflammation of the bladder. Cystitis most commonly occurs as a result of infection but can also accompany calculi, tumors, or other conditions. Bacteria usually enters the bladder through the urethra. Cystitis occurs more frequently in women than in men because the female urethra is shorter and closer to the anus (a source of bacteria) than in the male. Bladder infections are characterized by pelvic pain, an urge to urinate frequently, and hematuria. Two common forms of "nonbacterial" cystitis are *urethral syndrome* and *interstitial cystitis*. Urethral syndrome, which occurs most commonly in young women, has unknown causes but often develops into a bacterial infection. Interstitial cystitis is believed to be an autoimmune disease of the bladder.

Pyelonephritis

Nephritis (nef-RYE-tis) is a general term referring to kidney disease, especially inflammatory conditions. **Pyelonephritis** (pie-el-o-nef-RYE-tis) is literally "pelvis nephritis" and refers to inflammation of the renal pelvis and connective tissues of the kidney. As with cystitis, pyelonephritis is usually caused by bacterial infection but can also result from viral infection, mycosis, calculi, tumors, pregnancy, and other conditions.

Acute pyelonephritis develops rapidly and is characterized by fever, chills, pain in the sides (flank), nausea, and an urge to urinate frequently. It often results from the spread of infection from the lower urinary tract or through the blood from other organs. *Chronic pyelonephritis* is believed to be an autoimmune disease but is often preceded by a bacterial infection or urinary blockage.

GLOMERULAR DISORDERS

Glomerular disorders, collectively called **glomerulonephritis** (glo-mair-yoo-lo-nef-RYE-tis), result from damage to the glomerular-capsular membrane. This damage can be caused by immune mechanisms, heredity, or bacterial infections. Without successful treatment, glomerular disorders can progress to kidney failure.

Nephrotic Syndrome

Nephrotic (nef-ROT-ik) **syndrome** is a collection of signs and symptoms that accompany various glomerular disorders. This syndrome is characterized by the following:

1 Proteinuria (pro-ten-YOO-ree-ah)—presence of proteins (especially *albumin*) in the urine. Protein, normally absent from urine, filters through damaged glomerular-capsular membranes and is not reabsorbed by the kidney tubules.

2 Hypoalbuminemia (hye-po-al-byoo-min-EE-me-ah)—low albumin concentration in the blood, resulting from the loss of albumin from the blood

through holes in the damaged glomeruli. Albumin is the most abundant plasma protein. Because it normally cannot leave the blood vessels, it usually remains as a "permanent" solute in the plasma. This keeps plasma water concentration low and thus prevents osmosis of large amounts of water out of the blood and into tissue spaces. (You may wish to review the discussion of osmosis in Chapter 2 to help you understand this process.) In hypoalbuminemia, this function is lost and fluid leaks out of the blood vessels and into tissue spaces—causing widespread edema.

3 Edema—general tissue swelling caused by accumulation of fluids in the tissue spaces. Edema associated with nephrotic syndrome is caused by loss of plasma protein (albumin) and the resulting osmosis of fluid from the blood.

Acute Glomerulonephritis

Acute glomerulonephritis is the most common form of kidney disease. It is caused by a delayed immune response to streptococcal infection—the same mechanism that causes valve damage in rheumatic heart disease (see Chapter 12). For this reason, it is sometimes called *postinfectious glomerulonephritis*. Occurring 1 to 6 weeks after a streptococcal infection, this disorder is characterized by hematuria, oliguria, proteinuria, and edema. Antibiotic therapy and bed rest are the usual treatments for acute glomerulonephritis. Recovery is often complete but may progress to a chronic form of glomerulonephritis.

Chronic Glomerulonephritis

Chronic glomerulonephritis is the general name for a variety of noninfectious glomerular disorders characterized by progressive kidney damage leading to renal failure. Early stages of chronic glomerulonephritis are asymptomatic. As this disorder progresses, hematuria, proteinuria, oliguria, and edema develop. Immune mechanisms are believed to be the major causes of chronic glomerulonephritis. One immune mechanism involves antigen-antibody complexes that form in the blood when antibodies bind with foreign antigens (or possibly self-antigens). These antigen-antibody complexes lodge in the glomerular-capsular membrane and trigger an inflammation response. Less commonly, formation of antibodies that directly attack the glomerular-capsular membrane causes chronic glomerulonephritis.

ARTIFICIAL KIDNEY

The artificial kidney is a mechanical device that uses the principle of dialysis to remove or separate waste products from the blood. In the event of kidney failure, the process, appropriately called **hemodialysis** (Greek *haima*—"blood" and *lysis*—"separate"), is a reprieve from death for the patient. During a hemodialysis treatment, a semipermeable membrane is used to separate large (nondiffusible) particles such as blood cells from small (diffusible) ones such as urea and other wastes. Figure *A* shows blood from the radial artery passing through a porous (semipermeable) cellophane tube that is housed in a tanklike container. The tube is surrounded by a bath or dialyzing solution containing varying concentrations of electrolytes and other chemicals. The pores in the membrane are small and allow only very small molecules, such as urea, to escape into the surrounding fluid. Larger molecules and blood cells cannot escape and are returned through the tube to reenter the patient via a wrist vein. By constantly replacing the bath solution in the dialysis tank with freshly mixed solution, levels of waste materials can be kept at low levels. As a result, wastes such as urea in the blood rapidly pass into the surrounding wash solution. For a patient with complete kidney failure, two or three hemodialysis treatments a week are required. New dialysis

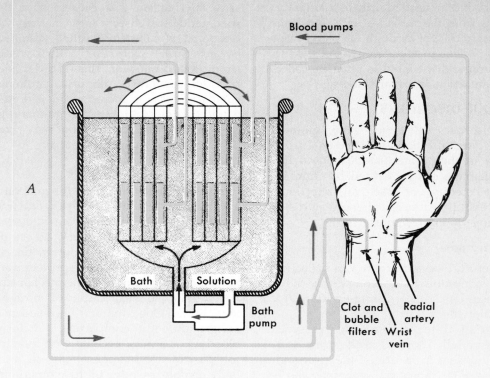

A

Blood pumps

Bath   Solution

Bath pump

Clot and bubble filters

Radial artery

Wrist vein

methods are now being developed, and dramatic advances in treatment are expected in the next few years.

Another technique used in the treatment of renal failure is called **continuous ambulatory peritoneal dialysis (CAPD).** In this procedure, 1 to 3 L of sterile dialysis fluid is introduced directly into the peritoneal cavity through an opening in the abdominal wall (Figure *B*). Peritoneal membranes in the abdominal cavity transfer waste products from the blood into the dialysis fluid, which is then drained back into a plastic container after about 2 hours. This technique is less expensive than hemodialysis and does not require the use of complex equipment.

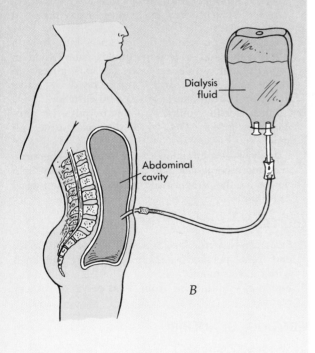

Dialysis fluid

Abdominal cavity

B

KIDNEY FAILURE

Kidney failure or **renal failure** is simply the failure of the kidney to properly process blood and form urine. Renal failure can be classified as either *acute* or *chronic*.

Acute Renal Failure

Acute renal failure is an abrupt reduction in kidney function characterized by oliguria and a sharp rise in nitrogenous compounds in the blood. The concentration of nitrogenous wastes in the blood are often assessed by the *blood urea nitrogen (BUN)* test; a high BUN result indicates failure of the kidneys to remove urea from the blood. Acute renal failure can be caused by a variety of factors that alter blood pressure or otherwise affect glomerular filtration. For example, hemorrhage, severe burns, acute glomerulonephritis or pyelonephritis, or obstruction of the lower urinary tract may progress to kidney failure. If the underlying cause of renal failure is treated, recovery is usually rapid and complete.

Chronic Renal Failure

Chronic renal failure is a slow, progressive condition resulting from the gradual loss of nephrons. There are dozens of diseases that may result in the gradual loss of nephron function, including infections, glomerulonephritis, tumors, systemic autoimmune disorders, and obstructive disorders. As kidney function is lost, the glomerular filtration rate (GFR) decreases, causing the blood urea nitrogen (BUN) and creatinine levels to climb. Chronic renal failure can be described as progressing through the three stages given here:

1 Stage 1—During the first stage, some nephrons are lost but the remaining healthy nephrons compensate by enlarging and taking over the function of the lost nephrons. BUN is kept within normal limits even though up to 75% of the nephrons are lost (as indicated by a 75% drop in GFR). This stage is often asymptomatic and may last for years, depending on the underlying cause.

2 Stage 2—The second stage is often called *renal insufficiency*. During this stage, the kidney can no longer adapt to the loss of nephrons. The remaining healthy nephrons cannot handle the urea load, and BUN levels climb dramatically.

Because the kidney's ability to concentrate urine is impaired, polyuria and dehydration may occur.

3 Stage 3—The final stage of chronic renal failure is called *uremia* or *uremic syndrome*. *Uremia* literally means "high blood urea" and is characterized by a very high BUN value caused by loss of kidney function. During this stage, low GFR causes low urine production and oliguria. Because fluids are retained by the body rather than being eliminated by the kidneys, edema and hypertension often occur. The uremic syndrome includes a long list of other symptoms caused directly or indirectly by the loss of kidney function. Unless an artificial kidney is used or a new kidney is transplanted (see Chapter 3), the progressive loss of kidney function eventually causes death.

OUTLINE SUMMARY

KIDNEYS

A Location—under back muscles, behind parietal peritoneum, just above waistline; right kidney usually a little lower than left (Figure 18-1)
B Internal structure (Figure 18-2)
   1 Cortex—outer layer of kidney substance
   2 Medulla—inner portion of kidney
   3 Pyramids—triangular divisions of medulla
   4 Papilla—narrow, innermost end of pyramid
   5 Pelvis—expansion of upper end of ureter; lies inside kidney
   6 Calyces—divisions of renal pelvis
C Microscopic structure—nephrons are microscopic units of kidneys; consist of following parts (Figure 18-3)
   1 Renal corpuscle—Bowman's capsule with its glomerulus
      a Bowman's capsule—the cup-shaped top of the nephron
      b Glomerulus—network of blood capillaries surrounded by Bowman's capsule
   2 Renal tubule
      a Proximal convoluted tubule—first segment of a renal tubule
      b Loop of Henle—extension of proximal tubule; consists of descending limb, loop, and ascending limb
      c Distal convoluted tubule—extension of ascending limb of loop of Henle
      d Collecting tubule—straight extension of distal tubule
D Functions
   1 Excretes toxins and nitrogenous wastes
   2 Regulates levels of many chemicals in blood
   3 Maintains water balance
   4 Helps regulate blood pressure via secretion of renin

FORMATION OF URINE (Figure 18-5)

A Occurs by a series of three processes that take place in successive parts of nephron
   1 Filtration—goes on continually in renal corpuscles; glomerular blood pressure causes water and dissolved substances to filter out of glomeruli into Bowman's capsule; normal glomerular filtration rate 125 ml per minute
   2 Reabsorption—movement of substances out of renal tubules into blood in peritubular capillaries; water, nutrients, and various ions are reabsorbed; water is reabsorbed by osmosis from proximal tubules
   3 Secretion—movement of substances into urine in the distal and collecting tubules from blood in peritubular capillaries; hydrogen ions, potassium ions, and certain drugs are secreted by active transport; ammonia is secreted by diffusion
B Control of urine volume—mainly by posterior pituitary hormone's ADH, which decreases urine volume
C Urinalysis—examination of the physical, chemical, and microscopic characteristics of urine; may help determine the presence and nature of a pathological condition

URETERS

A Structure—narrow long tubes with expanded upper end (renal pelvis) located inside kidney and lined with mucous membrane
B Function—drain urine from renal pelvis to urinary bladder

URINARY BLADDER

A Structure (Figure 18-7)
   1 Elastic muscular organ, capable of great expansion

2 Lined with mucous membrane arranged in rugae, like stomach mucosa

B Functions

  1 Storage of urine before voiding

  2 Voiding

URETHRA

A Structure

  1 Narrow tube from urinary bladder to exterior

  2 Lined with mucous membrane

  3 Opening of urethra to the exterior called *urinary meatus*

B Functions

  1 Passage of urine from bladder to exterior of the body

  2 Passage of male reproductive fluid (semen) from the body

MICTURITION

A Passage of urine from body (also called *urination* or *voiding*)

B Regulatory sphincters

  1 Internal urethral sphincter (involuntary)

  2 External urethral sphincter (voluntary)

C Bladder wall permits storage of urine with little increase in pressure

D Emptying reflex

  1 Initiated by stretch reflex in bladder wall

  2 Bladder wall contracts

  3 Internal sphincter relaxes

  4 External sphincter relaxes, and urination occurs

E Urinary retention—urine produced but not voided

F Urinary suppression—no urine produced but bladder is normal

G Incontinence—urine is voided involuntarily

  1 May be caused by spinal injury or stroke

  2 Retention of urine may cause cystitis

RENAL AND URINARY DISORDERS

A Obstructive disorders interfere with normal urine flow, possibly causing urine to back up and cause hydronephrosis or other kidney damage

  1 Renal calculi (kidney stones) may block ureters, causing intense pain called *renal colic*

  2 Neurogenic bladder—paralysis or abnormal function of the bladder, preventing normal flow of urine out of the body

  3 Tumors—renal cell carcinoma (kidney cancer) and bladder cancer are often characterized by hematuria (blood in the urine)

B Urinary tract infections (UTIs) are often caused by gram-negative bacteria

  1 Urethritis—inflammation of the urethra

  2 Cystitis—inflammation or infection of the urinary bladder

  3 Pyelonephritis—inflammation of the renal pelvis and connective tissues of the kidney; may be acute (infectious) or chronic (autoimmune)

C Glomerular disorders result from damage to the glomerular-capsular membrane of the renal corpuscles

  1 Nephrotic syndrome accompanies many glomerular disorders

    a Proteinuria—protein in the urine

    b Hypoalbuminemia—low plasma protein (albumin) level; caused by loss of proteins to urine

    c Edema—tissue swelling caused by loss of water from plasma due to hypoalbuminemia

  2 Acute glomerulonephritis is caused by delayed immune response to a streptococcal infection

  3 Chronic glomerulonephritis is a slow inflammatory condition caused by immune mechanisms and often leading to renal failure

D Kidney failure or renal failure occurs when the kidney fails to function

  1 Acute renal failure—abrupt reduction in kidney function that is usually reversible

  2 Chronic renal failure—slow, progressive loss of nephrons caused by a variety of underlying diseases

    a Early in this disorder, healthy nephrons often compensate for the loss of damaged nephrons

    b Loss of kidney function ultimately results in uremia (high BUN levels) and its life-threatening consequences

    c Complete kidney failure results in death unless a new kidney is transplanted or an artificial kidney substitute is used

NEW WORDS

atrial natriuretic hormone (ANH)	calyx	micturition	trigone
	catheterization	papilla	urination
Bowman's capsule	glomerulus	pyramid	voiding

Diseases and Other Clinical Terms

anuria	hypoalbuminemia	neurogenic bladder	renal colic
catheterization	incontinence	oliguria	renal failure
cystitis	lithotriptor	polyuria	uremia
glomerulonephritis	nephritis	proteinuria	urethritis
glycosuria	nephrotic syndrome	pyelonephritis	
hematuria			

CHAPTER TEST

1. Failure of the urinary system results in rapid accumulation of toxic wastes, a condition called _____.
2. Normally a little over _____ of the total blood pumped by the heart enters the kidneys.
3. The outer portion of the kidney is called the _____, and the inner portion is called the _____.
4. The functional unit of the kidney is called the _____.
5. The renal corpuscle is composed of the cup-shaped _____, and a network of blood capillaries called the _____.
6. The first segment of the renal tubule is called the _____ convoluted tubule.
7. Urine formation begins with the process of _____ of wastes from the blood.
8. Water is reabsorbed from the proximal convoluted tubules by _____.
9. Glucose is entirely reabsorbed from the _____ tubules.
10. Production of unusually large amounts of urine is called _____.
11. The basinlike upper end of the ureter located inside the kidney is called the renal _____.
12. Attacks of pain caused by the passage of a kidney stone are called _____ _____.
13. Urine passes from the bladder down the _____ and out its external opening, called the urinary _____.
14. Voiding urine involuntarily is called _____.
15. The medical term for bladder infection is _____.
16. Glomerular filtration normally occurs at the rate of about _____ ml per minute.

Select the most correct answer from Column B for each statement in Column A. (Only one answer is correct.)

COLUMN A

17. _____ High waste levels
18. _____ Contain renal corpuscles
19. _____ Functional unit of kidney
20. _____ Tuft of capillaries
21. _____ Glucose in urine
22. _____ Absence of urine
23. _____ Water-retaining hormone
24. _____ Kidney stones
25. _____ Urination
26. _____ Inflammation of the renal pelvis
27. _____ Inflammatory disease of the
glomerular-capsular   membrane
28. _____ Progressive loss of nephron
function
29. _____ Protein in urine

COLUMN B

a. Glycosuria
b. Glomerulus
c. ADH
d. Nephron
e. Uremia
f. Renal calculi
g. Anuria
h. Micturition
i. Cortex
j. Chronic renal failure
k. Glomerulonephritis
l. Proteinuria
m. Pyelonephritis

REVIEW QUESTIONS

1 What organs form the urinary system?
2 To operate on a kidney, does a surgeon have to cut through the peritoneum? Explain your answer.
3 Name the parts of a nephron.
4 What and where are the glomeruli and Bowman's capsules?
5 Explain briefly the functions of the glomeruli and Bowman's capsules.
6 Explain briefly the function of the renal tubules.
7 What kind of membrane lines the urinary tract?
8 Explain briefly the function of ADH. What is the full name of this hormone? What gland secretes it?
9 Suppose that ADH secretion increases noticeably. Would this increase or decrease urine volume? Why?
10 What hormone might appropriately be nicknamed the "water-retaining hormone"?

11 What hormone might appropriately be nicknamed the "salt- and water-retaining hormone"?
12 What hormone might be called the "salt- and water-losing hormone"?
13 What is the urinary meatus?
14 What and where are the ureters and the urethra?
15 Explain the process of micturition.
16 Name some conditions that obstruct the normal flow of urine. What medical consequences can such blockage have?
17 What is the difference between pyelonephritis and glomerulonephritis? What is urethritis?
18 How does loss of plasma proteins into the urine lead to edema of tissue throughout the body?
19 Describe the stages of chronic renal failure.

CLINICAL APPLICATIONS

1 Sue has *anorexia nervosa* (see Chapter 17). Her body fat content has decreased to a level that is far below normal. How might the changing structure of the body affect the position of Sue's kidneys? How can a change in the position of one or both kidneys lead to kidney failure?

2 Drugs called *thiazide diuretics* are sometimes prescribed to control hypertension (high blood pressure). These drugs act on kidney tubules to inhibit reabsorption of water. How does inhibition of water reabsorption by the kidney reduce high blood pressure? What effects would such drugs have on the volume of urine output?

3 Harriet is receiving *continuous ambulatory peritoneal dialysis (CAPD)*. As you may recall from the boxed essay earlier in this chapter, fluid is introduced into the peritoneal cavity and later withdrawn. Do you think that this dialysis fluid is hypertonic, isotonic, or hypotonic to normal blood plasma? Give reasons for your answer.

Fluid and Electrolyte Balance

C H A P T E R

19

Objectives

*After you have completed this chapter,
you should be able to:*

1 List, describe, and compare the body fluid compartments and their subdivisions.

2 Discuss avenues by which water enters and leaves the body and the mechanisms that maintain fluid balance.

3 Discuss the nature and importance of electrolytes in body fluids and explain the aldosterone mechanism of extracellular fluid volume control.

4 Explain the interaction between capillary blood pressure and blood proteins.

5 Give examples of common fluid imbalances.

*H*ave you ever wondered why you some- times excrete great volumes of urine and sometimes excrete almost none at all? Why sometimes you feel so thirsty that you can hardly get enough to drink and other times you want no liquids at all? These conditions and many more relate to one of the body's most important func- tions—that of maintaining its **fluid and elec- trolyte balance.**

The term *fluid balance* means several things. Of course, it means the same thing as homeostasis of fluids. To say that the body is in a state of fluid balance is to say that the total amount of water in the body is normal and that it remains relatively constant. Electrolytes are substances such as salts that dissolve or break apart in water solution. Health and sometimes even survival itself depend on maintaining proper balance of water and the electrolytes within it.

In this chapter you will find a discussion of body fluids and electrolytes, their normal values, the mechanisms that operate to keep them nor- mal, and some of the more common types of fluid and electrolyte imbalances.

Body Fluids

If you are a healthy young person and you weigh 55 kilograms (120 pounds), there is a good chance that, of the hundreds of compounds present in your body, one substance alone accounts for about 50% to 60% of your total weight. This, the body's most abundant compound, is water. It occupies three main locations known as **fluid compartments.** Look now at Figure 19-1. Note that the largest volume of water by far lies inside cells and that it is called, appropriately, **intracel- lular fluid (ICF).** Note, too, that the water outside of cells—**extracellular fluid (ECF)**—is located in two compartments: in the microscopic spaces between cells, where it is called **interstitial fluid (IF),** and in the blood vessels, where it is called **plasma.** Plasma is the liquid part of the blood, constituting a little more than half of the total blood volume (about 55%); blood cells make up the rest of the volume.

A normal body maintains fluid balance. The term *fluid balance* means that the volumes of ICF, IF, and plasma and the total volume of water in the body remain relatively constant. Of course,

not all bodies contain the same amount of water. The more a person weighs, the more water the body contains. This is true because, excluding fat or adipose tissue, about 55% of the body weight is water. Because fat is almost water free, the more fat present in the body, the less the total water content is per unit of weight. In other words, fat people have a lower water content per pound of body weight than slender people. The body of a slender adult man, for instance, typically consists of about 60% water. An obese male body, in con- trast, may consist of only 50% water or even less.

Sex and age also influence the amount of water in a body. Infants have more water compared with body weight than adults of either sex. In a newborn, water may account for up to 80% of total body weight. There is a rapid decline in the proportion of body water to body weight during the first year of life. Figure 19-2 illustrates the pro- portion of body weight represented by water in newborn infants, men, and women. The female body contains slightly less water per pound of weight because it contains slightly more fat than the male body. Age and the body's water content are inversely related. In general, as age increases, the amount of water per pound of body weight decreases.

Mechanisms that Maintain Fluid Balance

Under normal conditions, homeostasis of the total volume of water in the body is maintained or restored primarily by devices that adjust output (urine volume) to intake and secondarily by mechanisms that adjust fluid intake. There is no question about which of the two mechanisms is more important; the body's chief mechanism, by far, for maintaining fluid balance is to adjust its fluid output so that it equals its fluid intake.

Obviously, as long as output and intake are equal, the total amount of water in the body does not change. Figure 19-3 shows the three sources of fluid intake: the liquids we drink, the water in the foods we eat, and the water formed by catabolism of foods. Table 19-1 gives their normal volumes. However, these can vary a great deal and still be considered normal. Table 19-1 also indicates that fluid output from the body occurs through four organs: the kidneys, lungs, skin, and

FIGURE 19-1 Fluid Compartments of the Body. Percentages and volumes are given for young adults weighing 55 kg (120 lbs).

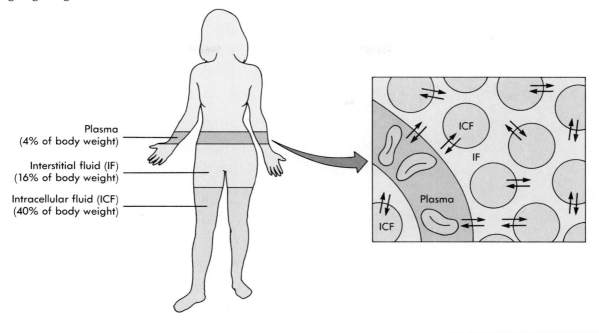

Plasma
(4% of body weight)

Interstitial fluid (IF)
(16% of body weight)

Intracellular fluid (ICF)
(40% of body weight)

ICF

IF

Plasma

ICF

FIGURE 19-2 Proportion of Body Weight Represented by Water.

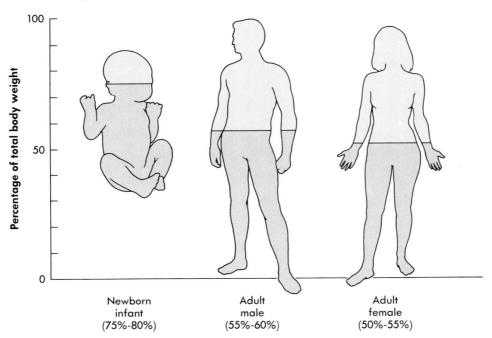

Percentage of total body weight

100

50

0

Newborn
infant
(75%-80%)

Adult
male
(55%-60%)

Adult
female
(50%-55%)

FIGURE 19-3 Sources of Fluid Intake and Output.

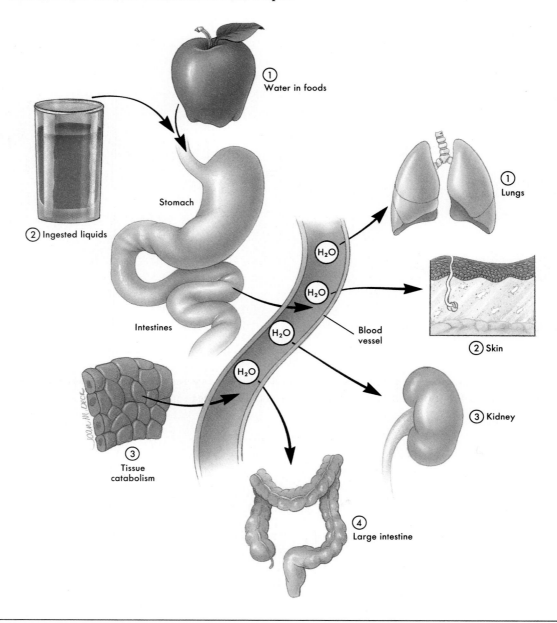

intestines. The fluid output that changes the most is that from the kidneys. The body maintains fluid balance mainly by changing the volume of urine excreted to match changes in the volume of fluid intake. Everyone knows this from experience. The more liquid one drinks, the more urine one excretes. Conversely, the less the fluid intake, the less the urine volume. How changes in urine volume come about was discussed on p. 459. This would be a good time to review these paragraphs.

It is important to remember from your study of the urinary system that the rate of water and salt resorption by the renal tubules is the most important factor in determining urine volume. Urine

TABLE 19-1 Typical Normal Values for Each Portal of Water Entry and Exit (24 Hours)

INTAKE	AMOUNT*	OUTPUT	AMOUNT*
Water in foods	700 ml	Lungs (water in expired air)	350 ml
Ingested liquids	1500 ml	Skin	
Water formed by catabolism	200 ml	By diffusion	350 ml
		By sweat	100 ml
		Kidneys (urine)	1400 ml
		Intestines (in feces)	200 ml
TOTALS	2400 ml		2400 ml

*Amounts vary widely.

volume is regulated chiefly by hormones secreted by the posterior lobe of the pituitary gland (antidiuretic hormone or ADH) and the adrenal cortex (aldosterone). Atrial natriuretic hormone (ANH) from the atrial wall of the heart also affects urine volume. See p. 460 for a review of the hormonal control of urine volume.

Several factors act as mechanisms for controlling plasma, IF, and ICF volumes. We shall limit our discussion to naming only three of these factors, stating their effects on fluid volumes, and giving some specific examples of them. Three of the main factors that influence ECF and ICF volumes are:

1 The concentration of electrolytes in ECF
2 The capillary blood pressure
3 The concentration of proteins in blood

REGULATION OF FLUID INTAKE

Physiologists disagree about the details of the mechanism for controlling and regulating fluid intake to compensate for factors that would lead to dehydration. In general it appears to operate in this way: when dehydration starts to develop—that is, when fluid loss from the body exceeds fluid intake—salivary secretion decreases, producing a "dry-mouth feeling" and the sensation of thirst. The individual then drinks water, thereby increasing fluid intake and compensating for previous fluid losses. This tends to restore fluid balance (Figure 19-4). If an individual takes nothing by mouth for days, can his fluid output decrease to zero? The answer—no—becomes obvious after reviewing the information in Table 19-1. Despite every effort of homeostatic mechanisms to compensate for the zero intake, some output (loss) of fluid occurs as long as life continues. Water is continually lost from the body through expired air and diffusion through the skin.

Although the body adjusts its fluid intake to some extent, factors that adjust fluid output, such as electrolytes and blood proteins, are far more important.

IMPORTANCE OF ELECTROLYTES IN BODY FLUIDS

The bonds that hold together the molecules of certain organic substances such as glucose are such that they do not permit the compound to break up or **dissociate** in water solution. Such compounds are called **nonelectrolytes.** Compounds such as ordinary table salt or sodium chloride (NaCl) that have molecular bonds that permit them to break up or dissociate in water solution into separate particles (Na^+ and Cl^-) are known as **electrolytes.** The dissociated particles of an electrolyte are called **ions** and carry an electrical charge. Positively charged particles (ions) such as Na^+ are called **cations** (CAT-eye-ons), and negatively charged particles such as Cl^- are called **anions** (AN-eye-ons).

Important cations include sodium (Na^+), calcium (Ca^{++}), potassium (K^+), and magnesium (Mg^{++}). Important anions include chloride (Cl^-), bicarbonate (HCO_3^-), phosphate (HPO_4^-), and

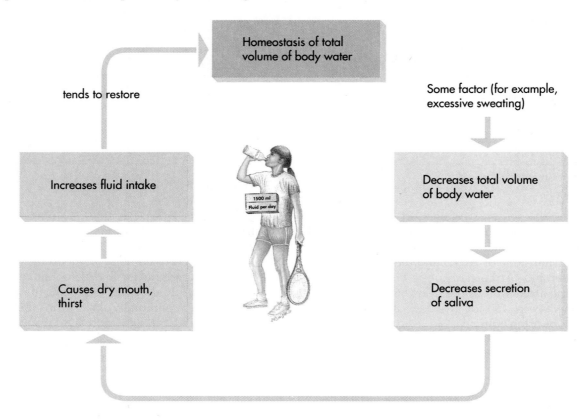

▼ **FIGURE 19-4 Homeostasis of the Total Volume of Body Water.** A basic mechanism for adjusting intake to compensate for excess output of body fluid is diagrammed.

▼ **TABLE 19-2** Common Electrolytes Found in Blood Plasma	
CATIONS	**ANIONS**
142 mEq Na$^+$	102 mEq Cl$^-$
4 mEq K$^+$	26 mEq HCO$_3^-$
5 mEq Ca^{++}	17 mEq protein$^-$
2 mEq Mg^{++}	6 mEq other$^-$
	2 mEq HPO$_4^-$
153 mEq/L plasma	153 mEq/L plasma

many proteins. Table 19-2 shows that although blood plasma contains a number of important electrolytes, by far the most abundant one is sodium chloride (ordinary table salt, Na$^+$ Cl$^-$).

A variety of anions and cations have important nutrient or regulatory roles in the body. Many ions are major or important "trace" elements in the body (see Appendix A). Iron, for example, is required for hemoglobin production, and iodine must be available for synthesis of thyroid hormones. Electrolytes are also required for many cellular activities such as nerve conduction and muscle contraction.

Additionally, electrolytes influence the movement of water between the three fluid compartments of the body. To remember how ECF

DIURETICS

The word **diuretic** is from the Greek word **diouretikos** meaning "causing urine." By definition a diuretic drug is a substance that promotes or stimulates the production of urine.

As a group, diuretics are among the most commonly used drugs in medicine. They are used because of their role in influencing water and electrolyte balance, especially sodium, in the body. Diuretics have their effect on tubular function in the nephron, and the differing types of diuretics are often classified according to their major site of action. Examples would include: (1) *proximal tubule diuretics* such as acetazolamide (Diamox), (2) *loop of Henle diuretics* such as ethacrynic acid (Edecrin) or furosemide (Lasix), and (3) *distal tubule diuretics* such as chlorothiazide (Diuril).

Classification can also be made according to the effect the drug has on the level or concentration of sodium (Na^+), chloride (Cl^-), potassium (K^+), and bicarbonate (HCO_3^-) ions in the tubular fluid. Using this classification, ethacrynic acid would be described as a diuretic that acts by inhibiting the reabsorption of chloride ions (in the loop of Henle). When chloride ion levels are increased in the tubular fluid, reabsorption of sodium ions is also blocked. The result is retention of NaCl, which must be excreted in the urine, carrying body water with it.

Nursing implications for patients receiving diuretics include keeping a careful record of fluid intake and output and assessing the patient for signs and symptoms of electrolyte and water imbalance.

---

electrolyte concentration affects fluid volumes, remember this one short sentence: where sodium goes, water soon follows. If, for example, the concentration of sodium in interstitial fluid spaces rises above normal, the volume of IF soon reaches abnormal levels, too—a condition called *edema*, which results in tissue swelling.

Figure 19-5 traces one mechanism that tends to maintain fluid homeostasis. Aldosterone, secreted by the adrenal cortex, increases Na^+ reabsorption by the kidney tubules. Water reabsorption also increases, causing an increase in ECF volume. Begin in the upper right of the diagram and follow, in sequence, each of the informational steps. In summary:

1 Overall fluid balance requires that fluid output equal fluid intake.
2 The type of fluid output that changes most is urine volume.
3 Renal tubule regulation of salt and water is the most important factor in determining urine volume.
4 Aldosterone controls sodium reabsorption in the kidney.

5 The presence of sodium forces water to move (where sodium goes, water soon follows).

The flow chart diagram in Figure 19-5 explains, in a very concise and brief way, the aldosterone mechanism that helps to restore normal ECF volume when it decreases below normal. Can you construct a similar diagram to show the effect of ADH secretion on ECF volume?

Although wide variations are possible, the average daily diet contains about 100 milliequivalents of sodium. The milliequivalent (mEq) (Table 19-2) is a unit of measurement related to reactivity. In a healthy individual, sodium excretion from the body by the kidney is about the same as intake. The kidney acts as the chief regulator of sodium levels in body fluids. It is important to know that many electrolytes such as sodium not only pass into and out of the body but also move back and forth between a number of body fluids during each 24-hour period. Figure 19-6 shows the large volumes of sodium-containing internal secretions produced each day. During a 24-hour

FIGURE 19-5 Aldosterone Mechanism. Aldosterone restores normal ECF volume when such levels decrease below normal. Excess aldosterone, however, leads to excess ECF volume—that is, excess blood volume (hypervolemia) and excess interstitial fluid volume (edema)—and also to an excess of the total Na$^+$ content of the body.

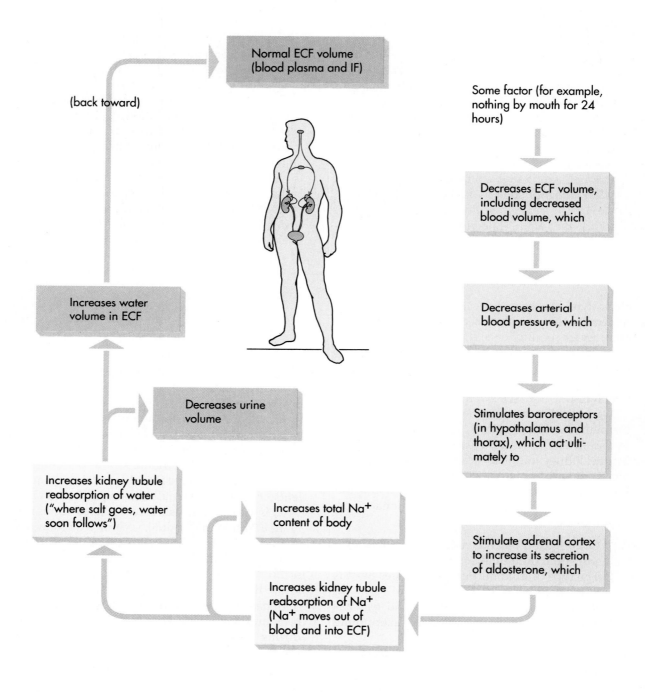

FIGURE 19-6 Sodium-Containing Internal Secretions. The total volume of these secretions may reach 8000 ml or more in 24 hours.

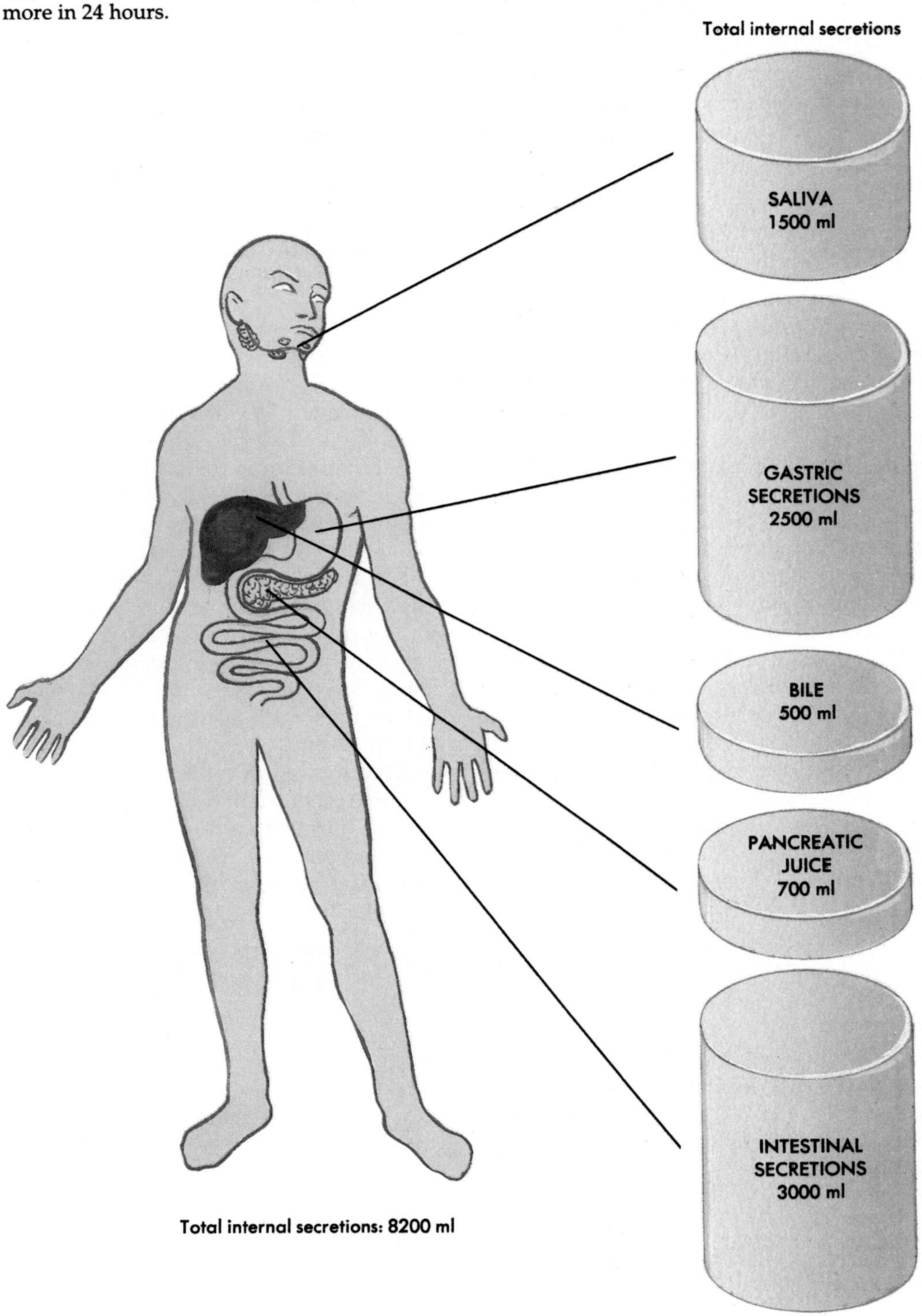

Total internal secretions

SALIVA
1500 ml

GASTRIC
SECRETIONS
2500 ml

BILE
500 ml

PANCREATIC
JUICE
700 ml

INTESTINAL
SECRETIONS
3000 ml

Total internal secretions: 8200 ml

period, over 8 liters of fluid containing 1000 to 1300 mEq of sodium are poured into the digestive system as part of saliva, gastric secretions, bile, pancreatic juice, and IF secretions. This sodium, along with most of that contained in the diet, is almost completely reabsorbed in the large intestine. Very little sodium is lost in the feces. Precise regulation and control of sodium levels are required for survival.

CAPILLARY BLOOD PRESSURE AND BLOOD PROTEINS

Capillary blood pressure is a "water-pushing" force. It pushes fluid out of the blood in capillaries into the IF. Therefore if capillary blood pressure increases, more fluid is pushed—filtered— out of blood into the IF. The effect of an increase in capillary blood pressure, then, is to transfer fluid from blood to IF. In turn this fluid shift, as it is called, changes blood and IF volumes. It decreases blood volume by increasing IF volume. If, on the other hand, capillary blood pressure decreases, less fluid filters out of blood into IF.

Water continually moves in both directions through the membranous walls of capillaries (Figure 19-1). The amount that moves out of capillary blood into IF depends largely on capillary blood pressure, a water-pushing force. The amount that moves in the opposite direction (that is, into blood from IF) depends largely on the concentration of proteins in blood plasma. Plasma proteins act as a water-pulling or water-holding force. They hold water in the blood and pull it into the blood from IF. If, for example, the con-

centration of proteins in blood decreases appreciably—as it does in some abnormal conditions such as dietary deficiency—less water moves into blood from IF. As a result, blood volume decreases and IF volume increases. Of the main body fluids, IF volume varies the most. Plasma volume usually fluctuates slightly and briefly. If a pronounced change in volume occurs, adequate circulation cannot be maintained.

Fluid Imbalances

Fluid imbalances are common ailments. They take several forms and stem from a variety of causes, but they all share a common characteristic—that of abnormally low or abnormally high volumes of one or more body fluids.

Dehydration is the fluid imbalance seen most often. In this potentially dangerous condition, IF volume decreases first, but eventually, if treatment has not been given, ICF and plasma volumes also decrease below normal levels. Either too small a fluid intake or too large a fluid output causes dehydration. Prolonged diarrhea or vomiting may result in dehydration due to the loss of body fluids. This is particularly true in infants where the total fluid volume is much smaller than it is in adults.

Overhydration can also occur but is much less common than dehydration. The grave danger of giving intravenous fluids too rapidly or in too large amounts is overhydration, which can put too heavy a burden on the heart.

OUTLINE SUMMARY

BODY FLUIDS

A Major locations—inside cells (ICF) and outside of cells in extracellular fluid (ECF) located between the cells as interstitial fluid (IF) and in blood vessels (plasma) (Figure 19-1)

B Percentage of body weight and volumes, assuming that weight is 55 kg:

ECF		
Plasma	4%	2.4 L
IF	16%	9.6 L
ICF	40%	24 L
Total body fluid	60%	36 L

C Variation in total body water related to:
1 Total body weight—the more a person weighs, the more water the body contains
2 Fat content of body—the more fat present, the less total water content per unit of weight (Fat is almost water free.)
3 Sex—proportion of body weight represented by water is about 10% less in women than in men (Figure 19-2)
4 Age—in newborn infant water may account for 80% of total body weight

MECHANISMS THAT MAINTAIN FLUID BALANCE

A Fluid output, mainly urine volume, adjusts to fluid intake; ADH from posterior pituitary gland acts to increase kidney tubule reabsorption of sodium and water from tubular urine into blood, thereby tending to increase ECF (and total body fluid) by decreasing urine volume (Figure 19-5)

B ECF electrolyte concentration (mainly Na^+ concentration) influences ECF volume; an increase in ECF Na^+ tends to increase ECF volume by increasing movement of water out of ICF and by increasing ADH secretion, which decreases urine volume, and this, in turn, increases ECF volume

C Capillary blood pressure pushes water out of blood, into IF; blood protein concentration pulls water into blood from IF; hence these two forces regulate plasma and IF volume under usual conditions

D Importance of electrolytes in body fluids
  1 Nonelectrolytes—organic substances that do not break up or dissociate when placed in water solution (for example, glucose)
  2 Electrolytes—compounds that break up or dissociate in water solution into separate particles called *ions* (for example, ordinary table salt or sodium chloride)
  3 Ions—the dissociated particles of an electrolyte that carry an electrical charge (for example, sodium ion [Na^+])

4 Cations—positively charged ions (for example, potassium [K^+] and sodium [Na^+])
5 Anions—negatively charged particles (ions) (for example, chloride [Cl^-] and bicarbonate [HCO_3^-])
6 Electrolyte composition of blood plasma—Table 19-2
7 Sodium—most abundant and important plasma cation
  a Normal plasma level—142 mEq/L
  b Average daily intake (diet)—100 mEq
  c Chief method of regulation—kidney
  d Aldosterone increases Na^+ reabsorption in kidney tubules (Figure 19-5)
  e Sodium containing internal secretions—Figure 19-6
E Capillary blood pressure and blood proteins

FLUID IMBALANCES

A Dehydration—total volume of body fluids smaller than normal; IF volume shrinks first, and then if treatment is not given, ICF volume and plasma volume decrease; dehydration occurs when fluid output exceeds intake for an extended period
B Overhydration—total volume of body fluids larger than normal; overhydration occurs when fluid intake exceeds output; various factors may cause this (for example, giving excessive amounts of intravenous fluids or giving them too rapidly may increase intake above output)

$\mathcal{N}$EW WORDS

anions	dissociate	extracellular fluid (ECF)	intracellular fluid (ICF)
cations	electrolyte	interstitial fluid (IF)	ions

Diseases and Other Clinical Terms

dehydration	diuretic	edema	overhydration

$\mathcal{C}$HAPTER TEST

1. The largest volume of water in the body is classified as _____ fluid.
2. ECF can be divided into fluid between cells, or _____ fluid, and fluid in blood, or _____.
3. Substances that do not break up or dissociate in water solution are called _____.
4. The dissociated particles of an electrolyte in water solution are called _____.
5. Positively charged particles in solution are _____; negatively charged particles are _____.
6. The largest quantity of water leaving the body exits as _____ produced by the kidneys.
7. The most abundant cation of blood plasma is _____, and the most abundant anion is _____.
8. A drug that promotes or stimulates the production of urine is called a _____.
9. Abnormally large amounts of fluid in the intercellular tissue spaces of the body is called _____.

Circle the "T" before each true statement and the "F" before each false statement.

T F 10. About 60% of total body weight is water.

T F 11. Blood plasma is one type of ECF.

T F 12. There is more IF in the body than ICF.

T F 13. The more fat in the body the more water per pound of weight.

T F 14. The chief mechanism for maintaining fluid balance is to adjust fluid output to equal fluid intake.

T F 15. Capillary blood pressure does not influence ECF or ICF volume.

T F 16. Water is continually lost from the body by way of expired air and diffusion through the skin.

T F 17. Electrolytes dissociate in solution to yield charged particles called *ions*.

T F 18. Edema is a serious side effect of diuretic drugs.

REVIEW QUESTIONS

1 Suppose a person who had never heard the term *fluid balance* were to ask you what it meant. How would you explain it briefly and simply?

2 Approximately what percentage of a slender man's body weight consists of water? How does that compare to the percentage of body weight consisting of water in a slender woman?

3 The volume of blood plasma in a normal-sized adult weighs approximately what percentage of body weight?

4 The proportion of body weight represented by water is about 10% higher in men than in women. Why?

5 ICF makes up approximately what percentage of adult body weight?

6 IF makes up approximately what percentage of adult body weight?

7 To maintain fluid balance, does output usually change to match intake or does intake usually adjust to output?

8 Explain in words and by a diagram how ADH functions to maintain fluid balance.

9 Define anion, cation, electrolyte, ion, and nonelectrolyte.

10 List the important anions and cations that are in blood plasma.

11 Use the phrase, *where sodium goes, water soon follows,* to explain how ECF electrolyte concentration affects fluid volumes.

12 List the sodium-containing internal secretions.

13 Explain by words or diagram how capillary blood pressure and blood protein concentration function to maintain fluid balance.

14 Suppose that an individual has suffered a hemorrhage and that, as a result, her capillary blood pressure has decreased below normal. What change would occur in blood and IF volumes as a result of this decrease in capillary blood pressure?

15 Suppose that an individual has a type of kidney disease that allows plasma proteins to be lost in the urine and that, as a result, his plasma protein concentration decreases. How would this tend to change blood and IF volumes?

16 If an individual becomes dehydrated, which fluid volume decreases first?

CLINICAL APPLICATIONS

1 Like most people in the United States, Tom ingests 20 to 30 times more sodium each day than he needs for survival. How does Tom's body compensate for this imbalance to restore homeostasis?

2 Jo has not eaten anything all day but has consumed an excessive amount of distilled water. Will this affect her urine output? What unusual characteristics are likely to appear in a urinalysis of Jo's urine?

3 Jo is overhydrated (see 2). This chapter states that *overhydration* can place a dangerously heavy burden on the heart. Explain how overhydration can tax the heart.

Acid-Base Balance

CHAPTER 20

Objectives

*After you have completed this chapter,
you should be able to:*

1 Discuss the concept of pH and define the phrase *acid-base balance*.

2 Define the term *buffer* and *buffer pair* and contrast strong and weak acids and bases.

3 Contrast the respiratory and urinary mechanisms of pH control.

4 Discuss compensatory mechanisms that may help return blood pH to near-normal levels in cases of pH imbalances.

5 Compare and contrast metabolic and respiratory types of pH imbalances.

One of the requirements for homeostasis and healthy survival is that the body maintain, or quickly restore, the **acid-base balance** of its fluids. Maintaining acid-base balance means keeping the concentration of hydrogen ions in body fluids relatively constant. This is of vital importance. If the hydrogen ion concentration veers away from normal even slightly, cellular chemical reactions cannot take place normally, and survival is thereby threatened.

pH of Body Fluids

Water and all water solutions contain **hydrogen ions (H^+)** and **hydroxide ions (OH^-)**. The term *pH* followed by a number indicates a solution's hydrogen ion concentration. More specifically, pH 7.0 means that a solution contains an equal concentration of hydrogen and hydroxide ions. Therefore pH 7.0 also means that a fluid is neutral in reaction (that is, neither acid nor alkaline) (Figure 20-1). The pH of water, for example, is 7.0. A pH higher than 7.0 indicates an alkaline or basic solution (that is, one with a lower concentration of hydrogen than hydroxide ions). The more alkaline a solution, the higher is its pH. A pH lower than 7.0 indicates an acid solution (that is, one with a higher hydrogen ion concentration than hydroxide ion concentration). The higher the hydrogen ion concentration, the lower the pH and the more acid a solution is. With a pH of about 1.6, gastric juice is the most acid substance in the body. Saliva has a pH of 7.7, on the alkaline side. Normally, the pH of arterial blood is about 7.45, and the pH of venous blood is about 7.35. By applying the information given in the last few sentences, you can deduce the answers to the following questions. Is arterial blood slightly acid or slightly alkaline? Is venous blood slightly acid or slightly alkaline? Which is a more accurate statement—venous blood is more acid than arterial blood or venous blood is less alkaline than arterial blood?

Arterial and venous blood are both slightly alkaline because both have a pH slightly higher than 7.0. Venous blood, however, is less alkaline than arterial blood because venous blood's pH of about 7.35 is slightly lower than arterial blood's pH of 7.45.

Mechanisms that Control pH of Body Fluids

The body has three mechanisms for regulating the pH of its fluids. They are (1) the buffer mechanism, (2) the respiratory mechanism, and (3) the urinary mechanism. Together, they constitute the complex pH homeostatic mechanism—the machinery that normally keeps blood slightly alkaline with a pH that stays remarkably constant. Its usual limits are very narrow, about 7.35 to 7.45.

The slightly lower pH of venous blood compared with arterial blood results primarily from carbon dioxide (CO_2) entering venous blood as a waste product of cellular metabolism. As carbon dioxide enters the blood, some of it combines with water (H_2O) and is converted into carbonic acid by **carbonic anhydrase,** an enzyme found in red blood cells. The following chemical equation represents this reaction. If you need to review chemical formulas and equations, please refer to Appendix A.

$$CO_2 + H_2O \xrightarrow{\text{carbonic anhydrase}} H_2CO_3$$

The lungs remove the equivalent of over 30 L of carbonic acid each day from the venous blood by elimination of CO_2. This almost unbelievable quantity of acid is so well buffered that a liter of venous blood contains only about $^1/_{100,000,000}$ g more H^+ than does 1 L of arterial blood. What incredible constancy! The pH homeostatic mechanism does indeed control effectively—astonishingly so.

BUFFERS

Buffers are chemical substances that prevent a sharp change in the pH of a fluid when an acid or base is added to it. Strong acids and bases, if added to blood, would "dissociate" almost completely and release large quantities of H^+ or OH^- ions. The result would be drastic changes in blood pH. Survival itself depends on protecting the body from such drastic pH changes.

More acids than bases are usually added to body fluids. This is because catabolism, a process that goes on continually in every cell of the body, produces acids that enter blood as it flows through tissue capillaries. Almost immediately, one of the salts present in blood—a buffer, that is—reacts with these relatively strong acids to change them to weaker acids. The weaker acids decrease blood pH only slightly, whereas the stronger acids formed by catabolism would have decreased it greatly if they were not buffered.

Buffers consist of two kinds of substances and are therefore often called **buffer pairs.** One of the main blood buffer pairs is ordinary baking soda (sodium bicarbonate or $NaHCO_3$) and carbonic acid (H_2CO_3).

Let us consider, as a specific example of buffer action, how the $NaHCO_3$–H_2CO_3 system works with a strong acid or base.

Addition of a strong acid, such as hydrochloric acid (HCl), to the $NaHCO_3$–H_2CO_3 buffer system would initiate the reaction shown in Figure 20-2. Note how this reaction between HCl and $NaHCO_3$ applies the principle of buffering. As a result of the buffering action of $NaHCO_3$, the weak acid, $H \cdot HCO_3$, replaces the very strong acid, HCl, and therefore the H^+ concentration of the blood increases much less than it would have if HCl were not buffered.

If, on the other hand, a strong base, such as sodium hydroxide (NaOH), were added to the same buffer system, the reaction shown in Figure 20-3 would take place. The H^+ of H_2CO_3 ($H \cdot HCO_3$), the weak acid of the buffer pair, combines with the OH^- of the strong base NaOH to form H_2O. Note what this accomplishes. It decreases

ACID-FORMING POTENTIAL OF FOODS

Although citrus fruits such as oranges and grapefruit contain citric acid and may have an acid taste, they are not acid forming when metabolized. The acid-forming potential of a food is determined largely by the chloride, sulfur, and phosphorus elements found in the "noncombustible" mineral residue or ash after metabolism of the food has occurred. Acids such as those found in citrus fruits are normally fully oxidized by the cells during metabolism and leave no mineral residue. As a result, they have little influence on acid-base balance.

FIGURE 20-1 The pH Range. The overall pH range is expressed numerically on what is called a *logarithmic scale* of 1 to 14. This means that a change of 1 pH unit represents a tenfold difference in actual concentration of hydrogen ions. Note that, as the concentration of H^+ ions increases, the solution becomes increasingly acidic and the pH value decreases. As OH^- concentration increases, the pH value also increases, and the solution becomes more and more basic or alkaline. A pH of 7 is neutral; a pH of 1 is very acidic, and a pH of 13 is very basic.

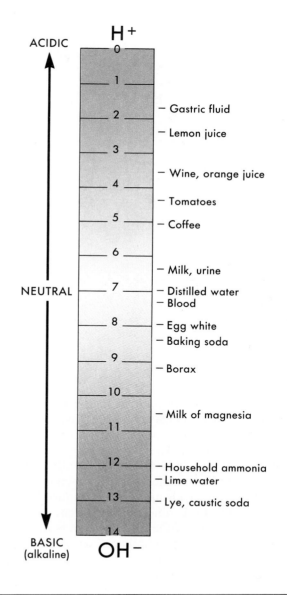

FIGURE 20-2 Buffering Action of Sodium Bicarbonate. Buffering of acid HCl by $NaHCO_3$. As a result of the buffer action, the strong acid (HCl) is replaced by a weaker acid (H · HCO_3). Note that HCl as a strong acid "dissociates" almost completely and releases more H^+ than H_2CO_3. Buffering decreases the number of H^+ in the system.

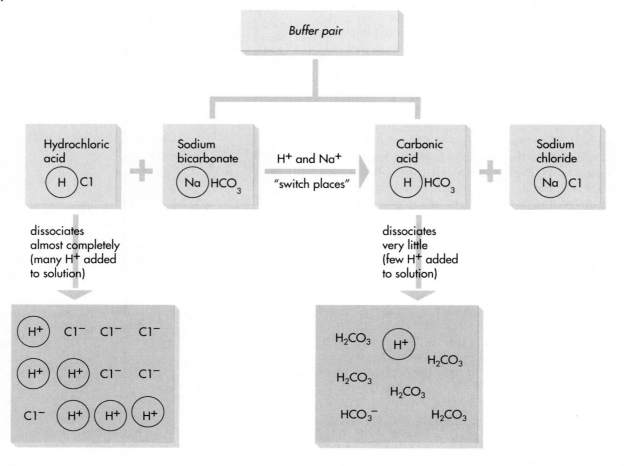

the number of OH^- added to the solution, and this in turn prevents the drastic rise in pH that would occur without buffering.

Figure 20-2 shows how a buffer system works with a strong acid. Although useful in demonstrating the principles of buffer action, HCl or similar strong acids are never introduced directly into body fluids under normal circumstances. Instead, the $NaHCO_3$ buffer system is most often called on to buffer a number of weaker acids produced during catabolism. Lactic acid is a good example. As a weak acid, it does not "dissociate"

as completely as HCl. Incomplete dissociation of lactic acid results in fewer hydrogen ions being added to the blood and a less drastic lowering of blood pH than would occur if HCl were added in an equal amount. Without buffering, however, lactic acid buildup results in significant H^+ accumulation over time. The resulting decrease of pH can produce serious acidosis. Ordinary baking soda (sodium bicarbonate or $NaHCO_3$) is one of the main buffers of the normally occurring "fixed" acids in blood. Lactic acid is one of the most abundant of the "fixed" acids (that is, acids

FIGURE 20-3 Buffering Action of Carbonic Acid. Buffering of base NaOH by H_2CO_3. As a result of buffer action, the strong base (NaOH) is replaced by $NaHCO_3$ and H_2O. As a strong base, NaOH "dissociates" almost completely and releases large quantities of OH^-. Dissociation of H_2O is minimal. Buffering decreases the number of OH^- in the system.

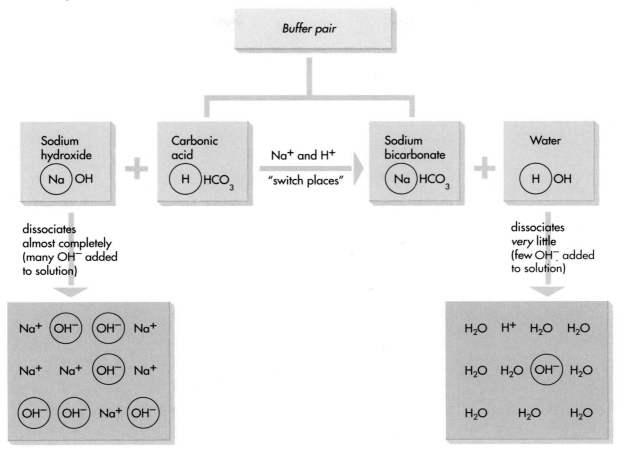

BICARBONATE LOADING

The buildup of lactic acid in the blood, released as a waste product from working muscles, has been blamed for the soreness and fatigue that sometimes accompanies strenuous exercise. Some athletes have adopted a technique called **bicarbonate loading,** ingesting large amounts of sodium bicarbonate ($NaHCO_3$) to counteract the effects of lactic acid buildup. Their theory is that fatigue is avoided because the $NaHCO_3$, a base, buffers the lactic acid. Unfortunately, the diarrhea that often results can trigger fluid and electrolyte imbalances. Long-term $NaHCO_3$ abuse can lead to disruption of acid-base balance and its disastrous effects.

FIGURE 20-4 Lactic Acid Buffered by Sodium Bicarbonate. Lactic acid (H · lactate) and other "fixed" acids are buffered by $NaHCO_3$ in the blood. Carbonic acid (H · HCO_3 or H_2CO_3, a weaker acid than lactic acid) replaces lactic acid. As a result, fewer H^+ are added to blood than would be if lactic acid were not buffered.

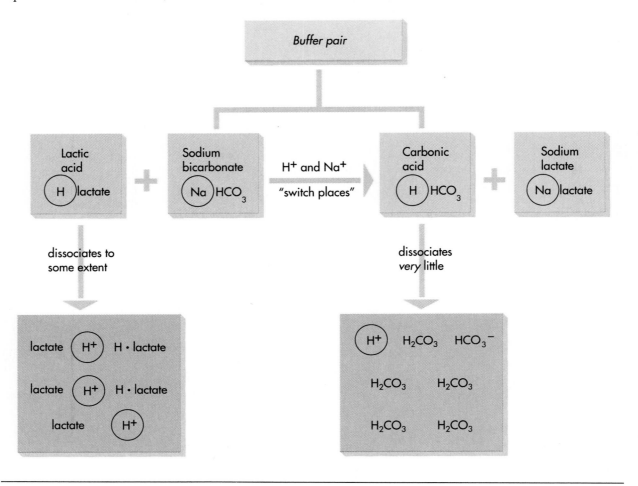

that do not break down to form a gas). Figure 20-4 shows the compounds formed by the buffering of lactic acid (a "fixed" acid), produced by normal body catabolism. The following changes in blood result from the buffering of fixed acids in the tissue capillaries:

1 The amount of H_2CO_3 in blood increases slightly because an acid (such as lactic acid) is converted to H_2CO_3.

2 The amount of bicarbonate in blood (mainly $NaHCO_3$) decreases because bicarbonate ions become part of the newly formed H_2CO_3. Normal arterial blood with a pH of 7.45 contains 20 times more $NaHCO_3$ than

H_2CO_3. If this ratio decreases, blood pH decreases below 7.45.

3 The H^+ concentration of blood increases slightly. H_2CO_3 adds hydrogen ions to blood, but it adds fewer of them than lactic acid would have because it is a weaker acid than lactic acid. In other words the buffering mechanisms do not totally prevent blood hydrogen ion concentration from increasing. It simply minimizes the increase.

4 Blood pH decreases slightly because of the small increase in blood concentration.

H_2CO_3 is the most abundant acid in body fluids because it is formed by the buffering of fixed acids and also because CO_2 forms it by combining with H_2O. Large amounts of CO_2, an end product of catabolism, continually pour into tissue capillary blood from cells. Much of the H_2CO_3 formed in blood diffuses into red blood cells where it is buffered by the potassium salt of hemoglobin. H_2CO_3 breaks down to form the gas, CO_2, and H_2O. This takes place in blood as it moves through the lung capillaries. Read the next paragraphs to find out how this affects blood pH.

RESPIRATORY MECHANISM OF pH CONTROL

Respirations play a vital part in controlling pH. With every expiration, CO_2 and H_2O leave the body in the expired air. The CO_2 has diffused out of the venous blood as it moves through the lung capillaries. Less CO_2 therefore remains in the arterial blood leaving the lung capillaries, so less of it is available for combining with water to form H_2CO_3. Hence arterial blood contains less H_2CO_3 and fewer hydrogen ions and has a higher pH (7.45) than does venous blood (pH 7.35).

Let us consider now how a change in respirations can change blood pH. Suppose you were to pinch your nose shut and hold your breath for a full minute or a little longer. Obviously, no CO_2 would leave your body by way of the expired air during that time and the blood's CO_2 content would necessarily increase. This would increase the amount of H_2CO_3 and the hydrogen-ion concentration of blood, which in turn would decrease blood pH. Here are two facts to remember. Anything that causes an appreciable decrease in respirations will in time produce **acidosis.** Conversely, anything that causes an excessive increase in respirations will in time produce **alkalosis.**

URINARY MECHANISM OF pH CONTROL

Most people know that the kidneys are vital organs and that life soon ebbs away if they stop functioning. One reason is that the kidneys are the body's most effective regulators of blood pH. They can eliminate much larger amounts of acid than can the lungs and, if it becomes necessary, they can also excrete excess base. The lungs cannot. In short, the kidneys are the body's last and best defense against wide variations in blood pH. If they fail, homeostasis of pH—acid-base balance—fails.

Because more acids than bases usually enter blood, more acids than bases are usually excreted by the kidneys. In other words, most of the time the kidneys acidify urine; that is, they excrete enough acid to give urine an acid pH—frequently as low as 4.8. (How does this compare with normal blood pH?) The distal tubules of the kidneys rid the blood of excess acid and at the same time conserve the base present in it by the two mechanisms illustrated by Figures 20-5 and 20-6. To

VOMITING

Vomiting, sometimes referred to as *emesis* (EM-e-sis), is the forcible emptying or expulsion of gastric and occasionally intestinal contents through the mouth. It can occur as a result of many stimuli, including foul odors or tastes, irritation of the stomach or intestinal mucosa, and some vomitive (emetic) drugs such as ipecac. A "vomiting center" in the brain regulates the many coordinated (but primarily involuntary) steps involved. Severe vomiting such as the pernicious vomiting of pregnancy or the repeated vomiting associated with pyloric obstruction in infants can be life threatening. One of the most frequent and serious complications of vomiting is metabolic alkalosis. The bicarbonate excess of metabolic alkalosis results because of the massive loss of chloride from the stomach as HCl. The loss of chloride causes a compensatory increase of bicarbonate in the extracellular fluid. The result is metabolic alkalosis. Therapy includes intravenous administration of chloride-containing solutions such as **normal saline.** The chloride ions of the solution replace the bicarbonate ions and thus help relieve the bicarbonate excess that is responsible for the imbalance.

▼ **FIGURE 20-5** Acidification of Urine and Conservation of Base by Distal Renal Tubule Secretion of H⁺ Ions.

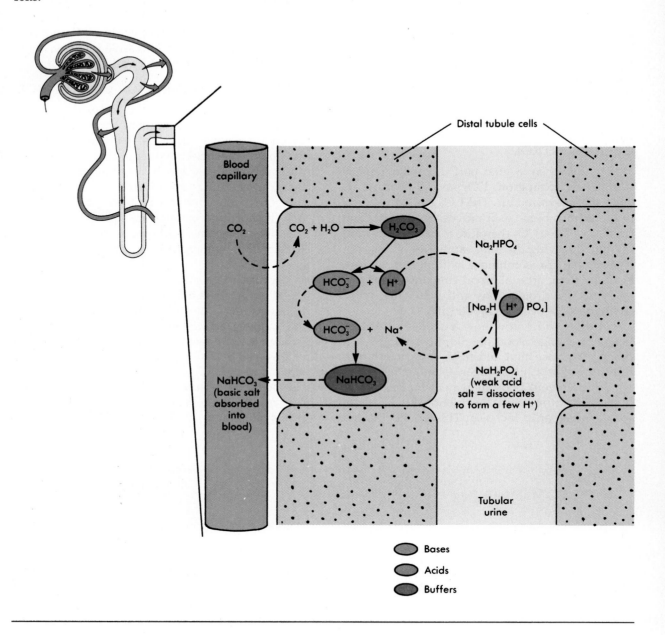

understand these figures fully, you need to have some grasp of basic chemistry. If necessary, refer to Appendix A before proceeding. Then look at Figure 20-5 and find the CO_2 leaving the blood (as it flows through a kidney capillary) and entering one of the cells that helps form the wall of a distal kidney tubule. Note that in this cell the CO_2 combines with water to form H_2CO_3. This occurs rapidly because the cell contains carbonic anhydrase, an enzyme that accelerates this reaction. As soon as H_2CO_3 forms, some of it dissociates to yield hydrogen ions and bicarbonate ions. Note

▼ **FIGURE 20-6 Acidification of Urine by Tubule Secretion of Ammonia (NH_3).** An amino acid (glutamine) moves into the tubule cell and loses an amino group (NH_2) to form ammonia, which is secreted into urine. In exchange, the tubule cell reabsorbs a basic salt (mainly $NaHCO_3$) into blood from urine.

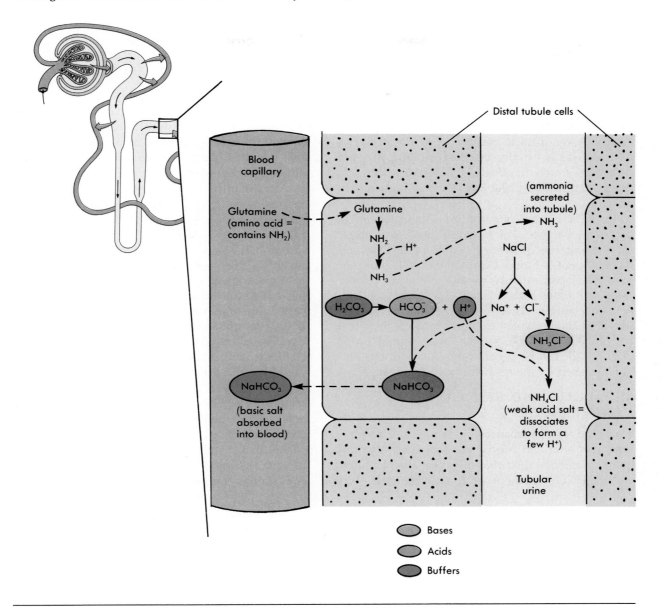

what happens to these ions. Hydrogen ions diffuse out of the tubule cell into the urine trickling down the tubule. There, it replaces one of the sodium ions (Na^+) in a salt (Na_2HPO_4) to form another salt (NaH_2PO_4), which leaves the body in the urine. Notice next that the Na^+ displaced from Na_2HPO_4 by the H^+ moves out of the tubular urine into a tubular cell. Here it combines with a bicarbonate (HCO_3^-) ion to form sodium bicarbonate, which then is resorbed into the blood. What this complex of reactions has accomplished is to add hydrogen ions to the urine—that is, acidify it—and to conserve NaHCO$_3$ by reabsorbing it into the blood.

Figure 20-6 illustrates another method of acidifying urine, as explained in the legend.

▼ pH Imbalances

Acidosis and **alkalosis** are the two kinds of pH or acid-base imbalance. In acidosis the blood pH falls as H$^+$ ion concentration increases. Only rarely does it fall as low as 7.0 (neutrality) and almost never does it become even slightly acid, because death usually intervenes before the pH drops this much. In alkalosis, which develops less often than acidosis, the blood pH is higher than normal.

From a clinical standpoint, disturbances in acid-base balance can be considered dependent on the relative quantities (ratio) of H$_2$CO$_3$ and NaHCO$_3$ in the blood. Components of this important buffer pair must be maintained at the proper ratio (20 times more NaHCO$_3$ than H$_2$CO$_3$) if acid-base balance is to remain normal. It is fortunate that the body can regulate both chemicals in the NaHCO$_3$–H$_2$CO$_3$ buffer system. Blood levels of NaHCO$_3$ can be regulated by the kidneys and H$_2$CO$_3$ levels by the respiratory system (lungs).

METABOLIC AND RESPIRATORY DISTURBANCES

Two types of disturbances, metabolic and respiratory, can alter the proper ratio of these components. Metabolic disturbances affect the bicarbonate (NaHCO$_3$) element of the buffer pair, and respiratory disturbances affect the H$_2$CO$_3$ element, as follows:

1 **Metabolic disturbances**
   a *Metabolic acidosis* (bicarbonate deficit)
   b *Metabolic alkalosis* (bicarbonate excess)
2 **Respiratory disturbances**
   a *Respiratory acidosis* (H$_2$CO$_3$ excess)
   b *Respiratory alkalosis* (H$_2$CO$_3$ deficit)

The *ratio* of NaHCO$_3$ to H$_2$CO$_3$ levels in the blood is the key to acid-base balance. If the normal ratio (20:1 NaHCO$_3$/H$_2$CO$_3$) can be maintained, the acid-base balance and pH remain normal despite changes in the absolute amounts of either component of the buffer pair in the blood.

As a clinical example, in a person suffering from untreated diabetes, abnormally large amounts of acids enter the blood. The normal 20:1 ratio is altered as the NaHCO$_3$ component of the buffer pair reacts with the acids. Blood levels of NaHCO$_3$ decrease rapidly in these patients. The result is a lower ratio of NaHCO$_3$ to H$_2$CO$_3$ (perhaps 10:1) and lower blood pH. The condition is called **uncompensated metabolic acidosis.** The body attempts to correct or *compensate* for the acidosis by altering the *ratio* of NaHCO$_3$ to H$_2$CO$_3$. Acidosis in a diabetic patient is often accompanied by rapid breathing or hyperventilation. This compensatory action of the respiratory system results in a "blow-off" of CO$_2$. Decreased blood levels of CO$_2$ result in lower H$_2$CO$_3$ levels. A new compensated ratio of NaHCO$_3$ to H$_2$CO$_3$ (perhaps 10:0.5) may result. In such individuals the blood pH returns to normal or near-normal levels. The condition is **compensated metabolic acidosis.**

OUTLINE SUMMARY

pH OF BODY FLUIDS

A Definition of pH—a number that indicates the hydrogen ion (H$^+$) concentration of a fluid; pH 7.0 indicates neutrality, pH higher than 7.0 indicates alkalinity, and pH less than 7.0 indicates acidity—see Figure 20-1
B Normal arterial blood pH—about 7.45
C Normal venous blood pH—about 7.35

MECHANISMS THAT CONTROL pH OF BODY FLUIDS

A Buffers
   1 Definition—substances that prevent a sharp change in the pH of a fluid when an acid or base is added to it—see Figures 20-2 and 20-3
   2 "Fixed" acids are buffered mainly by sodium bicarbonate (NaHCO$_3$)
   3 Changes in blood produced by buffering of "fixed" acids in the tissue capillaries

a Amount of carbonic acid (H_2CO_3) in blood increases slightly

b Amount of $NaHCO_3$ in blood decreases; ratio of amount of $NaHCO_3$ to the amount of H_2CO_3 does not normally change; normal ratio is 20:1

c H^+ concentration of blood increases slightly

d Blood pH decreases slightly below arterial level

B Respiratory mechanism of pH control—respirations remove some CO_2 from blood; as blood flows through lung capillaries, the amount of H_2CO_3 in blood is decreased and thereby its H^+ concentration is decreased, and this in turn increases blood pH from its venous to its arterial level

C Urinary mechanism of pH control—the body's most effective regulator of blood pH, kidneys usually acidify urine by the distal tubules secreting hydrogen ions and ammonia (NH_3) into the urine from blood in exchange for $NaHCO_3$ being reabsorbed into the blood

pH IMBALANCES

A Acidosis and alkalosis are the two kinds of pH or acid-base imbalances

B Disturbances in acid-base balance depend on relative quantities of $NaHCO_3$ and H_2CO_3 in the blood

C Body can regulate both of the components of the $NaHCO_3$–H_2CO_3 buffer system

1 Blood levels of $NaHCO_3$ regulated by kidneys

2 H_2CO_3 levels regulated by lungs

D Two basic types of pH disturbances, metabolic and respiratory, can alter the normal 20:1 ratio of $NaHCO_3$ to H_2CO_3 in blood

1 Metabolic disturbances affect the $NaHCO_3$ levels in blood

2 Respiratory disturbances affect the H_2CO_3 levels in blood

E Types of pH or acid-base imbalances

1 Metabolic disturbances

  a Metabolic acidosis—bicarbonate ($NaHCO_3$) deficit

  b Metabolic alkalosis—bicarbonate ($NaHCO_3$) excess

2 Respiratory disturbances

  a Respiratory acidosis (H_2CO_3 excess)

  b Respiratory alkalosis (H_2CO_3 deficit)

F In uncompensated metabolic acidosis, the normal ratio of $NaHCO_3$ to H_2CO_3 is changed; in compensated metabolic acidosis, the ratio remains at 20:1, but the total amount of $NaHCO_3$ and H_2CO_3 changes

$\mathcal{N}$EW WORDS

acid solution	base	buffer pairs	pH
alkaline solution	buffer	carbonic anhydrase	

Diseases and Other Clinical Terms

Metabolic acidosis	Respiratory acidosis	Metabolic alkalosis	Respiratory alkalosis

$\mathcal{C}$HAPTER TEST

1. A fluid having a pH of 7.0 would be described as _____ in reaction.
2. A pH higher than 7.0 indicates an _____ solution, and one with a pH lower than 7.0 indicates an _____ solution.
3. As CO_2 enters blood, some of it is converted into H_2CO_3 by an enzyme called _____ _____ found in red blood cells.
4. Substances that prevent a sharp change in pH of fluid when acid or base is added are _____.
5. The most abundant acid in the body is _____ acid.
6. Anything that causes an appreciable decrease in respirations will produce _____.
7. Acidification of urine can occur by renal tubule excretion of _____.
8. One of the most frequent complications of vomiting is metabolic _____.
9. A body deficit of $NaHCO_3$ results in metabolic _____.
10. The ability of the body to maintain a constant blood pH during changing conditions is an example of _____.

Select the most correct answer from Column B for each statement in Column A. (Only one answer is correct.)

COLUMN A

11. _____ pH of arterial blood
12. _____ Neutral in reaction
13. _____ Lost as CO_2
14. _____ Prevent pH changes
15. _____ "Fixed" acid
16. _____ Hydrogen ion concentration
17. _____ Acidification of urine
18. _____ Vomiting
19. _____ H_2CO_3 excess
20. _____ Compensates for acidosis

COLUMN B

a. Lactic acid
b. pH 7.0
c. Excretion of ammonia
d. Buffers
e. Metabolic alkalosis
f. pH 7.45
g. Respiratory acidosis
h. pH
i. Hyperventilation
j. H_2CO_3

REVIEW QUESTIONS

1 Explain briefly what the term *pH* means.
2 What is a typical normal pH for venous blood? For arterial blood?
3 When the body is in acid-base balance, arterial blood contains how many times more sodium bicarbonate than carbonic acid? In other words, what is the normal $NaHCO_3/H_2CO_3$ ratio in blood?
4 What are the functions of buffers?
5 Explain how respirations affect blood pH.
6 Explain how the kidneys maintain normal blood pH.
7 How does prolonged hyperventilation (abnormally increased respirations) affect blood pH?
8 Which is the most important for maintaining acid-base balance—buffers, respirations, or kidney functioning?
9 Briefly, how does compensated acidosis differ from uncompensated acidosis?
10 List the types of pH or acid-base imbalances.

CLINICAL APPLICATIONS

1 Compensated respiratory acidosis is commonly found in persons with chronic bronchitis, an obstructive respiratory disorder discussed in Chapter 15. Can you state what abnormal blood values a person should expect in such a case, and what factors produced them?

2 Larry is a diabetic who is suffering from metabolic acidosis. His breathing seems abnormally rapid. Is there a connection between Larry's acidosis and his rapid breathing? If so, can you explain the connection?

3 The hormone aldosterone affects kidney tubule function. One of its effects is to increase H^+ secretion by the kidney tubules. What effect does this action have on the pH of the internal environment (blood plasma)? What might occur if there is hypersecretion of aldosterone? Hyposecretion of aldosterone?

The Reproductive Systems

C H A P T E R

21

Objectives

*After you have completed this chapter,
you should be able to:*

1 List the essential and accessory organs of the
male and female reproductive systems and give
the generalized function of each.

2 Describe the gross and microscopic structure of
the gonads in both sexes and explain the developmental steps in spermatogenesis and oogenesis.

3 Discuss the primary functions of the sex hormones and identify the cell type or structure
responsible for their secretion.

4 Identify and describe the structures that constitute the external genitals in both sexes.

5 Identify and discuss the phases of the endometrial or menstrual cycle and correlate each
phase with its occurrence in a typical 28-day
cycle.

6 List the major disorders of the male and female
reproductive systems and briefly describe each.

7 Define the term *sexually transmitted disease* and
describe the major types.

Boxed Essays

Detecting Prostate Cancer
Ectopic Pregnancy
Hysterectomy
Fibrocystic Disease

The offspring of many one-celled plants and bacteria come from a single parent. These organisms are said to be *asexual* because they do not produce specialized reproductive or sex cells called **gametes** (GAM-eets). In humans, gametes, called **ova** and **sperm,** fuse during the process of fertilization to produce a cell called the **zygote** (ZYE-gote), which ultimately develops into the new individual. The zygote, which contains an intermingling of genetic messages from the sex cells of both parents, ultimately permits development of new human life. Reproduction in humans therefore is said to be *sexual.* As with all sexually produced offspring, new human life results from the equal contribution of not one but two parent cells—the female ovum and male sperm.

This chapter deals with the structure and function of the reproductive system in men and women. We are truly "fearfully and wonderfully made." Almost any one of the body's organ systems might have inspired this statement, but of them all, perhaps the reproductive systems best deserve such praise. Their awesome achievement is the creation of nature's most complex and beautiful structure—the human body; their ultimate goal is the survival of the human species. This chapter covers normal and pathological anatomy and physiology of the reproductive systems in both sexes and ends with a discussion of sexually transmitted diseases. Chapter 22 will cover the topic of human development—a process extending from fertilization until death.

Common Structural and Functional Characteristics Between the Sexes

Although the organs and specific functions of the male and female reproductive systems will be discussed separately, it is important to understand that a common general structure and function can be identified between the systems in both sexes and that both sexes contribute in uniquely important ways to overall reproductive success.

In both men and women, the organs of the reproductive system are adapted for the specific sequence of functions that permit development of sperm or ova followed by successful fertilization and then the normal development and birth of a baby. In addition, production of hormones that permits development of secondary sex characteristics, such as breast development in women and beard growth in men, occurs as a result of normal reproductive system activity.

As you study the specifics of each system keep in mind that the male organs function to produce, store, and ultimately introduce mature sperm into the female reproductive tract and that the female system is designed to produce ova, receive the sperm, and permit fertilization. In addition, the highly developed and specialized reproductive system in women permits the fertilized ovum to develop and mature until birth. The complex and cyclic control of reproductive functions in women are particularly crucial to overall reproductive success in humans. The production of sex hormones is required for development of the secondary sexual characteristics and for normal reproductive functions in both sexes.

Male Reproductive System

STRUCTURAL PLAN

So many organs make up the male reproductive system that we need to look first at the structural plan of the system as a whole. Reproductive organs can be classified as **essential** or **accessory.**

Essential Organs

The essential organs of reproduction in men and women are called the **gonads.** The gonads of men consist of a pair of main sex glands called the **testes** (TES-teez). The testes produce the male sex cells or **spermatozoa** (sper-ma-toe-ZO-ah).

Accessory Organs

The accessory organs of reproduction in men consist of the following structures:

1 A series of passageways or ducts that carry the sperm from the testes to the exterior
2 Additional sex glands that provide secretions that protect and nurture sperm
3 The external reproductive organs called the external genitals

Table 21-1 lists the names of the essential and accessory organs of reproduction in men, and Figure 21-1 shows the location of most of them. The table and the illustration are included very

FIGURE **21-1** Organization of the Male Reproductive Organs.

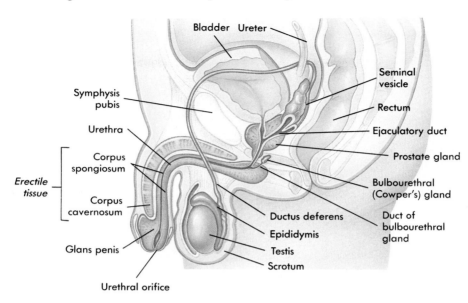

early in the chapter to provide a preliminary but important overview. Refer back to this table and illustration frequently as you learn about each organ in the pages that follow.

TESTES

Structure and Location

The paired **testes** are the gonads of men. They are located in the pouchlike **scrotum** (SKRO-tum), which is suspended outside of the body cavity behind the penis (Figure 21-1). This exposed location provides an environment about 1° C (3° F) cooler than normal body temperature, an important requirement for the normal production and survival of sperm. Each testis is a small, oval gland about 3.8 cm (1½ inches) long and 2.5 cm (1 inch) wide. The testis is shaped like an egg that has been flattened slightly from side to side. Note in Figure 21-2 that each testis is surrounded by a tough, whitish membrane called the **tunica** (TOO-ni-kah) **albuginea** (al-byoo-JIN-ee-ah). This membrane covers the testicle and then enters the gland to form the many septa that divide it into sections or lobules. As you can see in Figure 21-2, each lobule consists of a narrow but long and coiled

TABLE **21-1** Male Reproductive Organs

ESSENTIAL ORGANS	ACCESSORY ORGANS
Gonads: testes (right testis and left testis)	Ducts: epididymis (two), vas deferens (two), ejaculatory duct (two), and urethra
	Supportive sex glands; seminal vesicle (two), bulbourethral or Cowper's gland (two), and prostate gland
	External genitals: scrotum and penis

seminiferous (se-mi-NIF-er-us) **tubule.** These coiled structures form the bulk of the testicular tissue mass. Small, specialized cells lying near the septa that separate the lobules can be seen in Figure 21-3. These are the **interstitial cells** of the testes that secrete the male sex hormone **testosterone** (tes-TOS-te-rone).

FIGURE 21-2 Tubules of the Testis and Epididymis. The ducts and tubules are exaggerated in size. In the photograph, the testicle is the darker sphere in the center.

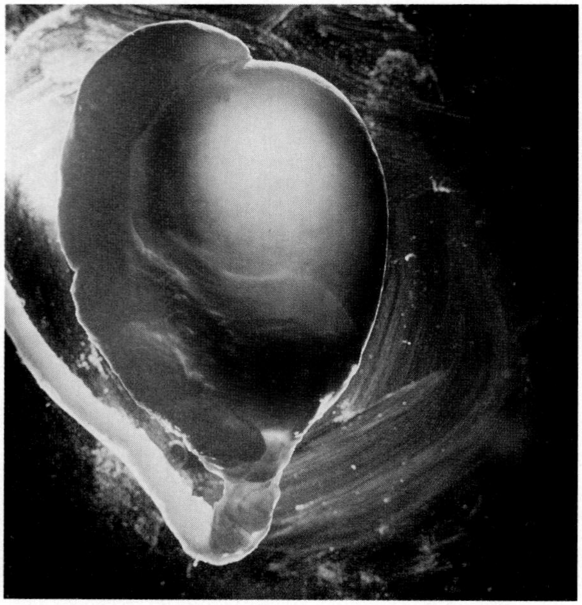

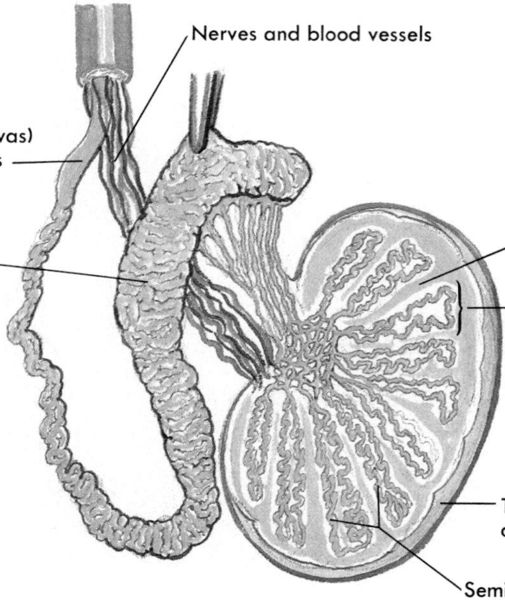

Nerves and blood vessels

Ductus (vas) deferens

Body of epididymis

Septum

Lobule

Tunica albuginea

Seminiferous tubules

Each seminiferous tubule is a long duct with a central lumen or passageway (Figure 21-3). Sperm develop in the walls of the tubule and are then released into the lumen and begin their journey to the exterior of the body.

Testis Functions

Spermatogenesis. Sperm production is called **spermatogenesis** (sper-ma-toe-JEN-e-sis). From puberty on, the seminiferous tubules continuously form spermatozoa or sperm. Although the number of sperm produced each day diminish with increasing age, most men continue to produce significant numbers throughout life.

The testes prepare for sperm production before puberty by increasing the numbers of sperm pre-

cursor (stem) cells called **spermatogonia** (sper-ma-toe-GO-nee-ah). These cells are located near the outer edge of each seminiferous tubule (Figure 21-4, *A*). Before puberty, spermatogonia increase in number by the process of mitotic cell division, which was described in Chapter 2. Recall that mitosis results in the division of a "parent" cell into two "daughter" cells, each identical to the parent and each containing a complete copy of the genetic material represented in the normal number of 46 chromosomes.

When a boy enters puberty, circulating levels of follicle-stimulating hormone (FSH) cause a spermatogonium to undergo a unique type of cell division. When the spermatogonium undergoes cell division and mitosis under the influence of

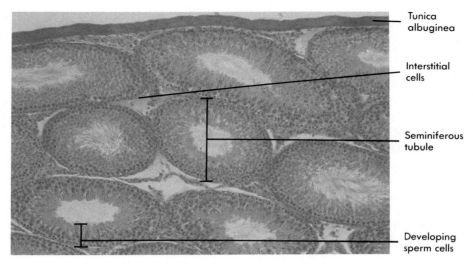

FIGURE 21-3 Testis Tissue. Several seminiferous tubules surrounded by septa containing interstitial cells are shown.

Tunica albuginea

Interstitial cells

Seminiferous tubule

Developing sperm cells

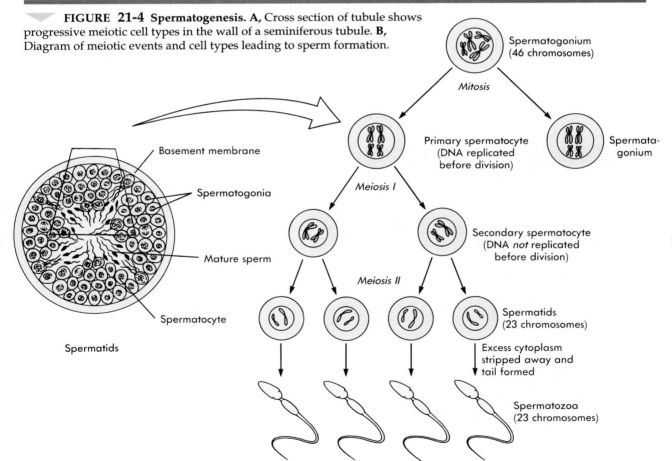

FIGURE 21-4 Spermatogenesis. A, Cross section of tubule shows progressive meiotic cell types in the wall of a seminiferous tubule. **B,** Diagram of meiotic events and cell types leading to sperm formation.

A

Basement membrane

Spermatogonia

Mature sperm

Spermatocyte

Spermatids

B

Spermatogonium (46 chromosomes)

Mitosis

Primary spermatocyte (DNA replicated before division)

Spermata-gonium

Meiosis I

Secondary spermatocyte (DNA *not* replicated before division)

Meiosis II

Spermatids (23 chromosomes)

Excess cytoplasm stripped away and tail formed

Spermatozoa (23 chromosomes)

FSH, it produces two daughter cells. One of these cells remains as a spermatogonium and the other forms another, more specialized cell called a **primary spermatocyte** (SPER-ma-toe-cyte). These primary spermatocytes then undergo a specialized type of division called **meiosis** (my-O-sis), which ultimately results in sperm formation. Note in Figure 21-4, *B*, that during meiosis two cell divisions occur (not one as in mitosis) and that four daughter cells (not two as in mitosis) are formed. The daughter cells are called **spermatids** (SPER-ma-tids). Unlike the two daughter cells that result from mitosis, the four spermatids, which will develop into spermatozoa, have only half the genetic material and half of the chromosomes (23) of other body cells.

In women, meiosis results in the production of a single ovum, which also has 23 chromosomes. This will be discussed in more detail later in the chapter.

Look again at the diagram of meiosis in Figure 21-4, *B*. It shows that each primary spermatocyte ultimately produces four sperm cells. Note that, in the portion of a seminiferous tubule shown in Figure 21-4, *A*, spermatogonia are found at the outer surface of the tubule, primary and secondary spermatocytes lie deeper in the tubule wall, and mature sperm are seen about to enter the lumen of the tube and begin their journey through the reproductive ducts to the exterior of the body.

Spermatozoa. Spermatozoa are among the smallest and most highly specialized cells in the body (Figure 21-5, *A*). All of the characteristics that a baby will inherit from its father at fertilization are contained in the condensed nuclear (genetic) material found in each sperm head. However, this genetic information from the father can fuse with genetic material contained in the mother's ovum only if successful fertilization occurs. Ejaculation of sperm into the female vagina during sexual intercourse is only one step in the long journey that these sex cells must make before they can meet and fertilize an ovum. To accomplish their task, these specialized packages of genetic information are equipped with a tail for motility and are designed to penetrate the outer membrane of the ovum when contact occurs.

The structure of a mature sperm is diagrammed in Figure 21-5, *B*. Note the sperm head

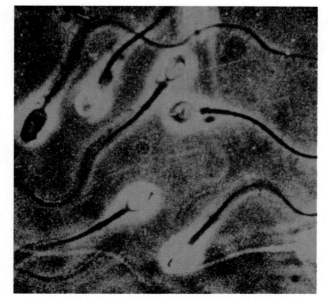

A

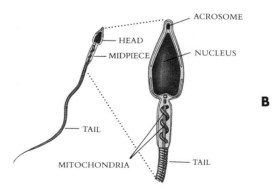

FIGURE 21-5 Human Sperm. A, Micrograph shows the heads and long, slender tails of several spermatozoa. **B,** Illustration shows the components of a mature sperm cell and an enlargement of a sperm head.

ACROSOME

HEAD

MIDPIECE

NUCLEUS

TAIL

MITOCHONDRIA

TAIL

B

containing the nucleus with its genetic material from the father. The nucleus is covered by the **acrosome** (AK-ro-sohm)—a specialized structure containing enzymes that enable the sperm to break down the covering of the ovum and permit entry if contact occurs. In addition to the head with its covering acrosome, each sperm has a midpiece and an elongated tail. Mitochondria in the midpiece break down adenosine triphosphate (ATP) to provide energy for the tail movements required to propel the sperm and allow them to "swim" for relatively long distances through the female reproductive ducts.

Production of testosterone. In addition to spermatogenesis, the other function of the testes is to secrete the male hormone, testosterone. This function is carried on by the interstitial cells of the testes, not by its seminiferous tubules. Testosterone serves the following general functions:

1 It masculinizes. The various characteristics that we think of as "male" develop because of testosterone's influence. For instance, when a young boy's voice changes, it is testosterone that brings this about.
2 It promotes and maintains the development of the male accessory organs (prostate gland, seminal vesicles, and so on).
3 It has a stimulating effect on protein anabolism. Testosterone thus is responsible for the greater muscular development and strength of the male.

A good way to remember testosterone's functions is to think of it as "the masculinizing hormone" and the "anabolic hormone."

REPRODUCTIVE DUCTS

The ducts through which sperm must pass after exiting from the testes until they reach the exterior of the body are important components of the accessory reproductive structures. The other two components included in the listing of accessory organs of reproduction in the male—the supportive sex glands and external genitals—will be discussed separately.

Sperm are formed within the walls of the seminiferous tubules of the testes. When they exit from these tubules within the testis, they enter and then pass, in sequence, through the epididymis, ductus (vas) deferens, ejaculatory duct,

and the urethra on their journey out of the human body.

Epididymis

Each **epididymis** (ep-i-DID-i-mis) consists of a single and very tightly coiled tube about 6 m (20 feet) in length. It is a comma-shaped structure (see Figure 21-2) that lies along the top and behind the testes inside the scrotum. Sperm mature and develop their ability to move or swim as they pass through the epididymis.

Ductus (Vas) Deferens

The **ductus** (DUK-tus) **deferens** (DEF-er-enz) or **vas deferens** is the tube that permits sperm to exit from the epididymis and pass from the scrotal sac upward into the abdominal cavity. Each ductus deferens is a thick, smooth, very muscular, and movable tube that can easily be felt or "palpated" through the thin skin of the scrotal wall. It passes through the inguinal canal into the abdominal cavity as part of the *spermatic cord,* a connective tissue sheath that also encloses blood vessels and nerves.

Once in the abdominal cavity, the ductus deferens extends over the top and down the posterior surface of the bladder, where it joins the duct from the seminal vesicle to form the **ejaculatory** (ee-JAK-yoo-lah-toe-ree) **duct.** Note in Figure 21-1 that the ejaculatory duct passes through the substance of the prostate gland and permits sperm to empty into the urethra, which eventually passes through the penis and opens to the exterior at the external urethral orifice.

ACCESSORY OR SUPPORTIVE SEX GLANDS

The term **semen** (SEE-men) or **seminal fluid** is used to describe the mixture of sex cells or sperm produced by the testes and the secretions of the accessory or supportive sex glands. The accessory glands, which contribute over 95% of the secretions to the gelatinous fluid part of the semen, include the two seminal vesicles, one prostate gland, and two bulbourethral (Cowper's) glands. In addition to the production of sperm, the seminiferous tubules of the testes contribute somewhat less than 5% of the seminal fluid volume. Usually 3 to 5 ml (about 1 teaspoon) of semen is ejaculated at one time; each milliliter contains about 100 million sperm. Semen is alkaline and protects sperm from the acidic environment of the female reproductive passageways.

Seminal Vesicles

The paired **seminal vesicles** are pouchlike glands that contribute about 60% of the seminal fluid volume. Their secretions are yellowish, thick, and rich in the sugar fructose. This fraction of the seminal fluid helps provide a source of energy for the highly motile sperm.

Prostate Gland

The **prostate gland** lies just below the bladder and is shaped like a doughnut. The urethra passes through the center of the prostate before traversing the penis to end at the external urinary orifice. The prostate secretes a thin, milk-colored fluid that constitutes about 30% of the total seminal fluid volume. This fraction of the ejaculate helps to activate the sperm and maintain their motility.

Bulbourethral Glands

Each of the two **bulbourethral** (BUL-bo-yoo-REE-thral) **glands** (also called *Cowper's glands*) resemble peas in size and shape. They are located just below the prostate gland and empty their secretions into the penile portion of the urethra. The mucuslike secretions of these glands lubricate the

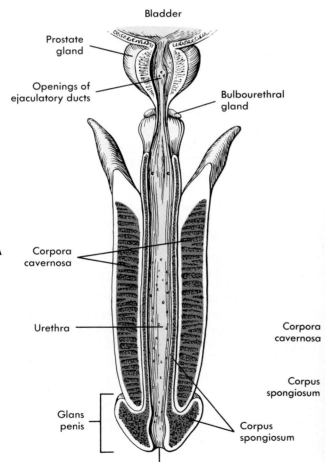

A

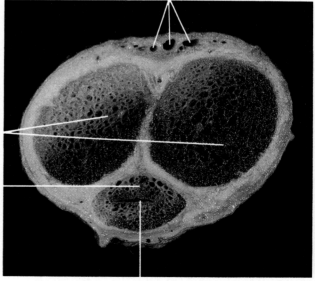

B

FIGURE 21-6 The Penis. A, In this sagittal section of the penis viewed from above, the urethra is exposed throughout its length and can be seen exiting from the bladder and passing through the prostate gland before entering the penis to end at the external urethral orifice. **B,** Photograph of a cross section of the shaft of the penis showing the three columns of erectile or cavernous tissue. Note the urethra within the substance of the corpus spongiosum.

terminal portion of the urethra and contribute less than 5% of the seminal fluid volume.

EXTERNAL GENITALS

The **penis** (PEE-nis) and **scrotum** constitute the external reproductive organs or **genitalia** (jen-i-TAL-ee-ah) of men. The penis (Figure 21-6) is the organ that, when made stiff and erect by the filling of its spongy or erectile tissue components with blood during sexual arousal, can enter and deposit sperm in the vagina during intercourse. The penis has three separate columns of erectile tissue in its shaft: one **corpus** (KOR-pus) **spongiosum** (spun-jee-O-sum), which surrounds the urethra, and two **corpora** (KOR-por-ah) **cavernosa** (kav-er-NO-sa), which lie above. The spongy nature of erectile tissue is apparent in Figure 21-6. At the distal end of the shaft of the penis is the enlarged **glands,** over which the skin is folded doubly to form a loose-fitting retractable casing called the **foreskin** or **prepuce** (PRE-pus). If the foreskin fits too tightly about the glans, a **circumcision** or surgical removal of the foreskin is usually performed to prevent irritation. The external urethral orifice is the opening of the urethra at the tip of the glans.

The scrotum is a skin-covered pouch suspended from the groin. Internally, it is divided into two sacs by a septum; each sac contains a testis, epididymis, the lower part of the ductus deferens, and the beginning of the spermatic cords.

Disorders of the Male Reproductive System

Several disorders of the male reproductive system cause **infertility.** Infertility is an abnormally low ability to reproduce. If there is a complete inability to reproduce, the condition is called **sterility.** Infertility or sterility involve an abnormally reduced capacity to deliver healthy sperm to the female reproductive tract. Reduced reproductive capacity may result from factors such as a decrease in the testes' production of sperm, structural abnormalities in the sperm, or obstruction of the reproductive ducts.

DISORDERS OF THE TESTES

Disruption of the sperm-producing function of the seminiferous tubules can result in decreased sperm production, a condition called **oligospermia** (ol-i-go-SPER-mee-ah). If the *sperm count* is too low, infertility may result. A large number of sperm is needed to ensure that many sperm will reach the ovum and dissolve its coating—allowing a single sperm to unite with the ovum. Oligospermia can result from factors such as infection, fever, radiation, malnutrition, and high temperature in the testes. In some cases, oligospermia is temporary—as in some acute infections. Oligospermia is a leading cause of infertility. Of course, total absence of sperm production results in sterility.

Early in fetal life the testes are located in the abdominal cavity near the kidneys but normally descend into the scrotum about 2 months before birth. Occasionally a baby is born with undescended testes, a condition called **cryptorchidism** (krip-TOR-ki-dizm), which is readily observed by palpation of the scrotum at delivery. The word *cryptorchidism* is from the Greek words *kryptikos* (hidden) and *orchis* (testis). Failure of the testes to descend may be caused by hormonal imbalances in the developing fetus or by a physical deficiency or obstruction. Regardless of cause, in the cryptorchid infant the testes remain "hidden" in the abdominal cavity. Because the higher temperature inside the body cavity inhibits spermatogenesis, measures must be taken to bring the testes down into the scrotum to prevent permanent sterility. Early treatment of this condition by surgery or by injection of testosterone, which stimulates the testes to descend, may result in normal testicular and sexual development.

Most testicular cancers arise from the sperm-producing cells of the seminiferous tubules. Malignancies of the testes are most common among men 25 to 35 years old. Besides age, this type of cancer is associated with genetic predisposition, trauma or infection of the testis, and cryptorchidism. Treatment of testicular cancer is most effective when the diagnosis is made early in the development of the tumor. Many physicians encourage male patients to perform regular self-examination of their testes, especially if they are in a high-risk group.

DISORDERS OF THE PROSTATE

A noncancerous condition called **benign** (bee-NINE) **prostatic hypertrophy** (hye-PER-tro-fee) is a common problem in older men. The condition is characterized by an enlargement or hypertrophy of the prostate gland. The fact that the urethra passes through the center of the prostate after exiting from the bladder is a matter of considerable clinical significance in this condition. As the prostate enlarges, it squeezes the urethra, frequently closing it so completely that urination becomes very difficult or even impossible. In such cases, surgical removal of a part or all of the gland, a procedure called **prostatectomy** (pros-ta-TEK-toe-me), may become necessary.

DISORDERS OF THE PENIS AND SCROTUM

The penis is subject to a number of sexually transmitted infections, as well as structural abnormalities. One such structural abnormality is **phimosis** (fi-MO-sis), a condition in which the foreskin fits so tightly over the glans that it cannot retract. As stated earlier in this chapter, the usual treatment for this condition is circumcision—a procedure in which the foreskin is cut along the base of the glans and removed. Severe phimosis can obstruct the flow of urine, possibly causing the death of an infant born with this condition. Milder phimosis can result in accumulation of dirt and organic matter under the foreskin, possibly causing severe infections.

Failure to achieve an erection of the penis is called **impotence** (IM-po-tens). Although impotence does not affect sperm production, it may cause infertility because normal intercourse may not be possible. Anxiety and psychological stress are often cited as causes of impotence. Impotence may also result from an abnormality in the erectile tissues of the penis or a failure of the nerve reflexes that control erection. Drugs and alcohol can cause temporary impotence by interfering with the nerves and blood vessels involved in producing an erection.

Swelling of the scrotum can be caused by a variety of conditions. One of the most common causes of scrotal swelling is an accumulation of fluid called a **hydrocele** (HYE-dro-seel). Hydroceles may be congenital, resulting from structural abnormalities present at birth. In adults, hydrocele often occurs when fluid produced by the serous membrane lining the scrotum is not absorbed properly. The cause of adult hydrocele is not always known, but in some cases, it can be linked to trauma or infection.

Swelling of the scrotum may also occur when the intestines push through the weak area of the abdominal wall that separates the abdominopelvic cavity from the scrotum. This condition is a form of **inguinal** (IN-gwi-nal) **hernia.** If the intestines protrude too far into the scrotum, the digestive tract may become obstructed—resulting in death. Inguinal hernia often occurs while lifting heavy objects, because of the high internal pressure generated by the contraction of abdominal muscles. Inguinal hernia may also be congenital. Small inguinal hernias may be treated with external supports that prevent organs from protruding into the scrotum; more serious hernias must be repaired surgically.

DETECTING PROSTATE CANCER

Many of the 32,000 men who die each year from prostate cancer—the most common cancer in American men—could be saved if the cancer was detected early enough for effective treatment. Several screening tests are now available for early detection of prostate cancer. For example, physicians can sometimes detect prostate cancer early by palpating the prostate through the wall of the rectum. Unfortunately, by the time prostate cancer can be palpated, it has usually spread to other organs. Many cancer experts believe prostate cancer screening can be more effective if rectal examinations are performed with a newer type of screening test called the *PSA test*. This test is a type of blood analysis that screens for *prostate specific antigen (PSA)*, a substance often found in the blood of men with prostate cancer. Because of the prevalence of this disease, adult men are encouraged to have regular prostate examinations and to report *any* urinary or sexual difficulty to their physicians.

Female Reproductive System

STRUCTURAL PLAN

The structural plan of the reproductive system in both sexes is similar in that organs are characterized as **essential** or **accessory.**

Essential Organs

The essential organs of reproduction in women, the **gonads,** are the paired **ovaries.** The female sex cells or **ova** are produced here.

Accessory Organs

The accessory organs of reproduction in women consist of the following structures:

1. A series of ducts or modified duct structures that extend from near the ovaries to the exterior
2. Additional sex glands, including the mammary glands, which have an important reproductive function only in women
3. The external reproductive organs or genitals

Table 21-2 lists the names of the essential and accessory organs of reproduction, and Figure 21-7 shows the location of most of them. Refer back to this table and illustration as you read about each structure in the pages that follow.

OVARIES

Structure and Location

The paired ovaries are the gonads of women. They have a puckered, uneven surface; each weighs about 3 g. They resemble large almonds in size and shape and are attached to ligaments in the pelvic cavity on each side of the uterus.

TABLE 21-2 Female Reproductive Organs	
ESSENTIAL ORGANS	**ACCESSORY ORGANS**
Gonads: ovaries (right ovary and left ovary)	Ducts: uterine tubes (two), uterus, vagina
	Accessory sex glands: Bartholin's glands (two), breasts (two)
	External genitals: vulva

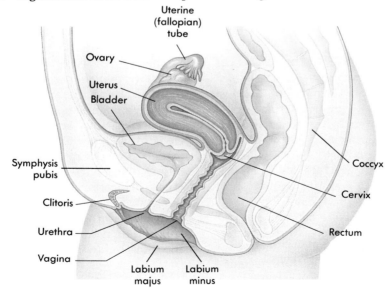

FIGURE 21-7 Organization of the Female Reproductive Organs.

FIGURE 21-8 Diagram of Ovary and Oogenesis. Cross section of mammalian ovary shows successive stages of ovarian (Graafian) follicle and ovum development. Begin with the first stage (primary follicle) and follow around clockwise to the final state (degenerating corpus luteum).

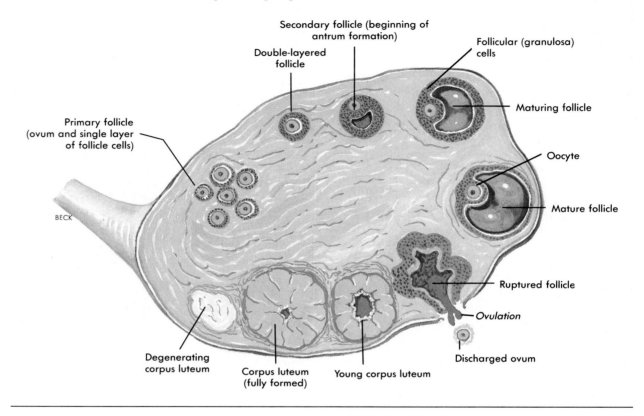

Embedded in a connective tissue matrix just below the outer layer of each ovary in a newborn baby girl are about 1 million **ovarian follicles;** each contains an **oocyte,** an immature stage of the female sex cell. By the time a girl reaches puberty, however, further development has resulted in the formation of a reduced number (about 400,000) of what are now called **primary follicles.** Each primary follicle has a layer of **granulosa cells** around the oocyte. During the reproductive lifetime of most women, only about 350 to 500 of these primary follicles fully develop into **mature follicles,** which ovulate and release an ovum for potential fertilization. Follicles that do not mature degenerate and are reabsorbed into the ovarian tissue. A mature ovum is sometimes called a **Graafian** (GRAHF-ee-an) **follicle,** in honor of the Dutch anatomist who discovered them 300 years ago.

The progression of development from primary follicle to ovulation is shown in Figure 21-8. As the thickness of the granulosa cell layer around the oocyte increases, a hollow chamber called an *antrum* (AN-trum) appears, and a **secondary follicle** is formed. Development continues, and, after ovulation, the ruptured follicle is transformed into a hormone-secreting glandular structure called the **corpus** (KOR-pus) **luteum** (LOO-tee-um), which is described later. Corpus luteum is from the Latin word meaning "yellow body," an appropriate name to describe the yellow appearance of this glandular structure.

Ovary Functions

Oogenesis. The production of female gametes or sex cells is called **oogenesis** (o-o-JEN-e-sis). The specialized type of cell division that results in sperm formation, meiosis, is also responsible for development of ova. During the developmental phases experienced by the female sex cell from its earliest stage to just after fertilization, two meiotic divisions occur. As a result of meiosis in the female sex cell, the number of chromosomes is reduced equally in each daughter cell to half the number (23) found in other body cells (46). However, the amount of cytoplasm is divided unequally. The result is formation of one large ovum and small daughter cells called *polar bodies* that degenerate.

The ovum with its large supply of cytoplasm and organelles is the body's largest cell and is uniquely designed to provide nutrients for rapid development of the embryo until implantation in the uterus occurs. At fertilization, the sex cells from both parents fuse, and the normal chromosome number (46) is achieved.

Production of estrogen and progesterone. The second major function of the ovary, in addition to oogenesis, is secretion of the sex hormones, **estrogen** and **progesterone**. Hormone production in the ovary begins at puberty with the cyclic development and maturation of the ovum. The granulosa cells around the oocyte in the growing and mature follicle secrete estrogen. The corpus luteum, which develops after ovulation, chiefly secretes progesterone but also some estrogen.

Estrogen is the sex hormone that causes the development and maintenance of the female *secondary sex characteristics* and stimulates growth of the epithelial cells lining the uterus. Some of the actions of estrogen include the following:

1 Development and maturation of female reproductive organs, including the external genitals
2 Appearance of pubic hair and breast development
3 Development of female body contours by deposition of fat below the skin surface and in the breasts and hip region
4 Initiation of the first menstrual cycle

Progesterone is produced by the corpus luteum, which is a glandular structure that develops from a ruptured mature follicle. If stimulated by the appropriate anterior pituitary hormone, the corpus luteum continues to produce progesterone for about 11 days after ovulation. Pro-

gesterone stimulates proliferation of the epithelial lining of the uterus and acts with estrogen to initiate the menstrual cycle in girls entering puberty.

REPRODUCTIVE DUCTS

Uterine Tubes

The two **uterine tubes,** also called **fallopian** (fal-LO-pee-an) **tubes** or **oviducts** (O-vi-dukts), serve as ducts for the ovaries, even though they are not attached to them. The outer end of each tube terminates in an expanded, funnel-shaped structure that has fringelike projections called **fimbriae** (FIM-bree-ee) along its edge. This part of the tube curves over the top of each ovary (Figure 21-9) and opens into the abdominal cavity. The inner end of each uterine tube attaches to the uterus, and the cavity inside the tube opens into the cavity in the uterus. Each tube is about 10 cm (4 inches) in length.

After ovulation the discharged ovum first enters the abdominal cavity and then finds its way into the uterine tube assisted by the wavelike movement of the fimbriae and the beating of the cilia on their surface. Once in the tube the ovum begins its journey to the uterus. Some ova never

---

ECTOPIC PREGNANCY

The term **ectopic** (ek-TOP-ic) **pregnancy** is used to describe a pregnancy resulting from the implantation of a fertilized ovum in any location other than the uterus. Occasionally, because the outer ends of the uterine tubes open into the pelvic cavity and are not actually connected to the ovaries, an ovum does not enter an oviduct but becomes fertilized and remains in the abdominal cavity. Although rare, if implantation occurs on the surface of an abdominal organ or on one of the mesenteries, development may continue to term. In such cases delivery by cesarean section is required. Most ectopic pregnancies involve implantation in the uterine tube and are therefore called *tubal pregnancies.* They result in tubal rupture and fetal death.

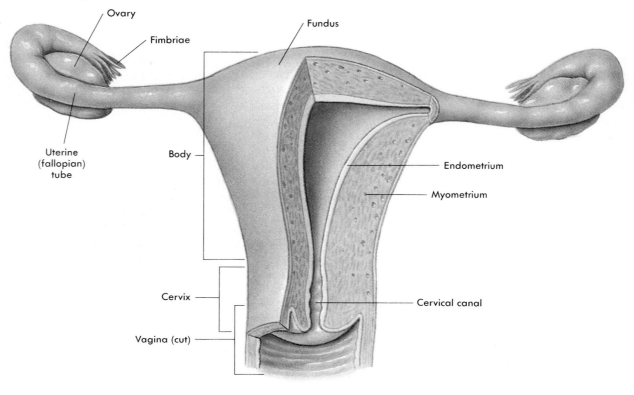

FIGURE 21-9 The Uterus. Sectioned view shows muscle layers of the uterus and its relationship to the ovaries and vagina.

find their way into the oviduct and remain in the abdominal cavity where they are reabsorbed. In Chapter 22 the details of fertilization, which normally occurs in the outer one third of the uterine tube, will be discussed.

The mucosal lining of the uterine tubes is directly continuous with the lining of the abdominal cavity on one end and with the lining of the uterus and vagina on the other. This is of great clinical significance because infections of the vagina or uterus such as gonorrhea may pass into the abdominal cavity, where they may become life threatening.

Uterus

The **uterus** (YOO-ter-us) is a small organ—only about the size of a pear—but it is extremely strong. It is almost all muscle or **myometrium** (my-o-ME-tree-um), with only a small cavity

inside. During pregnancy the uterus grows many times larger so that it becomes big enough to hold a baby and a considerable amount of fluid. The uterus is composed of two parts: an upper portion, the **body,** and a lower narrow section, the **cervix.** Just above the level where the uterine tubes attach to the body of the uterus, it rounds out to form a bulging prominence called the **fundus** (see Figure 21-9). Except during pregnancy, the uterus lies in the pelvic cavity just behind the urinary bladder. By the end of pregnancy, it becomes large enough to extend up to the top of the abdominal cavity. It then pushes the liver against the underside of the diaphragm—a fact that explains such a comment as "I can't seem to take a deep breath since I've gotten so big," made by many women late in their pregnancies.

The uterus functions in three processes—menstruation, pregnancy, and labor. The corpus luteum stops secreting progesterone and decreases its secretion of estrogens about 11 days after ovulation. About 3 days later, when the progesterone and estrogen concentrations in the blood are at their lowest, menstruation starts. Small pieces of the mucous membrane lining of the uterus, or the **endometrium** (en-doe-ME-tree-um) pull loose, leaving torn blood vessels underneath. Blood and bits of endometrium trickle out of the uterus into the vagina and out of the body. Immediately after menstruation the endometrium starts to repair itself. Hormonal stimulation resumes, and it again grows thick and becomes lavishly supplied with blood in preparation for pregnancy. If fertilization does not take place, the uterus once more sheds the lining made ready for a pregnancy that did not occur. Because these changes in the uterine lining continue to repeat themselves, they are spoken of as the **menstrual cycle** (see p. 513).

If fertilization occurs, pregnancy begins, and the endometrium remains intact. The events of pregnancy are discussed in Chapter 22.

Menstruation first occurs at puberty, often around the age of 12 years. Normally it repeats itself about every 28 days or thirteen times a year for some 30 to 40 years before it ceases at *menopause* (MEN-o-pawz), when a woman is somewhere around the age of 50 years.

Vagina

The **vagina** (vah-JYE-nah) is a distensible tube about 10 cm (4 inches) long made mainly of smooth muscle and lined with mucous membrane. It lies in the pelvic cavity between the urinary bladder and the rectum (Figure 21-7). As the part of the female reproductive tract that opens to the exterior, the vagina is the organ that sperm enter during their journey to meet an ovum, and it is also the organ from which a baby emerges to meet its new world.

ACCESSORY OR SUPPORTIVE SEX GLANDS

Bartholin's Glands

One of the small **Bartholin's** (BAR-toe-linz) or **greater vestibular** (ves-TIB-yoo-lar) **glands** lies to the right of the vaginal outlet and one lies to the left of it. Secretion of a mucuslike lubricating fluid is the function of Bartholin's glands. Their ducts

open into the space between the labia minora and the vaginal orifice called the **vestibule** (see Figure 21-11).

Breasts

The **breasts** lie over the pectoral muscles and are attached to them by connective tissue ligaments (of Cooper). Breast size is determined more by the amount of fat around the glandular (milk-secreting) tissue than by the amount of glandular tissue itself. Hence the size of the breast has little to do with its ability to secrete adequate amounts of milk after the birth of a baby.

Each breast consists of 15 to 20 divisions or lobes that are arranged radially (Figure 21-10). Each lobe consists of several lobules, and each lobule consists of milk-secreting glandular cells. The milk-secreting cells are arranged in grapelike clusters called *alveoli*. Small **lactiferous** (lak-TIF-er-us) **ducts** drain the alveoli and converge toward the nipple like the spokes of a wheel.

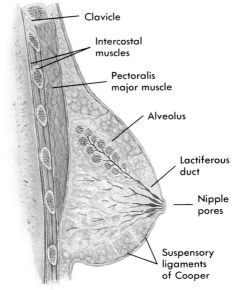

▽   **FIGURE 21-10 Lateral View of the Breast.** Sagittal section shows the gland fixed to the overlying skin and the pectoral muscles by the suspensory ligaments of Cooper. Each lobule of secretory tissue is drained by a lactiferous duct that opens through the nipple.

- Clavicle
- Intercostal muscles
- Pectoralis major muscle
- Alveolus
- Lactiferous duct
- Nipple pores
- Suspensory ligaments of Cooper

Only one lactiferous duct leads from each lobe to an opening in the nipple. The colored area around the nipple is the **areola** (ah-REE-o-lah).

A knowledge of the lymphatic drainage of the breast is important because cancerous cells from breast tumors often spread to other areas of the body through the lymphatic system. This lymphatic drainage is discussed in Chapter 14 (see also Figure 14-4).

EXTERNAL GENITALS

The **external genitalia** or **vulva** (VUL-vah) of women consist of the following:

1 Mons pubis
2 Clitoris
3 Orifice of urethra
4 Labia minora (singular, *labium minus*) (small lips)
5 Hymen

6 Orifice, duct of Bartholin's gland
7 Orifice of vagina
8 Labia majora (singular, *labium majus*) (large lips)

The **mons pubis** is a skin-covered pad of fat over the symphysis pubis. Hair appears on this structure at puberty and persists throughout life. Extending downward from the elevated mons pubis are the **labia** (LAY-bee-ah) **majora** (ma-JO-rah) or "large lips." These elongated folds, which are composed mainly of fat and numerous glands, are covered with pigmented skin and hair on the outer surface and are smooth and free from hair on the inner surface. The **labia minora** or "small lips" are located within the labia majora and are covered with modified skin. These two lips join anteriorly at the midline. The area between the labia minora is the **vestibule** (Figure 21-11). Several genital structures are located in the

▽ **FIGURE 21-11 External Genitals of the Female.**

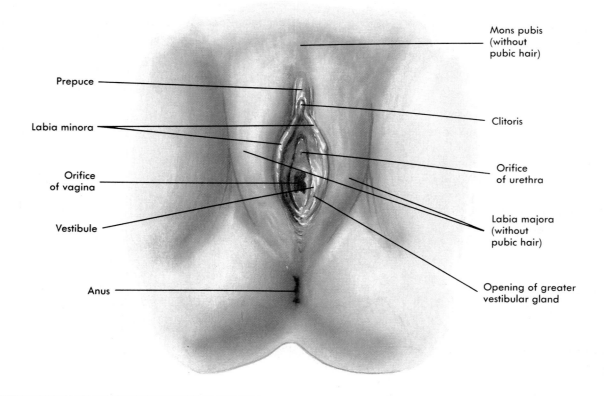

vestibule. The **clitoris** (KLIT-o-ris), which is composed of erectile tissue, is located just behind the anterior junction of the labia minora. Situated between the clitoris above and the vaginal opening below is the orifice of the urethra. The vaginal orifice is sometimes partially closed by a membranous **hymen** (HYE-men). The ducts of Bartholin's glands open on either side of the vaginal orifice inside the labia minora.

The term **perineum** (pair-i-NEE-um) is used to describe the area between the vaginal opening and anus. This area is sometimes cut in a surgical procedure called an **episiotomy** (e-piz-ee-OT-o-me) to prevent tearing of tissue during childbirth.

MENSTRUAL CYCLE

Phases and Events

The menstrual cycle consists of many changes—in the uterus, ovaries, vagina, and breasts and in the anterior pituitary gland's secretion of hormones. In the majority of women, these changes occur with almost precise regularity throughout their reproductive years. The first indication of changes comes with the first menstrual period. The first **menses** (MEN-seez) or menstrual flow is referred to as the **menarche** (me-NAR-kee).

A typical menstrual cycle covers a period of about 28 days. The length of the cycle varies among women. Some women, for example, may have a regular cycle that covers about 24 days. The length of the cycle also varies within one woman. Some women, for example, may have irregular cycles that range from 21 to 28 days, whereas the cycles of other women may be 2 to 3 months long. Each cycle consists of three phases. Although they have been called by several different names, we shall call them the *menstrual phase*, the *postmenstrual phase*, and the *premenstrual phase*. Examine Figure 21-12 to see the changes that take place in the lining of the uterus (endometrium) and in the ovaries during each phase of the menstrual cycle. Be sure that you do not overlook the event that occurs around day 14 of a 28-day cycle.

As a general rule, during the 30 or 40 years that a woman has periods, only one ovum matures each month. However, there are exceptions to this rule. Some months, more than one matures, and some months no ovum matures. Ovulation occurs 14 days before the next menstrual phase begins.

In a 28-day cycle, this means that ovulation occurs around day 14 of the cycle, as shown in Figure 21-12. (Note that the first day of the menstrual phase is considered the first day of the cycle.) In a 30-day cycle, however, ovulation would not occur on the fourteenth cycle day, but the sixteenth. And in a 25-day cycle, ovulation would occur the eleventh cycle day.

This matter of the time of ovulation has great practical importance. An ovum lives only a short time after it is ejected from its follicle, and sperm live only a short time after they enter the female body. Fertilization of an ovum by a sperm therefore can occur only around the time of ovulation. In other words, a woman's fertile period lasts only a few days each month.

Control of Menstrual Cycle Changes

The anterior pituitary gland plays a critical role in regulating the cyclic changes that characterize the functions of the female reproductive system (see Chapter 10). From day 1 to about day 7 of the menstrual cycle, the anterior pituitary gland secretes increasing amounts of FSH. A high blood concentration of FSH stimulates several immature ovarian follicles to start growing and secreting estrogens (Figure 21-13). As the estrogen content of blood increases, it stimulates the anterior pituitary gland to secrete another hormone, luteinizing hormone (LH). LH causes maturing of a follicle and its ovum, ovulation (rupturing of mature follicle with ejection of ovum), and luteinization (formation of a yellow body, the corpus luteum, from the ruptured follicle).

Which hormone—FSH or LH—would you call the "ovulating hormone"? Do you think ovulation could occur if the blood concentration of FSH remained low throughout the menstrual cycle? If you answered LH to the first question and no to the second, you answered both questions correctly. Ovulation cannot occur if the blood level of FSH stays low because a high concentration of this hormone is essential to stimulation of ovarian follicle growth and maturation. With a low level of FSH, no follicles start to grow, and therefore none become ripe enough to ovulate. Ovulation is caused by the combined actions of FSH and LH. Birth control pills that contain estrogen substances suppress FSH secretion. This indirectly prevents ovulation.

FIGURE **21-12** **The 28-day Menstrual Cycle.**

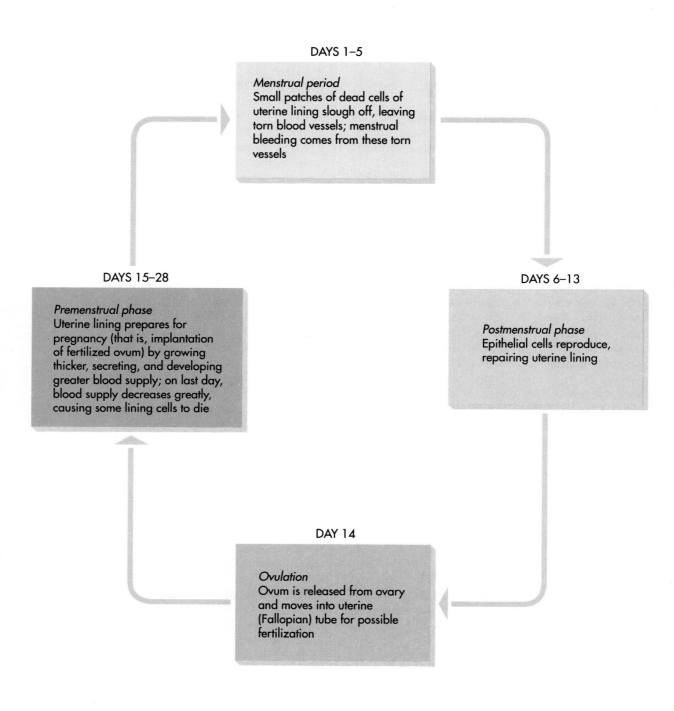

DAYS 1–5

Menstrual period
Small patches of dead cells of uterine lining slough off, leaving torn blood vessels; menstrual bleeding comes from these torn vessels

DAYS 6–13

Postmenstrual phase
Epithelial cells reproduce, repairing uterine lining

DAY 14

Ovulation
Ovum is released from ovary and moves into uterine (Fallopian) tube for possible fertilization

DAYS 15–28

Premenstrual phase
Uterine lining prepares for pregnancy (that is, implantation of fertilized ovum) by growing thicker, secreting, and developing greater blood supply; on last day, blood supply decreases greatly, causing some lining cells to die

FIGURE 21-13 The Human Menstrual Cycle. Diagram illustrates the interrelationship of pituitary, ovarian, and uterine functions throughout a usual 28-day cycle. A sharp increase in LH levels causes ovulation, whereas menstruation (sloughing off of the endometrial lining) is initiated by lower levels of progesterone.

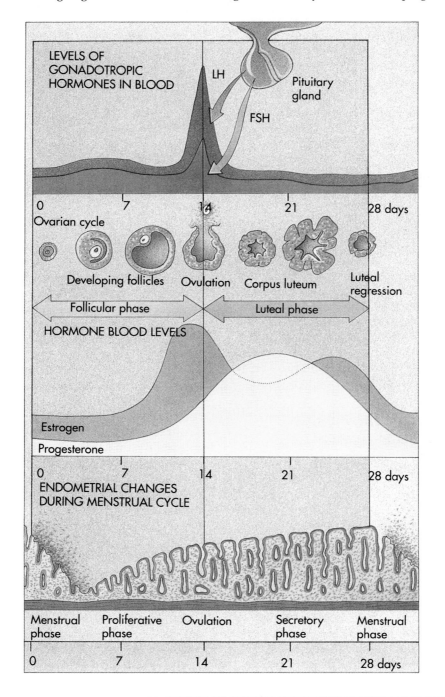

HYSTERECTOMY

The word **hysterectomy** (his-te-REK-toe-me) comes from the combination of two Greek words: *hystera*, meaning "uterus," and *ektome*, meaning "to cut out." By definition it is the surgical removal of the uterus. *Hysterectomy* is a term that is often misused, however, by incorrectly expanding its definition to include the removal of the ovaries or other reproductive structures. Only the uterus is removed in a hysterectomy. If the total uterus, including the cervix, is removed, the terms *total hysterectomy* or *panhysterectomy* may be used. If the cervical portion of the uterus is left in place and only the body of the organ is removed, the term *subtotal hysterectomy* is appropriate. The actual removal of the uterus may be performed through an incision made in the abdominal wall—*abdominal hysterectomy*—or through the vagina—*vaginal hysterectomy*. The term **oophorectomy** (o-off-o-REK-toe-me) is used to describe removal of the ovaries. Although the two surgical procedures may take place during the same operation—for a woman with uterine or ovarian cancer, for example—the terms used to describe them should not be used interchangeably.

Ovulation occurs, as we have said, because of the combined actions of the two anterior pituitary hormones, FSH and LH. The next question is: what causes menstruation? A brief answer is this: a sudden, sharp decrease in estrogen and progesterone secretion toward the end of the premenstrual period causes the uterine lining to break down and another menstrual period to begin.

Disorders of the Female Reproductive System

HORMONAL AND MENSTRUAL DISORDERS

Dysmenorrhea (dis-men-o-REE-ah) or painful menstruation can have several causes. *Primary dysmenorrhea* occurs in the absence of an associated pelvic disease, such as an infection or tumor. Symptoms include painful abdominal cramps as menstruation begins, sometimes accompanied by headache, backache, anorexia, and vomiting. These symptoms can last from several hours to several days. Primary dysmenorrhea is thought to be caused by an abnormally increased concentration of certain prostaglandins (tissue hormones) produced by the uterine lining. High concentrations of prostaglandin E_2 (PGE_2) and prostaglandin $F_{2\alpha}$ ($PGF_{2\alpha}$) cause painful spasms of uterine muscle. For this reason, prostaglandin-inhibiting drugs such as *ibuprofen* are sometimes used to relieve the symptoms of primary dysmenorrhea. Oral contraceptives, which are hormone preparations that inhibit development of the uterine lining, are sometimes prescribed for this condition. Application of heat, which often relieves the pain of muscle spasms, offers temporary relief in some cases. *Secondary dysmenorrhea* can result from a variety of conditions, including inflammatory conditions and cervical stenosis (narrow cervix). Treatment of secondary dysmenorrhea involves treating the underlying disorder.

Amenorrhea (a-men-o-REE-ah) is the absence of normal menstruation. *Primary amenorrhea* is the failure of menstrual cycles to begin and may be caused by a variety of factors, such as hormone imbalances, genetic disorders, brain lesions, or structural deformities of the reproductive organs. *Secondary amenorrhea* occurs when a woman who has previously menstruated slows to three or fewer cycles per year. Secondary amenorrhea may be a symptom of weight loss, pregnancy, lactation, menopause, or disease of the reproductive organs. Treatment of amenorrhea involves treating the underlying disorder or condition.

Dysfunctional uterine bleeding (DUB) is irregular or excessive uterine bleeding that results from a hormonal imbalance rather than from an infection or other disease condition. In DUB, hormonal imbalances may cause excessive growth (*hyperplasia*) of the endometrium—or abnormally frequent menstrual flows. Excessive uterine bleeding from any cause can result in life-threatening anemia because of the chronic loss of blood.

Premenstrual syndrome (PMS) is a condition that involves a collection of symptoms that regularly occur in many women during the premenstrual phase of their reproductive cycles. Symptoms include irritability, fatigue, nervousness, depression, and other problems that are often distressing enough to affect personal relationships. Because the cause of PMS is still unclear, current treatments focus on relieving the symptoms.

INFECTION AND INFLAMMATION

Infections of the female reproductive tract are often classified as *exogenous or endogenous*. Exogenous infections result from pathogenic organisms transmitted from another person, such as **sexually transmitted diseases (STDs).** Endogenous infections result from pathogens that normally inhabit the intestines, vulva, or vagina. You may recall from Chapter 4 that many areas of the body are normally inhabited by pathogenic microbes but that they cause infection only when there is a change in conditions or they are moved to a new area.

Pelvic inflammatory disease (PID) is an acute inflammatory condition caused by infection. PID can involve the uterus, uterine (fallopian) tubes, or ovaries. Uterine tube inflammation is termed **salpingitis** (sal-pin-JYE-tis), which simply means "tube inflammation." Inflammation of the ovaries is termed **oophritis** (o-o-FRY-tis). PID can be caused by several different pathogens, which usually spread upward from the vagina. Pelvic inflammatory disease may be accompanied by fever and pain or may have no symptoms at all. PID can lead to serious complications, including infertility resulting from obstruction or other damage to the reproductive tract. The infection may also spread to other tissues, including the blood—where it may cause septic shock and death (see Chapter 13).

You may recall earlier discussions of another infection of the female reproductive tract that may cause septic shock and death—*toxic shock syndrome (TSS)*. Although it can occur in men, TSS most often occurs in women who use super-absorbent tampons to absorb menstrual flow. These tampons provide ideal conditions for the sudden growth of staphylococcus and other types of bacteria. If a severe infection occurs, it may spread to the blood through tiny lesions in the vaginal wall often caused by tampons. After it is in the blood, the infection can cause circulatory (septic) shock and death. To avoid TSS, women are advised to use regular-absorbency tampons or sanitary napkins—and to change them frequently.

Vaginitis (vaj-in-EYE-tis) is inflammation or infection of the vaginal lining. Vaginitis most often results from STDs or from a "yeast infection." So-called yeast infections are usually opportunistic infections of the fungus *Candida albicans*, producing *candidiasis* (kan-did-EYE-as-is) (see Appendix B). Candidiasis infections are characterized by a whitish discharge—a symptom known as **leukorrhea** (loo-ko-REE-ah).

TUMORS AND RELATED CONDITIONS

Myoma (my-O-ma) or **fibromyoma** (fi-BRO-my-o-ma) is a benign tumor of smooth muscle and fibrous connective tissue that accounts for many uterine tumors. Also called *fibroids*, these tumors are usually small and often produce no symptoms. If they interfere with pregnancy or become large enough to cause a health risk, they may be removed surgically.

Ovarian cysts are benign enlargements on one or both ovaries. An ovarian cyst is a fluid-filled sac that develops from a follicle that fails to rupture completely or from a corpus luteum that fails to degenerate. Ovarian cysts rarely become dangerous and often disappear within a few months of their appearance.

Another benign condition that commonly affects the female reproductive tract is **endometriosis** (en-doe-me-tree-O-sis). Endometriosis is the presence of functioning endometrial tissue outside the uterus. The displaced endometrial tissue can occur in many different places throughout the body but is most often found in or on pelvic and abdominal organs. The tissue reacts to ovarian hormones in the same way as the normal endometrium—exhibiting a cycle of growth and sloughing off. Such tissues may also become the sites of ectopic pregnancies. Symptoms of endometriosis may include unusual bleeding, dysmenorrhea, and pain during intercourse.

FIBROCYSTIC DISEASE

The terms *fibrocystic disease* and *mammary dysplasia* are just two of the many names for a group of conditions characterized by benign lumps in one or both breasts. It is common in adult women before menopause, occurring in half of all women at some time, and is considered the most frequent breast lesion. The lumps that characterize fibrocystic disease are often painful, especially during the premenstrual phase of the reproductive cycle. Treatment is usually aimed at relieving pain or tenderness that may occur. Although it is commonly called a *disease*, most experts agree that fibrocystic disease is simply a collection of normal variations in breast tissue. Even though the lumps associated with fibrocystic disease are benign, any suspicious lump or other change in breast tissue should be regarded as possibly cancerous until determined otherwise by a physician.

Malignancies of reproductive and related organs, especially the breasts, account for the majority of cancer cases among women. Some studies show that 1 in 10 women eventually get breast cancer, often a form of adenocarcinoma. Treatment of breast cancer is often successful if the cancerous tumor is detected early. Because such tumors are often painless, most physicians recommend regular, frequent self-examination of breast tissue, as well as annual mammograms, for women (see Chapter 4). Treatments often involve surgery, chemotherapy, and radiation therapy. Surgeries can be very conservative, as in a simple lump removal or *lumpectomy*. If metastasis to surrounding tissue is suspected, a **radical mastectomy** (mas-TEK-toe-me) may be performed. In this procedure the entire breast, with nearby muscle tissues and lymph nodes, is removed.

Breast cancer often metastasizes to the ovaries, where it produces ovarian cancer. Of course, cancer can begin in the ovaries independent of breast cancer. Usually a type of adenocarcinoma, ovarian cancer is difficult to detect early—often not easily apparent until it has grown into a large mass. Regular pelvic examinations that include palpation of the ovaries may result in earlier detection. Risk factors for ovarian cancer include age (over 40), infertility, childlessness or few children, a history of miscarriages, and endometriosis. Ovarian cancer is often treated by surgical removal of the ovaries combined with radiation therapy and chemotherapy.

Cancer in the uterus can affect the body of the uterus or the cervix. Cancers of the uterine body most often involve the endometrium and mostly affect women beyond childbearing years. Risk factors for this type of cancer include obesity, prolonged estrogen therapy, infertility, and uterine bleeding. Cervical cancer occurs most frequently in women between the ages of 30 to 50. Cervical cancer is often diagnosed early, through screening tests such as the **Papanicolaou** (pap-a-nik-o-LAH-oo) **test** or *Pap smear*. In this test, cells swabbed from the cervix are smeared on a glass slide and examined microscopically to determine whether any abnormalities exist. The American Cancer Society recommends two Pap smears 1 year apart beginning at age 21. If these two Pap smears are negative (that is, revealing no abnormalities), subsequent Pap smears should occur every 1 to 3 years thereafter. Because early or frequent intercourse is a risk factor for cervical cancer, sexually active young women should have their first Pap smear much earlier—and have follow-ups done more often. Because screening tests and other early detection methods have been so successful, the death rates for uterine cancers have dropped dramatically over the last few decades.

INFERTILITY

Like in the male reproductive system, various disorders can disrupt normal function of the female reproductive tract so that successful reproduction is unlikely (infertility) or impossible (sterility). Infections, tumors, hormonal imbalances, and other factors can contribute to infertility or sterility in women. For example, inflammation or infection of the uterine tubes can result in scarring that blocks sperm from reaching the ovum or prevents the ovum from traveling to the uterus. Infections, cancer, or hormonal imbalances may inhibit the female reproductive cycle—preventing the production and release of a healthy ovum each month. Such conditions may also interfere with the development of the uterine lining that is essential for successful pregnancy.

Because sexual reproduction requires normal function of both male and female systems, infertility of a couple may result from the infertility of either partner. A couple is considered infertile if a pregnancy does not occur after a year of normal sexual intercourse (without contraception). When couples seek help for infertility problems, one of the first steps in diagnosis is to determine whether there is a problem in the male partner or the female partner—or both.

Summary of Male and Female Reproductive Systems

The reproductive systems in both sexes are centered around the production of highly specialized reproductive cells or gametes (sperm and ova), as well as mechanisms to ensure union of these two cells; the fusion of these cells enables transfer of parental genetic information to the next generation. Table 21-3 compares several analogous components of the reproductive systems in both

sexes. You can see that men and women have similar structures to accomplish complementary functions. In addition, the female reproductive system permits development and birth of the offspring—the first subject of our next chapter.

Sexually Transmitted Diseases

Sexually transmitted diseases (STDs) or *venereal diseases* are infections caused by communicable pathogens such as viruses, bacteria, fungi, and protozoans (see Appendix B). The factor that links all these diseases and gives this disease category its name is the fact that they can all be transmitted by sexual contact. The term *sexual contact* refers to normal intercourse in addition to any contact between the genitals of one person and the body of the another person. Diseases classified as STDs can be transmitted sexually but do not have to be. For example, *acquired immune deficiency syndrome (AIDS)* is a viral condition that can be spread through sexual contact but is also spread by transfusion of infected blood and use of contaminated medical instruments such as intravenous needles and syringes. Candidiasis, or yeast infection, is a common opportunistic infection, but it can be transmitted through sexual contact. Sexually transmitted diseases are the most common of all communicable diseases. Table 21-4 summarizes a few of the principal STDs.

TABLE 21-3 Analogous Features of the Reproductive Systems

FEATURE	FEMALE	MALE
Essential organs	Ovaries	Testes
Sex cells	Ova (eggs)	Sperm
Hormones	Estrogen and progesterone	Testosterone
Hormone-producing cells	Granulosa cells and corpus luteum	Interstitial cells
Duct systems	Uterine (fallopian) tubes, uterus, and vagina	Epididymis, urethra, and ductus (vas) deferens
External genitals	Clitoris and vulva	Penis and scrotum

TABLE 21-4   Examples of Sexually Transmitted Disease (STD)

DISEASE	PATHOGEN	DESCRIPTION
Acquired immune deficiency syndrome (AIDS)	*Virus:* Human immunodeficiency virus (HIV)	HIV is transmitted by direct contact of body fluids, often during sexual contact. After a latent period that sometimes lasts many years, HIV infection produces the condition known as AIDS. AIDS is characterized by damage to lymphocytes (T cells), resulting in immune system impairment. Death results from secondary infections or tumors. *AIDS-related complex (ARC)* is collection of milder symptoms that may precede full-blown AIDS.
Candidiasis	*Fungus: Candida albicans*	This yeast infection is characterized by a white discharge (leukorrhea), peeling of skin, and bleeding. Although it can occur as an ordinary opportunistic infection, it may be transmitted sexually.
Chancroid	*Bacterium: Haemophilus ducreyi*	A highly contagious STD, it characterized by papules on the skin of the genitals that eventually ulcerate. About 90% of cases are reported by men.
Genital herpes	*Virus:* Herpes simplex virus (HSV)	HSV causes blisters on the skin of the genitals. The blisters may disappear temporarily but may reappear occasionally, especially as a result of stress.
Genital warts	*Virus:* Human papillomavirus (HPV-6, HPV-7)	Genital warts are nipplelike neoplasms of skin covering the genitals.
Giardiasis	*Protozoan: Giardia lamblia*	This intestinal infection may be spread by sexual contact. Symptoms range from mild diarrhea to malabsorption syndrome, with about half of all cases being asymptomatic.
Gonorrhea	*Bacterium: Neisseria gonorrhoeae*	Gonorrhea primarily involves the genital and urinary tracts but can affect throat, conjunctiva, or lower intestines. It may progress to PID.
Hepatitis	*Virus:* Hepatitis B virus (HBV)	This acute-onset liver inflammation may develop into a severe chronic disease, perhaps ending in death.
Lymphogranuloma venereum (LGV)	*Bacterium: Chlamydia trachomatis*	This chronic STD is characterized by genital ulcers, swollen lymph nodes, headache, fever, and muscle pain. *C. trachomatis* infection may cause a variety of other syndromes, including conjunctivitis, urogenital infections, and systemic infections. *C. trachomatis* infections constitute the most common STD in the U.S.
Scabies	*Animal: Sarcoptes scabiei*	Scabies is caused by infestation by the *itch mite*, which burrows into the skin to lay eggs. About 2 to 4 months after initial contact, a hypersensitivity reaction occurs, causing a rash along each burrow that itches intensely. Secondary bacterial infection is possible.
Syphilis	*Bacterium: Treponema pallidum*	Although transmitted sexually, syphilis can affect any system. *Primary syphilis* is characterized by chancre sores on exposed areas of the skin. If untreated, *secondary syphilis* may appear 2 months after chancres disappear. The secondary stage occurs when the spirochete has spread through the body, presenting a variety of symptoms, and is still highly contagious—even through kissing. *Tertiary syphilis* may appear years later, possibly resulting in death.
Trichomoniasis	*Protozoan: Trichomonas vaginalis*	This urogenital infection is asymptomatic in most women and nearly all men. Vaginitis may occur, characterized by itching or burning and a foul-smelling discharge.

OUTLINE SUMMARY

COMMON STRUCTURAL AND FUNCTIONAL CHARACTERISTICS BETWEEN THE SEXES

A Common general structure and function can be identified between the systems in both sexes

B Systems adapted for development of sperm or ova followed by successful fertilization, development, and birth of offspring

C Sex hormones in both sexes important in development of secondary sexual characteristics and normal reproductive system activity

MALE REPRODUCTIVE SYSTEM

A Structural plan—organs classified as essential or accessory (Figure 21-1)
   1 Essential organs of reproduction are the gonads (testes), which produce sex cells (sperm)
   2 Accessory organs of reproduction
      a Ducts—passageways that carry sperm from testes to exterior
      b Sex glands—produce protective and nutrient solution for sperm
      c External genitals

B Testes—the gonads of men
   1 Structure and location (Figure 21-2)
      a Testes in scrotum—lower temperature
      b Covered by tunica albuginea, which divides testis into lobules containing seminiferous tubules
      c Interstitial cells produce testosterone
   2 Functions
      a Spermatogenesis is process of sperm production (Figure 21-4)
         (1) Sperm precursor cells called *spermatogonia*
         (2) Meiosis produces primary spermatocyte, which forms four spermatids with 23 chromosomes
         (3) Spermatozoa—highly specialized cell
            (a) Head contains genetic material
            (b) Acrosome contains enzymes to assist sperm in penetration of ovum
            (c) Mitochondria provide energy for movement
      b Production of testosterone by interstitial cells
         (1) Testosterone "masculinizes" and promotes development of male accessory organs
         (2) Stimulates protein anabolism and development of muscle strength

C Reproductive ducts—ducts through which sperm pass after exiting testes until they exit from the body

   1 Epididymis—single, coiled tube about 6 m in length; lies along the top and behind the testis in the scrotum
      a Sperm mature and develop the capacity for motility as they pass through epididymis
   2 Ductus (vas) deferens—receives sperm from the epididymis and transports them from scrotal sac through the abdominal cavity
      a Passes through inguinal canal
      b Joins duct of seminal vesicle to form the ejaculatory duct

D Accessory or supportive sex glands—semen: mixture of sperm and secretions of accessory sex glands. Averages 3 to 5 ml per ejaculation, with each milliliter containing about 100 million sperm
   1 Seminal vesicles
      a Pouchlike glands that produce about 60% of seminal fluid volume
      b Secretion is yellowish, thick, and rich in fructose to provide energy needed by sperm for motility
   2 Prostate gland
      a Shaped like a doughnut and located below bladder
      b Urethra passes through the gland
      c Secretion represents 30% of seminal fluid volume—is thin and milk-colored
      d Activates sperm and is needed for ongoing sperm motility
   3 Bulbourethral (Cowper's) glands
      a Resemble peas in size and shape
      b Secrete mucuslike fluid constituting less than 5% of seminal fluid volume

E External genitals
   1 Penis and scrotum called *genitalia*
   2 Penis has three columns of erectile tissue—two dorsal columns called *corpora cavernosa* and one ventral column surrounding urethra called *corpus spongiosum*
   3 Glans penis covered by foreskin
   4 Surgical removal of foreskin called *circumcision*

DISORDERS OF THE MALE REPRODUCTIVE SYSTEM

A May cause reduced reproductive ability (infertility) or total inability to reproduce (sterility)

B Disorders of the testes
   1 Oligospermia—low sperm production
   2 Cryptorchidism—undescended testes
   3 Testicular cancer—most common in young adult men of ages 25 to 35

C Disorders of the prostate
   1 Benign prostatic hypertrophy—enlargement of prostate common in older men
   2 Prostate cancer is a leading cause of cancer deaths in men over age 50
D Disorders of the penis and scrotum
   1 Phimosis—tight foreskin
   2 Impotence—failure to achieve erection of the penis
   3 Hydrocele—accumulation of watery fluid in the scrotum
   4 Inguinal hernia—protrusion of abdominopelvic organs, possibly into the scrotum

FEMALE REPRODUCTIVE SYSTEM

A Structural plan—organs classified as essential or accessory (Figure 21-7)
   1 Essential organs are gonads (ovaries), which produce sex cells (ova)
   2 Accessory organs of reproduction
     a Ducts or modified ducts—including oviducts, uterus, and vagina
     b Sex glands—including those in the breasts
     c External genitals
B Ovaries
   1 Structure and location
     a Paired glands weighing about 3 g each
     b Resemble large almonds
     c Attached to ligaments in pelvic cavity on each side of uterus
     d Microscopic structure (Figure 21-8)
       (1) Ovarian follicles—contain oocyte, which is immature sex cell (about 1 million at birth)
       (2) Primary follicles—about 400,000 at puberty are covered with granulosa cells
       (3) About 350 to 500 mature follicles ovulate during the reproductive lifetime of most women—sometimes called *Graafian follicles*
       (4) Secondary follicles have hollow chamber called *antrum*
       (5) Corpus luteum forms after ovulation
   2 Functions
     a Oogenesis—this meiotic cell division produces daughter cells with equal chromosome numbers (23) but unequal cytoplasm. Ovum is large; polar bodies are small and degenerate
     b Production of estrogen and progesterone
       (1) Granulosa cells surrounding the oocyte in the mature and growing follicles produce estrogen
       (2) Corpus luteum produces progesterone
       (3) Estrogen causes development and maintenance of secondary sex characteristics

       (4) Progesterone—stimulates secretory activity of uterine epithelium and assists estrogen in initiating menses
C Reproductive ducts
   1 Uterine (fallopian) tubes
     a Extend about 10 cm from uterus into abdominal cavity
     b Expanded distal end surrounded by fimbriae
     c Mucosal lining of tube is directly continuous with lining of abdominal cavity
   2 Uterus—composed of body, fundus, and cervix (Figure 21-9)
     a Lies in pelvic cavity just behind urinary bladder
     b Myometrium is muscle layer
     c Endometrium lost in menstruation
     d Menopause—end of repetitive menstrual cycles (about 45 years of age)
   3 Vagina
     a Distensible tube about 10 cm long
     b Located between urinary bladder and rectum in the pelvis
     c Receives penis during sexual intercourse and is birth canal for normal delivery of baby at termination of pregnancy
D Accessory or supportive sex glands
   1 Bartholin's (greater vestibular) glands
     a Secrete mucuslike lubricating fluid
     b Ducts open between labia minora
   2 Breasts (Figure 21-10)
     a Located over pectoral muscles of thorax
     b Size determined by fat quantity more than amount of glandular (milk-secreting) tissue
     c Lactiferous ducts drain at nipple, which is surrounded by pigmented areola
     d Lymphatic drainage important in spread of cancer cells to other body areas
E External genitals (Figure 21-11)
   1 Include mons pubis, clitoris, orifice of urethra, Bartholin's gland, and vagina, labia minora and majora, and hymen
   2 Perineum—area between vaginal opening and anus
     a Surgical cut during birth called *episiotomy*
F Menstrual cycle—involves many changes in the uterus, ovaries, vagina, and breasts (Figures 21-12 and 21-13)
   1 Length—about 28 days, varies from month to month in individuals and in the same individual
   2 Phases
     a Menstrual phase or menses—about the first 4 or 5 days of the cycle, varies somewhat; characterized by sloughing of bits of endometrium (uterine lining) with bleeding
     b Postmenstrual phase—days between the end of menses and ovulation; varies in length; the

shorter the cycle, the shorter the postmenstrual phase; the longer the cycle, the longer the postmenstrual phase; examples: in 28-day cycle, postmenstrual phase ends on day 13, but in 26-day cycle, it ends on the 11th day and in 32-day cycle, it ends on day 17; characterized by repair of endometrium

   c Premenstrual phase—days between ovulation and beginning of next menses; ovulation about 14 days before next menses; characterized by further thickening of endometrium and secretion by its glands in preparation for implantation of fertilized ovum; combined actions of the anterior pituitary hormones FSH and LH cause ovulation; sudden sharp decrease in estrogens and progesterone bring on menstruation if pregnancy does not occur.

DISORDERS OF THE FEMALE REPRODUCTIVE SYSTEM

  1 Hormonal and menstrual disorders
    a Dysmenorrhea—painful menstruation
    b Amenorrhea—absence of normal menstruation
    c Dysfunctional uterine bleeding (DUB)—irregular or excessive bleeding resulting from a hormonal imbalance
    d Premenstrual syndrome (PMS)—collection of symptoms that occur in many women before menstruation
  2 Infection and inflammation
    a Exogenous infections are often sexually transmitted; endogenous infections are caused by organisms already in or on the body
    b Pelvic inflammatory disease (PID)—acute inflammatory condition of the uterus, uterine tubes, or ovaries caused by infection
    c Toxic shock syndrome (TSS)—infection of the blood causing septic shock, it results from stapholococcal infections in the vagina
    d Vaginitis—infection of vaginal lining, it most often results from STDs or yeast infections

  3 Tumors and related conditions
    a Myoma or fibromyoma tumors (fibroids)—benign tumors of the uterus
    b Ovarian cysts—fluid-filled enlargements; usually benign
    c Endometriosis—presence of functioning endometrial tissue outside the uterus
    d Breast cancer is the most common type of cancer in women
    e Ovarian cancer can result from metastasis of breast cancer or can arise independently
    f Cervical cancer is often detected by a Papanicolaou test (Pap smear)
  4 Infertility can result from factors such as infection and inflammation, tumors, and hormonal imbalances

SUMMARY OF MALE AND FEMALE REPRODUCTIVE SYSTEMS

A In men and women the organs of the reproductive system are adapted for the specific sequence of functions that permit development of sperm or ova followed by the successful fertilization and then the normal development and birth of offspring (Table 21-3)

B The male organs produce, store, and ultimately introduce mature sperm into the female reproductive tract

C The female system produces ova, receives the sperm, and permits fertilization followed by fetal development and birth, with lactation afterward

D Production of sex hormones is required for development of secondary sex characteristics and for normal reproductive functions in both sexes

SEXUALLY TRANSMITTED DISEASES (STDs)

A STDs are transmitted sexually but can also be transmitted in other ways

B STDs are the most common of all communicable diseases

C STDs are caused by a variety of organisms (Table 21-4)

NEW WORDS

areola	gametes	ovulation	seminiferous tubule
clitoris	genitals	perineum	spermatogenesis
corpus luteum	gonads	polar body	spermatozoa
ejaculation	Graafian follicle	prepuce	testes
endometrium	meiosis	progesterone	testosterone
epididymis	menopause	scrotum	vulva
estrogen	menses	semen	zygote
fimbriae	oocyte		

Diseases and Other Clinical Terms

amenorrhea	endometriosis	myoma	premenstrual
benign prostatic	episiotomy	oligospermia	syndrome (PMS)
hypertrophy	fibromyoma	Papanicolaou test	prostatectomy
circumcision	hydrocele	pelvic inflammatory	radical mastectomy
cryptorchidism	inguinal hernia	disease (PID)	salpingitis
dysmenorrhea	leukorrhea	phimosis	vaginitis

CHAPTER TEST

1. The gonads of the male are the _____.
2. In men the sex cells are called _____.
3. The essential organs of reproduction in both sexes are called the _____.
4. Each lobule of the testis consists of a narrow but long and coiled tube called the _____ tubule.
5. Testosterone formation is the function of the _____ cells of the testis.
6. In men the urethra passes through the center of the doughnut-shaped _____ gland.
7. The narrow section of the uterus that opens into the vagina is called the _____.
8. The corpus luteum secretes _____.
9. The mucous membrane lining the uterus is called the _____.
10. The colored area around the nipple is called the _____.
11. The scientific name for the beginning of the menses is called _____.
12. The average length of a typical menstrual cycle is _____ days.

Select the most correct answer from Column B for each statement in Column A. (Only one answer is correct.)

COLUMN A

13. _____ Female gonads
14. _____ External genitals
15. _____ Secrete estrogen
16. _____ Corpus luteum
17. _____ Sperm formation
18. _____ "Ovulating hormone"
19. _____ Surgical removal of foreskin
20. _____ Feminizing hormone
21. _____ Female sex cell
22. _____ Male sex hormone
23. _____ Absence of menstruation
24. _____ Painful menstruation
25. _____ Undescended testes
26. _____ Presence of endometrium outside the uterus
27. _____ Low sperm count
28. _____ White discharge
29. _____ Inflammation of uterine (fallopian) tubes
30. _____ Removal of prostate tissue

COLUMN B

a. Ovarian follicles
b. Testosterone
c. LH
d. Secrete progesterone
e. Circumcision
f. Ovaries
g. Estrogen
h. Spermatogenesis
i. Vulva
j. Ovum
k. Amenorrhea
l. Cryptorchidism
m. Dysmenorrhea
n. Endometriosis
o. Leukorrhea
p. Oligospermia
q. Prostatectomy
r. Salpingitis

REVIEW QUESTIONS

1 Identify the essential and accessory organs of reproduction in men.
2 What organs are included in the *external genitals* of the male?
3 Discuss the structure of the testes. Explain the function of the seminiferous tubules and the interstitial cells of the testes.
4 What is the name of the masculinizing hormone? What secretes it? What are its general functions?
5 Discuss the anatomy of a sperm cell. Why is it motile? What parts of the sperm cell are designed to provide motility?
6 Trace a sperm cell from its point of formation in the testes through the male reproductive ducts to ejaculation.
7 How many sperm are normally present in one ejaculation of semen? What is the usual volume of semen ejaculated at one time?
8 Castration is an operation that removes the testes. Would this sterilize the male? Why? What other effects would you expect to see as a result of castration? Why?
9 Discuss the functions of the accessory glands in men.

10 Identify and describe the major disorders of the male reproductive tract and how they might lead to infertility or sterility.
11 Name the reproductive ducts in men and women.
12 Identify the feminizing hormone by its scientific name. What gland secretes it?
13 Identify the ovulating hormone by its scientific name. What glands secrete it?
14 Identify and locate the alveoli of breast, areola, cervix, clitoris, fundus of uterus, Graafian follicle, labia majora, mons pubis, ovum, and uterine tube.
15 What causes ovulation?
16 What is menstruation? What causes it?
17 How many female sex cells are usually formed each month? How does this compare with the number of male sex cells formed each month?
18 What is amenorrhea? dysmenorrhea? What factors may cause each of these conditions?
19 Name and describe the principal infections that involve the reproductive tract. What distinguishes an infection as an STD?
20 How can endometriosis lead to an ectopic pregnancy?

CLINICAL APPLICATIONS

1 As stated in a boxed essay, one procedure that has been commonly used to screen for prostate cancer is palpation of the prostate through the wall of the rectum. Explain why digital (finger) palpation of the prostate is the only way to examine this gland from the outside without special equipment. What other prostate disorders might be detected this way?

2 Liz and Zeke have come to their physician with a problem: after 2 years of trying, they cannot seem to have a child. According to the technical definition, is this couple infertile? On examination, Liz has been found to have pelvic inflammatory disease (PID). Could this condition have caused infertility?

3 Heather, age 22, briefly checks her breasts every few months and never detected anything abnormal. However, a routine examination by her physician has revealed some small, tender lumps in both breasts. Does she have breast cancer? Can you think of any reason that Heather did not detect this condition herself?

Growth and Development

Outline

Objectives

*After you have completed this chapter,
you should be able to:*

1 Discuss the concept of development as a biological process characterized by continuous modification and change.

2 Discuss the major developmental changes characteristic of the prenatal stage of life from fertilization to birth.

3 Discuss the three stages of labor that characterize a normal, vaginal birth.

4 Identify the three primary germ layers and several derivatives in the adult body that develop from each layer.

5 Identify and describe the major disorders associated with pregnancy.

6 List and discuss the major developmental changes characteristic of the four postnatal periods of life.

7 Discuss the effects of aging on the major body organ systems.

Boxed Essays

Fetal Alcohol Syndrome
In Vitro Fertilization
Antenatal Diagnosis and Treatment
Exercise and Aging

_M_any of your fondest and most vivid memories are probably associated with your birthdays. The day of birth is an important milestone of life. Most people continue to remember their birthday in some special way each year; birthdays serve as pleasant and convenient reference points to mark periods of transition or change in our lives. The actual day of birth marks the end of one phase of life called the **prenatal period** and the beginning of a second called the **postnatal period.** The prenatal period begins at conception and ends at birth; the postnatal period begins at birth and continues until death. Although important periods in our lives such as childhood and adolescence are often remembered as a series of individual and isolated events, they are in reality part of an ongoing and continuous process. In reviewing the many changes that occur during the cycle of life from conception to death, it is often convenient to isolate certain periods such as infancy or adulthood for study. It is important to remember, however, that life is not a series of stop-and-start events or individual and isolated periods of time. Instead, it is a biological process that is characterized by continuous modification and change.

This chapter discusses events and changes that occur in the development of the individual from conception to death. Study of development during the prenatal period is followed by a description of the birth process, disorders of pregnancy, and a review of development during infancy and adulthood; finally some important changes that occur in the individual organ systems of the body as a result of aging are discussed.

Prenatal Period

The **prenatal stage of development** begins at the time of conception or fertilization (that is, at the moment the female ovum and the male sperm cells unite) (Figure 22-1). The period of prenatal development continues until the birth of the child about 39 weeks later. The science of the development of the individual before birth is called **embryology** (em-bree-OL-o-jee). It is a story of miracles, describing the means by which a new human life is started and the steps by which a sin-

gle microscopic cell is transformed into a complex human being.

FERTILIZATION TO IMPLANTATION

After ovulation the discharged ovum first enters the abdominal cavity and then finds its way into the uterine (fallopian) tubes. Sperm cells "swim" up the uterine tubes toward the ovum. Look at the relationship of the ovary, the two uterine tubes, and the uterus in Figure 22-2. Recall from Chapter 21 that each uterine tube extends outward from the uterus for about 10 cm. It then ends in the abdominal cavity near the ovary, as you can see in Figure 22-2, in an opening surrounded by fringelike processes, the fimbriae. Using the uterus as a point of reference, anatomists divide each uterine tube into three parts. The innermost part of the tube actually extends through the uterine wall, the middle third extends out into the abdominal cavity, and the outer most third of the tube ends near the ovary in the dilated, funnel-shaped opening described above. Sperm cells that are deposited in the vagina must enter and "swim" through the uterus and then move out of the uterine cavity and through the uterine tube to meet the ovum. Fertilization most often occurs in the outer one third of the oviduct as shown in Figure 22-2. The fertilized ovum or **zygote** (ZYE-gote) is genetically complete; it represents a new single-celled individual. Time and nourishment are all that is needed for expression of characteristics such as sex, body build, and skin color that were determined at the time of fertilization. As you can see in the figure, the zygote immediately begins to divide, and in about 3 days a solid mass of cells called a **morula** (MOR-yoo-lah) is formed (Figure 22-2). The cells of the morula continue to divide, and by the time the developing embryo reaches the uterus, it is a hollow ball of cells called a **blastocyst** (BLAS-toe-sist).

During the 10 days from the time of fertilization to the time when the blastocyst is completely implanted in the uterine lining, no nutrients from the mother are available. The rapid cell division taking place up to the blastocyst stage occurs with no significant increase in total mass compared to the zygote (Figure 22-3). One of the specializations of the ovum is its incredible store of nutri-

FIGURE 22-1 Fertilization. Fertilization is a specific biological event. It occurs when the male and female sex cells fuse. After union between a sperm cell and the ovum has occurred, the cycle of life begins. The scanning electron micrograph shows spermatazoa attaching themselves to the surface of an ovum. Only one will penetrate and fertilize the ovum.

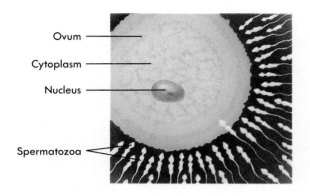
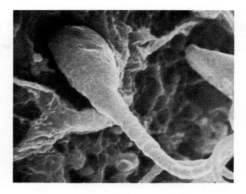

FIGURE 22-2 Fertilization and Implantation. At ovulation, an ovum is released from the ovary and begins its journey through the uterine tube. While in the tube, the ovum is fertilized by a sperm to form the single-celled zygote. After a few days of rapid mitotic division, a ball of cells called a *morula* is formed. After the morula develops into a hollow ball called a *blastocyst*, implantation occurs.

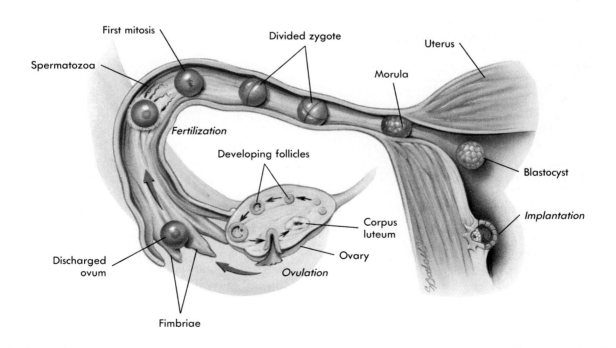

FIGURE 22-3 Early Stages of Human Development. A, Fertilized ovum or zygote. **B** to **D,** Early cell divisions produce more and more cells. The solid mass of cells shown in **D** forms the morula—an early stage in embryonic development.

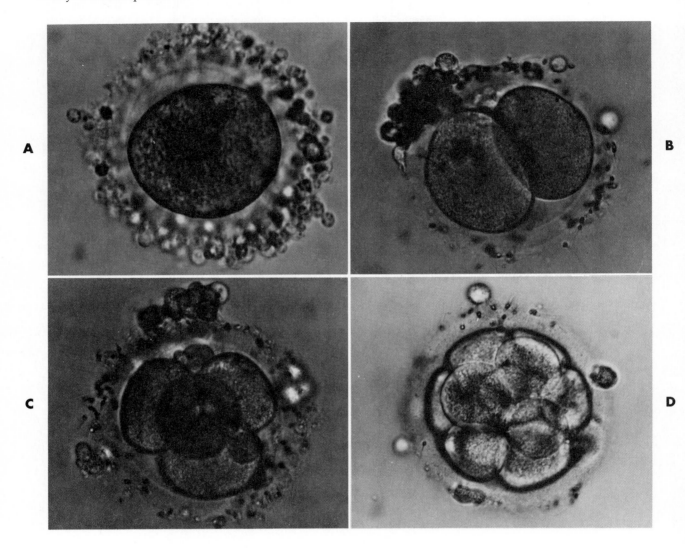

ents that support this embryonic development until implantation has occurred.

Note in Figure 22-4 that the blastocyst consists of an outer layer of cells and an inner cell mass. As the blastocyst continues to develop, it forms a structure with two cavities called the **yolk sac** and **amniotic** (am-nee-OT-ik) **cavity.** The yolk sac is most important in animals, such as birds, that depend heavily on yolk as a nutrient for the developing embryo. In these animals the yolk sac digests the yolk and provides nutrients to the embryo. Because uterine fluids provide nutrients to the developing embryo in humans until the placenta develops, the function of the yolk sac is not a nutritive one. Instead, it has other functions, including production of blood cells.

FIGURE 22-4 Implantation and Early Development. The hollow blastocyst implants itself in the uterine lining about 10 days after ovulation. Until the placenta is functional, nutrients are obtained by diffusion from uterine fluids. Notice the developing chorion and how the blastocyst eventually forms a yolk sac and amniotic cavity.

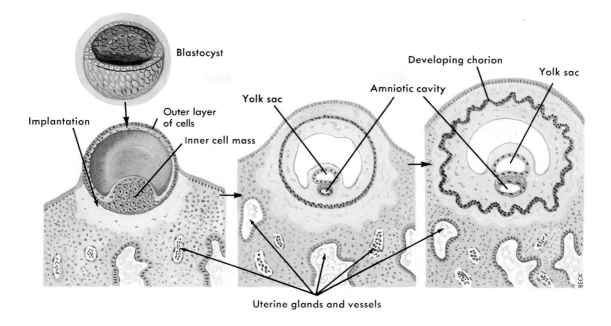

FETAL ALCOHOL SYNDROME

Consumption of alcohol during pregnancy can have tragic effects on a developing fetus. Educational efforts to inform pregnant women about the dangers of alcohol are now receiving national attention. Even very limited consumption of alcohol during pregnancy poses significant hazards to the developing baby because alcohol can easily cross the placental barrier and enter the fetal bloodstream.

When alcohol enters the fetal blood, the potential result, called **fetal alcohol syndrome (FAS)**, can cause tragic congenital abnormalities such as "small head" or microcephaly (my-kro-SEF-ah-lee), low birth weight, developmental disabilities such as mental retardation, and even fetal death.

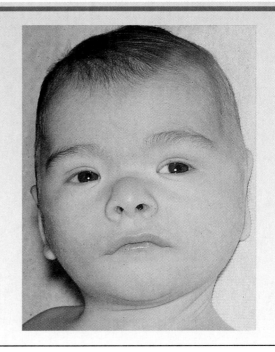

The amniotic cavity becomes a fluid-filled, shock-absorbing sac, sometimes called the *bag of waters*, in which the embryo floats during development. The **chorion** (KO-ree-on), shown in Figures 22-4 and 22-5, develops into an important fetal membrane in the **placenta** (plah-SEN-tah).

The *chorionic villi* shown in Figure 22-5 connect the blood vessels of the chorion to the placenta. The placenta (Figure 22-5) anchors the developing fetus to the uterus and provides a "bridge" for the exchange of nutrients and waste products between mother and baby.

FIGURE 22-5 Structural Features of the Placenta. The close placement of the fetal blood supply and the maternal blood supply permits diffusion of nutrients and other substances. It also forms a thin barrier to prevent diffusion of most harmful substances. No mixing of fetal and maternal blood occurs.

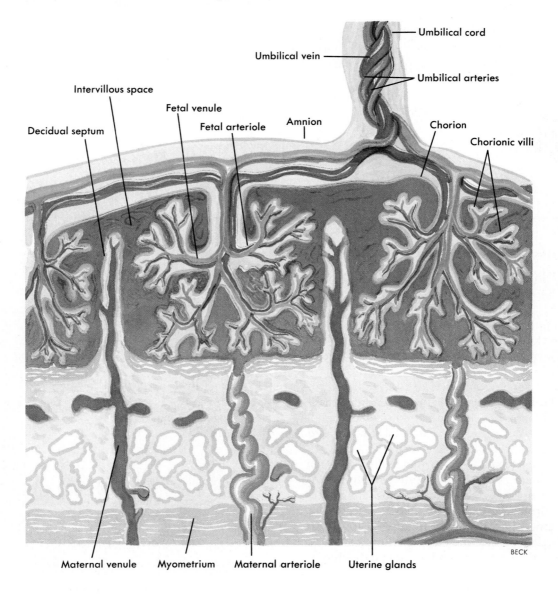

The placenta is a unique and highly specialized structure that has a temporary but very important series of functions during pregnancy. It is composed of tissues from mother and child and functions not only as structural "anchor" and nutritive bridge, but also as an excretory, respiratory, and endocrine organ (Figure 22-5).

Placental tissue normally separates the maternal and fetal blood supplies so that no intermixing occurs. The very thin layer of placental tissue

---

IN VITRO FERTILIZATION

The world's first "test tube" baby was born in Oldham, England, on July 25, 1978. This historic and highly publicized event represented a major medical breakthrough and set the stage for ongoing and sophisticated research in the area of reproductive physiology that continues in many fertility clinics and laboratories around the world.

Nine months before the historic birth, baby Louise Brown was conceived in vitro (in VEE-tro). The Latin term *in vitro* means, literally, "within a glass" and refers to the glass laboratory container where a mature ovum from her mother was fertilized by her father's sperm. The ovum was obtained from the mother using a specialized optical viewing tube called a **laparoscope** (LAP-ah-ro-skope) in a procedure sometimes called "belly button surgery." After 2½ days' growth in a temperature-controlled environment, the fertilized ovum (at the 8-cell stage) was returned by the physicians to the mother's uterus. Implantation was successful, and the subsequent pregnancy and birth were normal in every respect. (The figure shows Louise Brown at age 10.)

Current in vitro fertilization procedures also use a laparoscope. The newer and smaller fiberoptic instruments can be inserted through a very small incision in the woman's abdomen or through the bladder wall after being introduced through the urethra in a specially designed catheter. Once in the abdominal cavity the device allows the physician to not only see the ovary but also to puncture a mature follicle. The mature ovum that is released is sucked up and then transferred to and maintained in a specialized medium until it is fully mature and capable of being fertilized.

Most women who seek in vitro fertilization as a treatment for infertility have uterine tubes that either do not permit movement of sperm to reach and fertilize the ovum or that do not allow a developing zygote to reach the uterus after fertilization has occurred. Treatment begins with hormone treatment to induce simultaneous development and maturation of several ova, which are then collected and fertilized in vitro with spermatozoa that have been treated to bring about their full maturation.

After the developing zygotes have reached the 8- or 16-cell stage, they are then transferred to the mother's uterus. Under normal conditions, only an estimated 50% of fertilized ova successfully implant. In the most successful fertility clinics in the United States, 20% to 30% of in vitro fertilization cases result in implantation and a pregnancy that progresses to term.

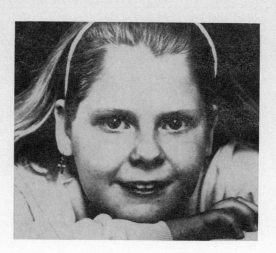

that separates maternal and fetal blood also serves as an effective "barrier" that can protect the developing baby from many harmful substances that may enter the mother's bloodstream. Unfortunately, toxic substances, such as alcohol and some infectious organisms, may penetrate this protective placental barrier and injure the developing baby. The virus responsible for German measles (rubella), for example, can easily pass through the placenta and cause tragic developmental defects in the fetus.

PERIODS OF DEVELOPMENT

The length of pregnancy (about 39 weeks)—called the *gestation period*—is divided into three 3-month segments called *trimesters*. A number of terms are used to describe development during these periods known as the first, second, and third trimesters of pregnancy.

During the first trimester or 3 months of pregnancy, a number of terms are used. *Zygote* is used to describe the ovum just after fertilization by a sperm cell. After about 3 days of constant cell division, the solid mass of cells, identified earlier as the *morula*, enters the uterus. Continued development transforms the morula into the hollow blastocyst, which then implants into the uterine wall.

The embryonic phase of development extends from fertilization until the end of the week 8 of gestation. During this period in the first trimester, the term *embryo* is used to describe the developing individual. The fetal phase is used to indicate development from week 18 to 39. During this period, the term *embryo* is replaced by *fetus*.

By day 35 of gestation (Figure 22-6, *A*), the heart is beating and, although the embryo is only 8 mm (about ³⁄₈ inch) long, the eyes and so-called limb buds, which ultimately form the arms and legs, are clearly visible. Figure 22-6, *C* shows the stage of development, now called a *fetus*, at the end of the first trimester of gestation. Body size is about 7 to 8 cm (3.2 inches) long. The facial features of the fetus are apparent, the limbs are complete, and gender can be identified. By month 4 (Figure 22-6, *D*), all organ systems are formed and functioning.

FORMATION OF THE PRIMARY GERM LAYERS

Early in the first trimester of pregnancy, three layers of specialized cells develop that embryologists call the **primary germ layers.** Each layer gives rise to definite structures such as the skin, nervous tissue, muscles, or digestive organs. Table 22-1 lists a number of structures derived from each primary germ layer called, respectively, **endoderm** (EN-doe-derm) or inside layer, **ectoderm** (EK-toe-derm) or outside layer, and **mesoderm** (MEZ-o-derm) or middle layer.

TABLE 22-1   Primary Germ Layer Derivatives

ENDODERM	ECTODERM	MESODERM
Lining of gastrointestinal tract	Epidermis of skin	Dermis of skin
Lining of lungs	Tooth enamel	Circulatory system
Lining of hepatic and pancreatic ducts	Lens and cornea of eye	Many glands
Kidney ducts and bladder	Outer ear	Kidneys
Anterior pituitary gland (adenohypophysis)	Nasal cavity	Gonads
Thymus gland	Facial bones	Muscle
Thyroid gland	Skeletal muscles in head	Bones (except facial)
Parathyroid gland	Brain and spinal cord	
Tonsils	Sensory neurons	
	Adrenal medulla	

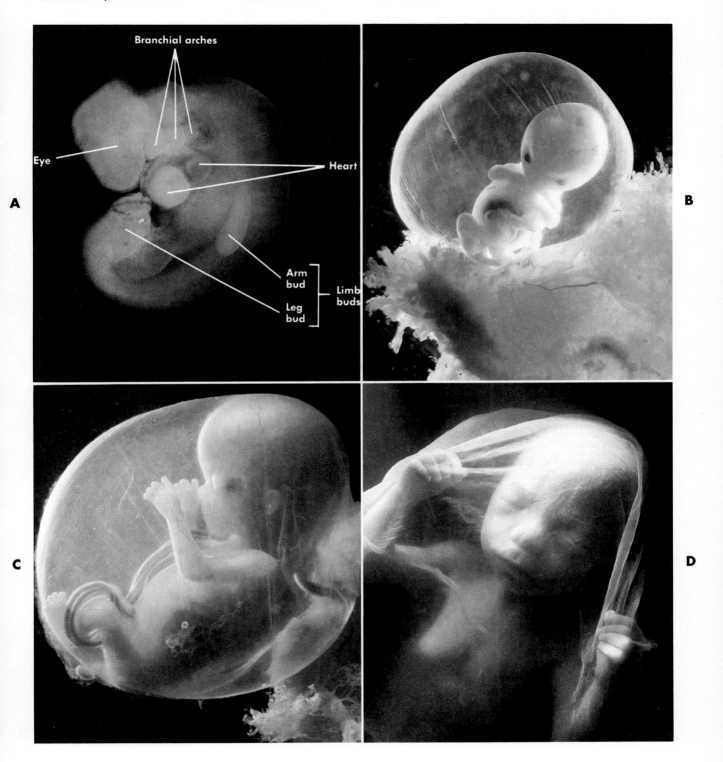

FIGURE 22-7 **Parturition. A,** The relation of the fetus to the mother. **B,** The fetus moves into the opening of the birth canal, and the cervix begins to dilate. **C,** Dilation of the cervix is complete. **D,** The fetus is expelled from the uterus. **E,** The placenta is expelled.

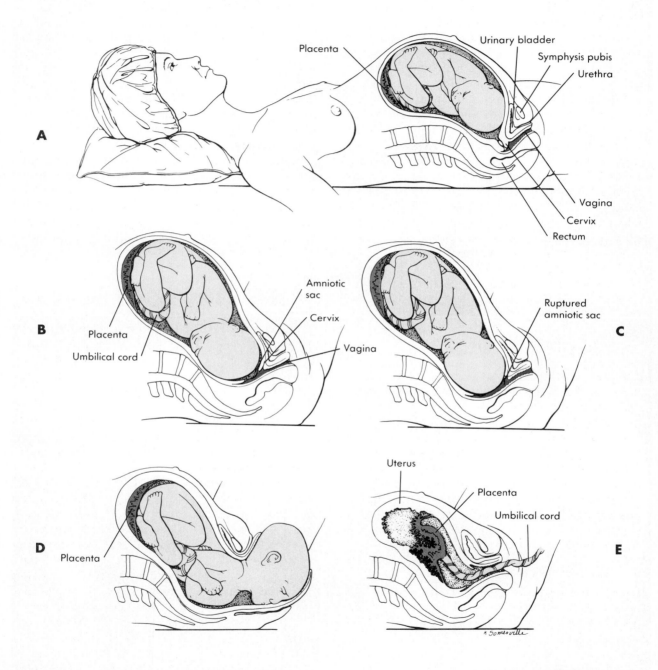

HISTOGENESIS AND ORGANOGENESIS

The study of how the primary germ layers develop into many different kinds of tissues is called **histogenesis** (his-toe-JEN-e-sis). The way that those tissues arrange themselves into organs is called **organogenesis** (or-ga-no-JEN-e-sis). The fascinating story of histogenesis and organogenesis in human development is long and complicated; its telling belongs to the science of embryology. But for the beginning student of anatomy, it seems sufficient to appreciate that life begins when two sex cells unite to form a single-celled zygote and that the new human body evolves by a series of processes consisting of cell differentiation, multiplication, growth, and rearrangement, all of which take place in definite, orderly sequence. Development of structure and function go hand in hand, and from 4 months of gestation, when every organ system is in place and functioning, until term (about 280 days), development of the fetus is mainly a matter of growth. Figure 22-7, *A*, shows the normal intrauterine placement of a fetus just before birth in a full-term pregnancy.

Birth or Parturition

The process of birth or **parturition** (par-too-RISH-un) is the point of transition between the prenatal and postnatal periods of life. As pregnancy draws to a close, the uterus becomes "irritable" and, ultimately, muscular contractions begin and cause the cervix to dilate or open, thus permitting the fetus to move from the uterus through the vagina or "birth canal" to the exterior. The process normally begins with the fetus taking a head-down position against the cervix (Figure 22-7, *A*). When contractions occur, the amniotic sac or "bag of waters" ruptures, and labor begins.

STAGES OF LABOR

Labor is the term used to describe the process that results in the birth of the baby. It is divided into three stages (Figure 22-7, *B* to *E*):

1 Stage one—period from onset of uterine contractions until dilation is complete

2 Stage two—period from the time of maximal cervical dilation until the baby exits through the vagina

3 Stage three—process of expulsion of the placenta through the vagina

The time required for normal vaginal birth varies widely and may be influenced by many variables, including whether the woman has previously had a child. In most cases, stage one of labor lasts from 6 to 24 hours, and stage two lasts from a few minutes to an hour. Delivery of the placenta (stage three) normally occurs within 15 minutes after the birth of the baby.

If abnormal conditions of the mother or fetus (or both) make normal vaginal delivery hazardous or impossible, physicians may suggest a **cesarean** (se-ZAIR-ee-an) **section.** Often called simply a *C-section*, it is a surgical procedure in which the newborn is delivered through an incision in the abdomen and uterine wall.

MULTIPLE BIRTHS

The term *multiple birth* refers to the birth of two or more infants from the same pregnancy. The birth of twins is more common than the birth of triplets, quadruplets, or quintuplets. Multiple-birth babies are often born prematurely, so they are at a greater than normal risk of complications in infancy. However, premature infants that have modern medical care available have a much lower risk of complications than without such care.

Twinning, or double births, can result from two different processes:

1 **Identical twins** result from the splitting of embryonic tissue from the same zygote early in development. As Figure 22-8, *A*, shows, identical twins usually share the same placenta but have separate umbilical cords. Because they develop from the same fertilized egg, identical twins have the same genetic code. Despite this, identical twins are not absolutely identical in terms of structure and function. Different environmental factors and personal experiences lead to individuality even in genetically identical twins.

2 **Fraternal twins** result from the fertilization of two different ova by two different spermatozoa (Figure 22-8, *B*). Fraternal twinning requires the

FIGURE 22-8 Multiple Births. A, Identical twins develop when embryonic tissue from a single zygote splits to form two individuals. Notice that the placenta and the part of the amnion separating the amniotic cavities are shared by the twins. **B,** Fraternal twins develop when two ova are fertilized at about the same time, producing two separate zygotes. Notice that each fraternal twin has its own placenta and amnion.

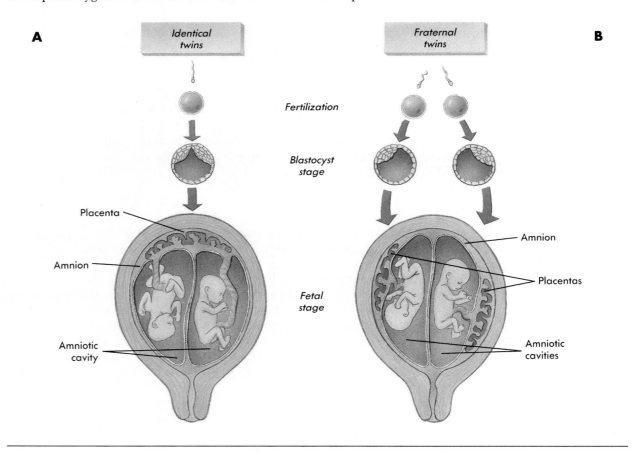

production of more than one mature ovum during a single menstrual cycle, a trait that is often inherited. Multiple ovulation may also occur in response to certain fertility drugs, especially the gonadotropin preparations. Fraternal twins are no more closely related genetically than any other brother-sister relationship. Because two separate fertilizations must occur, it is even possible for fraternal twins to have different biological fathers. Triplets, quadruplets, and other multiple births may be identical, fraternal, or any combination.

Disorders of Pregnancy
IMPLANTATION DISORDERS

A pregnancy has the best chance of a successful outcome, the birth of a healthy baby, if the blastocyst is implanted properly in the uterine wall. However, proper implantation does not always occur. Many offspring are lost before implantation occurs, often for unknown reasons. As mentioned in the previous chapter, implantation outside the uterus results in an ectopic pregnancy. If the blastocyst implants in a region of endometrio-

ANTENATAL DIAGNOSIS AND TREATMENT

Advances in **antenatal** (from the Latin *ante*, "before" and *natus*, "birth") **medicine** now permit extensive diagnosis and treatment of disease in the fetus much like any other patient. This new dimension in medicine began with techniques by which Rh$^+$ babies (see Figure 11-10) could be given transfusions before birth.

Current procedures using images provided by ultrasound equipment (see Figure 23-11) allow physicians to prepare for and perform, before the birth of a baby, corrective surgical procedures such as bladder repair. These procedures also allow physicians to monitor the progress of other types of treatment on a developing fetus. The first figure shows placement of the ultrasound transducer on the abdominal wall. The resulting image (see second figure), called an *ultrasonogram*, shows a 6-week embryo. The image plane is showing the head and trunk. The hollow cerebral ventricles are visible in the image.

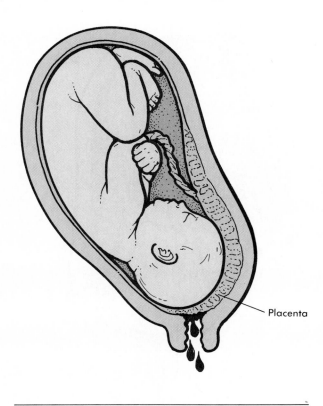

FIGURE 22-9 Placenta Previa.

Placenta

sis or normal peritoneal membrane, the pregnancy may be successful if there is room for the developing fetus to grow. Ectopic pregnancies that do succeed must be delivered by C-section rather than by normal vaginal birth. If an ectopic pregnancy occurs in a uterine tube, which cannot stretch to accommodate the developing offspring, the tube may rupture and cause life-threatening hemorrhaging. So-called **tubal pregnancies** are the most common type of ectopic pregnancy.

Occasionally, the blastocyst implants in the uterine wall near the cervix. This in itself may present no problem, but if the placenta grows too closely to the cervical opening a condition called **placenta previa** (PRE-vee-ah) results. The normal dilation and softening of the cervix that occurs in the third trimester often causes painless bleeding

as the placenta near the cervix separates from the uterine wall. The massive blood loss that may result can be life-threatening for both mother and offspring (Figure 22-9).

Separation of the placenta from the uterine wall can occur even when implantation occurs in the upper part of the uterus. When this occurs in a pregnancy of 20 weeks or more, the condition is called **abruptio placentae** (ab-RUP-chee-o pla-SEN-tay). Complete separation of the placenta causes the immediate death of the fetus. The severe hemorrhaging that often results, sometimes hidden in the uterus, may cause circulatory shock and death of the mother within minutes. A cesarean section and perhaps also a hysterectomy must be performed immediately to prevent blood loss and death (Figure 22-10).

▼ FIGURE 22-10 Abruptio Placentae.

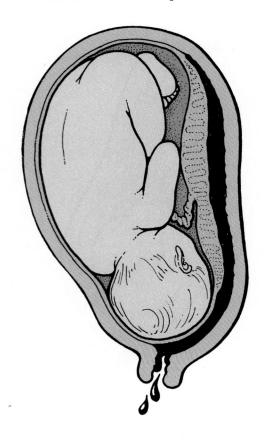

PREECLAMPSIA

Preeclampsia (pre-ee-KLAMP-see-ah), also called *toxemia of pregnancy,* is a serious disorder that occurs in about 1 in every 20 pregnancies. This disorder is characterized by the onset of acute hypertension after the twenty-fourth week, accompanied by proteinuria and edema. The causes of preeclampsia are largely unknown, despite intense research efforts. Preeclampsia can result in complications such as abruptio placentae, stroke, hemorrhage, fetal malnutrition, and low birth weight. This condition can progress to *eclampsia,* a life-threatening form of toxemia that causes severe convulsions, coma, kidney failure, and perhaps death of the fetus and mother.

FETAL DEATH

A *miscarriage* is the loss of an embryo or fetus before the twentieth week (or a fetus weighing less than 500 g or 1.1 lb). Technically known as a **spontaneous abortion,** the most common cause of such a loss is a structural or functional defect in the developing offspring. Abnormalities of the mother, such as hypertension, uterine abnormalities, and hormonal imbalances, can also cause spontaneous abortions. After 20 weeks, delivery of a lifeless infant is termed a **stillbirth.**

BIRTH DEFECTS

Birth defects, also called **congenital abnormalities,** include any structural or functional abnormality present at birth. Congenital defects may be inherited or may be acquired during gestation or delivery. Inherited defects will be discussed in the next chapter. Acquired defects result from agents called **teratogens** (TAIR-et-o-jenz) that disrupt normal histogenesis and organogenesis. Some teratogens are chemicals such as alcohol, antibiotics, and other drugs. Microbes, such as those that cause rubella (a viral infection), can also cross the placental barrier and disrupt normal embryonic development. Radiation and other physical factors can also cause birth defects. Some teratogens are also mutagens because they do their damage by changing the genetic code in cells of the developing embryo.

POSTPARTUM DISORDERS

Puerperal (poo-ER-per-al) **fever** or *childbed fever* is a syndrome of postpartum mothers characterized by bacterial infection that progresses to septicemia (blood infection) and possibly death. Until the 1930s, puerperal fever was the leading cause of maternal death—claiming the lives of more than 20% of postpartum women. Modern antiseptic techniques prevent most postpartum infections now. Puerperal infections that do occur are usually treated successfully by an immediate and intensive program of antibiotic therapy.

After a child is born, it needs the nourishment of milk to survive. However, a number of disorders of *lactation* (milk production) may occur to prevent a mother from nursing her infant. For example, anemia, malnutrition, emotional stress, and structural abnormalities of the breast can all interfere with normal lactation. **Mastitis** (mas-

TIE-tis), or breast inflammation often caused by infection, can result in lactation problems or production of milk contaminated with pathogenic organisms. In many cultures, the availability of other nursing mothers or breast-milk substitutes allows proper nourishment of the infant, even when lactation problems develop. Most breast-milk substitutes are formulations of milk from another mammal, such as the cow. Infants who lack the enzyme *lactase* may not be able to digest the lactose present in human or animal milk, resulting in a condition called **lactose intolerance.** Infants with lactose intolerance are sometimes given a lactose-free milk substitute made from soy or other plant products.

Postnatal Period

The **postnatal period** begins at birth and lasts until death. It is often divided into major periods for study, but people need to understand and appreciate the fact that growth and development are continuous processes that occur throughout the life cycle. Gradual changes in the physical appearance of the body as a whole and in the relative proportions of the head, trunk, and limbs are quite noticeable between birth and adolescence. Note in Figure 22-11 the obvious changes in the size of bones and in the proportionate sizes between different bones and body areas. The head, for example, becomes proportionately smaller. Whereas the infant head is approximately one fourth the total height of the body, the adult head is only about one eighth the total height. The facial bones also show several changes between infancy and adulthood. In an infant the face is one eighth of the skull surface, but in an adult the face is half of the skull surface. Another change in proportion involves the trunk and lower extremities. The legs become proportionately longer and the trunk proportionately shorter. In addition, the thoracic and abdominal contours change from round to elliptical.

FIGURE 22-11 Changes in the Proportions of Body Parts from Birth to Maturity. Note the dramatic differences in head size.

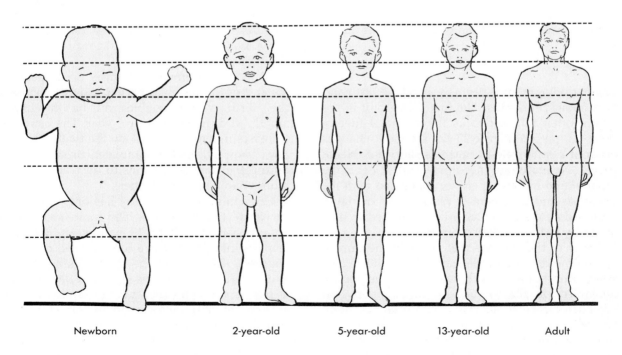

| Newborn | 2-year-old | 5-year-old | 13-year-old | Adult |

FIGURE **22-12** **The Neonate Infant.** The umbilical cord has been cut.

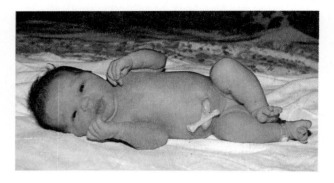

FIGURE **22-13** Normal Curvature of the Infant's Spine.

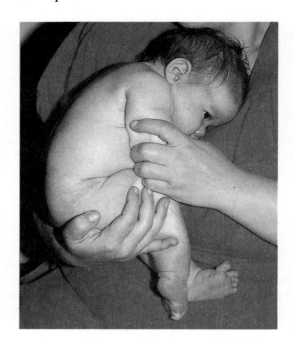

Such changes are good examples of the ever-changing and ongoing nature of growth and development. It is unfortunate that many of the changes that occur in the later years of life do not result in an increased function. These degenerative changes are certainly important, however, and will be discussed later in this chapter (see pp. 544-546). The following are the most common postnatal periods: (1) **infancy,** (2) **childhood,** (3) **adolescence** and **adulthood,** and (4) **older adulthood.**

INFANCY

The period of infancy begins abruptly at birth and lasts about 18 months. The first 4 weeks of infancy are often referred to as the **neonatal** (nee-o-NAY-tal) **period** (Figure 22-12). Dramatic changes occur at a rapid rate during this short but critical period. **Neonatology** (nee-o-nay-TOL-o-jee) is the medical and nursing specialty concerned with the diagnosis and treatment of disorders of the newborn. Advances in this area have resulted in dramatically reduced infant mortality.

Many of the changes that occur in the cardiovascular and respiratory systems at birth are necessary for survival. Whereas the fetus totally depended on the mother for life support, the newborn infant, to survive, must become totally self-supporting in terms of blood circulation and respiration immediately after birth. A baby's first breath is deep and forceful. The stimulus to breathe results primarily from the increasing amounts of carbon dioxide (CO_2) that accumulate in the blood after the umbilical cord is cut shortly after delivery.

Many developmental changes occur between the end of the neonatal period and 18 months of age. Birth weight doubles during the first 4 months and then triples by 1 year. The baby also increases in length by 50% by the twelfth month. The "baby fat" that accumulated under the skin during the first year begins to decrease, and the plump infant becomes leaner.

Early in infancy the baby has only one spinal curvature (Figure 22-13). The lumbar curvature appears between 12 and 18 months, and the once-helpless infant becomes a toddler who can stand (Figure 22-14). One of the most striking changes to occur during infancy is the rapid development of the nervous and muscular systems. This permits the infant to follow a moving object with the eyes (2 months); lift the head and raise the chest (3 months); sit when well supported (4 months); crawl (10 months); stand alone (12 months); and run, although a bit stiffly (18 months).

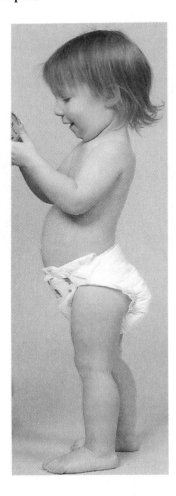

FIGURE 22-14 Normal Lumbar Curvature of a Toddler's Spine.

CHILDHOOD

Childhood extends from the end of infancy to sexual maturity or puberty—12 to 14 years in girls and 14 to 16 years in boys. Overall, growth during early childhood continues at a rather rapid pace, but month-to-month gains become less consistent. By the age of 6 the child appears more like a preadolescent than an infant or toddler. The child becomes less chubby, the potbelly becomes flatter, and the face loses its babyish look. The nervous and muscular systems continue to develop rapidly during the middle years of childhood; by 10 years of age the child has developed numerous motor and coordination skills.

The *deciduous teeth*, which began to appear at about 6 months of age, are lost during childhood, beginning at about 6 years of age. The *permanent teeth*, with the possible exception of the third molars or wisdom teeth, have all erupted by age 14.

ADOLESCENCE AND ADULTHOOD

The average age range of **adolescence** varies but generally the teenage years (13 to 19) are used. The period is marked by rapid and intense physical growth, which ultimately results in sexual maturity. Many of the developmental changes that occur during this period are controlled by the secretion of sex hormones and are classified as **secondary sex characteristics.** Breast development is often the first sign of approaching puberty in girls, beginning about age 10. Most girls begin to menstruate at 12 to 13 years of age. In boys the first sign of puberty is often enlargement of the testicles, which begins between 10 and 13 years of age. Both sexes show a spurt in height during adolescence. In girls the spurt in height begins between the ages of 10 and 12 and is nearly complete by 14 or 15. In boys the period of rapid growth begins between 12 and 13 and is generally complete by 16.

Many developmental changes that began early in childhood are not completed until the early or middle years of **adulthood.** Examples include the maturation of bone, resulting in the full closure of the growth plates, and changes in the size and placement of other body components such as the sinuses. Many body traits do not become apparent for years after birth. Normal balding patterns, for example, are determined at the time of fertilization by heredity but do not appear until maturity. As a general rule, adulthood is characterized by the maintenance of existing body tissues. With the passage of years the ongoing effort of maintenance and repair of body tissues becomes more and more difficult. As a result, degeneration begins. This is the process of aging, and it culminates in death.

OLDER ADULTHOOD

Most body systems are in peak condition and function at a high level of efficiency during the early years of adulthood. As a person grows older, a gradual but certain decline takes place in the functioning of every major organ system in the body. The study of aging is called *gerontology*.

FIGURE 22-15 **Some Physical Changes Associated with Maturity and Aging.** These include graying, thinning, or loss of hair *(left);* skin pigmentation changes *(center);* and barrel chest *(right).* Although physical capacity may decline with age, older adults often lead active, productive lives.

Unfortunately, the mechanisms and causes of aging are not well understood.

Some gerontologists believe that an important aging mechanism is the limit on cell reproduction. Laboratory experiments have shown that many types of human cells cannot reproduce more than 50 times—thus limiting the maximum lifespan. Cells die continually, no matter what a person's age, but in older adulthood, many dead cells are not replaced—causing degeneration of tissues. Perhaps the cells are not replaced because the surrounding cells have reached their limit of reproduction. Perhaps differences in each individual's aging process result from differences in the reproductive capacity of cells. This mechanism seems to operate in individuals with **progeria** (pro-JAIR-ee-ah), a rare, inherited condition in which a person appears to age rapidly.

A variety of factors that affect the rates of cell death and cell reproduction have been cited as causes of aging. Some gerontologists believe that nutrition, injury, disease, and other environmental factors affect the aging process. A few have even proposed that aging results from cellular changes caused by slow-acting "aging" viruses found in all living cells. Other gerontolists have proposed that aging is caused by "aging" genes—genes in which aging is "preprogrammed." Yet another proposed cause of aging is autoimmunity. You may recall from Chapter 14 that autoimmunity occurs when the immune system attacks a person's own tissues.

Although the causes and basic mechanisms of aging are yet to be understood, at least many of the signs of aging are obvious. The remainder of this chapter deals with a number of the more common degenerative changes that frequently characterize **senescence** (se-NES-ens) or older adulthood.

Effects of Aging

SKELETAL SYSTEM

In older adulthood, bones undergo changes in texture, degree of calcification, and shape. Instead of clean-cut margins, older bones develop indistinct and shaggy-appearing margins with spurs—a process called *lipping.* This type of degenerative change restricts movement because of the piling up of bone tissue around the joints. With advancing age, changes in calcification may result in reduction of bone size and in bones that are porous and subject to fracture. The lower cervical and thoracic vertebrae are the site of frequent fractures. The result is curvature of the spine and the shortened stature so typical of late adulthood. Degenerative joint diseases such as **osteoarthritis** (OS-tee-o-ar-THRYE-tis) are also common in elderly adults.

INTEGUMENTARY SYSTEM (SKIN)

With advancing age the skin becomes dry, thin, and inelastic. It "sags" on the body because of increased wrinkling and skinfolds. Pigmentation changes and the thinning or loss of hair are also common problems associated with the aging (Figure 22-15).

URINARY SYSTEM

The number of nephron units in the kidney decreases by almost 50% between the ages of 30 and 75. Also, because less blood flows through the kidneys as an individual ages, there is a reduction in overall function and excretory capacity or the ability to produce urine. In the bladder, significant age-related problems often occur because of diminished muscle tone. Muscle atrophy (wasting) in the bladder wall results in decreased capacity and inability to empty or void completely.

RESPIRATORY SYSTEM

In older adulthood the costal cartilages that connect the ribs to the sternum become hardened or calcified. This makes it difficult for the rib cage to expand and contract as it normally does during inspiration and expiration. In time the ribs gradually become "fixed" to the sternum, and chest movements become difficult. When this occurs the rib cage remains in a more expanded position, respiratory efficiency decreases, and a condition called "barrel chest" results (Figure 22-15). With advancing years a generalized atrophy or wasting of muscle tissue takes place as the contractile muscle cells are replaced by connective tissue. This loss of muscle cells decreases the strength of the muscles associated with inspiration and expiration.

CARDIOVASCULAR SYSTEM

Degenerative heart and blood vessel disease is one of the most common and serious effects of aging. Fatty deposits build up in blood vessel walls and narrow the passageway for the movement of blood, much as the buildup of scale in a water pipe decreases flow and pressure. The resulting condition, called **atherosclerosis** (ath-er-o-skle-RO-sis), often leads to eventual blockage of the coronary arteries and a "heart attack." If fatty accumulations or other substances in blood vessels calcify, actual hardening of the arteries or **arteriosclerosis** (ar-te-ree-o-skle-RO-sis) occurs. Rupture of a hardened vessel in the brain (stroke) is a frequent cause of serious disability or death in the older adult. **Hypertension** or high blood pressure is also more common.

SPECIAL SENSES

The sense organs, as a group, all show a gradual decline in performance and capacity as a person ages. Most people are farsighted by age 65 because eye lenses become hardened and lose elasticity; the lenses cannot become curved to accommodate for near vision. This hardening of the lens is called **presbyopia** (pres-bee-O-pee-ah), which means "old eye." Many individuals first notice the change at about 40 or 45 years of age, when it becomes difficult to do close-up work or read without holding printed material at arm's length. This explains the increased need, with advancing age, for bifocals or glasses that incorporate two lenses to automatically accommodate for near and distant vision. Loss of transparency of the lens or its covering capsule is another common age-related eye change. If the lens actually

EXERCISE AND AGING

A sound exercise program throughout life can reduce some of the common effects of aging. The improved cardiovascular condition associated with aerobic training can prevent or reduce the effects of several age-related circulatory problems. The loss of bone mass seen in older adulthood is less troublesome if one enters this period with a higher than average bone density developed through exercising. Lastly, the relatively high ratio of body fat to muscle seen in the elderly can be avoided entirely by a life-long fitness program.

becomes cloudy and significantly impairs vision, it is called a **cataract** (KAT-ah-rakt) and must be removed surgically. The incidence of **glaucoma** (glaw-KO-mah), the most serious age-related eye disorder, increases with age. Glaucoma causes an increase in the pressure within the eyeball and, unless treated, often results in blindness.

In many elderly people a very significant loss of hair cells in the organ of Corti (inner ear) causes a serious decline in the ability to hear certain frequencies. In addition, the eardrum and attached ossicles become more fixed and less able to transmit mechanical sound waves. Some degree of hearing impairment is universally present in the older adult.

The sense of taste is also decreased. This loss of appetite may be caused, at least in part, by the replacement of taste buds with connective tissue cells. Only about 40% of the taste buds present at age 30 remain in an individual at age 75.

REPRODUCTIVE SYSTEMS

Although most men and women remain sexually active throughout their later years, mechanisms of the sexual response may change, and fertility declines. In men, erection may be more difficult to achieve and maintain, and urgency for sex may decline. In women, lubrication of the vagina may decrease. Although men can continue to produce gametes as they age, women experience a cessation of reproductive cycling between the ages of 45 and 60—**menopause.** Menopause results from a decrease in the cyclic production of gonadotropins from the pituitary, which in turn reduces estrogen secretion. The decrease in estrogen accounts for the common symptoms of menopause: cessation of menstrual cycles, hot flashes, and thinning of the vaginal wall. The exact mechanism of hot flashes is not clearly understood, but it is related to the hormonal changes that occur during menopause. Rarely serious, hot flashes usually subside over time. Estrogen therapy is used to relieve menopause symptoms in extreme cases.

The decrease in estrogen levels associated with menopause may also contribute to *osteoporosis*. This condition, characterized by loss of bone mass (see Chapter 6), is often treated with estrogen therapy when it occurs in postmenopausal women.

OUTLINE SUMMARY

PRENATAL PERIOD

A Prenatal period begins at conception and continues until birth (about 39 weeks)

B Science of fetal growth and development called *embryology*

C Fertilization to implantation requires about 10 days

  1 Fertilization normally occurs in outer third of oviduct (Figure 22-2)

  2 Fertilized ovum called a *zygote;* zygote is genetically complete—all that is needed for expression of hereditary traits is time and nourishment

  3 After 3 days of cell division, the zygote has developed into a solid cell mass called a *morula*

  4 Continued cell divisions of the morula produce a hollow ball of cells called a *blastocyst*

  5 Blastocyst implants in the uterine wall about 10 days after fertilization

  6 Blastocyst forms the amniotic cavity and chorion of the placenta (Figure 22-4)

  7 Placenta provides for exchange of nutrients between the mother and fetus

D Periods of development

  1 Length of pregnancy or gestation period is about 39 weeks

  2 Embryonic phase extends from fertilization to the end of week 8 of gestation

  3 Fetal phase extends from week 8 to 39 of gestation

E Three primary germ layers appear in the developing embryo after implantation of the blastocyst (Table 22-1):

  1 Endoderm—inside layer

  2 Ectoderm—outside layer

  3 Mesoderm—middle layer

  4 All organ systems are formed and functioning by month 4 of gestation (Figure 22-6)

F Histogenesis and organogenesis

  1 Formation of new organs and tissues occurs from specific development of the primary germ layers

  2 Each primary germ layer gives rise to definite structures such as the skin and muscles

  3 Growth processes include cell differentiation, multiplication, growth, and rearrangement

4 From 4 months of gestation until delivery, the development of the baby is mainly a matter of growth

BIRTH OR PARTURITION

A Process of birth called *parturition* (Figure 22-7)
  1 At the end of week 39 of gestation, the uterus becomes "irritable"
  2 Fetus takes head-down position against the cervix
  3 Muscular contractions begin, and labor is initiated
  4 Amniotic sac ("bag of waters") ruptures
  5 Cervix dilates
  6 Fetus moves through vagina to exterior
B Stages of labor
  1 Stage one—period from onset of uterine contractions until dilation of the cervix is complete
  2 Stage two—period from the time of maximal cervical dilation until the baby exits through the vagina
  3 Stage three—process of expulsion of the placenta through the vagina
  4 Cesarean section (C-section)—surgical delivery, usually through an incision in the abdomen and uterine wall
C Multiple births—two or more infants from the same pregnancy
  1 Identical siblings result from the splitting of tissue from the same zygote, making them genetically identical
  2 Fraternal siblings develop from different ova that are fertilized separately

DISORDERS OF PREGNANCY

  1 Implantation disorders
    a Ectopic pregnancy—implantation outside the uterus (for example, tubal pregnancy)
    b Placenta previa—growth of the placenta at or near cervical opening, often resulting in separation of the placenta from the uterine wall
    c Abruptio placentae—separation of a normally-placed placenta from the uterine wall
  2 Preeclampsia (toxemia of pregnancy)—syndrome of pregnancy that includes hypertension, proteinuria, and edema; may progress to eclampsia, a severe toxemia that may result in death
  3 Fetal death
    a Spontaneous abortion (miscarriage)—loss before week 20 (or 500 g)
    b Stillbirth—loss after 20 weeks
  4 Birth defects; also called *congenital abnormalities*
    a May be inherited or acquired
    b Acquired defects are caused by teratogens (agents that disrupt normal development)

5 Postpartum disorders
  a Puerperal fever is caused by bacterial infection that may progress to septicemia and death; occurs in mothers after delivery (postpartum)
  b Lactation and thus infant nutrition can be disrupted by anemia, malnutrition, and other factors
    (1) Mastitis—inflammation or infection of the breast
    (2) Milk can be supplied by another nursing mother or by breast-milk substitutes
    (3) Lactose intolerance results from an infant's inability to digest lactose present in human or animal milk

POSTNATAL PERIOD

A Postnatal period begins at birth and lasts until death
B Divisions of postnatal period into isolated time frames can be misleading; life is a continuous process; growth and development are continuous
C Obvious changes in the physical appearance of the body—in whole and in proportion—occur between birth and maturity (Figure 22-11)
D Divisions of postnatal period
  1 Infancy
  2 Childhood
  3 Adolescence and adulthood
  4 Older adulthood
E Infancy
  1 First 4 weeks called *neonatal period* (Figure 22-13)
  2 Neonatology—medical and nursing specialty concerned with the diagnosis and treatment of disorders of the newborn
  3 Many cardiovascular changes occur at the time of birth; fetus is totally dependent on mother, whereas the newborn must immediately become totally self-supporting (respiration and circulation)
  4 Respiratory changes at birth include a deep and forceful first breath
  5 Developmental changes between the neonatal period and 18 months include:
    a Doubling of birth weight by 4 months and tripling by 1 year
    b 50% increase in body length by 12 months
    c Development of normal spinal curvature by 15 months (Figure 22-14)
    d Ability to raise head by 3 months
    e Ability to crawl by 10 months
    f Ability to stand alone by 12 months
    g Ability to run by 18 months
F Childhood
  1 Extends from end of infancy to puberty—13 years in girls and 15 in boys

2 Overall rate of growth remains rapid but decelerates

3 Continuing development of motor and coordination skills

4 Loss of deciduous or baby teeth and eruption of permanent teeth

G Adolescence and adulthood

1 Average age range of adolescence varies from 13 to 19 years

2 Period of rapid growth resulting in sexual maturity (adolescence)

3 Appearance of secondary sex characteristics regulated by secretion of sex hormones

4 Growth spurt typical of adolescence; begins in girls at about 10 and in boys at about 12

5 Growth plates fully close in adult; other structures such as the sinuses acquire adult placement

6 Adulthood characterized by maintenance of existing body tissues

7 Degeneration of body tissue begins in adulthood

H Older adulthood

1 Degenerative changes characterize older adulthood or sensecence

2 Every organ system of the body undergoes degenerative changes

3 Senescence culminates in death

4 The causes and mechanisms of aging are poorly understood

EFFECTS OF AGING

A Skeletal system

1 Aging causes changes in the texture, calcification, and shape of bones

2 Bone spurs develop around joints

3 Bones become porous and fracture easily

4 Degenerative joint diseases such as osteoarthritis are common

B Integumentary system (skin)

1 With age, skin "sags" and becomes:

  a Thin

  b Dry

  c Wrinkled

2 Pigmentation problems are common

3 Frequent thinning or loss of hair occurs

C Urinary system

1 Nephron units decrease in number by 50% between ages 30 and 75

2 Blood flow to kidney and therefore ability to form urine decrease

3 Bladder problems such as inability to void completely are caused by muscle wasting in the bladder wall

D Respiratory system

1 Calcification of costal cartilages causes rib cage to remain in expanded position—barrel chest

2 Wasting of respiratory muscles decreases respiratory efficiency

3 Respiratory membrane thickens; movement of oxygen from alveoli to blood is slowed

E Cardiovascular system

1 Degenerative heart and blood vessel disease are among the most common and serious effects of aging

2 Fat deposits in blood vessels (atherosclerosis) decreases blood flow to the heart and may cause complete blockage of the coronary arteries

3 Hardening of arteries (arteriosclerosis) may result in rupture of blood vessels, especially in the brain (stroke)

4 Hypertension or high blood pressure is common in older adulthood

F Special senses

1 All sense organs show a gradual decline in performance with age

2 Eye lenses become hard and cannot accommodate for near vision; result is farsightedness in many people by age 45 (presbyopia or "old eye")

3 Loss of transparency of lens or cornea is common (cataract)

4 Glaucoma (increase in pressure in eyeball) is often the cause of blindness in older adulthood

5 Loss of hair cells in inner ear produces frequency deafness in many older people

6 Decreased transmission of sound waves caused by loss of elasticity of eardrum and fixing of the bony ear ossicles is common in older adulthood

7 Some degree of hearing impairment is universally present in the aged

8 Only about 40% of the taste buds present at age 30 remain at age 75

G Reproductive system

1 Changes in the sexual response

  a Men—erection is more difficult to achieve and maintain; urgency for sex may decline

  b Women—lubrication during intercourse may decrease

2 Changes in fertility

  a Men—may continue to be fertile throughout later adult years

  b Women—experience menopause (cessation of reproductive cycling) between the ages of 45 and 60

NEW WORDS

blastocyst	histogenesis	parturition	senescence
chorion	implantation	primary germ layers	zygote
embryology	morula	endoderm	
fertilization	neonate	ectoderm	
gestation	organogenesis	mesoderm	

Diseases and Other Clinical Terms

abruptio placentae	lactose intolerance	preeclampsia	puerperal fever
cesarean section	mastitis	progeria	teratogen
congenital	placenta previa		

CHAPTER TEST

1. The prenatal period begins at _____ and ends at _____.
2. The postnatal period begins at _____ and continues until _____.
3. The science of the development of the individual before birth is called _____.
4. The fertilized ovum is also called a _____.
5. The fluid-filled, shock-absorbing sac in which the embryo floats during development is called the _____ cavity.
6. The "bridge" that permits exchange of nutrients between mother and baby before birth is called the _____.
7. The way tissues arrange themselves into organs during development is called _____.
8. The first 4 weeks of infancy are often referred to as the _____ period.
9. The developmental period that extends from the end of infancy to sexual maturity is called _____.
10. Hardening of the arteries is called _____.
11. If the lens of the eye becomes cloudy and impairs vision the condition is called a _____.

Circle the T before each true statement and the F before each false statement.

T   F   12. The fertilized ovum is called the *morula*.
T   F   13. There are 5 primary germ layers in the developing embryo.
T   F   14. In humans, all organ systems are formed and functioning by the month 4 of pregnancy.
T   F   15. The placenta develops in part from the fetal membrane called the *chorion*.
T   F   16. Neonatology is the medical and nursing specialty concerned with the diagnosis and treatment of disorders of the newborn.
T   F   17. Childhood extends from the end of infancy to sexual maturity.
T   F   18. The term *senescence* refers to older adulthood.
T   F   19. The kidney is unique in that it does not show degenerative changes with advancing age.
T   F   20. The term *arteriosclerosis* refers to "hardening of the arteries."

Select the most correct answer from Column B for each statement in column A. (Only one answer is correct)

COLUMN A

21. _____ Placenta at or near cervical opening
22. _____ Toxemia of pregnancy
23. _____ Causes birth defects
24. _____ Postpartum infection
25. _____ Cessation of menstrual cycle

COLUMN B

a. Menopause
b. Placenta previa
c. Preeclampsia
d. Puerperal fever
e. Teratogen

REVIEW QUESTIONS

1 What biological event separates the prenatal from the postnatal period of development?

2 Define embryology, gestation, fertilization, implantation, morula, and blastocyst.

3 What is the difference between an ovum and a zygote?

4 What is the difference between an embryo and a fetus?

5 Discuss the role of the placenta in fetal development.

6 What are the primary germ layers and how are they related to development of the fetus?

7 Briefly define histogenesis and organogenesis.

8 At what point in development of the fetus are all the organ systems formed and functioning?

9 What is the difference between identical twins and fraternal twins?

10 What disorders involve premature separation of the placenta from the uterine wall?

11 What is the difference between a miscarriage and a stillbirth?

12 What causes birth defects?

13 When does the postnatal period begin and how long does it last?

14 List the four subdivisions of the postnatal period.

15 What is the relationship of the neonatal period to infancy? What is neonatology?

16 What is it necessary for a baby's first breath to be deep and forceful? What is the primary stimulus that causes a newborn to take its first breath?

17 List five developmental changes that occur during infancy.

18 Is the period of childhood the same length for both boys and girls? Explain the reason for your answer.

19 What is the average age range of adolescence? Functionally, how would the reproductive system in an individual at the end of adolescence be described?

20 Define *senescence*.

21 Describe the degenerative changes that occur in the skeletal system as a result of aging.

22 How is the skin or integumentary system affected by aging?

23 What is the difference between atherosclerosis and arteriosclerosis?

24 Define presbyopia, glaucoma, and cataract.

25 What is menopause? Do both men and women remain fertile during the later adult years?

CLINICAL APPLICATIONS

1 Mary is pregnant with her first baby and is touring the maternity floor of a local hospital. She keeps asking about their precautions regarding aseptic technique in the "home-style" birthing rooms. She is concerned because her own grandmother died of an infection after delivering Mary's mother. Why should Mary be concerned? What might happen if aseptic techniques are not adhered to?

2 Two-year old Abe has always been small for his age and is now diagnosed as having developmental disabilities. Abe's grandmother believes that these disabilities are associated with the moderate drinking done by Abe's mother during her pregnancy. Is Abe's grandmother simply trying to justify her meddling—or could his mother's drinking have caused his problems?

3 Lactose intolerance is sometimes treated by removing lactose from the diet. For example, infants with lactose intolerance may be given a lactose-free milk substitute to avoid the indigestion, diarrhea, abdominal discomfort, and other symptoms of this condition. Older children and adults with lactose intolerance may avoid dairy products and other foods high in lactose. Your friend Aileen, who has lactose intolerance, takes a tablet with her favorite ice-cream (chocolate fudge) that helps her avoid any problems. What might this tablet contain?

Genetics and Genetic Diseases

C H A P T E R

23

Objectives

*After you have completed this chapter,
you should be able to:*

1 Explain how genes can cause disease.

2 Distinguish between dominant and recessive genetic traits.

3 Describe sex-linked inheritance.

4 List some important inherited diseases.

5 Describe how nondisjunction can result in trisomy or monosomy and list some disorders that result from it.

6 List some tools used in genetic counseling and how they are used to help clients.

7 Describe how genetic disorders can be treated.

*I*t seems that today we are hearing more and more about the relationship of **genetics,** the scientific study of inheritance, and human disease. Popular news magazines are running story after story on the revolution in treating fatal inherited disorders by using something called *gene therapy*. Health and science columns in newspapers keep us informed of the latest discoveries of disease-causing genes. Television programs outline the progress of the largest coordinated biological quest that anyone can remember: mapping the entire human genetic code. Clearly, a person cannot be informed about the mechanisms of human disease today without some knowledge of basic genetics. In this chapter, we will briefly review the essential concepts of genetics and explain how information in the genetic code can cause disease.

Genetics and Human Disease

History shows that humans have been aware of inheritance for thousands of years; however, it was not until the 1860s that the scientific study of inheritance—**genetics**—was born. At that time, a monk living in what is now Czechoslovakia was the first to discover the basic mechanism by which traits are transmitted from parents to offspring. That man, Gregor Mendel, proved that independent units (which we now call *genes*) are responsible for the inheritance of biological traits.

The science of genetics developed from Mendel's quest to explain how normal biological characteristics are inherited. As time went by, and more genetic studies were done, it became clear that certain diseases have a genetic basis. As you may recall from Chapter 4, some diseases are inherited directly. For example, the blood-clotting disorder called *hemophilia* is inherited by children from a parent who has the genetic code for hemophilia. Other diseases are only partly determined by genetics; that is, they involve genetic risk factors (Chapter 4). For example, certain forms of skin cancer are thought to have a genetic basis. A person who inherits the genetic code associated with skin cancer will develop the disease only if the skin is also heavily exposed to the ultraviolet radiation in sunlight.

Chromosomes and Genes

MECHANISM OF GENE FUNCTION

Mendel proposed that the genetic code is transmitted to offspring in discrete, independent units that we now call **genes** (jeans). Recall from Chapter 2 that each gene is a sequence of nucleotide bases in the deoxyribonucleic acid (DNA) molecule (also see Appendix A). Each gene contains a genetic code that the cell transcribes to a messenger RNA (mRNA) molecule. Each mRNA molecule moves to a ribosome, where the code is translated to form a specific protein molecule. Many of the protein molecules thus formed are *enzymes*, functional proteins that permit specific biochemical reactions to occur (see Appendix A). Because enzymes regulate the biochemistry of the body, they regulate the entire structure and function of the body. Thus genes determine the structure and function of the human body by producing a set of specific regulatory enzymes.

As described in Chapter 2, genes are simply segments of a DNA molecule. While the genetic codes of its genes are being actively transcribed, the DNA is in a threadlike form called *chromatin*. During cell division, each replicated strand of chromatin coils to form a compact mass called a *chromosome* (Figure 23-1). Thus each DNA molecule can be called either a *chromatin strand* or a *chromosome*. Throughout this chapter, we will use the term *chromosome* for DNA and the term *gene* for each distinct code within a DNA molecule.

DISTRIBUTION OF CHROMOSOMES TO OFFSPRING

Each cell of the human body contains 46 chromosomes. The only exceptions to this principle are the **gametes**—male *spermatozoa* and female *ova*. Recall from Chapter 21 that a special form of nuclear division called *meiosis* (Figure 23-2) produces gametes with only 23 chromosomes—exactly half the usual number. When a sperm (with its 23 chromosomes) unites with an ovum (with its 23 chromosomes) at conception, a *zygote* with 46 chromosomes is formed. Thus the zygote has the same number of chromosomes as each body cell in the parents.

As the photograph in Figure 23-1 shows, the 46 human chromosomes can be arranged in 23 pairs

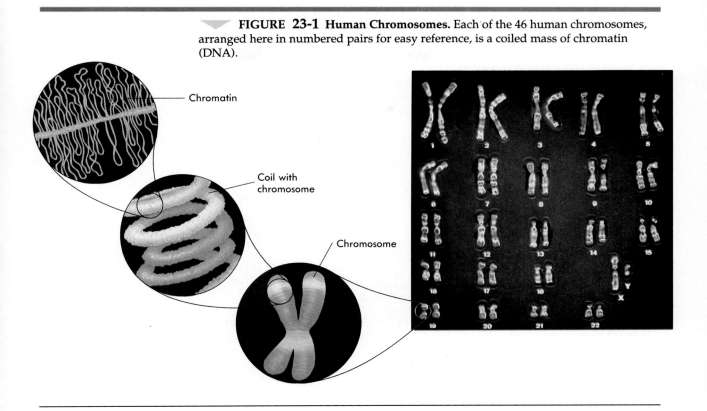

FIGURE 23-1 Human Chromosomes. Each of the 46 human chromosomes, arranged here in numbered pairs for easy reference, is a coiled mass of chromatin (DNA).

Chromatin

Coil with chromosome

Chromosome

according to size. One pair called the **sex chromosomes** may not match, but the remaining 22 pairs of **autosomes** (AW-toe-sohms) always appear to be nearly identical to each other.

Because half of an offspring's chromosomes are from the mother and half are from the father, a unique blend of inherited traits is formed. According to principles first discovered by Mendel, each chromosome assorts itself independently during meiosis (Figure 23-2). This means that as sperm are formed, chromosome pairs separate and the maternal and paternal chromosomes get mixed up and redistribute themselves independently of the other chromosome pairs. Thus each sperm is likely to have a *different* set of 23 chromosomes. Because ova are formed in the same manner, each ovum is likely to be genetically different from the ovum that preceded it. Independent assortment of chromosomes ensures that each offspring from a single set of parents is very likely to be genetically unique—a phenomenon known as *genetic variation*.

The principle of independent assortment also applies to individual genes or groups of genes. During one phase of meiosis, pairs of matching chromosomes line up along the equator of the cell and exchange genes. This process is called *crossing-over* because genes from a particular location cross over to the same location on the matching gene (Figure 23-3). Sometimes a whole group stays together and crosses over as a single unit—a phenomenon called *gene linkage*. Crossing-over introduces additional possibilities for genetic variation among offspring of a set of parents.

Gene Expression

HEREDITARY TRAITS

Mendel discovered that the genetic units we now call *genes* may be expressed differently among individual offspring. After rigorous experimentation with pea plants, he discovered that each inherited trait is controlled by two sets of similar

FIGURE 23-2 Meiosis. In meiosis, a series of two divisions results in the production of gametes with half the number of chromosomes of the original parent cell. In this figure, the original cell has 4 chromosomes and the gametes each have 2 chromosomes. During the first division of meiosis, pairs of similar chromosomes line up along the cell's equator for even distribution to daughter cells. Because different pairs assort independently of each other, any of four (2^2) different combinations of chromosomes may occur. Because human cells have 23 pairs of chromosomes, over 8 million (2^{23}) different combinations are possible.

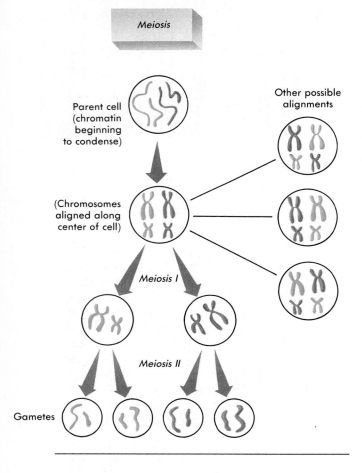

FIGURE 23-3 Crossing-Over. Genes (or linked groups of genes) from one chromosome are exchanged with matching genes in the other chromosome of a pair during meiosis.

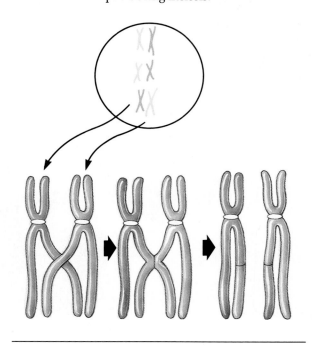

genes, one from each parent. We now know that each autosome in a pair matches its partner in the type of genes it contains. In other words, if one autosome has a gene for hair color, its partner will also have a gene for hair color—in the same location on the autosome. Although both genes specify hair color, they may not specify the *same* hair color. Mendel discovered that some genes are

dominant and some are **recessive.** A dominant gene is one whose effects are seen and that is capable of masking the effects of a recessive gene for the same trait.

Consider the example of **albinism** (AL-bin-izm), a total lack of melanin pigment in the skin and eyes. Because they lack dark pigmentation, people with this condition have difficulty with seeing and protecting themselves from burns in direct sunlight. The genes that cause albinism are recessive; the genes that cause normal melanin production are dominant. By convention, dominant genes are represented by uppercase letters and recessive genes by lowercase letters. One can represent the gene for albinism as *a* and the gene for normal skin pigmentation as *A*. A person with the gene combination *AA* has two dominant genes—and so will exhibit a normal skin color. Someone with the gene combination *Aa* will also have normal skin color because the normal gene *A* is dominant over the recessive albinism gene *a*. Only a person with the gene combination *aa* will

FIGURE 23-4 Inheritance of Albinism. Albinism is a recessive trait, producing abnormalities only in those with two recessive genes (a). Presence of the dominant gene (A) prevents albinism.

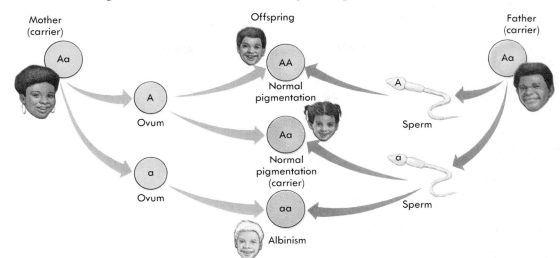

have albinism because there is no dominant gene to mask the effects of the two recessive genes.

In the example of albinism, a person with the gene combination of *Aa* is said to be a genetic **carrier** of albinism. This means that the person can transmit the albinism gene, *a*, to offspring. Thus two normal parents each having the gene combination *Aa* can produce both normal children and children that have albinism (Figure 23-4).

What happens if two different dominant genes occur together? Suppose there is a gene A^1 for light skin and a gene A^2 for dark skin. In a form of dominance called **co-dominance,** they will simply have equal effects and a person with the gene combination A^1A^2 will exhibit a skin color that is something between light and dark. Recall from Chapter 11 that the genes for sickle cell anemia behave this way. A person with two sickle cell genes will have *sickle cell anemia*, whereas a person with one normal gene and one sickle cell gene will have a milder form of the disease called *sickle-cell trait*.

SEX-LINKED TRAITS

Recall from our earlier discussion that, besides the 22 pairs of autosomes, there is one pair of sex chromosomes. Notice in Figure 23-1 that the chromosomes of this pair do not have matching structures. The larger sex chromosome is called the *X chromosome*, and the smaller one is called the *Y chromosome*. The X chromosome is sometimes called the *female chromosome* because it has genes that determine female sexual characteristics. If a person has only X chromosomes, she is genetically a female. The Y chromosome is often called the *male chromosome* because anyone possessing a Y chromosome is genetically a male. Thus all normal females have the sex chromosome combination XX and all normal males have the combination XY. Because men produce both X-bearing and Y-bearing sperm, any two parents can produce male or female children (Figure 23-5).

These large X chromosomes contain many genes besides those needed for female sexual traits. Genes for producing certain clotting factors, the photopigments in the retina of the eye, and many other proteins are also found on the X chromosome. The tiny Y chromosome, on the other hand, contains few genes other than those that determine the male sexual characteristics. Thus males and females need at least one normal X chromosome; otherwise the genes for clotting factors and other essential proteins would be missing. Nonsexual traits thus carried on sex

▼ **FIGURE 23-5 Sex Determination.** The presence of the Y chromosome specifies maleness. In the absence of a Y chromosome, an individual develops into a female.

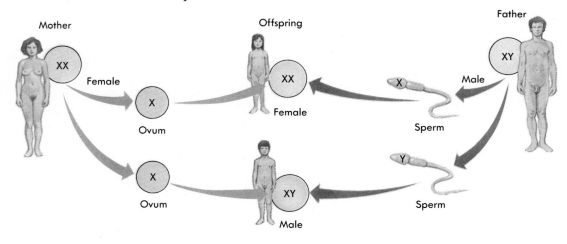

chromosomes are called **sex-linked traits.** Most sex-linked traits are called *X-linked traits* because they are determined by the genes in the large X chromosome.

Dominant X-linked traits appear in each person as one would expect for any dominant trait. In females, recessive X-linked genes are masked by dominant genes in the other X chromosome. Only females with two recessive X-linked genes can exhibit the recessive trait. Because males inherit only one X chromosome (from the mother), the presence of only one recessive X-linked gene is enough to produce the recessive trait. In short, in males, there are no matching genes in the Y chromosome to mask recessive genes in the X chromosome. For this reason, X-linked recessive traits appear much more frequently in males than in females.

An example of a recessive X-linked condition is *red-green color blindness,* which involves a deficiency of photopigments in the retina (see Chapter 9). In this condition, male children of a parent who carries the recessive abnormal gene on an X chromosome may be color blind (Figure 23-6). A female can inherit this form of color blindness only if her father is color blind *and* her mother is color blind or is a color-blindness carrier. The X chromosome has been studied in great detail by many researchers, and general locations for genes

causing at least 59 distinct X-linked diseases have been identified.

GENETIC MUTATIONS

The term *mutation* simply means "change." A *genetic mutation* is a change in the genetic code. Mutations may occur spontaneously, without the influence of factors outside the DNA itself. However, most genetic mutations are believed to be caused by **mutagens**—agents that cause mutations. Genetic mutagens include chemicals, some forms of radiation, and even viruses. Some mutations involve a change in the genetic code within a single gene, perhaps a slight rearrangement of the nucleotide sequence. Other mutations involve damage to a portion of a chromosome or a whole chromosome. For example, a portion of a chromosome may completely break away.

Beneficial mutations allow organisms to adapt to their environments. Because such mutant genes benefit survival, they tend to spread throughout a population over the course of several generations. Harmful mutations inhibit survival, so they are not likely to spread through the population. Most harmful mutations kill the organism in which they occur or at least prevent successful reproduction—and so are never passed to offspring. If a harmful mutation is only mildly harmful, it may persist in a population over many generations.

FIGURE 23-6 Sex-Linked Inheritance. Some forms of color blindness involve recessive X-linked genes. In this case, a female carrier of the abnormal gene can produce male children who are color-blind.

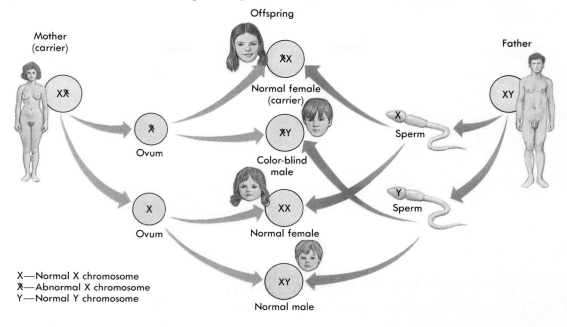

X—Normal X chromosome
X̶—Abnormal X chromosome
Y—Normal Y chromosome

Genetic Diseases

MECHANISMS OF GENETIC DISEASE

Genetic diseases are diseases produced by an abnormality in the genetic code. Many genetic diseases are caused by individual mutant genes that are passed from one generation to the next—making them *single-gene diseases*. In single-gene diseases, the mutant gene may make an abnormal product that causes disease, or it may fail to make a product required for normal function. Some disease conditions result from the combined effects of inheritance and environmental factors. Because they are not solely caused by genetic mechanisms, such conditions are not genetic diseases in the usual sense of the word; they are instead said to involve a *genetic predisposition*.

Some genetic diseases do not result from an abnormality in a single gene. Instead, these diseases result from chromosome breakage or from the abnormal presence or absence of entire chromosomes. For example, a condition called **trisomy** (TRY-so-me) may occur where there is a triplet of autosomes rather than a pair. Trisomy results from a mistake in meiosis called *nondisjunction* when a pair of chromosomes fails to separate. This produces a gamete with two autosomes that are "stuck together" instead of the usual one. When this abnormal gamete joins with a normal gamete to form a zygote, the zygote has a triplet of autosomes (Figure 23-7). Usually trisomy of any autosome pair is fatal. However, if trisomy occurs in autosome pair 13, 15, 18, 21, or 22, a person may survive for a time—but not without profound developmental defects. **Monosomy** (MAHN-o-so-me), the presence of only one autosome instead of a pair, may also result from conception involving a gamete produced by nondisjunction (Figure 23-7). Like trisomy, monosomy may produce life-threatening abnormalities. Because most trisomic and monosomic individuals die before they can reproduce, these conditions are not usually passed from generation to generation. Trisomy and monosomy are congenital conditions that are sometime referred to as *chromosomal genetic diseases*.

FIGURE 23-7 Effects of Nondisjunction. Nondisjunction, failure of a chromosome pair to separate during gamete production, may result in trisomy or monosomy in the offspring.

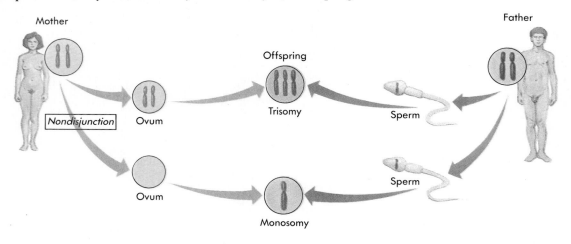

MAPPING THE GENOME

All genetic material in each cell of the human body is called the **genome** (JEE-nome). The typical human genome consists of 46 individual chromosomes. An intense, coordinated effort is currently underway to map all the gene locations in the human genome and read the different genetic codes possible at each location. Scientists predict that all genes will be identified within the next few years. Already, the location of about 2000 genes that may cause disease have been identified (see Appendix B). Much of the work is being done under the leadership of James Watson, one of a team of researchers that originally discovered the structure of DNA in the early 1950s.

SINGLE-GENE DISEASES

There are many single-gene diseases. A few are discussed here and summarized in Table 23-1.

Cystic fibrosis (SIS-tik fi-BRO-sis) **(CF)** is caused by recessive genes in chromosome pair 7. The primary effect of the recessive genes is impairment of chloride ion transport across cell membranes. This disruption causes exocrine cells to secrete thick mucus and sweat. The thickened mucus is especially troublesome in the gastrointestinal and respiratory tracts, where it can cause obstruction that leads to death.

Phenylketonuria (feen-il-kee-toe-NOO-ree-ah) **(PKU)** is caused by recessive genes that fail to produce the enzyme *phenylalanine hydroxylase*. This is needed to convert the amino acid phenylalanine into another amino acid, tyrosine. Thus phenylalanine absorbed from ingested food accumulates—resulting in the abnormal presence of phenylketone in the urine (hence the name *phenylketonuria*). A high concentration of phenylalanine destroys brain tissue, so babies born with this condition are at risk of progressive mental retardation and perhaps death. Many PKU victims are identified at birth by state-mandated tests. Once identified, PKU victims are put on diets low in phenylalanine, thus avoiding a toxic accumulation of it. You may be familiar with the printed warning for phenylketonurics commonly seen on products that contain aspartame (Nutrasweet) or other substances made from phenylalanine.

CHROMOSOMAL DISEASES

Some genetic disorders are not inherited but result from nondisjunction during gametes formation. As Figure 23-7 shows, nondisjunction results in trisomy or monosomy.

TABLE 23-1   Examples of Single-Gene Disorders

DISORDER	DOMINANCE	DESCRIPTION
Hemophilia (some forms)	Recessive (X-linked)	Group of blood clotting disorders caused by a failure to form clotting factors VIII, IX, or XI
Albinism	Recessive	Lack of the dark brown pigment *melanin* in the skin and eyes, resulting in vision problems and susceptibility to sunburn and skin cancer
Sickle-cell anemia and Sickle-cell trait	Codominant	Blood disorder in which abnormal hemoglobin is produced, causing red blood cells to deform into a sickle shape
Red-green color blindness	Recessive (X-linked)	Inability to distinguish red and green light resulting from a deficiency of photopigments in the cone cells of the retina.
Cystic fibrosis (CF)	Recessive	Condition characterized by excessive secretion of thick mucus and concentrated sweat, often causing obstruction of the GI or respiratory tracts.
Phenylketonuria (PKU)	Recessive	Excess of phenylketone in the urine, which is caused by accumulation of phenylalanine in the tissues and may cause brain injury and death
Tay-Sachs disease	Recessive	Fatal condition in which abnormal lipids accumulate in the brain and cause damage that leads to death by age 4
Osteogenesis imperfecta	Dominant	Group of connective tissue disorders characterized by imperfect skeletal development that produces brittle bones
Multiple neurofibromatosis	Dominant	Disorder characterized by multiple, sometimes disfiguring, benign tumors of the Schwann cells (neuroglia) that surround nerve fibers
Duchenne muscular dystrophy (DMD)	Recessive (X-linked)	Muscle disorder characterized by progressive atrophy of skeletal muscle without nerve involvement
Hypercholesterolemia	Dominant	High blood cholesterol that may lead to atherosclerosis and other cardiovascular problems
Huntington's disease (HD)	Dominant	Degenerative brain disorder characterized by chorea (purposeless movements) progressing to severe dementia and death by age 55
Severe combined immune deficiency (SCID)	Recessive	Failure of the lymphocytes to develop properly, in turn causing failure of the immune system's defense of the body; usually caused by adenosine deaminase (ADA) deficiency

MITOCHONDRIAL INHERITANCE

Mitochondria are tiny, bacteria-like organelles in every cell (see Chapter 2). Like a bacterium, each mitochondrion has its own DNA molecule, sometimes called *mitochondrial DNA (mDNA)*. Inheritance of mDNA occurs only through one's mother because sperm do not contribute mitochondria to the ovum during fertilization. Because mDNA contains the only genetic code for several important enzymes, it can carry mutations that produce disease. Mitochondrial inheritance is known to transmit genes for several degenerative nerve and muscle disorders. One such disease is *Leber's hereditary optic neuropathy*. In this disease, young adults lose their eyesight as the optic nerve degenerates—resulting in total blindness by age 30. Some medical researchers believe that *Parkinson's disease*, a nervous disorder that produces muscle tremors, may also be inherited through mitochondrial DNA.

The most famous chromosomal disorder is *trisomy 21,* which produces a group of symptoms called **Down syndrome.** As Figure 23-8, *A,* shows, in this condition, there is a triplet of chromosome 21 rather than the usual pair. In the general population, trisomy 21 occurs in only 1 of every 600 or so live births. After age 35, however, a mother's chances of producing a trisomic child increase dramatically—to as high as 1 in 80 births by age 40. Down syndrome results from trisomy 21 and rarely from other genetic abnormalities (which can be inherited). This syndrome is characterized by mental retardation (ranging from mild to severe) and multiple defects that include distinctive facial appearance (Figure 23-8, *B*), enlarged tongue, short hands and feet with stubby digits, congenital heart disease, and susceptibility to acute leukemia. People with Down syndrome have a shorter-than-average life expectancy but can survive to old age.

Klinefelter's syndrome is another genetic disorder resulting from nondisjunction of chromosomes. This disorder occurs in males with a Y chromosome and at least two X chromosomes, typically the XXY pattern. Characteristics of Klinefelter's syndrome include long legs, enlarged breasts, low intelligence, small testes, sterility, and chronic pulmonary disease.

Turner's syndrome, sometimes called *XO syndrome,* occurs in females with a single sex chromosome, X. Like the conditions described above, it results from nondisjunction during gamete formation. Turner syndrome is characterized by failure of the ovaries and other sex organs to mature (causing sterility), cardiovascular defects, dwarfism or short stature, a webbed neck, and possible learning disorders. Symptoms of Turner's syndrome can be reduced by hormone therapy using estrogens and growth hormone. Cardiovascular defects may be repaired surgically.

▼ *Prevention and Treatment of Genetic Diseases*

GENETIC COUNSELING

The term *genetic counseling* refers to professional consultations with families regarding genetic dis-

▼ **FIGURE 23-8 Down Syndrome. A,** Down syndrome is usually associated with trisomy of chromosome 21. **B,** A child with Down syndrome sitting on his father's knee. Notice the distinctive anatomical features: "oriental" folds around the eyes, flattened nose, round face, and small hands with short fingers.

A

B

eases. Trained genetic counselors may help a family determine the risk of producing children with genetic diseases. Parents with a high risk of producing children with genetic disorders may decide to avoid having children. Genetic counselors may also help evaluate whether any offspring already have a genetic disorder and offer advice on treatment or care. A growing list of tools are available to genetic counselors, some of which are described below.

Pedigree

A **pedigree** is chart that illustrates genetic relationships in a family over several generations (Figure 23-9). Using medical records and family histories, the genetic counselors assemble the chart, beginning with the client and moving backward through as many generations as are known. Squares represent males; circle represent females. Fully shaded symbols represent affected individuals, whereas unshaded symbols represent normal individuals. Partially-shaded symbols represent carriers of a recessive trait. A horizontal line

between symbols designates a sexual relationship that produced offspring.

The pedigree is useful in determining the possibility of producing offspring with certain genetic disorders. It also may tell a person whether he or she might have a genetic disorder that appears late in life, such as Huntington's disease. In either case, a family can prepare emotionally, financially, and medically before a crisis occurs.

Punnett Square

The Punnett (PUN-et) square, named after the English geneticist Reginald Punnett, is a grid used to determine the *probability* of inheriting genetic traits. As Figure 23-10, *A* shows, genes in the mother's gametes are represented along one axis of the grid, and genes in the father's gametes are along the other axis. The ratio of different gene combinations in the offspring predicts their probability of occurrence in the next generation. Thus offspring produced by two carriers of PKU (a recessive disorder) have a one in four (25%) chance of inheriting this recessive condition

FIGURE 23-9 Pedigree. Pedigrees chart the genetic history of family lines. Squares represent males, and circles represent females. Fully shaded symbols indicate affected individuals; partly-shaded symbols indicate carriers, and unshaded symbols indicate unaffected noncarriers. Roman numerals indicate the order of generations. This pedigree reveals the presence of an X-linked recessive trait.

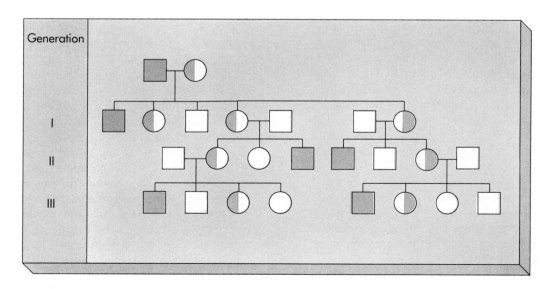

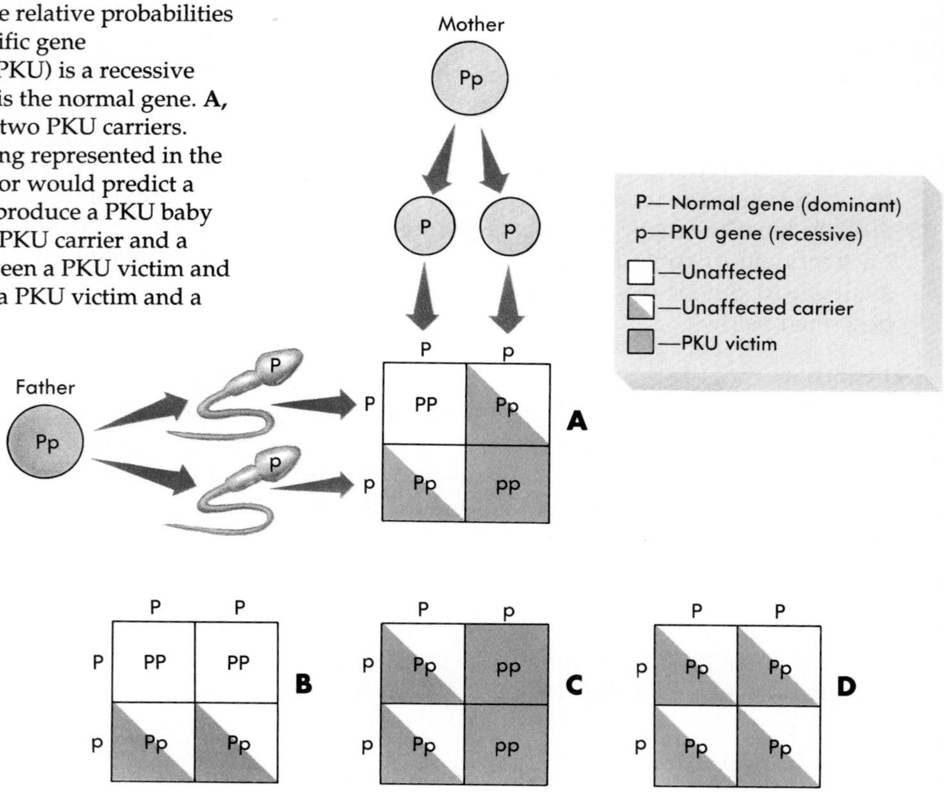

FIGURE 23-10 Punnett Square. The Punnett square is a grid used to determine relative probabilities of producing offspring with specific gene combinations. Phenylketonuria (PKU) is a recessive disorder caused by the gene *p*. *P* is the normal gene. **A,** Possible results of cross between two PKU carriers. Because one in four of the offspring represented in the grid have PKU, a genetic counselor would predict a 25% chance that this couple will produce a PKU baby at each birth. **B,** Cross between a PKU carrier and a normal noncarrier. **C,** Cross between a PKU victim and a PKU carrier. **D,** Cross between a PKU victim and a normal noncarrier.

GENETIC BASIS OF CANCER

Recall from Chapter 4 that some forms of cancer are thought to be caused, at least in part, by abnormal genes called *oncogenes*. One hypothesis states that most normal cells contain such cancer-causing genes. However, it is uncertain how these genes become activated and produce cancer. Perhaps oncogenes can transform a cell into a cancer cell only when certain environmental conditions occur.

Another hypothesis states that normal cells contain another class of genes, *tumor suppressor genes*. According to this hypothesis, such genes regulate cell division so that it proceeds normally. When a tumor suppressor gene is nonfunctional because of a genetic mutation, it then allows cells to divide abnormally—possibly producing cancer.

Another possible genetic basis for cancer relates to the genes that govern the cell's ability to repair damaged DNA. As mentioned in Chapter 5, a rare genetic disorder called *xeroderma pigmentosum* is characterized by the inability of skin cells to repair genetic damage caused by ultraviolet radiation in sunlight. Individuals with this condition almost always develop skin cancer when exposed to direct sunlight. The genetic abnormality does not cause skin cancer directly but inhibits the cell's cancer-preventing mechanisms.

Cancer researchers are now working feverishly to determine the exact role various genes play in the development of cancer. The more we understand the genetic basis of cancer, the more likely it is that we will find effective treatments—or even cures.

(Figure 23-10, *A*). There is a two in four (50%) chance that an individual child produced will be a PKU carrier. Figure 23-10, *B* shows that off-spring between a carrier and a noncarrier cannot inherit PKU. What is the chance of an individual offspring being a PKU carrier in this case? Figure 23-10, *C* shows the probability of producing an affected offspring when a PKU victim and a PKU carrier have children. Figure 23-10, *D* shows the genetic probability when a PKU victim and a non-carrier produce children.

Karyotype

Disorders that involve trisomy (extra chromosomes), monosomy (missing chromosomes), and broken chromosomes can be detected after a **karyotype** (KAIR-ee-o-type) is produced. The first step in producing a karyotype is getting a sample of cells from the individual to be tested. This can be done by scraping cells from the lining of the cheek or from a blood sample containing WBCs. Fetal tissue can be collected by **amniocentesis** (am-nee-o-sen-TEE-sis), a procedure in which fetal cells floating in the amniotic fluid are collected with a syringe (Figure 23-11). **Chorionic villus sampling** is a newer procedure in which cells from chorionic villi that surround a young embryo (see Chapter 22) are collected through the opening of the cervix. Collected cells are grown in a special culture medium and allowed to reproduce. Cells in metaphase (when the chromosomes are most distinct) are stained and photographed using a microscope. The chromosomes are cut out of the photo and pasted on a chart in pairs according to size, like in Figures 23-1 and 23-8, *A*. Genetic counselors then examine the karyotype, looking for chromosome abnormalities. What chromosome abnormality is visible in Figure 23-8, *A*? Is this a male or female karyotype?

TREATING GENETIC DISEASES

Until recently, the only hope of treating genetic diseases was to treat the symptoms. In some diseases, such as PKU, this works well. If PKU victims simply avoid large amounts of phenylalanine in their diets, especially during critical stages of development, severe complications can be avoided. In Klinefelter's and Turner's syndromes, hormone therapy and surgery can alleviate some

symptoms. However, there are no effective treatments for the vast majority of genetic disorders. Fortunately, medical science now offers us some hope of treating genetic disorders through **gene therapy.**

In a therapy called *gene replacement,* genes that specify production of abnormal, disease-causing proteins are replaced by normal or "therapeutic" genes. To get the therapeutic genes to cells that need them, researchers have proposed using genetically altered viruses as carriers. Recall from Chapter 1 and Chapter 4 that viruses are easily capable of inserting new genes into the human genome. If the therapeutic genes behave as

FIGURE 23-11 Amniocentesis. In amniocentesis, a syringe is used to collect amniotic fluid. Ultrasound imaging is used to guide the tip of the syringe needle to prevent damage to the placenta and fetus. Fetal cells in the collected amniotic fluid can then be chemically tested or used to produce a karyotype of the developing baby.

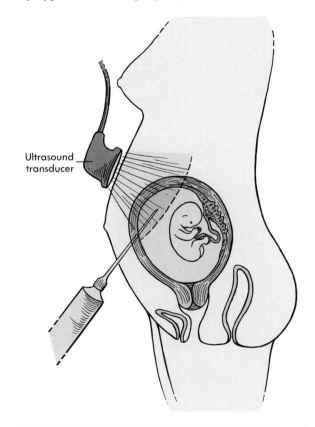

Ultrasound transducer

DNA ANALYSIS

As a result of the intense efforts underway to map the entire human genome, new techniques have been developed to analyze the genetic makeup of individuals. Automated machines can now chemically analyze chromosomes and "read" their sequence of nucleotides—the genetic code. One method by which this is done is called **electrophoresis** (el-ek-tro-fo-REE-sis), which means "electric separation" *(A)*. In electrophoresis, DNA fragments are chemically processed then placed in a thick fluid or *gel*. An electric field in the gel causes the DNA fragments to separate into groups according to their relative sizes. The resulting pattern *(B)* that results represents the sequence of codons in the DNA fragment. This process is also the basis for so-called *DNA fingerprinting*. Like a fingerprint pattern, each person's DNA sequence is unique. After the exact sequences for specific diseases have been discovered, genetic counselors will be able to provide more details about the genetic makeup of their clients.

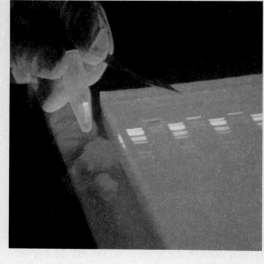

A

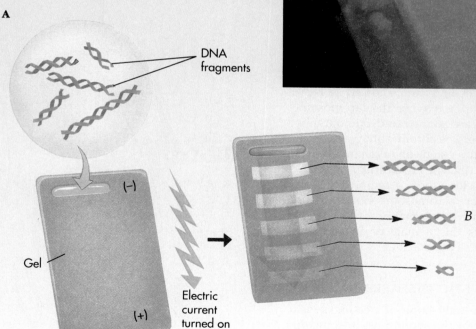

A

DNA fragments

Gel

(−)

(+)

Electric current turned on

B

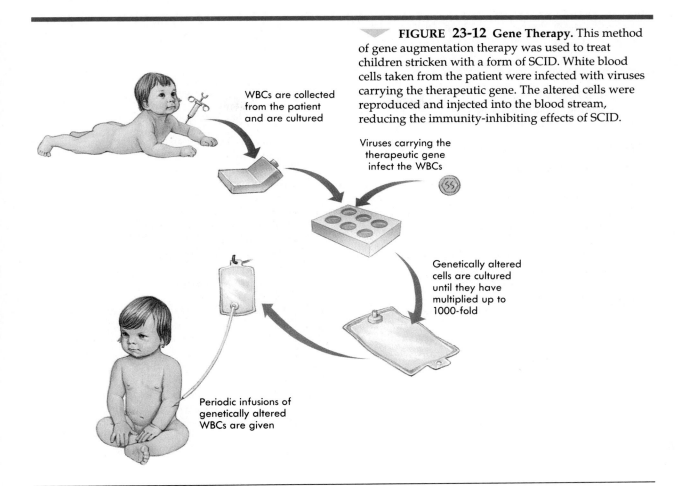

FIGURE 23-12 Gene Therapy. This method of gene augmentation therapy was used to treat children stricken with a form of SCID. White blood cells taken from the patient were infected with viruses carrying the therapeutic gene. The altered cells were reproduced and injected into the blood stream, reducing the immunity-inhibiting effects of SCID.

WBCs are collected from the patient and are cultured

Viruses carrying the therapeutic gene infect the WBCs

Genetically altered cells are cultured until they have multiplied up to 1000-fold

Periodic infusions of genetically altered WBCs are given

expected, a cure may result. Thus the goal of gene replacement therapy is to genetically alter existing body cells in the hope of eliminating the cause of a genetic disease.

In a therapy called *gene augmentation,* normal genes are introduced with the hope that they will augment (add to) the production of the needed protein. In one form of gene augmentation, virus-altered cells are injected into the blood or implanted under the skin of a patient to produce adequate amounts of the missing protein. Gene augmentation attempts to add genetically altered cells to the body, rather than change existing body cells as in gene replacement therapy.

The use of genetic therapy began in 1990 with a group of young children having **adenosine deaminase (ADA) deficiency.** In this rare recessive dis-

order, the gene for producing the enzyme ADA is missing from both autosomes in pair 20. Deficiency of ADA results in *severe combined immune deficiency (SCID),* making its victims highly susceptible to infection (see Chapter 14). As Figure 23-12 shows, white blood cells from each patient were collected and infected with viruses carrying therapeutic genes. After reproducing a 1000-fold, the genetically altered white blood cells were then injected into the patient. The first attempts at gene therapy in humans met with some success, but many technical problems are yet to be overcome before it can be used widely. It is too early to say for sure, but there may soon come a time when many genetic diseases are treated—or even cured—with gene therapy.

OUTLINE SUMMARY

GENETICS AND HUMAN DISEASE

A Genetics, begun by Mendel over 100 years ago, is the scientific study of inheritance

B Inherited traits can produce disease (see Chapter 4)

CHROMOSOMES AND GENES

A Mechanism of gene function

  1 Gene—independent genetic units (DNA segments) that carry the genetic code

  2 Genes dictate the production of enzymes and other molecules, which in turn dictate the structure and function of a cell

  3 Genes are active in the chromatin (strand) form and inactive when DNA is in the chromosome (compact) form (Figure 23-1)

B Distribution of chromosomes to offspring

  1 Meiotic cell division produces gametes with 23 chromosomes each (Figure 23-2)

  2 At conception, two gametes join and produce a zygote with 46 chromosomes—the complete human genome

  3 Twenty-two pairs of chromosomes are called *autosomes;* each member of a pair resembles its partner

  4 The remaining pair of chromosomes (pair 23) are called *sex chromosomes*

  5 Genetic variation among offspring is increased by:

    a Independent assortment of chromosomes during gamete formation (Figure 23-2)

    b Crossing-over of genes or linked groups of genes between chromosome partners during meiosis (Figure 23-3)

GENE EXPRESSION

A Hereditary traits have been rigorously studied

  1 Dominant genes have effects that appear in the offspring (dominant forms of a gene are often represented by uppercase letters)

  2 Recessive genes have effects that do not appear in the offspring when they are masked by a dominant gene (recessive forms of a gene are represented by lowercase letters)

B Sex-linked traits (Figures 23-5 and 23-6)

  1 The large X chromosome ("female chromosome") contains genes for female sexual characteristics and many other traits

  2 The small Y chromosome ("male chromosome") contains only genes for male sexual characteristics

  3 Normal males have XY as pair 23; normal females have XX as pair 23

  4 Nonsexual traits carried on sex chromosomes are sex-linked traits; most are X-linked traits

C Genetic mutations can result in abnormalities in the genetic code that cause disease

GENETIC DISEASES

A Mechanisms of genetic disease

  1 Single-gene diseases result from individual mutant genes (or groups of genes) that are passed from generation to generation

  2 Chromosomal diseases result from chromosome breakage or from nondisjunction (failure of a chromosome pair to separate during gamete formation) (Figure 23-7)

    a Trisomy—a chromosome triplet (instead of the usual pair)

    b Monosomy—a single chromosome (instead of a pair)

B Examples of single-gene diseases (Table 23-1)

  1 Cystic fibrosis—recessive autosomal condition characterized by excessive secretion of mucus and sweat, often causing obstruction of the gastrointestinal or respiratory tracts

  2 Phenylketonuria (PKU)—recessive autosomal condition characterized by excess phenylketone in urine, caused by accumulation of phenylalanine in tissues; may cause brain injury and death

C Examples of chromosomal diseases (Table 23-2)

  1 Down syndrome—usually caused by trisomy of chromosome 21; characterized by mental retardation and multiple structural defects (Figure 23-8)

  2 Klinefelter's syndrome—caused by the presence of two or more X chromosomes in a male (usually trisomy XXY); characterized by long legs, enlarged breasts, low intelligence, small testes, sterility, chronic pulmonary disease

  3 Turner's syndrome—caused by monosomy of the X chromosome (XO); characterized by immaturity of sex organs (resulting in sterility), short stature, webbed neck, cardiovascular defects, and learning disorders

PREVENTION AND TREATMENT OF GENETIC DISEASES

A Genetic counseling—professional consultations with families regarding genetic diseases

  1 Pedigree—chart illustrating genetic relationships over several generations (Figure 23-9)

  2 Punnett square—grid used to determine the probability of inheriting genetic traits (Figure 23-10)

  3 Karyotype—arrangement of chromosome photographs used to detect abnormalities

    a Amniocentesis—involves collection of fetal cells floating in the amniotic fluid (via a syringe through the uterine wall) (Figure 23-11)

b Chorionic villus sampling—involves collection of embryonic cells from outside of chorionic tissue (via tube through cervical opening)

B Treating genetic diseases

1 Most current treatments for genetic diseases are based on relieving or avoiding symptoms rather than attempting a cure

2 Gene therapy manipulates genes to cure genetic problems (Figure 23-12); most forms of gene therapy have not yet been proven effective in humans

1 Gene replacement therapy—abnormal genes in existing body cells are replaced by therapeutic genes

2 Gene augmentation therapy—cells carrying normal genes are introduced into the body to augment production of a needed protein

NEW WORDS

autosomes
carrier
codominance

dominant
genetics

genome
recessive

sex chromosomes
sex-linked trait

Diseases and Other Clinical Terms

adenosine deaminase
(ADA) deficiency
albinism
amniocentesis

chorionic villus
sampling
Down syndrome
gene therapy

karyotype
Klinefelter's syndrome
monosomy
pedigree

phenylketonuria (PKU)
Punnett square
trisomy
Turner's syndrome

CHAPTER TEST

1. The scientific study of inheritance is called _____.
2. The DNA molecule in its compact form is called the _____.
3. There are 22 pairs of _____, or matching chromosomes, in the human genome.
4. A _____ gene is one whose effects appear in an individual, masking the effects of a _____ gene for the same trait.
5. A _____ is a person who has a recessive gene that is not expressed.
6. Nonsexual traits governed by genes in sex chromosomes are called _____ traits.
7. Nondisjunction of chromosomes during meiosis may result in offspring with three matching chromosomes instead of the usual pair, a condition called _____.
8. _____ _____ is a group of dominant, inherited bone disorders characterized by brittleness of the skeleton.
9. A _____ is a chart that illustrates genetic relationships over several generations.
10. Amniocentesis can be used to collect cells so that their chromosomes can be photographed and arranged in an orderly pattern called a _____.

Circle the T before each true statement and the F before each false statement.

T F 11. Cancer can be caused by genes.
T F 12. Monosomy can result from nondisjunction during gamete formation.
T F 13. Chorionic villus sampling can be used to collect cells needed to produce a karyotype.
T F 14. In gene replacement therapy, normal genes are introduce in the hope that they will augment production of a needed protein.
T F 15. Genes determine biological traits by specifying which proteins a cell will synthesize.
T F 16. Red-green color blindness is an example of a dominant trait.
T F 17. Females are more likely to exhibit recessive X-linked traits that are males.
T F 18. Down syndrome is a dominant trait that is often passed from generation to generation.

Select the most correct answer from Column B for each statement in column A. (Only one answer is correct)

COLUMN A

19. _____ Abnormal accumulation of phenylalanine
20. _____ Characterized by accumulation of thick mucus
21. _____ Abnormal lipids in the brain cause death by age 4
22. _____ Produces tumors of tissue surrounding nerve fibers
23. _____ Lack of melanin pigment in skin and eyes
24. _____ Occurs in males with more than one X chromosome
25. _____ Usually results from trisomy 21

COLUMN B

a. Albinism
b. Cystic fibrosis
c. Down syndrome
d. Klinefelter's syndrome
e. Multiple neurofibromatosis
f. Phenylketonuria
g. Tay-Sachs disease

REVIEW QUESTIONS

1 How do genes cause disease?
2 What is the difference between an autosome and a sex chromosome?
3 What mechanisms cause genetic variation among offspring?
4 What is a dominant genetic trait? A recessive trait?
5 What is codominance? Why is sickle cell anemia considered to be an example of codominance?
6 What is X-linked inheritance? Using hemophilia as an example, explain why X-linked recessive traits affect men more often than women.
7 How to mutagens affect the genetic code?
8 How are single-gene conditions such as cystic fibrosis different from chromosomal diseases such as Down syndrome?
9 What is nondisjunction? How can it result in offspring that exhibit trisomy? How does it cause monosomy in offspring?

10 Briefly describe cystic fibrosis and PKU.
11 Briefly describe Down, Klinefelter's, and Turner's syndromes.
12 What is genetic counseling?
13 What is a pedigree? How is it used by genetic counselors?
14 What is a Punnett square? How is it used to predict the probability of inheriting specific disorders?
15 How is a karyotype prepared? What information can be revealed in a karyotype?
16 What is the basic strategy currently used to treat most genetic disorders?
17 What is gene replacement therapy? How has gene therapy been used to treat severe combined immune deficiency?

CLINICAL APPLICATIONS

1 Quentin's family physician suspects that Quentin may have Klinefelter's syndrome. Quentin's long limbs, small testes, and enlarged breasts seem to support the diagnosis. What test might the physician order to confirm the syndrome? What test results would be expected? What causes the genetic abnormality that produces Klinefelter's syndrome?

2 A young infant just born at Memorial Hospital has parents that both report a history of Tay-Sachs disease in their families. Assuming both parents have the Tay-Sachs gene, what is the probability that this infant will develop this deadly disease?

Chemistry of Life

APPENDIX

A

L ife is chemistry. It's not quite that simple, but the more we learn about human structure and function, the more we realize that it all boils down to interactions among chemicals. The digestion of food, the formation of bone tissue, and the contraction of a muscle are all chemical processes. Thus, the basic principles of anatomy and physiology are ultimately based on principles of chemistry. A whole field of science, **biochemistry,** is devoted to studying the chemical aspects of life. To truly understand the human body, it is important to understand a few basic facts about biochemistry, the chemistry of life.

Levels of Chemical Organization

Matter is anything that occupies space and has mass. Biochemists classify matter into several levels of organization for easier study. In the body, most chemicals are in the form of **molecules.** Molecules are particles of matter that are composed of one or more smaller units called **atoms.** Atoms in turn are composed of several kinds of *subatomic particles:* **protons, electrons,** and **neutrons.** (Refer to the Mini-Glossary on p. A-14).

ATOMS

Atoms are units that until recently could not be seen by scientists. New instruments, including *tunneling microscopes* and *atomic force microscopes,* produce pictures of atoms that confirm current models of how atoms are put together. At the core of each atom is a **nucleus** composed of positively charged protons and uncharged neutrons. The number of protons in the nucleus is an atom's **atomic number.** The number of protons and neutrons combined is the atom's **atomic mass.**

Negatively charged electrons surround the nucleus at a distance. In an electrically neutral atom, there is one electron for every proton. Electrons move about within certain limits called **orbitals.** Each orbital can hold two electrons. Orbitals are arranged into **energy levels** (shells), depending on their distance from the nucleus. The farther an orbital extends from the nucleus, the higher its energy level. The energy level closest to the nucleus has one orbital, so it can hold two electrons. The next energy level has up to four orbitals, so it can hold eight electrons. Figure 1 shows a carbon (C) atom. Notice that the first energy level (the innermost shell) contains two electrons and the outer energy level contains four electrons. The outer energy level of a carbon atom

FIGURE 1 A Model of the Atom. The nucleus—protons (+) and neutrons—is at the core. Electrons inhabit outer regions called *energy levels.* This is a carbon atom, a fact that is determined by the number of its protons. All carbon atoms (and only carbon atoms) have six protons. (One proton in the nucleus is not visible in this illustration.)

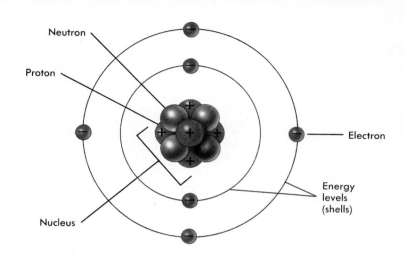

TABLE 1 Important Elements in the Human Body

	NAME	SYMBOL	NUMBER OF ELECTRONS IN OUTER SHELL*
Major elements (over 96% of body weight)	Oxygen	O	6
	Carbon	C	4
	Hydrogen	H	1
	Nitrogen	N	5
Trace elements (examples of more than 20 trace elements found in the body)	Calcium	Ca	2
	Phosphorus	P	5
	Sodium (Latin *natrium*)	Na	1
	Potassium (Latin *kalium*)	K	1
	Chlorine	Cl	7
	Iodine	I	7

*Maximum is eight, except for hydrogen. The maximum for that element is two.

could hold up to four more electrons (for a total of eight). The number of electrons in the outer energy level of an atom determines how it behaves chemically (that is, how it may unite with other atoms). This behavior, called *chemical bonding,* will be discussed later.

ELEMENTS, MOLECULES, AND COMPOUNDS

Substances can be classified as **elements** or **compounds.** Elements are pure substances, composed of only one of over a hundred types of atoms that exist in nature. Only four kinds of atoms (**oxygen,**

RADIOACTIVE ISOTOPES

Each element is unique because of the number of protons it has. In short, each element has its own *atomic number*. However, atoms of the same element can have different numbers of neutrons. Two atoms that have the same atomic number but different atomic masses are **isotopes** of the same element. An example is hydrogen. Hydrogen has three isotopes: ^1H (the most common isotope), ^2H, and ^3H. The figure shows that each different isotope has only one proton but different numbers of neutrons.

Some isotopes have unstable nuclei that radiate (give off) particles. Radiation particles include protons, neutrons, electrons, and altered versions of these normal subatomic particles. An isotope that emits radiation is called a **radioactive isotope.**

Radioactive isotopes of common elements are sometimes used to evaluate the function of body parts. Radioactive iodine (^{125}I) put into the body and taken up by the thyroid gland gives off radiation that can be easily measured. Thus the rate of thyroid activity can be determined. Images of internal organs can be formed by radiation scanners that plot out the location of injected or ingested radioactive isotopes. For example, radioactive technetium (^{99}Tc) is commonly used to image the liver and spleen. The radioactive isotopes ^{13}N, ^{15}O, and ^{11}C are often used to study the brain in a technique called the *PET scan.*

Radiation can damage cells. Exposure to high levels of radiation may cause cells to develop into cancer cells. Higher levels of radiation completely destroy tissues, causing *radiation sickness.* Low doses of radioactive substances are sometimes given to cancer patients to destroy cancer cells. The side effects of these treatments result from the unavoidable destruction of normal cells with the cancer cells.

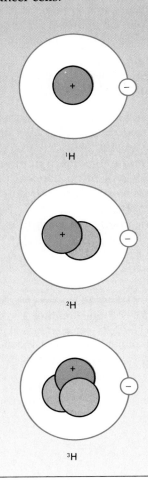

^1H

^2H

^3H

carbon, hydrogen,** and **nitrogen**) make up about 96% of the human body. There are traces of about 20 other elements in the body. Table 1 lists some of the elements in the body. Table 1 also gives for each element its universal chemical *symbol*—the abbreviation used by chemists worldwide.

Atoms usually unite with each other to form larger chemical units called **molecules.** Some molecules are made of several atoms of the same element. *Compounds* are substances whose molecules have more than one element in them. The *formula* for a compound contains symbols for the elements in each molecule. The number of atoms of each element in the molecule is expressed as a subscript after the elemental symbol. For example, each molecule of the compound **carbon dioxide** has one carbon (C) atom and two oxygen (O) atoms; thus its molecular formula is CO_2.

> *Matter is composed of molecules, which are composed of atoms. Atoms are composed of protons, electrons, and neutrons. Elements are pure substances, composed of only one kind of atom. Compounds are composed of molecules having more than kind of atom.*

Chemical Bonding

Chemical bonds form to make atoms more stable. An atom is said to be chemically stable when its outer energy level is "full" (that is, when its energy shells have the maximum number of electrons they can hold). All but a handful of atoms have room for more electrons in their outermost energy level. A basic chemical principle states that atoms react with one another in ways to make their outermost energy level full. To do this, atoms can share, donate, or borrow electrons.

For example, a hydrogen atom has one electron and one proton. Its single energy shell has one electron but *can* hold two—so it's not full. If two hydrogen atoms "share" their single electrons with each other, then both will have full energy shells, making them more stable *as a molecule* than either would be as an atom. This is one example of how atoms **bond** to form molecules. Other atoms may donate or borrow electrons until the outermost energy level is full.

IONIC BONDS

One common way in which atoms make their outermost energy level full is to form **ionic bonds** with other atoms. Such a bond forms between an atom that has only one or two electrons in the outermost level (that would normally hold eight) and an atom that needs only one or two electrons to fill its outer level. The atom with one or two electrons simply "donates" its outer shell electrons to the one that needs one or two.

For example, as you can see in Table 1, the sodium (Na) atom has one electron in its outer

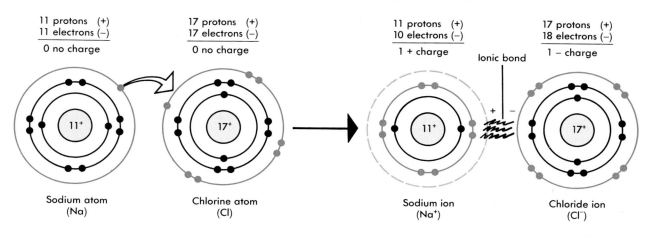

FIGURE 2 Ionic Bonding. The sodium atom donates the single electron in its outer energy level to a chlorine atom having seven electrons in its outer level. Now both have eight electrons in their outer shells. Because the electron/proton ratio changes, the sodium atom becomes a positive sodium ion. The chlorine atom becomes a negative chloride ion. The positive-negative attraction between these oppositely charged ions is called an *ionic bond*.

11 protons (+)	17 protons (+)	11 protons (+)	17 protons (+)
11 electrons (−)	17 electrons (−)	10 electrons (−)	18 electrons (−)
0 no charge	0 no charge	1 + charge	1 − charge

Ionic bond

| Sodium atom (Na) | Chlorine atom (Cl) | Sodium ion (Na^+) | Chloride ion (Cl^-) |

level and the chlorine (Cl) atom has seven. Both need to have eight electrons in their outer shell. Figure 2 shows how sodium and chlorine form an ionic bond when sodium "donates" the electron in its outer shell to chlorine. Now both atoms have full outer shells (although sodium's outer shell is now one energy level lower). Because the sodium atom lost an electron, it now has one more proton that it has electrons. This makes it a positive **ion,** an electrically charged atom. Chlorine has "borrowed" an electron to become a negative ion called the *chloride* ion. Because oppositely charged particles attract one another, the sodium and chloride ions are drawn together to form a sodium chloride (NaCl) molecule—common table salt. The molecule is held together by an *ionic bond.*

Ionic molecules usually dissolve easily in water because water molecules wedge between the ions and force them apart. When this happens, we say the molecules **dissociate** (dis-SO-see-ayt) to form free ions. Molecules that form ions when dissolved in water are called **electrolytes** (el-EK-tro-lites). Chapter 19 describes mechanisms that maintain the homeostasis of electrolytes in the body. Table 2 lists some of the more important ions present in body fluids.

The formula of an ion always shows its charge by a superscript after the chemical symbol. Thus the sodium ion is Na^+, and the chloride ion is

FIGURE 3 Covalent Bonding. Two hydrogen atoms move together, overlapping their energy levels. Although neither gains nor loses an electron, the atoms share the electrons, forming a covalent bond.

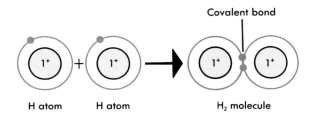

Cl^-. Calcium (Ca) atoms lose two electrons when they form ions, so the formula for the calcium ion is Ca^{++}.

COVALENT BONDS

Atoms may also fill their energy levels by *sharing* electrons rather than donating or receiving them. When atoms share electrons, a **covalent** (ko-VAY-lent) **bond** forms. For example, Figure 3 shows how two hydrogen atoms may move together closely so that their energy levels overlap. Each energy level contributes its one electron to the sharing relationship. This way, both outer levels have access to both electrons. Because atoms involved in a covalent bond must stay close to each other, it is not surprising that covalent bonds are not easily broken. Covalent bonds normally do not break apart in water.

Ionic bonding occurs when one atom donates an electron to another atom and the resulting ions attract each other. In covalent bonding, atoms share electrons.

Inorganic Chemistry

In living organisms, there are two kinds of compounds: **organic** and **inorganic.** Organic compounds are composed of molecules that contain carbon-carbon (C-C) covalent bonds or carbon-

TABLE 2 Important Ions in Human Body Fluids	
NAME	**SYMBOL**
Sodium	Na^+
Chloride	Cl^-
Potassium (Latin *kalium*)	K^+
Calcium	Ca^{++}
Hydrogen	H^+
Magnesium	Mg^{++}
Hydroxide	OH^-
Phosphate	$PO_4^{\equiv}$
Bicarbonate	HCO_3^-

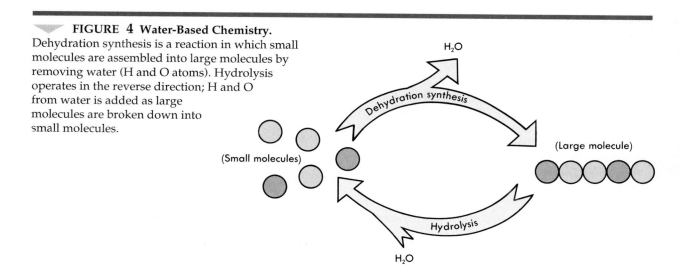

▼ **FIGURE 4 Water-Based Chemistry.**
Dehydration synthesis is a reaction in which small molecules are assembled into large molecules by removing water (H and O atoms). Hydrolysis operates in the reverse direction; H and O from water is added as large molecules are broken down into small molecules.

hydrogen (C-H) covalent bonds—or both kinds of bonds. Few inorganic compounds have carbon atoms in them and none have C-C or C-H bonds. Organic molecules are generally larger and more complex than inorganic molecules. The human body has both kinds of compounds because both are equally important to the chemistry of life. We will discuss the chemistry of inorganic compounds first, then move on to some of the important types of organic compounds.

WATER

One of the compounds that is most essential to life, water, is an inorganic compound. Water is the most abundant compound in the body, found in and around each cell. It is the **solvent** in which most other compounds or **solutes** are dissolved. When water is the solvent for a *mixture* (a blend of two or more kinds of molecules), the mixture is called an **aqueous solution.** An aqueous solution containing common salt (NaCl) and other molecules forms the "internal sea" of the body. Water molecules not only compose the basic internal environment of the body, but they also participate in many important *chemical reactions.* Chemical reactions are interactions among molecules in which atoms regroup into new combinations.

A common type of chemical reaction in the body is **dehydration synthesis.** In any kind of

synthesis reaction, the **reactants** combine to form a larger **product.** In dehydration synthesis, reactants combine only after hydrogen (H) and oxygen (O) atoms are removed. These leftover H and O atoms come together, forming H_2O or water. As Figure 4 shows, the result is both the large product molecule and a water molecule. Just as dehydration of a cell is a loss of water from the cell and dehydration of the body is loss of fluid from the entire internal environment, dehydration synthesis is a reaction in which water is lost from the reactants.

Another common reaction in the body, **hydrolysis** (hye-DROL-i-sis), also involves water. In this reaction, water *(hydro-)* disrupts the bonds in large molecules, causing them to be broken down into smaller molecules *(lysis).* Hydrolysis is virtually the reverse of dehydration synthesis, as Figure 4 shows.

Chemical reactions always involve energy transfers. Energy is required to build the molecules. Some of that energy is stored as potential energy in the chemical bonds. The stored energy can then be released when the chemical bonds in the molecule are later broken apart. For example, a molecule called **adenosine triphosphate (ATP)** breaks apart in the muscle cells to yield the energy needed for muscle contraction (see Figure 17-2).

Chemists often use a *chemical equation* to represent a chemical reaction. In a chemical equation, the reactants are separated from the products by an arrow ($\rightarrow$) showing the "direction" of the reaction. Reactants are separated from each other and products are separated from each other by addition signs (+). Thus the reaction *potassium and chloride combine to form potassium chloride* can be expressed as the equation:

$$K^+ + Cl^- \rightarrow KCl$$

The single arrow ($\rightarrow$) is used for equations that occur in only one direction. For example, when hydrochloric acid (HCl) is dissolved in water, *all* of it dissociates to form H^+ and Cl^-.

$$HCl \rightarrow H^+ + Cl^-$$

The double arrow ($\leftrightarrows$) is used for reactions that happen in "both directions" at the same time. When carbonic acid (H_2CO_3) dissolves in water, *some* of it dissociates into H^+ (hydrogen ion) and HCO_3^- (bicarbonate) but not all of it. As additional ions dissociate, previously dissociated ions bond together again forming H_2CO_3.

$$H_2CO_3 \leftrightarrows H^+ + HCO_3^-$$

In short, the double arrow indicates that at any instant in time both reactants and products are present in the solution at the same time.

ACIDS, BASES, AND SALTS

Besides water, many other inorganic compounds are important in the chemistry of life. For example, **acids** and **bases** are compounds that profoundly affect chemical reactions in the body. As explained in more detail at the beginning of Chapter 20, a few water molecules dissociate to form the H^+ ion and the OH^- (hydroxide) ion:

$$H_2O \leftrightarrows H^+ + OH^-$$

In pure water, the balance between these two ions is equal. However, when an acid such as hydrochloric acid (HCl) dissociates into H^+ and Cl^-, it shifts this balance in favor of excess H^+ ions. In the blood, carbon dioxide (CO_2) forms carbonic acid (H_2CO_3) when it dissolves in water. Some of the carbonic acid then dissociates to form H^+ ions and HCO_3^- (bicarbonate) ions, producing an excess of H^+ ions in the blood. Thus high CO_2 levels in the blood make the blood acidic.

Bases or **alkaline** compounds, on the other hand, shift the balance in the opposite direction. For example, sodium hydroxide (NaOH) is a base that forms OH^- ions but no H^+ ions. In short, acids are compounds that produce an excess of H^+ ions, and bases are compounds that produce an excess of OH^- ions (or a decrease in H^+).

The relative H^+ concentration is a measure of how acidic or basic a solution is. The H^+ concentration is usually expressed in units of **pH.** The formula used to calculate pH units gives a value of 7 to pure water. A higher pH value indicates a low relative concentration of H^+—a base. A lower pH value indicates a higher H^+ concentration—an acid. Figure 5 shows a scale of pH from 0 to 14. Notice that when the pH of a solution is less than 7, the scale "tips" toward the side marked "high H^+." When the pH is more than 7, the scale "tips" toward the side marked "low H^+." pH units increase or decrease by factors of 10. Thus a pH 5 solution has ten times the H^+ concentration of a pH 6 solution. A pH 4 solution has a hundred times the H^+ concentration of a pH 6 solution.

A *strong acid* is an acid that completely, or almost completely, dissociates to form H^+ ions. A *weak acid*, on the other hand, dissociates very little and therefore produces few excess H^+ in solution.

When a strong acid and a strong base mix, excess H^+ ions may combine with the excess OH^- ions to form water. That is, they may *neutralize* each other. The remaining ions usually form neutral ionic compounds called **salts.** For example:

$$\underset{\text{acid}}{HCl} + \underset{\text{base}}{NaOH} \rightarrow H^+ + Cl^- + Na^+ + OH^- \rightarrow \underset{\text{water}}{H_2O} + \underset{\text{salt}}{NaCl}$$

The pH of body fluids affects body chemistry so greatly that normal body function can be maintained only within a narrow range of pH. The body can remove excess H^+ by excreting them in the urine (see Chapter 18). Another way to remove acid is by increasing the loss of CO_2 (an acid) by the respiratory system (see Chapter 15). A third way to adjust the body's pH is the use of **buffers**—chemicals in the blood that maintain pH. Buffers maintain pH balance by preventing sudden changes in the H^+ ion concentration. Buffers do this by forming a chemical system that neutralizes acids and bases as they are added to a solution. The mechanisms by which the body maintains pH homeostasis, or acid-base balance, are discussed further in Chapter 20.

▼ **FIGURE 5 The pH Scale.** The H^+ concentration is balanced with the OH^- concentration at pH 7. At values above 7 (low H^+), the scale tips in the basic direction. At values below 7 (high H^+), the scale tips toward the acid side.

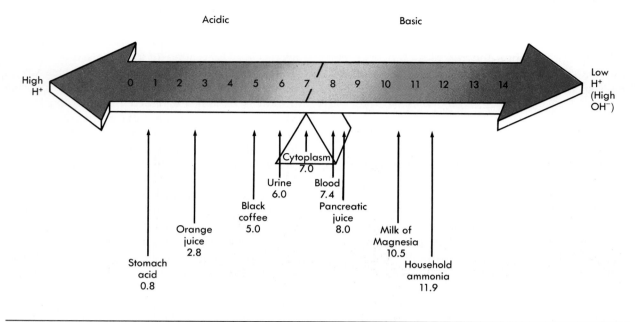

Organic compounds always contain carbon (forming carbon-carbon or carbon-hydrogen bonds); few inorganic compounds even contain carbon. Biological reactions take place in aqueous solutions. In hydrolysis, chemical bonds are broken. In dehydration synthesis, bonds are formed. The pH (relative H^+ concentration) of body fluids affects chemical reactions.

▼ Organic Chemistry

Organic compounds are much more complex than inorganic compounds. In this section, we will describe the basic structure and function of each major type of organic compound found in the body: **carbohydrates, lipids** (fats), **proteins,** and **nucleic acids.** Table 3 summarizes the structure and the function of each type. Refer to this table as you read through the descriptions that follow.

CARBOHYDRATES

The name *carbohydrate* literally means "carbon (C) and water (H_2O)," signifying the types of atoms that form carbohydrate molecules. The basic unit of carbohydrate molecules is called a *monosaccharide* (mah-no-SAK-ah-ride) (Figure 6). Glucose (dextrose) is an important monosaccharide in the body; cells use it as their primary source of energy (see Chapter 17). A molecule made of two saccharide units is a double sugar or *disaccharide*. The disaccharides sucrose (table sugar) and lactose (milk sugar) are important dietary carbohydrates. After they are eaten, the body digests them to form monosaccharides that can be used as cellular fuel. Many saccharide units joined together form *polysaccharides*. Examples of polysaccharides are **glycogen** (GLY-ko-jen) and *starch*. Each glycogen molecule is a chain of glucose molecules joined together. Liver cells and muscle cells form glycogen when there is an excess of glucose in the blood, thus putting them into "storage" for later use.

TABLE 3   Major Types of Organic Compounds

EXAMPLE	COMPONENTS	FUNCTIONS
Carbohydrate		
Monosaccharide (glucose, galactose, fructose)	Single monosaccharide unit	Used as source of energy; used to build other carbohydrates
Disaccharide (sucrose, lactose, maltose)	Two monosaccharide units	Can be broken into monosaccharides
Polysaccharide (glycogen, starch)	Many monosaccharide units	Used to store monosaccharides (thus to store energy)
Lipid		
Triglyceride	One glycerol, three fatty acids	Stores energy
Phospholipid	Phosphorus-containing unit, two fatty acids	Forms cell membranes
Cholesterol	Four carbon rings at core	Transports lipids; is basis of steroid hormones
Protein		
Structural proteins (fibers)	Amino acids	Form structures of the body
Functional proteins (enzymes, hormones)	Amino acids	Facilitate chemical reactions; send signals; regulate functions
Nucleic Acid		
Deoxyribonucleic acid (DNA)	Nucleotides (contain deoxyribose)	Contains information (genetic code) for making proteins
Ribonucleic acid (RNA)	Nucleotides (contain ribose)	Serves as a copy of a portion of the genetic code

Carbohydrates have potential energy stored in their bonds. When the bonds are broken in cells, the energy is released and then trapped by the cell's chemistry to do work. Chapter 17 explains more about the process by which the body extracts energy from carbohydrates and other food molecules.

LIPIDS

Lipids are fats and oils. Fats are lipids that are solid at room temperature, such as the fat in butter and lard. Oils, such as corn oil and olive oil, are liquid at room temperature. There are several important types of lipids in the body:

1 **Triglycerides** (try-GLIS-er-ides) are lipid molecules formed by a *glycerol* unit joined to

FIGURE 6 Carbohydrates. Monosaccharides are single carbohydrate units joined by dehydration synthesis to form disaccharides and polysaccharides.

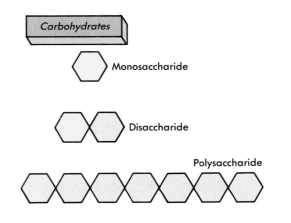

three *fatty acids* (Figure 7). Like carbohydrates, their bonds can be broken apart to yield energy (see Chapter 17). Thus triglycerides are useful in storing energy in cells for later use.

2 **Phospholipids** are similar to triglycerides but have phosphorus-containing units in them, as their name implies. The phosphorus-containing unit in each molecule forms a "head" that attracts water. Two fatty acid "tails" repel water. Figure 8, *A* shows the head and tail of the phospholipid molecule. This structure allows them to form a stable *bilayer* in water that forms the foundation for the cell membrane. In Figure 8, *B,* the water-attracting heads face the water and the water-repelling tails face away from the water (and toward each other).

3 **Cholesterol** is a *steroid* lipid that performs several important functions in the body. It combines with phospholipids in the cell membrane to help stabilize its bilayer structure. The body also uses cholesterol as a starting point in making steroid hormones such as estrogen, testosterone, and cortisone (see Chapter 10).

PROTEINS

Proteins are very large molecules composed of basic units called **amino acids.** In addition to car-

bon, hydrogen, and oxygen, amino acids contain nitrogen (N). By means of a process described fully in Chapter 2, a particular sequence of amino acids is strung together and held by **peptide bonds.** Positive-negative attractions between different atoms in the long amino acid strand cause it to coil on itself and maintain its shape. The complex, three-dimensional molecule that results is a protein molecule (Figure 9).

The shape of a protein molecule determines its role in body chemistry. **Structural proteins** are shaped in ways that allow them to form essential structures of the body. Collagen, a protein with a fiber shape, holds most of the body tissues together. Keratin, another structural protein, forms a network of water-proof fibers in the outer layer of the skin. **Functional proteins** participate in chemical processes of the body. Functional proteins include some of the hormones, growth factors,

BLOOD LIPOPROTEINS

A lipid such as cholesterol can travel in the blood only after it has attached to a protein molecule—forming a lipoprotein. Some of these molecules are called *high-density lipoproteins (HDLs)* because they have a high density of protein (more protein than lipid). Another type of molecule contains less protein (and more lipid), so it is called *low-density lipoprotein (LDL).*

The cholesterol in LDLs is often called "bad" cholesterol because high blood levels of LDL are associated with **atherosclerosis,** a life-threatening blockage of arteries. LDLs carry cholesterol *to cells,* including the cells that line blood vessels. HDLs, on the other hand, carry so-called "good" cholesterol *away from cells* and toward the liver for elimination from the body. A high proportion of HDL in the blood is associated with a low risk of developing atherosclerosis. Factors such as cigarette smoking decrease HDL levels and thus contribute to risk of atherosclerosis. Factors such as exercise increase HDL levels and thus decrease the risk of atherosclerosis.

▽ **FIGURE 7 Triglyceride.** Each triglyceride is composed of three fatty acid units attached to a glycerol unit.

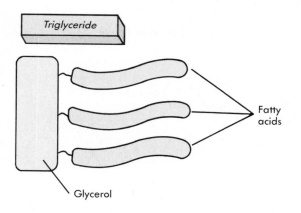

FIGURE 8 Phospholipids. A, Each phospholipid molecule has a phosphorus-containing "head" that attracts water and a lipid "tail" that repels water. **B,** Because the tails repel water, phospholipid molecule often arrange themselves so that their tails face away from water. The stable structure that results is a bilayer sheet forming a small bubble.

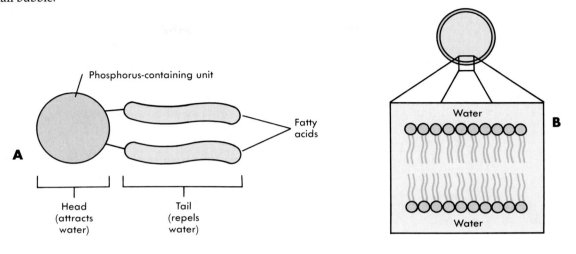

FIGURE 9 Protein. Protein molecules are large, complex molecules formed by a twisted and folded strand of amino acids. Each amino acid is connected to the next amino acid by covalent peptide bonds.

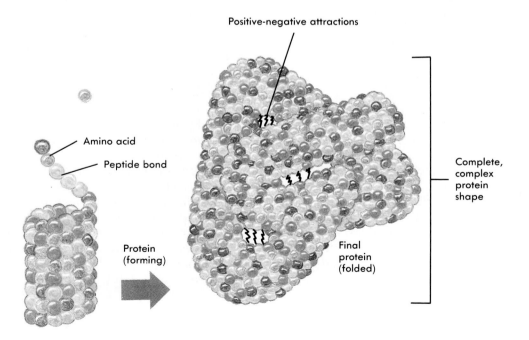

FIGURE 10 Enzyme Action. Enzymes are functional proteins whose molecular shape allows them to catalyze chemical reactions. Molecules *A* and *B* are brought together by the enzyme to form a larger molecule, *AB*.

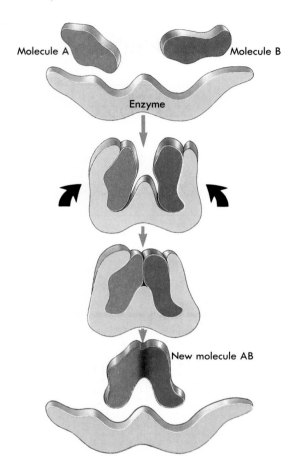

unless the specific enzymes needed for that reaction are present. Figure 10 illustrates how shape is important to the function of enzyme molecules. Each enzyme has a shape that "fits" the specific molecules it works on—much as a key fits specific locks. This explanation of enzyme action is sometimes called the **lock-and-key model.**

Proteins can bond with other organic compounds and form "mixed" molecules. For example, *glycoproteins* are proteins with sugars attached. *Lipoproteins* are lipid-protein combinations.

NUCLEIC ACIDS

The two forms of nucleic acid are **deoxyribonucleic acid (DNA)** and **ribonucleic acid (RNA).** As outlined in Chapter 2, the basic building blocks of nucleic acids are called **nucleotides.** Each nucleotide consists of a *phosphate unit*, a sugar (*ribose* or *deoxyribose*), and a *nitrogen base.* DNA nucleotide bases include **adenine, thymine, guanine,** and **cytosine.** RNA uses the same set of bases, except for the substitution of **uracil** for thymine. See Table 4.

Nucleotides bind to one another to form strands or other structures. In the DNA molecule, nucleotides are arranged in a twisted, double strand called a **double helix** (Figure 11).

The sequence of different nucleotides along the DNA double helix is the "master code" for assembling proteins and other nucleic acids. *Messenger RNA (mRNA)* molecules have a sequence that forms a temporary "working copy" of the DNA code. The code in nucleic acids ultimately directs the entire symphony of living chemistry.

Carbohydrates are composed of monosaccharide units and can be broken apart to yield energy. Lipids are fat molecules composed mainly of glycerol and fatty acids and have many functions. Proteins are complex strings of amino acids. Proteins comprise body structures or regulate body functions. Nucleic acids DNA and RNA are composed of nucleotides. The sequence of nucleotides serves as a code for assembling proteins.

cell membrane channels and receptors, and enzymes.

Enzymes are chemical catalysts. This means that they help a chemical reaction occur but are not reactants or products themselves. They participate in chemical reactions but are not changed by the reactions. Enzymes are vital to body chemistry. No reaction in the body occurs fast enough

FIGURE 11 DNA. Deoxyribonucleic acid (DNA), like all nucleic acids, is composed of units called *nucleotides*. Each nucleotide has a phosphate, a sugar, and a nitrogen base. In DNA, the nucleotides are arranged in a double helix formation.

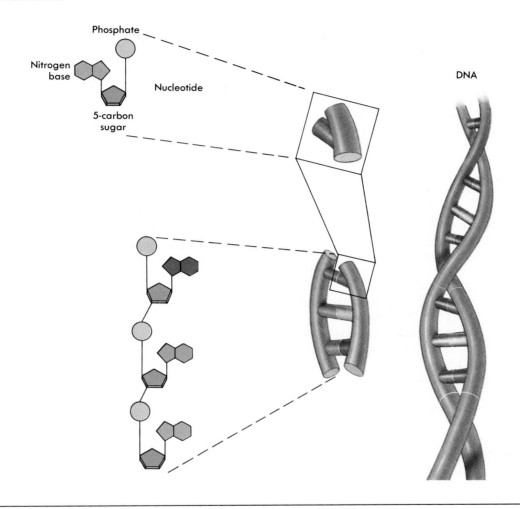

TABLE 4 Components of Nucleotides

NUCLEOTIDE	DNA	RNA
Sugar	Deoxyribose	Ribose
Phosphate	Phosphate	Phosphate
Nitrogen base	Cytosine	Cytosine
	Guanine	Guanine
	Adenine	Adenine
	Thymine	Uracil

Mini-Glossary

acid any substance that, when dissolved in water, contributes to an excess of H^+ ions

aqueous (AY-kwee-us) solution liquid mixture in which water is the solvent; for example, saltwater is an aqueous solution because water is the solvent

atom smallest particle of a pure substance (element) that still has the chemical properties of that substance; composed of protons, electrons, and neutrons (subatomic particles)

atomic mass combined total number of protons and neutrons in an atom

atomic number total number of protons in an atom's nucleus; atoms of each element have a characteristic atomic number

base alkaline; any substance that, when dissolved in water, contributes to an excess of OH^- ions

carbohydrate (KAR-bo-HYE-drate) organic molecule composed of one or more monosaccharides (containing C, H, and O in a 1:2:1 ratio)

compound (KOM-pound) substance whose molecules have more than one kind of element in them

covalent (ko-VAY-lent) bond chemical bond formed when atoms share electrons by overlapping their energy levels (electron shells)

dehydration (dee-hye-DRAY-shun) synthesis (SIN-the-sis) chemical reaction in which large molecules are formed by removing water from smaller molecules and joining them together

dissociation (dis-so-see-AY-shun) separation of ions as they dissolve in water

double helix (HE-lix) shape of DNA molecules; a double spiral

electrolyte (el-EK-tro-lite) compound whose molecules dissolve in water to form ions

electron (e-LEK-tron) negatively charged particle orbiting the nucleus of an atom

element (EL-e-ment) pure substance, composed of only one type of atom

energy level limited region surrounding the nucleus of a atom at a certain distance containing electrons; also called a shell

enzyme (EN-zime) functional protein that catalyzes chemical reactions (helps them occur more rapidly)

glycogen (GLYE-ko-jen) polysaccharide consisting of a chain of glucose (monosaccharide) molecules

hydrolysis (hye-DRO-li-sis) chemical reaction in which water is added to a large molecule causing it to break apart into smaller molecules

inorganic compounds compounds whose molecules generally do not contain carbon

ionic (eye-ON-ic) bond chemical bond formed by the positive-negative attraction between two ions

lipid organic molecule usually composed of glycerol and fatty acid units; types include triglycerides, phospholipids, and cholesterol

matter any substance that occupies space and has mass

molecule (MOL-e-kyool) particle of matter composed of one or more smaller units called atoms

neutron (NOO-tron) electrically neutral particle within the nucleus of an atom

nucleic (noo-KLEE-ic) acid complex organic molecule composed of units called nucleotides that each include a phosphate, a five carbon sugar, and a nitrogen base

nucleus (of an atom) central core of the atom; contains protons and neutrons (hydrogen atoms have a single proton and no neutrons)

organic (or-GAN-ic) compounds compounds whose large molecules contain carbon, which forms C-C bonds and/or C-H bonds

peptide bond covalent bond linking amino acids within a protein molecule

pH unit expressing relative H+ concentration (acidity); pH value higher than 7 is basic, pH value less than 7 is acidic, pH value equal to 7 is neutral

product any substance formed as a result of a chemical reaction

protein (PRO-teen) nitrogen-containing organic compound composed of a folded strand of amino acids

proton (PRO-ton) positively charged particle within the nucleus of an atom

reactant (ree-AK-tant) any substance entering (and being changed by) a chemical reaction

solute (SOL-yoot) substance that dissolves into another substance; for example, in saltwater the salt is the solute dissolved in water

solvent (SOL-vent) substance in which other substances are dissolved; for example, in saltwater the water is the solvent for salt

Examples of Pathological Conditions

APPENDIX

B

TABLE 1   Leading Health Problems*

CONDITION	CHAPTER REFERENCE
Diseases of the heart and blood vessels	(Chapters 12 and 13)
Cancer	(Chapter 4)
Accidents	—
Chronic obstructive pulmonary disease (COPD)	(Chapter 15)
Pneumonia and influenza	(Chapters 4 and 15)
Diabetes mellitus	(Chapter 10)
Cirrhosis of the liver	(Chapter 16)

*Principal causes of death in the United States ranked by number of deaths caused by each condition.

TABLE 2   Viral Conditions

DISEASE	VIRUS	DESCRIPTION
Acquired immune deficiency syndrome (AIDS)	Human immunodeficiency virus (HIV)	Identified in the west in 1981, it may have existed in Africa for many years. HIV is transmitted by direct contact with body fluids, perhaps within WBCs in blood or semen. AIDS is characterized by T-lymphocyte damage, resulting in immune dysfunction. Death results from secondary infections or tumors.
Acute T-cell lymphocytic leukemia (ATLL)	Human T-lymphotropic virus 1 (HTLV-1)	This form of cancer in adults can be caused by the *oncovirus* ("cancer virus") HTLV-1. This disease is one of many forms of leukemia and does not appear until at least 30 years after initial infection. HTLV-1 is transmitted in the same manner as HIV.
Chickenpox (varicella) and shingles (herpes zoster)	Varicella zoster virus (VZV)	Chickenpox is usually a childhood infection involving typical blisters and fever. Herpes zoster or shingles occurs later (in adulthood) in those who already had the varicella infection. Shingles often involves a rash along a single dermatome on one side of the body accompanied by severe pain.
Common cold and upper respiratory infections (URIs)	Rhinoviruses	This mild, contagious infection is characterized by nasal inflammation, weakness, cough, and low fever. Dozens of different rhinoviruses have been typed.
Fever blisters and herpes	Herpes simplex 1 and 2	This virus causes blisters on the hands or face (fever blisters) or genitals (genital herpes). The blisters may disappear temporarily but may reappear, especially as a result of stress.
Hepatitis (serum)	Hepatitis B virus	This acute-onset liver inflammation may develop into a severe chronic disease, perhaps ending in death.
Hepatitis (infectious)	Hepatitis A virus	Liver inflammation is characterized by slow onset and complete recovery. This virus is spread by direct contact or contaminated food or water.
Infectious mononucleosis	Epstein-Barr virus (EBV)	This acute infection is characterized by fever, sore throat, increased count and abnormal shape of lymphocytes, and liver, spleen, or lymph node swelling.
Influenza	Influenza A, B, C, etc.	This highly contagious respiratory infection is characterized by sore throat, fever, cough, muscle pain, and weakness. New strains of viruses A, B, and C appear at intervals—usually originating in the Orient.
Measles	*Morbillivirus*	This acute, contagious respiratory infection is characterized by fever, headache, and the measles rash.
Mumps	*Paramyxovirus*	An acute infection characterized by swollen parotid salivary glands, fever, and in adult males, swollen testes, mumps is most common in children but can occur at any age.
Poliomyelitis	Poliovirus 1, 2, and 3	This acute infection has several different forms (depending on extent of infection): asymptomatic, mild, nonparalyzing, and paralyzing. It is no longer common in the United States because of successful vaccination programs.

TABLE 2—cont'd   Viral Conditions

DISEASE	VIRUS	DESCRIPTION
Rabies	Rabies virus	This fatal infection of the central nervous system is usually transmitted through the bites of infected animals.
Rubella (German measles)	Rubella virus	This contagious infection is characterized by upper respiratory inflammation, swollen lymph nodes, joint pain, and measle-like rash. In pregnant women, it can spread to the fetus and cause congenital defects.
Viral encephalitis	(many)	*Viral encephalitis* is a general term for any brain inflammation caused by a virus. Brain damage may occur, perhaps causing death. Many different forms exist because many different viruses may infect the brain (for example, St. Louis encephalitis, California encephalitis, and equine encephalitis). Most encephalitis viruses are transmitted by mosquitoes.
Warts and genital warts	Human papillomaviruses (HPV)	Warts are nipple-like neoplasms of the skin. Forty-six HPV types have been identified. HPV types 6 and 11 cause genital warts, a common sexually transmitted disease (STD).

TABLE 3   Bacterial Conditions

DISEASE	ORGANISM	DESCRIPTION
Acute bacterial conjunctivitis	*Staphylococcus, Haemophilus, Proteus* and other organisms	This acute inflammation of the conjunctiva covering the eye is characterized by a discharge of mucous pus; it is highly contagious (compare with trachoma)
Anthrax	*Bacillus anthracis*	Usually transmitted from farm animals, this infection is characterized by a reddish-brown skin lesion but can also infect the respiratory tract. It can be fatal.
Botulism	*Clostridium botulinum* (bacillus)	This is a possibly fatal food poisoning resulting from ingestion of food contaminated with toxins produced by *C. botulinum*.
Brucellosis	*Brucella* species (bacilli)	Also called *undulant fever*, it is transmitted from farm animals and is characterized by chills, fever, weight loss, and weakness. Serious complications can occur if it is not treated.
Cholera	*Vibrio cholerae* (curved)	This acute intestinal infection is characterized by diarrhea, vomiting, cramps, dehydration, and electrolyte imbalance caused by bacterial toxins. It can be fatal if untreated. It spreads through contaminated food or water.
Dental caries	*Streptococcus mutans* (coccus) and other organisms	Tooth demineralization is caused by acids formed when nutrients on the tooth's surface are metabolized by bacteria. It can progress to a bacterial invasion of the tooth's pulp cavity and beyond.

Continued.

TABLE 3—cont'd   Bacterial Conditions

DISEASE	ORGANISM	DESCRIPTION
Diphtheria	*Corynebacterium diphtheriae* (bacillus)	Diptheria is an acute, contagious disease characterized by systemic poisoning by bacterial toxins and development of a "false membrane" lining of the throat that may obstruct breathing. Untreated, it may be fatal.
External otitis (swimmer's ear)	*Pseudomonas aeruginosa,* *Staphylococcus aureus,* *Streptococcus pyogenes,* etc.	Inflammation of the external ear canal is usually caused by bacteria but can also result from herpes infections, allergy, and other factors.
Gastroenteritis	(many)	*Gastroenteritis* is a general term for any inflammation of the gastrointestinal tract. Many different bacterial infections can cause this condition. (See salmonellosis).
Gonorrhea	*Neisseria gonorrhoeae* (coccus)	A common STD. It primarily infects the genital and urinary tracts but can affect the throat, conjunctiva, or lower intestines. It may progress to pelvic inflammatory disease (see later entry).
Legionnaires' disease	*Legionella pneumophila* (bacillus)	This is a type of pneumonia characterized by influenza-like symptoms followed by high fever, muscle pain, and headache—possible progressing to dry cough and pleurisy. It is spread by moist environmental sources (for example, air conditioning cooling units and soil) rather than person-to-person contact.
Lyme disease	*Borrelia burgdorferi* (spirochete)	Although the first cases were known only near Lyme (Conn), this tick-borne disease is now endemic over much of the United States. It usually first presents as a "bull's-eye" rash but later may cause chronic nerve, heart, and joint problems.
Lymphogranuloma venereum (LGV)	*Chlamydia trachomatis* (small)	This is a chronic STD characterized by genital ulcers, swollen lymph nodes, headache, fever, and muscle pain. *C. trachomatis* infection may cause a variety of other syndromes, including conjunctivitis, urogenital infections, and systemic infections. *C. trachomatis* infections constitute the most common STD in the United States.
Meningitis	*Streptococcus pneumoniae,* *Neisseria meningitidis,* *Haemophilus influenzae,* and other organisms	Meningitis is any inflammation of the meninges covering the brain and spinal cord. Several different bacteria can infect the meninges, as can several fungi. Can be mild but if severe, it can cause death.
Parrot fever (psittacosis)	*Chlamydia psittaci* (small)	Also called ornithosis, this pneumonia-like infection is transmitted by parrots and other birds. Characterized by cough, fever, loss of appetite, and severe headache.
Pelvic inflammatory disease (PID)	*Neisseria gonorrhoeae* (coccus), *Mycoplasma hominis* (small free-living), and other organisms	PID refers to any extensive inflammation of the female pelvic structures. Chronic inflammation associated with PID can cause tissue damage that leads to sterility.
Pertussis (whooping cough)	*Bordetella pertussis* (bacillus)	Pertussis is an acute, contagious infection of the respiratory tract characterized by coughs that end with "whooping" respirations.
Pneumonia	*Streptococcus pnueumoniae* (coccus) and other organisms	An acute lung infection that commonly follows the flu or other condition that prevents clearance of the lungs. It is characterized by blockage of the pulmonary airways.

TABLE 3—cont'd   Bacterial Conditions

DISEASE	ORGANISM	DESCRIPTION
Q fever	*Coxiella burnetti* (small)	Q (for "query") fever usually involves respiratory infection and is characterized by fever, headache, and muscle pain. Acute and chronic forms may develop after exposure to infected animals or animal products. A rickettsial disease.
Rheumatic fever	Group A beta-hemolytic Streptococci (cocci)	Inflammatory disease resulting from a delayed reaction to "strep" infection. May affect heart, brain, joints, or skin.
Rocky Mountain spotted fever (RMSF)	*Rickettsia rickettsii* (small)	Sometimes fatal, tick-borne disease characterized by fever, chills, headache, muscle pain, rash, constipation, and hemorrhagic lesions. May progress to shock and renal failure.
Salmonellosis	*Salmonella* species (bacilli)	This type of bacterial gastroenteritis is caused by ingestion of contaminated food.
Shigellosis (Shigella dysentery and bacillary dysentery)	*Shigella* species (bacilli)	This common disease is characterized by bloody, mucous diarrhea, cramps, fevere, and fatigue. It can cause dehydration, electrolyte imbalance, and acidosis if not treated. Antibiotic resistant strains of *Shigella* organisms make this condition a serious health threat—especially in areas with poor santitation.
Staphylococcal infections	*Staphylococcus* species (cocci)	This bacterial infection is characterized by abscesses. Includes staphylococcal scalded skin syndrome (SSSS), a skin disorder of infants.
Syphilis	*Treponema pallidum* (spirochete)	This sexually transmitted disease can affect any system. Primary syphilis is characterized by chancre sores on exposed areas of the skin. Untreated, secondary syphilis may appear 2 months after chancres disappear. The secondary stage occurs when the spirochete has spread through the body, presenting a variety of symptoms, and is still highly contagious—even through kissing. Tertiary syphilis may occur years later, possibly resulting in death.
Tetanus	*Clostridium tetani* (bacillus)	In this acute, sometimes fatal central nervous system infection, the bacteria usually enter a wound then produce a toxin that causes headache, fever, and painful muscle spasms.
Toxic shock syndrome (TSS)	*Staphylococcus aureus* strains (cocci)	This acute, severe toxic infection is associated with the use of highly absorbent tampons but can occur under a variety of circumstances. It begins as a high fever, headache, sore throat, etc., and may progress to renal failure, liver failure, and possibly death.
Trachoma (chlamydial conjunctivitis)	*Chlamydia trachomatis* (small)	This chronic infection of the conjunctiva covering the eye is characterized by painful inflammation, photophobia (light sensitivity), and excessive production of tears; if untreated, it will progress to form granular lesions that eventually affect the cornea and cause blindness
Typhoid fever	*Salmonella typhi* (bacillus)	Also called *enteric fever*, it is characterized by fever, headache, cough, diarrhea, and rash. It is transmitted through contaminated food or water.

▼ **TABLE 4** Mycotic (Fungal) Conditions

DISEASE	ORGANISM	DESCRIPTION
Aspergillosis	*Aspergillus* species (mold)	Uncommon, opportunistic mold infection by any of a number of different species has many different forms. It often affects the ear but can affect any organ, where it produces characteristic "fungus ball" lesions. If the infection becomes widespread, it can be fatal.
Blastomycosis	*Blastomyces dermatitidis* (mold*)	As is histoplasmosis, most cases of blastomycosis are asymptomatic. The most common symptomatic forms are skin ulcers and bone lesions, but the infection may spread to the lungs, kidneys, or nervous system.
Candidiasis	*Candida albicans* and other species (yeasts)	An opportunistic yeast infection characterized by a white discharge, peeling, and bleeding, candidiasis has several forms, depending on the severity and where it occurs: thrush (skin), diaper rash (skin), vaginitis, endocarditis, etc. It can be transmitted sexually, making it a sexually transmitted disease (STD).
Coccidioidomycosis (San Joaquin fever)	*Coccidioides immitis* (mold*)	Also called *desert fever*, it is endemic to dry regions of the United States southwest, Central and South America. It is characterized by cold- or influenza-like symptoms. A small number of cases develop into a more serious infection.
Histoplasmosis	*Histoplasma capsulatum* (mold*)	Histoplasmosis is a fungal infection most common in the midwestern United States, where it is spread through contaminated soil. In most cases, it is asymptomatic, but acute pneumonia may develop in a few cases.
Mycosis	(Many)	*Mycosis* is a general term used to describe any disease caused by fungi. *Mycoses* is the plural form.
Tinea	*Epidermophyton, Microsporum,* and *Trichophyton* species (molds)	Examples of opportunistic cutaneous mycoses include tinea pedis (athlete's foot), tinea cruris (jock itch), tinea corporis (body ringworm), tinea capitis (scalp ringworm), and tinea ungunum (nail fungus). All are characterized by inflammation accompanied by itching, scaling, and (occasionally) painful lesions.

*These molds are normally multicellular but transform to a unicellular phase when they infect humans.

▽ **TABLE 5**   Conditions Caused by Protozoa

DISEASE	ORGANISM	DESCRIPTION
Amebiasis and amebic dysentery	*Entamoeba histolytica, Entamoeba polecki*, and other organisms (ameba)	Usually acquired through contaminated food and water, this is an amebic infection of the intestine or liver. Mild cases are asymptomatic. More severe forms are characterized by diarrhea, abdominal pain, jaundice, and weight loss.
Balantidiasis	*Balantidium coli* (ciliate)	*B. coli* can be carried asymptomatically in the gastrointestinal tract. The disease is characterized by abdominal pain, nausea, and diarrhea. It may progress to intestinal ulceration and subsequent secondary infections.
Giardiasis (traveler's diarrhea)	*Giardia lamblia* (flagellate)	Intestinal infection is spread through contaminated food or through person-to-person contact. Symptoms range from mild diarrhea to malabsorption syndrome, with about half of all cases being asymptomatic.
Isosporiasis	*Isospora belli* (sporozoan)	Transmitted through contaminated food or oral-anal sexual contact, isosporiasis is an intestinal infection that may be asymptomatic. Symptomatic manifestations range from mild to severe, resembling giardiasis.
Malaria	*Plasmodium* species (sporozoa)	This serious disease is caused by blood-cell parasites that require two hosts: mosquitoes and humans (or other animals). Malaria is characterized by fever, anemia, swollen spleen, and possible relapse months or years later.
Toxoplasmosis	*Toxoplasma gondii* (sporozoan)	A common infection of blood and other tissue cells, it is often asymptomatic. It is transmitted through cat feces and undercooked meat. It is characterized by fever, lymphatic involvement, headache, fatigue, nervous disorders, and heart problems. If transmitted from mother to fetus, it can cause congenital defects that often lead to death.
Trichomoniasis	*Trichomonas vaginalis* (flagellate)	This urogenital infection is asymptomatic in most female patients and nearly all male patients. Vaginitis may occur, characterized by itching or burning and a foul-smelling discharge. It is usually spread through sexual contact.

▽   **TABLE 6**   Conditions Caused by Pathogenic Animals

DISEASE	ORGANISM	DESCRIPTION
Ascariasis (roundworm infestation)	*Ascaris lumbricoides* (nematode)	It is transmitted through contaminated food or contact with contaminated surfaces (such as hands). Eggs hatch in the small intestine, and the larvae travel to the lungs, where they cause coughing and fever. Intestinal and liver involvement may also be serious.
Bites and stings	Arachnida and Insecta	Symptoms of bites and stings usually result from mechanical injury and the release of toxins at the injury site. Some individuals may be hypersensitive to certain toxins and thus exhibit an allergic reaction, perhaps even anaphylaxis and death. Bites and stings may also transmit pathogens when the culprit is a vector.
Enterobiasis (pinworm infestation)	*Enterobius vermicularis* (nematode)	It is a common parasite infestation in which eggs can be transmitted by contaminated hands (a common cause of reinfection) or on inhaled dust particles. The infestation is localized in the large intestine. The adult female lays eggs around the outside of the anus, causing itching and possibly insomnia.
Fish tapeworm infestation	*Diphyllobothrium latum* (platyhelminth)	Spread by eating undercooked, contaminated fish, it is usually asymptomatic but can cause pernicious anemia if too much vitamin B_{12} is absorbed from the host.
Liver fluke infestation	*Fasciola hepatica, Opisthorchis sinensis,* and other organisms (platyhelminths)	Transmitted through watercress contaminated by infected snails, especially in sheep-raising regions, this infestation causes inflammation and swelling of the liver. The symptoms may progress to include hepatitis, bile duct obstruction, and secondary infections.
Pork and beef tapeworm infestation	*Taenia solium* (pork tapeworm) and *Taenia saginata* (beef tapeworm) (platyhelminths)	It is spread by eating undercooked, contaminated pork or beef. Adult tapeworms mature in the gastrointestinal tract, usually producing mild symptoms of diarrhea and weight loss. Larvae may spread to other tissue, sometimes causing serious infections.
Schistosomiasis (snail fever)	*Schistosoma mansoni, Schistosoma japonicum,* and *Schistosoma haematobium* (platyhelminths)	It is a parasitic condition transmitted in the form of skin-penetrating parasites released by freshwater snails in water contaminated by human feces. Characteristics of the disease depend on the organs involved and the species of fluke.
Trichinosis (threadworm infestation)	*Trichinella spiralis* (nematode)	It is an infestation characterized by diarrhea, nausea, and fever, possibly progressing to muscle pain and fatigue. In severe cases, the heart, lungs, and brain may become involved, sometimes resulting in death. The parasite is transmitted through undercooked pork, bear, and other meats.

▽ **TABLE 7** Conditions Caused by Physical Agents

CONDITION	PHYSICAL AGENT	DESCRIPTION
Bone fracture	Mechanical injury (for example, intense pressure, blow to the body, and abnormal turn while bearing weight)	Complete or incomplete break of hard bone tissue in one or more localized areas is often characterized by pain, swelling, and limited motion; compound fractures break the skin and may thus allow infection.
Burn	Chemical agents (for example, acids and bases) Intense heat Ionizing radiation (for example, x and gamma rays) Nonionizing radiation (for example, ultraviolet) Electricity	This is an injury to tissues caused by the factors listed in which the extent of the injury is proportional to exposure to the causative agent; it is characterized by tissue destruction, "burning" pain, and resulting inflammation response. Untreated or severe burns may become infected and may cause severe fluid loss.
Cancer	Mechanical injury Ionizing radiation (for example, x and gamma rays) Nonionizing radiation (for example, ultraviolet) Chemical agents (for example, irritants and carcinogens)	Malignant neoplasm (abnormal tissue growth) is characterized by invasion of surrounding tissue and metastasis (spread) to other parts of the body; it often progresses to death if not treated.
Chronic obstructive pulmonary disease (COPD)	Chemical pollutants (in air) Airborne particulates	This group of disorders is characterized by progressive, irreversible obstruction of air flow in the lungs; it includes bronchitis, emphysema, asthma. The incidence in the U.S. population has increased with exposure to air pollutants, including cigarette smoke.
Contusion	Mechanical injury (for example, blow to the body and intense pressure)	A contusion is a localized tissue lesion characterized by breakage of blood vessels and surrounding tissue cells without external bleeding; it is sometimes called a *bruise*.
Crush syndrome	Mechanical pressure (intense)	This severe, life-threatening condition is characterized by massive destruction of muscle and bone, hemorrhage, fluid loss, hypovolemic shock, hematuria (bloody urine), and kidney failure—often progressing to coma.
Diarrhea	Chemical agents (ingested) Ionizing radiation (for example, x and gamma rays)	Frequent passing of loose, watery feces (stools) results from increased peristalsis (motility) of the colon, in this case resulting from irritation by physical agents; the resulting fluid and electrolyte imbalance may cause dehydration or another life-threatening condition.
Headache	Mechanical injury (for example, blow to the head) Chemical pollutants (for example, inhaled organic compounds)	Pain in the head in this case results from injury by the agents listed.
Hearing impairment	High-volume (intensity) sound (for example, noise pollution)	Chronic exposure to loud noise causes hearing loss proportional to exposure—resulting from damage to the organ of Corti.

Continued.

TABLE 7—cont'd   Conditions Caused by Physical Agents

CONDITION	PHYSICAL AGENT	DESCRIPTION
Hypersensitivity reaction and physical allergy	Chemical substances in environment Light (as in photosensitivity) Temperature (as in cold or heat sensitivity)	Inappropriate, intense immune reaction to otherwise harmless physical agents is characterized by urticaria (hives), edema, and other allergy symptoms; specific antigens are usually associated with the reaction.
Laceration	Mechanical injury (sharp-edged object)	This is a mechanical injury in which tissue is cut or torn, often characterized by bleeding; if untreated, it may become infected.
Nausea	Chemical agents (ingested) Ionizing radiation (for example, x and gamma rays)	This is an unpleasant sensation of the gastrointestinal tract that commonly precedes the urge to vomit (that is, "upset stomach").
Pneumonia	Inhaled substances	This abnormal condition is characterized by acute inflammation of the lungs (in this case, triggered by irritation caused by inhaled substance) in which alveoli and bronchial passages become plugged with thick fluid (exudate).
Poisoning	Naturally-occurring toxins Synthetic toxins Drugs (for example, abuse, overdose, toxic interaction) Environmental pollutants (for example, air, water)	This condition results from exposure to a poison or toxin—a substance that impairs health or destroys life; effects may be local or systemic. Sometimes antidotes reverse toxicity, but sometimes the condition is irreversible. The toxin may be ingested, injected, inhaled, or absorbed through skin or may enter the body in some other way.
Radiation sickness	Ionizing radiation (for example, x and gamma rays)	Depending on the length, intensity, and location of exposure to radiation, this condition may be mild (headache, nausea, vomiting, anorexia, and diarrhea) to severe (sterility, fetal injury, cancer, alopecia, and cataracts); excessive radiation exposure may cause death.
Visual impairment	Mechanical injury (for example, blow to the head) Intense light (for example, direct sunlight and laser) Ionizing radiation (for example, x and gamma rays) Nonionizing radiation (for example, ultraviolet)	Blow to the head may cause detachment of retina; intense light or other radiation may damage retinal tissue. Radiation may also cloud lens or cornea, producing cataracts.
Windburn and abrasion burn	Abrasives (for example, windblown particles and rough surfaces)	This injury is similar to a heat or chemical burn but caused by mechanical abrasion of the skin or other tissues.

▽ **TABLE 8** Endocrine Conditions		
CONDITION	**MECHANISM**	**DESCRIPTION**
Acromegaly	Hypersecretion of growth hormone (GH) during adulthood	This is a chronic metabolic disorder characterized by gradual enlargement or elongation of facial bones and extremities.
Addison's disease	Hyposecretion of adrenal cortical hormones *(adrenal cortical insufficiency)*	Caused by tuberculosis, autoimmunity, or other factors, this life-threatening condition is characterized by weakness, anorexia, weight loss, nausea, irritability, decreased cold tolerance, dehydration, increased skin pigmentation, and emotional disturbance; it may lead to an acute phase (adrenal crisis) characterized by circulatory shock.
Aldosteronism	Hypersecretion of aldosterone	Often caused by adrenal hyperplasia, this condition is characterized by sodium retention and potassium loss—producing Conn's syndrome: severe muscle weakness, hypertension (high blood pressure), kidney dysfunction, and cardiac problems.
Cretinism	Hyposecretion of thyroid hormone during early development	This congenital condition is characterized by dwarfism, retarded mental development, facial puffiness, dry skin, umbilical hernia, and lack of muscle coordination.
Cushing's disease	Hypersecretion of adrenocorticotropic hormone (ACTH)	Caused by adenoma of the anterior pituitary, increased ACTH causes hypersecretion of adrenocortical hormones, producing *Cushing's syndrome.*
Cushing's syndrome	Hypersecretion (or injection) of glucocorticoids	This metabolic disorder is characterized by fat deposits on upper back, striated pad of fat on chest and abdomen, rounded "moon" face, muscular atrophy, edema, hypokalemia (low blood potassium), and possible abnormal skin pigmentation; it occurs in *Cushing's disease.*
Diabetes insipidus	Hyposecretion of (or insensitivity to) antidiuretic hormone (ADH)	This metabolic disorder is characterized by extreme polyuria (excessive urination) and polydipsia (excessive thirst) due to a decrease in the kidney's retention of water.
Gestational diabetes mellitus (GDM)	Temporary decrease in blood levels of insulin during pregnancy	This carbohydrate-metabolism disorder occurs in some pregnant women; it is characterized by polydipsia, polyuria, overeating, weight loss, fatigue, and irritability.
Gigantism	Hypersecretion of GH before age 25	This condition is characterized by extreme skeletal size caused by excess protein anabolism during skeletal development.
Graves' disease	Hypersecretion of thyroid hormone	This inherited, possibly autoimmune disease is characterized by hyperthyroidism
Hashimoto's disease	Autoimmune damage to thyroid causing hyposecretion of thyroid hormone	Enlargement of thyroid (goiter) is sometimes accompanied by hypothyroidism, typically occurring between ages 30 and 50; it is 20 times more common in females than males.

Continued.

TABLE 8—cont'd   Endocrine Conditions

CONDITION	MECHANISM	DESCRIPTION
Hyperparathyroidism	Hypersecretion of parathyroid hormone (PTH)	This condition is characterized by increased reabsorption of calcium from bone tissue and kidneys and increased absorption by the gastrointestinal tract; it produces hypercalcemia, resulting in confusion, anorexia, abdominal pain, muscle pain, and fatigue, possibly progressing to circulatory shock, kidney failure, and death.
Hypothyroidism (adult)	Hyposecretion of thyroid hormone	This condition, characterized by sluggishness, weight gain, skin dryness, constipation, arthritis and general slowing of body function, may lead to myxedema, coma, or death of untreated.
Hyperthyroidism (adult)	Hypersecretion of thyroid hormone	This condition, characterized by nervousness, exophthalmos (protruding eyes), tremor, weight loss, excessive hunger, fatigue, heat intolerance, heart arrhythmia, and diarrhea, is caused by a general acceleration of body function.
Insulin shock	Hypersecretion (or overdose injection) of insulin, decreased food intake, and excessive exercise	Hypoglycemic (low blood glucose) shock is characterized by nervousness, sweating and chills, irritability, hunger, and pallor—progressing to convulsion, coma, and death if untreated.
Insulin-dependent diabetes mellitus (IDDM) (type I diabetes mellitus)	Hyposecretion of insulin	This inherited condition with sudden childhood onset is characterized by polydipsia, polyuria, overeating, weight loss, fatigue, and irritability resulting from the inability of cells to secure and metabolize carbohydrates.
Myxedema	Extreme hyposecretion of thyroid hormone during adulthood	This is a severe form of adult hypothyroidism characterized by edema of the face and extremities, often progressing to coma and death.
Non-insulin-dependent diabetes mellitus (NIDDM) (type II diabetes mellitus)	Insensitivity of target cells to insulin	This carbohydrate-metabolism disorder with slow adulthood onset is thought to be caused by a combination of genetic and environmental factors and characterized by polydipsia, polyuria, overeating, weight loss, fatigue, and irritability.
Osteoporosis	Hyposecretion of estrogen in postmenopausal women	This bone disorder is characterized by loss of minerals and collagen from bone matrix, producing holes or porosities that weaken the skeleton.
Pituitary dwarfism	Hyposecretion of GH before age 25	This condition is characterized by reduced skeletal size caused by decreased protein anabolism during skeletal development.
Simple goiter	Lack of iodine in diet	Enlargement of thyroid tissue results from the inability of the thyroid to make thyroid hormone because of a lack of iodine; a positive-feedback situation develops in which low thyroid hormone levels trigger hypersecretion of thyroid-stimulating hormone (TSH) by pituitary—which stimulates thyroid growth.
Sterility	Hyposecretion of sex hormones	This is a loss of reproductive function.
Winter (seasonal) depression	Hypersecretion of (or hypersensitivity to) melatonin	This abnormal emotional state is characterized by sadness and melancholy resulting from exaggerated melatonin effects; melatonin levels are inhibited by sunlight so they increase when day length decreases during winter.

▼ **TABLE 9** Autoimmune Diseases

DISEASE	POSSIBLE SELF-ANTIGEN	DESCRIPTION
Addison's disease	Surface antigens on adrenal cells	Hyposecretion of adrenal hormones results in weakness, reduced blood sugar, nausea, loss of appetite, and weight loss.
Cardiomyopathy	Cardiac muscle	Disease of cardiac muscle (that is, the myocardium) results in a loss of pumping efficiency (heart failure).
Diabetes mellitus (insulin-dependent)	Pancreatic islet cells, insulin, and insulin receptors	Hyposecretion of insulin by the pancreas results in extremely elevated blood glucose levels (in turn causing a host of metabolic problems, even death if untreated).
Glomerulonephritis	Blood antigens that form immune complexes that are deposited in kidney	Disease of the filtration apparatus of the kidney (renal corpuscle) results in fluid and electrolyte imbalance and possibly total kidney failure and death.
Hemolytic anemia	Surface antigens on red blood cells (RBCs)	Condition of low RBC count in the blood results from excessive destruction of mature RBCs (hemolysis).
Graves' disease (type of hyperthyroidism)	Thyroid-stimulating hormone (TSH) receptors on thyroid cells	Hypersecretion of thyroid hormone results in increase in metabolic rate.
Multiple sclerosis	Antigens in myelin sheaths of nervous tissue	Progressive degeneration of myelin sheaths results in widespread impairment of nerve function (especially muscle control).
Myasthenia gravis	Antigens at neuromuscular junction	Muscle disorder is characterized by progressive weakness and chronic fatigue.
Myxedema	Antigens in thyroid cells	Hyposecretion of thyroid hormone in adulthood causes decreased metabolic rate; it is characterized by reduced mental and physical vigor, weight gain, hair loss, and edema.
Pernicious anemia	Antigens on parietal cells and intrinsic factor	Abnormally low RBC count results from the inability to absorb vitamin B_{12}, a substance critical to RBC production.
Reproductive infertility	Antigens on sperm or tissue surround ovum (egg)	This is an inability to produce offspring (in this case, resulting from destruction of gametes).
Rheumatic fever	Cardiac cell membranes (cross-reaction with Group A streptococcal antigen)	This causes rheumatic heart disease and inflammatory cardiac damage (especially to the endocardium or valves).
Rheumatoid arthritis	Collagen	Inflammatory joint disease is characterized by synovial inflammation that spreads to other fibrous tissues.
Systemic lupus erythematosus	Numerous	Chronic inflammatory disease has widespread effects and is characterized by arthritis, a red rash on the face, and other signs.
Ulcerative colitis	Mucous cells of colon	Chronic inflammatory disease of the colon is characterized by watery diarrhea containing blood, mucus, and pus.

TABLE 10 Deficiency Diseases

CONDITION	DEFICIENT SUBSTANCE*	DESCRIPTION
Avitaminosis K	Vitamin K	This occurs almost exclusively in children and is characterized by an impaired blood-clotting ability.
Beriberi	Vitamin B_1 (thiamine)	Peripheral nerve condition is characterized by diarrhea, fatigue, anorexia, edema, heart failure, and limb paralysis leading to muscle atrophy.
Folate-deficiency anemia	Vitamin B_9 (folic acid)	Blood disorder is characterized by a decrease in red blood cell (RBC) count.
Iron-deficiency anemia	Iron (Fe)	Blood disorder is characterized by a decrease in size and pigmentation of RBC that causes fatigue and pallor.
Kwashiorkor	Protein and calories	This form of protein-calorie malnutrition is characterized by wasting of muscle and subcutaneous tissue, dehydration, lethargy, edema and ascites, and retarded growth; it is caused by deficiency of proteins in the presence of adequate caloric intake (see marasmus).
Marasmus	Protein and calories	This form of protein-calorie malnutrition is characterized by progressive wasting of muscle and subcutaneous tissue accompanied by fluid and electrolyte imbalances; it is caused by deficiency of both protein and calories (see kwashiorkor).
Night blindness (nyctalopia)	Vitamin A	Relative inability to see in dim light results from failure to produce sufficient photopigment in the rods of the retina.
Osteomalacia	Vitamin D, calcium (Ca), and/or phosphorus (P)	Adult form of rickets is characterized by reduced mineralization of bone tissue accompanied by weakness, pain, anorexia, and weight loss.

▽   **TABLE 10**—cont'd   Deficiency Diseases

CONDITION	DEFICIENT SUBSTANCE*	DESCRIPTION
Pellagra	Vitamin B_3 (niacin) or tryptophan (an amino acid)	Disease is characterized by sun-sensitive scaly dermatitis, inflammation of mucosa, diarrhea, confusion, and depression.
Pernicious anemia	Vitamin B_{12}	Blood disorder is characterized by a reduced number of RBCs, causing weakness, pallor, tingling of the extremities, and anorexia.
Protein-calorie malnutrition (PCM)	Protein and calories	Abnormal condition resulting from dietary deficiency of calories in general and protein in particular; its forms include kwashiorkor and marasmus.
Rickets	Vitamin D, calcium (Ca), and/or phosphorus (P)	Juvenile form of osteomalacia is characterized by weakness and abnormal skeletal formation resulting from reduced mineralization of bone tissue.
Scurvy	Vitamin C	Reduced manufacture and maintenance of collagen and other functions results in weakness, anemia, edema, weakness of gingiva and loosening of teeth, and hemorrhaging (especially in skin and mucous membranes).
Simple goiter	Iodine (I)	Enlargement of thyroid tissue results from inability of thyroid to make thyroid hormone because of lack of iodine; positive-feedback situation develops: low thyroid hormone levels trigger hypersecretion of thyroid-stimulating hormone (TSH) by pituitary, which stimulates thyroid growth
Zinc deficiency	Zinc (Zn)	Condition is characterized by fatigue, decreased alertness, retarded growth, decreased smell and taste sensitivity, impaired healing, immunity.

*Deficiency may be caused by dietary deficiency or an inability to absorb or chemically process the listed substances.

▼ **TABLE 11**   Genetic Conditions

CHROMOSOME LOCATION	DISEASE	DESCRIPTION
Single-Gene Inheritance (Nuclear DNA)		
Dominant		
7, 17	Osteogenesis imperfecta	Group of connective tissue disorders is characterized by imperfect skeletal development that produces brittle bones.
17	Multiple neurofibromatosis	Disorder is characterized by multiple, sometimes disfiguring benign tumors of the Schwann cells (neuroglia) that surround nerve fibers.
5	Hypercholesterolemia	High blood cholesterol may lead to atherosclerosis and other cardiovascular problems.
4	Huntington's disease (HD)	Degenerative brain disorder is characterized by chorea (purposeless movements) progressing to severe dementia and death by age 55.
Co-dominant		
11	Sickle cell anemia Sickle cell trait	Blood disorder in which abnormal hemoglobin is produced causes red blood cells (RBCs) to deform into a sickle shape; sickle cell anemia is the severe form, and sickle cell trait the milder form.
11, 16	Thalassemia	Group of inherited hemoglobin disorders is characterized by production of hypochromic, abnormal RBCs.
Recessive (Autosomal)		
7	Cystic fibrosis (CF)	Condition is characterized by excessive secretion of thick mucus and concentrated sweat, often causing obstruction of the gastrointestinal or respiratory tracts.
15	Tay-Sachs disease	Fatal condition in which abnormal lipids accumulate in the brain and cause tissue damage leads to death by age 4.
12	Phenylketonuria (PKU)	Excess of phenylketone in the urine is caused by accumulation of phenylalanine in the tissues; it may cause brain injury and death if phenylalanine (amino acid) intake is not managed properly.
11	Albinism (total)	Lack of the dark brown pigment *melanin* in the skin and eyes results in vision problems and susceptibility to sunburn and skin cancer.
20	Severe combined immune deficiency (SCID)	Failure of the lymphocytes to develop properly causes failure of the immune system's defense of the body; it is usually caused by adenosine deaminase (ADA) deficiency.
Recessive (X-Linked)		
23 (X)	Hemophilia	Group of blood clotting disorders is caused by a failure to form clotting factors VIII, IX, or XI.

▽   **TABLE 11**—cont'd   Genetic Conditions

CHROMOSOME LOCATION	DISEASE	DESCRIPTION
Recessive (X-linked)—cont'd		
23 (X)	Duchenne muscular dystrophy (DMD)	Muscle disorder is characterized by progressive atrophy of skeletal muscle without nerve involvement.
23 (X)	Red-green color blindness	Inability to distinguish red and green light results from a deficiency of photopigments in the cone cells of the retina.
23 (X)	Fragile X syndrome	Mental retardation results from breakage of X chromosome in males.
23 (X)	Ocular albinism	Form of albinism in which the pigmented layers of the eyeball lack melanin results in hypersensitivity to light and other problems.
23 (X)	Androgen insensitivity	Inherited insensitivity to androgens (steroid sex hormones associated with maleness) results in reduced effects of these hormones.
23 (X)	Cleft palate (X-linked form)	One form of a congenital deformity in which the skull fails to develop properly, it is characterized by a gap in the palate (plate separating mouth from nasal cavity).
23 (X)	Retinitis pigmentosa	Condition causes blindness, characterized by clumps of melanin in retina of eyes.
Single-Gene Inheritance (Mitochondrial DNA)		
mDNA	Leber's hereditary optic neuropathy	Optic nerve degeneration in young adults results in total blindness by age 30.
mDNA	Parkinson's disease (?)	Nervous disorder is characterized by involuntary trembling and muscle rigidity.
Chromosomal Abnormalities		
Trisomy		
21	Down syndrome	Condition is characterized by mental retardation and multiple structural defects.
23	Klinefelter's syndrome	Condition is caused by the presence of two or more X chromosomes in a male (XXY); it is characterized by long legs, enlarged breasts, low intelligence, small testes, sterility, and chronic pulmonary disease.
Monosomy		
23	Turner's syndrome	Condition is caused by monosomy of the X chromosome (XO); it is characterized by immaturity of sex organs (causing sterility), webbed neck, cardiovascular defects, and learning disorders.

Medical Terminology

*I*f you are unfamiliar with it, medical and scientific terminology can seem overwhelming. The length and apparent complexity of many medical terms often scares people who have not had any training or practice in scientific terminology. Although it requires knowledge of some basic word parts and a few rules for using them, medical terminology is not as difficult as it seems. This appendix provides what you need to get you started. First, there are a handful of hints to help you learn and use medical terms. Second, there are several tables containing many of the most commonly used word parts and examples of how they are used. This appendix does not attempt to teach you the entire field of medical terminology—but with the information given here, and a little practice, you will soon become comfortable with the basics.

Hints for Learning and Using Medical Terms

1 Many medical terms are derived from the Latin and Greek languages. This is so because many of the anatomists, physiologists, and physicians who discovered the basic principles of modern life science used these languages themselves—so they could communicate with each other without having to learn dozens of native languages. Thus Latin and Greek have become the "universal" language of scientific terminology. Not only are many words derived from these classical languages, but also some of their rules of usage. The more useful of those rules are given later in this section.

2 One set of rules for using Latin and Greek is essential to understanding medical terminology. Both of these languages rely on the ability to combine word parts to make new words. Thus almost all medical terms are constructed by combining smaller word elements to make a meaningful term. Because of this combining technique, many medical terms appear at first glance to be long and complex. However, if you read a new term as a *series of word elements* rather than a single word, you will find it less imposing. One of the easiest ways to learn medical terminology is to develop the ability to *instantly analyze* new terms to discover the word parts that compose them. There are different kinds of word parts, depending on exactly how they fit with other word parts to form a complete term.

A **prefix** is a word part that is added to the beginning of an existing word to alter its meaning. We use prefixes in English, as well: the meaning of *sense* changes when we add the prefix *non-* to make the word *nonsense*.

A **suffix** is a word part that is added to the end of an existing word to alter its meaning. Once again, suffixes are also sometimes used in English. For example, the meaning of *sense* changes when we add the suffix *-less* to make the word *senseless*. A complex term can have a series of suffixes, a series of prefixes, or both. For example, the word *senselessness* has two suffixes: *-less* and *-ness*.

A **root** is a word part that serves as the starting point for forming a new term. In the previous examples from English, the word *sense* was the root to which was added a prefix or a suffix. Word parts commonly used as roots can also be used as suffixes or prefixes in forming a new term. Also, several roots are sometimes combined to form a larger root to which suffixes or prefixes can be added.

Combining vowels are vowels (*a, e, i, o, u, y*) that are used to link word parts—often to make pronunciation easier. For example, to link the suffix *-tion* to the root *sense*, we must use the combining vowel *-a-* to form the new term *sensation*. Using the *-e* that is already there would make the term difficult to pronounce. A root and a combining vowel together, such as *sensa-*, is often called the **combining form** of the word part.

3 Another set of rules for using Latin and Greek terms that you will find useful pertain to pluralization. To form a plural in English, we often simply add *-s* or *-es* to a word. For example, the plural for *sense* is *senses*. Because we have adopted medical terms into English to form sentences, we often simply use the pluralization rules of English and add the *-s* or *-es*. Often, however, you will run across a term that has been plural-

ized according to Latin or Greek rules. This brief list will help you distinguish between many plural and singular forms:

SINGULAR	PLURAL	EXAMPLE
-a	*-ae*	*Ampulla, ampullae*
-ax	*-aces*	*Thorax, thoraces*
-en	*-ena*	*Lumen, lumena*
-en	*-ina*	*Foramen, foramina*
-ex	*-ices*	*Cortex, cortices*
-is	*-es*	*Neurosis, neuroses*
-ix	*-ices*	*Appendix, appendices*
-on	*-a*	*Mitochondrion, mitochondria*
-um	*-a*	*Datum, data*
-ur	*-ora*	*Femur, femora*
-us	*-i*	*Villus, villi*
-yx	*-yces*	*Calyx, calyces*
-ma	*-mata*	*Lymphoma, lymphomata*

4 Correct spelling of medical terms is essential to their meanings. This is especially true of terms that are very close in spelling but very different in meaning. For example, the *perineum* is the region of the trunk around the genitals and anus, whereas the *peritoneum* is a membrane that lines the abdominal cavity and covers abdominal organs. Even a mistake in one letter can change the meaning of a word, as in the case of *ilium* (part of the pelvis) and *ileum* (part of the small intestine).

5 Because many medical terms are spoken, correct pronunciation is as important as correct spelling. Medical terms can usually be pronounced phonetically—by sounding out each letter sound of each syllable. It is best to check the pronunciation keys given in each chapter and in the glossary.

6 As you know, practice makes perfect. Practice using the medical terms in this or another book until you become comfortable with medical terminology. It won't take long—and you'll probably have fun doing it.

▼ TABLE 1   Word Parts Commonly Used as Prefixes

WORD PART	MEANING	EXAMPLE	MEANING OF EXAMPLE
a-	Without, not	Apnea	Cessation of breathing
af-	Toward	Afferent	Carrying toward
an-	Without, not	Anuria	Absence of urination
ante-	Before	Antenatal	Before birth
anti-	Against; resisting	Antibody	Unit that resists foreign substances
auto-	Self	Autoimmunity	Self-immunity
bi-	Two; double	Bicuspid	Two-pointed
circum-	Around	Circumcision	Cutting around
co-, con-	With; together	Congenital	Born with
contra-	Against	Contraceptive	Against conception
de-	Down from, undoing	Defibrillation	Stop fibrillation
dia-	Across, through	Diarrhea	Flow through (intestines)
dipl-	Twofold, double	Diploid	Two sets of chromosomes
dys-	Bad; disordered; difficult	Dysplasia	Disordered growth
ectop-	Displaced	Ectopic pregnancy	Displaced pregnancy
ef-	Away from	Efferent	Carrying away from
em-, en-	In, into	Encyst	Enclose in a cyst
endo-	Within	Endocarditis	Inflammation of heart lining
epi-	Upon	Epimysium	Covering of a muscle
ex-, exo-	Out of, out from	Exophthalmos	Protruding eyes
extra-	Outside of	Extraperitoneal	Outside the peritoneum
eu-	Good	Eupnea	Good (normal) breathing
hapl-	Single	Haploid	Single set of chromosomes
hem-, hemat-	Blood	Hematuria	Bloody urine
hemi-	Half	Hemiplegia	Paralysis in half the body
hom(e)o-	Same; equal	Homeostasis	Standing the same
hyper-	Over; above	Hyperplasia	Excessive growth
hypo-	Under; below	Hypodermic	Below the skin
infra-	Below, beneath	Infraorbital	Below the (eye) orbit
inter-	Between	Intervertebral	Between vertebrae
intra-	Within	Intracranial	Within the skull
iso-	Same, equal	Isometric	Same length
macro-	Large	Macrophage	Large eater (phagocyte)
mega-	Large; million(th)	Megakaryocyte	Cell with large nucleus

Continued.

TABLE 1—cont'd   Word Parts Commonly Used as Prefixes

WORD PART	MEANING	EXAMPLE	MEANING OF EXAMPLE
mes-	Middle	Mesentery	Middle of intestine
meta-	Beyond, after	Metatarsal	Beyond the tarsals (ankle bones)
micro-	Small; millionth	Microcytic	Small-celled
milli-	Thousandth	Milliliter	Thousandth of a liter
mono-	One (single)	Monosomy	Single chromosome
neo-	New	Neoplasm	New matter
non-	Not	Nondisjunction	Not disjoined
oligo-	Few, scanty	Oliguria	Scanty urination
ortho-	Straight; correct, normal	Orthopnea	Normal breathing
para-	By the side of; near	Parathyroid	Near the thyroid
per-	Through	Permeable	Able to go through
peri-	Around; surrounding	Pericardium	Covering of the heart
poly-	Many	Polycythemia	Condition of many blood cells
post-	After	Postmortem	After death
pre-	Before	Premenstrual	Before menstruation
pro-	First; promoting	Progesterone	Hormone that promotes pregnancy
quadr-	Four	Quadriplegia	Paralysis in four limbs
re-	Back again	Reflux	Backflow
retro-	Behind	Retroperitoneal	Behind the peritoneum
semi-	Half	Semilunar	Half-moon
sub-	Under	Subcutaneous	Under the skin
super-, supra-	Over, above, excessive	Superior	Above
trans-	Across; through	Transcutaneous	Through the skin
tri-	Three; triple	Triplegia	Paralysis of three limbs

▼ **TABLE 2** Word Parts Commonly Used as Suffixes

WORD PART	MEANING	EXAMPLE	MEANING OF EXAMPLE
-al, -ac	Pertaining to	Instestinal	Pertaining to the intestines
-algia	Pain	Neuralgia	Nerve pain
-aps, -apt	Fit; fasten	Synapse	Fasten together
-arche	Beginning; origin	Menarche	First menstruation
-ase	Signifies an enzyme	Lipase	Enzyme that acts on lipids
-blast	Sprout; make	Osteoblast	Bone maker
-centesis	A piercing	Amniocentesis	Piercing the amniotic sac
-cide	To kill	Fungicide	Fungus killer
-clast	Break; destroy	Osteoclast	Bone breaker
-crine	Release; secrete	Endocrine	Secrete within
-ectomy	A cutting out	Appendectomy	Removal of the appendix
-emia	Refers to blood condition	Hypercholesterolemia	High blood cholesterol level
-emesis	Vomiting	Hematemesis	Vomiting blood
-flux	Flow	Reflux	Backflow
-gen	Creates; forms	Lactogen	Milk producer
-genesis	Creation, production	Oogenesis	Egg production
-gram*	Something written	Electroencephalogram	Record of brain's electrical activity
-graph(y)*	To write, draw	Electrocardiograph	Apparatus that records heart's electrical activity
-hydrate	Containing H_2O (water)	Dehydration	Loss of water
-ia, -sia	Condition; process	Arthralgia	Condition of joint pain
-iasis	Abnormal condition	Giardiasis	Giardia infestation
-ic, -ac	Pertaining to	Cardiac	Pertaining to the heart
-in	Signifies a protein	Renin	Kidney protein
-ism	Signifies "condition of"	Gigantism	Condition of gigantic size
-itis	Signifies "inflammation of"	Gastritis	Stomach inflammation
-lemma	Rind; peel	Neurilemma	Covering of a nerve fiber
-lepsy	Seizure	Epilepsy	Seizure upon seizure
-lith	Stone; rock	Lithotripsy	Stone-crushing
-logy	Study of	Cardiology	Study of the heart
-lunar	Moon; moonlike	Semilunar	Half-moon
-malacia	Softening	Osteomalacia	Bone softening
-megaly	Enlargement	Splenomegaly	Spleen enlargement

*A term ending in -*graph* refers to an apparatus that results in a visual and/or recorded representation of biological phenomena, whereas a term ending in -*graphy* is the technique or process of using the apparatus. A term ending in -*gram* is the record itself. Example: In electrocardio*graphy*, an electrocardio*graph* is used in producing an electrocardio*gram*.   *Continued.*

TABLE 2—cont'd   Word Parts Commonly Used as Suffixes

WORD PART	MEANING	EXAMPLE	MEANING OF EXAMPLE
-metric, -metry	Measurement, length	Isometric	Same length
-oid	Like; in the shape of	Sigmoid	S-shaped
-oma	Tumor	Lipoma	Fatty tumor
-opia	Vision, vision condition	Myopia	Nearsightedness
-ose	Signifies a carbohydrate (Especially sugar)	Lactose	Milk sugar
-osis	Condition, process	Dermatosis	Skin condition
-oscopy	Viewing	Laparoscopy	Viewing the abdominal cavity
-ostomy	Formation of an opening	Tracheostomy	Forming an opening in the trachea
-otomy	Cut	Lobotomy	Cut of a lobe
-philic	Loving	Hydrophilic	Water-loving
-penia	Lack	Leukopenia	Lack of white (cells)
-phobic	Fearing	Hydrophobic	Water-fearing
-phragm	Partition	Diaphragm	Partition separating thoracic and abdominal cavities
-plasia	Growth, formation	Hyperplasia	Excessive growth
-plasm	Substance, matter	Neoplasm	New matter
-plasty	Shape; make	Rhinoplasty	Reshaping the nose
-plegia	Paralysis	Triplegia	Paralysis in three limbs
-pnea	Breath, breathing	Apnea	Cessation of breathing
-(r)rhage, -(r)rhagia	Breaking out, discharge	Hemorrhage	Blood discharge
-(r)rhaphy	Sew, suture	Meningeorrhaphy	Suturing of meninges
-(r)rhea	Flow	Diarrhea	Flow through (intestines)
-some	Body	Chromosome	Stained body
-tensin, -tension	Pressure	Hypertension	High pressure
-tonic	Pressure, tension	Isotonic	Same pressure
-tripsy	Crushing	Lithotripsy	Stone-crushing
-ule	Small, little	Tubule	Small tube
-uria	Refers to urine condition	Proteinuria	Protein in the urine

▽ **TABLE 3** Word Parts Commonly Used as Roots

WORD PART	MEANING	EXAMPLE	MEANING OF EXAMPLE
acro-	Extremity	Acromegaly	Enlargement of extremities
aden-	Gland	Adenoma	Tumor of glandular tissue
alveol-	Small hollow, cavity	Alveolus	Small air sac in the lung
angi-	Vessel	Angioplasty	Reshaping a vessel
arthr-	Joint	Arthritis	Joint inflammation
asthen-	Weakness	Myasthenia	Condition of muscle weakness
bar-	Pressure	Baroreceptor	Pressure receptor
bili-	Bile	Bilirubin	Orange-yellow bile pigment
brachi-	Arm	Brachial	Pertaining to the arm
brady-	Slow	Bradycardia	Slow heart rate
bronch-	Air passage	Bronchitis	Inflammation of pulmonary passages (bronchi)
calc-	Calcium; limestone	Hypocalcemia	Low blood calcium level
capn-	Smoke	Hypercapnia	Elevated blood CO_2 level
carcin-	Cancer	Carcinogen	Cancer-producer
card-	Heart	Cardiology	Study of the heart
cephal-	Head, brain	Encephalitis	Brain inflammation
cerv-	Neck	Cervicitis	Inflammation of (uterine) cervix
chem-	Chemical	Chemotherapy	Chemical treatment
chol-	Bile	Cholecystectomy	Removal of bile (gall) bladder
chondr-	Cartilage	Chondroma	Tumor of cartilage tissue
chrom-	Color	Chromosome	Stained body
corp-	Body	Corpus luteum	Yellow body
cortico-	Pertaining to cortex	Corticosteroid	Steroid secreted by (adrenal) cortex
crani-	Skull	Intracranial	Within the skull
crypt-	Hidden	Cryptorchidism	Undescended testis
cusp-	Point	Tricuspid	Three-pointed
cut(an)-	Skin	Transcutaneous	Through the skin
cyan-	Blue	Cyanosis	Condition of blueness
cyst-	Bladder	Cystitis	Bladder inflammation
cyt-	Cell	Cytotoxin	Cell poison
dactyl-	Fingers, toes (digits)	Syndactyly	Joined digits
dendr-	Tree; branched	Oligodendrocyte	Branched nervous tissue cell
dent-	Tooth	Dentalgia	Toothache
derm-	Skin	Dermatitis	Skin inflammation
diastol-	Relax; stand apart	Diastole	Relaxation phase of heart beat

Continued.

▽   **TABLE 3**—cont'd   Word Parts Commonly Used as Roots

WORD PART	MEANING	EXAMPLE	MEANING OF EXAMPLE
dips-	Thirst	Polydipsia	Excessive thirst
ejacul-	To throw out	Ejaculation	Expulsion (of semen)
electr-	Electrical	Electrocardiogram	Record of electrical activity of heart
enter-	Intestine	Enteritis	Intestinal inflammation
eryth(r)-	Red	Erythrocyte	Red (blood) cell
esthe-	Sensation	Anesthesia	Condition of no sensation
febr-	Fever	Febrile	Pertaining to fever
gastr-	Stomach	Gastritis	Stomach inflammation
gest-	To bear, carry	Gestation	Pregnancy
gingiv-	Gums	Gingivitis	Gum inflammation
glomer-	Wound into a ball	Glomerulus	Rounded tuft of vessels
gloss-	Tongue	Hypoglossal	Under the tongue
gluc-	Glucose, sugar	Glucosuria	Glucose in urine
glutin-	Glue	Agglutination	Sticking together (of particles)
glyc-	Sugar (carbohydrate); glucose	Glycolipid	Carbohydrate-lipid combination
hepat-	Liver	Hepatitis	Liver inflammation
hist-	Tissue	Histology	Study of tissues
hydro-	Water	Hydrocephalus	Water on the brain
hyster-	Uterus	Hysterectomy	Removal of the uterus
iatr-	Treatment	Podiatry	Foot treatment
kal-	Potassium	Hyperkalemia	Elevated blood potassium level
kary-	Nucleus	Karyotype	Array of chromosomes from nucleus
kerat-	Cornea	Keratotomy	Cutting of the cornea
kin-	To move; divide	Kinesthesia	Sensation of body movement
lact-	Milk; milk production	Lactose	Milk sugar
lapar-	Abdomen	Laparoscopy	Viewing the abdominal cavity
leuk-	White	Leukorrhea	White flow (discharge)
lig-	To tie, bind	Ligament	Tissue that binds bones
lip-	Lipid (fat)	Lipoma	Fatty tumor
lys-	Break apart	Hemolysis	Breaking of blood cells
mal-	Bad	Malabsorption	Improper absorption
melan-	Black	Melanin	Black protein
men-, mens-, (menstru-)	Month (monthly)	Amenorrhea	Absence of monthly flow
metr-	Uterus	Endometrium	Uterine lining

	TABLE 3—cont'd Word Parts Commonly Used as Roots		
WORD PART	**MEANING**	**EXAMPLE**	**MEANING OF EXAMPLE**
muta-	Change	Mutagen	Change-maker
my-, myo-	Muscle	Myopathy	Muscle disease
myel-	Marrow	Myeloma	(Bone) marrow tumor
myc-	Fungus	Mycosis	Fungal condition
myx-	Mucus	Myxedema	Mucous edema
nat-	Birth	Neonatal	Pertaining to newborns (infants)
natr-	Sodium	Natriuresis	Elevated sodium in urine
nephr-	Nephron, kidney	Nephritis	Kidney inflammation
neur-	Nerve	Neuralgia	Nerve pain
noct-, nyct-	Night	Nocturia	Urination at night
ocul-	Eye	Binocular	Two-eyed
odont-	Tooth	Periodontitis	Inflammation (of tissue) around the teeth
onco-	Cancer	Oncogene	Cancer gene
ophthalm-	Eye	Ophthalmology	Study of the eye
orchid-	Testis	Orchiditis	Testis inflammation
osteo-	Bone	Osteoma	Bone tumor
oto-	Ear	Otosclerosis	Hardening of ear tissue
ov-, oo-	Egg	Oogenesis	Egg production
oxy-	Oxygen	Oxyhemoglobin	Oxygen-hemoglobin combination
path-	Disease	Neuropathy	Nerve disease
ped-	Children	Pediatric	Pertaining to treatment of children
phag-	Eat	Phagocytosis	Cell eating
pharm-	Drug	Pharmacology	Study of drugs
phleb-	Vein	Phlebitis	Vein inflammation
photo-	Light	Photopigment	Light-sensitive pigment
physio-	Nature (function) of	Physiology	Study of biological function
pino-	Drink	Pinocytosis	Cell drinking
plex-	Twisted; woven	Nerve plexus	Complex of interwoven nerve fibers
pneumo-	Air, breath	Pneumothorax	Air in the thorax
pneumon-	Lung	Pneumonia	Lung condition
pod-	Foot	Podocyte	Cell with feet
poie-	Make; produce	Hemopoiesis	Blood cell production
pol-	Axis, having poles	Bipolar	Having two ends
presby-	Old	Presbyopia	Old vision

Continued.

▽ **TABLE 3**—cont'd   Word Parts Commonly Used as Roots

WORD PART	MEANING	EXAMPLE	MEANING OF EXAMPLE
proct-	Rectum	Proctoscope	Instrument for viewing the rectum
pseud-	False	Pseudopodia	False feet
psych-	Mind	Psychiatry	Treatment of the mind
pyel-	Pelvis	Pyelogram	Image of the kidney pelvis
pyo-	Pus	Pyogenic	Pus-producing
pyro-	Heat; fever	Pyrogen	Fever producer
ren-	Kidney	Renocortical	Referring to the cortex of the kidney
rhino-	Nose	Rhinoplasty	Reshaping the nose
rigor-	Stiffness	Rigor mortis	Stiffness of death
sarco-	Flesh; muscle	Sarcolemma	Muscle fiber membrane
scler-	Hard	Scleroderma	Hard skin
semen-, semin-	Seed; sperm	Seminiferous tubule	Sperm-bearing tubule
sept-	Contamination	Septicemia	Contamination of the blood
sigm-	Greek Σ or Roman S	Sigmoid colon	S-shaped colon
sin-	Cavity; recess	Paranasal sinus	Cavity near the nasal cavity
son-	Sound	Sonography	Imaging using sound
spiro-, -spire	Breathe	Spirometry	Measurement of breathing
stat-, stas-	A standing, stopping	Homeostasis	Staying the same
syn-	Together	Syndrome	Signs appearing together
systol-	Contract; stand together	Systole	Contraction phase of the heart beat
tachy-	Fast	Tachycardia	Rapid heart rate
therm-	Heat	Thermoreceptor	Heat receptor
thromb-	Clot	Thrombosis	Condition of abnormal blood clotting
tom-	A cut; a slice	Tomography	Image of a slice or section
tox-	Poison	Cytotoxin	Cell poison
troph-	Grow; nourish	Hypertrophy	Excessive growth
tympan-	Drum	Tympanum	Eardrum
varic-	Enlarged vessel	Varicose vein	Enlarged vein
vas-	Vessel, duct	Vasoconstriction	Vessel narrowing
vesic-	Bladder; blister	Vesicle	Blister
vol-	Volume	Hypovolemic	Characterized by low volume

Clinical and Laboratory Values

APPENDIX

D

TABLE 1 Blood, Plasma, or Serum Values

TEST	NORMAL VALUES*	SIGNIFICANCE OF A CHANGE
Acid phosphatase	*Women:* 0.01-0.56 sigma U/ml *Men:* 0.13-0.63 sigma U/ml	↑ in prostate cancer ↑ in kidney disease ↑ after trauma and in fever
Alkaline phosphatase	*Adult:* 13-39 IU/l *Child:* up to 104 IU/l	↑ in bone disorders ↑ in liver disease ↑ during pregnancy ↓ in hypothyroidism
Bicarbonate	22-26 mEq/L	↑ in metabolic alkalosis ↑ in respiratory alkalosis ↓ in metabolic acidosis ↓ in respiratory alkalosis
Blood urea nitrogen (BUN)	5-25 mg/100 ml	↑ with increased protein intake ↓ in kidney failure
Blood volume	*Women:* 65 ml/kg body weight *Men:* 69 ml/kg body weight	↓ during hemorrhage
Calcium	8.4-10.5 mg/100 ml	↑ in hypervitaminosis D ↑ in hyperparathyroidism ↑ in bone cancer and other bone diseases ↓ in severe diarrhea ↓ in hypoparathyroidism ↓ in avitaminosis D (rickets and osteomalacia)

*Values vary with the analysis method used; 100 ml = 1 dl.

Continued.

▽ **TABLE 1**—cont'd   Blood, Plasma, or Serum Values

TEST	NORMAL VALUES*	SIGNIFICANCE OF A CHANGE
Carbon dioxide content	24-32 mEq/L	↑ in severe vomiting ↑ in respiratory disorders ↑ in obstruction of intestines ↓ in acidosis ↓ in severe diarrhea ↓ in kidney disease
Chloride	96 to 107 mEq/L	↑ in hyperventilation ↑ in kidney disease ↑ in Cushing's syndrome ↓ in diabetic acidosis ↓ in severe diarrhea ↓ in severe burns ↓ in Addison's disease
Clotting time	5-10 minutes	↓ in hemophilia ↓ (occasionally) in other clotting disorders
Copper	100-200 μg/100 ml	↑ in some liver disorders
Creatine phosphokinase (CPK)	*Women:* 5-35 mU/ml *Men:* 5-55 mU/ml	↑ in Duchenne muscular dystrophy ↑ during myocardial infarction ↑ in muscle trauma
Creatinine	0.6-1.5 mg/100 ml	↑ in some kidney disorders
Glucose	70-110 mg/100 ml (fasting)	↑ in diabetes mellitus ↑ in kidney disease ↑ in liver disease ↑ during pregnancy ↑ in hyperthyroidism ↓ in hypothyroidism ↓ in Addison's disease ↓ in hyperinsulinism
Hematocrit (packed cell volume)	*Women:* 38-47% *Men:* 40-54%	↑ in polycythemia ↑ in severe dehydration ↓ in anemia ↓ in leukemia ↓ in hyperthyroidism ↓ in cirrhosis of liver
Hemoglobin	*Women:* 12-16 g/100 ml *Men:* 13-18 g/100 ml *Newborn:* 14-20 g/100 ml	↑ in polycythemia ↑ in chronic obstructive pulmonary disease ↑ in congestive heart failure ↓ in anemia ↓ in hyperthyroidism ↓ in cirrhosis of liver
Iron	50-150 μg/100 ml (can be higher in male)	↑ in liver disease ↑ in anemia (some forms) ↓ in iron-deficiency anemia
Lactic dehydrogenase (LDH)	60-120 U/ml	↑ during myocardial infarction ↑ in anemia (several forms) ↑ liver disease ↑ in acute leukemia and other cancers

▽ **TABLE 1**—cont'd   Blood, Plasma, or Serum Values

TEST	NORMAL VALUES*	SIGNIFICANCE OF A CHANGE
Lipids—total	450-1,000 mg/100 ml	↑ (total) in diabetes mellitus
Cholesterol—total	120-220 mg/100 ml	↑ (total) in kidney disease
High-density		↑ (total) in hypothyroidism
lipoprotein (HDL)	> 40 mg/100 ml	↓ (total) in hyperthyroidism
Low-density		↑ in inherited hypercholesterolemia
lipoprotein (LDL)	< 180 mg/100 ml	↑ (cholesterol) in chronic hepatitis
Triglycerides	40-150 mg/100 ml	↓ (cholesterol) in acute hepatitis
Phospholipids	145-200 mg/100 ml	↑ (HDL) with regular exercise
Fatty acids	190-420 mg/100 ml	
Mean corpuscular volume	82-98 μl	↑ or ↓ in various forms of anemia
Osmolality	285-295 mOsm/L	↑ or ↓ in fluid and electrolyte imbalances
P_{CO_2}	35-43 mm Hg	↑ in severe vomiting
		↑ in respiratory disorders
		↑ in obstruction of intestines
		↓ in acidosis
		↓ in severe diarrhea
		↓ in kidney disease
pH	7.35-7.45	↑ during hyperventilation
		↑ in Cushing's syndrome
		↓ during hypoventilation
		↓ in acidosis
		↓ in Addison's disease
Phosphorus	2.5-4.5 mg/100 ml	↑ in hypervitaminosis D
		↑ in kidney disease
		↑ in hypoparathyroidism
		↑ in acromegaly
		↓ hyperparathyroidism
		↓ in hypovitaminosis D (rickets and osteomalacia)
Plasma volume	*Women:* 40 ml/kg body weight	↑ or ↓ in fluid and electrolyte imbalances
	Men: 39 ml/kg body weight	↓ during hemorrhage
Platelet count	150,000-400,000/mm³	↑ in heart disease
		↑ in cancer
		↑ in cirrhosis of liver
		↑ after trauma
		↓ in anemia (some forms)
		↓ during chemotherapy
		↓ in some allergies
P_{O_2}	75-100 mm Hg	↑ in polycythemia
	(breathing standard air)	↓ in anemia
		↓ in chronic obstructive pulmonary disease
Potassium	3.8 to 5.1 mEq/L	↑ in hypoaldosteronism
		↑ in acute kidney failure
		↓ in vomiting or diarrhea
		↓ in starvation
Protein—total	6-8.4 g/100 ml	↑ (total) in severe dehydration
Albumin	3.5-5 g/100 ml	↓ (total) during hemorrhage
Globulin	2.3-3.5 g/100 ml	↓ (total) in starvation

Continued.

TABLE 1—cont'd   Blood, Plasma, or Serum Values

TEST	NORMAL VALUES*	SIGNIFICANCE OF A CHANGE
Red blood cell count	*Women:* 4.2-5.4 million/mm^3 *Men:* 4.5-6.2 million/mm^3	↑ in polycythemia ↑ in dehydration ↓ in anemia (several forms) ↓ in Addison's disease ↓ in systemic lupus erythematosus
Reticulocyte count	25,000-75,000/mm^3 (0.5%-1.5% of RBC count)	↑ in hemolytic anemia ↑ in leukemia and metastatic carcinoma ↓ pernicious anemia ↓ in iron-deficiency anemia ↓ during radiation therapy
Sodium	136-145 mEq/L	↑ in dehydration ↑ in trauma or disease of the central nervous system ↑ or ↓ in kidney disorders ↓ in excessive sweating, vomiting, diarrhea ↓ in burns (sodium shift into cells)
Specific gravity	1.058	↑ or ↓ in fluid imbalances
Transaminase	10-40 U/ml	↑ during myocardial infarction ↑ in liver disease
Uric acid	*Women:* 2.5-7.5 mg/100 ml *Men:* 3-9 mg/100 ml	↑ in gout ↑ in toxemia of pregnancy ↑ during trauma
Viscosity	1.4-1.8 times the viscosity of water	↑ in polycythemia ↑ in dehydration

White blood cell count
  Total
  Neutrophils
  Eosinophils
  Basophils
  Lymphocytes
  Monocytes
	
	4,500-11,000/mm^3
	60-70% of total
	2-4% of total
	0.5-1% of total
	20-25% of total
	3-8% of total
		↑ (total) in acute infections
		↑ (total) in trauma
		↑ (total) some cancers
		↓ (total) in anemia (some forms)
		↓ (total) during chemotherapy
		↑ (neutrophil) in acute infection
		↑ (eosinophil) in allergies
		↓ (basophil) in severe allergies
		↑ (lymphocyte) during antibody reactions
		↑ (monocyte) in chronic infections

TABLE 2   Urine Components

TEST	NORMAL VALUES*	SIGNIFICANCE OF A CHANGE
Routine Urinalysis		
Acetone and acetoacetate	None	↑ during fasting ↑ in diabetic acidosis
Albumin	None to trace	↑ in hypertension ↑ in kidney disease ↑ after strenuous exercise (temporary)
Ammonia	20-70 mEq/L	↑ in liver disease ↑ in diabetes mellitus
Bile and bilirubin	—	↑ during obstruction of the bile ducts
Calcium	< 150 mg/day	↑ in hyperparathyroidism ↓ in hypoparathyroidism
Color	Transparent yellow, straw-colored, or amber	Abnormal color or cloudiness may indicate: blood in urine, bile, bacteria, drugs, food pigments, or high solute concentration
Odor	Characteristic slight odor	Acetone odor in diabetes mellitus (diabetic ketosis)
Osmolality	500-800 mOsm/L	↑ in dehydration ↑ in heart failure ↓ in diabetes insipidus ↓ in aldosteronism
pH	4.6-8.0	↑ in alkalosis ↑ during urinary infections ↓ in acidosis ↓ in dehydration ↓ in emphysema
Potassium	25-100 mEq/L	↑ dehydration ↑ in chronic kidney failure ↓ in diarrhea or vomiting ↓ in adrenal insufficiency
Sodium	75-200 mg/day	↑ in starvation ↑ in dehydration ↓ acute kidney failure ↓ in Cushing syndrome
Creatinine clearance	100-140 ml/min	↑ in kidney disease
Creatinine	1-2 g/day	↑ in infections ↓ in some kidney diseases ↓ in anemia (some forms)
Glucose	0	↑ in diabetes mellitus ↑ in hyperthyroidism ↑ in hypersecretion of adrenal cortex
Urea clearance	> 40 ml blood cleared per min	↑ in some kidney diseases

*Values vary with the analysis method used.

Continued.

▽ **TABLE 2**—cont'd   Urine Components

TEST	NORMAL VALUES*	SIGNIFICANCE OF A CHANGE
Routine Urinalysis—cont'd		
Urea	25-35 g/day	↑ in some liver diseases ↑ in hemolytic anemia ↓ during obstruction of bile ducts ↓ in severe diarrhea
Uric acid	0.6-1.0 g/day	↑ in gout ↓ in some kidney diseases
Microscopic Examination		
Bacteria	< 10,000/ml	↑ during urinary infections
Blood cells (RBC)	0-trace	↑ in pyelonephritis ↑ from damage by calculi ↑ in infection ↑ in cancer
Blood cells (WBC)	0-trace	↑ in infections
Blood cell casts (RBC)	0-trace	↑ in pyelonephritis
Blood cell casts (WBC)	0-trace	↑ in infection
Crystals	0-trace	↑ in urinary retention Very large crystalline masses are calculi
Epithelial casts	0-trace	↑ in some kidney disorders ↑ in heavy metal toxicity
Granular casts	0-trace	↑ in some kidney disorders
Hyaline casts	0-trace	↑ in some kidney disorders ↑ in fever

Common Abbreviations and Symbols

Scientists and health care professionals must use a variety of technical terms to keep accurate records and to communicate with each other effectively. Over the years, a kind of shorthand for many of these terms has evolved. Some terms can be represented by a symbol. For example, nearly everyone knows that ♀ represents the term *female*. Other terms are simply shortened to a recognizable, abbreviated form. For example, the abbreviation for the blood protein *hemoglobin* is *Hb* or *Hgb*. Phrases or very long words are often represented with an acronym— an abbreviation composed of the first letter of each word or word part. For example, the disease *insulin-dependent diabetes mellitus* can be shorted to *IDDM*.

The following is a list of commonly used symbols, acronyms, and other abbreviations commonly used in discussing the human body in health and disease. The list has been subdivided into groups for easier reference. As you use this list, be careful to watch for multiple meanings. For example, the abbreviation *DC* can mean "discharge," "discontinue," or "doctor of chiropractic," depending on the context. Conversely, a single term may have several commonly used abbreviations. For example, the term *electrocardiogram* can be shortened to either *ECG* or *EKG*. If you can learn to recognize all or most of the abbreviations listed here you will be well on your way to mastering the language of the health professions.

DISEASES AND INFECTIOUS AGENTS

AIDS   acquired immune deficiency syndrome
ARC   AIDS-related complex
ARDS   adult respiratory distress syndrome
ATLL   acute T-cell lymphocyte leukemia
CF   cystic fibrosis
CHF   congestive heart failure
COPD   chronic obstructive pulmonary disease
CVA   cerebrovascular accident
DJD   degenerative joint disease
DM   diabetes mellitus
DMD   Duchenne muscular dystrophy
EBV   Epstein-Barr virus
EP   ectopic pregnancy
FAS   fetal alcohol syndrome
GDM   gestational diabetes mellitus
HBV   hepatitis B virus
HD   Huntington's disease
HIV   human immunodeficiency virus
HPV   human papillomavirus
IDDM   insulin-dependent diabetes mellitus
IM   infectious mononucleosis

IRDS infant respiratory distress syndrome
LE lupus erythematosus
LGV lymphogranuloma venereum
MI myocardial infarction
MS multiple sclerosis
NIDDM non-insulin-dependent diabetes mellitus
PCM protein-calorie malnutrition
PID pelvic inflammatory disease
PKU phenylketonuria
PMS premenstrual syndrome
RhA rheumatoid arthritis
RMSF Rocky Mountain spotted fever
SCID severe combined immune deficiency
SIDS sudden infant death syndrome
STD sexually transmitted disease
T.b., TB tuberculosis
TSS toxic shock syndrome
URI upper respiratory infection
VD, VDRL venereal disease
VZV varicella zoster virus

CHARTING TERMS

a.c. before meals
a.m.a. against medical advice
āā of each
ABC aspiration biopsy cytology
ABC airway, breathing, circulation
ABG arterial blood gas
ad lib. as much as desired
ADL activities of daily living
alb. albumin
AM before noon
amt. amount
ante before
aq. water
Av. average
b.i.d. twice a day
b.m. bowel movement
BE barium enema
Bib drink
BMR basal metabolic rate
BP blood pressure
BPM beats per minute
BRP bathroom priveleges
BSA body surface area
BUN blood urea nitrogen
c/o complains of
CA cancer, carcinoma
CAD coronary artery disease

cap capsule
CAT computed axial tomography
CATH catheter
CBC complete blood count
CBI continuous bladder irrigation
CCU coronary care unit
CCU cardiac care unit
CDC Centers for Disease Control
CPR cardiopulmonary respiration
CSR Cheyne-Stokes respiration
CT computed tomography
c̄ with
d/c discontinue
D & C dilation and curettage
D_5W 5% dextrose in water
DC discharge
DC discontinue
DC doctor of chiropractic
decub decubitus ulcer (bedsore)
DNR do not resuscitate
DO doctor of osteopathy
DOA dead on arrival
DRNG drainage
DRSG dressing
dx, Dx diagnosis
ECG electrocardiogram
EDC expected date of confinement
EEG electroencephalogram
EENT ear, eye, nose, throat
EKG electrocardiogram
EMS emergency medical service
ENT ear, nose, throat
EP evoked potential
ER emergency room
ESR erythrocyte sedimentation rate
EVP evoked potential
FBS fasting blood sugar
FMH family medical history
FP false positive
FUO fever of undetermined origin
fx, Fx fracture
GI gastrointestinal
GP general practitioner
GTT glucose tolerance test
GU genitourinary
H & P history and physical
h.s. at bedtime
HCT, Hct hematocrit
I & O intake and output
ICU intensive care unit

IM   intramuscular
IND   investigational new drug
IUD   intrauterine device
IV   intravenous
IVP   intravenous pyelogram
IVU   intravenous urogram
s̄   without
s̄s̄   one half
KUB   kidney, ureter, and bladder
KVO   keep vein open
LLQ   left lower quadrant
LMP   last menstrual period
LOC   loss of consciousness
LP   lumbar puncture
LPN   licensed practical nurse
LUQ   left upper quadrant
MD   medical doctor
MRI   magnetic resonance imaging
NMR   nuclear magnetic resonance
non rep.   do not repeat
NPO   nothing by mouth
NYD   not yet diagnosed
OB-GYN   obstetrics and gynecology
OD   overdose
OD   right eye
od   once daily
OR   operating room
OS   left eye
OTA   open to air
OTC   over-the-counter
p.c.   after meals
p.r.n.   as needed
P-A, PA, P/A   posterior-anterior
P_{CO_2} carbon dioxide pressure
PCV   packed cell volume
per   by
PET   positron emission tomography
PH   past history
Phar., Pharm.   pharmacy
PI   previous illness
PM   postmortem
PM   after noon
PO   by mouth
P_{O_2} oxygen pressure
PT   prothrombin time
PT   physical therapy
PTX, Px   pneumothorax
PX   physical examination
Px   prognosis
q.   every

q.d.   every day
q.h.   every hour
q.i.d.   four times a day
q.n.s.   quantity not sufficient
q.o.d.   every other day
q.s.   quantity sufficient
R/O   rule out
RLQ   right lower quadrant
RN   registered nurse
ROM   range of motion
RUQ   right upper quadrant
SC   subcutaneous
SOB   shortness of breath
sos   if necessary
sp. gr.   specific gravity
SPECT   single-photon emission computed tomography
stat.   at once, immediately
STD   skin test dose
sub Q   subcutaneous
T & A   tonsillectomy and adenoidectomy
T   temperature
t.i.d.   three times a day
TIA   transient ischemic attack
TPN   total parenteral nutrition
TPR   temperature, pulse, respiration
TUR   transurethral resection
UA   urinalysis
UTI   urinary tract infection
VS   vital signs
VSS   vital signs stable
w/c   wheel chair

SYMBOLS

α   alpha
@   at
β   beta
Δ   change
↓   decrease
°   degree
=   equal
♀   female
γ   gamma
>   greater than
≥   greater than or equal to
↑   increase
≤   less than or equal to
<   less than
♂   male
μ   micro

± plus or minus
℞ prescription
1° primary
∝ proportional
2° secondary
σ sigma
3° tertiary

UNITS OF MEASUREMENT

Å Angstrom
°C degrees Celsius (Centigrade)
cc cubic centimeter (= milliliter)
cm centimeter
d. day
dl deciliter
°F degrees Fahrenheit
g, Gm, gm gram
gr grain
gtt drops
h. hour
hr. hour
kg kilogram
l, ℓ, L liter
m meter
mcg microgram
mg milligram
ml milliliter (= cubic centimeter)
mm Hg millimeters of mercury
mm millimeter
msec millisecond
μg microgram
μm micrometer
pH hydrogen ion concentration (acidity; alkalinity)
s., sec. second

GENERAL SCIENTIFIC ABBREVIATIONS

A-V, AV, A/V atrioventricular
ADH antidiuretic hormone
Ba barium
BBB blood-brain barrier
Ca calcium
CNS central nervous system
CO carbon monoxide
CO cardiac output
Co cobalt
CO_2 carbon dioxide
CSF cerebrospinal fluid
CT calcitonin
DNA deoxyribonucleic acid
EM electron micrograph
FSH follicle-stimulating hormone
GH growth hormone
H_2O water
Hb hemoglobin
HCG human chorionic gonadotropin
Hg mercury
Hgb hemoglobin
HGH human growth hormone
HR heart rate
I iodine
K potassium
LH luteinizing hormone
O_2 oxygen (gas)
PDL periodontal ligament
RBC red blood cell
RNA ribonucleic acid
SEM scanning electron micrograph
Tc technetium
TEM transmission electron micrograph
WBC white blood cell

Chapter Test and Clinical Application Answers

CHAPTER 1
Chapter Test
1. anatomy; physiology; pathology
2. tissue
3. lateral
4. anterior; posterior
5. axial; appendicular
6. homeostasis
7. h
8. e
9. a
10. j
11. b
12. d
13. c
14. f
15. i
16. g
17. a
18. d
19. c
20. c

Clinical Applications

1. Mrs. Miller's mole is on the anterior (front) surface of her trunk, inferior to (below) the rib cage, and superior to (above) and left of the navel (Figure 1-6). With the choices given, it is best to ask Mrs. Miller to assume a supine (belly-up) position. Mrs. Miller's occipital mole is on the back of her head (Figure 1-8). This is best examined if she rolls to a prone (belly-down) position or sits up and tilts her head forward.

2. Mr. Sanchez's injury was at the end of a finger. A typical response to a sudden drop in blood pressure is to increase pumping of blood by the heart. Under normal circumstances, this response would bring the blood pressure back up to its average value. In this case, however, increased pumping by the heart will increase blood loss—causing a further drop in blood pressure and threatening Mr. Sanchez's homeostatic balance even more. If the blood loss is not stopped soon, Mr. Sanchez could lose a large volume of blood. Because this response amplifies the drop in blood pressure rather than returning it to its normal value, it is a case of positive feedback.

CHAPTER 2
Chapter Test
1. a
2. b
3. c
4. d
5. a
6. b
7. d
8. c
9. a
10. f
11. e
12. d
13. b
14. h
15. g
16. ribosomes
17. active
18. prophase
19. epithelial
20. epithelial

Clinical Applications

1. Glycogen-digesting enzymes are likely to be found within *lysosomes*. Recall that lysosomes are tiny sacs that contain enzymes used to digest food molecules within cells. The nonfunctional enzymes responsible for Pompe's disease were synthesized at ribosomes with information obtained from each cell's DNA—information that was incorrect. The incorrect genetic information in DNA originally came from the genetic code in the DNA

passed from parent to offspring in the sperm and egg cells. A more detailed explanation of DNA inheritance is given in Chapter 23.

2. Drugs that interfere with formation of spindle fibers halt mitosis by preventing chromosomes from lining up during metaphase and from separating in an orderly fashion during anaphase. Mitosis cannot proceed without spindles to which chromosomes can attach. Drugs that prevent DNA synthesis affect cells before mitosis even begins—during interphase when DNA replication should be occurring. If DNA is not replicated, the cell will not enter prophase.

3. CF causes thickened secretions in many areas of the body. Thickened mucus secretions in the airways of the lungs often interferes with breathing and promotes recurring infections. A traditional treatment for youngsters with CF is to put them over a pillow or other support, head down, and deliver a series of sharp slaps to the back of the thorax. This promotes drainage of the thick mucus out of the airways, making breathing easier and reducing the likelihood of pulmonary infection.

CHAPTER 3
Chapter Test

1. skin
2. bones
3. communication; integration; control; recognition
4. hormones
5. lymphatic
6. urine
7. accessory
8. alveoli
9. d
10. g
11. b
12. h
13. a
14. c
15. e
16. f

Clinical Applications

1. The kidneys are part of the urinary system, so it is this system that is *primarily* involved in Tommy's condition. However, because the kidneys clear metabolic wastes from the blood and help maintain homeostasis of the entire internal environment, all systems will be affected if these functions are lost. Based on information given in this chapter, Tommy's physicians may recommend use of a dialysis machine or similar procedure (see Chapter 18) for short-term use. Because the kidney failure is permanent, a kidney transplant may be recommended as a long-term solution.

2. The male urethra serves two functions: it conducts urine from the bladder to the outside of the body and it conducts sperm out of the body during the male sexual response. Thus both urinary and reproductive functions may be impaired by Mr. Davidson's condition.

CHAPTER 4
Chapter Test

1. idiopathic
2. pathogenesis
3. etiology
4. pathophysiology
5. microbe
6. bacilli
7. vector
8. vaccine
9. antibiotics
10. neoplasm
11. cancer
12. carcinogens
13. oncogene
14. biopsy
15. mediator
16. b
17. j
18. c
19. f
20. i
21. h
22. d
23. g
24. a
25. e

Clinical Applications

1. An *epidemiologist* would probably label this outbreak of bacterial infection an *epidemic*. Bacterial pathogens can be transmitted by person-to-person contact (which can occur rapidly in crowded conditions or close-knit communities), by environmental contamination (such as unsanitary drinking water or improper waste disposal), by transmission by a vector (such as a mosquito), and by other means.

2. So-called "staph" infections are caused by *Staphylococcus* bacteria, round (coccus) bacteria that adhere to one another to form clusters of bacteria. Antibiotics are usually derived from living organisms such as fungi and bacteria and are thus classified as "natural" rather than synthetic. Chemists are able to produce synthetic copies or variants of natural compounds, so it is *possible* that the antibiotic was manufactured synthetically.

3. Fred's symptoms are classic symptoms of the inflammatory response. See Figures 4-16 and 4-17 for the details of how these symptoms are produced. In normal inflammatory responses, the body's defense mechanisms are called into play—thus taking care of the damage (including possible infection). If this normal response is inhibited, the body's defenses will be interrupted. This might increase the likelihood of a severe infection and will slow the healing process. Antiinflammatory agents are normally only used when the inflammation response occurs at inappropriate times or is so great that it poses a threat of damage itself (as in allergic reactions or inflammatory disease).

CHAPTER 5
Chapter Test

1. epithelial; connective tissue
2. parietal; visceral
3. synovial
4. skin
5. cutaneous
6. epidermis; dermis
7. keratin
8. melanin
9. root
10. pressure
11. eccrine; apocrine
12. sebaceous
13. burns
14. decubitus ulcers
15. squamous cell
16. d
17. g
18. e
19. a
20. c
21. b
22. f
23. i
24. j
25. h

Clinical Applications

1. Dana can use the rule of nines illustrated in Figure 5-11. Each arm has about 9% of the total surface area of the body (4.5% in front, 4.5% in back). Both arms have about 18% of the total. If half of both arms are burned, and half of 18 is 9, then the patient has about 9% of the total skin surface affected by the burn.

2. Although one cannot be positive in this case, the characteristics and location of the lesion indicate that it is likely to be basal cell carcinoma. Aunt Gina might find some comfort in knowing that this form of cancer is less likely to spread (metastasize) that other types of skin cancer. She might also appreciate knowing that, because she is dark-skinned, she is in a lower risk group for developing this type of cancer (see Chapter 4).

3. The ringlike pattern, redness, and scaling of the skin lesion make it likely that the patient has *ringworm*, a type of fungal infection (mycosis) called *tinea*. Even after treatment with antifungal agents, some of the fungi responsible for the infection may remain—possibly causing a later opportunistic infection. Recurrence of tinea infections can be prevented by keeping the skin dry because dry conditions inhibit the growth of tinea-causing fungi.

CHAPTER 6
Chapter Test

1. trabeculae
2. Haversian systems
3. osteoblasts; osteoclasts
4. diaphysis
5. axial; appendicular
6. sinuses

7. cervical
8. scapula; clavicle
9. radius; ulna
10. olecranon
11. carpal; phalanges
12. ilium; ischium; pubis
13. patella; tibia
14. feet
15. diarthrotic
16. scoliosis
17. osteoporosis
18. osteomyelitis
19. open (or compound)
20. osteoarthritis
21. F
22. T
23. F
24. F
25. T
26. F
27. T
28. T
29. F
30. T

Clinical Applications

1. The term *greenstick fracture* refers to a type of incomplete fracture in which a bone bends and then breaks only along the outside curve of the bend. This type of fracture is so named because it imitates the incomplete break commonly seen when one tries to snap a green (fresh) wooden stick into two pieces. Greenstick fractures and other types of incomplete fractures heal more rapidly than other fractures because the broken edges of bone tissue remain close to one another after the injury—facilitating the repair process.

2. As Figure 6-11, *C*, shows, such an abnormality may result from *scoliosis*—abnormal lateral curvature of the vertebral column. Common among adolescent women, scoliosis can be treated in a variety of ways. The text mentions braces (for example, the Milwaukee brace), transcutaneous muscle stimulation, and spinal fusion surgery as possible treatments. Other treatments you may know include prescribed exercises, traction (pulling on the vertebral column to straighten it), body casts, or other therapies.

3. Osteoporosis, literally "bone porosity," is characterized by the loss of bone volume as open

spaces or pores develop in bone tissue. Such bone loss causes the skeleton to become brittle and thus easily broken. Sometimes the bone weakness is so profound that seemingly normal stresses cause fractures—often called *spontaneous fractures* because they seem to have no direct cause. Fractured bone, as any broken tissue, may become infected —especially if proper medical treatment is delayed. *Osteomyelitis* is infection of bone tissue and may result from infection secondary to a fracture.

CHAPTER 7
Chapter Test

1. b
2. a
3. a
4. b
5. b
6. d
7. a
8. smooth or involuntary
9. tendons
10. myosin
11. sliding-filament
12. synergists
13. posture
14. oxygen debt
15. isotonic
16. plantar
17. atrophy
18. hamstrings
19. muscle strain
20. fibromyositis
21. poliomyelitis
22. muscular dystrophy
23. myasthenia gravis
24. c
25. a
26. d
27. b
28. g
29. e
30. f

Clinical Applications

1. Pseudohypertrophic muscular dystrophy or *Duchenne muscular dystrophy (DMD)* is an X-linked inherited disorder, so the cause of Tom's condition is inheritance of a defective gene on the X chromosome. More information on X-linked

traits is found in Chapter 23. It is unlikely that Geri will develop DMD because females rarely exhibit X-linked disorders. Your risk of developing DMD depend on several different factors. First, if you are female you are not likely to develop DMD. Second, Tom's defective X chromosome had to have been inherited from his mother (see Chapter 23)—so if you are related to Tom's mother you are more likely to have the defective gene than if you are unrelated to her. Third, because DMD typically strikes by age 3 you are unlikely to develop the disease now.

2. Your friend has sustained a tear or overstretching of muscle fibers in a major lower leg muscle located in the "calf" region. Elena is likely suffering from myalgia (muscle pain) and myositis (muscle inflammation). The irritation of the injury and the resulting inflammation response may also cause muscle cramps. The gastrocnemius muscle plantar flexes the foot, causing the toes to point downward. Almost any movement of the lower leg is likely to disturb the injured tissue, so Elena should avoid leg movements as recommended by her physician. Walking or even standing are most likely to worsen Elena's injury.

3. Robert may have some increase, or hypertrophy, of his upper body muscles. However, because racquetball primarily involves endurance or "aerobic" exercise, the increase will not be great. Massive muscle hypertrophy, or "body building," results from strength training or "anaerobic" exercise such as weight-lifting. Robert seems out of breath because the aerobic exercise undertaken in a game of racquetball causes an *oxygen debt* that must be "repaid" after exercise by continued heavy breathing—perhaps making it difficult for Robert to speak to you.

CHAPTER 8
Chapter Test
1. central; peripheral
2. meninges
3. hydrocephalus
4. neurons; neuroglia
5. oligodendroglia
6. away; axon
7. synapse
8. neurotransmitters
9. dermatome
10. endorphins; enkephalins
11. hypothalamus; thalamus
12. 12; 31
13. cardiac muscle; smooth muscle; glandular epithelium
14. sympathetic; parasympathetic
15. sympathetic
16. c
17. h
18. a
19. b
20. e
21. i
22. f
23. g
24. d
25. T
26. F
27. T
28. T
29. F
30. T

Clinical Applications
1. Many epileptics experience very mild seizures which may appear to Tony's teachers as "daydreaming" or "not paying attention." Tony may indeed be simply daydreaming, but it is *possible* that he has epilepsy and his physician would be well advised to test him—perhaps by ordering an EEG (electroencephalogram). If Tony experiences an episode while the EEG machine is recording his brain waves, a sudden burst of increased, abnormal activity may be detected. Such a burst indicates a seizure.

2. Although the information given in this item is limited, the description of Angela's condition fits that of *multiple neurofibromatosis*. This disease is characterized by tumors of the Schwann cells that wrap around the axons of cutaneous nerves—nerves near the surface of the skin—as well as other nerves. Thus a disorder of nerves in the skin can cause skin lesions. Similarly, the lesions of shingles are caused by viral (herpes) infection of cutaneous nerves—usually along a single dermatome.

CHAPTER 9
Chapter Test
1. general
2. rods; cones
3. sclera; choroid; retina
4. iris
5. aqueous humor
6. cataracts
7. choroid
8. mechanoreceptors
9. external; middle; inner (internal)
10. malleus
11. perilymph
12. Corti
13. chemoreceptors
14. gustatory
15. g
16. a
17. i
18. c
19. b
20. h
21. d
22. f
23. e
24. F
25. T
26. T
27. F
28. F
29. T
30. T

Clinical Applications

1. Roger's condition, *diabetic retinopathy*, involves progressive degeneration of the retina. Early in this disease, small hemorrhages of retinal vessels disrupts the normal supply of blood to retinal cells. Later, the retina begins to form abnormal, thick vessels that block vision and may cause retinal detachment. The damage involved starts at the edges of the retina and progresses inward toward the optic nerve. Roger is *legally* blind, meaning that his vision is worse than 20-200. Roger may have seen the other pedestrian stumble in front him, if only as a very fuzzy image. Even without any vision, some blind individuals have learned to use their other senses—especially hearing—to detect changes in their environment.

2. Obviously, Mrs. Stark has a form of color blindness, or *color deficiency*. Because Mrs. Stark did not have difficulty in distinguishing colors until late adulthood, this could not have been one of the common, inherited forms of color blindness such as red-green color blindness. Being female, it is unlikely that she would have been born with such an X-linked condition anyway. Additionally, inherited forms of color blindness typically cause problems with specific ranges of color—not all colors. Considering her age and the suddenness of onset, it is likely that Mrs. Stark suffered a cerebrovascular accident (stroke) that damaged the color-perception centers of her brain. If damage was limited to that area only, she may not have any other symptoms besides color blindness. This condition has sometimes been called *acquired cortical color blindness*.

3. Otitis media is inflammation of the middle ear resulting from infection by bacterial or viral pathogens. A pathogen has entered the air-filled middle ear, possibly through the auditory (eustachian) tube, and infected the mucous lining. Inflammation and accumulation of exudate has likely occurred as a result. The swelling and fluid accumulation may affect your hearing—creating a conduction impairment. If left untreated, your body's immune system may be able to fight off the infection, and your ear tissues may repair themselves. On the other hand, the infection may become severe and spread to surrounding tissues. If the inner ear or vestibulocochlear nerve becomes involved, the infection may cause permanent hearing loss—a type of nerve impairment. In an extreme case, the infection may spread to yet other tissues, perhaps to the blood or to the brain, causing coma or death.

CHAPTER 10
Chapter Test
1. hormones
2. exocrine
3. second messenger
4. negative
5. prostaglandins
6. hypersecretion
7. thyroid stimulating
8. anterior pituitary
9. antidiuretic

10. thyroxine
11. goiter
12. glomerulosa
13. medulla
14. glucagon
15. diabetes mellitus
16. progesterone
17. testosterone
18. thymosin
19. pineal
20. estrogen
21. F
22. T
23. T
24. F
25. T
26. F
27. T
28. F
29. T
30. F
31. T
32. T
33. T
34. T
35. F

Clinical Applications

1. George's excessive level of activity and his abnormal heart rhythm (atrial fibrillation) are common characteristics of hyperthyroidism. Hypersecretion of thyroid hormone (T_3 and T_4) causes a general increase in the metabolism of all cells, and therefore increases activity in all organs. The heart is no exception, so many sufferers of hyperthyroidism experience heart problems such as atrial fibrillation. George's physicians will probably recommend surgical removal or destruction (via radiation) of some of the thyroid tissue in an attempt to reduce thyroid hormone secretion to normal levels.

2. George's surgeons will, as in any surgical procedure, be careful to avoid unnecessary injury to local blood vessels, nerves, and other tissues. Not only must George's surgeons be careful to avoid damaging George's trachea and larynx, they will probably be careful to avoid damaging or removing his *parathyroid glands*. As you know, these glands necessary because they produce parathyroid hormone (PTH), which is essential to the vital calcium balance of the body.

3. Lynn's condition, type I diabetes mellitus, is also known as *insulin-dependent diabetes mellitus (IDDM)* as described in the text and Appendix B. This name indicates the most common form of treatment: insulin therapy. Usually, several small doses of insulin are injected into a person's body each day. There are a number of additional measures, such as diet control and exercise, that often supplement insulin therapy. Tissue grafts and other new methods of treating IDDM are also being explored. IDDM results from hyposecretion of insulin, the hormone that allows glucose to enter cells. The hyperglycemia associated with IDDM results from accumulation of nutrients (glucose) that would otherwise have entered the cells for catabolism. Thus without sufficient insulin the excess glucose really isn't available for cell use, and the body's cells literally "starve in the midst of plenty."

CHAPTER 11
Chapter Test

1. b
2. b
3. d
4. a
5. c
6. d
7. a
8. b
9. plasma
10. red bone marrow
11. polycythemia
12. hemoglobin
13. differential
14. hemophilia
15. embolus
16. antigen
17. AB
18. Rh

Clinical Applications

1. Angelo's physician will probably order a test in which the concentration of certain enzymes in Angelo's blood plasma are abnormally elevated. In this chapter, you learned that *transaminase* is a plasma enzyme that increases after a heart attack.

2. Of course, the wisest course of action usually involves seeking and following professional medical advice rather than acting on one's own initiative. In this case, Yvonne's actions are not

likely to help her because pernicious anemia is more often caused by a lack of *intrinsic factor* needed to absorb B_{12} into the blood. If this is true in Yvonne's case, no matter how many B_{12} tablets she takes, she will probably not absorb enough B_{12} to reverse her condition. Her physician would likely recommend an intramuscular injection of B_{12}—a method that bypasses the absorption problem.

3. The actual diagnosis can only be made by a qualified professional, but based on the results given and the data in Table 11-2, a good guess is a type of hemolytic anemia. Most types of hemolytic anemia, such as sickle-cell anemia and thalassemia, are inherited, so it is possible that you inherited the same defective gene or genes as your brother. Depending on the exact type, however, it is unlikely that you would have gotten this far in life without developing some symptoms already.

CHAPTER 12
Chapter Test

1. atria; ventricles
2. myocardium
3. pericardium
4. atrioventricular
5. tricuspid
6. vena cava
7. pulmonary
8. angina pectoris
9. pacemaker
10. electrocardiogram (ECG)
11. ventricles
12. g
13. a
14. e
15. c
16. d
17. b
18. f
19. T
20. T
21. F
22. F
23. T
24. T
25. T

Clinical Applications

1. The large spikes (tall waves) on the ECG monitor are probably the QRS complexes, which represent depolarization of the ventricular myocardium—the point at which your friend's ventricles are about to pump blood out of the heart during each cardiac cycle. The observed change probably resulted from the electrical "noise" produced by a muscular movement made by the patient. If so, the ECG pattern will return to normal when she stops moving. However, the sudden change *may* be due to ventricular fibrillation, a judgement best made by a trained professional. As with any medical emergency, the first thing to do after determining that there is a problem is to summon help. First aid for such a problem may involve CPR (have you had your CPR refresher course this year?). When they arrive, the hospital staff may use a defibrillator or other treatments to correct your friend's problem.

2. Vivian's *mitral valve prolapse (MVP)* is a condition in which the left atrioventricular (mitral) valve billows backward into the left atrium (see Figure 12-4). Because of this defect, the edges of the mitral valve may not meet to form a tight seal. Thus blood may leak back into the left atrium during contraction of the left ventricle. The severity of Vivian's condition depends largely on the amount of blood leakage that occurs.

3. Your uncle's heart probably has blockage of some of the major coronary arteries. This blockage could be due to a number of factors, the most likely of which is *atherosclerosis*. In this condition, fatty deposits form in the wall of arteries—decreasing the diameter of the lumen and reducing blood flow. Without correction, your uncle may suffer a myocardial infarction resulting from oxygen deprivation of heart muscle supplied by the affected arteries. The triple-bypass surgery will graft new vessels into your uncle's coronary circulation to route blood around the blocked areas (Figure 12-7).

CHAPTER 13
Chapter Test

1. arteries
2. tunica media
3. capillaries
4. arteries; veins
5. systemic
6. capillary
7. foramen ovale
8. aneurysm
9. shock

10. systolic; diastolic
11. g
12. a
13. i
14. h
15. c
16. d
17. e
18. b
19. f

Clinical Applications

1. In Appendix A, in the boxed essay dealing with cholesterol, it states that exercise increases the ratio of "good" cholesterol, thus decreasing the ratio of the "bad" cholesterol that causes atherosclerosis. Atherosclerosis, which develops into hardening of the arteries (arteriosclerosis), may block vessels and cause myocardial infarction, ischemia or necrosis of other tissues, aneurysms, CVAs, and other serious problems.

2. Advanced atherosclerosis of a leg artery may reduce flow to skeletal muscles in the leg enough to make it difficult for them to get oxygen during walking. The muscles will use anaerobic respiration, which increases lactic acid levels and causes a burning pain often associated with muscle fatigue. Leo's physician has many choices of treatment. One choice would be to use vasodilator drugs, which will relax and expand the affected artery and thus improve blood flow. Another choice is angioplasty, in which the obstruction is mechanically altered to improve blood flow.

3. To reach the anterior tibial artery, the tip of the catheter must proceed through the femoral artery, then through the popliteal artery (see Figure 13-2). To reach the mitral (left AV) valve from the same point, the tip must pass superiorly through the femoral artery, through the external iliac artery, through the abdominal aorta, through the thoracic aorta, through the arch of the aorta, past the aortic valve and through the left atrium (Figure 13-2).

CHAPTER 14

Chapter Test

1. thoracic duct
2. lymph nodes
3. artificial
4. antibodies; complement
5. antigens

6. humoral
7. antibodies
8. lymphocytes
9. neutrophils; monocytes
10. lymphocytes
11. humoral
12. cell-mediated
13. T
14. F
15. T
16. F
17. F
18. T
19. T
20. T
21. T
22. F
23. F
24. e
25. g
26. b
27. d
28. c
29. f
30. a

Clinical Applications

1. One possible explanation is that the infection in the lymph nodes of the groin (area between the legs) produced scarring of the lymph nodes and/or the lymphatic vessels connected to the nodes. Scarring often occurs as a result of damage caused by infection. Such scarring may have blocked lymphatic drainage from the leg, causing tissue fluid to accumulate in the leg and produce swelling—a condition called *lymphedema*. If the scarring only occurred on one side, only one leg would be affected this way.

2. Although the spleen performs some important functions, other organs can take over these functions well enough that Keith can survive without a spleen. This is fortunate because the usual treatment recommended for a ruptured spleen is immediate *splenectomy* or surgical removal of the spleen. Without such treatment, Keith will lose a great deal of blood. He has probably already lost quite a bit of blood through hemorrhaging from the rupture because the spleen itself can hold over a pint of blood.

3. Children with SCID are unable to defend themselves against infection by pathogens. There-

fore such children are defenseless against the pathogenic bacteria found throughout the natural environment. Children without SCID have immune defenses that protect them from infection by these common pathogens. The "bubble" strategy was used to protect the boy from exposure to these pathogens. Recently, injections of antibodies have afforded partial, temporary immune protection to people with SCID. The boy in question died at age 12 after leaving the bubble to receive a bone marrow transplant. More recently, bone marrow transplants have been successful in treating some cases of SCID by replacing abnormal stem cells with normal donor cells.

CHAPTER 15
Chapter Test
1. diffusion
2. surfactant
3. lower
4. oropharynx
5. nasopharynx
6. larynx
7. epiglottis
8. pneumothorax
9. external
10. oxyhemoglobin
11. vital capacity
12. medulla
13. alveoli
14. atelectasis
15. infant respiratory distress syndrome
16. rhinitis
17. pneumonia
18. asthma
19. e
20. b
21. g
22. d
23. f
24. c
25. a

Clinical Applications
1. Because Curtis is obviously in "respiratory distress," he is probably suffering from *adult respiratory distress syndrome (ARDS)*. ARDS can be caused by impairment of the surfactant lining the alveoli of the lungs resulting from accidental inhalation of water in a swimming pool. The water has probably diluted the surfactant and thus increased the surface tension of the fluid lining the lungs. Many of Curtis's alveoli have probably collapsed as a result, restricting his breathing.

2. Aspergillosis is a *mycotic* or fungal condition in which a type of mold typically produces "fungus ball" lesions that can obstruct airways when they occur in the lung. Partial blockage of the bronchi, the principal airways leading into the lungs, constitutes an *obstructive disorder*. Because obstructive disorders typically do not reduce lung volumes or capacities but obstruct normal inspiration and expiration. Thus a spirometry test should show an increased time for forced maximal inspiration or expiration.

3. Considering the circumstances (swallowing food) and the fact that your friend can't speak indicate that the food has lodged in the larynx or upper trachea. The Heimlich maneuver (see the boxed essay) is the recommended first aid in this situation. Many people in such circumstances run to the restroom for privacy during their distress—often to their detriment because there may not be anyone to help them. Medical personnel may elect to perform an emergency tracheostomy if first aid procedures don't remove the obstruction. In a tracheostomy, an incision in the trachea (a tracheotomy) is made and a hollow tube inserted into the airway, allowing the subject to breathe.

CHAPTER 16
Chapter Test
1. mechanical; chemical
2. feces
3. gastroenterology
4. enteritis
5. accessory
6. lumen
7. mucosa
8. palates
9. crown
10. mastication
11. amylase
12. esophagus
13. chyme
14. peristalsis
15. jejunum
16. d
17. a
18. f

19. b
20. h
21. c
22. e
23. g
24. j
25. i
26. T
27. F
28. F
29. T
30. T
31. F
32. F
33. F
34. T
35. T
36. T

Clinical Applications

1. Improper care of the teeth and gums may lead to gingivitis (gum inflammation). If untreated, the gingivitis may progress to periodontitis, which causes loosening of the teeth. Teeth may also be lost through a failure to treat caries (tooth decay), a condition that also results from poor dental hygiene.

2. Famotidine, like cimetidine, decreases secretion of stomach acid. Because hypersecretion of stomach acid is the usual cause of progressive ulceration of the gastrointestinal lining, this treatment should slow or stop the damage and allow healing of the ulcer. Because it is more potent that cimetidine, smaller, less frequent doses are required. Sucralfate helps in the treatment of ulcers by sticking to and protecting the areas of the gastrointestinal tract that have a damaged mucus coat—specifically the areas that already have damage (that is, the ulcer itself). The antipepsin effect prevents the stomach enzyme pepsin from digesting the proteins in exposed tissues lining the gastrointestinal tract and possibly perforating the gastrointestinal wall.

3. The barium swallow test, also called an *upper GI study*, stretches (fills) the stomach and makes it easily visible in x-ray photography. In a hiatal hernia, the bottom of the esophagus may be abnormally stretched and the cardiac sphincter dilated. An x-ray photograph would likely show the barium backing up into the lower esophagus.

If stomach acid backs up this way, it would produce the heartburn symptom experienced by Fred. Backflow, or *reflux*, would more likely occur when bending or reclining because gravity would pull the stomach contents toward the esophagus. When standing up, gravity pulls the stomach contents *away from* the esophagus.

CHAPTER 17
Chapter Test
1. catabolism
2. anabolism
3. hepatic portal vein
5. glycolysis
6. glycogenesis
7. fats
8. vitamins
9. ATP
10. basal metabolic rate (BMR)
11. hypothalamus
12. heat stroke
13. frostbite
14. F
15. T
16. T
17. F
18. F
19. T
20. T

Clinical Applications

1. Your friend is probably suffering from *heat exhaustion.* Because of the loss of fluid and electrolytes while sweating, her body's internal environment has become imbalanced, producing the symptom of nausea and muscle cramps (called *heat cramps* in this case). First aid for your friend should include rest in a cool environment—perhaps in an air-conditioned building or in the shade—and plenty of isotonic fluids (for example, Gatorade).

2. Scurvy, an abnormal condition resulting from a deficiency of vitamin C certainly can cause the loss of teeth, as well as other forms of degeneration or injury. Without sufficient vitamin C, the body cannot maintain the collagen fibers that hold together most of the body. The collagen forming the periodontal ligament (PDL) that holds each tooth in its socket can loosen enough to make tooth loss unavoidable (see Chapter 16).

In the past, sailors were particularly vulnerable to scurvy because they often did not have sufficient fresh vitamin-C–containing foods available to them on long voyages.

3. The body requires an assortment of 20 amino acids to synthesize proteins needed for normal body function. These proteins include functional proteins such as enzymes, neurotransmitters, and protein hormones, as well as structural proteins such as collagen and keratin that hold the body together. Some of these 20 can be manufactured by the body from compounds already present, so they are considered *nonessential* in the diet. Some of them, however, cannot be made by the body so they are *essential* in the diet. Without sufficient essential amino acids in her diet, Andrea may suffer consequences such as an inability to produce one or more important proteins. She need not be overly concerned, however, because the proper combination of vegetables can easily supply all the essential amino acids Andrea needs.

CHAPTER 18
Chapter Test
1. uremia
2. 10%
3. cortex; medulla
4. nephron
5. Bowman's capsule; glomerulus
6. proximal
7. filtration
8. osmosis
9. proximal
10. polyuria
11. pelvis
12. renal colic
13. urethra; meatus
14. incontinence
15. cystitis
16. 125
17. e
18. i
19. d
20. b
21. a
22. g
23. c
24. f
25. h
26. m
27. k
28. j
29. l

Clinical Applications
1. The loss of the fatty pad that surrounds and supports the kidneys may cause them to drop from their normal positions, a condition called *nephroptosis*. In nephroptosis, the ureters may become kinked and cause urine to backflow into the kidneys and produce hydronephrosis. Hydronephrosis, in turn, may prevent normal filtration of fluids into nephrons and thus cause kidney failure.

2. Thiazide diuretics lower blood pressure by inhibiting water reabsorption in the kidney, thus reducing the amount of water retained by the blood. As you may recall from Chapter 13, such a reduction in blood volume reduces the overall blood pressure. Because the water is prevented from re-entering the blood stream, it remains as urine, greatly increasing the volume of urine output. The term *diuretic* is applied to any agent that increases the volume of urine output.

3. The fluid used in CAPD is *isotonic* to normal body fluids, including blood plasma. If the fluid is significantly hypertonic to plasma, it will draw water osmotically from the blood and cause dehydration of the body. If the fluid is significantly hypotonic to plasma, it will lose water osmotically to the blood, causing the blood cells to swell and burst. The dialyzing fluid is isotonic to plasma but contains a different mixture of solutes than plasma, a mixture devoid of metabolic wastes such as urea.

CHAPTER 19
Chapter Test
1. intracellular
2. interstitial; plasma
3. nonelectrolytes
4. ions
5. cations; anions
6. urine
7. sodium; chloride
8. diuretic
9. edema
10. T
11. T
12. F

13. F
14. T
15. F
16. T
17. T
18. F

Clinical Applications

1. Tom's body can deal with an excessive input of sodium by excreting an increased amount of sodium in the urine. The specific mechanism involves decreased reabsorption of sodium ions in the kidney tubules (see Chapter 18).

2. If Jo has consumed a large amount of distilled water, especially without also consuming salts, she is likely to produce an excessive amount of urine. This occurs because the body attempts to maintain its homeostasis of fluid volume (so it gets rid of excess fluid) and its homeostasis of electrolyte and water concentration (so it gets rid of excess water to maintain the normal osmotic balance). A urinalysis might reveal a decreased specific gravity (density), perhaps one that is closer to the density of pure water (1.000). Because of the increased water content in Jo's urine, the remaining components (solutes) will be diluted. Thus the color may appear lighter than usual and the concentration of each solute will be less than normal.

3. Until and unless the body can reverse overhydration by increasing urinary output (see 2), the presence of excess water in the internal environment makes the blood volume greater than normal. Increased blood volume means increased blood pressure. If peripheral blood pressure increases, the heart must pump harder to exceed that pressure and thus maintain blood flow. (Recall from Chapter 13 that the heart must generate a pressure higher than peripheral vessels to maintain the pressure gradient that allows blood to flow through the circulatory system.)

CHAPTER 20
Chapter Test

1. neutral
2. alkaline; acid
3. carbonic anhydrase
4. buffers
5. carbonic
6. acidosis
7. ammonia
8. alkalosis
9. acidosis
10. homeostasis
11. f
12. b
13. j
14. d
15. a
16. h
17. c
18. e
19. g
20. i

Clinical Applications

1. Chronic obstruction of the airways, as in chronic bronchitis, may reduce the respiratory system's rate of CO_2 excretion. This elevates the concentration of carbonic acid (H_2CO_3) in the blood and thus decreases pH—respiratory acidosis. In such chronic disorders, the body *compensates* for this pH imbalance by increasing sodium bicarbonate ($NaHCO_3^-$) levels to increase the pH to normal levels. One would expect to see blood pH within the normal range (because this is *compensated* acidosis) but an elevated level of H_2CO_3 and $NaHCO_3$.

2. There is indeed a connection; Larry's body is attempting to compensate for the reduced blood pH characteristic of metabolic acidosis. By hyperventilating, Larry is increasing the rate of CO_2 excretion by the lungs and thus reducing his blood CO_2 level. This in turn decreases his blood H_2CO_3 level. This sequence of events thus lowers the acid content of Larry's blood and increases blood pH toward the normal range.

3. If aldosterone increases secretion of H^+ into kidney tubules (that is, into the urine) from the blood, then the urine is acidified (gaining H^+) and the blood becomes less acid (losing H^+). Thus the blood pH increases. In hypersecretion of aldosterone, the blood pH may rise dramatically. In hyposecretion, there is less aldosterone present and thus the blood will retain the H^+ it would have otherwise lost, decreasing the blood pH.

CHAPTER 21
Chapter Test

1. testes
2. spermatozoa
3. gonads

4. seminiferous
5. interstitial
6. prostate
7. cervix
8. progesterone
9. endometrium
10. areola
11. menarche
12. 28
13. f
14. j
15. a
16. d
17. h
18. c
19. e
20. g
21. i
22. b
23. a
24. c
25. b
26. d
27. f
28. e
29. h
30. g

Clinical Applications

1. Considering the anatomical position of the prostate under the urinary bladder (see Figure 21-1), the only way to palpate or "feel" the organ with the hands or fingers from the outside of the body is by a digital rectal examination. The prostate is just anterior to the rectum, so any swelling or other physical abnormality will probably be felt easily through the thin rectal wall. Benign prostatic hypertrophy, or any other disorder involving a physical change in the prostate, might be detected this way.

2. According to definition, a couple that does not conceive after 1 year of normal sex without contraception is infertile; therefore Liz and Zeke could be said to be an infertile couple. Liz's PID could be a cause of infertility. PID can cause scarring or other damage that might impair ovulation, obstruct the uterine tubes, or prevent the uterine lining from sustaining a pregnancy. Only one partner needs to be infertile for the couple to be infertile. This couple's physician will likely recommend treatment of the PID.

3. Heather obviously did not follow recommended procedures for breast self-examination because her self-examinations were "brief" and infrequent. She should have asked her physician or another health professional for directions on how to examine her breasts properly. Perhaps Heather would still have missed the lumps, but proper self-examination often reveals lesions that improper examinations do not. Whether Heather has breast cancer cannot be deduced from the information given. The presence of lumps is a sign of *possible* breast cancer but is more likely to be a form of *fibrocystic disease*. Fibrocystic disease is a name for the extremely common, normal changes that sometimes occur in breast tissue. Although the lumps are *probably* benign, they should be studied thoroughly to rule out cancer—they *could be* cancerous, after all, and thus require prompt treatment.

CHAPTER 22
Chapter Test
1. conception; birth
2. birth; death
3. embryology
4. zygote
5. amniotic
6. placenta
7. organogenesis
8. postnatal
9. childhood
10. arteriosclerosis
11. cataract
12. F
13. F
14. T
15. T
16. T
17. T
18. T
19. F
20. T
21. b
22. c

23. e
24. d
25. a

Clinical Applications

1. Until the 1930s, *puerperal fever* (childbed fever) was a leading cause of death associated with childbirth. Caused by infection during the birthing process, it can be avoided by using aseptic or clean technique in and around the delivery area. Mary is justified in her concern, given her family's sad experience and her knowledge of puerperal fever's cause. She was probably assured by the hospital staff that they do take precautions in safeguarding against infections before, during, and after the delivery procedure.

2. Although Abe's grandmother may indeed use this as an excuse for meddling in the affairs of her daughter, it is possible that even moderate drinking during pregnancy can cause congenital abnormalities. Alcohol is a teratogen that easily crosses the placental barrier between the mother's blood and that of the developing offspring. *Fetal alcohol syndrome (FAS)* is a pronounced collection of signs caused by maternal alcohol consumption. FAS may include microcephaly, low birth weight, and mental retardation or other developmental disabilities.

3. Lactose intolerance is also called *lactase deficiency*, a name that implies the mechanism of this condition. Deficiency of the lactose-digesting enzyme *lactase* causes congenital and adult-onset forms of this condition. Affecting a large segment of the human population (as high as 90% in some regions), this condition can be treated by avoiding lactose in the diet. Aileen has chosen to take a tablet that probably contains a powdered form of lactase to help her digest ice cream and thus avoid the unpleasant symptoms of her condition.

CHAPTER 23
Chapter Test

1. genetics
2. chromosome
3. autosomes
4. dominant; recessive
5. carrier
6. sex-linked
7. trisomy
8. osteogenesis imperfecta
9. pedigree
10. karyotype
11. T
12. T
13. T
14. F
15. T
16. F
17. F
18. F
19. f
20. b
21. g
22. e
23. a
24. d
25. c

Clinical Applications

1. Quentin's physician might order a karyotype of Quentin. Blood or other tissue will be taken and cultured. Cells in metaphase will be stained and photographed, producing an image of Quentin's chromosomes. Because Klinefelter's syndrome occurs only when there is more than one X chromosome in a male, a result of XXY, XXXY, or a similar multiple-X result, will confirm that the signs exhibited by Quentin do indeed constitute Klinefelter's syndrome. Such an abnormality usually results from nondisjunction during formation of gametes produced by Quentin's parents.

2. Tay-Sachs disease is a recessive disorder caused by an abnormal gene thought to be located on chromosome 15. Let T represent the normal, dominant gene and let t represent the recessive Tay-Sachs gene. Each parent must have the gene combination Tt, because each has the gene but does not have the disease. If either parent had the combination tt, they would have Tay-Sachs disease and would have died by age 4. Using a Punnett square, as in Figure 23-10, *A*, to determine probability, one would expect a 25% probability (1 in 4 chance) that the infant will exhibit Tay-Sachs disease.

Glossary

abdomen (AB-doe-men) body area between the diaphragm and pelvis

abdominal cavity (ab-DOM-i-nal KAV-i-tee) the cavity containing the abdominal organs

abdominal muscles (ab-DOM-i-nal MUS-els) muscles supporting the anterior aspect of the abdomen

abdominal quadrants (ab-DOM-i-nal KWOD-rants) health professionals divide the abdomen (through the navel) into four areas to help locate specific organs

abdominal regions (ab-DOM-i-nal REE-juns) anatomists have divided the abdomen into nine regions to identify the location of organs

abdominopelvic cavity (ab-DOM-i-no-PEL-vik KAV-i-tee) term used to describe the single cavity containing the abdominal and pelvic organs

abduction (ab-DUK-shun) moving away from the midline of the body, opposite motion of adduction

abruptio placentae (ab-RUP-shee-o plah-SEN-tah) separation of normally positioned placenta from the uterine wall; may result in hemorrhage and death of the fetus and/or mother

absorption (ab-SORP-shun) passage of a substance through a membrane such as skin or mucosa, into blood

accessory organ (ak-SES-o-ree OR-gan) an organ that assists other organs in accomplishing their functions

acetabulum (as-e-TAB-yoo-lum) socket in the hip bone (ox coxa or innominate bone) into which the head of the femur fits

acetylcholine (as-e-til-KO-lean) chemical neurotransmitter

acid-base balance (AS-id base BAL-ans) maintaining the concentration of hydrogen ions in body fluids

acidosis (as-i-DOE-sis) condition in which there is an excessive proportion of acid in the blood

acne vulgaris (AK-nee vul-GAR-is) inflammatory skin condition affecting sebaceous gland ducts; see *comedones*

acquired immune deficiency syndrome (AIDS) (ah-KWIRED i-MYOON de-FISH-en-see SIN-drome) disease in which the human immunodeficiency virus attacks T cells, thereby compromising the body's immune system

acquired immunity (ah-KWIRED i-MYOO-ni-tee) immunity that is obtained after birth through the use of injections or exposure to a harmful agent

acromegaly (ak-ro-MEG-ah-lee) condition caused by hypersecretion of growth hormone after puberty, resulting in enlargement of facial features (for example, jaw, nose), fingers, and toes

acrosome (AK-ro-sohm) specialized structure on the sperm containing enzymes that break down the covering of the ovum to allow entry

actin (AK-tin) contractile protein found in the thin myofilaments of skeletal muscle

action potential (AK-shun po-TEN-shal) nerve impulse

active transport (AK-tiv TRANS-port) movement of a substance into and out of a living cell requiring the use of cellular energy

acute (ah-KYOOT) intense; short in duration—as in *acute disease*

Addison's disease (AD-i-sons di-ZEEZ) disease of the adrenal gland resulting in low blood sugar, weight loss, and weakness

adduction (ah-DUK-shun) moving toward the midline of the body, opposite motion of abduction

adenocarcinoma (ad-e-no-kar-si-NO-mah) cancer of glandular epithelium

adenohypophysis (ad-e-no-hye-POF-i-sis) anterior pituitary gland, which has the structure of an endocrine gland

adenoid (AD-e-noyd) literally, gland-like; adenoids, or pharyngeal tonsils, are paired lymphoid structures in the nasopharynx

adenoma (ad-e-NO-mah) benign tumor of glandular epithelium

adenosine deaminase (ADA) deficiency (ah-DEN-o-sen dee-AM-i-nase de-FISH-en-see) rare, inherited condition in which production of the enzyme adenosine deaminase is deficient, resulting in severe combined immune deficiency (SCID); was the first human disorder treated by gene therapy

adenosine diphosphate (ADP) (ah-DEN-o-sen dye-FOS-fate) molecule similar to adenosine triphosphate but containing only two phosphate groups

adenosine triphosphate (ATP) (ah-DEN-o-sen try-FOS-fate) chemical compound that provides energy for use by body cells

adipose (AD-i-pose) fat tissue

adolescence (ad-o-LES-ens) period between puberty and adulthood

adrenal cortex (ah-DREE-nal KOR-teks) outer portion of adrenal gland that secretes hormones called corticoids

adrenal gland (ah-DREE-nal gland) glands that rest on the top of the kidneys, made up of the cortex and medulla

adrenal medulla (ah-DREE-nal me-DUL-ah) inner portion of adrenal gland that secretes epinephrine and norepinephrine

adrenergic fibers (a-dre-NER-jik FYE-bers) axons whose terminals release norepinephrine and epinephrine

adrenocorticotropic hormone (ACTH) (ah-dree-no-kor-te-ko-TRO-pic HOR-mone) hormone that stimulates the adrenal cortex to secrete larger amounts of hormones

aerobic (air-O-bik) requiring oxygen

adulthood (ah-DULT-hood) period after adolescence

aerobic respiration (air-O-bik res-pi-RAY-shun) the stage of cellular respiration requiring oxygen

aerobic training (air-O-bik TRAIN-ing) continuous vigorous exercise requiring the body to increase its consumption of oxygen and develop the muscles' ability to sustain activity over a long period of time

afferent (AF-fer-ent) carrying or conveying toward the center (for example, an afferent neuron carries nerve impulses toward the central nervous system)

agglutinate (ah-GLOO-tin-ate) antibodies causing antigens to clump or stick together

aging process (AJ-ing PROS-es) the gradual degenerative changes that occur after young adulthood as a person ages

AIDS-related complex (ARC) (AIDS ree-LAY-ted KOM-pleks) a more mild form of AIDS that produces fever, weight loss, and swollen lymph nodes

albinism (AL-bi-nizm) recessive, inherited condition characterized by a lack of the dark brown pigment *melanin* in the skin and eyes, resulting in vision problems and susceptibility to sunburn and skin cancer; ocular albinism is a lack of pigment in the layers of the eyeball

aldosterone (AL-doe-ste-rone) hormone that stimulates the kidney to retain sodium ions and water

alimentary canal (al-e-MEN-tar-ee kah-NAL) the digestive tract as a whole

alkaline phosphatase (AL-kah-lin FOS-fah-tase) enzyme present in blood plasma in high concentration during certain liver and malignant bone marrow disorders

alkalosis (al-kah-LO-sis) condition in which there is an excessive proportion of alkali in the blood, opposite of acidosis

allergen (AL-er-jen) harmless environmental antigen that stimulates an allergic reaction (hypersensitivity reaction) in a susceptible, sensitized person

allergy (AL-er-jee) hypersensitivity of the immune system to relatively harmless environmental antigens

all or none when stimulated, a muscle fiber will contract fully or not at all. Whether a contraction occurs depends on whether the stimulus reaches the required threshold

alopecia (al-o-PEE-she-ah) clinical term referring to hair loss

alpha cell (AL-fah sell) pancreatic cell that secretes glucagon

alveolar duct (al-VEE-o-lar dukt) airway that branches from the smallest bronchioles; alveolar sacs arise from alveolar ducts

alveolar sac (al-VEE-o-lar sak) each alveolar duct ends in several sacs that resemble a cluster of grapes

alveolus (al-VEE-o-lus) literally, a small cavity; alveoli of lungs are microscopic saclike dilations of terminal bronchioles

Alzheimer's disease (ALZ-hye-merz di-ZEEZ) brain disorder of the middle and late adult years characterized by loss of memory and dementia

ameba (ah-MEE-bah) protozoan of changing shape capable of causing infection

amenorrhea (ah-men-o-REE-ah) absence of normal menstruation

amino acid (ah-MEE-no AS-id) structural units from which proteins are built

amniocentesis (am-nee-o-sen-TEE-sis) procedure in which a sample of amniotic fluid is removed with a syringe for use in genetic testing, perhaps to produce a karyotype of the fetus

amniotic cavity (am-nee-OT-ik KAV-i-tee) cavity within the blastocyst that will become a fluid-filled sac in which the embryo will float during development

amphiarthrosis (am-fee-ar-THRO-sis) slightly movable joint such as the joint joining the two pubic bones

amylase (AM-i-lase) enzyme that digests carbohydrates

anabolic steroid (a-nah-BOL-ik STE-royd) hormones that stimulate the building of large molecules, specifically proteins in muscle and bones

anabolism (ah-NAB-o-lizm) cells making complex molecules (for example, hormones) from simpler compounds (for example, amino acids), opposite of catabolism

anaerobic (an-air-O-bik) requiring the absence of oxygen

anal canal (AY-nal kah-NAL) terminal portion of the rectum

anaphase (AN-ah-faze) stage of mitosis; duplicate chromosomes move to poles of dividing cell

anaphylactic shock (an-ah-fi-LAK-tik shock) circulatory failure (shock) caused by a type of severe allergic reaction characterized by blood vessel dilation; may be fatal

anaplasia (an-ah-PLAY-zee-ah) growth of abnormal (undifferentiated) cells, as in a tumor or neoplasm

anatomical position (an-ah-TOM-i-kal po-ZISH-un) the reference position for the body, which gives meaning to directional terms

anatomy (ah-NAT-o-mee) the study of the structure of an organism and the relationships of its parts

androgen (AN-dro-jen) male sex hormone

anemia (ah-NEE-mee-ah) deficient number of red blood cells or deficient hemoglobin

anesthesia (an-es-THEE-zee-ah) loss of sensation

aneurysm (AN-yoo-rizm) abnormal widening of the arterial wall; aneurysms promote the formation of thrombi and also tend to burst

angina pectoris (an-JYE-nah PECK-tor-is) severe chest pain resulting when the myocardium is deprived of sufficient oxygen

angioplasty (AN-jee-o-plas-tee) medical procedure in which vessels occluded by arteriosclerosis are opened (that is, the channel for blood flow is widened)

Angstrom (ANG-strom) unit that is 0.1 mm (1/10,000,000,000 of a meter or about 1/250,000,000 of an inch)

anion (AN-eye-on) negatively charged particle

anorexia (an-o-REK-see-ah) loss of appetite (a symptom, rather than a distinct disorder)

anorexia nervosa (an-o-REK-see-ah ner-VO-sah) behavioral eating disorder characterized by chronic refusal to eat, often related to an abnormal fear of becoming obese

antagonist muscle (an-TAG-o-nist MUS-el) those having opposing actions; for example, muscles that flex the upper arm are antagonists to muscles that extend it

antebrachial (an-tee-BRAY-kee-al) refers to the forearm

antecubital (an-tee-KYOO-bi-tal) refers to the elbow

antenatal medicine (an-tee-NAY-tal MED-i-sin) prenatal medicine

anterior (an-TEER-ee-or) front or ventral; opposite of posterior or dorsal

antibiotic (an-ti-bye-OT-ik) compound usually produced by living organisms that destroys or inhibits microbes

antibody (AN-ti-bod-ee) substance produced by the body that destroys or inactivates a specific substance (antigen) that has entered the body

antibody-mediated immunity (AN-ti-bod-ee MEE-dee-ate-ed i-MYOO-ni-tee) immunity that is produced when antibodies make antigens unable to harm the body

antidiuretic hormone (ADH) (an-ti-dye-yoo-RET-ik HOR-mone) hormone produced in the posterior pituitary gland to regulate the balance of water in the body by accelerating the reabsorption of water

antigen (AN-ti-jen) substance that, when introduced into the body, causes formation of antibodies against it

antrum (AN-trum) cavity

anuria (ah-NOO-ree-ah) absence of urine

anus (AY-nus) distal end or outlet of the rectum

aorta (ay-OR-tah) main and largest artery in the body

aortic body (ay-OR-tik BOD-ee) small cluster of chemosensitive cells that respond to carbon dioxide and oxygen levels

aortic semilunar valve (ay-OR-tic sem-i-LOO-nar valve) valve between the aorta and left ventricle that prevents blood from flowing back into the ventricle

apex (A-peks) pointed end of a conical structure

aplastic anemia (ah-PLAS-tik ah-NEE-mee-ah) blood disorder characterized by a low red blood cell count, caused by destruction of myeloid tissue in the bone marrow

apnea (AP-nee-ah) temporary cessation of breathing

apocrine (AP-o-krin) sweat glands located in the axilla and genital regions; these glands enlarge and begin to function at puberty

appendicitis (ah-pen-di-SYE-tis) inflammation of the vermiform appendix

appendage (ah-PEN-dij) something that is attached

appendicular (a-pen-DIK-yoo-lar) refers to the upper and lower extremities of the body

appendicular skeleton (a-pen-DIK-yoo-lar SKEL-e-ton) the bones of the upper and lower extremities of the body

aqueous humor (AY-kwee-us HYOO-mor) watery fluid that fills the anterior chamber of the eye, in front of the lens

arachnoid (ah-RAK-noyd) delicate, weblike middle membrane covering the brain, the meninges

areola (ah-REE-o-lah) small space; the pigmented ring around the nipple

areolar (ah-REE-o-lar) a type of connective tissue consisting of fibers and a variety of cells embedded in a loose matrix of soft, sticky gel

arrector pili (ah-REK-tor PYE-lie) smooth muscles of the skin, which are attached to hair follicles; when contraction occurs, the hair stands up, resulting in "goose flesh"

arrhythmia (ah-RITH-mee-ah) term referring to any abnormality of cardiac rhythm

arteriole (ar-TEER-ee-ole) small branch of an artery

arteriosclerosis (ar-tee-ree-o-skle-RO-sis) hardening of arteries; materials such as lipids (as in atherosclerosis) accumulate in arterial walls, often becoming hardened via calcification

artery (AR-ter-ee) vessel carrying blood away from the heart

arthritis (ar-THRY-tis) inflammatory joint disease, characterized by inflammation of the synovial membrane and a variety of systemic signs or symptoms

arthropod (AR-thro-pod) type of animal capable of infesting or parasitizing humans

articular cartilage (ar-TIK-yoo-lar KAR-ti-lij) cartilage covering the joint ends of bones

articulation (ar-tik-yoo-LAY-shun) joint

artificial kidney (ar-ti-FISH-al KID-nee) mechanical device that removes wastes from the blood that would normally be removed by the kidney

artificial pacemaker (ar-ti-FISH-al PAYS-may-ker) an electrical device that is implanted into the heart to treat a heart block

aseptic technique (ay-SEP-tik tek-NEEK) approach to limiting the spread of infection by preventing or reducing contacts with contaminated surfaces

asexual (a-SEKS-yoo-al) one-celled plants and bacteria that do not produce specialized sex cells

assimilation (ah-sim-i-LAY-shun) when food molecules enter the cell and undergo chemical changes

asthma (AZ-mah) obstructive pulmonary disorder characterized by recurring spasms of muscles in bronchial walls accompanied by edema and mucus production, making breathing difficult

astigmatism (ah-STIG-mah-tizm) irregular curvature of the cornea or lens that impairs refraction of a well-focused image in the eye

astrocyte (AS-tro-site) a neuroglial cell

atelectasis (at-e-LEK-tah-sis) total or partial collapse of the alveoli of the lung

atherosclerosis (ath-er-o-skle-RO-sis) type of "hardening of the arteries" in which lipids and other substances build up on the inside wall of blood vessels.

atrial natriuretic hormone (ANH) (AY-tree-al na-tree-yoo-RET-ik HOR-mone) hormone secreted by the heart cells that regulates fluid and electrolyte homeostasis

atrioventricular valves (ay-tree-o-ven-TRIK-yoo-lar valves) two valves that separate the atrial chambers from the ventricles

atrium (AY-tree-um) chamber or cavity; for example, atrium of each side of the heart

atrophy (AT-ro-fee) wasting away of tissue; decrease in size of a part; sometimes referred to as disuse atrophy

auditory tube (AW-di-toe-ree tube) tube that connects the throat with the middle ear

auricle (AW-ri-kul) part of the ear attached to the side of the head; earlike appendage of each atrium of the heart

autoimmunity (aw-toe-i-MYOO-ni-tee) process in which a person's immune system attacks the person's own body tissues—the underlying cause of several diseases

autonomic effector (aw-toe-NOM-ik ef-FEK-tor) tissues to which autonomic neurons conduct impulses

autonomic nervous system (ANS) (aw-toe-NOM-ik NER-vus SIS-tem) division of the human nervous system that regulates involuntary actions

autonomic neuron (aw-toe-NOM-ik NOO-ron) motor neurons that make up the autonomic nervous system

autosome (AW-toe-sohm) one of the 44 (22 pairs) chromosomes in the human genome besides the two sex chromosomes; means "same body," referring to the fact that each member of a pair of autosomes match in size and other structural features

AV bundle (A V BUN-dul) fibers in the heart that relay a nerve impulse from the AV node to the ventricles; also known as the bundle of HIS

avitaminosis (ay-vye-tah-mi-NO-sis) general name for any condition resulting from a vitamin deficiency

axial (AK-see-al) refers to the head, neck, and torso, or trunk of the body

axial skeleton (AK-see-al SKEL-e-ton) the bones of the head, neck, and torso

axilla (AK-sil-ah) refers to the armpit

axon (AK-son) nerve cell process that transmits impulses away from the cell body

B cells (B sells) a lymphocyte; activated B cells develop into plasma cells, which secrete antibodies into the blood

bacillus (bah-SIL-us) rod-shaped bacterium

bacterium (bak-TEE-ree-um) microbe capable of causing disease, it is a primitive, single-celled organism without membranous organelles

bartholinitis (bar-toe-lin-NYE-tis) inflammation of the Bartholin's glands, accessory organs of the female reproductive tract

Bartholin's glands (BAR-toe-lins glands) gland, located on either side of the vaginal outlet, that secretes mucuslike lubricating fluid; also know as greater vestibular glands

basal ganglia (BAY-sal GANG-glee-ah) islands of gray matter located in the cerebral cortex that are responsible for automatic movements and postures

basal metabolic rate (BMR) (BAY-sal met-ah-BOL-ik rate) number of calories of heat that must be produced per hour by catabolism to keep the body alive, awake, and comfortably warm

basement membrane (BASE-ment MEM-brane) the connective tissue layer of the serous membrane that holds and supports the epithelial cells

basophil (BAY-so-fil) white blood cell that stains readily with basic dyes

Bell's palsy (bellz PAWL-zee) temporary or permanent paralysis of facial features caused by damage to the seventh cranial nerve (facial nerve)

benign (be-NINE) refers to a tumor or neoplasm that does not metastasis or spread to different tissues

benign prostatic hypertrophy (be-NINE pros-TAT-ik hye-PER-tro-fee) benign enlargement of the prostate, a condition common in older males

benign tumor (be-NINE TOO-mer) a relatively harmless neoplasm

beta cell (BAY-tah sell) pancreatic islet cell that secretes insulin

bicarbonate loading (bye-KAR-bo-nate LOHD-ing) ingesting large amounts of sodium bicarbonate to counteract the effects of lactic acid buildup, thereby reducing fatigue; however, there are potential dangerous side effects

biceps brachii (BYE-seps BRAY-kee-eye) the primary flexor of the forearm

biceps femoris (BYE-seps FEM-o-ris) powerful flexor of the lower leg

bicuspid (bye-KUS-pid) premolars

bicuspid valve (bye-KUS-pid valve) one of the two AV valves, it is located between the left atrium and ventricle and is sometimes called the mitral valve

bile (bile) substance that reduces large fat globules into smaller droplets of fat that are more easily broken down

bile duct (bile dukt) duct that drains bile into the small intestine and is formed by the union of the comman hepatic and cystic ducts

biological filtration (bye-o-LOJ-e-kal fil-TRAY-shun) process in which cells alter the contents of the filtered fluid

biopsy (BYE-op-see) procedure in which living tissue is removed from a patient for laboratory examination, as in determining the presence of cancer cells

blackhead (BLACK-hed) when sebum accumulates, darkens, and enlarges some of the ducts of the sebaceous glands; also known as a comedo

bladder (BLAD-der) a sac, usually referring to the urinary bladder

blastocyst (BLAS-toe-sist) postmorula stage of developing embryo; hollow ball of cells

blister (BLIS-ter) fluid-filled skin lesion; see *vesicle*

blood-brain barrier (blud brane BAR-ee-er) structural and functional barrier formed by astrocytes and blood vessel walls in the brain, it prevents some substances from diffusing from the blood into brain tissue

blood doping (blud DOE-ping) a practice used to improve athletic performance by removing red blood cells weeks before an event and then reinfusing them just before competition to increase the oxygen-carrying capacity of the blood

blood pressure (blud PRESH-ur) pressure of blood in the blood vessels, expressed as systolic pressure/diastolic pressure (for example, 120/80 mmHg)

blood pressure gradient (blud PRESH-ur GRAY-dee-ent) the difference between two blood pressures in the body

blood types (blud tipes) the different types of blood that are identified by certain antigens in red blood cells (A, B, AB, O and Rh-negative or Rh-positive)

body (BOD-ee) unified and complex assembly of structurally and functionally interactive components

body composition (BOD-ee com-po-ZISH-un) assessment to identify the percentage of the body that is lean tissue and the percentage that is fat

bolus (BO-lus) a small, rounded mass of masticated food to be swallowed

bone (bone) highly specialized connective tissue whose matrix is hard and calcified

bone marrow (bone MAR-o) soft material that fills cavities of the bones; red bone marrow is vital to blood cell formation; yellow bone marrow is inactive fatty tissue

bone marrow transplant (bone MAR-o TRANS-plant) treatment in which healthy blood-forming marrow tissue from a donor is intravenously introduced into a recipient

bony labyrinth (BONE-ee LAB-i-rinth) the fluid-filled complex maze of three spaces (the vestibule, semicircular canals, and cochlea) in the temporal bone

Bowman's capsule (BO-mens KAP-sul) the cup-shaped top of a nephron that surrounds the glomerulus

brachial (BRAY-kee-al) pertaining to the arm

bradycardia (bray-de-KAR-dee-ah) slow heart rhythm (below 60 beats/minute)

breast (brest) anterior aspect of the chest; in females, also an accessory sex organ

bronchi (BRONG-ki) the branches of the trachea

bronchiole (BRONG-kee-ole) small branch of a bronchus

bronchitis (brong-KYE-tis) inflammation of the bronchi of the lungs, characterized by edema and excessive mucus production that causes coughing and difficulty in breathing (especially expiration)

buccal (BUK-al) pertaining to the cheek

buffer (BUF-er) compound that combines with an acid or with a base to form a weaker acid or base, thereby lessening the change in hydrogen-ion concentration that would occur without the buffer

buffer pairs (BUF-er pairs) two kinds of chemical substances that prevent a sharp change in the pH of a fluid; for example, sodium bicarbonate ($NaHCO_3$) and carbonic acid (H_2CO_3)

bulbourethral gland (BUL-bo-yoo-REE-thral gland) small glands located just below the prostate gland whose mucuslike secretions lubricate the terminal portion of the urethra and contribute less than 5% of the seminal fluid volume; also known as Cowper's glands

bulimia (bu-LEEM-ee-ah) behavioral eating disorder characterized by an alternating pattern of overeating followed by self-denial (and perhaps purging of GI contents)

bundle of His (BUN-dul of his) see *AV bundle*

burn (bern) an injury to tissues resulting from contact with heat, chemicals, electricity, friction, or radiant and electromagnetic energy; classified into three categories, depending on the number of tissue layers involved

bursae (BER-see) small, cushionlike sacs found between moving body parts, making movement easier

bursitis (ber-SYE-tis) inflammation of a bursa

cachexia (kah-KEK-see-ah) syndrome associated with cancer and other chronic diseases that involves loss of appetite, weight loss, and general weakness

calcaneus (kal-KAY-nee-us) heel bone; largest tarsal in the foot

calcitonin (kal-si-TOE-nin) a hormone secreted by the thyroid that decreases calcium in the blood

calculi (KAL-kyoo-lie) hard, crystalline stones that form in the lumen of hollow organs such as the gall bladder or liver (biliary calculi) or renal passages (renal calculi)

callus (KAL-us) bony tissue that forms a sort of collar around the broken ends of fractured bone during the healing process

calorie (c) (KAL-or-ree) heat unit; the amount of heat needed to raise the temperature of 1 g of water 1° C

Calorie (C) (KAL-or-ree) heat unit; kilocalorie; the amount of heat needed to raise the temperature of 1 kilogram of water 1° C

calyx (KAY-liks) cup-shaped division of the renal pelvis

canaliculi (kan-ah-LIK-yoo-lie) an extremely narrow tubular passage of channel in compact bone

canine tooth (KAY-nine tooth) the tooth with the longest crown and the longest root, which is located lateral to the second incisor

capillary (KAP-i-lair-ee) tiny vessels that connect arterioles and venules

capillary blood pressure (KAP-i-lair-ee blud PRESH-ur) the blood pressure found in the capillary vessels

capsule (KAP-sul) found in diarthrotic joints, holds the bones of joints together while allowing movement; made of fibrous connective tissue lined with a smooth slippery synovial membrane

carbaminohemoglobin (kar-bam-ee-no-hee-mo-GLO-bin) compound formed by the union of carbon dioxide with hemoglobin

carbohydrate (kar-bo-HYE-drate) organic compounds containing carbon, hydrogen, and oxygen in certain specific proportions; for example, sugars, starches, and cellulose

carbohydrate loading (kar-bo-HYE-drate LOHD-ing) a method used by athletes to increase the stores of muscle glycogen, allowing more sustained aerobic exercise

carbonic anhydrase (kar-BON-ik an-HYE-drays) the enzyme that converts carbon dioxide into carbonic acid

carbuncle (KAR-bung-kl) a mass of connected boils, pus-filled lesions associated with hair follicle infections; see *furuncle*

carcinogen (kar-SIN-o-jen) substance that promotes the development of cancer

cardiac (KAR-dee-ak) refers to the heart

cardiac cycle (KAR-dee-ak MUS-el) each complete heart beat, including contraction and relaxation of the atria and ventricles

cardiac muscle (KAR-dee-ak SYE-kul) the specialized muscle that makes up the heart

cardiac output (KAR-dee-ak OUT-put) volume of blood pumped by one ventricle per minute

cardiac sphincter (KAR-dee-ak SFIN-GK-ter) a ring of muscle between the stomach and esophagus that prevents food from reentering the esophagus when the stomach contracts

cardiac tamponade (KAR-dee-ak tam-po-NAHD) compression of the heart caused by fluid buildup in the pericardial space, as in pericarditis or mechanical damage to the pericardium

cardiogenic shock (kar-dee-o-JEN-ik shock) circulatory failure (shock) caused by heart failure; literally "heart-caused" shock

cardiologist (kar-dee-OL-o-jist) physician or researcher who specializes in the structure and function of the heart and associated structures

cardiomyopathy (kar-dee-o-my-OP-ah-thee) general term for disease of the myocardium (heart muscle)

cardiopulmonary resuscitation (CPR) (kar-dee-o-PUL-mo-nair-ee ree-sus-i-TAY-shun) combined external cardiac (heart) massage and artificial respiration

cardiovascular (kar-dee-o-VAS-kyoo-lar) pertaining to the heart and blood vessels

caries (KARE-eez) decay of teeth or of bone

carotid body (kah-ROT-id BOD-ee) chemoreceptor located in the carotid artery that detects changes in oxygen, carbon dioxide, and blood acid levels

carpal (KAR-pal) pertaining to the wrist

carpal tunnel syndrome (KAR-pal TUN-el SIN-drome) muscle weakness, pain, and tingling in the radial side (thumb side) of the wrist, hand, and fingers—perhaps radiating to the forearm and shoulder; caused by compression of the median nerve within the carpal tunnel (a passage along the ventral concavity of the wrist)

carrier (KARE-ee-er) in genetics, a person who possesses the gene for a recessive trait, but who does not actually exhibit the trait

cartilage (KAR-ti-lij) a specialized, fibrous connective tissue that has the consistency of a firm plastic or gristle-like gel

catabolism (kah-TAB-o-lizm) breakdown of food compounds or cytoplasm into simpler compounds; opposite of anabolism, the other phase of metabolism

catalyst (KAT-ah-list) chemical that speeds up reactions without being changed itself

cataract (KAT-ah-rakt) opacity of the lens of the eye

catecholamines (kat-e-kol-AM-eens) norepinephrine and epinephrine

catheterization (kath-e-ter-i-ZAY-shun) passage of a flexible tube (catheter) into the bladder through the urethra for the withdrawal of urine (urinary catheterization)

cation (KAT-eye-on) positively charged particle

cavity (KAV-i-tee) hollow place or space in a tooth; dental caries

cecum (SEE-kum) blind pouch; the pouch at the proximal end of the large intestine

cell (sell) the basic biological and structural unit of the body consisting of a nucleus surrounded by cytoplasm and enclosed by a membrane

cell body (sell BOD-ee) the main part of a neuron from which the dendrites and axons extend

cell-mediated immunity (sell MEE-dee-ate-ed i-MYOO-ni-tee) resistance to disease organisms resulting from the actions of cells; chiefly sensitized T cells

cellular respiration (SELL-yoo-lar res-pi-RAY-shun) enzymes in the mitochondrial wall and matrix using oxygen to break down glucose and other nutrients to release energy needed for cellular work

centimeter (SEN-ti-mee-ter) 1/100 of a meter; approximately 2.5 cm equal 1 inch

central nervous system (CNS) (SEN-tral NER-vus SIS-tem) the brain and spinal cord

central venous pressure (SEN-tral VEE-nus PRESH-ur) venous blood pressure within the right atrium that influences the pressure in the large peripheral veins

centriole (SEN-tree-ol) one of a pair of tiny cylinders in the centrosome of a cell; believed to be involved with the spindle fibers formed during mitosis

centromere (SEN-tro-meer) a beadlike structure that attaches one chromatid to another during the early stages of mitosis

cephalic (se-FAL-ik) refers to the head

cerebellum (sair-e-BELL-um) the second largest part of the human brain that plays an essential role in the production of normal movements

cerebral cortex (se-REE-bral KOR-teks) a thin layer of gray matter made up of neuron dentrites and cell bodies that compose the surface of the cerebrum

cerebral palsy (se-REE-bral PAWL-zee) abnormal condition characterized by permanent, nonprogressive paralysis (usually spastic paralysis) of one or more extremities caused by damage to motor control areas of the brain before, during, or shortly after birth

cerebrospinal fluid (CSF) (se-ree-bro-SPY-nal FLOO-id) fluid that fills the subarachnoid space in the brain and spinal cord and in the cerebral ventricles

cerebrovascular accident (CVA) (se-ree-bro-VAS-kyoo-lar AK-si-dent) a hemorrhage or cessation of blood flow through cerebral blood vessels resulting in destruction of neurons; commonly called a stroke

cerebrum (se-REE-brum) the largest and uppermost part of the human brain that controls consciousness, memory, sensations, emotions, and voluntary movements

cerumen (se-ROO-men) ear wax

ceruminous gland (se-ROO-mi-nus gland) gland that produces a waxy substance called cerumen (ear wax)

cervical (SER-vi-kal) refers to the neck

cervicitis (ser-vi-SYE-tis) inflammation of the cervix of the uterus

cervix (SER-viks) neck; any necklike structure

cesarean section (se-SAIR-ee-an SEK-shun) surgical removal of a fetus, often through an incision of the skin and uterine wall; also called C-section

chemoreceptors (kee-mo-ree-SEP-tors) receptors that respond to chemicals and are responsible for taste and smell

chemotaxis (kee-mo-TAK-sis) process in which white blood cells move toward the source of inflammation mediators

chemotherapy (kee-mo-THER-ah-pee) technique of using chemicals to treat disease (for example, infections, cancer)

chest (chest) thorax

Cheyne-Stokes respiration (CSR) (chain stokes res-pi-RAY-shun) pattern of breathing associated with critical conditions such as brain injury or drug overdose and characterized by cycles of apnea and hyperventilation

childhood (CHILD-hood) from infancy to puberty

chlamydia (klah-MID-ee-ah) small bacterium that infects human cells as an obligate parasite

cholecystitis (koh-lee-sis-TIE-tis) inflammation of the gallbladder

cholecystokinin (CCK) (ko-le-sis-toe-KYE-nin) hormone secreted from the intestinal mucosa of the duodenum that stimulates the contraction of the gall bladder, resulting in bile flowing into the duodenum

cholelithiasis (koh-lee-li-THYE-ah-sis) condition of having gall (bile) stones, hard mineral deposits that may form and collect in the gallbladder

chondrocyte (KON-dro-site) cartilage cell

chondroma (kon-DRO-mah) benign tumor of cartilage

chondrosarcoma (kon-dro-sar-KOH-mah) cancer of cartilage tissue

chordae tendineae (KOR-dee ten-DIN-ee) stringlike structures that attach the AV valves to the wall of the heart

chorion (KO-ree-on) develops into an important fetal membrane in the placenta

chorionic gonadotropins (ko-ree-ON-ik go-na-doe-TRO-pins) hormones that are secreted as the uterus develops during pregnancy

chorionic villi (ko-ree-ON-ik VIL-eye) connect the blood vessels of the chorion to the placenta

chorionic villus sampling (ko-ree-ON-ik VIL-lus SAM-pling) procedure in which a tube is inserted through the (uterine) cervical opening and a sample of the chorionic tissue surrounding a developing embryo is removed for genetic testing; compare **amniocentesis**

choroid (KO-royd) middle layer of the eyeball that contains a dark pigment to prevent the scattering of incoming light rays

choroid plexus (KO-royd PLEK-sus) a network of brain capillaries that are involved with the production of cerebrospinal fluid

chromatids (KRO-mah-tids) a chromosome strand

cholinergic fiber (ko-lin-ER-jik FYE-ber) axons whose terminals release acetylcholine

chromatin granules (KRO-mah-tin GRAN-yools) deep-staining substance in the nucleus of cells; divides into chromosomes during mitosis

chromosome (KRO-mo-sohm) DNA molecule that has coiled to form a compact mass during mitosis or meiosis; each chromosome is composed of regions called genes, each of which transmits hereditary information

chronic (KRON-ik) long-lasting, as in *chronic disease*

chronic obstructive pulmonary disease (COPD) (KRON-ik ob-STRUK-tiv PUL-mo-nair-ee di-ZEEZ) general term referring to a group of disorders characterized by progressive, irreversible obstruction of air flow in the lungs; see *bronchitis, emphysema, asthma*

chyme (kime) partially digested food mixture leaving the stomach

cilia (SIL-ee-ah) hairlike projections of cells

ciliate (SIL-ee-at) type of protozoan having cilia

circulatory shock (SER-kyoo-lah-tor-ee shock) failure of the circulatory system to deliver adequate oxygen to the tissues of the body

circulatory system (SER-kyoo-lah-tor-ee SIS-tem) the system that supplies transportation for cells of the body

circumcision (ser-kum-SIZH-un) surgical removal of the foreskin or prepuce

cirrhosis (sir-RO-sis) degeneration of liver tissue characterized by the replacement of damaged liver tissue with fibrous or fatty connective tissue

cisterna chyli (sis-TER-nah KYE-lye) an enlarged pouch on the thoracic duct that serves as a storage area for lymph moving towards its point of entry in to the venous system

citric acid cycle (SIT-rik AS-id SYE-kul) the second series of chemical reactions in the process of glucose metabolism; it is an aerobic process

clavicle (KLAV-i-kul) collar bone, connects the upper extremity to the axial skeleton

cleavage furrow (KLEEV-ij FUR-o) appears at the end of anaphase and begins to divide the cell into two daughter cells

clitoris (KLIT-o-ris) erectile tissue located within the vestibule of the vagina

clone (klone) any of a family of many identical cells descended from a single "parent" cell

closed fracture (closed FRAK-chur) simple fracture; a bone fracture in which the skin is not pierced by bone fragments

coccus (KOK-us) spherical bacterial cell

cochlea (KOKE-lee-ah) snail shell or structure of similar shape

cochlear duct (KOKE-lee-ar dukt) membranous tube within the bony cochlea

co-dominance (ko-DOM-i-nance) in genetics, a form of dominance in which two dominant versions of a trait are both expressed in the same individual

colitis (ko-LIE-tis) any inflammatory condition of the colon and/or rectum

collagen (KOL-ah-jen) principle organic constituent of connective tissue

collecting tubule (ko-LEK-ting TOO-byool) a straight part of a renal tubule formed by distal tubules of several nephrons joining together

colloid (KOL-oyd) dissolved particles with diameters of 1 to 100 millimicrons (1 millimicron equals about 1/25,000,000 of an inch)

colon (KO-lon) intestine

color blindness (KUL-or BLIND-ness) X-linked inherited condition in which one or more photopigments in the cones of the retina are abnormal or missing

colorectal cancer (kol-o-REK-tal KAN-ser) common form of cancer, usually adenocarcinoma, associated with advanced age, low-fiber/high-fat diet, and genetic predisposition

columnar (ko-LUM-nar) shape in which cells are higher than they are wide

combining sites (kom-BINE-ing sites) antigen-binding sites, antigen receptor regions on antibody molecule; shape of each combining site is complementary to shape of a specific antigen

comedones (kom-e-DONS) (singular *comedo*) inflamed lesions associated with early stages of acne formed when sebaceous gland ducts become blocked

comminuted fracture (KOM-i-noo-ted FRAK-chur) bone fracture characterized by many bone fragments

communicable (ko-MYOO-ni-kah-bl) able to spread from one individual to another

compact bone (kom-PAKT bone) dense bone

compensated metabolic acidosis (KOM-pen-say-ted met-ah-BOL-ik as-i-DOE-sis) when metabolic acidosis occurs and the body is able to adjust to return the blood pH to near normal levels

complement (KOM-ple-ment) any of several inactive enzymes normally present in blood, which when activated kill foreign cells by dissolving them

complementary base pairing (kom-ple-MEN-ta-ree base PAIR-ing) bonding purines and pyridimes in DNA; adenine always binds with thymine, and cytosine always binds with guanine

complement fixation (KOM-ple-ment fik-SAY-shun) highly specialized antigen-antibody complexes are formed to destroy a foreign cell

complete fracture (kom-PLEET FRAK-chur) bone fracture characterized by complete separation of bone fragments

computed tomography (CT) (kom-PYOO-ted to-MOG-rah-fee) radiographic imaging technique in which a patient is scanned with x-rays and a computer constructs an image that appears to be a cut section of the person's body

concave (KON-kave) a rounded, somewhat depressed surface

concentric lamella (kon-SEN-trik lah-MEL-ah) ring of calcified matrix surrounding the Haversian canal

conchae (KONG-kee) shell-shaped structure; for example, bony projections into the nasal cavity

conduction (kon-DUK-shun) transfer of heat energy to the skin and then the external environment

cone (cone) receptor cell located in the retina that is stimulated by bright light

congenital (kon-JEN-i-tal) term that refers to a condition present at birth; congenital conditions may be inherited or may be acquired in the womb or during delivery

congestive heart failure (CHF) (kon-JES-tiv hart FAIL-yoor) left heart failure; inability of the left ventricle to pump effectively, resulting in congestion in the systemic and pulmonary circulations

conjunctiva (kon-junk-TIE-vah) mucous membrane that lines the eyelids and covers the sclera (white portion)

conjunctivitis (kon-junk-ti-VYE-tis) inflammation of the conjunctiva, usually due to irritation, infection, or allergy

connective tissue (ko-NEK-tiv TISH-yoo) most abundant and widely distributed tissue in the body and has numerous functions

connective tissue membrane (ko-NEK-tiv TISH-yoo MEM-brane) one of the two major types of body membranes; composed exclusively of various types of connective tissue

constipation (kon-sti-PAY-shun) condition caused by decreased motility of the large intestine, resulting in the formation of small, hard feces and difficulty in defecation

contact dermatitis (KON-takt der-mah-TIE-tis) a local skin inflammation lasting a few hours or days after being exposed to an antigen

continuous ambulatory peritoneal dialysis (CAPD) (kon-TIN-yoo-us AM-byoo-lah-tor-ee pair-i-toe-NEE-al dye-AL-i-sis) an alternative form of treatment for renal failure rather than the more complex and expensive hemodialysis

contractile unit (kon-TRAK-til YOO-nit) the sarcomere, the basic functional unit of skeletal muscle

contractility (kon-TRAK-til-i-tee) ability to contract a muscle

contraction (kon-TRAK-shun) ability of muscle cells to shorten or contract

contusion (kon-TOO-zhun) local injury caused by mechanical trauma characterized by limited hemorrhaging under the skin, as in a muscle contusion or skin contusion caused by a blow to the body; a bruise

convection (kon-VEK-shun) transfer of heat energy to air that is flowing away from the skin

convex (KON-veks) a rounded, somewhat elevated surface

coronal (ko-RO-nal) literally "like a crown"; a coronal plane divides the body or an organ into anterior and posterior regions

coronary artery (KOR-o-nair-ee AR-ter-ee) the first artery to branch off the aorta, supplies blood to the myocardium (heart muscle)

coronary bypass surgery (KOR-o-nair-ee BYE-pass SER-jer-ee) surgery to relieve severely restricted coronary blood flow; veins are taken from other parts of the body to bypass the partial blockage

coronary circulaton (KOR-o-nair-ee ser-kyoo-LAY-shun) delivery of oxygen and removal of waste product from the myocardium (heart muscle)

coronary embolism (KOR-o-nair-ee EM-bo-lizm) blocking of a coronary blood vessel by a clot

coronary heart disease (KOR-o-nair-ee hart di-ZEEZ) disease (blockage or other deformity) of the vessels that supply the myocardium (heart muscle); one of the leading causes of death among adults in the United States

coronary sinus (KOR-o-nair-ree SYE-nus) area that receives deoxygenated blood from the coronary veins and empties into the right atrium

coronary thrombosis (KOR-o-nair-ree throm-BO-sis) formation of a blood clot in a coronary blood vessel

coronary vein (KOR-o-nair-ee vane) any vein that carries blood from the myocardial capillary beds to the coronary sinus

cor pulmonale (kor pul-mo-NA-lee) failure of the right atrium and ventricle to pump blood effectively, resulting from obstruction of pulmonary blood flow

corpora cavernosa (KOR-por-ah kav-er-NO-sah) two columns of erectile tissue found in the shaft of the penis

corpus callosum (KOR-pus kal-LO-sum) where the right and left cerebral hemispheres are joined

corpus luteum (KOR-pus LOO-tee-um) a hormone-secreting glandular structure transformed after ovulation from a ruptured follicle; it secretes chiefly progesterone with some estrogen secreted as well

corpus spongiosum (KOR-pus spun-jee-O-sum) a column of erectile tissue surrounding the urethra in the penis

cortex (KOR-teks) outer part of an internal organ; for example, the outer part of the cerebrum and of the kidneys

corticoids (KOR-ti-koyds) hormones secreted by the three cell layers of the adrenal cortex

cramps (kramps) painful muscle spasms (involuntary twitches) that result from irritating stimuli, as in mild inflammation, or from ion imbalances

cranial (KRAY-nee-al) toward the head

cranial cavity (KRAY-nee-al KAV-i-tee) space inside the skull that contains the brain

cranial nerve (KRAY-nee-al nerv) any of twelve pairs of nerves that attach to the undersurface of the brain and conduct impulses between the brain and structures in the head, neck, and thorax

craniosacral (kray-nee-o-SAY-kral) pertaining to parasympathetic nerves

cranium (KRAY-nee-um) bony vault made up of eight bones that encases the brain

crenation (kre-NAY-shun) abnormal notching in an erythrocyte due to shrinkage after suspension in a hypertonic solution

cretinism (KREE-tin-izm) dwarfism caused by hyposecretion of the thyroid gland

crista ampullaris (KRIS-tah am-pyoo-LAIR-is) a specialized receptor located within the semicircular canals that detects head movements

crown (krown) topmost part of an organ or other structure

crural (KROOR-al) refers to the leg

crust (krust) scab; area of the skin covered by dried blood or exudate

cryptorchidism (krip-TOR-ki-dizm) undescended testicles

cubital (KYOO-bi-tal) refers to the elbow

cuboid (KYOO-boyd) resembling a cube

cuboidal (KYOO-boyd-al) cell shape resembling a cube

Cushing's syndrome (KOOSH-ings SIN-drome) condition caused by the hypersecretion of glucocorticoids fom the adrenal cortex

cuspid (KUS-pid) canine tooth, serves to pierce or tear food being eaten

cutaneous (kyoo-TANE-ee-us) pertaining to the skin

cutaneous membrane (ku-TANE-ee-us MEM-brane) primary organ of the integumentary system; the skin

cuticle (KYOO-ti-kul) skin fold covering the root of the nail

cyanosis (sye-ah-NO-sis) condition in which light-skinned individuals exhibit a bluish coloration resulting from relatively low oxygen content in the arterial blood; literally "blue condition"

cyclic AMP (SIK-lik A M P) one of several second messengers that delivers information inside the cell and thus regulates the cell's activity

cystic duct (SIS-tik dukt) joins with the common hepatic duct to form the common bile duct

cystic fibrosis (SIS-tik fye-BRO-sis) inherited disease involving abnormal chloride ion (Cl⁻) transport; causes secretion of abnormally thick mucus and other problems

cystitis (sis-TIE-tis) inflammation or infection of the urinary bladder

cytoplasm (SYE-toe-plazm) the gel-like substance of a cell exclusive of the nucleus and other organelles

deciduous (de-SID-yoo-us) temporary; shedding at a certain stage of growth; for example, deciduous teeth, which are commonly referred to as baby teeth

decubitus ulcer (de-KYOO-bi-tus UL-ser) pressure sore that often develops when lying in one position for prolonged periods

deep (deep) farther away from the body's surface

deglutition (deg-loo-TISH-un) swallowing

dehydration (dee-hye-DRAY-shun) clinical term that refers to an abnormal loss of fluid from the body's internal environment

deltoid (DEL-toyd) triangular; for example, the deltoid muscle

dementia (de-MEN-she-ah) syndrome of brain abnormalities including loss of memory, shortened attention span, personality changes, reduced intellectual capacity, and motor dysfunction

dendrite (DEN-drite) branching or treelike; a nerve cell process that transmits impulses toward the body

dense bone (dense bone) bone that has a hard, dense outer layer

deoxyribonucleic acid (DNA) (dee-ok-see-rye-bo-NOO-klee-ik AS-id) genetic material of the cell that carries the chemical "blueprint" of the body

depilatories (de-PIL-ah-toe-rees) hair removers

depolarization (dee-po-lar-i-ZAY-shun) the electrical activity that triggers a contraction of the heart muscle

dermal-epidermal junction (DER-mal-EP-i-der-mal JUNK-shun) junction between the thin epidermal layer of the skin and the dermal layer providing support for the epidermis

dermal papillae (DER-mal pah-PIL-ee) upper region of the dermis that forms part of the dermal-epidermal junction and forms the ridges and grooves of fingerprints

dermatitis (der-mah-TIE-tis) general term referring to any inflammation of the skin

dermatomes (DER-mah-tohms) skin surface areas supplied by a single spinal nerve

dermatosis (der-mah-TOE-sis) general term meaning "skin condition"

dermis (DER-mis) the deeper of the two major layers of the skin, composed of dense fibrous connective tissue interspersed with glands, nerve endings, and blood vessels; sometimes called the "true skin"

developmental process (de-vel-op-MEN-tal PROSS-es) changes and functions occurring during a human's early years as the body becomes more efficient and more effective

deviated septum (DEE-vee-ay-ted SEP-tum) abnormal condition in which the nasal septum is far from its normal position, possibly obstructing normal nasal breathing

diabetes insipidus (dye-ah-BEE-teez in-SIP-i-dus) condition resulting from hyposecretion of ADH in which large volumes of urine are formed and, if left untreated, may cause serious health problems

diabetes mellitus (dye-ah-BEE-teez mell-EYE-tus) a condition resulting when the pancreatic islets secrete too little insulin, resulting in increased levels of blood glucose

dialysis (dye-AL-i-sis) separation of smaller (diffusible) particles from larger (nondiffusable) particles through a semipermeable membrane

diaphragm (DYE-ah-fram) membrane or partition that separates one thing from another; the flat muscular sheet that separates the thorax and abdomen and is a major muscle of respiration

diaphysis (dye-AF-i-sis) shaft of a long bone

diarthroses (dye-ar-THRO-sis) freely movable joint

diarrhea (dye-ah-REE-ah) defecation of liquid feces

diastole (dye-AS-toe-lee) relaxation of the heart, interposed between its contractions; opposite of systole

diastolic pressure (dye-ah-STOL-ik PRESH-ur) blood pressure in arteries during diastole (relaxation) of the heart

diencephalon (dye-en-SEF-ah-lon) "between" brain; parts of the brain between the cerebral hemispheres and the mesencephalon or midbrain

differential WBC count (dif-fer-EN-shal WBC count) special type of white blood cell count in which proportions of each type of white blood cell are reported as percentages of the total count

diffusion (di-FYOO-shun) spreading; for example, scattering of dissolved particles

digestion (di-JEST-chun) the breakdown of food materials either mechanically (that is, chewing) or chemically (that is, digestive enzymes)

digestive system (di-JEST-tiv SIS-tem) organs that work together to ensure proper digestion and absorption of nutrients

digital (DIJ-i-tal) refers to fingers and toes

discharging chambers (dis-CHARJ-ing CHAM-bers) the two lower chambers of the heart called ventricles

disease (di-ZEEZ) any significant abnormality in the body's structure or function that disrupts a person's vital function or physical, mental, or social well-being

dissection (di-SEK-shun) cutting technique used to separate body parts for study

dissociate (di-SO-see-ate) when a compound breaks apart in solution

distal (DIS-tal) toward the end of a structure; opposite of proximal

distal convoluted tubule (DIS-tal KON-vo-loo-ted TOO-byool) the part of the tubule distal to the ascending limb of the loop of Henle in the kidney

disuse atrophy (DIS-yoos AT-ro-fee) when prolonged inactivity results in the muscles getting smaller in size

diuretic (dye-yoo-RET-ik) a substance that promotes or stimulates the production of urine; diuretic drugs are among the most commonly used drugs in medicine

diverticulitis (dye-ver-tik-yoo-LIE-tis) inflammation of diverticula (abnormal outpouchings) of the large intestine, possibly causing constipation

DNA replication (DNA rep-li-KAY-shun) the unique ability of DNA molecules to make copies of themselves

dominant (DOM-i-nant) in genetics, the term dominant refers to genes that have effects that appear in the offspring (dominant forms of a gene are often represented by upper case letters); compare *recessive*

dopamine (DOE-pah-meen) chemical neurotransmitter

dorsal (DOR-sal) referring to the back; opposite of ventral; in humans, the posterior is dorsal

dorsal body cavity (DOR-sal BOD-ee KAV-i-tee) includes the cranial and spinal cavities

dorsiflexion (dor-si-FLEK-shun) when the top of the foot is elevated (brought toward the front of the lower leg) with the toes pointing upward

Down syndrome (down SIN-drome) group of symptoms usually caused by trisomy of chromosome 21; characterized by mental retardation and multiple structural defects, including facial, skeletal, and cardiovascular abnormalities

ductless gland (DUKT-less gland) specialized gland that secretes hormones directly into the blood

ductus arteriosus (DUK-tus ar-teer-ee-O-sus) connects the aorta and the pulmonary artery, allowing most blood to bypass the fetus' developing lungs

ductus deferens (DUK-tus DEF-er-ens) a thick, smooth, muscular tube that allows sperm to exit from the epididymis and pass from the scrotal sac into the abdominal cavity; also known as the vas deferens

ductus venosus (DUK-tus ve-NO-sus) a continuation of the umbilical vein that shunts blood returning from the placenta past the fetus' developing liver directly into the inferior vena cava

duodenal papillae (doo-o-DEE-nal pah-PIL-ee) ducts located in the middle third of the duodenum that empty pancreatic digestive juices and bile from the liver into the small intestine; there are two ducts, the major duodenal papillae and the minor papillae

duodenum (doo-o-DEE-num) the first subdivision of the small intestine where most chemical digestion occurs

dura mater (DOO-rah MAH-ter) literally "strong or hard mother"; outermost layer of the meninges

dust cells (dust sells) macrophages that ingest particulate matter in the small air sacs of the lungs

dwarfism (DWARF-izm) condition of abnormally small stature, sometimes resulting from hyposecretion of growth hormone

dysfunctional uterine bleeding (DUB) (dis-FUNK-shun-al YOO-ter-in BLEED-ing) irregular or excessive bleeding from the uterus resulting from a hormonal imbalance

dysmenorrhea (dis-men-o-REE-ah) painful menstruation

dyspnea (DISP-nee-ah) difficult or labored breathing

eardrum (EAR-drum) the tympanic membrane that separates the external ear and middle ear

eccrine (EK-rin) small sweat glands distributed over the total body surface

echocardiography (ek-o-kar-dee-OG-rah-fee) heart imaging technique in which ultrasound echoes back from heart tissues to form a continuous recording of heart structure movement during a series of cardiac cycles

ectoderm (EK-toe-derm) the innermost of the primary germ layers that develops early in the first trimester of pregnancy

ectopic pregnancy (ek-TOP-ik PREG-nan-see) a pregnancy in which the fertilized ovum implants some place other than in the uterus

eczema (EK-ze-mah) inflammatory skin condition associated with a variety of diseases and characterized by erythema, papules, vesicles, and crusts

edema (e-DEE-mah) accumulation of fluid in a tissue, as in inflammation; swelling

effector (ef-FEK-tor) responding organ; for example, voluntary and involuntary muscle, the heart, and glands

efferent (EF-fer-ent) carrying from, as neurons that transmit impulses from the central nervous system to the periphery; opposite of afferent

ejaculation (ee-jak-yoo-LAY-shun) sudden discharging of semen from the body

ejaculatory duct (ee-JAK-yoo-lah-toe-ree dukt) duct formed by the joining of the ductus deferens and the duct from the seminal vesicle that allows sperm to enter the urethra

electrocardiogram (ECG) (e-lek-tro-KAR-dee-o-gram) graphic record of the heart's action potentials

electroencephalogram (e-lek-tro-en-SEF-lo-gram) graphic representation of voltage changes in brain tissue used to evaluate nerve tissue function

electrolyte (e-LEK-tro-lite) substance that ionizes in solution, rendering the solution capable of conducting an electric current

electrolyte balance (e-LEK-tro-lite BAL-ans) homeostasis of electrolytes

electrophoresis (e-lek-tro-fo-REE-sis) laboratory procedure in which different types of molecules are separated according to molecular weight by passing a weak electric current through their liquid medium

elephantiasis (el-e-fan-TIE-ah-sis) extreme lymphedema (swelling due to lymphatic blockage) in the limbs caused by a parasitic worm infestation, so called because the limbs swell to "elephant proportions"

embolism (EM-bo-lizm) obstruction of a blood vessel by foreign matter carried in the bloodstream

embolus (EM-bo-lus) a blood clot or other substance (bubble of air) that is moving in the blood and may block a blood vessel

embryo (EM-bree-o) animal in early stages of intrauterine development; in humans, the first 3 months after conception

embryology (em-bree-OL-o-gee) study of the development of an individual from conception to birth

embryonic phase (em-bree-ON-ik faze) the period extending from fertilization until the end of the eigth week of gestation; during this phase the term embryo is used

emesis (EM-e-sis) vomiting

emphysema (em-fi-SEE-mah) abnormal condition characterized by trapping of air in alveoli of the lung that causes them to rupture and fuse to other alveoli

emptying reflex (EMP-tee-ing REE-fleks) the reflex that causes the contraction of the bladder wall and relaxation of the internal sphincter to allow urine to enter the urethra, which is followed by urination if the external sphincter is voluntarily relaxed

emulsify (e-MUL-se-fye) in digestion, when bile breaks up fats

endemic (en-DEM-ik) refers to a disease native to a local region of the world

endocarditis (en-doe-kar-DYE-tis) inflammation of the lining of the heart

endocardium (en-doe-KAR-dee-um) thin layer of very smooth tissue lining each chamber of the heart

endochondral ossification (en-doe-KON-dral os-i-fi-KAY-shun) the process in which most bones are formed from cartilage models

endocrine (EN-doe-krin) secreting into the blood or tissue fluid rather than into a duct; opposite of exocrine

endocrine glands (EN-doe-krin glands) ductless glands that are part of the endocrine system and secrete hormones into intercellular spaces

endocrine system (EN-doe-krin SIS-tem) the series of ductless glands that are found in the body

endoderm (EN-doe-derm) the outermost layer of the primary germ layers that develops early in the first trimester of pregnancy

endometriosis (en-doe-mee-tree-O-sis) presence of functioning endometrial tissue outside the uterus

endometrium (en-doe-MEE-tree-um) mucous membrane lining the uterus

endoneurium (en-doe-NOO-ree-um) the thin wrapping of fibrous connective tissue that surrounds each axon in a nerve

endoplasmic reticulum (ER) (en-doe-PLAS-mik re-TIK-yoo-lum) network tubules and vesicles in cytoplasm

endorphins (en-DOR-fins) chemical in central nervous system that influences pain perception; a natural painkiller

endosteum (en-DOS-tee-um) a fibrous membrane that lines the medullary cavity

endothelium (en-doe-THEE-lee-um) squamous epithelial cells that line the inner surface of the entire circulatory system and the vessels of the lymphatic system

endurance training (en-DOOR-ance TRAIN-ing) continuous vigorous exercise requiring the body to increase its consumption of oxygen and developing the muscles' ability to sustain activity over a prolonged period of time

enkephalins (en-KEF-ah-lins) peptide chemical in the central nervous system that acts as a natural painkiller

enzyme (EN-zime) biochemical catalyst allowing chemical reactions to take place

eosinophil (ee-o-SIN-o-fils) white blood cell that is readily stained by eosin

epicardium (ep-i-KAR-dee-um) the inner layer of the pericardium that covers the surface of the heart; it is also called the visceral pericardium

epidemic (ep-i-DEM-ik) refers to a disease that occurs in many individuals at the same time

epidemiology (EP-i-dee-mee-OL-o-jee) study of the occurrence, distribution, and transmission of diseases in human populations

epidermis (ep-i-DER-mis) "false" skin; outermost layer of the skin

epididymis (ep-i-DID-i-mis) tightly coiled tube that lies along the top and behind the testes where sperm mature and develop the ability to swim

epididymitis (ep-i-did-i-MY-tis) inflammation of the epididymis

epiglottis (ep-i-GLOT-is) lidlike cartilage overhanging the entrance to the larynx

epilepsy (EP-i-lep-see) a seizure disorder characterized by recurring seizures

epinephrine (ep-i-NEF-rin) adrenaline; secretion of the adrenal medulla

epineurium (ep-i-NOO-ree-um) a tough fibrous sheath that covers the whole nerve

epiphyseal fracture (ep-i-FEEZ-ee-al FRAK-cher) when the epiphyseal plate is separated from the epiphysis or diaphysis; this type of fracture can disrupt the normal growth of the bone

epiphyseal plate (ep-i-FEEZ-ee-al plate) the cartilage plate that is between the epiphysis and the diaphysis and allows growth to occur; sometimes referred to as a growth plate

epiphyses (e-PIF-i-sees) ends of a long bone

episiotomy (e-piz-ee-OT-o-mee) a surgical procedure used during birth to prevent a laceration of the mother's perineum or the vagina

epistaxis (ep-i-STAK-sis) clinical term referring to a bloody nose

epithelial membrane (ep-i-THEE-lee-al MEM-brane) membrane composed of epithelial tissue with an underlying layer of specialized connective tissue

epithelial tissue (ep-i-THEE-lee-al TISH-yoo) covers the body and its parts; lines various parts of the body; forms continuous sheets that contain no blood vessels; classified according to shape and arrangement

erythroblastosis fetalis (e-rith-ro-blas-TOE-sis fee-TAL-is) condition of a fetus or infant caused by the mother's Rh antibodies reacting with the baby's Rh-positive RBCs, characterized by massive agglutination of the blood and resulting life-threatening circulatory problems

erythrocytes (e-RITH-ro-sites) red blood cells

esophagus (e-SOF-ah-gus) the muscular, mucus-lined tube that connects the pharynx with the stomach; also known as the foodpipe

essential organs (ee-SEN-shal OR-gans) reproductive organs that must be present for reproduction to occur and are known as gonads

estrogen (ES-tro-jen) sex hormone secreted by the ovary that causes the development and maintenance of the female secondary sex characteristics and stimulates growth of the epithelial cells lining the uterus

etiology (e-tee-OL-o-jee) theory, or study, of the factors involved in causing a disease

eupnea (YOOP-nee-ah) normal respiration

eustachian tube (yoo-STAY-shun toob) tube extending from inside the ear to the throat to equalize air pressure

evaporation (ee-vap-o-RAY-shun) heat being lost from the skin by sweat being vaporized

excoriation (eks-ko-ree-AY-shun) skin lesion in which epidermis has been removed, as in a scratch wound

exhalation (eks-hah-LAY-shun) moving air out of the lungs; also known as expiration

exocrine (EK-so-krin) secreting into a duct; opposite of endocrine

exocrine gland (EK-so-krin glands) glands that secrete their products into ducts that empty onto a surface or into a cavity; for example, sweat glands

exophthalmos (ek-sof-THAL-mos) condition of abnormally protruding eyeballs, occurring in a form of hyperthyroidism called Graves disease; also called *exophthalmia*

expiration (eks-pi-RAY-shun) moving air out of the lungs; also known as exhalation

expiratory center (eks-PYE-rah-tor-ee SEN-ter) one of the two most important respiratory control centers, located in the medulla

expiratory muscles (eks-PYE-rah-tor-ee MUS-els) muscles that allow more forceful expiration to increase the rate and depth of ventilation; the internal intercostals and the abdominal muscles

expiratory reserve volume (ERV) (eks-PYE-rah-tor-ee re-ZERV VOL-yoom) the amount of air that can be forcibly exhaled after expiring the tidal volume (TV)

extension (ek-STEN-shun) increasing the angle between two bones at a joint

external auditory canal (eks-TER-nal AW-di-toe-ree kah-NAL) a curved tube (approximately 2.5 cm) extending from the auricle into the temporal bone, ending at the tympanic membrane

external ear (eks-TER-nal ear) the outer part of the ear that is made up of the auricle and the external auditory canal

external genitalia (eks-TER-nal jen-i-TAIL-yah) external reproductive organs

external intercostals (eks-TER-nal inter KOS-tals) inspiratory muscles that enlarge the thorax, causing the lungs to expand and air to rush in

external nares (eks-TER-nal NAY-reez) nostrils

external oblique (eks-TER-nal o-BLEEK) the outermost layer of the anterolateral abdominal wall

external otitis (eks-TER-nal o-TIE-tis) a common infection of the external ear; also known as swimmer's ear

external respiration (eks-TER-nal respi-RAY-shun) the exchange of gases between air in the lungs and in the blood

extracellular fluid (ECF) (eks-trah-SELL-yoo-lar FLOO-id) the water found outside of cells located in two compartments between cells (interstitial fluid) and in the blood (plasma)

facial (FAY-shal) referring to the face

fallen arch (fallen arch) when the tendons and ligaments of the foot weaken, allowing the normally curved arch to flatten out

fallopian tubes (fal-LO-pee-an toobs) the pair of tubes that conduct the ovum from the ovary to the uterus

false ribs (fawls ribs) the eighth, ninth, and tenth pairs of ribs, which are attached to the cartilage of the seventh ribs rather than the sternum

fasciculus (fah-SIK-yoo-lus) little bundle

fat (fat) one of the three basic food types; primarily a source of energy

fatigue (fah-TEEG) loss of muscle power; weakness

fat tissue (fat TISH-yoo) adipose tissue; specialized to store lipids

feces (FEE-seez) waste material discharged from the intestines

feedback control loop (FEED-bak kon-TROL loop) a highly complex and integrated communication control network, classified as negative or positive. Negative feedback loops are the most important and numerous homeostatic control mechanisms

femoral (FEM-or-al) referring to the thigh

femur (FEE-mur) the thigh bone, which is the longest bone in the body

fertilization (FER-ti-li-ZAY-shun) the moment the female's ovum and the male's sperm cell unite

fetal alcohol syndrome (FAS) (FEE-tal AL-ko-hol SIN-drome) a condition that may cause congenital abnormalities in a baby; it results from a woman consuming alcohol during pregnancy

fetal phase (FEE-tal faze) period extending from the eighth to the thirty-ninth week of gestation; during this phase the term *fetus* is used

fetus (FEE-tus) unborn young, especially in the later stages; in human beings, from the third month of the intrauterine period until birth

fibers (FYE-bers) threadlike structures; for example, nerve fibers

fibrillation (fi-bri-LAY-shun) condition in which individual muscle fibers, or small groups of fibers, contract asynchronously (out of time) with other muscle fibers in an organ, producing no effective movement

fibrin (FYE-brin) insoluble protein in clotted blood

fibrinogen (fye-BRIN-o-jen) soluble blood protein that is converted to insoluble fibrin during clotting

fibrous connective tissue (FYE-brus ko-NEK-tiv TISH-yoo) strong, nonstretchable, white collagen fibers that compose tendons

fibromyoma (fye-bro-my-O-mah) benign tumor of smooth muscle and fibrous connective tissue commonly occurring in the uterine wall, where it is often called *fibroid*; see *myoma*

fibromyositis (fye-bro-my-o-SYE-tis) inflammation of muscle tissue accompanied by inflammation of nearby tendon tissue

fibrosarcoma (fye-bro-sar-KO-mah) cancer of fibrous connective tissue

fibula (FIB-yoo-lah) the slender non-weight-bearing bone located on the lateral aspect of the leg

fight or flight syndrome (fite or flite SIN-drome) the changes produced by increased sympathetic impulses allowing the body to deal with any type of stress

filtration (fil-TRAY-shun) movement of water and solutes through a membrane by a higher hydrostatic pressure on one side

fimbriae (FIM-bree-ee) fringe

flagellate (FLAJ-e-lat) protozoan possessing flagella

flagellum (flah-JEL-um) single projection extending from the cell surface; only example in human is the "tail" of the male sperm

flat bone (flat bone) one of the four types of bone; the frontal bone is an example of a flat bone

flat feet (flat feet) when the tendons and ligaments of the foot weaken, allowing the normally curved arch to flatten out

flatulence (FLAT-yoo-lens) presence of air or other gases in the lumen of the gastrointestinal tract

flexion (FLEK-shun) act of bending; decreasing the angle between two bones at the joint

floating ribs (FLOW-ting ribs) the eleventh and twelfth pairs of ribs, which are only attached to the thoracic vertebrae

fluid balance (FLOO-id BAL-ans) homeostasis of fluids; the volumes of interstitial fluid, intracellular fluid, and plasma and total volume of water remain relatively constant

fluid compartments (FLOO-id kom-PART-ments) the areas in the body where the fluid is located; for example, interstitial fluid

folate-deficiency anemia (FO-late de-FISH-en-see ah-NEE-mee-ah) blood disorder characterized by a decrease in the red blood cell count, caused by a deficiency of folic acid (vitamin B_9) in the diet (as in malnourished individuals)

follicles (FOL-li-kuls) specialized structures required for hair growth

follicle-stimulating hormone (FSH) (FOL-li-kul STIM-yoo-lay-ting HOR-mone) hormone present in males and females; in males, FSH stimulates the production of sperm; in females, FSH stimulates the ovarian follicles to mature and follicle cells to secrete estrogen

fontanels (FON-tah-nels) "soft spots" on the infant's head; unossified areas in the infant skull

foramen (fo-RAY-men) small opening; for example, the vertebral foramen, which allows the spinal cord to pass through the vertebral canal

foramen ovale (fo-RAY-men o-VAL-ee) shunts blood from the right atrium directly into the left atrium allowing most blood to bypass the baby's developing lungs

foreskin (FORE-skin) a loose-fitting, retractable casing located over the glans of the penis; also known as the prepuce

fractal geometry (FRAK-tul jee-OM-e-tree) the study of surfaces with a seemingly infinite area, such as the lining of the small intestine

free nerve endings (free nerv END-ings) specialized receptors in the skin that respond to pain

frenulum (FREN-yoo-lum) the thin membrane that attaches the tongue to the floor of the mouth

frontal (FRON-tal) lengthwise plane running from side to side, dividing the body into anterior and posterior portions

frontal muscle (FRON-tal MUS-el) one of the muscles of facial expression; it moves the eyebrows and furrows the skin of the forehead

frontal sinusitis (FRON-tal sye-nyoo-SYE-tis) inflammation in the frontal sinus

frostbite (FROST-bite) local tissue damage caused by extreme cold

fungus (FUNG-gus) organism similar to plants but lacking chlorophyll and capable of producing mycotic (fungal) infections

furuncle (FUR-un-kl) boil; a pus-filled cavity formed by some hair follicle infections

gangrene (GANG-green) tissue death (necrosis) that involves decay of tissue

gastroenteritis (gas-tro-en-ter-EYE-tis) inflammation of the stomach and intestines

gene (jean) one of many segments of a chromosome (DNA molecule); each gene contains the genetic code for synthesizing a protein molecule such as an enzyme or hormone

gene therapy (jeen THER-ah-pee) manipulation of genes to cure genetic problems; most forms of gene therapy have not yet been proven effective in humans

genetics (je-NET-iks) scientific study of heredity and the genetic code

genitalia (jen-i-TAIL-yah) reproductive organs

genome (JEE-nome) entire set of chromosomes in a cell; the *human genome* refers to the entire set of human chromosomes

gerontology (jair-on-TAHL-o-jee) study of the aging process

gestation (jes-TAY-shun) the length of pregnancy, approximately 9 months in humans

gigantism (jye-GAN-tizm) a condition produced by hypersecretion of growth hormone during the early years of life; results in a child who grows to gigantic size

gingivitis (jin-ji-VYE-tis) inflammation of the gum (gingiva), often as a result of poor oral hygiene

gland (gland) secreting structure

glandular epithelium (GLAN-dyoo-lar ep-i-THEE-lee-um) cells that are specialized for secreting activity

glans (glans) the distal end of the shaft of the penis

glaucoma (glaw-KO-mah) disorder characterized by elevated pressure in the eye

glioma (glye-O-mah) one of the most common types of brain tumors

glomerulonephritis (glo-mer-yoo-lo-ne-FRY-tis) inflammatory disease of the glomerular-capsular membranes of the kidney

glomerulus (glo-MARE-yoo-lus) compact cluster; for example, capillaries in the kidneys

glottis (GLOT-is) the space between the vocal cords

glucagon (GLOO-kah-gon) hormone secreted by alpha cells of the pancreatic islets

glucocorticoids (GC) (gloo-ko-KOR-ti-koyds) hormones that influence food metabolism; secreted by the adrenal cortex

gluconeogenesis (gloo-ko-nee-o-JEN-e-sis) formulation of glucose or glycogen from protein or fat compounds

glucose (GLOO-kose) monosaccharide or simple sugar; the principal blood sugar

gluteal (GLOO-tee-al) of or near the buttocks

gluteus maximus (GLOO-tee-us MAX-i-mus) major extensor of the thigh and also supports the torso in an erect position

glycerol (GLIS-er-ol) product of fat digestion

glycogen (GLYE-ko-jen) polysaccharide; animal starch

glycogen loading (GLYE-ko-jen LOHD-ing) see carbohydrate loading

glycogenesis (glye-ko-JEN-e-sis) formation of glycogen from glucose or from other monosaccharides, fructose, or galactose

glycogenolysis (glye-ko-je-NOL-i-sis) hydrolysis of glycogen to glucose-6-phosphate or to glucose

glycolysis (glye-KOL-i-sis) the first series of chemical reactions in glucose metabolism; changes glucose to pyruvic acid in a series of anaerobic reactions

glycosuria (glye-ko-SOO-ree-ah) glucose in the urine; a sign of diabetes mellitus

goblet cells (GOB-let sells) specialized cells found in simple columnar epithelium that produce mucus

goiter (GOY-ter) enlargement of the thyroid gland

golgi apparatus (GOL-jee ap-ah-RA-tus) small sacs stacked on one another near the nucleus that makes carbohydrate compounds, combines them with protein molecules, and packages the product in a globule

golgi tendon receptors (GOL-jee TEN-don ree-SEP-tors) sensors that are responsible for proprioception

gonads (GO-nads) sex glands in which reproductive cells are formed

gout (gowt) abnormal condition in which excess uric acid is deposited in joints and other tissues as sodium urate crystals—the crystals produce inflammation or *gout arthritis*

Graafian follicle (GRAF-ee-an FOL-li-kul) a mature ovum in its sac

gradient (GRAY-dee-ent) a slope or difference between two levels; for example, blood pressure gradient: a difference between the blood pressure in two different vessels

gram (gram) the unit of measure in the metric system on which mass is based (approximately 454 grams equals 1 pound)

granulosa cell (gran-yoo-LO-sah sell) cell layer surrounding the oocyte

Graves disease (gravz di-ZEEZ) inherited, possibly immune endocrine disorder characterized by hyperthyroidism accompanied by exophthalmos (protruding eyes)

gray matter (gray MAT-er) tissue comprising cell bodies and unmyelinated axons and dendrites

greater omentum (GRATE-er o-MEN-tum) a pouchlike extension of the visceral peritoneum

growth hormone (growth HOR-mone) hormone secreted by the anterior pituitary gland that controls the rate of skeletal and visceral growth

gustatory cell (GUS-tah-tor-ee sell) cells of taste

gyrus (JYE-rus) convoluted ridge

hair follicle (hair FOL-li-kul) a small tube where hair growth occurs

hair papilla (hair pah-PIL-ah) a small, cap-shaped cluster of cells located at the base of the follicle where hair growth begins

hamstring muscles (HAM-string MUS-els) powerful flexors of the hip made up of the semimenbranosus, semitendinosis, and biceps femoris muscles

Haversian canal (ha-VER-shun kah-NAL) the canal in the Haversian system that contains a blood vessel

Haversian system (hah-VER-shun SIS-tem) the circular arrangements of calcified matrix and cells that give bone its characteristic appearance

health (helth) physical, mental, and social well-being; the absence of disease

heart block (hart blok) a blockage of impulse conduction from atria to ventricles so that the heart beats at a slower rate than normal

heart disease (hart di-ZEEZ) any of a group of cardiac disorders that together constitute the leading cause of death in the United States

heart failure (hart FAYL-yoor) inability of the heart to pump returned blood sufficiently

heartburn (HART-bern) esophageal pain caused by backflow of stomach acid into esophagus

heat exhaustion (heet eg-ZAWS-chun) condition caused by fluid loss resulting from activity of thermoregulatory mechanisms in a warm external environment

heat stroke (heet stroke) life-threatening condition characterized by high body temperature; failure of thermoregulatory mechanisms to maintain homeostasis in a very warm external environment

Heimlich maneuver (HIME-lik mah-NOO-ver) lifesaving technique used to free the trachea of objects blocking the airway

hematocrit (he-MAT-o-krit) volume percent of blood cells in whole blood

hematuria (hem-ah-TOO-ree-ah) symptom of blood in the urine, often the result of a renal or urinary disorder

hemiplegia (hem-e-PLEE-jee-ah) paralysis (lack of voluntary muscle control) of one entire side of the body

hemodialysis (hee-mo-dye-AL-i-sis) use of dialysis to separate waste products from the blood

hemoglobin (hee-mo-GLO-bin) iron-containing protein in red blood cells

hemolytic anemia (hee-mo-LIT-ik ah-NEE-mee-ah) any of a group of blood disorders characterized by deficient or abnormal hemoglobin that causes deformation and fragility of red blood cells (for example, sickle-cell anemia, thalassemia)

hemophilia (hee-mo-FIL-ee-ah) any of a group of X-linked inherited blood clotting disorders caused by a failure to form clotting factors VIII, IX, or XI

hemopoiesis (hee-mo-poy-EE-sis) blood cell formation

hemopoietic tissue (hee-mo-poy-ET-ik TISH-yoo) specialized connective tissue that is responsible for the formation of blood cells and lymphatic system cells; found in red bone marrow, spleen, tonsils, and lymph nodes

hemorrhoid (HEM-o-royd) varicose vein in the rectum; hemorrhoids are also called piles

hemothorax (hee-mo-THO-raks) abnormal condition in which blood is present in the pleural space surrounding the lung, possibly causing collapse of the lung

heparin (HEP-ah-rin) substance obtained from the liver; inhibits blood clotting

hepatic colic flexure (he-PAT-ik KOL-ik FLEK-sher) the bend between the ascending colon and the transverse colon

hepatic ducts (he-PAT-ik dukts) drain bile out of the liver

hepatic portal circulation (he-PAT-ik POR-tal ser-kyoo-LAY-shun) the route of blood flow through the liver

hepatic portal vein (he-PAT-ik POR-tal vane) delivers blood directly from the gastrointestinal tract to the liver

hepatitis (hep-ah-TIE-tis) inflammation of the liver; may be caused by toxins, viruses (for example, hepatitis A, hepatitis B), bacteria, or parasites

herpes zoster (HER-peez ZOS-ter) "shingles," viral infection that affects the skin of a single dermatone

hiatal hernia (hye-AY-tal HER-nee-ah) condition in which a portion of the stomach is pushed through the hiatus (opening) of the diaphragm, often weakening or expanding the cardiac sphincter at the inferior end of the esophagus

hiccup (HIK-up) involuntary spasmodic contraction of the diaphragm

hip (hip) the joint connecting the legs to the trunk

histogenesis (his-toe-JEN-e-sis) formation of tissues from primary germ layers of embryo

Hodgkin's disease (HOJ-kinz di-ZEEZ) type of lymphoma (malignant lymph tumor) characterized by painless swelling of lymph nodes in the neck, progressing to other regions

homeostasis (ho-me-o-STAY-sis) relative uniformity of the normal body's internal environment

homeostatic mechanism (ho-mee-o-STAT-ik MEK-ah-nizm) a system that maintains a constant environment enabling body cells to function effectively

hormone (HOR-mone) substance secreted by an endocrine gland

human immunodeficiency virus (HYOO-man i-myoo-no-de-FISH-en-see VYE-rus) the retrovirus that causes acquired immune deficiency syndrome (AIDS)

human lymphocyte antigen (HLA) (HYOO-man LIM-fo-site AN-ti-jen) type of self-antigen that the immune system uses to distinguish one's own tissue from that of a foreign entity

humerus (HYOO-mer-us) the second longest bone in the body; the long bone of the arm

humoral immunity (HYOO-mor-al i-MYOO-ni-tee) antibody-mediated immunity

Huntington's disease (HD) (HUNT-ing-tunz di-ZEEZ) degenerative, inherited brain disorder characterized by chorea (purposeless movements) progressing to severe dementia and death by age 55

hybridoma (hye-brid-O-ma) fused or hybrid cells that continue to produce the same antibody as the original lymphocyte

hydrocele (HYE-dro-seel) abnormal accumulation of watery fluid, as can occur in the scrotum

hydrocephalus (hye-dro-SEF-ah-lus) abnormal accumulation of cerebrospinal fluid; "water on the brain"

hydrocortisone (hye-dro-KOR-ti-zone) a hormone secreted by the adrenal cortex; cortisol; compound F

hydrogen ion (HYE-dro-jen eye-on) found in water and water solutions; produces an acidic solution; H^1

hydrostatic pressure (hye-dro-STAT-ik PRESH-ur) the force of a fluid pushing against some surface

hydroxide ion (hye-DROK-side eye-on) found in water and water solutions; produces an alkaline solution; H^2

hymen (HYE-men) Greek for "membrane"; mucous membrane that may partially or entirely occlude the vaginal outlet

hyperacidity (hye-per-a-SID-i-tee) excessive secretion of acid; an important factor in the formation of ulcers

hypercalcemia (hye-per-kal-SEE-mee-ah) a condition in which there is harmful excesses of calcium in the blood

hypercholesterolemia (hye-per-ko-lester-ol-EE-mee-ah) condition of high blood cholesterol content

hyperglycemia (hye-per-glye-SEE-mee-ah) higher than normal blood glucose concentration

hyperopia (hye-per-O-pee-ah) refractive disorder of the eye caused by a shorter than normal eyeball; farsightedness

hyperplasia (hye-per-PLAY-zee-ah) growth of an abnormally large number of cells at a local site, as in a neoplasm or tumor

hypersecretion (hye-per-se-KREE-shun) too much of a substance is being secreted

hypersensitivity (hye-per-SEN-si-tiv-i-tee) inappropriate or excessive response of the immune system

hypertension (hye-per-TEN-shun) abnormally high blood pressure

hyperthyroidism (hye-per-THYE-royd-izm) oversecretion of thyroid hormones, which increases metabolic rate resulting in loss of weight, increased appetite, and nervous irritability

hypertonic (hye-per-TON-ik) a solution containing a higher level of salt (NaCl) than is found in a living red blood cell (above 0.9% NaCl)

hypertrophy (hye-PER-tro-fee) increased size of a part caused by an increase in the size of its cells

hyperventilation (hye-per-ven-ti-LAY-shun) very rapid, deep respirations

hypervitaminosis (hye-per-vye-tah-mi-NO-sis) general name for any condition resulting from an abnormally high intake of vitamins

hypoalbuminemia (hye-po-al-boo-min-EE-mee-ah) condition of low albumin (protein) in the blood plasma; it often results from renal disorders or malnutrition; loss of plasma protein usually causes edema of the tissue spaces.

hypoglycemia (hye-po-glye-SEE-mee-ah) lower-than-normal blood glucose concentration

hyposecretion (hye-po-se-KREE-shun) too little of a substance is being secreted

hypothalamus (hye-po-THAL-ah-mus) portion of the floor lateral wall of the third ventricle of the brain

hypothermia (hye-po-THER-mee-ah) failure of thermoregulatory mechanisms to maintain homeostasis in a very cold external environment

hypothesis (hye-POTH-e-sis) a proposed explanation of an observed phenomena

hypothyroidism (hye-po-THYE-royd-izm) undersecretion of thyroid hormones; early in life results in cretinism; later in life results in myxedema

hypotonic (hye-po-TON-ik) a solution containing a lower level of salt (NaCl) than is found in a living red blood cell (below 0.9% NaCl)

hypoventilation (hye-po-ven-ti-LAY-shun) slow and shallow respirations

hypovolemic shock (hye-po-vo-LEE-mik shock) circulatory failure (shock) caused by a drop in blood volume that causes blood pressure (and blood flow) to drop; literally "low volume" shock

hysterectomy (his-te-REK-toe-mee) surgical removal of the uterus

idiopathic (id-ee-o-PATH-ik) refers to a disease of undetermined cause

ileocecal valve (il-ee-o-SEE-kal valv) the sphincterlike structure between the end of the small intestine and the beginning of the large intestine

ileum (IL-ee-um) the distal portion of the small intestine

iliac crest (IL-ee-ak krest) the superior edge of the illium

iliopsoas (il-ee-op-SO-as) a flexor of the thigh and an important stabilizing muscle for posture

ilium (IL-ee-um) one of the three separate bones that forms the ox coxa

immune deficiency (i-MYOON de-FISH-en-see) general term for complete or relative failure of the immune system to defend the internal environment of the body

immune system (i-MYOON SIS-tem) the body's defense system against disease

immunization (i-myoo-ni-ZAY-shun) deliberate artificial exposure to disease to produce acquired immunity

immunosuppressive drugs (i-myoo-no-soo-PRES-iv drugs) compounds that suppress, or reduce the capacity of, the immune system; such drugs are sometimes used to prevent rejection of transplanted tissues

immunotherapy (i-myoo-no-THER-ah-pee) therapeutic technique that bolsters a person's immune system in an attempt to control a disease

impetigo (im-pe-TIE-go) a highly contagious bacterial skin infection that occurs most often in children

implantation (im-plan-TAY-shun) when a fertilized ovum implants in the uterus

impotence (IM-po-tense) failure to achieve erection of the penis resulting in an inability to reproduce

inborn immunity (IN-born i-MYOO-ni-tee) immunity to disease that is inherited

incompetent (cardiac) valves (in-KOM-pe-tent (KAR-dee-ak) valvs) cardiac valves that "leak," allowing some blood to flow back into the chamber from which it came

incomplete fracture (in-kom-PLEET FRAK-chur) bone fracture in which the bone fragments remain partially joined

incontinence (in-KON-ti-nens) when an individual voids urine involuntarily

incubation (in-kyoo-BAY-shun) early, latent stage of an infection, during which an infection has begun but signs or symptoms have not yet developed

incus (IN-kus) the anvil, the middle ear bone that is shaped like an anvil

induced abortion (in-DOOST ah-BOR-shun) purposeful termination of a pregnancy before the fetus is able to survive outside the womb

infancy (IN-fan-see) from birth to about 18 months of age

infant respiratory distress syndrome (IN-fant re-SPY-rah-toe-ree di-STRESS SIN-drome) leading cause of death in premature babies, due to a lack of surfacant in the alveolar air sacs

inferior (in-FEER-ee-or) lower; opposite of superior

inferior vena cava (in-FEER-ee-or VEE-nah KAY-vah) one of two large veins carrying blood into the right atrium

infertility (in-fer-TIL-i-tee) lower-than-normal ability to reproduce

inflammation (in-flah-MAY-shun) group of responses to a tissue irritant marked by signs of redness, heat, swelling, and pain

inflammation mediators (in-flah-MAY-shun MEE-dee-ay-tors) chemicals (for example, prostaglandins, histamine, kinins) released by irritated tissues that promote the events of the inflammation response

inflammatory exudate (in-FLAM-ah-to-ree EKS-yoo-date) fluid that accumulates in inflamed tissues as a result of increased permeability of blood vessels

inguinal (ING-gwi-nal) of the groin

inguinal hernia (ING-gwi-nal HER-nee-ah) protrusion of abdominopelvic organs through the inguinal canal and into the scrotum

inhalation (in-hah-LAY-shun) inspiration or breathing in; opposite of exhalation or expiration

inherited immunity (in-HAIR-i-ted i-MYOO-ni-tee) inborn immunity

inhibiting hormone (in-HIB-i-ting HOR-mone) hormone produced by the hypothalamus that slows the release of anterior pituitary hormones

insertion (in-SER-shun) attachment of a muscle to the bone that it moves when contraction occurs (as distinguished from its origin)

inspiration (in-spi-RAY-shun) inhalation, moving air into the lungs; same as inhalation, opposite of exhalation or expiration

inspiratory muscle (in-SPY-rah-tor-ee MUS-el) the muscles that increase the size of the thorax, including the diaphragm and external intercostals, and allow air to rush into the lungs

inspiratory center (in-SPY-rah-tor-ee SEN-ter) one of the two most important control centers located in the medulla; the other is the expiratory center

inspiratory reserve volume (IRV) (in-SPY-rah-tor-ee re-SERV VOL-yoom) the amount of air that can be forcibly inspired over and above a normal respiration

insulin (IN-suh-lin) hormone secreted by the pancreatic islets

integument (in-TEG-yoo-ment) refers to the skin

integumentary system (in-teg-yoo-MEN-tar-ee SIS-tem) the skin; the largest and most important organ in the body

intercalated disks (in-TER-kah-lay-ted disks) cross striations and unique dark bands that are found in cardiac muscle fibers

intercostal muscle (in-ter-KOS-tal MUS-el) the respiratory muscles located between the ribs

interferon (in-ter-FEER-on) small proteins produced by the immune system that inhibits virus multiplication

internal oblique (in-TER-nal o-BLEEK) the middle layer of the anterolateral abdominal walls

internal respiration (in-TER-nal re-spi-RAY-shun) the exchange of gases that occurs between the blood and cells of the body

interneuron (in-ter-NOO-ron) nerves that conduct impulses from sensory neurons to motor neurons

interphase (IN-ter-faze) the phase immediately before the visible stages of cell division when the DNA of each chromosome replicates itself

interstitial cell (in-ter-STISH-al sell) small specialized cells in the testes that secrete the male sex hormone, testosterone

interstitial cell-stimulating hormone (ICSH) (in-ter-STISH-al sell STIM-yoo-lay-ting HOR-mone) the previous name for luteinizing hormone in males; causes testes to develop and secrete testosterone

interstitial fluid (in-ter-STISH-al FLOO-id) fluid located in the microscopic spaces between the cells

intestine (in-TES-tin) the part of the digestive tract that is after the stomach; separated into two segments, the small and the large

intestinal gland (in-TES-ti-nal gland) thousands of glands found in the mucous membrane of the mucosa of the small intestines; secrete intestinal digestive juices

intracellular fluid (ICF) (in-tra-SELL-yoo-lar FLOO-id) fluid located within the cells; largest fluid compartment

in vitro (in VEE-tro) refers to the glass laboratory container where a mature ovum is fertilized by a sperm

involuntary muscle (in-VOL-un-tare-ee MUS-el) smooth muscles that are not under conscious control and are found in organs such as the stomach and small intestine

involution (in-vo-LOO-shun) return of an organ to its normal size after an enlargement; also retrograde or degenerative change

ion (EYE-on) electrically charged atom or group of atoms

iron deficiency anemia (EYE-ern de-FISH-en-see ah-NEE-mee-ah) when there are inadequate levels of iron in the diet so that less hemoglobin is produced; results in extreme fatigue

ischemia (is-KEE-me-ah) reduced flow of blood to tissue resulting in impairment of cell function

ischium (IS-kee-um) one of three separate bones that forms the ox coxa

isoimmunity (eye-so-i-MYOO-ni-tee) immune response to antigens of another human, as in transplanted (grafted) tissues; in some cases it is called rejection syndrome

isometric (eye-so-MET-rik) type of muscle contraction in which muscle does not shorten

isotonic (eye-so-TON-ik) of the same tension or pressure

jaundice (JAWN-dis) abnormal yellowing of skin, mucous membranes, and white of eyes

jejunum (je-JOO-num) the middle third of the small intestine

joints (joynts) articulation

Kaposi's sarcoma (KAH-po-seez sar-KO-mah) rare malignant neoplasm of the skin that often spreads to lymph nodes and internal organs; Kaposi's sarcoma is often found in AIDS patients

karyotype (KARE-ee-o-type) ordered arrangement of photographs of chromosomes from a single cell used in genetic counseling to identify chromosomal disorders such as trisomy or monosomy

keloid (KEE-loid) an unusually thick fibrous scar on the skin

keratin (KARE-ah-tin) protein substance found in hair, nails, outer skin cells, and horny tissues

kidney (KID-nee) organ that cleanses the blood of waste products continually produced by metabolism

kilocalorie (Kcal) (KIL-o-kal-o-ree) 1000 calories

kinesthesia (kin-es-THEE-zee-ah) "muscle sense"; that is, sense of position and movement of body parts

Klinefelter's syndrome (KLINE-fel-ter SIN-drome) genetic disorder caused by the presence of two or more X chromosomes in a male (typically trisomy XXY); characterized by long legs, enlarged breasts, low intelligence, small testes, sterility, chronic pulmonary disease

Krause's end bulb (KROWZ end bulb) skin receptor that detects sensations of cold

Kupffer cell (KOOP-fer sell) macrophage found in spaces between liver cells

kyphosis (kye-FO-sis) abnormally exaggerated thoracic curvature of the vertebral column

labia majora (LAY-bee-ah ma-JO-rah) "large lips" of the vulva

labia minora (LAY-bee-ah mi-NO-rah) "small lips" of the vulva

labor (LAY-bor) the process that results in the birth of the baby

lacrimal gland (LAK-ri-mal gland) the glands that produce tears, located in the upper lateral portion of the orbit

lacteal (LAK-tee-al) a lymphatic vessel located in each villus of the intestine; serves to absorb fat materials from the chyme passing through the small intestine

lactiferous duct (lak-TIF-er-us dukt) the duct that drains the grapelike cluster of milk-secreting glands in the breast

lactose intolerance (LAK-tose in-TOL-er-ans) lack of the enzyme lactase, resulting in an inability to digest lactose (a disaccharide present in milk and dairy products)

lacuna (lah-KOO-nah) space or cavity; for example, lacunae in bone contain bone cells

lambdoidal suture (LAM-doyd-al SOO-chur) the immovable joint formed by the parietal and occipital bones

lamella (lah-MEL-ah) thin layer, as of bone

lanugo (lah-NOO-go) the extremely fine and soft hair found on a newborn infant

laparoscope (LAP-ah-ro-skope) specialized optical viewing tube

laryngitis (lar-in-JYE-tis) inflammation of the mucous tissues of the larynx (voicebox)

laryngopharynx (lah-ring-go-FAIR-inks) the lowest part of the pharynx

larynx (LAIR-inks) the voice box located just below the pharynx; the largest piece of cartilage making up the larynx is the thyroid cartilage, commonly known as the Adam's apple

lateral (LAT-er-al) of or toward the side; opposite of medial

latissimus dorsi (la-TIS-i-mus DOR-si) an extensor of the upper arm

lens (lenz) the refracting mechanism of the eye that is located directly behind the pupil

lesion (LEE-zhun) any objective abnormality in a body structure

leukemia (loo-KEE-mee-ah) blood cancer characterized by an increase in white blood cells

leukocyte (LOO-ko-site) white blood cells

leukocytosis (loo-ko-SYE-toe-sis) abnormally high white blood cell numbers in the blood

leukopenia (loo-ko-PEE-nee-ah) abnormally low white blood cell numbers in the blood

leukoplakia (loo-ko-PLAY-kee-ah) white patches in the mouth, commonly seen in chronic cigarette smokers; may lead to mouth cancer

leukorrhea (loo-ko-REE-ah) whitish discharge from the urogenital tract

ligament (LIG-ah-ment) bond or band connecting two objects; in anatomy a band of white fibrous tissue connecting bones

linear fracture (LIN-ee-ar FRAK-chur) bone fracture characterized by a fracture line that is parallel to the bone's long axis

lipase (LYE-pase) fat-digesting enzymes

lipoma (li-PO-mah) benign tumor of adipose (fat) tissue

lithotriptor (LITH-o-trip-tor) a specialized ultrasound generator that is used to pulverize kidney stones

liver glycogenolysis (LIV-er glye-ko-je-NOL-i-sis) chemical process by which liver glycogen is converted to glucose

lobectomy (lo-BEK-toe-mee) surgical removal of a single lobe of an organ, as in the removal of one lobe of a lung

longitudinal arch (lon-ji-TOO-di-nal arch) two arches, the medial and lateral, that extend lengthwise in the foot

loop of Henle (loop of HEN-lee) extension of the proximal tubule of the kidney

lordosis (lor-DOE-sis) abnormally exaggerated lumbar curvature of the vertebral column

lumbar (LUM-bar) lower back, between the ribs and pelvis

lumbar puncture (LUM-bar PUNK-chur) when some cerebrospinal fluid is withdrawn from the subarachnoid space in the lumbar region of the spinal cord

lumen (LOO-men) the hollow space within a tube

lung (lung) organ of respiration; the right lung has three lobes and the left lung has two lobes

lunula (LOO-nyoo-lah) crescent-shaped white area under the proximal nail bed

luteinization (loo-te-ni-ZAY-shun) the formation of a golden body (corpus luteum) in the ruptured follicle

luteinizing hormone (LH) (LOO-te-nye-zing HOR-mone) acts in conjunction with follicle-stimulating hormone (FSH) to stimulate follicle and ovum maturation and release of estrogen and ovulation; known as the ovulating hormone; in males, causes testes to develop and secrete testosterone

lymph (limf) specialized fluid formed in the tissue spaces that returns excess fluid and protein molecules to the blood

lymph node (limf node) performs biological filtration of lymph on its way to the circulatory system

lymphangitis (lim-fan-JYE-tis) inflammation of lymph vessels, usually due to infection, characterized by fine red streaks extending from the site of infection; may progress to *septicemia* (blood infection)

lymphatic capillaries (lim-FAT-ik CAP-i-lair-ees) tiny, blind-ended tubes distributed in the tissue spaces

lymphatic duct (lim-FAT-ik dukt) terminal vessel into which lymphatic vessels empty lymph; the duct then empties the lymph into the circulatory system

lymphatic system (lim-FAT-ik SIS-tem) a system that plays a critical role in the functioning of the immune system, moves fluids and large molecules from the tissue spaces and fat-related nutrients from the digestive system to the blood

lymphatic tissue (lim-FAT-ik TISH-yoo) tissue that is responsible for manufacturing lymphocytes and moncytes; found mostly in the lymph nodes, thymus, and spleen

lymphatic vessels (lim-FAT-ik VES-els) vessels that carry lymph to its eventual return to the circulatory system

lymphedema (lim-fe-DEE-mah) swelling (edema) of tissues due to partial or total blockage of the lymph vessels that drain the affected tissue

lymphocytes (LIM-fo-sites) one type of white blood cell

lymphoma (lim-FO-mah) cancer of lymphatic tissue

lyse (lize) disintegration of a cell

lysosome (LYE-so-sohm) membranous organelles containing various enzymes that can dissolve most cellular compounds; hence called digestive bags or suicide bags of cells

macrophage (MAK-ro-faje) phagocytic cells in the immune system

macule (MAK-yool) a flat skin lesion distinguished from the surrounding tissue by a difference in coloration

magnetic resonance imaging (MRI) (mag-NET-ik REZ-o-nans IM-ah-jing) scanning technique that uses a magnetic field to induce tissues to emit radio waves that can be used by computer to construct a sectional view of a patient's body

malabsorption syndrome (mal-ab-SORP-shun SIN-drome) group of symptoms associated with the failure to absorb food properly: anorexia, ascites, cramps, anemia, fatigue

malignant (mah-LIG-nant) refers to a tumor or neoplasm that is capable of metastasizing or spreading to new tissues (that is, cancer)

malleus (MAL-ee-us) hammer; the tiny middle ear bone that is shaped like a hammer

mammary gland (MAM-er-ee gland) breasts; classified as external accessory sex organs in females

mastication (mas-ti-KAY-shun) chewing

mastitis (mas-TIE-tis) inflammation or infection of the breast

mastoiditis (mas-toy-DYE-tis) inflammation of the air cells within the mastoid portion of the temporal bone; usually caused by infection

matrix (MAY-triks) the intracellular substance of a tissue; for example, the matrix of bone is calcified, whereas that of blood is liquid

mature follicle (mah-CHUR FOL-li-kul) graafian follicle

maximum oxygen consumption (Vo$_{2max}$) (MAX-i-mum OKS-i-jen kon-SUMP-shun) the maximum amount of oxygen taken up by the lungs, transported to the tissues, and used to do work

mechanoreceptor (mek-an-o-ree-SEP-tor) receptors that are mechanical in nature; for example, equilibrium and balance sensors in the ears

medial (MEE-dee-al) of or toward the middle; opposite of lateral

mediastinum (mee-dee-as-TI-num) a subdivision in the midportion of the thoracic cavity

medulla (me-DUL-ah) Latin for "marrow"; hence the inner portion of an organ in contrast to the outer portion or cortex

medulla oblongata (me-DUL-ah ob-long-GAH-tah) the lowest part of the brain stem; an enlarged extension of the spinal cord; the vitals centers are located within this area

medullary cavity (MED-yoo-lair-ee KAV-i-tee) hollow area inside the diaphysis of the bone that contains yellow bone marrow

meiosis (my-O-sis) nuclear division in which the number of chromosomes are reduced to half their original number; produce gametes

Meissner's corpuscle (MIZS-ners KOR-pus-ul) a sensory receptor located in the skin close to the surface that detects light touch

melanin (MEL-ah-nin) brown skin pigment

melanocyte (me-LAN-o-site) specialized cells in the pigment layer that produce melanin

melanocyte-stimulating hormone (MSH) (me-LAN-o-site STIM-yoo-lay-ting HOR-mone) responsible for a rapid increase in the synthesis and dispersion of melanin granules in specialized skin cells

melanoma (mel-ah-NO-mah) cancer of pigmented epithelial cells

melatonin (mel-ah-TOE-nin) important hormone produced by the pineal gland; it is believed to regulate the onset of puberty and the menstrual cycle; also referred to as the third eye because it responds to levels of light and is thought to be involved with the body's internal clock

membrane (MEM-brane) thin layer or sheet

membranous labyrinth (MEM-brah-nus LAB-i-rinth) a membranous sac that follows the shape of the bony labyrinth and is filled with endolymph

memory cell (MEM-o-ree sell) cells that remain in reserve in the lymph nodes until their ability to secrete antibodies is needed

menarche (me-NAR-kee) beginnings of the menstrual function

Meniere's disease (men-ee-AIRZ di-ZEEZ) chronic inner ear disorder characterized by tinnitus, progressive nerve deafness, and vertigo

meninges (me-NIN-jeez) fluid-containing membranes surrounding the brain and spinal cord

meningitis (men-in-JYE-tis) inflammation of the meninges caused by a variety of factors including bacterial infection, mycosis, viral infection, and tumors

menopause (MEN-o-pawz) termination of menstrual cycles

menses (MEN-seez) menstrual flow

menstrual cycle (MEN-stroo-al SYE-kul) the cyclical changes in the uterine lining

mesentary (MEZ-en-tair-ee) a large double fold of peritoneal tissue that anchors the loops of the digestive tract to the posterior wall of the abdominal cavity

mesoderm (MEZ-o-derm) the middle layer of the primary germ layers

messenger RNA (mRNA) (MES-en-jer RNA) a duplicate copy of a gene sequence on the DNA that passes from the nucleus to the cytoplasm

metabolic acidosis (met-ah-BOL-ik as-i-DOE-sis) a disturbance affecting the bicarbonate element of the bicarbonate-carbonic acid buffer pair; bicarbonate deficit

metabolic alkalosis (met-ah-BOL-ik al-kah-LO-sis) disturbance affecting the bicarbonate element of the bicarbonate-carbonic acid buffer pair; bicarbonate excess

metabolism (me-TAB-o-lizm) complex process by which food is used by a living organism

metacarpal (met-ah-KAR-pal) the part of the hand between the wrist and fingers

metaphase (MET-ah-faze) second stage of mitosis, during which the nuclear membrane and nucleolus disappear

metastasis (me-TAS-tah-sis) process by which malignant tumor cells fall off a primary tumor then migrate to a new tissue to colonize a secondary tumor

metatarsal arch (met-ah-TAR-sal arch) the arch that extends across the ball of the foot; also called the transverse arch

meter (MEE-ter) a measure of length in the metric system; equal to about 39.5 inches

microbe (MY-krobe) term that refers to any microscopic organism

microcephaly (my-kro-SEF-ah-lee) a congenital abnormality in which an infant is born with a small head

microglia (my-KROG-lee-ah) one type of connective tissue found in the brain and spinal cord

micron (MY-kron) 1/1000 millimeter; 1/25,000 inch

microvilli (my-kro-VIL-eye) the brush-like border made up of epithelial cells found on each villus in the small intestine; increases the surface area for absorption of nutrients

micturition (mik-too-RISH-un) urination, voiding

midbrain (MID-brain) one of the three parts of the brain stem

middle ear (MID-ul eer) a tiny and very thin epithelium-lined cavity in the temporal bone that houses the ossicles; in the middle ear, sound waves are amplified

midsagittal (mid-SAJ-i-tal) a cut or plane that divides the body or any of its parts into two equal halves

minerals (MIN-er-als) inorganic elements or salts found naturally in the earth that are vital to the proper functioning of the body

mineralocorticoid (MC) (min-er-al-o-KOR-ti-koyd) hormone that influences mineral salt metabolism; secreted by adrenal cortex; aldosterone is the chief mineralocorticoid

mitochondria (my-toe-KON-dree-ah) threadlike structures

mitosis (my-TOE-sis) indirect cell division involving complex changes in the nucleus

mitral valve (MY-tral valv) also known as the bicuspid valve; located between the left atrium and ventricle

mitral valve prolapse (MVP) (MY-tral valv PRO-laps) condition in which the bicuspid (mitral) valve extends into the left atrium, causing incompetence (leaking) of the valve

mold (mold) large fungus (compared to a yeast, which is a small fungus)

monoclonal antibody (mon-o-KLONE-al AN-ti-bod-ee) specific antibody produced from a population of identical cells

monocyte (MON-o-site) a phagocyte

monosomy (MON-o-so-me) abnormal genetic condition in which cells have only one chromosome where there should be a pair; usually caused by nondisjunction (failure of chromosome pairs to separate) during gamete production

mons pubis (monz PYOO-bis) skin-covered pad of fat over the symphysis pubis in the female

morbidity (mor-BID-i-tee) illness or disease; the rate of incidence of a specific illness or disease in a specific population

mortality (mor-TAL-i-tee) death; the rate of deaths caused by a specific disease within a specific population

morula (MOR-yoo-lah) a solid mass of cells formed by the divisions of a fertilized egg

motor neuron (MO-tor NOO-ron) transmits nerve impulses from the brain and spinal cord to muscles and glandular epithelial tissues

motor unit (MO-tor YOO-nit) a single motor neuron with the muscle cells it innervates

mucocutaneous junction (myoo-ko-kyoo-TAY-nee-us JUNK-shun) the transitional area where the skin and mucous membrane meet

mucosa (myoo-KO-sah) mucous membrane

mucous membrane (MYOO-kus MEM-brane) epithelial membranes that line body surfaces opening directly to the exterior and secrete a thick, slippery material called mucus

mucus (MYOO-kus) thick, slippery material that is secreted by the mucous membrane and keeps the membrane moist

multiple neurofibromatosis (MUL-ti-pul noo-ro-fye-bro-mah-TOE-sis) disorder characterized by multiple, sometimes disfiguring, benign tumors of the Schwann cells (neuroglia) that surround nerve fibers

multiple sclerosis (MS) (MUL-ti-pul skle-RO-sis) the most common primary disease of the central nervous system; a myelin disorder

muscle fiber (MUS-el FYE-ber) the specialized contractile cells of muscle tissue that are grouped together and arranged in a highly organized way

muscle strain (MUS-el strane) muscle injury resulting from overexertion or trauma and involving overstretching or tearing of muscle fibers

muscular dystrophy (MUS-kyoo-lar DIS-tro-fee) a group of muscle disorders characterized by atrophy of skeletal muscle without nerve involvement; Duchenne muscular dystrophy (DMD) is the most common type

muscular system (MUS-kyoo-lar SIS-tem) the muscles of the body

muscularis (mus-kyoo-LAIR-is) two layers of muscle surrounding the digestive tube that produce wavelike, rhythmic contractions, called peristalsis, which move food material

mutagen (MYOO-tah-jen) agent capable of causing mutation (alteration) of DNA

myalgia (my-AL-jee-ah) general term referring to the symptom of pain in muscle tissue

myasthenia gravis (my-as-THEE-nee-ah GRA-vis) autoimmune muscle disorder characterized by progressive weakness and chronic fatigue

mycotic infection (my-KO-tik in-FEK-shun) fungal infection

myelin (MY-e-lin) lipoid substance found in the myelin sheath around some nerve fibers

myelinated fiber (MY-e-li-nay-ted FYE-ber) axons outside the central nervous system that are surrounded by a segmented wrapping of myelin

myeloid (MY-e-loyd) pertaining to bone marrow

myeloma (my-e-LO-mah) malignant tumor of bone marrow

myocardial infarction (my-o-KAR-dee-al in-FARK-shun) death of cardiac muscle cells resulting from inadequate blood supply, as in coronary thrombosis

myocardium (my-o-KAR-dee-um) muscle of the heart

myofilaments (my-o-FIL-ah-ments) ultramicroscopic, threadlike structures found in myofibrils

myoma (my-O-mah) benign tumor of smooth muscle commonly occurring in the uterine wall; see also *fibromyoma*

myometrium (my-o-MEE-tree-um) muscle layer in the uterus

myopathy (my-OP-ah-thee) general term referring to any muscle disease

myopia (my-O-pee-ah) refractive disorder of the eye caused by an elongated eyeball; nearsightedness

myosin (MY-o-sin) contractile protein found in the thick filaments of skeletal muscle

myositis (my-o-SYE-tis) general term referring to muscle inflammation, as in infection or injury

myxedema (mik-se-DEE-mah) condition caused by deficiency of thyroid hormone in adults

nail body (nail BOD-ee) the visible part of the nail

nail root (nail root) the part of the nail hidden by the cuticle

nanometer (NAN-o-mee-ter) a measure of length in the metric system; one billionth of a meter

nares (NAY-reez) nostrils

nasal cavity (NAY-zal KAV-i-tee) the moist, warm cavities lined by mucosa located just beyond the nostrils; olfactory receptors are located in the mucosa

nasal septum (NAY-zal SEP-tum) a partition that separates the right and left nasal cavities

nasopharynx (nay-zo-FAIR-inks) the uppermost portion of the tube just behind the nasal cavities

nausea (NAW-see-ah) unpleasant sensation of the gastrointestinal tract that commonly precedes the urge to vomit; upset stomach

necrosis (ne-KRO-sis) death of cells in a tissue, often resulting from ischemia (reduced blood flow)

nematode (NEM-ah-tode) roundworms—large parasites capable of infesting humans

neonatology (nee-o-nay-TOL-o-jee) diagnosis and treatment of disorders of the newborn infant

neoplasm (NEE-o-plazm) an abnormal mass of proliferating cells that may be either benign or malignant

nephritis (ne-FRY-tis) general term referring to inflammatory or infectious conditions of renal (kidney) tissue

nephron (NEF-ron) anatomical and functional unit of the kidney, consisting of the renal corpuscle and the renal tubule

nephropathy (NEF-ro-path-ee) kidney disease

nephrotic syndrome (ne-FROT-ik SIN-drome) group of symptoms and signs that often accompany glomerular disorders of the kidney: proteinuria, hypoalbuminemia, and edema

nerve (nerv) collection of nerve fibers

nerve impulse (nerv IM-puls) signals that carry information along the nerves

nervous tissue (NER-vus TISH-yoo) consists of neurons and neuroglia and provides rapid communication and control of body function

nucleic acids (noo-KLEE-ik AS-ids) the two nucleic acids are ribonucleic acid, found in the cytoplasm, and deoxyribonucleic acid, found in the nucleus

neuralgia (noo-RAL-jee-ah) general term referring to nerve pain

neurilemma (noo-ri-LEM-mah) nerve sheath

neuritis (noo-RYE-tis) general term referring to nerve inflammation

neuroblastoma (noo-roe-blas-TOE-mah) malignant tumor of sympathetic nervous tissue, found mainly in young children

neurogenic bladder (noo-ro-JEN-ik BLAD-der) condition in which the nervous control of the urinary bladder is impaired, causing abnormal or obstructed flow of urine from the body

neurogenic shock (noo-ro-JEN-ik shock) circulatory failure (shock) caused by a nerve condition that relaxes (dilates) blood vessels and thus reduces blood flow; literally "nerve-caused" shock

neuroglia (noo-ROG-lee-ah) supporting cells of nervous tissue

neurohypophysis (noo-ro-hye-POF-i-sis) posterior pituitary gland

neuroma (noo-RO-mah) general term for nervous tissue tumors

neuromuscular junction (noo-ro-MUS-kyoo-lar JUNK-shun) the point of contact between the nerve endings and muscle fibers

neuron (NOO-ron) nerve cell, including its processes (axons and dendrites)

neurotransmitter (noo-ro-trans-MIT-ter) chemicals by which neurons communicate

neutrophil (NOO-tro-fils) white blood cell that stains readily with neutral dyes

nevus (NEE-vus) small, pigmented benign tumor of the skin (for example, a mole)

nodes of Ranvier (nodes of rahn-vee-AY) indentations found between adjacent Schwann cells

non-Hodgkin's lymphoma (non-HOJ-kinz lim-FO-mah) type of lymphoma (malignant lymph tumor) characterized by swelling of lymph nodes and progressing to other areas

norepinephrine (nor-ep-i-NEF-rin) hormone secreted by adrenal medulla; released by sympathetic nervous system

nose (noze) respiratory organ

nuclear membrane (NOO-klee-ar MEM-brane) membrane that surrounds the cell nucleus

nucleolus (noo-KLEE-o-lus) critical to protein formation because it "programs" the formation of ribosomes in the nucleus

nucleoplasm (NOO-klee-o-plazm) a special type of cytoplasm found in the nucleus

nucleus (NOO-klee-us) spherical structure within a cell; a group of neuron cell bodies in the brain or spinal cord

nutrition (noo-TRI-shun) food, vitamins, and minerals that are ingested and assimilated into the body

nyctalopia (nik-tah-LO-pee-ah) condition caused by retinal degeneration or avitaminosis A and characterized by the relative inability to see in dim light; night blindness

obesity (o-BEES-i-tee) condition characterized by abnormally high proportion of body fat

oblique fracture (o-BLEEK FRAK-chur) bone fracture characterized by a fracture line that is diagonal to the long axis of the broken bone

old age (old age) see senescence

olecranon fossa (o-LEK-rah-non FOS-ah) a large depression on the posterior surface of the humerus

olecranon process (o-LEK-rah-non PROSS-es) the large bony process of the ulna; commonly referred to as the tip of the elbow

olfaction (ol-FAK-shun) sense of smell

olfactory receptor (ol-FAK-tor-ee ree-SEP-tor) chemical receptors responsible for the sense of smell; located in the epithelial tissue in the upper part of the nasal cavity

oligodendroglia (ol-i-go-den-DROG-lee-ah) holds nerve fibers together and, more important, produces the myelin sheath around axons in the CNS

oligospermia (ol-i-go-SPER-me-ah) low sperm production

oliguria (ol-i-GOO-ree-ah) scanty amounts of urine

oncogene (ON-ko-jeen) gene (DNA segment) thought to be responsible for the development of a cancer

oocyte (O-o-site) immature stage of the female sex cell

oogenesis (o-o-JEN-e-sis) production of female gametes

oophorectomy (o-off-o-REK-toe-mee) surgical procedure to remove the ovaries

open fracture (O-pen FRAK-chur) compound fracture; bone fracture in which bone fragments pierce the skin

opposition (op-o-ZISH-un) moving the thumb to touch the tips of the fingers; the movement used to hold a pencil to write

optic disc (OP-tic disk) the area in the retina where the optic nerve fibers exit and there are no rods or cones; also known as a blind spot

oral cavity (OR-al KAV-i-tee) mouth

orbicularis oculi (or-bik-yoo-LAIR-is OK-yoo-lie) facial muscle that causes a squint

orbicularis oris (or-bik-yoo-LAIR-is O-ris) facial muscle that puckers the lips

orchitis (or-KYE-tis) inflammation of the testes, often due to infection

organ (OR-gan) group of several tissue types that performs a special function

organelle (or-gah-NELL) cell organ; for example, the ribosome

organism (OR-gah-nizm) an individual living thing

organ of Corti (OR-gan of KOR-tee) the organ of hearing located in the cochlea and filled with endolymph

organogenesis (or-ga-no-JEN-e-sis) formation of organs from the primary germ layers of the embryo

origin (OR-i-jin) the attachment of a muscle to the bone, which does not move when contraction occurs, as distinguished from insertion

oropharynx (o-ro-FAIR-inks) the portion of the pharynx that is located behind the mouth

orthopnea (or-THOP-nee-ah) dyspnea (difficulty in breathing) that is relieved after moving into an upright or sitting position

osmosis (os-MO-sis) movement of a fluid through a semipermeable membrane

ossicles (OS-si-kls) little bones; found in the ears

osteoarthritis (os-tee-o-ar-THRY-tis) degenerative joint disease; a noninflammatory disorder of a joint characterized by degeneration of articular cartilage

osteoblast (OS-tee-o-blast) bone-forming cell

osteoclast (OS-tee-o-klast) bone-absorbing cell

osteocyte (OS-tee-o-site) bone cell

osteogenesis imperfecta (os-tee-o-JEN-e-sis im-per-FEK-tah) dominant, inherited disorder of connective tissue characterized by imperfect skeletal development, resulting in brittle bones

osteoma (os-tee-O-mah) benign bone tumor

osteomalacia (os-tee-o-mah-LAY-she-ah) bone disorder usually caused by vitamin D deficiency and characterized by loss of mineral in the bone matrix; the adult form of rickets

osteomyelitis (os-tee-o-my-e-LIE-tis) bacterial (usually staphylococcus) infection of bone tissue

osteoporosis (os-tee-o-po-RO-sis) bone disorder characterized by loss of minerals and collagen from bone matrix, reducing the volume and strength of skeletal bone

osteosarcoma (os-tee-o-sar-KO-mah) bone cancer

otitis (o-TIE-tis) general term referring to inflammation or infection of the ear; otitis media involves the middle ear

otitis media (o-TIE-tis MEE-dee-ah) a middle ear infection

otosclerosis (o-toe-skle-RO-sis) inherited bone disorder involving structural irregularities of the stapes in the middle ear and characterized by tinnitus progressing to deafness

ova (O-vah) female sex cells (singular ovum)

oval window (O-val WIN-doe) a small, membrane-covered opening that separates the middle and inner ear

ovarian follicles (o-VARE-ee-an FOL-i-kuis) contain oocytes

ovaries (O-var-ees) female gonads that produce ova (sex cells)

overhydration (o-ver-hye-DRAY-shun) too large a fluid input that can put a burden on the heart

oviducts (O-vi-dukts) uterine or fallopian tubes

os coxae (os KOK-see) hip bones

oxygen debt (OK-si-jen det) continued increased metabolism that occurs in a cell to remove excess lactic acid that resulted from exercise

oxyhemoglobin (ok-see-hee-mo-GLO-bin) hemoglobin combined with oxygen

oxytocin (ok-see-TOE-sin) hormone secreted by the posterior pituitary gland before and after delivering a baby; thought to initiate and maintain labor and it causes the release of breast milk into ducts for the baby to suck

pacemaker (PASE-may-ker) see **sino-atrial node**

pacinian corpuscle (pah-SIN-ee-an KOR-pus-ul) a receptor found deep in the dermis that detects pressure on the skin surface

Paget's disease (PAJ-ets di-ZEEZ) osteitis deformans; an often mild metabolic bone disorder characterized by replacement of normal spongy bone with disorganized bone matrix

palate (PAL-let) the roof of the mouth; made up of the hard (anterior portion of the mouth) and soft (posterior portion of the mouth) palates

palmar (PAHL-mar) palm of the hand

pancreas (PAN-kree-as) endocrine gland located in the abdominal cavity; contains pancreatic islets that secrete glucagon and insulin

pancreatic islets (pan-kree-AT-ik eye-LETS) endocrine portion of the pancreas; made up of alpha and beta cells, among others

pandemic (pan-DEM-ik) refers to a disease that affects many people worldwide

Papanicolaou test (pap-ah-nik-o-LA-oo test) cancer-screening test in which cells swabbed from the uterine cervix are smeared on a glass slide and examined for abnormalities; also called Pap smear or Pap test

papillae (pah-PIL-ee) small, nipple-shaped elevations

papilloma (pap-i-LO-mah) benign skin tumor characterized by fingerlike projections (for example, a wart)

papule (PAP-yool) raised, firm skin lesion less than 1 cm in diameter

paralysis (pah-RAL-i-sis) loss of the power of motion, especially voluntary motion

paranasal sinus (pair-ah-NAY-sal SYE-nus) four pairs of sinuses that have openings into the nose

paraplegia (par-ah-PLEE-jee-ah) paralysis (loss of voluntary muscle control) of both legs

parasite (PAR-ah-syte) any organism that lives in or on another organism (a *host*) to obtain its nutrients; parasites may be harmless to the host, or they may disrupt normal body functions of the host and thus cause disease

parasympathetic nervous system (PNS) (par-ah-sim-pah-THE-tic NER-vus SIS-tem) part of the autonomic nervous system; ganglia are connected to the brain stem and the sacral segments of the spinal cord; controls many visceral effectors under normal conditions

parathyroid glands (pair-ah-THYE-royd glands) endocrine glands located in the neck on the posterior aspect of the thyroid gland; secrete parathyroid hormone

parathyroid hormone (PTH) (pair-ah-THYE-royd HOR-mone) hormone released by the parathryroid gland that increases the concentration of calcium in the blood

parietal (pah-RYE-i-tal) of the walls of an organ or cavity

parietal pericardium (pah-RYE-i-tal pair-i-KAR-dee-um) pericardium surrounding the heart like a loose-fitting sack to allow the heart enough room to beat

parietal portion (pah-RYE-i-tal POR-shun) serous membrane that lines the walls of a body cavity

Parkinson's disease (PAR-kin-sunz di-ZEEZ) nervous disorder characterized by abnormally low levels of the neurotransmitter dopamine in parts of the brain that control voluntary movement—patients usually exhibit involuntary trembling and muscle rigidity

parturition (par-too-RISH-un) act of giving birth

patella (pah-TEL-ah) small, shallow pan; the kneecap

pathogenesis (path-o-JEN-e-sis) pattern of a disease's development

pathology (pah-THOL-o-jee) the scientific study of disease

pathophysiology (path-o-fiz-ee-OL-o-jee) study of the underlying physiological aspects of disease

pectoral girdle (PEK-toe-ral GIR-dul) shoulder girdle; the scapula and clavicle

pectoralis major (pek-tor-RAL-is MAY-jor) major flexor of the upper arm

pedal (PEED-al) foot

pedigree (PED-i-gree) chart used in genetic counseling to illustrate genetic relationships over several generations

pelvic cavity (PEL-vik KAV-i-tee) the lower portion of the ventral cavity; the distal portion of the abdominopelvic cavity

pelvic girdle (PEL-vik GIR-dul) connects the legs to the trunk

pelvic inflammatory disease (PID) (PEL-vik in-FLAM-ah-tor-ee di-ZEEZ) acute inflammatory condition of the uterus, fallopian tubes, and/or ovaries —usually the result of an STD or other infection

pelvis (PEL-vis) basin- or funnel-shaped structure

penis (PEE-nis) forms part of the male genitalia; when sexually aroused, becomes stiff to enable it to enter and deposit sperm in the vagina

pepsinogen (pep-SIN-o-jen) component of gastric juice that is converted into pepsin by hydrochloric acid

pericarditis (pair-i-kar-DYE-tis) condition in which the pericardium becomes inflamed

pericardium (pair-i-KAR-dee-um) membrane that surrounds the heart

periodontitis (per-ee-o-don-TIE-tis) inflammation of the periodontal membrane (periodontal ligament) that anchors teeth to jaw bone; common cause of tooth loss among adults

perilymph (PAIR-i-limf) a watery fluid which fills the bony labyrinth of the ear

perineal (pair-i-NEE-al) area between the anus and genitals; the perineum

perineum (pair-i-NEE-um) see perineal

periosteum (pair-i-OS-tee-um) tough, connective tissue covering the bone

peripheral (pe-RIF-er-al) pertaining to an outside surface

peripheral nervous system (PNS) (pe-RIF-er-al NER-vus SIS-tem) the nerves connecting the brain and spinal cord to other parts of the body

peristalsis (pair-i-STAL-sis) wavelike, rhythmic contractions of the stomach and intestines that move food material along the digestive tract

peritoneal space (pair-i-toe-NEE-al space) small, fluid-filled space between the visceral and parietal layers that allows the layers to slide over each other freely in the abdominopelvic cavity

peritoneum (pair-i-toe-NEE-um) large, moist, slippery sheet of serous membrane that lines the abdominopelvic cavity (parietal layer) and its organs (visceral layer)

peritonitis (pair-i-toe-NYE-tis) inflammation of the serous membranes in the abdominopelvic cavity; sometimes a serious complication of an infected appendix

permeable membrane (PER-mee-ah-bul MEM-brane) a membrane that allows passage of substances

permease system (PER-mee-ase SIS-tem) a specialized cellular component that allows a number of active transport mechanisms to occur

pernicious anemia (per-NISH-us ah-NEE-mee-ah) deficiency of red blood cells due to a lack of vitamin B_{12}

peroneal muscles (per-o-NEE-al MUS-els) plantar flexors and evertors of the foot; the peroneus longus forms a support arch for the foot

perspiration (per-spi-RAY-shun) transparent, watery liquid released by glands in the skin that eliminates ammonia and uric acid and helps maintain body temperature; also known as sweat

phagocytes (FAG-o-sites) white blood cells that engulf microbes and digest them

phagocytosis (fag-o-sye-TOE-sis) ingestion and digestion of articles by a cell

phalanges (fah-LAN-jeez) the bones that make up the fingers and toes

pharyngitis (fair-in-JYE-tis) sore throat; inflammation or infection of the pharynx

pharynx (FAIR-inks) organ of the digestive and respiratory system; commonly called the throat

phenylketonuria (PKU) (fen-il-kee-toe-NOO-ree-ah) recessive, inherited condition characterized by excess of phenylketone in the urine, caused by accumulation of phenylalanine (an amino acid) in the tissues; may cause brain injury and death if phenylalanine intake is not managed properly

phimosis (fi-MO-sis) abnormal condition in which the prepuce (foreskin) fits tightly over the glans of the penis

phospholipid (fos-fo-LIP-id) phosphate-containing fat molecule

photopigments (fo-toe-PIG-ments) chemicals in retinal cells that are sensitive to light

phrenic nerve (FREN-ik nerv) the nerve that stimulates the diaphragm to contract

physiology (fiz-ee-OL-o-jee) the study of body function

pia mater (PEE-ah MAH-ter) the vascular innermost covering (meninx) of the brain and spinal cord

pigment layer (PIG-ment LAY-er) the layer of the epidermis that contains the melanocytes that produce melanin to give skin its color

pineal gland (PI-nee-al gland) endocrine gland located in the third ventricle of the brain; produces melatonin

pinocytosis (pin-o-sye-TOE-sis) the active transport mechanism used to transfer fluids or dissolved substances into cells

pituitary gland (pi-TOO-i-tair-ee gland) endocrine gland located in the skull, made up of the adenohypophysis and the neurohypophysis

placenta (plah-SEN-tah) anchors the developing fetus to the uterus and provides a "bridge" for the exchange of nutrients and waste products between the mother and developing baby

placenta previa (plah-SEN-tah PRE-vee-ah) abnormal condition in which a blastocyst implants in the lower uterus, developing a placenta that approaches or covers the cervical opening; placenta previa involves the risk of placental separation and hemorrhage

plantar (PLAN-tar) pertaining to the sole of the foot

plantar flexion (PLAN-tar FLEK-shun) the bottom of the foot is directed downward; this motion allows a person to stand on their tip-toes

plaque (plak) raised skin lesion greater than 1 cm in diameter

plasma (PLAZ-mah) the liquid part of the blood

plasma cells (PLAZ-mah sells) cells that secrete copious amounts of antibody into the blood

plasma membrane (PLAZ-mah MEM-brane) membrane that separates the contents of a cell from the tissue fluid; encloses the cytoplasm and forms the outer boundary of the cell

platelet factors (PLATE-let FAK-tors) substances that are formed when platelets break up; combine with other substances to form thrombin

platyhelminth (plat-ee-HEL-minth) flatworm or fluke—animal parasite capable of infesting humans

pleura (PLOOR-ah) the serous membrane in the thoracic cavity

pleural cavity (PLOOR-al KAV-i-tee) a subdivision of the thorax

pleural space (PLOOR-al space) the space between the visceral and parietal plueras filled with just enough fluid to allow them to glide effortlessly with each breath

pleurisy (PLOOR-i-see) inflammation of the pleura

plica (PLYE-kah) multiple circular folds

pneumocystitis (noo-mo-sis-TYE-tis) a protozoan infection, most likely to invade the body when the immune system has been compromised

pneumonectomy (noo-mo-NEK-toe-mee) surgical procedure in which an entire lung is removed

pneumonia (noo-MO-nee-ah) abnormal condition characterized by acute inflammation of the lungs in which alveoli and bronchial passages become plugged with thick fluid (exudate)

pneumothorax (noo-mo-THO-raks) abnormal condition in which air is present in the pleural space surrounding the lung, possibly causing collapse of the lung

poliomyelitis (po-lee-o-my-e-LYE-tis) viral disease that damages motor nerves, often progressing to paralysis of skeletal muscles

polycythemia (pol-ee-sye-THEE-me-ah) an excessive number of red blood cells

polyuria (pol-ee-YOO-ree-ah) unusually large amounts of urine

pons (ponz) the part of the brain stem between the medulla oblongata and the midbrain

popliteal (pop-li-TEE-al) behind the knee

pore (pore) pinpoint-size openings on the skin that are outlets of small ducts from the eccrine sweat glands

posterior (pos-TEER-ee-or) located behind; opposite of anterior

posterior pituitary gland (pos-TEER-ee-or pi-TOO-i-tair-ee gland) neurohypophysis; hormones produced are ADH and oxytocin

posterior root ganglion (pos-TEER-ee-or GANG-lee-on) ganglion located near the spinal cord; where the neuron cell body of the dendrites of the sensory neuron is located

postganglionic neurons (post-gang-glee-ON-ik NOO-rons) autonomic neurons that conduct nerve impulses from a ganglion to cardiac or smooth muscle or glandular epithelial tissue

postnatal period (POST-nay-tal PEER-ee-od) the period after birth and ending at death

postsynaptic neuron (post-si-NAP-tik NOO-ron) a neuron situated distal to a synapse

posture (POS-chur) position of the body

precapillary sphincter (pree-CAP-pi-lair-ee SFINGK-ter) smooth muscle cells that guard the entrance to the capillary

preeclampsia (pree-e-KLAMP-see-ah) syndrome of abnormal conditions in pregnancy of uncertain cause; syndrome includes hypertension, proteinuria, and edema; also called toxemia of pregnancy, it may progress to eclampsia, severe toxemia that may cause death

preganglionic neurons (pree-gang-glee-ON-ik NOO-rons) autonomic neurons that conduct nerve impulses between the spinal cord and a ganglion

premature (cardiac) contractions (pree-mah-TUR (KAR-dee-ak) kon-TRAK-shun) contractions of the heart wall that occur before expected; extrasystoles

premenstrual syndrome (PMS) (pree-MEN-stroo-al SIN-drome) syndrome of psychological changes (such as irritability) and physical changes (localized edema) that occur before menstruation in many women

prenatal period (PREE-nay-tal PEER-i-od) the period after conception until birth

presbycusis (pres-be-KYOO-sis) progressive hearing loss associated with advanced age

presbyopia (pres-bee-O-pee-ah) farsightedness of old age

presynaptic neuron (pree-si-NAP-tik NOO-ron) a neuron situated proximal to a synapse

primary follicles (PRYE-mare-ee FOL-i-kuls) the follicles present at puberty; covered with granulosa cells

primary germ layers (PRYE-mar-ee jerm LAY-ers) three layers of specialized cells that give rise to definite structures as the embryo develops

primary spermatocyte (SPER-mah-toe-site) specialized cell that undergoes meiosis to ultimately form sperm

prime mover (prime MOO-ver) the muscle responsible for producing a particular movement

progesterone (pro-JES-ter-ohn) hormone produced by the corpus luteum; stimulates secretion of the uterine lining; with estrogen, helps to initiate the menstrual cycle in girls entering puberty

progeria (pro-JEE-ree-ah) rare, inherited condition in which a person appears to age rapidly due to abnormal, widespread degeneration of tissues; adult and childhood forms exist, with the childhood form resulting in death by age 20 or so

prolactin (pro-LAK-tin) hormone secreted by the anterior pituitary gland during pregnancy to stimulate the breast development needed for lactation

pronate (PRO-nate) to turn the palm downward

prone (prone) used to describe the body lying in a horizontal position facing downward

prophase (PRO-faze) first stage of mitosis during which chromosomes become visible

proprioceptors (pro-pree-o-SEP-tors) receptors located in the muscles, tendons, and joints; allows the body to recognize its position

prostaglandins (PG) (pross-tah-GLAN-dins) a group of naturally occurring fatty acids that affect many body functions

prostatectomy (pross-tah-TEK-toe-mee) surgical removal of part or all of the prostate gland

prostate gland (PROSS-tate gland) lies just below the bladder; secretes a fluid that constitutes about 30% of the seminal fluid volume; helps activate sperm and helps them maintain motility

protease (PRO-tee-ase) protein-digesting enzyme

protein (PRO-teen) one of the basic nutrients needed by the body; usually involved with anabolism

protein hormone (PRO-teen HOR-mone) first messenger, provides communication between endocrine glands and target organs; triggers second messengers to affect the cell's activity

proteinuria (pro-teen-YOO-ree-ah) presence of abnormally high amounts of plasma protein in the urine; usually an indicator of kidney disease

prothrombin (pro-THROM-bin) a protein present in normal blood that is required for blood clotting

protozoa (pro-toe-ZO-ah) single-celled organisms with nuclei and other membranous organelles that can infect humans

proximal (PROK-si-mal) next or nearest; located nearest the center of the body or the point of attachment of a structure

proximal convoluted tubule (PROK-si-mal kon-vo-LOO-ted TOOB-yool) the first segment of a renal tubule

pseudo (SOO-doe) false

psoriasis (so-RYE-ah-sis) chronic, inflammatory skin disorder characterized by cutaneous inflammation and scaly plaques

pubis (PYOO-bis) joint in the midline between the two pubic bones

puerperal fever (pu-ER-per-al FEE-ver) condition caused by bacterial infection in a woman after delivery of an infant, possibly progressing to septicemia and death; also called childbed fever

pulmonary artery (PUL-mo-nair-ee AR-ter-ee) artery that carries deoxygenated blood from the right ventricle to the lungs

pulmonary circulation (PUL-mo-nair-ee ser-kyoo-LAY-shun) venous blood flow from the right atrium to the lung and returning to the left atrium

pulmonary embolism (PUL-mo-nair-ee EM-bo-lizm) blockage of the pulmonary circulation by a thrombus or other matter; may lead to death if blockage of pulmonary blood flow is significant

pulmonary semilunar valve (PUL-mo-nair-ee sem-i-LOO-nar valv) valve located at the beginning of the pulmonary artery

pulmonary vein (PUL-mo-nair-ee vanes) any vein that carries oxygenated blood from the lungs to the left atrium

pulmonary ventilation (PUL-mo-nair-ee ven-ti-LAY-shun) breathing; process that moves air in and out of the lungs

Punnett square (PUN-it skwair) grid used in genetic counseling to determine the probability of inheriting genetic traits

pupil (PYOO-pil) the opening in the center of the iris that regulates the amount of light entering the eye

purkinje fibers (pur-KIN-jee FYE-bers) specialized cells located in the walls of the ventricles; relay nerve impulses from the AV node to the ventricles causing them to contract

pus (pus) accumulation of white blood cells, dead bacterial cells, and damaged tissue cells at the site of an infection

pustule (PUS-tyool) small, raised skin lesion filled with pus

P wave (P wave) deflection on an ECG that occurs with depolarization of the atria

pyelonephritis (pye-e-lo-ne-FRY-tis) infectious condition characterized by inflammation of the renal pelvis and connective tissues of the kidney

pyloric sphincter (pye-LOR-ik SFIN-GK-ter) sphincter that prevents food from leaving the stomach and entering the duodenum

pyloric stenosis (pye-LOR-ik ste-NO-sis) anatomical abnormality in which the opening through the pylorus or pyloric sphincter is unusually narrow

pylorus (pye-LOR-us) the small narrow section of the stomach that joins the first part of the small intestine

pyramids (PEER-ah-mids) triangular-shaped divisions of the medulla of the kidney

QRS complex (QRS KOM-pleks) deflection on an ECG that occurs as a result of depolarization of the ventricles

quadriceps femoris (KWOD-re-seps fe-MOR-is) extensor of the lower leg

quadriplegia (kwod-ri-PLEE-jee-ah) paralysis (loss of voluntary muscle control) in all four limbs

quickening (KWIK-en-ing) when a pregnant woman first feels recognizable movements of the fetus

radiation (ray-dee-AY-shun) flow of heat waves away from the blood

radical mastectomy (RAD-i-kal mas-TEK-toe-mee) surgical procedure in which a cancerous breast is removed along with nearby muscle tissue and lymph nodes

radiography (ray-de-OG-rah-fee) imaging technique using x-rays that pass through certain tissues more easily than others, allowing an image of tissues to form on a photographic plate or other sensitive surface

radius (RAY-dee-us) one of the two bones in the forearm; located on the thumb side of the forearm

reabsorption (ree-ab-SORP-shun) process of absorbing again that occurs in the kidneys

receiving chambers (ree-SEE-ving CHAM-bers) atria of the heart; receive blood from the superior and inferior vena cava

receptor (ree-SEP-tor) peripheral beginning of a sensory neuron's dendrite

recessive (ree-SES-iv) in genetics, the term recessive refers to genes that have effects that do not appear in the offspring when they are masked by a dominant gene (recessive forms of a gene are represented by lower case letters); compare *dominant*

rectum (REK-tum) distal portion of the large intestine

rectus abdominis (REK-tus ab-DOM-i-nis) muscle that runs down the middle of the abdomen; protects the abdominal viscera and flexes the spinal column

reflex (REE-fleks) involuntary action

reflex arc (REE-fleks ark) allows an impulse to travel in only one direction

refraction (ree-FRAK-shun) bending of a ray of light as it passes from a medium of one density to one of a different density

regeneration (ree-jen-er-AY-shun) the process of replacing missing tissue with new tissue by means of cell division

releasing hormones (ree-LEE-sing HOR-mones) hormone produced by the hypothalamus gland that causes the anterior pituitary gland to release its hormones

remission (ree-MISH-un) stage of a disease during which a temporary recovery from symptoms occurs

renal calculi (REE-nal KAL-kyoo-lie) kidney stones

renal colic (REE-nal KOL-ik) pain caused by the passage of a kidney stone

renal corpuscle (REE-nal KOR-pus-ul) the part of the nephron located in the cortex of the kidney

renal failure (REE-nal FAIL-yoor) acute or chronic loss of kidney function; acute kidney failure is often reversible, but chronic kidney failure slowly progresses to total loss of renal function (and death if kidney function is not restored through a kidney transplant or use of artificial kidneys)

renal pelvis (REE-nal PEL-vis) basin-like upper end of the ureter that is located inside the kidney

renal ptosis (REE-nal TOE-sis) condition in which one or both kidneys descend, often because of loss of the fat pad that surrounds each kidney

renal tubule (REE-nal TOOB-yool) one of the two principal parts of the nephron

renin (RE-nin) enzyme produced by the kidney that catalyzes the formation of angiotensin, a substance that increases blood pressure

repolarization (ree-po-lah-ri-ZAY-shun) begins just before the relaxation phase of cardiac muscle activity

reproductive system (ree-pro-DUK-tiv SIS-tem) produces hormones that permit the development of sexual characteristics and the propagation of the species

residual volume (RV) (re-ZID-yoo-al VOL-yoom) the air that remains in the lungs after the most forceful expiration

respiratory acidosis (re-SPY-rah-tor-ee as-i-DOE-sis) a respiratory disturbance that results in a carbonic acid excess

respiratory alkalosis (re-SPY-rah-tor-ee al-kah-LO-sis) a respiratory disturbance that results in a carbonic acid deficit

respiratory arrest (re-SPY-rah-tor-ee ah-REST) cessation of breathing without resumption

respiratory control centers (re-SPY-rah-tor-ee kon-TROL SEN-ters) centers located in the medulla and pons that stimulate the muscles of respiration

respiratory distress syndrome (re-SPY-rah-tor-ee di-STRESS SIN-drome) difficulty in breathing caused by absence or failure of the surfactant in fluid lining the alveoli of the lung; IRDS is infant respiratory distress syndrome; ARDS is adult respiratory distress syndrome

respiratory membrane (re-SPY-rah-tor-ee MEM-brane) the single layer of cells that makes up the wall of the alveoli

respiratory mucosa (re-SPY-rah-tor-ee myoo-KO-sah) mucus-covered membrane that lines the tubes of the resiratory tree

respiratory muscles (re-SPY-rah-tor-ee MUS-els) muscles responsible for the changing shape of the thoracic cavity that allows air to move in and out of the lungs

respiratory system (re-SPY-rah-tor-ee SIS-tem) the organs that allow the exchange of oxygen from the air with the carbon dioxide from the blood

respiratory tract (re-SPY-rah-tor-ee trakt) the two divisions of the respiratory system are the upper and lower respiratory tracts

reticular formation (re-TIK-yoo-lar for-MAY-shun) located in the medulla where bits of gray and white matter mix intricately

retina (RET-i-nah) innermost layer of the eyeball; contains rods and cones and continues posteriorly with the optic nerve

retroperitoneal (re-tro-pair-i-toe-NEE-al) area outside of the peritoneum

rheumatic heart disease (roo-MAT-ik hart di-ZEEZ) cardiac damage (especially to the endocardium, including the valves) resulting from a delayed inflammatory response to streptococcal infection

rheumatoid arthritis (ROO-mah-toid ar-THRY-tis) an autoimmune inflammatory joint disease characterized by synovial inflammation that spreads to other tissue

rhinitis (rye-NYE-tis) inflammation of the nasal mucosa often caused by nasal infections

Rh-negative (RH NEG-ah-tiv) red blood cells that do not contain the antigen called Rh factor

RhoGAM (RO-gam) an injection of a special protein given to an Rh-negative woman who is pregnant to prevent her body from forming anti-Rh antibodies, which may harm an Rh-positive baby

Rh-positive (RH POZ-i-tiv) red blood cells that contain an antigen called Rh factor

ribonucleic acid (RNA) (rye-bo-noo-KLEE-ik AS-id) a nucleic acid found in the cytoplasm that is crucial to protein synthesis

ribosome (RYE-bo-sohm) organelle in the cytoplasm of cells that synthesizes proteins; also known as a protein factory

rickets (RIK-ets) childhood form of osteomalacia, a bone-softening condition caused by vitamin D deficiency

rickettsia (ri-KET-see-ah) small bacterium that infects human cells as an obligate parasite

rigor mortis (RIG-or MOR-tis) literally "stiffness of death," the permanent contraction of muscle tissue after death caused by the depletion of ATP during the actin-myosin reaction—preventing myosin from releasing actin to allow relaxation of the muscle

risk factor (risk FAK-tor) predisposing condition; factor that puts one at a higher than usual risk for developing a disease

rods (rods) receptors located in the retina that are responsible for night vision

rotation (ro-TAY-shun) movement around a longitudinal axis; for example, shaking your head no

rugae (ROO-gee) wrinkles or folds; singular, ruga (ROO-gah)

"rule of nines" a frequently used method to determine the extent of a burn injury; the body is divided into 11 areas of 9% each to help estimate the amount of skin surface burned in an adult

sagittal (SAJ-i-tal) longitudinal; like an arrow

salivary amylase (SAL-i-vair-ee AM-i-lase) digestive enzyme found in the saliva that begins the chemical digestion of carbohydrates

salpingitis (sal-pin-JYE-tis) inflammation of the fallopian tubes

saltatory conduction (SAL-tah-tor-ee kon-DUK-shun) when a nerve impulse encounters myelin and "jumps" from one node of Ranvier to the next

sarcomere (SAR-ko-meer) contractile unit of muscle; length of a myofibril between two **Z** bands

scapula (SKAP-yoo-lah) shoulder blade

scar (skahr) thickened mass of tissue, usually fibrous connective tissue, that remains after a damaged tissue has been repaired

Schwann cells (shwon sells) large nucleated cells that form myelin

sclera (SKLE-rah) white outer coat of the eyeball

scleroderma (skle-ro-DER-mah) rare disorder affecting the vessels and connective tissue of skin and other tissues, characterized by tissue hardening

scoliosis (sko-lee-O-sis) abnormal lateral (side-to-side) curvature of the vertebral column

scotoma (sko-TOE-mah) loss of the central visual field due to nerve degeneration, it sometimes occurs with neuritis associated with multiple sclerosis

scrotum (SKRO-tum) pouchlike sac that contains the testes

sebaceous gland (se-BAY-shus gland) oil-producing glands found in the skin

sebum (SEE-bum) secretion of sebaceous glands

secondary infection (SEK-on-dair-ee in-FEK-shun) infection that occurs as a consequence of the weakened state or damage caused by a previously existing disease

secondary sexual characteristics (SEK-on-dair-ee SEK-shoo-al kair-ak-ter-IS-tiks) external physical characteristics of sexual maturity resulting from the action of the sex hormones; they include male and female patterns of body hair and fat distribution, as well as development of the external genitals

second messenger (SEK-und MES-en-jer) provide communication within a hormone's target cell; for example, cyclic AMP

sella turcica (SEL-lah TER-si-kah) small depression of the sphenoid bone that contains the pituitary gland

semen (SEE-men) male reproductive fluid

semicircular canals (sem-i-SIR-kyoo-lar kah-NALS) located in the inner ear; contain a specialized receptor called crista ampullaris that generates a nerve impulse on movement of the head

semilunar valves (sem-i-LOO-nar valvs) valves located between the two ventricular chambers and the large arteries that carry blood away from the heart; valves found in the veins

seminal fluid (SEM-i-nal FLOO-id) semen

seminal vesicle (SEM-i-nal VES-i-kul) paired, pouchlike glands that contribute about 60% of the seminal fluid volume; rich in fructose, which is a source of energy for sperm

seminiferous tubule (se-mi-NIF-er-us TOOB-yool) long, coiled structure that forms the bulk of the testicular mass

senescence (se-NES-enz) older adulthood; aging

sensory neurons (SEN-sor-ee NOO-rons) neurons that transmit impulses to the spinal cord and brain from all parts of the body

septic shock (SEP-tik shock) circulatory failure (shock) resulting from complications of septicemia (toxins in blood resulting from infection)

serosa (se-RO-sah) outermost covering of the digestive tract; composed of the parietal pleura in the abdominal cavity

serotonin (sair-o-TOE-nin) a neurotransmitter that belongs to a group of compounds called catecholamines

serous membrane (SE-rus MEM-brane) a two-layer epithelial membrane that lines body cavities and covers the surfaces of organs

serum (SEER-um) blood plasma minus its clotting factors, still contains antibodies

severe combined immune deficiency (SCID) (SE-veer kom-BIND i-MYOON de-FISH-en-see) nearly complete failure of the lymphocytes to develop properly, in turn causing failure of the immune system's defense of the body; very rare congenital immune disorder

sex chromosomes (seks KRO-mo-soms) pair of chromosomes in the human genome that determine gender; normal males have one X chromosome and one Y chromosome (XY), while normal females have two X chromosomes (XX)

sex-linked trait (seks-linked trait) nonsexual, inherited trait governed by genes located in a sex chromosome (X or Y); most known sex-linked traits are X-linked

sexually transmitted disease (STD) (SEKS-yoo-al-ee trans-MI-ted di-ZEEZ) any communicable disease that is commonly transmitted through sexual contact

shingles (SHING-guls) see **herpes zoster**

sickle cell anemia (SIK-ul sell ah-NEE-mee-ah) severe, possible fatal, hereditary disease due to an abnormal type of hemoglobin

sickle cell trait (SIK-ul sell trate) when only one defective gene is inherited and only a small amount of hemoglobin that is less soluble than usual is produced

sigmoid colon (SIG-moyd KO-lon) S-shaped segment of the large intestine that terminates in the rectum

sign (sine) objective deviation from normal that marks the presence of a disease

sinoatrial (SA) node (sye-no-AY-tree-al node) the heart's pacemaker; where the impulse conduction of the heart normally starts; located in the wall of the right atrium near the opening of the superior vena cava

sinus (SYE-nus) a space or cavity inside some of the cranial bones

sinus arrhythmia (SYE-nus ah-RITH-me-ah) variation in the rhythm of heart rate during the breathing cycle (inspiration and expiration)

sinusitis (sye-nyoo-SYE-tis) sinus infections

skeletal muscle (SKEL-e-tal MUS-el) also known as voluntary muscle; muscles under willed or voluntary control

skeletal system (SKEL-e-tal SIS-tem) the bones, cartilage, and ligaments that provide the body with a rigid framework for support and protection

smooth muscle (smooth MUS-el) muscles that are not under conscious control; also known as involuntary or visceral; forms the walls of blood vessels and hollow organs

solute (SOL-yoot) dissolved particles in water

somatic nervous system (so-MA-tik NER-vus SIS-tem) the motor neurons that control the voluntary actions of skeletal muscles

spastic paralysis (SPAS-tik pah-RAL-i-sis) loss of voluntary muscle control characterized by involuntary contractions of affected muscles

specific immunity (spe-SI-fik i-MYOON-i-tee) the protective mechanisms that provide specific protection against certain types of bacteria or toxins

sperm (sperm) the male spermatozoon; sex cell

spermatids (SPER-mah-tids) the resulting daughter cells from the primary spermatocyte undergoing meiosis; these cells have only half the genetic material and half the chromosomes of other body cells

spermatogenesis (sper-mah-toe-JEN-e-sis) production of sperm cells

spermatogonia (sper-mah-toe-GO-nee-ah) sperm precursor cells

spermatozoa (sper-mah-tah-ZO-ah) sperm cells (singular: spermatozoon)

sphincter (SFINGK-ter) ring-shaped muscle

sphygmomanometer (sfig-mo-mah-NOM-e-ter) device for measuring blood pressure in the arteries of a limb

spinal cavity (SPY-nal KAV-i-tee) the space inside the spinal column through which the spinal cord passes

spinal nerves (SPY-nal nervs) nerves that connect the spinal cord to peripheral structures such as the skin and skeletal muscles

spinal tracts (spy-nal tracts) the white columns of the spinal cord that provide two-way conduction paths to and from the brain; ascending tract carries information to the brain, whereas descending tracts conduct impulses from the brain

spindle fiber (SPIN-dul FYE-ber) a network of tubules formed in the cytoplasm between the centrioles as they are moving away from each other

spirometer (spi-ROM-e-ter) an instrument used to measure the amount of air exchanged in breathing

spleen (spleen) largest lymphoid organ; filters blood, destroys worn out red blood cells, salvages iron from hemoglobin, and serves as a blood reservoir

splenectomy (splen-NEK-toe-mee) surgical removal of the spleen

splenic flexure (SPLEEN-ik FLEK-shur) where the descending colon turns downward on the left side of the abdomen

splenomegaly (sple-no-MEG-ah-lee) condition of enlargement of the spleen

spongy bone (SPUN-jee bone) porous bone in the end of the long bone, which may be filled with marrow

spontaneous abortion (spon-TAY-nee-us ah-BOR-shun) miscarriage; loss of an embryo or fetus before the twentieth week of gestation (or under a weight of 500 g)

spore (spor) nonreproducing form of a bacterium that resists adverse environmental conditions; spores revert to the active multiplying form when conditions improve

sporozoa (spo-rah-ZO-ah) coccidia; parasitic protozoan that enters a host cell during one phase of a two-part life cycle

sprain (sprain) traumatic injury of ligaments forming a skeletal joint, sometimes also involving injury (strain) of skeletal muscle or tendon tissue

squamous (SKWAY-mus) scalelike

squamous suture (SKWAY-mus SOO-chur) the immovable joint between the temporal bone and the sphenoid bone

stapes (STAY-peez) tiny, stirrup-shaped bone in the middle ear

Stensen's ducts (STEN-sens dukts) the ducts of the parotid gland as they enter the mouth

sternoclavicular joint (ster-no-klah-VIK-yoo-lar joynt) the direct point of attachment between the bones of the upper extremity and the axial skeleton

sternocleidomastoid (stern-o-klye-doe-MAS-toyd) the "strap" muscle located on the anterior aspect of the neck

stenosed (cardiac) valves (ste-NOST KAR-dee-ak valvs) valves that are narrower than normal, slowing blood flow from a heart chamber.

sterility (ste-RIL-i-tee) as applied to humans, the inability to reproduce

steroid hormones (STE-royd HOR-mones) lipid-soluble hormones that pass intact through the cell membrane of the target cell and influence cell activity by acting on specific genes

stillbirth (STIL-berth) delivery of a dead fetus (after 20th week of gestation; before 20 weeks it is termed a spontaneous abortion)

stimulus (STIM-yoo-lus) agent that causes a change in the activity of a structure

stomach (STUM-ak) an expansion of the digestive tract between the esophagus and small intestine

strabismus (strah-BIS-mus) abnormal condition in which lack of coordination of, or weakness in, the muscles that control the eye cause improper focusing of images on the retina, thus making depth perception difficult

stratum corneum (STRA-tum KOR-nee-um) the tough outer layer of the epidermis; cells are filled with keratin

stratum germinativum (STRA-tum JER-mi-nah-tiv-um) the innermost of the tightly packed epithelial cells of the epidermis; cells in this layer are able to reproduce themselves

strength training (strength TRAIN-ing) contracting muscles against resistance to enhance muscle hypertrophy

striated muscle (STRYE-ay-ted MUS-el) see skeletal muscle

stroke volume (stroke VOL-yoom) the amount of blood that is ejected from the ventricles of the heart with each beat

subcutaneous tissue (sub-kyoo-TAY-nee-us TISH-yoo) tissue below the layers of skin; made up of loose connective tissue and fat

submucosa (sub-myoo-KO-sah) connective tissue layer containing blood vessels and nerves in the wall of the digestive tract

sudoriferous gland (soo-doe-RIF-er-us gland) glands that secrete sweat; also referred as sweat glands

sulcus (SUL-kus) furrow or groove

superficial (soo-per-FISH-al) near the body surface

superior (soo-PEER-ee-or) higher, opposite of inferior

superior vena cava (soo-PEER-ee-or VEE-nah KAY-vah) one of two large veins returning deoxygenated blood to the right atrium

supinate (SOO-pi-nate) to turn the palm of the hand upward; opposite of pronate

supine (SOO-pine) used to describe the body lying in a horizontal position facing upward

supraclavicular (soo-prah-cla-VIK-yoo-lar) area above the clavicle

surfactant (sur-FAK-tant) a substance covering the surface of the respiratory membrane inside the alveolus; it reduces surface tension and prevents the alveoli from collapsing

suture (SOO-chur) immovable joint

sweat (swet) transparent, watery liquid released by glands in the skin that eliminates ammonia and uric acid and helps maintain body temperature; also known as perspiration

sympathetic nervous system (sim-pah-THE-tik NER-vus SIS-tem) part of the autonomic nervous system; ganglia are connected to the thoracic and lumbar regions of the spinal cord; functions as an emergency system

sympathetic postganglionic neurons (sim-pah-THE-tik post-gang-glee-ON-ik NOO-rons) dendrites and cell bodies are in sympathetic ganglia and axons travel to a variety of visceral effectors

sympathetic preganglionic neurons (sim-pah-THE-tik pree-gang-glee-ON-ik NOO-rons) dendrites and cell bodies are located in the gray matter of the thoracic and lumbar segments of the spinal cord; leaves the cord through an anterior root of a spinal nerve and terminates in a collateral ganglion

symptom (SIMP-tum) subjective deviation from normal that marks the presence of a disease; sometimes used synonymously with *sign*

synapse (SIN-aps) junction between adjacent neurons

synaptic cleft (si-NAP-tik kleft) the space between a synaptic knob and the plasma membrane of a postsynaptic neuron

synaptic knob (si-NAP-tik nob) a tiny bulge at the end of a terminal branch of a presynaptic neuron's axon that contains vesicles with neurotransmitters

synarthrosis (sin-ar-THRO-sis) a joint in which fibrous connective tissue joins bones and holds them together tightly; commonly called sutures

syndrome (SIN-drome) collection of signs or symptoms, usually with a common cause that defines or gives a clear picture of a pathological condition

synergist (SIN-er-jist) muscle that assists a prime mover

synovial fluid (si-NO-vee-al FLOO-id) the thick, colorless lubricating fluid secreted by the synovial membrane

synovial membrane (si-NO-vee-al MEM-brane) connective tissue membrane lining the spaces between bones and joints that secretes synovial fluid

system (SIS-tem) group of organs arranged so that the group can perform a more complex function than any one organ can perform alone

systemic circulation (sis-TEM-ik ser-kyoo-LAY-shun) blood flow from the left ventricle to all parts of the body and back to the right atrium

systemic lupus erythematosus (SLE) (sis-TEM-ik LOO-pus er-i-them-ah-TOE-sus) chronic inflammatory disease caused by widespread attack of self-antigens by the immune system (autoimmunity) characterized by a red rash on the face and other signs

systole (SIS-toe-lee) contraction of the heart muscle

tachycardia (tak-e-KAR-dee-ah) rapid heart rhythm (above 100 beats/minute)

target organ cell (TAR-get OR-gan sell) organ or cell acted on by a particular hormone and responding to it

tarsals (TAR-sals) seven bones of the heel and back part of the foot; the calcaneus is the largest

taste buds (taste buds) chemical receptors that generate nerve impulses, resulting in the sense of taste

Tay-Sachs disease (TAY-saks di-ZEEZ) recessive, inherited condition in which abnormal lipids accumulate in the brain and cause tissue damage that leads to death by age 4

telophase (TEL-o-faze) last stage of mitosis in which the cell divides

temporal (TEM-po-ral) muscle that assists the masseter in closing the jaw

tendons (TEN-dons) bands or cords of fibrous connective tissue that attach a muscle to a bone or other structure

tendon sheath (TEN-don sheeth) tube-shaped structure lined with synovial membrane that encloses certain tendons

tenosynovitis (ten-o-sin-o-VYE-tis) inflammation of a tendon sheath

teratogen (TER-ah-toe-jen) physical or chemical agent that disrupts normal embryonic development and thus causes congenital defects

testes (TES-teez) male gonads that produce the male sex cells or sperm

testosterone (tes-TOS-te-rone) male sex hormone produced by the interstitial cells in the testes; the "masculinizing hormone"

tetanic contraction (te-TAN-ik kon-TRAK-shun) sustained contraction

tetanus (TET-ah-nus) sustained muscular contraction

thalamus (THAL-ah-mus) located just above the hypothalamus; its functions are to help produce sensations, associates sensations with emotions and plays a part in the arousal mechanism

thalassemia (thah-las-sah-NEE-me-ah) any of a group of inherited hemoglobin disorders characterized by production of hypochromic, abnormal red blood cells

thermoregulation (ther-mo-reg-yoo-LAY-shun) maintaining homeostasis of body temperature

thoracic duct (thor-AS-ik dukt) largest lymphatic vessel in the body

thorax (THOR-aks) chest

threshold stimulus (THRESH-hold STIM-yoo-lus) minimal level of stimulation required to cause a muscle fiber to contract

thrombin (THROM-bin) protein important in blood clotting

thrombocytes (THROM-bo-sites) also called platelets; play a role in blood clotting

thrombocytopenia (throm-bo-sye-toe-PEE-nee-ah) general term referring to an abnormally low blood platelet count

thrombophlebitis (throm-bo-fle-BYE-tis) vein inflammation (phlebitis) accompanied by clot formation

thrombosis (throm-BO-sis) formation of a clot in a blood vessel

thrombus (THROM-bus) stationary blood clot

thymosin (THY-mo-sin) hormone produced by the thymus that is vital to the development and functioning of the body's immune system

thymus gland (THY-mus gland) endocrine gland located in the mediastinum; vital part of the body's immune system

thyroid gland (THY-royd gland) endocrine gland located in the neck that stores its hormones until needed; thyroid hormones regulate cellular metabolism

thyroid-stimulating hormone (TSH) (THY-royd STIM-yoo-lay-ting HOR-mone) a tropic hormone secreted by the anterior pituitary gland that stimulates the thyroid gland to increase its secretion of thyroid hormone

thyroxine (T$_4$) (thy-ROK-sin) thyroid hormone that stimulates cellular metabolism

tibia (TIB-ee-ah) shinbone

tibialis anterior (tib-ee-AL-is an-TEER-ee-or) dorsiflexor of the foot

tidal volume (TV) (TIE-dal VOL-yoom) amount of air breathed in and out with each breath

tinnitus (ti-NYE-tus) abnormal sensation of ringing or buzzing in the ear

tissue (TISH-yoo) group of similar cells that perform a common function

tissue fluid (TISH-yoo FLOO-id) a dilute salt water solution that bathes every cell in the body

tissue hormone (TISH-yoo HOR-mone) prostaglandins; produced in a tissue and only diffuses a short distance to act on cells within the tissue

tissue typing (TISH-yoo TIE-ping) a procedure used to identify tissue compatability before an organ transplant

T lymphocytes (T LIM-fo-sites) cells that are critical to the function of the immune system; produce cell-mediated immunity

tonic contraction (TON-ik kon-TRAK-shun) special type of skeletal muscle contraction used to maintain posture

tonsillectomy (ton-si-LEK-toe-mee) surgical procedure used to remove the tonsils

tonsillitis (ton-si-LIE-tis) inflammation of the tonsils, usually due to infection

tonsils (TON-sils) masses of lymphoid tissue; protect against bacteria; three types: palatine tonsils, located on each side of the throat; pharyngeal tonsils (adenoids), near the posterior opening of the nasal cavity; and lingual tonsils, near the base of the tongue

total metabolic rate (TMR) (TOE-tal met-ah-BOL-ik rate) total amount of energy used by the body per day

trabeculae (trah-BEK-yoo-lee) needle-like threads of spongy bone that surround a network of spaces

trachea (TRAY-kee-ah) the windpipe; the tube extending from the larynx to the bronchi

trachoma (trah-KO-mah) chronic infection of the conjunctiva covering the eye caused by the bacterium *Chlamydia trachomatis*; also called *chlamydial conjunctivitis*

transaminase (trans-AM-i-nase) enzyme released from damaged tissues; high blood concentration may indicated a heart attack or other pathological event

transcription (trans-KRIP-shun) when the double stranded DNA molecule unwind and form mRNA

translation (trans-LAY-shun) the synthesis of a protein by ribosomes

transplant (trans-PLANT) tissue or organ graft; procedure in which a tissue (skin, bone marrow, etc.) or an organ (kidney, liver, etc.) from a donor is surgically implanted into a recipient

transverse arch (TRANS-vers arch) see metatarsal arch

transverse fracture (TRANS-vers FRAK-chur) bone fracture characterized by a fracture line that is at a right angle to the long axis of the bone

transversus abdominis (trans-VER-sus ab-DOM-i-nis) the innermost layer of the anterolateral abdominal wall

trapezium (trah-PEE-zee-um) the carpal bone of the wrist that forms the saddle joint that allows the opposition of the thumb

trapezius (trah-PEE-zee-us) triangular muscle in the back that elevates the shoulder and extends the head backwards

triceps brachii (TRY-seps BRAY-kee-eye) extensor of the elbow

tricuspid valve (try-KUS-pid valve) the valve located between the right atrium and ventricle

trigeminal neuralgia (tri-JEM-i-nal noo-RAL-jee-ah) pain in one or more (of three) branches of the fifth cranial nerve (trigeminal nerve) that runs along the face; also called *tic douloureux*

trigone (TRY-gon) triangular area on the wall of the urinary bladder

triiodothyronine (**T₃**) (try-eye-o-doe-THY-ro-nine) thyroid hormone that stimulates cellular metabolism

triplegia (try-PLEE-jee-ah) paralysis (loss of voluntary muscle control) in three limbs, often two legs and one arm

trisomy (TRY-so-me) abnormal genetic condition in which cells have three chromosomes (a triplet) where there should be a pair; usually caused by nondisjunction (failure of chromosome pairs to separate) during gamete production

tropic hormone (TRO-pik HOR-mone) hormone that stimulates another endocrine gland to grow and secrete its hormones

true ribs (troo ribs) the first seven pairs of ribs, which are attached to the sternum

tubal pregnancy (TOO-bal PREG-nan-see) ectopic pregnancy that occurs in a fallopian tube

tuberculosis (TB) (too-ber-kyoo-LO-sis) chronic bacterial (bacillus) infection of the lungs or other tissues caused by *Mycobacterium tuberculosis* organisms

tumor (TOO-mer) growth of tissues in which cell proliferation is uncontrolled and progressive

tunica albuginea (TOO-ni-kah al-byoo-JIN-ee-ah) a tough, whitish membrane that surrounds each testis and enters the gland to divide it into lobules

tunica externa (TOO-ni-kah eks-TER-nah) the outermost layer found in blood vessels

tunica interna (TOO-ni-kah in-TER-nah) endothelium that lines the blood vessels

tunica media (TOO-ni-kah MEE-dee-ah) the muscular middle layer found in blood vessels; the tunica media of arteries is more muscular than that of veins

Turner syndrome (TUR-ner SIN-drome) genetic disorder caused by monosomy of the X chromosome (XO) in females; characterized by immaturity of sex organs (causing sterility), webbed neck, cardiovascular defects, and learning disorders

T wave (T wave) deflection on an electrocardiogram that occurs with repolarization of the ventricles

twitch (twitch) a quick, jerky response to a single stimulus

tympanic (tim-PAN-ik) drumlike

ulcer (UL-ser) a necrotic open sore or lesion

ulna (UL-nah) one of the two forearm bones; located on the little finger side

ultrasonogram (ul-tra-SOHN-o-gram) a technique using sound to produce images

ultrasonography (ul-tra-son-OG-rah-fee) imaging technique in which high-frequency sound waves are reflected off tissue to form an image

umbilical artery (um-BIL-i-kul AR-ter-ee) two small arteries that carry oxygen-poor blood from the developing fetus to the placenta

umbilical cord (um-BIL-i-kul cord) flexible structure connecting the fetus with the placenta, which allows the umbilical arteries and vein to pass

umbilical vein (um-BIL-i-kul vane) a large vein carrying oxygen rich blood from the placenta to the developing fetus

urea (yoo-REE-ah) nitrogen-containing waste product

uremia (yoo-REE-mee-ah) condition in which blood urea concentration is abnormally elevated, expressed as a high BUN (blood urea nitrogen) value; uremia is often caused by renal failure; also called uremic syndrome

uremic poisoning (yoo-REE-mik POY-zon-ing) see *uremia*

urethra (yoo-REE-thrah) passageway for elimination of urine; in males, also acts as a genital duct that carries sperm to the exterior

urethritis (yoo-re-THRY-tis) inflammation or infection of the urethra

urinary meatus (YOOR-i-nair-ee mee-AY-tus) external opening of the urethra

urinary system (YOOR-i-nair-ee SIS-tem) system responsible for excreting liquid waste from the body

urination (yoor-i-NAY-shun) passage of urine from the body; emptying of the bladder

urine (YOOR-in) fluid waste excreted by the kidneys

urticaria (er-ti-KAIR-ee-ah) hives; an allergic or hypersenstivity response characterized by raised red lesions

uterus (YOO-ter-us) hollow, muscular organ where a fertilized egg implants and grows

uvula (YOO-vyoo-lah) cone-shaped process hanging down from the soft palate that helps prevent food and liquid from entering the nasal cavities

vaccine (VAK-seen) application of killed or attenuated (weakened) pathogens (or portions of pathogens) to a patient to stimulate immunity against that pathogen

vagina (vah-JYE-nah) internal tube from uterus to vulva

vaginitis (vaj-i-NYE-tis) inflammation of the vagina

varicose vein (VAIR-i-kose vane) enlarged vein in which blood pools; also called *varix* (plural, *varices*)

varix (VAIR-iks) varicose vein; plural, *varices*

vas deferens (vas DEF-er-enz) see ductus deferens

vastus (VAS-tus) wide; of great size

vector (VEK-tor) arthropod that carries an infectious pathogen from one organism to another

vein (vane) vessel carrying blood toward the heart

ventral (VEN-tral) of or near the belly; in humans, front or anterior; opposite of dorsal or posterior

ventricles (VEN-tri-kuls) small cavities

venule (VEN-yool) small blood vessels that collect blood from the capillaries and join to form veins

vermiform appendix (VERM-i-form ah-PEN-diks) a tubular structure attached to the cecum composed of lymphatic tissue

vertebrae (VER-te-bray) bones that make up the spinal column

vertebral column (ver-TEE-bral KOL-um) the spinal column, made up of a series of separate vertebrae that form a flexible, curved rod

vertigo (VER-ti-go) abnormal sensation of spinning; dizziness

vesicle (VES-i-kl) a clinical term referring to blisters, fluid-filled skin lesions

vestibular nerve (ves-TIB-yoo-lar nerv) a division of the vestibulocochlear nerve (the eighth cranial nerve)

vestibule (VES-ti-byool) located in the inner ear; the portion adjacent to the oval window between the semicircular canals and the cochlea

villi (VIL-eye) fingerlike folds covering the plicae of the small intestines

Vincent's infection (VIN-sent in-FEK-shun) bacterial (spirochete) infection of the gum, producing gingivitis; also called *Vincent's angina* and *trench mouth*

virus (VYE-rus) microscopic, parasitic entity consisting of a nucleic acid bound by a protein coat and sometimes a lipoprotein envelope

visceral pericardium (VIS-er-al pair-i-KAR-dee-um) the pericardium that covers the heart

visceral portion (VIS-er-al POR-shun) serous membrane that covers the surface of organs found in the body cavity

vital capacity (VC) (VYE-tal kah-PAS-i-tee) largest amount of air that can be moved in and out of the lungs in one inspiration and expiration

vitamins (VYE-tah-mins) organic molecules needed in small quantities to help enzymes operate effectively

vitreous humor (VIT-ree-us HYOO-mor) the jellylike fluid found in the eye, posterior to the lens

voiding (VOYD-ing) emptying of the bladder

volar (VO-lar) palm or sole

voluntary muscle (VOL-un-tair-ee MUS-el) see **skeletal muscle**

vulva (VUL-vah) external genitals of the female

vulvitis (vul-VYE-tis) inflammation of the vulva (the external female genitals)

white matter (wite MAT-ter) nerves covered with white myelin

withdrawl reflex (with-DRAW-al REE-fleks) a reflex that moves a body part away from an irritating stimulus

yeast (yeest) single-celled fungus (compared to mold, which is a multicellular fungus)

yolk sac (yoke sak) in humans, involved with the production of blood cells in the developing embryo

zona fasciculata (ZO-nah fas-sic-yoo-LAY-tah) middle zone of the adrenal cortes that secretes glucocorticoids

zona glomerulosa (ZO-nah glo-mare-yoo-LO-sah) outer zone of the adrenal cortex that secretes mineralocorticoids

zona reticularis (ZO-nah re-tik-yoo-LAIR-is) inner zone of the adrenal cortex that secretes small amounts of sex hormones

zygomaticus (zye-go-MAT-ik-us) muscle that elevates the corners of the mouth and lips; also know as the smiling muscle

zygote (ZYE-gote) a fertilized ovum

Chapter One: 1-1, 1-3, E.W. Beck; 1-2, 1-4, 1-7, 1-8, 1-9, Joan Beck.

Chapter Two: 2-1, 2-2A, William Ober; 2-2B, K.G. Murti/Visuals Unlimited; 2-3, Lennart Nilsson; Tonicity Box, Barbara Cousins; 2-4, 2-7, Rolin Graphics; 2-5, 2-8, 2-10, Joan Beck; 2-6, courtesy of Ingalls and Salerno (From *Maternal & Child Health Nursing*, ed. 6, 1987, Times-Mirror, Mosby College Publishing); 2-9, Raychel Ciemma; 2-11, 2-12A, 2-14A, 2-15A, 2-19A, 2-20A, 2-21A, Ed Reschke; 2-13A, 2-16A, 2-17A, 2-18A, 2-22A, Carolina Biological Supply; 2-24, courtesy of Thomas P. Habif (from *Clinical Dermatology*, ed. 2, The C.V. Mosby Co.).

Chapter Three: 3-2 through 3-13, Joan Beck; 3-14, 3-15, Rolin Graphics.

Chapter Four: 4-1, Centers for Disease Control, Atlanta, Georgia; 4-2, 4-5, 4-7, 4-9, 4-11, 4-13, 4-14, Rolin Graphics; 4-3, Lennart Nilsson; 4-4 courtesy of Raven & Johnson (from *Biology*, ed. 2, 1989, Times-Mirror, Mosby College Publishing); 4-6, 4-8, David M. Phillips/Visuals Unlimited; 4-10, courtesy of Stanley Erlandson, University of Minnesota; 4-12, John Cunningham/Visuals Unlimited; Lab ID box, courtesy of B.C. Stratford (from *An Atlas of Medical Microbiology*, 1977, Blackwell Scientific Publications); 4-15A courtesy of Dr. A.L. Le Treut (from *Mammography*, 1991, Mosby-Year Book, Inc.); 4-15B, courtesy of Drs. Alan L. Williams and Victor M. Haughton (from *Cranial Computed Tomography*, 1985, The C.V. Mosby Co.); 4-15C, courtesy of Dr. Val M. Runge (from *Enhanced Magnetic Resonance Imaging*, 1989, The C.V. Mosby Co.); 4-15D, courtesy of Dr. Sandra L. Hagen-Ansert (from *Textbook of Diagnostic Ultrasonography*, 1989, The C.V. Mosby Co.); 4-17 Michael P. Schenck.

Chapter Five: 5-1, 5-2, 5-8, Rolin Graphics; 5-4, 5-10, 5-12C&E, 5-13, 5-14, 5-15, courtesy of Thomas P. Habif (from *Clinical Dermatology*, ed. 2, The C.V. Mosby Co.); 5-5, Visuals Unlimited; 5-6, 5-9, 5-11, Joan Beck; 5-7, courtesy of Stephen B. Tucker, MD, Dept. of Dermatology, University of Texas Health Science Center at Houston; Table 5-1 figures, George J. Wassilchenko/David P. O'Connor; 5-12A, courtesy of Antoinette Hood, MD, Dept. of Dermatology, School of' Medicine, Johns Hopkins University, Baltimore; 5-12D, courtesy of Jaime A. Tschen, MD, Dept. of Dermatology, Baylor College of Medicine, Houston.

Chapter Six: 6-1, 6-5, 6-11, 6-20, 6-25, Joan Beck; 6-2, Laurie O'Keefe/John Daugherty; 6-3, courtesy of Dr. McMullen, Dept. of Biology, South Dakota State University; 6-6, 6-7, 6-8, 6-9, 6-10A, 6-13A&B, 6-14A, 6-15A&B, 6-16, 6-17, 6-18, 6-19, E.W. Beck; 6-10B, Ron Edwards; 6-12, 6-14B, 6-15C, Branislav Vidic; Palpable Bony Landmarks box, Terry Cockerham/Synapse Media Production; Knee Joint box photo, Allsport Photography; 6-21, David J. Mascaro & Associates; Knee Joint box figure, 6-22, Rolin Graphics; Arthroscopy box, courtesy of Dr. Johnson (from *Diagnostic & Surgical Arthroscopy*, ed. 2, 1981, The C.V. Mosby Co.); 6-23, courtesy of H.E. Hilt & S.B. Cogburn (from *Manual of Orthopedics*, 1980, The C.V. Mosby Co.); 6-24, David J. Mascaro & Associates; 6-26, Paul R. Manske.

Chapter Seven: 7-1, Christine Oleksyk; 7-2, Barbara Stackhouse; 7-3A, E.W. Beck; 7-3B, courtesy of Dr. H.E. Huxley; 7-4, Laurie O'Keefe/John Daugherty; 7-5, Occupational Health Problems box, 7-9A&C, 7-10A&C, Rolin Graphics; 7-6, 7-7, 7-8, John V. Hagen; 7-9B, 7-10B, 7-11A,C,D, Terry Cockerham/Synapse Media Production; 7-12, courtesy of Rob Williams (from *Athletic Injury Assessment*, ed. 2, 1989, Times-Mirror, Mosby College Publishing).

Chapter Eight: 8-2A&B, Blood Brain Barrier box, 8-25, Rolin Graphics; 8-2C, Ed Reschke; 8-3, 8-6B, 8-17, 8-18, Scott Bodell; 8-4, courtesy of Thomas P. Habif (from *Clinical Dermatology*, ed. 2, The C.V. Mosby Co.); 8-5, 8-8, Laurie O'Keefe/John Daugherty; 8-5 (inset) courtesy of K.L. McCance (from *Pathophysiology*, 1990, The C.V. Mosby Co.); 8-6, Barbara Cousins; 8-7A, 8-10, 8-24, Joan Beck; 8-9, George J. Wassilchenko/David P. O'Connor; 8-11A, 8-12, 8-14, William Ober; 8-11B, 8-12 (inset), 8-21, Branislav Vidic; Brain Studies box, courtesy of Dr. Nolte (from *The Human Brain*, The C.V. Mosby Co.; 8-13A, Rebecca S. Montgomery; 8-13B, courtesy of D.N. Markand, MD; 8-15, Christine Oleksyk; 8-16, 8-17 (inset), 8-19, 8-20, 8-22, E.W. Beck; 8-23, Raychel Ciemma.

Chapter Nine: 9-1, George J. Wassilchenko/David P. O'Connor; 9-2, Christine Oleksyk; 9-4, from Newell, 1986; 9-5, from Stein and Slatt, 1988; 9-7, 9-9, 9-10A, Marsha J. Dohrmann; 9-8, Cochlear Implant box, 9-12, Rolin Graphics; 9-10B&C, Kathy Mitchell Grey; 9-11, Joan Beck.

Chapter Ten: 10-1, Joan Beck; 10-2, 10-3, 10-11, 10-16, Rolin Graphics; 10-5, Kevin Sommerville; 10-6, 10-12, E.W. Beck; 10-9, Lester Bergman & Associates; 10-14, courtesy of Gower Medical Publishers.

Chapter Eleven: 11-1, 11-4, courtesy of G. Bevelander, and J.A. Ramaley (from *Essentials of Histology*, ed. 8, The C.V. Mosby Co.); 11-2, 11-9, Rolin Graphics; 11-3, Barbara Stackhouse; 11-5, Pagecrafters/Nadine

Sokol; 11-6, courtesy of Robert L. Callentine; 11-7, courtesy of Dr. A. Arlan Hinchee; 11-8, Laurie O'Keefe/John Daugherty; 11-10, Molly Babich/John Daugherty.

Chapter Twelve: 12-1, E.W. Beck; 12-2, Rusty Jones; 12-2 (inset), Barbara Stackhouse; 12-3A&B, 12-5, Christine Oleksyk; 12-3C, Branislav Vidic; 12-4, courtesy of K.L. McCance (from *Pathophysiology*, 1990, The C.V. Mosby Co.); 12-6, Rusty Jones; 12-8, 12-9, Lisa Shoemaker/Joan Beck; 12-10, Visuals Unlimited; 12-11, courtesy of Patricia Kane, Radiology Dept., Indiana University Medical Center.

Chapter Thirteen: 13-1, 13-9, E.W. Beck; 13-2, 13-3, George J. Wassilchenko; 13-4, Ronald J. Ervin; 13-6, 13-7, 13-8, Rolin Graphics; 13-8, Karen Waldo; Blood Pressure box, Joan Beck/Donna Odle; 13-11, G. David Brown.

Chapter Fourteen: 14-1, 14-3, Joan Beck; 14-4, George J. Wassilchenko/David P. O'Connor; 14-7, Laurie O'Keefe/John Daugherty; 14-8, courtesy of Robert L. Callentine; 14-9, courtesy of Emma Shelton; 14-13, courtesy of James T. Barrett; Immunization box, Rolin Graphics; 14-15, courtesy of Thomas P. Habif (from *Clinical Dermatology*, ed. 2, The C.V. Mosby Co.).

Chapter Fifteen: 15-1, Joan Beck; 15-2, Joan Beck/Lisa Shoemaker; 15-4, 15-14, Margaret Gerrity; 15-5, 15-6A, 15-9, E.W. Beck; 15-6B, Christine Oleksyk; 15-6C, Custom Medical Stock Photo; 15-7, Lisa Shoemaker/Joan Beck; 15-8, 15-11, Barbara Stackhouse; 15-10, courtesy of R.H. Daffner (from *Introduction to Clinical Radiology*, 1978, The C.V. Mosby Co.); 15-13, John Daugherty; 15-13 (insets), Barbara Cousins; Oxygen Therapy box, 15-17, Rolin Graphics; 15-18, courtesy of Dr. Andrew P. Evans, Indiana University.

Chapter Sixteen: 16-1, Lisa Shoemaker; 16-2, 16-5, 16-6, 16-7, 16-10, 16-11, 16-14, E.W. Beck; 16-3, 16-4, 16-15, 16-18, Rolin Graphics; 16-8, G. David Brown; 16-9, Table 16-2 figure, Laurie O'Keefe/John Daugherty; 16-12, from Kissane, 1985; 16-13, Branislav Vidic; 16-16, Science Photo Library/Photo Researchers; 16-17, Michael P. Schenck.

Chapter Seventeen: 17-1, 17-2, 17-3, Rolin Graphics; 17-5, Barabara Stackhouse.

Chapter Eighteen: 18-1A, 18-2, 18-3, 18-4, 18-7, E.W. Beck; 18-1B, courtesy of Patricia Kane, Indiana University Medical Center; 18-3 (inset), courtesy of Cleveland P. Hickman; 18-5, Rolin Graphics; 18-6, courtesy of G. Bevelander, and J.A. Ramaley (from *Essentials of Histology*, ed. 8, The C.V. Mosby Co.).

Chapter Nineteen: 19-1, 19-2, Rolin Graphics; 19-3, Joan Beck; 19-4 (inset), Barbara Stackhouse.

Chapter Twenty: 20-5 (inset), 20-6 (inset), Rolin Graphics.

Chapter Twenty-One: 21-1, 21-7, Barbara Cousins; 21-2 figure, 21-8, E.W. Beck; 21-2 photo, Lennart Nilsson; 21-4, Rolin Graphics; 21-5A, Carolyn B. Coulam/John A. McIntyre; 21-5B, 21-9, William Ober; 21-6A, 21-13, Kevin A. Somerville; 21-6B, Branislav Vidic; 21-10, George J. Wassilchenko/David P. O'Connor; 21-11, David J. Mascaro & Associates.

Chapter Twenty-Two: 22-1 figure, 22-8, Rolin Graphics; 22-1 photo, 22-6, Lennart Nilsson; 22-2, Scott Bodell; 22-3, Lucinda L. Veeck, Jones Institute for Reproductive Medicine, Norfolk, Virginia; 22-4, 22-5, E.W. Beck; FAS box, courtesy of Claus Simon/Michael Janner; In Vitro Fertilization box, Wide World Photos; 22-7, Kevin A. Somerville; Antenatal Diagnosis box figure, Rebecca S. Montgomery, photo courtesy of Carolyn B. Coulam /John A. McIntyre; 22-10, courtesy of Marjorie M. Pyle, Lifecircle, Costa Mesa, California; 22-11, Ron Edwards; 22-12, George J. Wassilchenko/David P. O'Connor; 22-13A, TIB, The Image Bank; 22-13B, Joel Gordon Photography; 22-13C, Visuals Unlimited.

Chapter Twenty-Three: 23-1 figure, 23-2, 23-4, 23-5, 23-6, 23-7, 23-10, 23-12, Rolin Graphics; 23-1 photo, CNRI/Science Photo Library; 23-8A, courtesy of Louis McGavran/Denver Children's Hospital; 23-8B, Richard Hutching/Photo Researchers; 23-9, Pagecrafters/Nadine Sokol; 23-11, Rebecca S. Montgomery; DNA Analysis box photo, BioProducts, Rockland, ME; DNA Analysis box figure, Barbara Cousins.

Appendix A: A-1, A-10, A-11, Ronald J. Ervin; A-5, courtesy of Trent Stephens; A-9, Rolin Graphics.

Index

A

Abdomen, *10*
  arteries of, 332t
  horizontal section of, *419*
  quadrants of, *8, 9*
Abdominal, *10*, 11t
Abdominal aorta, *330, 338, 419*
Abdominal cavity, 4, 6
  organs of, 5, 6
Abdominal hysterectomy, 516
Abdominal regions, 8
Abdominopelvic cavity, 4, 6
Abducens nerve, *220*, 221t
Abduction, 182, *183*
Abnormal body temperature, 445-446
ABO system of blood typing, 298-301
Abortion, spontaneous, 540
Abruptio placentae, 540
Absorption, 404, 426-427
  from large intestine, 420
Abuse
  steroid, 264
  substance, causing protein-calorie
    malnutrition, 443t
Accessory digestive organs, 62
Acetabulum, 149
Acetazolamide, 479
Acetylcholine, 204
  release of, 228
Achilles tendon, 181
Acid phosphatase, 298
Acid
  deoxyribonucleic, 26, 30
  nucleic, 30
  ribonucleic, 30
Acid-base balance, 485-496
Acid-forming potential of foods, 487
Acidification of urine
  by distal renal tubule secretion of
    hydrogen ions, *492*
  by tubule secretion of ammonia, *493*
Acidity, 486
Acidosis, 298, 494
  metabolic, 494
  respiratory, 494
Acne, 111
Acne vulgaris, 111

Acoustic nerve, *220*, 221t, *245*
Acquired cortical color blindness, 244
Acquired immune deficiency, 366
Acquired immune deficiency syndrome,
  24, 32, 79t, 85, 213, 295, 366, 519,
  520t
  viral codes and, 32
Acquired immunity, 355
Acromegaly, 266
Acromion process of scapula, *135, 146t,*
  184
  palpation for, 144
Acrosome, 503
ACTH; *see* Adrenocorticotropic hormone
Actin, 169
Action potentials, 199
Activated B cells, 361
Active immunity, 355
Active transport processes of cell, 26, 28-
  29
Acyclovir, 87
Adam's apple; *see* Thyroid cartilage
Addison's disease, 277
Adduction, 182, *183*
Adductor group of muscles, 175t, *176,*
  183t
Adductor longus muscle, 175t, *176*
Adductor magnus muscle, *177*
Adductor muscles, 181
Adenine, 30
Adenocarcinoma, 89
  of colon, 422
  of pancreas, 420
Adenofibroma, 89
Adenohypophysis, 265; *see also* Anterior
  pituitary gland
Adenoids, 354, *377*
Adenoma, 88
Adenosine deaminase deficiency, 565
Adenosine diphosphate, 436
Adenosine monophosphate, cyclic, 258,
  263
Adenosine triphosphate, 26
  in carbohydrate metabolism, 434-435
  and heat production, 170
  and muscle fatigue, 170
  structure of, *436*

Adenovirus, *79*
ADH; *see* Antidiuretic hormone
Adipose cell, nucleus of, *41*
Adipose connective tissue, 37t, 41
Adipose tissue, *378*
Adolescence, 543
ADP; *see* Adenosine diphosphate
Adrenal cortex, 272-274
  hormones of, functions of, 261t
  tumors of, 276-277
Adrenal glands, 60, *61, 259,* 272-277, *453*
Adrenal medulla, 274-277
  hormones of, functions of, 261t
Adrenaline
  functions of, 261t
  secretion of, *224*
Adrenergic fibers, 228
Adrenocorticotropic hormone, 265, *267*
  functions of, 260t
Adult respiratory distress syndrome,
  383
Adulthood, 543
Aerobic bacteria, 80
Aerobic respiration, 25
Aerobic training, 174
Afferent arteriole, *456, 457*
Afferent lymph vessels, 352
Afferent neurons, 194
Age
  and body functions, 14
  and cancer, 89-90
  as risk factor, 77
  skin changes with, 107
Age-related macular degeneration, 243
Agglutination, 299
Aging
  effects of, 544-546
  reproductive system changes in, 546
"Aging" genes, 544
Aging processes, 14
AIDS-related complex, 366
Air sac, *38*
Albinism, 554-555, 559t
Aldosterone, 273
  and control of urine volume, 460
  functions of, 261t
  and urine volume, 477

Italic type indicates mention only in figure; t indicates table mention only.

I-1